Real and Complex Analysis

A Program of Monographs, Textbooks, and Lecture Notes

Recent Titles

Christof Eck, Jiri Jarusek, and Miroslav Krbec, Unilateral Contact Problems: Variational Methods and Existence Theorems (2005)

M. M. Rao, Conditional Measures and Applications, Second Edition (2005)

A. B. Kharazishvili, Strange Functions in Real Analysis, Second Edition (2006)

Vincenzo Ancona and Bernard Gaveau, Differential Forms on Singular Varieties: De Rham and Hodge Theory Simplified (2005)

Santiago Alves Tavares, Generation of Multivariate Hermite Interpolating Polynomials (2005)

Sergio Macías, Topics on Continua (2005)

Mircea Sofonea, Weimin Han, and Meir Shillor, Analysis and Approximation of Contact Problems with Adhesion or Damage (2006)

Marwan Moubachir and Jean-Paul Zolésio, Moving Shape Analysis and Control: Applications to Fluid Structure Interactions (2006)

Alfred Geroldinger and Franz Halter-Koch, Non-Unique Factorizations: Algebraic, Combinatorial and Analytic Theory (2006)

Kevin J. Hastings, Introduction to the Mathematics of Operations Research with *Mathematica*®, Second Edition (2006)

Robert Carlson, A Concrete Introduction to Real Analysis (2006)

John Dauns and Yiqiang Zhou, Classes of Modules (2006)

N. K. Govil, H. N. Mhaskar, Ram N. Mohapatra, Zuhair Nashed, and J. Szabados, Frontiers in Interpolation and Approximation (2006)

Luca Lorenzi and Marcello Bertoldi, Analytical Methods for Markov Semigroups (2006)

M. A. Al-Gwaiz and S. A. Elsanousi, Elements of Real Analysis (2006)

Theodore G. Faticoni, Direct Sum Decompositions of Torsion-Free Finite Rank Groups (2007)

R. Sivaramakrishnan, Certain Number-Theoretic Episodes in Algebra (2006)

Aderemi Kuku, Representation Theory and Higher Algebraic K-Theory (2006)

Robert Piziak and P. L. Odell, Matrix Theory: From Generalized Inverses to Jordan Form (2007)

Norman L. Johnson, Vikram Jha, and Mauro Biliotti, Handbook of Finite Translation Planes (2007)

Lieven Le Bruyn, Noncommutative Geometry and Cayley-smooth Orders (2008)

Fritz Schwarz, Algorithmic Lie Theory for Solving Ordinary Differential Equations (2008)

Jane Cronin, Ordinary Differential Equations: Introduction and Qualitative Theory, Third Edition (2008)

Su Gao, Invariant Descriptive Set Theory (2009)

Christopher Apelian and Steve Surace, Real and Complex Analysis (2010)

Real and Complex Analysis

Christopher Apelian
Steve Surace

with Akhil Mathew

CRC Press is an imprint of the
Taylor & Francis Group, an **informa** business
A CHAPMAN & HALL BOOK

CRC Press
Taylor & Francis Group
6000 Broken Sound Parkway NW, Suite 300
Boca Raton, FL 33487-2742

First issued in paperback 2019

ISBN-13: 978-1-58488-806-2 (hbk)
ISBN-13: 978-0-367-38478-4 (pbk)

Library of Congress Cataloging-in-Publication Data

Apelian, Christopher.
Real and complex analysis / Christopher Apelian, Steve Surace.
p. cm. -- (Pure and applied mathematics. A program of monographs, textbooks, and lecture notes)
Includes bibliographical references and index.
ISBN 978-1-58488-806-2 (hardcover : alk. paper)
1. Mathematical analysis--Textbooks. 2. Functions--Textbooks. 3. Numbers, Complex--Textbooks. I. Surace, Steve. II. Title. III. Series.

QA300.A5685 2010
515--dc22 2009039342

Visit the Taylor & Francis Web site at
http://www.taylorandfrancis.com

and the CRC Press Web site at
http://www.crcpress.com

To my wife Paula.
For the sacrifices she made while this book was being written.

And to Ellie.
For being exactly who she is.

- CA

To family and friends.

- SS

To my wife Paula,

For the [illegible] made while this book was being written

And to [illegible]

For being exactly [illegible]

LA

To [illegible]

[illegible]

CONTENTS

PREFACE

The last thing one knows when writing a book is what to put first.

Blaise Pascal

This is a text for a two-semester course in analysis at the advanced undergraduate or first-year graduate level, especially suited for liberal arts colleges. Analysis is a very old and established subject. While virtually none of the content of this work is original to the authors, we believe its organization and presentation are unique. Unlike other undergraduate level texts of which we are aware, this one develops both the real and complex theory together. Our exposition thus allows for a unified and, we believe, more elegant presentation of the subject. It is also consistent with the recommendations given in the Mathematical Association of America's 2004 Curriculum Guide, available online at `www.maa.org/cupm`.

We believe that learning real and complex analysis separately can lead students to compartmentalize the two subjects, even though they—like all of mathematics—are inextricably interconnected. Learning them together shows the connections at the outset. The approach has another advantage. In small departments (such as ours), a combined development allows for a more streamlined sequence of courses in real and complex function theory (in particular, a two-course sequence instead of the usual three), a consideration that motivated Drew University to integrate real and complex analysis several years ago. Since then, our yearly frustration of having to rely on two separate texts, one for real function theory and one for complex function theory, ultimately led us to begin writing a text of our own.

We wrote this book with the student in mind. While we assume the standard background of a typical junior or senior undergraduate mathematics major or minor at today's typical American university, the book is largely self-contained. In particular, the reader should know multivariable calculus and the basic notions of set theory and logic as presented in a "gateway" course for mathematics majors or minors. While we will make use of matrices, knowledge of linear algebra is helpful, but not necessary.

We have also included over 1,000 exercises. The reader is encouraged to do all of the embedded exercises that occur within the text, many of which are necessary to understand the material or to complete the development of a particular topic. To gain a stronger understanding of the subject, however, the serious student should also tackle some of the supplementary exercises at the end of each chapter. The supplementary exercises include both routine skills problems as well as more advanced problems that lead the reader to new results not included in the text proper.

Both students and instructors will find this book's website, `http://users.drew.edu/capelian/rcanalysis.html`, helpful. Partial solutions or hints to selected exercises, supplementary materials and problems, and recommendations for further reading will appear there. Questions or comments can be sent either to capelian@drew.edu, or to ssurace@drew.edu.

This text was typeset using LaTeX® 2ε and version 5.5 of the WinEdt® editing environment. All figures were created with Adobe Illustrator®. The mathematical computing package *Mathematica*® was also used in the creation of Figures 6.1, 8.2, 8.19, 8.22, 10.1, 10.10, and 10.13. The quotes that open each chapter are from the Mathematical Quotation Server, at `http://math.furman.edu/~mwoodard/mqs/mquot.shtml`.

We sincerely hope this book will help make your study of analysis a rewarding experience.

CA
SS

ACKNOWLEDGMENTS

Everything of importance has been said before by somebody who did not discover it.

Alfred North Whitehead

We are indebted to many people for helping us produce this work, not least of whom are those authors who taught us through their texts, including Drs. Fulks, Rudin, Churchill, Bartle, and many others listed in the bibliography. We also thank and credit our instructors at Rutgers University, New York University, and the Courant Institute, whose lectures have inspired many examples, exercises, and insights. Our appreciation of analysis, its precision, power, and structure, was developed and nurtured by all of them. We are forever grateful to them for instilling in us a sense of its challenges, importance, and beauty. We would also like to thank our colleagues at Drew University, and our students, who provided such valuable feedback during the several years when drafts of this work were used to teach our course sequence of real and complex analysis. Most of all, we would like to extend our heartfelt gratitude to Akhil Mathew, whose invaluable help has undoubtedly made this a better book. Of course, we take full responsibility for any errors that might yet remain. Finally, we thank our families. Without their unlimited support, this work would not have been possible.

The Authors

Christopher Apelian completed a Ph.D. in mathematics in 1993 at New York University's Courant Institute of Mathematical Sciences and then joined the Department of Mathematics and Computer Science at Drew University. He has published papers in the applications of probability and stochastic processes to the modeling of turbulent transport. His other interests include the foundations and philosophy of mathematics and physics.

Steve Surace joined Drew University's Department of Mathematics and Computer Science in 1987 after earning his Ph.D. in mathematics from New York University's Courant Institute. His mathematical interests include analysis, mathematical physics and cosmology. He is also the Associate Director of the New Jersey Governor's School in the Sciences held at Drew University every summer.

1

THE SPACES $\mathbb{R}$, $\mathbb{R}^k$, AND $\mathbb{C}$

We used to think that if we knew one, we knew two, because one and one are two. We are finding that we must learn a great deal more about "and."

Sir Arthur Eddington

We begin our study of analysis with a description of those number systems and spaces in which all of our subsequent work will be done. The real spaces $\mathbb{R}$ and $\mathbb{R}^k$ for $k \geq 2$, and the complex number system $\mathbb{C}$, are all vector spaces and have many properties in common.[1] Among the real spaces, since each space $\mathbb{R}^k$ is actually a Cartesian product of k copies of $\mathbb{R}$, such similarities are not surprising. It might seem worthwhile then to consider the real spaces together, leaving the space $\mathbb{C}$ as the only special case. However, the set of real numbers $\mathbb{R}$ and the set of complex numbers $\mathbb{C}$ both distinguish themselves from the other real spaces $\mathbb{R}^k$ in a significant way: $\mathbb{R}$ and $\mathbb{C}$ are both *fields*. The space $\mathbb{R}$ is further distinguished in that it is an *ordered field*. For this reason, the real numbers $\mathbb{R}$ and the complex numbers $\mathbb{C}$ each deserve special attention. We begin by formalizing the properties of the real numbers $\mathbb{R}$. We then describe the higher-dimensional real spaces $\mathbb{R}^k$. Finally, we introduce the rich and beautiful complex number system $\mathbb{C}$.

1 THE REAL NUMBERS $\mathbb{R}$

In a course on the foundations of mathematics, or even a transition course from introductory to upper-level mathematics, one might have seen the development of number systems. Typically, such developments begin with the natural numbers $\mathbb{N} = \{1, 2, 3, \ldots\}$ and progress constructively to the real numbers $\mathbb{R}$. The first step in this progression is to supplement the natural numbers with an additive identity, 0, and each natural number's additive inverse to arrive at the integers $\mathbb{Z} = \{\ldots, -2, -1, 0, 1, 2, \ldots\}$. One then

[1]Throughout our development $\mathbb{R}^k$ will be our concise designation for the higher-dimensional real spaces, i.e., $k \geq 2$.

considers ratios of integers, thereby obtaining the rational number system $\mathbb{Q} = \left\{ \frac{p}{q} : p, q \in \mathbb{Z}, q \neq 0 \right\}$. The real number system is then shown to include $\mathbb{Q}$ as well as elements not in $\mathbb{Q}$, the so-called irrational numbers $\mathbb{I}$. That is, $\mathbb{R} = \mathbb{Q} \cup \mathbb{I}$, where $\mathbb{Q} \cap \mathbb{I} = \varnothing$. In fact, $\mathbb{N} \subset \mathbb{Z} \subset \mathbb{Q} \subset \mathbb{R}$. While it is reasonable, and even instructive, to "build up" the real number system from the simpler number systems in this way, this approach has its difficulties. For convenience, therefore, we choose instead to assume complete familiarity with the real number system $\mathbb{R}$, as well as its geometric interpretation as an infinite line of points. Our purpose in this section is to describe the main properties of $\mathbb{R}$. This summary serves as a valuable review, but it also allows us to highlight certain features of $\mathbb{R}$ that are typically left unexplored in previous mathematics courses.

1.1 Properties of the Real Numbers $\mathbb{R}$

The Field Properties

The set of real numbers $\mathbb{R}$ consists of elements that can be combined according to two binary operations: *addition* and *multiplication*. These operations are typically denoted by the symbols + and $\cdot$, respectively, so that given any two elements x and y of $\mathbb{R}$, their *sum* is denoted by $x + y$, and their *product* is denoted by $x \cdot y$, or more commonly, by xy (the $\cdot$ is usually omitted). Both the sum and the product are themselves elements of $\mathbb{R}$. Owing to this fact, we say that $\mathbb{R}$ is *closed* under addition and multiplication. The following algebraic properties of the real numbers are also known as *the field properties*, since it is exactly these properties that define a *field* as typically described in a course in abstract algebra.

1. (Addition is commutative) $x + y = y + x$ for all $x, y \in \mathbb{R}$.
2. (Addition is associative) $(x + y) + z = x + (y + z)$ for all $x, y, z \in \mathbb{R}$.
3. (Additive identity) There exists a unique element, denoted by $0 \in \mathbb{R}$, such that $x + 0 = x$ for all $x \in \mathbb{R}$.
4. (Additive inverse) For each $x \in \mathbb{R}$, there exists a unique element $-x \in \mathbb{R}$ such that $x + (-x) = 0$.
5. (Multiplication is commutative) $xy = yx$ for all $x, y \in \mathbb{R}$.
6. (Multiplication is associative) $(xy)z = x(yz)$ for all $x, y, z \in \mathbb{R}$.
7. (Multiplicative identity) There exists a unique element, denoted by $1 \in \mathbb{R}$, such that $1x = x$ for all $x \in \mathbb{R}$.
8. (Multiplicative inverse) For each nonzero $x \in \mathbb{R}$, there exists a unique element $x^{-1} \in \mathbb{R}$ satisfying $xx^{-1} = 1$.
9. (Distributive property) $x(y + z) = xy + xz$ for all $x, y, z \in \mathbb{R}$.

With the above properties of addition and multiplication, and the notions of additive and multiplicative inverse as illustrated in properties 4 and 8, we can easily define the familiar operations of subtraction and division. In particular, *subtraction* of two real numbers, a and b, is denoted by the relation $a - b$, and is given by $a - b = a + (-b)$. Similarly, *division* of two real numbers a and b is denoted by the relation a/b where it is necessary that $b \neq 0$, and is given by $a/b = ab^{-1}$. The notions of subtraction and division are conveniences only, and are not necessary for a complete theory of the real numbers.

The Order Properties

In an intuitive, geometric sense, the *order properties* of $\mathbb{R}$ are simply a precise way of expressing when one real number lies to the left (or right) of another on the real line. More formally, an *ordering* on a set S is a relation, generically denoted by $\prec$, between any two elements of the set satisfying the following two rules:

1. For any $x, y \in S$, exactly one of the following holds: $x \prec y,\ x = y,\ y \prec x$.
2. For any $x, y, z \in S$, $x \prec y$ and $y \prec z \ \Rightarrow\ x \prec z$.

If the set S is also a field, as $\mathbb{R}$ is, the ordering might also satisfy two more properties:

3. For any $x, y, z \in S$, $x \prec y \ \Rightarrow\ x + z \prec y + z$.
4. If $x, y \in S$ are such that $0 \prec x$ and $0 \prec y$, then $0 \prec xy$.

In the case where all four of the above properties are satisfied, S is called an *ordered field*.

With any ordering $\prec$ on S, it is convenient to define the relation $\succ$ according to the following convention. For x and y in S, we write $y \succ x$ if and only if $x \prec y$.

For students of calculus, the most familiar ordering on $\mathbb{R}$ is the notion of "less than" denoted by $<$. Because the ordering $<$ on $\mathbb{R}$ satisfies properties 1 through 4 above, we refer to $\mathbb{R}$ as an *ordered field*. As we will soon see, not every field with an ordering is an ordered field. That is, there are fields with orderings that do not satisfy properties 3 and 4.

The following list of facts summarizes some other useful order properties of $<$ on $\mathbb{R}$, each of which can be proved using properties 1 through 4 above. Suppose x, y, z, and w are real numbers. Then,

a) $x < y$ if and only if $0 < y - x$.
b) If $x < y$ and $w < z$, then $x + w < y + z$.
c) If $0 < x$ and $0 < y$, then $0 < x + y$.
d) If $0 < z$ and $x < y$, then $xz < yz$.
e) If $z < 0$ and $x < y$, then $yz < xz$.

f) If $0 < x$ then $0 < x^{-1}$.

g) If $x \neq 0$ then $0 < x^2$.

h) If $x < y$, then $x < \frac{(x+y)}{2} < y$.

► **1.1** *Prove properties a) through h) using order properties 1 through 4 and the field properties.*

As in the general case described above, it is convenient to define the relation $>$ on $\mathbb{R}$ according to the following rule. For x and y in $\mathbb{R}$ we will write $y > x$ if and only if $x < y$. Students will recognize $>$ as the "greater than" ordering on $\mathbb{R}$. Note also that when a real number x satisfies $0 < x$ or $x = 0$, we will write $0 \leq x$ and may refer to x as *nonnegative.* Similarly, when x satisfies $x < 0$ or $x = 0$, we will write $x \leq 0$ and may refer to x as *nonpositive.* In general, we write $x \leq y$ if either $x < y$ or $x = y$. We may also write $y \geq x$ when $x \leq y$. These notations will be convenient in what follows.

► **1.2** *A* **maximal element** *or* **maximum** *of a set $A \subset \mathbb{R}$ is an element $x \in A$ such that $a \leq x$ for all $a \in A$. Likewise, a* **minimal element** *or* **minimum** *of a set A is an element $y \in A$ such that $y \leq a$ for all $a \in A$. When they exist, we will denote a maximal element of a set A by $max\, A$, and a minimal element of a set A by $min\, A$. Can you see why a set $A \subset \mathbb{R}$ need not have a maximal or minimal element? If either a maximal or a minimal element exists for $A \subset \mathbb{R}$, show that it is unique.*

The Dedekind Completeness Property

Of all the properties of real numbers with which calculus students work, the completeness property is probably the one most taken for granted. In certain ways it is also the most challenging to formalize. After being introduced to the set $\mathbb{Q}$, one might have initially thought that it comprised the whole real line. After all, between any two rational numbers there are infinitely many other rational numbers (to see this, examine fact *(viii)* from the above list of properties carefully). However, as you have probably already discovered, $\sqrt{2}$ is a real number that isn't rational, and so the elements in $\mathbb{Q}$, when lined up along the real line, don't fill it up entirely. The irrational numbers "complete" the rational numbers by filling in these gaps in the real line. As simple and straightforward as this sounds, we must make this idea more mathematically precise. What we will refer to as the *Dedekind completeness property* accomplishes this. We state this property below.

The Dedekind Completeness Property *Suppose A and B are nonempty subsets of $\mathbb{R}$ satisfying the following:*

(i) $\mathbb{R} = A \cup B$,

(ii) $a < b$ *for every* $a \in A$ *and every* $b \in B$.

Then either there exists a maximal element of A, or there exists a minimal element of B.

This characterization of the completeness property of $\mathbb{R}$ is related to what is known more formally as a *Dedekind cut.*[2] Because of (ii) and the fact that the sets A and B referred to in the *Dedekind completeness property* form a partition of $\mathbb{R}$ (see the exercise below), we will sometimes refer to A and B as an *ordered partition* of $\mathbb{R}$. It can be shown that such an ordered partition of $\mathbb{R}$ uniquely defines (or is uniquely defined by) a single real number x, which serves as the dividing point between the subsets A and B. With respect to this dividing point x, the Dedekind completeness property of $\mathbb{R}$ is equivalent to the statement that x satisfies the following: $a \le x \le b$ for all $a \in A$ and for all $b \in B$, and x belongs to exactly one of the sets A or B.

► **1.3** *A* **partition** *of a set S is a collection of nonempty subsets $\{A_\alpha\}$ of S satisfying $\bigcup A_\alpha = S$ and $A_\alpha \cap A_\beta = \varnothing$ for $\alpha \ne \beta$. Prove that any two sets $A, B \subset \mathbb{R}$ satisfying the Dedekind completeness property also form a partition of $\mathbb{R}$.*

We illustrate the Dedekind completeness property in the following example.

Example 1.1 *Consider the sets A and B defined by the following.*

$$\begin{array}{lll} a) & A = \{x \in \mathbb{R} : x \le 0\}, & B = \{x \in \mathbb{R} : 0 < x\}. \\ b) & A = \{x \in \mathbb{R} : x < 1/2\}, & B = \{x \in \mathbb{R} : 1/2 \le x\}. \\ c) & A = \{x \in \mathbb{R} : x < 0\}, & B = \{x \in \mathbb{R} : 3/4 < x\}. \end{array}$$

The examples in a) and b) each define an ordered partition with dividing point at $x = 0$ and $x = 1/2$, respectively. In a), the set A has a maximal element, namely 0, while the set B does not have a minimal element. (How do we know that B does not have a minimal element? Suppose that it did, and call it y. Then, since y is in B, we know $0 < y$. But $0 < y/2$, and so $y/2$ is in B as well. Since $y/2 < y$, this contradicts the claim that y is the minimal element in B. Therefore, B has no minimal element.) In b), the set B has a minimal element, namely $1/2$, while the set A does not have a maximal element. Example c) does not define an ordered partition since the sets A and B exhibited there do not satisfy property (i) of our statement of the Dedekind completeness property, and therefore do not form a partition of $\mathbb{R}$. ◄

► **1.4** *In part b) of the above example, show that the set A does not have a maximal element.*

Related to the completeness property is the question of how the rational and irrational numbers are distributed on the real line. In order to investigate this idea a bit further, we take as given the following two properties. They will be of great use to us in establishing several important results.

The Archimedean Property of $\mathbb{R}$ *Given any real number $x \in \mathbb{R}$, there exists a natural number $n \in \mathbb{N}$ such that $x < n$.*

[2] So named after Richard Dedekind (1831–1916), who invented it as a way of "completing" the rational number system. Our description is not, strictly speaking, the same as Dedekind's, but it is equivalent in that it accomplishes the same end.

The Well-Ordered Property of $\mathbb{N}$ *If S is a nonempty subset of $\mathbb{N}$, then S has a minimal element.*

The meaning of each of these properties is rather subtle. On first blush, it seems that the Archimedean property is making the uncontroversial claim that there is no largest natural number. In fact, it is claiming a very significant characteristic of the *real* numbers, namely, that each real number is *finite.* When turned on its head, it also implies that there are no nonzero "infinitesimal" real numbers.[3] The meaning of the well-ordered property of $\mathbb{N}$ is also deep. In fact, although we do not prove it here, it is equivalent to the principle of mathematical induction.

► **1.5** *Establish the following:*

a) *Show that, for any positive real number x, there exists a natural number n such that $0 < \frac{1}{n} < x$. This shows that there are rational numbers arbitrarily close to zero on the real line. It also shows that there are real numbers between 0 and x.*

b) *Show that, for any real numbers x, y with x positive, there exists a natural number n such that $y < nx$.*

► **1.6** *Use the Archimedean property and the well-ordered property to show that, for any real number x, there exists an integer n such that $n - 1 \leq x < n$. To do this, consider the following cases in order: $x \in \mathbb{Z}$, $\{x \in \mathbb{R} : x > 1\}$, $\{x \in \mathbb{R} : x \geq 0\}$, $\{x \in \mathbb{R} : x < 0\}$.*

We will now use the results of the previous exercises to establish an important property of the real line. We will show that between any two real numbers there is a rational number. More specifically, we will show that for any two real numbers x and y satisfying $x < y$, there exists a rational number q such that $x < q < y$. To establish this, we will find $m \in \mathbb{N}$ and $n \in \mathbb{Z}$ such that $mx < n < my$. In a geometric sense, the effect of multiplying x and y by m is to "magnify" the space between x and y on the real line to ensure that it contains an integer value, n. To begin the proof, note that $0 < y - x$, and so by the Archimedean property there exists $m \in \mathbb{N}$ such that $0 < \frac{1}{m} < y - x$. This in turn implies that $1 < my - mx$, or $1 + mx < my$. According to a previous exercise, there exists an integer n such that $n - 1 \leq mx < n$. The right side of this double inequality is the desired $mx < n$. Adding 1 to the left side of this double inequality obtains $n \leq 1 + mx$. But we have already established that $1 + mx < my$, so we have the desired $n < my$. Overall, we have shown that there exists $m \in \mathbb{N}$ and $n \in \mathbb{Z}$ such that $mx < n < my$. Dividing by m yields $x < n/m < y$, and the result is proved.

Not only have we shown that between any two real numbers there is a rational number, it is not hard to see how our result actually implies there are in-

[3] Loosely speaking, if ξ and x are positive real numbers such that $\xi < x$, the number ξ is called *infinitesimal* if it satisfies $n\,\xi < x$ for all $n \in \mathbb{N}$. Sets with such infinitesimal elements are called non-Archimedean and are not a part of modern standard analysis. This is so despite the erroneous use of this term and a version of its associated concept among some mathematicians during the early days of analysis.

finitely many rational numbers between any two real numbers. It also proves that there was nothing special about zero two exercises ago. That is, there are rational numbers arbitrarily close to *any* real number.

▶ **1.7** *Show that if ξ is irrational and $q \neq 0$ is rational, then $q\xi$ is irrational.*

▶ **1.8** *In this exercise, we will show that for any two real numbers x and y satisfying $x < y$, there exists an irrational number ξ satisfying $x < \xi < y$. The result of this exercise implies that there are infinitely many* irrational *numbers between any two real numbers, and that there are* irrational *numbers arbitrarily close to any real number. To begin, consider the case $0 < x < y$, and make use of the previous exercise.*

1.2 The Absolute Value

We have seen several properties of real numbers, including algebraic properties, order properties, and the completeness property, that give us a fuller understanding of what the real numbers are. To understand $\mathbb{R}$ even better, we will need a tool that allows us to measure the *magnitude* of any real number. This task is accomplished by the familiar *absolute value.*

Definition 1.2 For any $x \in \mathbb{R}$, the **absolute value** of x is the nonnegative real number denoted by $|x|$ and defined by

$$|x| = \begin{cases} -x & \text{for } x < 0 \\ x & \text{for } x \geq 0 \end{cases}.$$

It is worth noting that for any pair of real numbers x and y, the difference $x - y$ is also a real number, so we may consider the magnitude $|x - y|$. Geometrically, $|x|$ can be interpreted as the "distance" from the point x to the origin, and similarly, $|x - y|$ can be interpreted as the "distance" from the point x to the point y. The idea of distance between two points in a space is an important concept in analysis, and the geometric intuition it affords will be exploited throughout what follows. The important role played by the absolute value in $\mathbb{R}$ as a way to quantify magnitudes and distances between points will be extended to $\mathbb{R}^k$ and $\mathbb{C}$ in the next chapter.

Note that a convenient, equivalent formula for computing the absolute value of a real number x is given by $|x| = \sqrt{x^2}$. To see the equivalence, consider the two cases $x \geq 0$ and $x < 0$ separately. If $x \geq 0$, then according to our definition, $|x| = x$. Also, $\sqrt{x^2} = x$, and so $|x| = \sqrt{x^2}$ for $x \geq 0$. If $x < 0$, then our definition yields $|x| = -x$, while $\sqrt{x^2} = \sqrt{(-x)^2} = -x$. Therefore, $|x| = \sqrt{x^2}$ for $x < 0$. This means for computing $|x|$ is particularly useful in proving many of the following properties of the absolute value, which we state as a proposition.

Proposition 1.3 *Absolute Value Properties*

a) $|x| \geq 0$ *for all* $x \in \mathbb{R}$, *with equality if and only if* $x = 0$.
b) $|-x| = |x|$ *for all* $x \in \mathbb{R}$.
c) $|x - y| = |y - x|$ *for all* $x, y \in \mathbb{R}$.
d) $|x\,y| = |x|\,|y|$ *for all* $x, y \in \mathbb{R}$.
e) $\left|\frac{x}{y}\right| = \frac{|x|}{|y|}$ *for all* $x, y \in \mathbb{R}$ *such that* $y \neq 0$.
f) If $c \geq 0$, *then* $|x| \leq c$ *if and only if* $-c \leq x \leq c$.
g) $-|x| \leq x \leq |x|$ *for all* $x \in \mathbb{R}$.

PROOF

a) Clearly $|x| = \sqrt{x^2} \geq 0$ for all $x \in \mathbb{R}$. Note that $|x| = 0$ if and only if $\sqrt{x^2} = 0$ if and only if $x^2 = 0$ if and only if $x = 0$.

b) $|-x| = \sqrt{(-x)^2} = \sqrt{x^2} = |x|$ for all $x \in \mathbb{R}$.

c) For all $x, y \in \mathbb{R}$, we have that $|x - y| = |-(x - y)|$ by part b), and $|-(x - y)| = |y - x|$.

d) For all $x, y \in \mathbb{R}$, we have $|xy| = \sqrt{(x\,y)^2} = \sqrt{x^2\,y^2} = \sqrt{x^2}\,\sqrt{y^2} = |x|\,|y|$.

e) $\left|\frac{x}{y}\right| = \sqrt{\left(\frac{x}{y}\right)^2} = \sqrt{\frac{x^2}{y^2}} = \frac{\sqrt{x^2}}{\sqrt{y^2}} = \frac{|x|}{|y|}$.

f) Note that

$$\begin{aligned}\{x \in \mathbb{R} : |x| \leq c\} &= \{x \in \mathbb{R} : |x| \leq c\} \cap (\{x \in \mathbb{R} : x \geq 0\} \cup \{x \in \mathbb{R} : x < 0\}) \\ &= \{x \in \mathbb{R} : x \geq 0, |x| \leq c\} \cup \{x \in \mathbb{R} : x < 0, |x| \leq c\} \\ &= \{x \in \mathbb{R} : x \geq 0, x \leq c\} \cup \{x \in \mathbb{R} : x < 0, -x \leq c\} \\ &= \{x \in \mathbb{R} : 0 \leq x \leq c\} \cup \{x \in \mathbb{R} : -c \leq x < 0\} \\ &= \{x \in \mathbb{R} : -c \leq x \leq c\}.\end{aligned}$$

g) We consider the proof casewise. If $x \geq 0$, then $-|x| \leq 0 \leq x = |x| \leq |x|$. If $x < 0$, then $-|x| \leq -|x| = x < 0 \leq |x|$. ♦

► **1.9** *Show that if* $c > 0$, *then* $|x| < c$ *if and only if* $-c < x < c$.

The following corollary to the previous result is especially useful. It can serve as a convenient way to prove that a real number is, in fact, zero.

Corollary 1.4 *Suppose* $x \in \mathbb{R}$ *satisfies* $|x| < \epsilon$ *for all* $\epsilon > 0$. *Then* $x = 0$.

► **1.10** *Prove the above corollary. Show also that the conclusion still holds when the condition $|x| < \epsilon$ is replaced by $|x| \le \epsilon$.*

► **1.11** *If $x < \epsilon$ for every $\epsilon > 0$, what can you conclude about x? Prove your claim. (Answer: $x \le 0$.)*

The inequalities in the following theorem are so important that they are worth stating separately from Proposition 1.3.

Theorem 1.5

a) $|x \pm y| \le |x| + |y|$ *for all* $x, y \in \mathbb{R}$. *The triangle inequality*

b) $\big|\, |x| - |y| \,\big| \le |x \pm y|$ *for all* $x, y \in \mathbb{R}$. *The reverse triangle inequality*

The results of the above theorem are sometimes stated together as

$$\big|\, |x| - |y| \,\big| \le |x \pm y| \le |x| + |y| \quad \text{for all } x, y \in \mathbb{R}.$$

PROOF We prove the "+" case of the stated "±," in both *a)* and *b)*, leaving the "−" case to the reader. First, we establish $|x + y| \le |x| + |y|$. We prove this casewise.

Case 1: If $x + y \ge 0$, then $|x + y| = x + y \le |x| + |y|$.

Case 2: If $x + y < 0$, then $|x + y| = -(x + y) = -x - y \le |x| + |y|$.

Now we establish that $\big|\, |x| - |y| \,\big| \le |x + y|$. Note by part *a)* that

$$|x| = |x + y - y| \le |x + y| + |y|,$$

which upon rearrangement gives $|x| - |y| \le |x + y|$. Similarly, we have that

$$|y| = |y + x - x| \le |y + x| + |x| = |x + y| + |x|,$$

and so $|x| - |y| \ge -|x + y|$. These results stated together yield

$$-|x + y| \le |x| - |y| \le |x + y|.$$

Finally, part *f)* of Proposition 1.3 gives $\big|\, |x| - |y| \,\big| \le |x + y|$. ◆

► **1.12** *Finish the proof of the triangle inequality. That is, show that*

$$\big|\, |x| - |y| \,\big| \le |x - y| \le |x| + |y| \quad \textit{for all } x, y \in \mathbb{R}.$$

► **1.13** *Show that* $|x - z| \le |x - y| + |y - z|$ *for all* x, y, *and* $z \in \mathbb{R}$.

A natural extension of the triangle inequality to a sum of n terms is often useful.

Corollary 1.6 *For all $n \in \mathbb{N}$ and for all $x_1, x_2, \ldots, x_n \in \mathbb{R}$,*

$$|x_1 + x_2 + x_3 + \cdots + x_n| \le |x_1| + |x_2| + |x_3| + \cdots + |x_n|.$$

► **1.14** *Prove the above corollary.*

1.3 Intervals in $\mathbb{R}$

To close out our discussion of the basic properties of the real numbers, we remind the reader of an especially convenient type of subset of $\mathbb{R}$, and one that will be extremely useful to us in our further development of analysis. Recall that an *interval* $I \subset \mathbb{R}$ is the set of all real numbers lying between two specified real numbers a and b where $a < b$. The numbers a and b are called the *endpoints* of the interval I. There are three types of intervals in $\mathbb{R}$:

1. *Closed.* Both endpoints are included in the interval, in which case we denote the interval as $I = [a, b]$.
2. *Open.* Both endpoints are excluded from the interval, in which case we denote the interval as $I = (a, b)$.
3. *Half-Open or Half-Closed.* Exactly one of the endpoints is excluded from the interval. If the excluded endpoint is a we write $I = (a, b]$. If the excluded endpoint is b we write $I = [a, b)$.

In those cases where one wishes to denote an *infinite interval* we will make use of the symbols $-\infty$ and ∞. For example, $(-\infty, 0)$ represents the set of negative real numbers, also denoted by $\mathbb{R}^-$, $(-\infty, 2]$ represents all real numbers less than or equal to 2, and $(5, \infty)$ represents the set of real numbers that are greater than 5. Note here that the symbols $-\infty$ and ∞ are not meant to indicate elements of the real number system; only that the collection of real numbers being considered has no left or right endpoint, respectively. Because $-\infty$ and ∞ are not actually real numbers, they should never be considered as included in the intervals in which they appear. For this reason, $-\infty$ and ∞ should always be accompanied by parentheses, not square brackets. The open interval $(0, 1)$ is often referred to as the *open unit interval*, and the closed interval $[0, 1]$ as the *closed unit interval*. We will also have need to refer to the *length* of an interval $I \subset \mathbb{R}$, a positive quantity we denote by $\ell(I)$. If an interval $I \subset \mathbb{R}$ has endpoints $a < b \in \mathbb{R}$, we define the length of I by $\ell(I) = b - a$. An infinite interval is understood to have infinite length.

2 The Real Spaces $\mathbb{R}^k$

Just as for the real numbers $\mathbb{R}$, we state the basic properties of elements of $\mathbb{R}^k$. Students of calculus should already be familiar with $\mathbb{R}^2$ and $\mathbb{R}^3$. In analogy with those spaces, each space $\mathbb{R}^k$ is geometrically interpretable as that space formed by laying k copies of $\mathbb{R}$ mutually perpendicular to one another with the origin being the common point of intersection. Cartesian product notation may also be used to denote $\mathbb{R}^k$, e.g., $\mathbb{R}^2 = \mathbb{R} \times \mathbb{R}$. We will see that these higher-dimensional spaces inherit much of their character from the copies of $\mathbb{R}$ that are used to construct them. Even so, they will not inherit *all* of the properties and structure possessed by $\mathbb{R}$.

Notation

Each real k-dimensional space is defined to be the set of all k-tuples of real numbers, hereafter referred to as k-dimensional *vectors*. That is,

$$\mathbb{R}^k = \Big\{(x_1, x_2, \ldots, x_k) : x_j \in \mathbb{R} \text{ for } 1 \leq j \leq k\Big\}.$$

For a given vector $(x_1, x_2, \ldots, x_k) \in \mathbb{R}^k$, the real number x_j for $1 \leq j \leq k$ is referred to as the vector's *jth coordinate* or as its *jth component*. For convenience, when the context is clear, we will denote an element of $\mathbb{R}^k$ by a single letter in bold type, such as $\mathbf{x}$, or by $[x_j]$ when reference to its coordinates is critical to the discussion. Another element of $\mathbb{R}^k$ different from $\mathbf{x}$ might be $\mathbf{y} = (y_1, y_2, \ldots, y_k)$, and so on. When there can be no confusion and we are discussing only a single element of $\mathbb{R}^2$, we may sometimes refer to it as (x, y) as is commonly done in calculus. When discussing more than one element of $\mathbb{R}^2$, we may also use subscript notation to distinguish the coordinates of distinct elements, e.g., (x_1, y_1) and (x_2, y_2).

In each space $\mathbb{R}^k$, the unit vector in the direction of the positive x_j-coordinate axis is the vector with 1 as the jth coordinate, and 0 as every other coordinate. Each such vector will be denoted by $\mathbf{e}_j$ for $1 \leq j \leq k$. (In three-dimensional Euclidean space, $\mathbf{e}_1$ is the familiar $\hat{\imath}$, for example.)

Students of linear algebra know that vectors are sometimes to be interpreted as column vectors rather than as row vectors, and that the distinction between column vectors and row vectors can be relevant to a particular discussion. If $\mathbf{x} = (x_1, x_2, \ldots, x_k)$ is a row vector in $\mathbb{R}^k$, its *transpose* is typically denoted by $\mathbf{x}^T$ or $(x_1, x_2, \ldots, x_k)^T$, and represents the column vector in $\mathbb{R}^k$. (Similarly, the transpose of a column vector is the associated row vector having the same correspondingly indexed components.) However, we prefer to avoid the notational clutter the transpose operator itself can produce. For this reason, we dispense with transpose notation and allow the context to justify each vector's being understood as a column vector or a row vector in each case.

Given these new higher-dimensional spaces and their corresponding elements, we now describe some of their basic properties.

2.1 Properties of the Real Spaces $\mathbb{R}^k$

The Algebraic Properties of $\mathbb{R}^k$

The set of points in $\mathbb{R}^k$ can be combined according to an *addition* operation similar to that of the real numbers $\mathbb{R}$. In fact, for all $\mathbf{x}, \mathbf{y} \in \mathbb{R}^k$, addition in $\mathbb{R}^k$ is defined componentwise by

$$\mathbf{x} + \mathbf{y} = (x_1, x_2, \ldots, x_k) + (y_1, y_2, \ldots, y_k) = (x_1 + y_1,\ x_2 + y_2, \ldots, x_k + y_k) \in \mathbb{R}^k.$$

Notice that the sum of two elements of $\mathbb{R}^k$ is itself an element of $\mathbb{R}^k$. That is, $\mathbb{R}^k$ is *closed* under this addition operation.

We can also define *scalar multiplication* in $\mathbb{R}^k$. For all $c \in \mathbb{R}$, and for all $\mathbf{x} \in \mathbb{R}^k$,

$$c\,\mathbf{x} = c\,(x_1, x_2, \ldots, x_k) = (c\,x_1, c\,x_2, \ldots, c\,x_k) \in \mathbb{R}^k.$$

The real number c in this product is referred to as a *scalar* in contexts involving $\mathbf{x} \in \mathbb{R}^k$. Note also that

$$c\,\mathbf{x} = c\,(x_1, x_2, \ldots, x_k) = (c\,x_1, c\,x_2, \ldots, c\,x_k) = (x_1\,c, x_2\,c, \ldots, x_k\,c) \equiv \mathbf{x}\,c \in \mathbb{R}^k$$

allows us to naturally define scalar multiplication in either order, "scalar times vector" or "vector times scalar." It should be noted that while the above property is true, namely, that $c\,\mathbf{x} = \mathbf{x}\,c$, rarely is such a product written in any other way than $c\,\mathbf{x}$.

There are some other nice, unsurprising algebraic properties associated with these two binary operations, and we list these now without proof.

1. (Addition is commutative) $\mathbf{x} + \mathbf{y} = \mathbf{y} + \mathbf{x}$ for all $\mathbf{x}, \mathbf{y} \in \mathbb{R}^k$.
2. (Addition is associative) $(\mathbf{x} + \mathbf{y}) + \mathbf{z} = \mathbf{x} + (\mathbf{y} + \mathbf{z})$ for all $\mathbf{x}, \mathbf{y}, \mathbf{z} \in \mathbb{R}^k$.
3. (Additive identity) $\mathbf{0} = (0, 0, \ldots, 0) \in \mathbb{R}^k$ is the unique element in $\mathbb{R}^k$ such that $\mathbf{x} + \mathbf{0} = \mathbf{x}$ for all $\mathbf{x} \in \mathbb{R}^k$.
4. (Additive inverse) For each $\mathbf{x} = (x_1, x_2, \ldots, x_k) \in \mathbb{R}^k$, there exists a unique element $-\mathbf{x} = (-x_1, -x_2, \ldots, -x_k) \in \mathbb{R}^k$ such that $\mathbf{x} + (-\mathbf{x}) = \mathbf{0}$.
5. (A type of commutativity) $c\mathbf{x} = \mathbf{x}c$ for all $c \in \mathbb{R}$ and for all $\mathbf{x} \in \mathbb{R}^k$.
6. (A type of associativity) $(cd)\mathbf{x} = c(d\mathbf{x})$ for all $c, d \in \mathbb{R}$ and for all $\mathbf{x} \in \mathbb{R}^k$.
7. (Scalar multiplicative identitiy) $1 \in \mathbb{R}$ is the unique element in $\mathbb{R}$ such that $1\mathbf{x} = \mathbf{x}$ for all $\mathbf{x} \in \mathbb{R}^k$.
9. (Distributive properties) For all $c, d \in \mathbb{R}$ and for all $\mathbf{x}, \mathbf{y} \in \mathbb{R}^k$, $c(\mathbf{x} + \mathbf{y}) = c\mathbf{x} + c\mathbf{y} = \mathbf{x}c + \mathbf{y}c = (\mathbf{x} + \mathbf{y})c$, and $(c + d)\,\mathbf{x} = c\mathbf{x} + d\mathbf{x} = \mathbf{x}\,(c + d)$.

► **1.15** *Prove the above properties.*

The numbering here is meant to mirror that of the analogous section in our discussion of addition and multiplication in $\mathbb{R}$, and it is here that we stress a very important distinction between $\mathbb{R}$ and $\mathbb{R}^k$ thus far: scalar multiplication as defined on $\mathbb{R}^k$ is not comparable to the multiplication operation previously defined on $\mathbb{R}$. Not only is the analog to property 8 missing, but this multiplication is not of the same type as the one in $\mathbb{R}$ in another, more fundamental way. In particular, multiplication as defined on $\mathbb{R}$ is an operation that combines two elements of $\mathbb{R}$ to get another element of $\mathbb{R}$ as a result. *Scalar* multiplication in $\mathbb{R}^k$, on the other hand, is a way to multiply one element of $\mathbb{R}$ with one element of $\mathbb{R}^k$ to get a new element of $\mathbb{R}^k$; this multiplication does *not* combine two elements of $\mathbb{R}^k$ to get another element of $\mathbb{R}^k$. Because of this, and the consequent lack of property 8 above, $\mathbb{R}^k$ *is not a field.*

Our set of elements with its addition and scalar multiplication as defined above does have some structure, however, even if it isn't as much structure as is found in a field. Any set of elements having the operations of addition and scalar multiplication with the properties listed above is called a *vector space.* So $\mathbb{R}^k$, while not a field, is a vector space. A little thought will convince you that $\mathbb{R}$ with its addition and multiplication (which also happens to be a scalar multiplication) is a vector space too, as well as a field.

Of course, the familiar notion of "subtraction" in $\mathbb{R}^k$ is defined in terms of addition in the obvious manner. In particular, for any pair of elements $\mathbf{x}$ and $\mathbf{y}$ from $\mathbb{R}^k$, we define their difference by $\mathbf{x}-\mathbf{y} = \mathbf{x}+(-\mathbf{y})$. Scalar multiplication of such a difference distributes as indicated by $c(\mathbf{x}-\mathbf{y}) = c\mathbf{x} - c\mathbf{y} = (\mathbf{x}-\mathbf{y})c$. And, unsurprisingly, $(c-d)\,\mathbf{x} = c\mathbf{x} - d\mathbf{x} = \mathbf{x}c - \mathbf{x}d = \mathbf{x}\,(c-d)$.

Order Properties of $\mathbb{R}^k$

As in $\mathbb{R}$, we can define a relation on the higher-dimensional spaces $\mathbb{R}^k$ that satisfies properties 1 and 2 listed on page 3, and so higher-dimensional spaces can possess an ordering. An example of such an ordering for $\mathbb{R}^2$ is defined as follows. Consider any element $\mathbf{y} = (y_1, y_2) \in \mathbb{R}^2$. Then any other element $\mathbf{x} = (x_1, x_2)$ from $\mathbb{R}^2$ must satisfy exactly one of the following:

1. $x_1 < y_1$
2. $x_1 = y_1$ and $x_2 < y_2$
3. $x_1 = y_1$ and $y_2 = x_2$
4. $x_1 = y_1$ and $y_2 < x_2$
5. $y_1 < x_1$

If $\mathbf{x}$ satisfies either 1 or 2 above, we will say that $\mathbf{x} < \mathbf{y}$. If $\mathbf{x}$ satisfies 3 we will say that $\mathbf{x} = \mathbf{y}$. And if $\mathbf{x}$ satisfies either 4 or 5, we will say that $\mathbf{y} < \mathbf{x}$. Clearly this definition satisfies property 1 of the ordering properties on page 3. We leave it to the reader to verify that it also satisfies property 2 there, and therefore is an ordering on $\mathbb{R}^2$.

▶ **1.16** *Verify property 2 on page 3 for the above defined relation on* $\mathbb{R}^2$*, confirming that it is, in fact, an ordering on* $\mathbb{R}^2$*. Draw a sketch of* $\mathbb{R}^2$ *and an arbitrary point* $\mathbf{y}$ *in it. What region of the plane relative to* $\mathbf{y}$ *in your sketch corresponds to those points* $\mathbf{x} \in \mathbb{R}^2$ *satisfying* $\mathbf{x} < \mathbf{y}$*? What region of the plane relative to* $\mathbf{y}$ *corresponds to those points* $\mathbf{x} \in \mathbb{R}^2$ *satisfying* $\mathbf{y} < \mathbf{x}$*?*

Note that since none of the higher-dimensional real spaces is a field, properties 3 and 4 listed on page 3 will not hold even if an ordering *is* successfully defined.

The Completeness Property of $\mathbb{R}^k$

Since $\mathbb{R}^k$ is a Cartesian product of k copies of $\mathbb{R}$, each of which is complete as described in the previous section, it might come as no surprise that each $\mathbb{R}^k$ is also complete. However, there are difficulties associated with the notion of Dedekind completeness for these higher-dimensional spaces. Fortunately, there is another way to characterize completeness for the higher-dimensional spaces $\mathbb{R}^k$. This alternate description of completeness, called the *Cauchy completeness property*, will be shown to be equivalent to the Dedekind version in the case of the real numbers $\mathbb{R}$. Once established, it will then serve as a more general description of completeness that will apply to all the spaces $\mathbb{R}^k$, and to $\mathbb{R}$ as well. For technical reasons, we postpone a discussion of the Cauchy completeness property until Chapter 3.

2.2 Inner Products and Norms on $\mathbb{R}^k$

Inner Products on $\mathbb{R}^k$

The reader will recall from a previous course in linear algebra, or even multivariable calculus, that there is a kind of product in each $\mathbb{R}^k$ that combines two elements of that space and yields a real number. The familiar "dot product" from calculus is an example of what is more generally referred to as an *inner product* (or *scalar product*). An arbitrary inner product is typically denoted by $\langle\cdot,\cdot\rangle$, so that if $\mathbf{x}$ and $\mathbf{y}$ is a pair of vectors from $\mathbb{R}^k$, the inner product of $\mathbf{x}$ and $\mathbf{y}$ is denoted by $\langle\mathbf{x},\mathbf{y}\rangle$. When an inner product has been specified for a given $\mathbb{R}^k$, the space is often called a *real inner product space*. Within a given $\mathbb{R}^k$, inner products comprise a special class of real-valued functions on pairs of vectors from that space. We begin by defining this class of functions.[4]

Definition 2.1 An **inner product** on $\mathbb{R}^k$ is a real (scalar) valued function of pairs of vectors in $\mathbb{R}^k$ such that for any elements $\mathbf{x}$, $\mathbf{y}$, and $\mathbf{z}$ from $\mathbb{R}^k$, and for c any real number (scalar), the following hold true:

1. $\langle\mathbf{x},\mathbf{x}\rangle \geq 0$, with equality if and only if $\mathbf{x} = \mathbf{0}$.
2. $\langle\mathbf{x},\mathbf{y}\rangle = \langle\mathbf{y},\mathbf{x}\rangle$.
3. $\langle\mathbf{x},\mathbf{y}+\mathbf{z}\rangle = \langle\mathbf{x},\mathbf{y}\rangle + \langle\mathbf{x},\mathbf{z}\rangle$.
4. $c\langle\mathbf{x},\mathbf{y}\rangle = \langle\mathbf{x},c\mathbf{y}\rangle$.

► **1.17** *Suppose $\langle\cdot,\cdot\rangle$ is an inner product on $\mathbb{R}^k$. Let $\mathbf{x}$ and $\mathbf{y}$ be arbitrary elements of $\mathbb{R}^k$, and c any real number. Establish the following.*

a) $c\langle\mathbf{x},\mathbf{y}\rangle = \langle c\mathbf{x},\mathbf{y}\rangle$ *b)* $\langle\mathbf{x}+\mathbf{y},\mathbf{z}\rangle = \langle\mathbf{x},\mathbf{z}\rangle + \langle\mathbf{y},\mathbf{z}\rangle$

► **1.18** *Define inner products on $\mathbb{R}$ in an analogous way to that of $\mathbb{R}^k$. Verify that ordinary multiplication in $\mathbb{R}$ is an inner product on the space $\mathbb{R}$, and so $\mathbb{R}$ with its usual multiplication operation is an inner product space.*

Many different inner products can be defined for a given $\mathbb{R}^k$, but the most common one is the inner product that students of calculus refer to as "the

[4]Even though functions will not be covered formally until Chapter 4, we rely on the reader's familiarity with functions and their properties for this discussion.

dot product." We will adopt the dot notation from calculus for this particular inner product, and we will continue to refer to it as "the dot product" in our work. In fact, *unless specified otherwise, "the dot product" is the inner product that is meant whenever we refer to "the inner product" on any space* $\mathbb{R}^k$, *as well as on* $\mathbb{R}$. It is usually the most natural inner product to use. Recall that for any $\mathbf{x} = (x_1, x_2, \dots, x_k)$ and $\mathbf{y} = (y_1, y_2, \dots, y_k)$ in $\mathbb{R}^k$, the *dot product of* $\mathbf{x}$ *and* $\mathbf{y}$ is denoted by $\mathbf{x} \cdot \mathbf{y}$ and is defined by

$$\mathbf{x} \cdot \mathbf{y} \equiv \sum_{j=1}^{k} x_j\, y_j.$$

The *dot product* is a genuine *inner product.* To verify this, we must verify that the five properties listed in Definition 2.1 hold true. We will establish property 4 and leave the rest as exercises for the reader. To establish 4, simply note that

$$\mathbf{x} \cdot (\mathbf{y} + \mathbf{z}) = \sum_{j=1}^{k} x_j\,(y_j + z_j) = \sum_{j=1}^{k} x_j\, y_j + \sum_{j=1}^{k} x_j\, z_j = \mathbf{x} \cdot \mathbf{y} + \mathbf{x} \cdot \mathbf{z}.$$

► **1.19** *Verify the other properties of Definition 2.1 for the dot product.*

A key relationship for any inner product in $\mathbb{R}^k$ is the Cauchy-Schwarz inequality. We state and prove this important result next.

Theorem 2.2 (The Cauchy-Schwarz Inequality)
For all $\mathbf{x}, \mathbf{y} \in \mathbb{R}^k$, *if* $\langle \cdot, \cdot \rangle$ *is an inner product on* $\mathbb{R}^k$, *then*

$$|\langle \mathbf{x}, \mathbf{y} \rangle| \le \langle \mathbf{x}, \mathbf{x} \rangle^{1/2}\, \langle \mathbf{y}, \mathbf{y} \rangle^{1/2}.$$

PROOF Clearly if either $\mathbf{x}$ or $\mathbf{y}$ is $\mathbf{0}$ the result is true, and so we consider only the case where $\mathbf{x} \neq \mathbf{0}$ and $\mathbf{y} \neq \mathbf{0}$. For every $\lambda \in \mathbb{R}$,

$$\langle \mathbf{x} - \lambda\, \mathbf{y},\, \mathbf{x} - \lambda\, \mathbf{y} \rangle = \langle \mathbf{x}, \mathbf{x} \rangle - 2\lambda\, \langle \mathbf{x}, \mathbf{y} \rangle + \lambda^2 \langle \mathbf{y}, \mathbf{y} \rangle. \tag{1.1}$$

The seemingly arbitrary choice[5] of $\lambda = \langle \mathbf{x}, \mathbf{y} \rangle / \langle \mathbf{y}, \mathbf{y} \rangle$ yields for the above quadratic expression

$$\langle \mathbf{x} - \lambda \mathbf{y},\, \mathbf{x} - \lambda \mathbf{y} \rangle = \langle \mathbf{x}, \mathbf{x} \rangle - 2\frac{\langle \mathbf{x}, \mathbf{y} \rangle^2}{\langle \mathbf{y}, \mathbf{y} \rangle} + \frac{\langle \mathbf{x}, \mathbf{y} \rangle^2}{\langle \mathbf{y}, \mathbf{y} \rangle} = \langle \mathbf{x}, \mathbf{x} \rangle - \frac{\langle \mathbf{x}, \mathbf{y} \rangle^2}{\langle \mathbf{y}, \mathbf{y} \rangle}. \tag{1.2}$$

Since the left side of expression (1.2) is nonnegative, we must have

$$\langle \mathbf{x}, \mathbf{x} \rangle \ge \frac{\langle \mathbf{x}, \mathbf{y} \rangle^2}{\langle \mathbf{y}, \mathbf{y} \rangle},$$

which in turn implies the result $|\langle \mathbf{x}, \mathbf{y} \rangle| \le \langle \mathbf{x}, \mathbf{x} \rangle^{1/2}\, \langle \mathbf{y}, \mathbf{y} \rangle^{1/2}$. Note that the

[5]This choice actually minimizes the quadratic expression in (1.1).

only way we have $|\langle \mathbf{x}, \mathbf{y}\rangle| = \langle \mathbf{x}, \mathbf{x}\rangle^{1/2} \langle \mathbf{y}, \mathbf{y}\rangle^{1/2}$ is when $\mathbf{x} - \lambda \mathbf{y} = \mathbf{0}$, i.e., when $\mathbf{x} = \lambda \mathbf{y}$ for the above scalar λ. ♦

Norms on $\mathbb{R}^k$

A *norm* provides a means for measuring magnitudes in a vector space and allows for an associated idea of "distance" between elements in the space. More specifically, a *norm* on a vector space is a nonnegative real-valued function that associates with each vector in the space a particular real number interpretable as the vector's "length." More than one such function might exist on a given vector space, and so the notion of "length" is norm specific. Not just any real-valued function will do, however. In order to be considered a *norm*, the function must satisfy certain properties. The most familiar norm to students of calculus is the absolute value function in $\mathbb{R}$, whose properties were reviewed in the last section. With the familiar example of the absolute value function behind us, we now define the concept of norm more generally in the higher-dimensional real spaces $\mathbb{R}^k$.

Definition 2.3 A **norm** $|\cdot|$ on $\mathbb{R}^k$ is a nonnegative real-valued function satisfying the following properties for all $\mathbf{x}, \mathbf{y} \in \mathbb{R}^k$.

1. $|\mathbf{x}| \geq 0$, with equality if and only if $\mathbf{x} = \mathbf{0}$.
2. $|c\mathbf{x}| = |c|\,|\mathbf{x}|$ for all $c \in \mathbb{R}$.
3. $|\mathbf{x} + \mathbf{y}| \leq |\mathbf{x}| + |\mathbf{y}|$.

▶ **1.20** *For a given norm* $|\cdot|$ *on* $\mathbb{R}^k$ *and any two elements* $\mathbf{x}$ *and* $\mathbf{y}$ *in* $\mathbb{R}^k$, *establish* $|\mathbf{x} - \mathbf{y}| \leq |\mathbf{x}| + |\mathbf{y}|$, *and* $\big||\mathbf{x}| - |\mathbf{y}|\big| \leq |\mathbf{x} \pm \mathbf{y}|$. *These results together with part 3 of Definition 2.3 are known as* the triangle inequality *and* the reverse triangle inequality, *as in the case for* $\mathbb{R}$ *given by Theorem 1.5 on page 9.*

▶ **1.21** *Suppose* $\mathbf{x} \in \mathbb{R}^k$ *satisfies* $|\mathbf{x}| < \epsilon$ *for all* $\epsilon > 0$. *Show that* $\mathbf{x} = \mathbf{0}$.

Once a norm is specified on a vector space, the space may be referred to as a *normed vector space.* While it is possible to define more than one norm on a given space, it turns out that if the space is an inner product space, there is a natural choice of norm with which to work. Fortunately, since the real spaces $\mathbb{R}$ and $\mathbb{R}^k$ are inner product spaces, this natural choice of norm is available to us. The norm we are referring to is called an *induced norm,* and we define it below.

Definition 2.4 For an inner product space with inner product $\langle \cdot, \cdot \rangle$, the **induced norm** is defined for each $\mathbf{x}$ in the inner product space by

$$|\mathbf{x}| = \langle \mathbf{x}, \mathbf{x}\rangle^{1/2}.$$

We now show that the induced norm is actually a norm. The first three properties are straightforward and hence left to the reader. We prove the fourth one. Let $\mathbf{x}, \mathbf{y} \in \mathbb{R}^k$. We have, by the Cauchy-Schwarz inequality:

$$\begin{aligned} |\mathbf{x}+\mathbf{y}|^2 &= \langle \mathbf{x}+\mathbf{y}, \mathbf{x}+\mathbf{y} \rangle \\ &= \langle \mathbf{x}, \mathbf{x} \rangle + 2 \langle \mathbf{x}, \mathbf{y} \rangle + \langle \mathbf{y}, \mathbf{y} \rangle \\ &\le |\mathbf{x}|^2 + 2 |\mathbf{x}| \, |\mathbf{y}| + |\mathbf{y}|^2 \\ &= (|\mathbf{x}| + |\mathbf{y}|)^2. \end{aligned}$$

Now take square roots to get the desired inequality.

▶ **1.22** *Verify the other norm properties for the induced norm.*

With the dot product as the inner product, which will usually be the case in our work, the induced norm is given by

$$|x| = \sqrt{x^2} \quad \text{for all } x \in \mathbb{R}, \tag{1.3}$$

and

$$|\mathbf{x}| = \sqrt{x_1^2 + x_2^2 + \ldots + x_k^2} \quad \text{for all } \mathbf{x} = (x_1, x_2, \ldots, x_k) \in \mathbb{R}^k. \tag{1.4}$$

If a norm is the norm induced by the dot product it will hereafter be denoted by $|\cdot|$, with other norms distinguished notationally by the use of subscripts, such as $|\cdot|_1$, for example, or by $\|\cdot\|$.[6] We now restate the Cauchy-Schwarz inequality on $\mathbb{R}^k$ in terms of the dot product and its induced norm.

Theorem 2.5 (The Cauchy-Schwarz Inequality for the Dot Product)
For all $\mathbf{x}, \mathbf{y} \in \mathbb{R}^k$, the dot product and its induced norm satisfy

$$\big| \, \mathbf{x} \cdot \mathbf{y} \big| \le |\mathbf{x}| \, |\mathbf{y}|.$$

▶ **1.23** *Verify that the absolute value function in $\mathbb{R}$ is that space's induced norm, with ordinary multiplication the relevant inner product.*

The following example illustrates another norm on $\mathbb{R}^2$, different from the dot product induced norm.

Example 2.6 *Consider the real valued function $|\cdot|_1 : \mathbb{R}^2 \to \mathbb{R}$ defined by $|\mathbf{x}|_1 = |x_1| + |x_2|$ for each $\mathbf{x} = (x_1, x_2) \in \mathbb{R}^2$. We will show that this function is, in fact, a norm on $\mathbb{R}^2$ by confirming that each of the three properties of Definition 2.3 hold true. Consider that*

$$|\mathbf{x}|_1 = |x_1| + |x_2| = 0 \ \textit{if and only if} \ \ \mathbf{x} = \mathbf{0},$$

and so property 1 is clearly satisfied. Note that

$$|c\,\mathbf{x}|_1 = |c\,x_1| + |c\,x_2| = |c|\big(|x_1| + |x_2|\big) = |c| \, |\mathbf{x}|_1,$$

[6]Such specific exceptions can be useful to illustrate situations or results that are norm dependent.

which establishes property 2. Finally, for $\mathbf{x}, \mathbf{y} \in \mathbb{R}^2$, *note that*

$$|\mathbf{x}+\mathbf{y}|_1 = \left| x_1 + y_1 \right| + \left| x_2 + y_2 \right| \le |x_1| + |y_1| + |x_2| + |y_2| = |\mathbf{x}|_1 + |\mathbf{y}|_1,$$

which is property 3. Note that the collection of vectors in $\mathbb{R}^2$ *with length* 1 *according to the norm* $|\cdot|_1$ *are those whose tips lie on the edges of a square (make a sketch), a somewhat different situation than under the usual norm in* $\mathbb{R}^2$. ◀

2.3 Intervals in $\mathbb{R}^k$

In $\mathbb{R}^k$, the notion of an interval $\mathcal{I}$ is the natural Cartesian product extension of that of an interval of the real line. In general we write

$$\mathcal{I} = I_1 \times I_2 \times \cdots \times I_k,$$

where each I_j for $1 \le j \le k$ is an interval of the real line as described in the previous section. An example of a closed interval in $\mathbb{R}^2$ is given by $[0, 1] \times [-1, 2] = \left\{(x, y) \in \mathbb{R}^2 : 0 \le x \le 1, \, -1 \le y \le 2\right\}$. In $\mathbb{R}^k$ a closed interval would be given more generally as

$$\mathcal{I} = [a_1, b_1] \times [a_2, b_2] \times \cdots \times [a_k, b_k] \text{ with } a_j < b_j,$$

for $1 \le j \le k$. To be considered *closed*, an interval in $\mathbb{R}^k$ must contain all of its endpoints a_j, b_j for $j = 1, 2, \ldots, k$. If all the endpoints are excluded, the interval is called *open*. All other cases are considered *half open* or *half closed*. Finally, if any I_j in the Cartesian product expression for $\mathcal{I}$ is infinite as described in the previous section, then $\mathcal{I}$ is also considered infinite. Whether an infinite interval is considered as open or closed depends on whether all the finite endpoints are excluded or included, respectively.

Example 2.7

a) $[0,1] \times [0,1]$ *is the closed unit interval in* $\mathbb{R}^2$, *sometimes referred to as the* closed unit square.

b) $(-\infty, \infty) \times [0, \infty)$ *is the closed upper half-plane, including the* x*-axis, in* $\mathbb{R}^2$.

c) $(0, \infty) \times (0, \infty) \times (0, \infty)$ *is the open first octant in* $\mathbb{R}^3$.

d) *The set given by* $\left\{(x, y) \in \mathbb{R}^2 : x \in [-1, 3], \, y = 0\right\}$ *is not an interval in* $\mathbb{R}^2$. ◀

Finally, we would like to characterize the "extent" of an interval in the real spaces $\mathbb{R}^k$ in a way that retains the natural notion of "length" in the case of an interval in $\mathbb{R}$. To this end, we define the *diameter* of an interval in $\mathbb{R}^k$ as follows. Let $\mathcal{I} = I_1 \times I_2 \times \cdots \times I_k$ be an interval in $\mathbb{R}^k$, with each I_j for $1 \le j \le k$ having endpoints $a_j < b_j$ in $\mathbb{R}$. We define the real number $\operatorname{diam}(\mathcal{I})$ by

$$\text{diam}(\mathcal{I}) \equiv \sqrt{(b_1 - a_1)^2 + (b_2 - a_2)^2 + \cdots + (b_k - a_k)^2}.$$

For example, the interval $\mathcal{I} = [-1, 2] \times (0, 4] \times (-3, 2) \subset \mathbb{R}^3$ has diameter given by $\text{diam}(\mathcal{I}) = \sqrt{3^2 + 4^2 + 5^2} = \sqrt{50}$. If the interval $\mathcal{I} \subset \mathbb{R}^k$ is an infinite interval, we define its diameter to be infinite. Note that this definition is, in fact, consistent with the concept of length of an interval $I \subset \mathbb{R}$, since for such an interval I having endpoints $a < b$, we have $\text{diam}(I) = \sqrt{(b-a)^2} = b - a = \ell(I)$.

3 The Complex Numbers ℂ

Having described the real spaces and their basic properties, we consider one last extension of our number systems. Real numbers in $\mathbb{R}$, and even real k-tuples in $\mathbb{R}^k$, are not enough to satisfy our mathematical needs. None of the elements from these number systems can satisfy an equation such as $x^2 + 1 = 0$. It is also a bit disappointing that the first extension of $\mathbb{R}$ to higher dimensions, namely $\mathbb{R}^2$, loses some of the nice properties that $\mathbb{R}$ possessed (namely the field properties). Fortunately, we can cleverly modify $\mathbb{R}^2$ to obtain a new number system, which remedies some of these deficiencies. In particular, the new number system will possess solutions to equations such as $x^2 + 1 = 0$, and, in fact, any other polynomial equation. It will also be a field (but not an ordered field). The new number system is the system of *complex numbers*, or more geometrically, *the complex plane* ℂ.

3.1 An Extension of $\mathbb{R}^2$

We take as our starting point the space $\mathbb{R}^2$ and all of its properties as outlined previously. One of the properties that this space lacked was a multiplication operator; a way to multiply two elements from the space and get another element from the space as a result. We now define such a multiplication operation on ordered pairs (x_1, y_1) and (x_2, y_2) from $\mathbb{R}^2$.

Definition 3.1 For any two elements (x_1, y_1) and (x_2, y_2) from $\mathbb{R}^2$, we define their product according to the following rule called **complex multiplication**:

$$(x_1,\ y_1) \cdot (x_2,\ y_2) \equiv (x_1\, x_2\ -\ y_1\, y_2,\ x_1\, y_2\ +\ x_2\, y_1).$$

Complex multiplication is a well-defined multiplication operation on the space $\mathbb{R}^2$. From now on, we will denote the version of $\mathbb{R}^2$ that includes complex multiplication by ℂ. We will distinguish the axes of ℂ from those of $\mathbb{R}^2$ by referring to the horizontal axis in ℂ as the *real axis* and referring to the vertical axis in ℂ as the *imaginary axis*.

Note that

$$(x,\ 0) + (y,\ 0) = (x + y,\ 0)$$

and

$$(x,\ 0)(y,\ 0) = (xy,\ 0),$$

and so the subset of $\mathbb{C}$ defined by $\{\ (x,\ 0) : x \in \mathbb{R}\}$ is isomorphic to the real numbers $\mathbb{R}$. In this way, the real number system can be algebraically identified with this special subset of the complex number system, and geometrically identified with the real axis of $\mathbb{C}$. Hence, not only is $\mathbb{C}$ a generalization of $\mathbb{R}^2$, it is also a natural generalization of $\mathbb{R}$. At this point, it might appear that $\mathbb{C}$ generalizes $\mathbb{R}$ in a merely notational way, through the correspondence $x \leftrightarrow (x,\ 0)$, but there is more to the generalization than mere notation.

In fact, one new to the subject might ask "If $\mathbb{R}^2$ already generalizes $\mathbb{R}$ geometrically, why bother with a new system like $\mathbb{C}$ at all?" It might initially appear that $\mathbb{R}^2$ is as good an option as $\mathbb{C}$ for generalizing $\mathbb{R}$. After all, both sets consist of ordered pairs of real numbers that seem to satisfy some of the same algebraic properties. For example, note that for any (x, y) in $\mathbb{C}$, the element $(0, 0)$ has the property that $(x, y) + (0, 0) = (x, y)$. That is, the element $(0, 0)$ is an *additive identity* in $\mathbb{C}$, just as in $\mathbb{R}^2$. Also, for each $(x, y) \in \mathbb{C}$ there exists an element $-(x, y) \in \mathbb{C}$ (namely, $-(x, y) = (-x, -y)$) such that $(x, y) + (-(x, y)) = (0, 0)$. That is, every element in $\mathbb{C}$ has an *additive inverse,* just as in $\mathbb{R}^2$. Finally, elements of the form $(x, 0) \in \mathbb{R}^2$ *seem* to generalize any $x \in \mathbb{R}$ just as well as $(x, 0) \in \mathbb{C}$, so what does $\mathbb{C}$ offer that $\mathbb{R}^2$ doesn't? Much. Recall that $\mathbb{R}^2$ is algebraically deficient in that it lacks a multiplication operation. The set $\mathbb{C}$ has such a multiplication, and like $\mathbb{R}$, has some extra algebraic structure associated with it. In fact, we will show that $\mathbb{C}$ is a field, just as $\mathbb{R}$ is. In particular, for each $(x, y) \in \mathbb{C}$ the element $(1, 0)$ has the property that $(x, y)\,(1, 0) = (x, y)$. That is, $(1, 0)$ is a *multiplicative identity* in $\mathbb{C}$. Also, for each (x, y) in $\mathbb{C}$ such that $(x, y) \neq (0, 0)$, we will see that there exists a unique $(x, y)^{-1} \in \mathbb{C}$, called the *multiplicative inverse* of (x, y), which has the property that $(x, y)\,(x, y)^{-1} = (1, 0)$. This, along with the other properties of complex multiplication, makes $\mathbb{C}$ a *field.* While the space $\mathbb{R}^2$ extends $\mathbb{R}$ geometrically, it is not a field. The space $\mathbb{C}$ extends $\mathbb{R}$ geometrically as $\mathbb{R}^2$ does, and retains the field properties that $\mathbb{R}$ originally had. This, we will soon discover, makes quite a difference.

Notation

As mentioned above, for any $(x,\ y) \in \mathbb{C}$, we have

$$(1,\ 0) \cdot (x,\ y) = (x\ -\ 0,\ y\ +\ 0) = (x,\ y),$$

and so $(1,\ 0)$ is called the *multiplicative identity* in $\mathbb{C}$.

Another element in $\mathbb{C}$ worth noting is $(0,\ 1)$. In fact, it is not hard to verify that $(0, 1)^2 = (-1, 0) = -(1, 0)$, an especially interesting result.

It is useful to introduce an alternative notation for these special elements of

$\mathbb{C}$. In particular, we denote the element $(1,\ 0)$ by 1, and the element $(0,\ 1)$ by i. Any element $(x,\ y)$ of $\mathbb{C}$ can then be written as a linear combination of 1 and i as follows:[7]

$$(x,\ y) = (x,\ 0) + (0,\ y) = x\,(1,\ 0) + y\,(0,\ 1) = x \cdot 1 + y \cdot i = x \cdot 1 + i \cdot y.$$

Of course, the 1 in this last expression is usually suppressed (as are the multiplication dots), and so for any $(x,\ y) \in \mathbb{C}$ we have that $(x,\ y) = x + i\,y$. Note that in this new notation our result $(0,\ 1)^2 = -(1,\ 0)$ becomes $i^2 = -1$. In fact, multiplication of two complex numbers in this new notation is easier to carry out than as described in the definition, since $(x_1,\ y_1) \cdot (x_2,\ y_2)$ becomes $(x_1 + i\,y_1) \cdot (x_2 + i\,y_2)$, which, by ordinary real multiplication and application of the identity $i^2 = -1$, quickly yields the result

$$\begin{aligned}(x_1 + i\,y_1) \cdot (x_2 + i\,y_2) &= (x_1\,x_2 - y_1\,y_2) + i\,(x_1\,y_2 + x_2\,y_1)\\ &= (x_1\,x_2 - y_1\,y_2,\ x_1\,y_2 + x_2\,y_1),\end{aligned}$$

thus matching the definition.

As a further convenience in working with complex numbers, we will often denote elements of $\mathbb{C}$ by a single letter. An arbitrary element of $\mathbb{C}$ is usually denoted by z. In this more compact notation we have $z = x + i\,y = (x,\ y)$, and we make the following definition.

Definition 3.2 For $z \in \mathbb{C}$, where $z = x + i\,y$, we define x to be $\mathrm{Re}(z)$, the **real part** of z, and we write $\mathrm{Re}(z) = x$. Likewise, we define y to be $\mathrm{Im}(z)$, the **imaginary part** of z, and we write $\mathrm{Im}(z) = y$.

Note in the definition above that $\mathrm{Re}(z)$ and $\mathrm{Im}(z)$ are real numbers. As we have already seen, the real numbers can be viewed as a subset of $\mathbb{C}$. Likewise, the elements $z \in \mathbb{C}$ with $\mathrm{Re}(z) = 0$ and $\mathrm{Im}(z) \neq 0$ are often referred to as *pure imaginary numbers*.

3.2 Properties of Complex Numbers

For convenience, we summarize those algebraic properties alluded to above; in particular, the properties inherited from $\mathbb{R}^2$ and supplemented by those associated with complex multiplication. In all that follows, z is an element of $\mathbb{C}$, which we assume has real and imaginary parts given by x and y, respectively. In cases where several elements of $\mathbb{C}$ are needed within the same statement, we will sometimes use subscripts to distinguish them. In general then, z_j will represent an element of $\mathbb{C}$ having real and imaginary parts given by x_j and y_j, respectively.

[7]This should not be a surprise, since a field is also a vector space. The elements $1 \equiv (1,\ 0)$ and $i \equiv (0,\ 1)$ form a basis of the two-dimensional vector space $\mathbb{C}$ over the field of real numbers as scalars.

Addition of two complex numbers z_1 and z_2 is defined as in $\mathbb{R}^2$. Namely,

$$z_1 + z_2 = (x_1 + i\,y_1) + (x_2 + i\,y_2) = (x_1 + x_2) + i\,(y_1 + y_2).$$

Note that $\mathbb{C}$ is closed under the addition operation.

Subtraction of two complex numbers z_1 and z_2 is also defined as in $\mathbb{R}^2$. That is,

$$z_1 - z_2 = (x_1 + i\,y_1) - (x_2 + i\,y_2) = (x_1 + i\,y_1) + (-x_2 - i\,y_2) = (x_1 - x_2) + i\,(y_1 - y_2).$$

Note that $\mathbb{C}$ is closed under the subtraction operation.

Scalar multiplication is defined in $\mathbb{C}$ in the same way as it is in $\mathbb{R}^2$. Namely, for $c \in \mathbb{R}$, we have

$$c\,z = c\,(x + i\,y) = c\,x + i\,c\,y.$$

Note that the result of scalar multiplication of an element of $\mathbb{C}$ is an element of $\mathbb{C}$. But this scalar multiplication is merely a special case of complex multiplication, which follows.

Complex multiplication is defined in $\mathbb{C}$ as described in the previous subsection. Since a scalar $c \in \mathbb{R}$ can be considered as an element of $\mathbb{C}$ having real part c and imaginary part 0, the notion of scalar multiplication is not really necessary, and we will no longer refer to it in the context of complex numbers.

Complex division will be described immediately after our next topic, the field properties of $\mathbb{C}$.

The Field Properties of $\mathbb{C}$

Given the operations of *addition* and *complex multiplication* as described above, we summarize their properties below.

1. (Addition is commutative) $z_1 + z_2 = z_2 + z_1$ for all $z_1, z_2 \in \mathbb{C}$.
2. (Addition is associative) $(z_1 + z_2) + z_3 = z_1 + (z_2 + z_3)$ for all $z_1, z_2, z_3 \in \mathbb{C}$.
3. (Additive identity) There exists a unique element $0 = (0 + i\,0) \in \mathbb{C}$ such that $z + 0 = z$ for all $z \in \mathbb{C}$.
4. (Additive inverse) For each $z = (x + i\,y) \in \mathbb{C}$, there exists a unique element $-z = (-x - i\,y) \in \mathbb{C}$ such that $z + (-z) = 0$.
5. (Multiplication is commutative) $z_1 z_2 = z_2 z_1$ for all $z_1, z_2 \in \mathbb{C}$.
6. (Multiplication is associative) $(z_1 z_2) z_3 = z_1 (z_2 z_3)$ for all $z_1, z_2, z_3 \in \mathbb{C}$.
7. (Multiplicative identity) There exists a unique element $1 = (1 + i\,0) \in \mathbb{C}$ such that $1\,z = z$ for all $z \in \mathbb{C}$.
8. (Multiplicative inverse) For each nonzero $z \in \mathbb{C}$, there exists a unique element $z^{-1} \in \mathbb{C}$ such that $z\,z^{-1} = 1$.
9. (Distributive property) $z_1(z_2 + z_3) = z_1 z_2 + z_1 z_3$ for all $z_1, z_2, z_3 \in \mathbb{C}$.

► **1.24** *Establish the above properties by using the definitions.*

With these properties, we see that $\mathbb{C}$ is a *field* just as $\mathbb{R}$ is. But it is worth elaborating on property 8 a bit further. Suppose $z = a + i\,b$, and $z \neq 0$. Then, writing z^{-1} as $z^{-1} = c + i\,d$ for some pair of real numbers c and d, one can easily show that $z\,z^{-1} = 1$ implies

$$c = \frac{a}{a^2 + b^2} \quad \text{and} \quad d = \frac{-b}{a^2 + b^2}.$$

That is,

$$z^{-1} = \frac{a}{a^2 + b^2} - i\,\frac{b}{a^2 + b^2}.$$

► **1.25** *Verify the above claim, and that* $z\,z^{-1} = 1$.

Division of two complex numbers is then naturally defined in terms of complex multiplication as follows,

$$\frac{z_1}{z_2} \equiv z_1\,z_2^{-1} \quad \text{for } z_2 \neq 0.$$

From this, we may conclude that $z^{-1} = \frac{1}{z}$ for $z \neq 0$.

Finally, *integer powers of complex numbers* will be of almost immediate importance in our work. To remain consistent with the corresponding definition of integer powers of real numbers, we define $z^0 \equiv 1$ for nonzero $z \in \mathbb{C}$. To make proper sense of such expressions as z^n and z^{-n} for any $n \in \mathbb{N}$, we naturally define z^n, by $z^n \equiv zz \cdots z$, where the right-hand side is the product of n copies of the complex number z. We then readily define z^{-n} for $z \neq 0$ by the expression $z^{-n} \equiv \left(z^{-1}\right)^n$.

► **1.26** *Suppose z and w are elements of $\mathbb{C}$. For $n \in \mathbb{Z}$, establish the following:*
a) $(zw)^n = z^n w^n$ *b)* $(z/w)^n = z^n/w^n$ *for* $w \neq 0$

The Order Properties of $\mathbb{C}$

Just like $\mathbb{R}^2$ from which $\mathbb{C}$ inherits its geometric character, there exist orderings on $\mathbb{C}$. In fact, any ordering on $\mathbb{R}^2$ is also an ordering on $\mathbb{C}$. However, unlike $\mathbb{R}^2$, the set $\mathbb{C}$ is also a field, and so we should determine whether properties 3 and 4 on page 3 also hold. In fact, they do not, and so even though $\mathbb{C}$ is a field with an ordering, it is *not* an ordered field. This fact might leave one somewhat dissatisfied with $\mathbb{C}$, as falling short of our goal of extending all the nice properties of $\mathbb{R}$ to a new, higher-dimensional system. However, it can be shown that any two complete ordered fields are isomorphic to each other. Therefore, if $\mathbb{C}$ had this property, it would really be $\mathbb{R}$ in disguise with nothing new to discover! We leave the details of establishing that $\mathbb{C}$ is not an ordered field as an exercise.

► **1.27** *Show that any ordering on $\mathbb{C}$ which satisfies properties 3 and 4 on page 3 cannot satisfy properties 1 and 2 on page 3. (Hint: Exploit the fact that $i^2 = -1$.)*

The Completeness Property of $\mathbb{C}$

$\mathbb{C}$ is complete, as inherited from $\mathbb{R}^2$. We say that $\mathbb{C}$ is a *complete field.* Justifying this fact more properly is postponed until the description of the *Cauchy completeness property* in Chapter 3, just as for the case of $\mathbb{R}^2$.

3.3 A Norm on $\mathbb{C}$ and the Complex Conjugate of z

A Norm on $\mathbb{C}$

As in $\mathbb{R}^k$, there are many possible norms one could define on $\mathbb{C}$, including norms induced by inner products. However, up to this point we haven't specified an inner product on $\mathbb{C}$, and so induced norms such as those associated with the dot product in $\mathbb{R}^k$ are not evident in this case.[8] Despite this, there is a natural choice of norm on $\mathbb{C}$. Recall that while $\mathbb{C}$ is algebraically different from $\mathbb{R}^2$, it is geometrically identical to it. For this reason, we will use a norm in $\mathbb{C}$ that is consistent with our choice of norm in $\mathbb{R}^2$. For each $z = x + i\,y \in \mathbb{C}$, we define the *norm* (sometimes also called the *modulus*) of z to be the real number given by

$$|z| = \sqrt{x^2 + y^2}\,. \tag{1.5}$$

Again, this choice of norm on $\mathbb{C}$ is consistent with the fact that $\mathbb{C}$ is a geometric copy of $\mathbb{R}^2$. That is, the norm of z is the length of the vector represented by the ordered pair $z = (x, y) = x + iy$. It is also worth noting that if z is real, i.e., $z = (x, 0)$ for some $x \in \mathbb{R}$, this norm on $\mathbb{C}$ yields the same result as the ordinary absolute value on $\mathbb{R}$.

▶ **1.28** *Suppose* $z \in \mathbb{C}$ *is such that* $|z| < \epsilon$ *for all* $\epsilon > 0$. *Show that* $z = 0$.

The Complex Conjugate of z

Complex conjugation will prove to be a convenient tool in various calculations, as well as in characterizing certain functions later in our studies. We formally define the complex conjugate of $z \in \mathbb{C}$ below.

Definition 3.3 For $z = x + i\,y \in \mathbb{C}$, the **complex conjugate** of z, denoted by $\overline{z}$, is an element of $\mathbb{C}$. It is given by the formula

$$\overline{z} = x - i\,y.$$

The complex conjugate $\overline{z}$ is z with its imaginary part "flipped" in sign. Geometrically this amounts to $\overline{z}$ being the reflection of z across the real axis. The

[8]It should be noted that the definition of inner product that we provided in Definition 2.1, while the most general one for the real spaces $\mathbb{R}$ and $\mathbb{R}^k$, is not the most general (or even the most common) definition of inner product on $\mathbb{C}$. Inner products on $\mathbb{C}$ are typically defined over the field of complex numbers as scalars, and the defining properties of the inner product must appropriately reflect this. In particular, properties 3 and 5 in Definition 2.1 must be properly generalized.

basic properties of complex conjugation are listed below. We leave the proof of each to the exercises.

a) $\overline{\overline{z}} = z$.

b) $\overline{z} = z \Leftrightarrow z$ is real.

c) $\overline{z} = -z \Leftrightarrow \mathrm{Re}(z) = 0$.

d) $\overline{z_1 \pm z_2} = \overline{z_1} \pm \overline{z_2}$.

e) $\overline{z_1\, z_2} = \overline{z_1}\, \overline{z_2}$.

f) $\overline{\left(\frac{z_1}{z_2}\right)} = \frac{(\overline{z_1})}{(\overline{z_2})}$ for $z_2 \neq 0$.

► **1.29** *Prove each of the above properties.*

There is another formula for our choice of norm in $\mathbb{C}$ that is often convenient to use. While equivalent to equation (1.5), this alternative formula does not refer to the real and imaginary parts of z explicitly, and is given by

$$|z| = \sqrt{z\,\overline{z}}. \tag{1.6}$$

► **1.30** *Verify the equivalence of (1.6) with (1.5).*

We list some useful properties of our adopted norm on $\mathbb{C}$ in the following theorem. Many of them are analogous to those listed in Proposition 1.3.

Proposition 3.4 *Norm Properties*

a) $|z| \geq 0$ *for all* $z \in \mathbb{C}$, *with equality if and only if* $z = 0$.

b) $|-z| = |z|$ *for all* $z \in \mathbb{C}$.

c) $|z_1 - z_2| = |z_2 - z_1|$ *for all* $z_1, z_2 \in \mathbb{C}$.

d) $|z_1 z_2| = |z_1|\,|z_2|$ *for all* $z_1, z_2 \in \mathbb{C}$.

e) $\left|\frac{z_1}{z_2}\right| = \frac{|z_1|}{|z_2|}$, *for all* z_1, *and nonzero* $z_2 \in \mathbb{C}$.

f) $|\overline{z}| = |z|$ *for all* $z \in \mathbb{C}$.

g) $z\,\overline{z} = |z|^2$ *for all* $z \in \mathbb{C}$.

h) $z^{-1} = \frac{\overline{z}}{|z|^2}$ *for all nonzero* $z \in \mathbb{C}$.

i) $\mathrm{Re}(z) = \frac{z+\overline{z}}{2}$ *and* $\mathrm{Im}(z) = \frac{z-\overline{z}}{2i}$ *for all* $z \in \mathbb{C}$.

j) $|\mathrm{Re}(z)| \leq |z|$ *and* $|\mathrm{Im}(z)| \leq |z|$ *for all* $z \in \mathbb{C}$.

► **1.31** *Prove the properties in the above proposition.*

The reader should recognize the first five norm properties in this result as the natural extensions of those that hold for the more familiar norm used in $\mathbb{R}$, i.e., the absolute value. One more norm property is important enough to be stated on its own. It is the complex version of Theorem 1.5 on page 9.

Theorem 3.5

a) $\left| z_1 \pm z_2 \right| \le |z_1| + |z_2|$ *for all* $z_1, z_2 \in \mathbb{C}$. *The triangle inequality*

b) $\big| |z_1| - |z_2| \big| \le \left| z_1 \pm z_2 \right|$ *for all* $z_1, z_2 \in \mathbb{C}$. *The reverse triangle inequality*

Sometimes the results of the above theorem are stated together as

$$\Big| |z_1| - |z_2| \Big| \;\le\; \left| z_1 \pm z_2 \right| \;\le\; |z_1| + |z_2| \quad \text{for all } z_1, z_2 \in \mathbb{C}.$$

PROOF We first establish the "forward" triangle inequality given by the expression $\left| z_1 \pm z_2 \right| \le |z_1| + |z_2|$. Note that

$$\begin{aligned}
\left| z_1 + z_2 \right|^2 &= (z_1 + z_2)\,\overline{(z_1 + z_2)} \\
&= (z_1 + z_2)\,(\overline{z_1} + \overline{z_2}) \\
&= z_1\,\overline{z_1} + z_1\,\overline{z_2} + \overline{z_1}\,z_2 + z_2\,\overline{z_2} \\
&= |z_1|^2 + |z_2|^2 + z_1\,\overline{z_2} + \overline{(z_1\,\overline{z_2})} \\
&= |z_1|^2 + |z_2|^2 + 2\,\mathrm{Re}\,(z_1\,\overline{z_2}) \\
&\le |z_1|^2 + |z_2|^2 + 2\,|z_1\,\overline{z_2}| \\
&= |z_1|^2 + |z_2|^2 + 2\,|z_1|\,|z_2| \\
&= \big(\,|z_1| + |z_2|\big)^2.
\end{aligned}$$

Taking the square root of each side yields the result $\left| z_1 + z_2 \right| \le |z_1| + |z_2|$. For the case $\left| z_1 - z_2 \right| \le |z_1| + |z_2|$, replace z_2 with $(-z_2)$. To prove the "reverse" triangle inequality, $\big| |z_1| - |z_2| \big| \le \left| z_1 \pm z_2 \right|$, consider that

$$|z_1| = \left| (z_1 + z_2) - z_2 \right| \le \left| z_1 + z_2 \right| + |z_2|,$$

and

$$|z_2| = \left| (z_2 + z_1) - z_1 \right| \le \left| z_2 + z_1 \right| + |z_1|.$$

These two inequalities, respectively, yield $|z_1| - |z_2| \le \left| z_1 + z_2 \right|$ and $|z_2| - |z_1| \le \left| z_1 + z_2 \right|$. Together, these last two inequalities are equivalent to

$$\big| |z_1| - |z_2| \big| \le \left| z_1 + z_2 \right|.$$

For the case of $\big| |z_1| - |z_2| \big| \le \left| z_1 - z_2 \right|$, replace z_2 with $(-z_2)$. ♦

▶ **1.32** *Describe how intervals in $\mathbb{C}$ should be defined. Do so in a manner that is consistent with what has already been described for intervals in $\mathbb{R}^k$.*

3.4 Polar Notation and the Arguments of z

The norm and the complex conjugate will turn out to be especially useful concepts. In fact, they will simplify the analysis of various problems, as well as aid in the conceptual understanding of certain aspects of the complex plane. To these ends, it is worth pointing out yet another means of describing a complex number z. We've already seen two ways of depicting z (as an ordered pair, and via the i notation). We now present a notation that employs the norm in a fundamental way.

Polar Notation

As all students of calculus have learned, the use of polar coordinates in situations involving circular symmetry in the plane can greatly simplify one's work, and one's understanding. This is true whether one is working in the real plane $\mathbb{R}^2$, or the complex plane $\mathbb{C}$. For a nonzero complex number $z = x + iy$ in $\mathbb{C}$, we may consider z geometrically as a point in $\mathbb{R}^2$. Such a point has polar coordinates (r, θ) where $r = \sqrt{x^2 + y^2} = |z|$, and $\tan\theta = y/x$. Of course, not just any angle θ satisfying $\tan\theta = y/x$ will do, since the equations $x = r\cos\theta$, and $y = r\sin\theta$ must be satisfied as well. Nonetheless, for a given θ that works, note that θ is not unique because $\theta + 2\pi k$ for any $k \in \mathbb{Z}$ serves just as well. For a specified $z = x + iy \neq 0$, we refer to the set of allowable polar angles for representing z as *the argument of* z, denote it by $\arg z$, and define it as follows:

$$\arg z \equiv \{\theta \in \mathbb{R} : x = r\cos\theta,\ y = r\sin\theta\}.$$

Using any one of the allowable angles in the set $\arg z$, we have

$$z = x + i\,y = r\cos\theta + i\,r\sin\theta = r\,(\cos\theta + i\,\sin\theta). \tag{1.7}$$

To make the notation more compact, we define $e^{i\,\theta}$ and $e^{-i\,\theta}$ for $\theta \in \mathbb{R}$ as follows:

$$e^{i\,\theta} \equiv (\cos\theta + i\,\sin\theta) \quad \text{and} \quad e^{-i\,\theta} \equiv e^{i\,(-\theta)} = \cos\theta - i\,\sin\theta. \tag{1.8}$$

Note that both of these expressions yield the value one expects to obtain when $\theta = 0$. The first identity in (1.8) is commonly known as *Euler's formula.* With it, expression (1.7) becomes

$$z = re^{i\,\theta}, \quad \text{where } r = |z| \text{ and } \theta \in \arg z. \tag{1.9}$$

We will hereafter refer to the expression (1.9) as a *polar form* of z. From the above discussion, we see that any nonzero $z \in \mathbb{C}$ has exactly one possible value of r, but infinitely many valid values of θ. So there are actually infinitely many valid polar forms for any given $z \neq 0$, each corresponding to a different value of θ from the infinite set $\arg z$. This polar form notation will be used frequently in what follows.

► **1.33** *Why doesn't $z = 0$ have a polar form?*

► **1.34** *Show that if θ and ϕ are both elements of* $\arg z$ *for some nonzero $z \in \mathbb{C}$, then $\phi = \theta + 2\pi k$ for some $k \in \mathbb{Z}$.*

It is useful at this point to derive certain results involving complex numbers in polar form, which will simplify our later work. These "basic exponent rules" are the natural ones that one would hope (and expect) to be true for complex numbers in polar form, and they follow rather naturally from the exponential notation adopted via Euler's formula. The verification of many of these properties is left to the exercises.

Proposition 3.6 *In the following list of identities, θ and ϕ are elements of $\mathbb{R}$.*

a) $e^{i\theta}e^{i\phi} = e^{i(\theta+\phi)}$.

b) $\left(e^{i\theta}\right)^n = e^{in\theta}$ *for* $n \in \mathbb{N}$.

c) $\left(e^{i\theta}\right)^{-1} = e^{-i\theta}$.

d) $\left(e^{i\theta}\right)^{-n} = e^{-in\theta}$ *for* $n \in \mathbb{N}$.

e) $e^{i\theta} = e^{i\phi}$ *if and only if* $\theta = \phi + 2\pi k$ *for some* $k \in \mathbb{Z}$.

f) $\left|e^{i\theta}\right| = 1$.

g) $\overline{e^{i\theta}} = e^{-i\theta}$.

h) $\cos\theta = \frac{1}{2}(e^{i\theta} + e^{-i\theta})$, *and* $\sin\theta = \frac{1}{2i}(e^{i\theta} - e^{-i\theta})$.

PROOF We prove the first four properties and leave the rest to the reader. To establish *a)*, note that

$$\begin{aligned} e^{i\theta}e^{i\phi} &= (\cos\theta + i\,\sin\theta)\,(\cos\phi + i\,\sin\phi) \\ &= (\cos\theta\cos\phi - \sin\theta\sin\phi) + i\,(\cos\theta\sin\phi + \sin\theta\cos\phi) \\ &= \cos(\theta+\phi) + i\,\sin(\theta+\phi) \\ &= e^{i\,(\theta+\phi)}. \end{aligned}$$

For *b)*, we will prove the result inductively. The $n = 1$ case is obvious. Assume that $(e^{i\theta})^n = e^{in\theta}$. Then,

$$(e^{i\theta})^{n+1} = (e^{i\theta})^n e^{i\theta} = e^{in\theta}e^{i\theta} = e^{i\,(n\theta+\theta)} = e^{i\,(n+1)\theta}.$$

To establish *c)*, use *a)* to readily verify that $(e^{i\theta})(e^{-i\theta}) = e^{i(\theta-\theta)} = e^0 = 1$, or, since $e^{i\theta} \neq 0$,

$$(e^{i\theta})^{-1} = \frac{1}{e^{i\theta}} = \frac{1}{e^{i\theta}}\left(\frac{e^{-i\theta}}{e^{-i\theta}}\right) = \frac{e^{-i\theta}}{e^0} = e^{-i\theta}.$$

To establish *d)*, note that

$$(e^{i\theta})^{-n} = ((e^{i\theta})^{-1})^n = (e^{-i\theta})^n = (e^{i(-\theta)})^n = (e^{in(-\theta)}) = e^{-in\theta}.$$

Proofs of parts *e)* through *h)* are left to the reader. ♦

▶ **1.35** *Prove the remaining properties in the above result.*

It is worth noting that the second and fourth properties in the above list are often written as

$$\left(\cos\theta + i\,\sin\theta\right)^n = \cos(n\theta) + i\sin(n\theta) \;\text{ for } n \in \mathbb{Z}, \tag{1.10}$$

a result commonly known as *de Moivre's Formula.*

The Arguments of z : $\arg z$ and $\mathrm{Arg}(z)$

A significant aspect of polar notation which we now address more carefully is the seeming multivalued nature of $\arg z$. As described above, for any given $z = x + iy \neq 0$ located in the complex plane, if the angle subtended as measured from the positive real axis to the ray from the origin through $x + iy$ can be recorded as θ, then z may be represented in polar notation as

$$z = re^{i(\theta+2\pi\,k)} \;\text{ for any }\; k \in \mathbb{Z}, \;\text{ where }\; r = \sqrt{x^2 + y^2} = |z|.$$

For a given nonzero $z \in \mathbb{C}$ then, it might seem that $\arg z$ is a multivalued "function" of z. In fact, as previously discussed, $\arg z$ for a specified nonzero $z \in \mathbb{C}$ represents an infinite set of values. Sometimes we refer to the values in such a set as being members of the same *equivalence class.* When we refer to $\arg z$, therefore, we are referring to all of the members of this equivalence class. In order to extract a true, single-valued *function* from $\arg z$, we make the following definition.

Definition 3.7 For nonzero $z = re^{i\theta} \in \mathbb{C}$, the **principal value** or **principal branch** of $\arg z$ is denoted by $\mathrm{Arg}(z)$ and is defined to be the unique value of $\arg z$ satisfying the following condition:

$$\mathrm{Arg}(z) \equiv \Theta \in \arg z \quad \text{such that} \quad -\pi < \Theta \leq \pi.$$

By restricting $\arg z$ in the above definition to those angles within $(-\pi, \pi]$, we have made $\mathrm{Arg}(z)$ a single-valued function of z. However, our choice of restriction was clearly not the only choice. For example, restricting the argument values to $[0, 2\pi)$ would work just as well. In fact, the restriction of the argument to any particular half-open interval of length 2π is referred to as a *branch* of $\arg z$. We will always assume the *principal branch*, $\mathrm{Arg}(z)$, as defined above unless specified otherwise. Note too that the principal branch value of the argument of z is denoted by a capital Θ.

We now investigate a key distinction in how to properly handle $\arg z$ and $\mathrm{Arg}(z)$ in certain computations. Consider two complex numbers, $z_1 \neq 0$ and $z_2 \neq 0$, given by

$$z_1 = r_1 e^{i\theta_1} \quad \text{and} \quad z_2 = r_2 e^{i\theta_2},$$

where $\theta_1 \in \arg z_1$ and $\theta_2 \in \arg z_2$. Denoting the product of z_1 and z_2 by z, we have then

$$z = z_1 z_2 = \left(r_1 e^{i\theta_1}\right)\left(r_2 e^{i\theta_2}\right) = r_1 r_2 e^{i(\theta_1+\theta_2)}.$$

Note here that $|z| = |z_1 z_2| = r_1 r_2$, and $(\theta_1 + \theta_2)$ is a particular element of $\arg z = \arg z_1 z_2$. Adding the exponents that represent arguments of z_1 and z_2 to arrive at an argument for the product $z = z_1 z_2$ is valid. In fact, any choice of argument for z from $\arg z = \arg z_1 z_2 = \{(\theta_1 + \theta_2) + 2\pi n \text{ for any } n \in \mathbb{Z}\}$ would be a valid choice, and we have conveniently selected the choice corresponding to $n = 0$. In this sense, "$\arg z_1 z_2 = \arg z_1 + \arg z_2$," if we interpret + in this last expression as a setwise addition operation. That is,

$$\arg z_1 + \arg z_2 \equiv \left\{\phi_1 + \phi_2 : \phi_1 \in \arg z_1,\ \phi_2 \in \arg z_2\right\}.$$

The moral here is that when one needs an argument for a product $z = z_1 z_2$ of complex numbers, one can conveniently compute an argument of the product z by adding arguments of the factors z_1 and z_2. Unfortunately, this same "convenient" addition should not be used when dealing with *the principal branch* of the argument $\text{Arg}(z)$ (or any branch of the argument for that matter). The reason should be clear from the following example.

Example 3.8 *Let z be the product of $z_1 = -1 + i$ and $z_2 = -1$. Note then that*

$$\arg z_1 = \left\{\tfrac{3\pi}{4} + 2\pi n \text{ for } n \in \mathbb{Z}\right\} \text{ and } \text{Arg}(z_1) = \tfrac{3\pi}{4},$$

while $\arg z_2 = \{\pi + 2\pi m \text{ for } m \in \mathbb{Z}\}$ and $\text{Arg}(z_2) = \pi$. As described above, we may conveniently compute the product z as

$$z = z_1 z_2 = \left(\sqrt{2}e^{i3\pi/4}\right)\left(e^{i\pi}\right) = \sqrt{2}e^{i7\pi/4},$$

where $\frac{7\pi}{4} \in \arg z$. Note, however, that $\text{Arg}(z) = -\frac{\pi}{4}$ but $\text{Arg}(z_1) + \text{Arg}(z_2) = \frac{3\pi}{4} + \pi = \frac{7\pi}{4}$, and so $\text{Arg}(z) = \text{Arg}(z_1 z_2) \neq \text{Arg}(z_1) + \text{Arg}(z_2)$. ◀

3.5 Circles, Disks, Powers, and Roots

Circles and Disks

Polar notation gives us an especially easy way to describe circular regions in the complex plane, $\mathbb{C}$. In particular, consider the equation of a circle of radius R centered at the origin in $\mathbb{C}$. This is given in terms of complex coordinates by the simple equation $|z| = R$. In fact, recalling that $|z| = \sqrt{x^2 + y^2}$ for $z = x + iy$ makes this explicitly clear. Generalizing to a circle centered at $z_0 = x_0 + iy_0$ isn't difficult. The circle of radius R centered at $z_0 = x_0 + iy_0$ is given by ordered pairs (x, y) in the plane satisfying the equation $(x - x_0)^2 + (y - y_0)^2 = R^2$, or $\sqrt{(x - x_0)^2 + (y - y_0)^2} = R$. It is easy to verify that this is equivalent to

$$|z - z_0| = R.$$

Of course, any point $z \in \mathbb{C}$ that satisfies this equation also satisfies $z - z_0 = R\,e^{i\theta}$ for some angle θ, and so the points z on this circle are just those points satisfying

$$z = z_0 + R\,e^{i\theta},$$

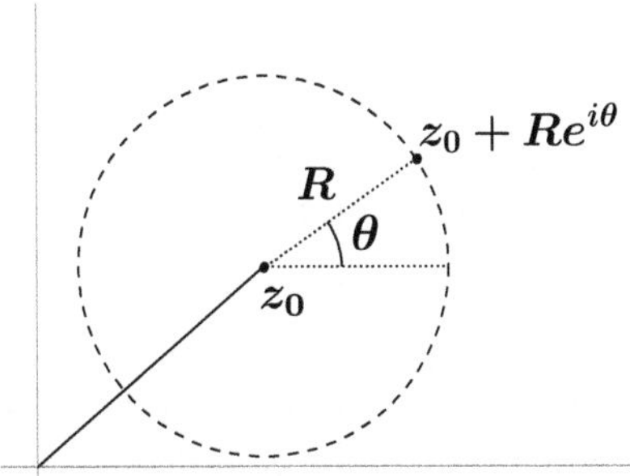

Figure 1.1 *The points z on the circle with center z_0 and radius R.*

as Figure 1.1 clearly illustrates.

Circular *regions* in the complex plane are easy to describe as well. The open disk of radius R centered at z_0 is given by the set of points z satisfying

$$|z - z_0| < R,$$

and the closed disk of radius R centered at z_0 is given by the set of points z satisfying

$$|z - z_0| \leq R.$$

Powers and Roots

With this circular geometry in mind, consider the effect of multiplying a given complex number $z = r\,e^{i\theta}$ by itself. Clearly we obtain $z^2 = r^2 e^{i2\theta}$, and so the resulting product complex number lies on a circle of radius r^2 at an angle 2θ as measured from the positive real axis. Similarly, any higher power $z^n = r^n e^{in\theta}$ for integer $n > 0$ lies on a circle of radius r^n at an angle $n\theta$ as measured from the positive real axis. If one considers the sequence of values, $z, z^2, z^3, \ldots$, one can visualize them as discrete points on a spiral that is spiraling away from the origin in the case where $r > 1$, spiraling toward the origin in the case where $r < 1$, or on a circle, circling the origin in the case where $r = 1$.

The formula z^n is just as easy to understand for negative integer powers. Suppose we wish to raise z to the $-n$ power where n is taken to be a positive integer. Recalling that z^{-n} is defined as $\left(z^{-1}\right)^n$, we have for $z = re^{i\theta}$ that $z^{-1} = \left(re^{i\theta}\right)^{-1} = r^{-1}e^{-i\theta}$, giving us

$$z^{-n} = \left(z^{-1}\right)^n = \left(r^{-1}e^{-i\theta}\right)^n = r^{-n}e^{-in\theta}.$$

The roots of a complex number are also of interest to us. How do we find them? In particular, how do we find the nth roots of a complex number z_0? It would seem natural to denote such roots as those complex numbers z satisfying

$$z = z_0^{1/n}.$$

Stated a bit differently, we seek complex numbers z satisfying

$$z^n = z_0.$$

We start by considering the special case where $z_0 = 1$. This case is referred to as finding *the nth roots of unity,* and it goes as follows. We wish to find all complex numbers z which satisfy the equation

$$z^n = 1.$$

For z having the form $z = re^{i\theta}$ this leads to

$$z^n = r^n e^{in\theta} = 1.$$

Clearly to satisfy this equation, we must have

$$r^n = 1 \text{ and } n\theta = 2\pi k \text{ for any integer } k.$$

Solving for r and θ, and considering only values of k which lead to *geometrically distinct* values of θ, we must have

$$r = 1 \text{ and } \theta = \tfrac{2\pi k}{n} \text{ for } k = 0, 1, 2, \ldots, n-1.$$

Denoting the kth nth root of unity by u_n^k, we have

$$u_n^k = e^{i\frac{2\pi k}{n}} \text{ for } k = 0, 1, 2, \ldots, n-1.$$

Note that $k = n$ corresponds to $\theta = 2\pi$, which is geometrically equivalent to $\theta = 0$, and the cycle would start over again. Therefore, there are *n nth roots of unity.* They are conveniently arranged around the unit circle, starting at $z = 1$ on the real axis and stepped along, counterclockwise, at $\frac{2\pi}{n}$ sized angular intervals as shown in Figure 1.2 for the case $n = 8$. In fact, since $e^{i\frac{2\pi}{n}} = u_n^1$, to get to the next nth root of unity from 1 we can think of multiplying 1 by u_n^1. Likewise, to get to the next root from there, we multiply u_n^1 by u_n^1 to get u_n^2. To get to the next root we again multiply by u_n^1 to get u_n^3, and so on. The sequence of roots so generated is

$$1, u_n^1, u_n^2, u_n^3, \ldots, u_n^{n-1}.$$

Each factor of u_n^1 rotates the complex number in the plane by $\frac{2\pi}{n}$ in the counterclockwise direction while leaving its magnitude equal to 1. We stop at u_n^{n-1} in our sequence since another rotation of $\frac{2\pi}{n}$ would take us back to 1.

We can generalize this idea to finding the nth roots of any complex number $z_0 = r_0 e^{i\theta_0}$. We seek all complex numbers $z = re^{i\theta}$ satisfying the equation

$$z = z_0^{1/n},$$

or

$$z^n = z_0.$$

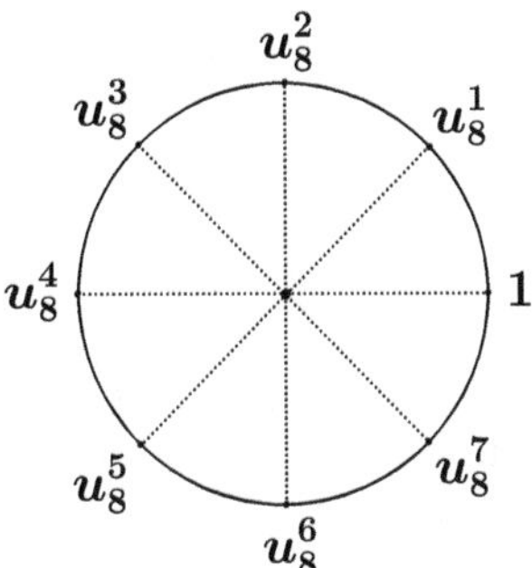

Figure 1.2 *The eighth roots of unity.*

This last equation amounts to

$$r^n e^{in\theta} = r_0 e^{i\theta_0},$$

which gives us the following two conditions to satisfy:

$$r^n = r_0, \quad \text{and} \quad n\theta = (\theta_0 + 2\pi k), \quad \text{for } k = 0, 1, 2, \ldots, n-1,$$

or

$$r = r_0^{1/n}, \quad \text{and} \quad \theta = \left(\tfrac{\theta_0}{n} + \tfrac{2\pi k}{n}\right), \quad \text{for } k = 0, 1, 2, \ldots, n-1.$$

If we denote the kth nth root of z_0 by c_n^k, we then have

$$c_n^k = r_0^{1/n} \exp\left[i\left(\tfrac{\theta_0}{n} + \tfrac{2\pi k}{n}\right)\right], \quad \text{for } k = 0, 1, 2, \ldots, n-1.$$

Example 3.9 *Consider the 6th roots of $z_0 = (1 + i\sqrt{3})$. First we convert the given z_0 to polar form, computing $z_0 = r_0 e^{i\theta_0}$, where $r_0 = \sqrt{1^2 + (\sqrt{3})^2} = 2$, and $\theta_0 = \tan^{-1}(\sqrt{3}) = \pi/3$, to obtain $z_0 = 2e^{i\frac{\pi}{3}}$. The kth 6th root of this particular z_0 is then given by*

$$c_6^k = 2^{1/6} \exp\left[i\left(\tfrac{\pi/3}{6} + \tfrac{2\pi k}{6}\right)\right] = 2^{1/6} e^{i\left(\frac{\pi}{18} + \frac{6\pi k}{18}\right)} \quad \textit{for } k = 0, 1, 2, 3, 4, 5.$$ ◀

► **1.36** *Compute* $\left(1 + i\sqrt{3}\right)^6$.

► **1.37** *Show that the n complex numbers that comprise the nth roots of z_0 are just the nth roots of unity each multiplied by the same complex factor. That is, the nth roots of z_0 are located on a circle of fixed radius centered at the origin in the complex plane.*

► **1.38** *Show that the kth nth root of z_0 is just the $(k-1)^{st}$ nth root of z_0 multiplied by u_n^1, the first complex nth root of unity. That is $c_n^k = c_n^{(k-1)} u_n^1$ for $k = 1, 2, \ldots, (n-1)$. What does this mean geometrically?*

► **1.39** *Show that every quadratic equation of the form* $a\,z^2+b\,z+c=0$ *where* $a, b, c \in \mathbb{C}$ *has at least one solution in* $\mathbb{C}$. *Also show that if* $b^2-4\,a\,c \neq 0$, *then the equation has two solutions in* $\mathbb{C}$.

3.6 Matrix Representation of Complex Numbers

So far we have seen several different methods of representing a complex number, $z \in \mathbb{C}$. These include denoting z as an ordered pair (or two-dimensional vector), using i notation, and polar notation. We present in this section yet another interpretation of z. While lacking in practical, computational benefits, it has a very nice conceptual upside. Complex numbers can be viewed as matrices.

As we saw in the previous section, multiplication of one complex number by another is geometrically equivalent to a scaling (stretching or shrinking) and a rotation in the complex plane. If we consider this process from the point of view of transformations or mappings from $\mathbb{C}$ to $\mathbb{C}$ (or $\mathbb{R}^2$ to $\mathbb{R}^2$), the complex number we use as the "multiplier," say, $z_0 = x_0 + iy_0$, takes complex numbers to complex numbers (or two-dimensional vectors to two-dimensional vectors). We already know such multiplication is a *linear* operation, since

$$z_0\,(z_1+z_2) = z_0z_1 + z_0z_2, \quad \text{and} \quad z_0\,(kz_1) = kz_0z_1, \quad \text{for } k \text{ a scalar.}$$

Therefore, this multiplication operation is a linear transformation from $\mathbb{C}$ to $\mathbb{C}$. Denote this linear transformation by $T_{z_0} : \mathbb{C} \to \mathbb{C}$ where $T_{z_0}(z) = z_0z$, and recall from linear algebra that any linear transformation from one two-dimensional vector space to another can be represented by a 2×2 matrix $\boldsymbol{A}$. In fact, given a specified basis for the vector spaces, this matrix representation is unique. In particular, consider the standard basis of $\mathbb{C}$ given by

$$\{1, i\} = \left\{ \begin{pmatrix}1\\0\end{pmatrix}, \begin{pmatrix}0\\1\end{pmatrix} \right\}.$$

Notice that we have expressed the complex number basis elements 1 and i by their corresponding two-dimensional column vectors, a notation with which students of linear algebra should be familiar. Corresponding to this standard basis, the unique standard matrix representation of T_{z_0} is given by

$$\boldsymbol{A} = \begin{bmatrix} \uparrow & \uparrow \\ T_{z_0}(1) & T_{z_0}(i) \\ \downarrow & \downarrow \end{bmatrix}.$$

Therefore, the first column of $\boldsymbol{A}$ is given by

$$T_{z_0}(1) = z_01 = x_0 + iy_0 = \begin{pmatrix}x_0\\y_0\end{pmatrix},$$

and the second column is given similarly by

$$T_{z_0}(i) = z_0i = -y_0 + ix_0 = \begin{pmatrix}-y_0\\x_0\end{pmatrix}.$$

Overall we have

$$A = \begin{bmatrix} x_0 & -y_0 \\ y_0 & x_0 \end{bmatrix}.$$

This is the unique matrix representation corresponding to T_{z_0}, and therefore to z_0, in the standard basis. This establishes a one-to-one correspondence between the complex numbers and the set of 2×2 matrices with real entries of the form $\begin{bmatrix} a & -b \\ b & a \end{bmatrix}$, i.e., we have

$$z = a + ib \quad \longleftrightarrow \quad \begin{bmatrix} a & -b \\ b & a \end{bmatrix}.$$

We will refer to the collection of such 2×2 matrices as $\mathcal{M}$.

► **1.40** *Show that for $z_0 = x_0 + iy_0 \in \mathbb{C}$, the product $z_0 z$ can be computed via the matrix product $\begin{bmatrix} x_0 & -y_0 \\ y_0 & x_0 \end{bmatrix} \begin{pmatrix} x \\ y \end{pmatrix}$ for any $z = x + iy \in \mathbb{C}$.*

► **1.41** *Verify that any matrix in $\mathcal{M}$ represents a scaling and rotation of the vector it multiplies.*

► **1.42** *Show that $\mathcal{M}$ forms a two-dimensional vector space, and that the mapping that takes elements of $\mathbb{C}$ to elements of $\mathcal{M}$ is a one-to-one correspondence. This shows that the class $\mathcal{M}$ of 2×2 matrices is* isomorphic *to the vector space of complex numbers.*

In this chapter we have focused primarily on the basic algebraic properties of the spaces $\mathbb{R}$, $\mathbb{R}^k$, and $\mathbb{C}$, with some extra attention paid to the basic geometry of complex numbers as well. In the next chapter, we set out to expand on certain shared, geometric features among all these spaces. In particular, each of these spaces is a normed space. We will see that within each such space the norm can be used to measure the "distance" between any pair of points. Since all norms share certain properties, what arises is a common framework for exploring what is referred to as the *topology* of such normed spaces. This in turn will provide the foundation for exploring certain distance-related properties of interest to us in our development of analysis, such as *open* and *closed* sets, *bounded* and *unbounded* sets, *convergence*, and *continuity*, to name just a few. While this motivation for the study of the topology of spaces is not the most general, it is the most natural one for us to pursue in our study of analysis on the Euclidean spaces of interest to us.

4 Supplementary Exercises

1. *Show that $\mathbb{Z}$ is not a field.*

2. *Consider a subset A of a field F. If A is itself a field with the addition and multiplication inherited from F, then A is called a* subfield *of F. Show that $\mathbb{Q}$ is a subfield*

of $\mathbb{R}$, and that there are no proper subfields of $\mathbb{Q}$. This means that $\mathbb{Q}$ is the smallest subfield of $\mathbb{R}$.

3. *Two fields F and G are said to be isomorphic if there exists a mapping $f : F \to G$ that is one-to-one and onto such that for all $a, b \in F$,*

(i) $f(a + b) = f(a) + f(b)$,
(ii) $f(ab) = f(a)f(b)$.

That is, f is a one-to-one correspondence between the elements of F and the elements of G that also preserves the algebraic field properties. Note that the + on the left side of the equality in (i) is the addition on F while the + on the right side is the addition on G, and similarly for the multiplications on each side of (ii), respectively. Show that $\mathbb{C}$ is not isomorphic to $\mathbb{R}$ by considering the following:

a) *Assume $f : \mathbb{C} \to \mathbb{R}$ possesses the above isomorphism properties.*

b) *Determine $f(0)$.*

c) *Determine $f(1)$.*

d) *Use the fact that $i^2 = -1$ to obtain a contradiction.*

4. *Show that for all $a, b, c \in \mathbb{R}$,*

a) $-(-a) = a$
b) $(a^{-1})^{-1} = a$ for $a \neq 0$
c) $a(-b) = (-a)(b) = -(ab)$
d) $(-a)(-b) = ab$
e) $ab = 0 \Rightarrow a = 0$ or $b = 0$
f) $a + b = a + c \Rightarrow b = c$
g) $a + b = a \Rightarrow b = 0$
h) $a + b = 0 \Rightarrow b = -a$
i) $a \neq 0$ and $ab = ac \Rightarrow b = c$
j) $a \neq 0$ and $ab = a \Rightarrow b = 1$
k) $a \neq 0$ and $ab = 1 \Rightarrow b = a^{-1}$
l) $0a = 0$

5. *Find the multiplicative inverse for each of the following complex numbers.*

a) i b) $7i$ c) $3 + i$ d) $5 - i$ e) $\sqrt{2} - \frac{\pi}{2}i$

6. *Express each of the following complex numbers in $x + i\,y$ form.*

a) $\frac{1}{3+i}$ b) $\frac{1+i}{\sqrt{2}-\frac{\pi}{2}i}$ c) $\frac{2i}{5-i}$ d) $\frac{e-\pi i}{7i}$

e) $(1 + 2i)(\pi i)$ f) $(7 + \pi i)(\pi + i)$ g) $(1 + 4i)(4 - 2i)$

7. *Suppose a and b are real numbers such that $0 \leq a \leq b$.*

a) *Show that $a^2 \leq b^2$.* b) *Show that $\sqrt{a} \leq \sqrt{b}$.*

8. *Can you define another ordering for $\mathbb{R}^2$ different from the one described on page 13? How about one defined in terms of polar coordinates?*

9. *Suppose you were to attempt to define an ordering on $\mathbb{R}^2$ in the following way: "Given two vectors $\mathbf{x}$ and $\mathbf{y}$ in $\mathbb{R}^2$, if $x_1 < y_1$ and $x_2 < y_2$ we will write $\mathbf{x} < \mathbf{y}$ and say that $\mathbf{x}$ is "less than" $\mathbf{y}$." What is wrong with this attempt?*

10. *Define an ordering for $\mathbb{R}^3$. How about more generally for $\mathbb{R}^k$?*

11. *Is there a "well-ordered property of $\mathbb{R}$" ? What about for $\mathbb{R}^k$? How about $\mathbb{C}$? (Answer: There is none. Explain this.)*

12. *Show that $\mathbb{Q}$ is not complete according to the Dedekind completeness property.*

13. *Suppose $q \in \mathbb{Q}$. If $\xi \in \mathbb{R}$ is irrational, show that $q + \xi$ is irrational.*

14. *Suppose a and b are two arbitrary real numbers. Simplify the expressions $\frac{1}{2}\big(a + b + |a - b|\big)$ and $\frac{1}{2}\big(a + b - |a - b|\big)$.*

15. *Suppose a square root of a complex number z is a complex number w such that $w^2 = z$. Find the two square roots of $1+i$. How do you know there are only two square roots? Find the only complex number having only one square root.*

16. *Find the complex solutions to the equation $z^2+z+1 = 0$. Show that every quadratic equation $az^2 + bz + c = 0$ where $a, b, c \in \mathbb{C}$ has a solution if $a \neq 0$. How is this different from the real case $ax^2 + bx + c = 0$ with $a, b, c \in \mathbb{R}$?*

17. *Define what it means for $w \in \mathbb{C}$ to be a fifth root of z. How many fifth roots does 32 have if complex numbers are allowed as roots? Graph the fifth roots of 32. What geometrical shape is associated with your graph? Why?*

18. *Express the following in $x + i\,y$ form: a) $(1 + i)^{100}$ b) $(\sqrt{3} - i)^{50}$*

19. *Suppose $z \in \mathbb{C}$ is on the circle of radius two centered at the origin. Show that $13 \leq \big|\,\overline{z} + \frac{1}{2}z^5 + 1\,\big| \leq 19$.*

20. *Let $z_i \in \mathbb{C}$ for $i = 1, 2, \ldots, m$. Show that $\big|\sum_{i=1}^m z_i\big| \leq \sum_{i=1}^m |z_i|$.*

21. *If $z \in \mathbb{C}$ is such that $|z| = 1$, compute $|1 + z|^2 + |1 - z|^2$.*

22. *If $\langle \cdot, \cdot \rangle$ is an inner product on $\mathbb{R}^k$, show that $\langle \mathbf{0}, \mathbf{x} \rangle = 0$ for all $\mathbf{x} \in \mathbb{R}^k$.*

23. *Show that for any real numbers a, b, c, d, the following is true:*

$$-\sqrt{(a^2 + b^2)(c^2 + d^2)} \leq ac - bd \leq \sqrt{(a^2 + b^2)(c^2 + d^2)}.$$

(Hint: Use the Cauchy-Schwarz inequality.)

24. *Let $\langle \cdot, \cdot \rangle$ be an inner product on $\mathbb{R}^k$, and let $B = \big\{\mathbf{x} \in \mathbb{R}^k : |\mathbf{x}| < 1\big\}$ where $|\cdot|$ is the induced norm. Show that B is convex, i.e., that for all $\mathbf{x}, \mathbf{y} \in B$, the line segment in $\mathbb{R}^k$ that joins $\mathbf{x}$ to $\mathbf{y}$ is contained in B.*

25. *Show that for all x and y in $\mathbb{R}$,*

$$|x + y|^2 + |x - y|^2 = 2\,|x|^2 + 2\,|y|^2,$$

and

$$x\,y = \tfrac{1}{4}\left[|x + y|^2 - |x - y|^2\right].$$

The first equality is called the parallelogram equality. *The second is called* the polarization identity.

26. *Consider* $\mathbb{R}^k$ *with the dot product as inner product. Show that for any* $\mathbf{x}, \mathbf{y} \in \mathbb{R}^k$,

$$|\mathbf{x}+\mathbf{y}|^2 + |\mathbf{x}-\mathbf{y}|^2 = 2\,|\mathbf{x}|^2 + 2\,|\mathbf{y}|^2$$

and

$$\mathbf{x}\cdot\mathbf{y} = \tfrac{1}{4}\Big[\,|\mathbf{x}+\mathbf{y}|^2 - |\mathbf{x}-\mathbf{y}|^2\Big].$$

These are the parallelogram equality *and* the polarization identity, *respectively, for the dot product and its induced norm on* $\mathbb{R}^k$. *The significance of these two equalities lies in the following fact, described more generally in terms of an arbitrary norm. If a norm on* $\mathbb{R}^k$ *satisfies the parallelogram equality, then an inner product that induces that norm can be defined according to the polarization identity (just replace* $\mathbf{x}\cdot\mathbf{y}$ *in the above equality with* $\langle\mathbf{x},\mathbf{y}\rangle$*). If the norm fails to satisfy the parallelogram equality, then no such inducing inner product exists for that norm.*

27. *Suppose* $\langle\cdot,\cdot\rangle$ *is an arbitrary inner product on* $\mathbb{R}^k$, *and let* $|\cdot|$ *be the associated induced norm. Show that the polarization identity and the parallelogram equality hold, i.e., for all* $\mathbf{x}, \mathbf{y} \in \mathbb{R}^k$,

$$|\mathbf{x}+\mathbf{y}|^2 + |\mathbf{x}-\mathbf{y}|^2 = 2|\mathbf{x}|^2 + 2|\mathbf{y}|^2, \text{ and } \langle\mathbf{x},\mathbf{y}\rangle = \tfrac{1}{4}\Big[\,|\mathbf{x}+\mathbf{y}|^2 - |\mathbf{x}-\mathbf{y}|^2\Big].$$

28. *For all* $\mathbf{x} = (x_1, x_2) \in \mathbb{R}^2$, *consider the norm on* $\mathbb{R}^2$ *given by* $|\mathbf{x}|_1 = |x_1| + |x_2|$. *Does this norm satisfy the parallelogram equality on* $\mathbb{R}^2$?

29. *Suppose* $|\cdot|$ *is an arbitrary norm on* $\mathbb{R}^k$ *satisfying the parallelogram equality,*

$$|\mathbf{x}+\mathbf{y}|^2 + |\mathbf{x}-\mathbf{y}|^2 = 2\,|\mathbf{x}|^2 + 2\,|\mathbf{y}|^2$$

for all $\mathbf{x}, \mathbf{y} \in \mathbb{R}^k$. *Show that the inner product defined by*

$$\langle\mathbf{x},\mathbf{y}\rangle \equiv \tfrac{1}{4}\Big[\,|\mathbf{x}+\mathbf{y}|^2 - |\mathbf{x}-\mathbf{y}|^2\Big]$$

for all $\mathbf{x}, \mathbf{y} \in \mathbb{R}^k$ *is a valid inner product that induces this norm. To do so, show*

a) $0 \le \langle\mathbf{x},\mathbf{x}\rangle = |\mathbf{x}|^2$ *for all* $\mathbf{x} \in \mathbb{R}^k$.

b) $\langle\mathbf{y},\mathbf{x}\rangle = \langle\mathbf{x},\mathbf{y}\rangle$ *for all* $\mathbf{x}, \mathbf{y} \in \mathbb{R}^k$.

c) $\langle\mathbf{x}+\mathbf{y},\mathbf{z}\rangle = \langle\mathbf{x},\mathbf{z}\rangle + \langle\mathbf{y},\mathbf{z}\rangle$ *for all* $\mathbf{x}, \mathbf{y}, \mathbf{z} \in \mathbb{R}^k$.

d) $\langle\mathbf{x},c\mathbf{y}\rangle = c\,\langle\mathbf{x},\mathbf{y}\rangle$ *for all* $\mathbf{x}, \mathbf{y} \in \mathbb{R}^k$ *and for all* $c \in \mathbb{R}$. *To establish this, consider the following special cases, in order:* $c = n \in \mathbb{Z}^+$, $c = -n \in \mathbb{Z}^-$, $c = \frac{1}{m}$ *for nonzero* $m \in \mathbb{Z}$, $c = \frac{n}{m}$ *for nonzero* $m \in \mathbb{Z}$ *and* $n \in \mathbb{Z}$, $c = r \in \mathbb{Q}$. *This step-by-step procedure proves the result for rational* c. *Finally, to establish the result for any real* c, *show that* $\big|\langle\mathbf{x},c\mathbf{y}\rangle - c\,\langle\mathbf{x},\mathbf{y}\rangle\big| < \epsilon$ *for all* $\epsilon > 0$. *To do this, consider that*

$$\big|\langle\mathbf{x},c\mathbf{y}\rangle - c\,\langle\mathbf{x},\mathbf{y}\rangle\big| = \Big|\langle\mathbf{x},(c-r)\mathbf{y}\rangle + \langle\mathbf{x},r\mathbf{y}\rangle - r\,\langle\mathbf{x},\mathbf{y}\rangle + (r-c)\,\langle\mathbf{x},\mathbf{y}\rangle\Big|,$$

where $r \in \mathbb{Q}$ *and* $|c-r|$ *can be made as small as you like.*

e) *Show that* $|\cdot|$ *is induced by this inner product.*

30. *Is the norm* $|\cdot|_1$ *described in Example 2.6 on page 17 induced by an inner product in* $\mathbb{R}^2$?

31. *What is the geometrical significance of the* parallelogram equality *in* $\mathbb{R}^2$?

32. *Consider the dot product induced norm* $|\cdot|$ *and* $|\cdot|_1$ *from Example 2.6 on page 17, both defined on* $\mathbb{R}^2$. *Show that*

$$\tfrac{1}{2}\,|\mathbf{x}|_1 \le |\mathbf{x}| \le 2\,|\mathbf{x}|_1 \quad \textit{for every } \mathbf{x} \in \mathbb{R}^2.$$

In general, any two norms $|\cdot|_1$ *and* $|\cdot|_2$ *that satisfy* $\beta^{-1}|\mathbf{x}|_1 \le |\mathbf{x}|_2 \le \beta\,|\mathbf{x}|_1$ *for* $\beta > 0$ *are called* **equivalent.** *It can be shown that any two norms on* $\mathbb{R}^k$ *are equivalent. This fact might surprise one at first. We will see in Chapter 3 that it is a very significant, and convenient truth.*

33. *For* $z, w \in \mathbb{C}$, *define the product* $\langle\cdot,\cdot\rangle : \mathbb{C}\times\mathbb{C} \to \mathbb{C}$ *by*

$$\langle z, w\rangle \equiv z\,\overline{w}.$$

Recall that the inner product properties in Definition 2.1 on page 14 were specified for real vector spaces. Is the product we define in this exercise an inner product on $\mathbb{C}$ *according to Definition 2.1 of the last subsection? If so, verify it. If not, can you think of a way to extend Definition 2.1 so as to include the product of this exercise while remaining consistent with the original definition when the vectors used are, in fact, real vectors?*

34. *Recall that* $\mathbb{C}$ *is a geometric copy of* $\mathbb{R}^2$. *Verify that the norm* $|\cdot|$ *in* $\mathbb{C}$ *defined in equation (1.6) on page 25 is just another way of writing the induced norm from* $\mathbb{R}^2$ *under the* $\mathbb{R}^2$ *dot product. Is this complex norm induced by an associated complex inner product? If so, what inner product on* $\mathbb{C}$ *induces it?*

35. *For the norm* $|\cdot|$ *in* $\mathbb{C}$ *defined by (1.6) on page 25, and the complex inner product given by* $\langle z, w\rangle \equiv z\,\overline{w}$, *show the following:*

a) $\big|\,z+w\big|^2 + \big|\,z-w\big|^2 + \big|\,z+iw\big|^2 + \big|\,z-iw\big|^2 = 4\,|z|^2 + 4\,|w|^2$

b) $\mathrm{Re}\big(\,\langle z,w\rangle\big) = \frac{1}{4}\big(\,|z+w|^2 - |z-w|^2\big)$

c) $\mathrm{Im}\big(\,\langle z,w\rangle\big) = \frac{1}{4}\big(\,|z+iw|^2 - |z-iw|^2\big)$

d) $\langle z,w\rangle = \frac{1}{4}\big(\,|z+iw|^2 - |z-iw|^2\big) + i\,\frac{1}{4}\big(\,|z+iw|^2 - |z-iw|^2\big)$

Part a) is the parallelogram equality, and part d) is the polarization identity for this complex inner product.

36. *Let* $\mathbb{C}^2 = \big\{\,(z,w) : z, w \in \mathbb{C}\big\}$, *and define*

$$\Big\langle (z_1, w_1), (z_2, w_2)\Big\rangle \equiv z_1\overline{w_1} + z_2\overline{w_2}.$$

Show that $\langle\cdot,\cdot\rangle$ *is an inner product on* $\mathbb{C}^2$ *with the complex numbers as scalars. What is the norm induced by this inner product?*

37. *Consider* $\mathbb{C}^2$ *with the inner product described in the last exercise. Find an associated Cauchy-Schwarz inequality.*

38. *Find* $\arg z$ *and* $\mathrm{Arg}(z)$ *for the following:*

a) $z = -i$ b) $z = -1 + i$ c) $z = 1 + i\sqrt{3}$

39. *For the following pairs of complex numbers, show that* $\arg z_1 z_2 = \arg z_1 + \arg z_2$.

a) $z_1 = -3,\ z_2 = 7i$ b) $z_1 = 1 - i,\ z_2 = 1 + i\sqrt{3}$ c) $z_1 = 1 - i,\ z_2 = 1 + i$

40. *For parts a), b), and c) of the previous exercise, determine whether or not* $\operatorname{Arg}(z_1 z_2) = \operatorname{Arg}(z_1) + \operatorname{Arg}(z_2)$.

41. *Suppose* $n \in \mathbb{N}, n > 1$. *Prove that*

$$\sum_{j=0}^{n-1} e^{\frac{2\pi i}{n} j} = 0.$$

(Hint: Set the value of that sum to be s*, and consider* $e^{\frac{2\pi i}{n}} s$*. Or use the geometric series.)*

2

POINT-SET TOPOLOGY

Mathematics would certainly have not come into existence if one had known from the beginning that there was in nature no exactly straight line, no actual circle, no absolute magnitude.

Friedrich Nietzsche

To someone unfamiliar with the subject, the *topology* of spaces such as $\mathbb{R}$, $\mathbb{R}^k$, or $\mathbb{C}$ can be a difficult concept to describe. Our desire in this introductory note is only to motivate our study of certain topological topics as they relate to analysis. The chapter itself will serve to make these ideas more precise. Also, it is worth emphasizing that many topological results applicable to more general spaces (that is, to spaces other than $\mathbb{R}$, $\mathbb{R}^k$, and $\mathbb{C}$) belong more properly to the subject of topology itself, not analysis, and so we will not pursue them. With these goals in mind, our motivation for the study of topology stems from the following fact: It will be useful in our work to characterize the points of a space in certain ways. In particular, if A is any subset of one of the spaces $\mathbb{R}$, $\mathbb{R}^k$, or $\mathbb{C}$, we will characterize the points of the space relative to A. That is, we will define a collection of properties that describe in some way how the points of A are configured relative to each other, and to the points of the space outside of A. We will collectively refer to the study of such properties as *point-set topology*. It is both significant and convenient that within the spaces $\mathbb{R}$, $\mathbb{R}^k$, and $\mathbb{C}$, these properties can be described in relation to the distances between points. It is just as significant and convenient that each of these spaces is a normed space, and that within each of them we can specify the distances between points using the norm on each space. This shared trait will allow us to efficiently characterize the topological features of these spaces in terms that are common to all of them. For this reason, we will use the symbol $\mathbb{X}$ throughout our discussion to represent any one of the spaces $\mathbb{R}$, $\mathbb{R}^k$, or $\mathbb{C}$ when the result to which we refer applies to each of them. If a result is unique to one of the spaces $\mathbb{R}$, $\mathbb{R}^k$, or $\mathbb{C}$ or does not apply to one of them, we will make explicit mention of that fact.

1 BOUNDED SETS

We begin by developing the concept of *distance* between elements of a space $\mathbb{X}$. We remind the reader that, as noted in the introduction to this chapter, we will generally limit our attention to the sets $\mathbb{R}$, $\mathbb{R}^k$, and $\mathbb{C}$. As we saw in the previous chapter, each of these spaces is a normed space. Since most of the results of interest to us will apply to all of them, we will let $\mathbb{X}$ be our concise notation for any (or all) of these spaces in statements that apply to them all. With this convention in mind, we note that the norm associated with a space is a convenient means for measuring distances between points in that space.[1] This motivates the following definition:

$$\textit{distance between } x \textit{ and } y \textit{ in } \mathbb{X} \equiv |\, x - y\,|.$$

Here, $|\cdot|$ is the usual norm associated with the space $\mathbb{X}$ unless otherwise specified. This distance function inherits many of the norm properties, and we list these now: For all $x, y \in \mathbb{X}$,

$$\begin{array}{ll} (i) & |\, x-y\,| \geq 0 \text{ with equality if and only if } x = y. \\ (ii) & |\, x-y\,| = |\, y-x\,|. \\ (iii) & |\, x-y\,| \leq |\, x-z\,| + |\, z-y\,| \text{ for any } z \in \mathbb{X}. \end{array}$$

To better suggest the geometric intuition associated with the notion of "distance" on normed spaces such as $\mathbb{X}$, the elements of $\mathbb{X}$ are often referred to as *points*.

1.1 Bounded Sets in $\mathbb{X}$

We will now develop certain concepts and their associated language for describing collections of points in $\mathbb{X}$. We start by defining what it means for a subset $A \subset \mathbb{X}$ to be *bounded* in $\mathbb{X}$.

Definition 1.1 The set $A \subset \mathbb{X}$ is called **bounded** if there exists a nonnegative real number M such that $|\, x\,| \leq M$ for each $x \in A$.

In the above definition, it sometimes helps to remember that $|\, x\,| = |\, x - 0\,|$ and is the distance from x to 0 in $\mathbb{X}$. With this in mind, a bounded set can be thought of as one that can be entirely contained inside a disk centered at the origin and of large enough finite radius. The following examples illustrate the above definition in the spaces of interest to us.

Example 1.2 *Let* $A = \{x \in \mathbb{R} : |\, x-4\,| < 2\}$. *Clearly* A *is bounded, since for*

[1] The function that measures distance in a given space can be defined in other ways as well. It is convenient when the distance function is induced by a norm in the space, but it is not necessary. The reader can see another kind of distance function in the supplementary exercises at the end of this chapter.

each $x \in A$ we have $|\,x\,| \le |\,x-4\,| + 4 < 6$. Letting $M = 6$ in Definition 1.1 gives the result. The set A consists of those points within $\mathbb{R}$ that are within a distance 2 of the point 4. ◀

Example 1.3 *Let $A = \left\{\frac{5n}{2n-1} : n \in \mathbb{N}\right\} \subset \mathbb{R}$. We will show that A is bounded. Note that each $x \in A$ satisfies*

$$|\,x\,| = \left|\frac{5n}{2n-1}\right| = \frac{5n}{2n-1} \le \frac{5n}{n} = 5.$$

Letting $M = 5$ in Definition 1.1 gives the result. ◀

Example 1.4 *Consider the first quadrant $Q_1 \subset \mathbb{R}^2$. We will show that Q_1 is not bounded using Definition 1.1 and a proof by contradiction. To this end, assume there exists a real number $M \ge 0$ such that*

$$|\,\mathbf{x}\,| \le M \quad \textit{for all } \mathbf{x} \in Q_1.$$

Choosing $\mathbf{x} = (n, n)$ where $n \in \mathbb{N}$, we then have

$$|\,\mathbf{x}\,| = \sqrt{2}\,n \le M \quad \textit{for all } n \in \mathbb{N}.$$

This is clearly a contradiction. (Why?) ◀

Example 1.5 *Let $A = \{z \in \mathbb{C} : |\,z\,| + |\,1-z\,| = 5\}$. We will show that A is bounded. Let z be an arbitrary element of A. Then $|\,z\,| + |\,1-z\,| = 5$. But $|\,z-1\,| \ge |\,z\,| - 1$ (Why?), and so we have*

$$|\,z\,| + |\,z\,| - 1 \le |\,z\,| + |\,1-z\,| = 5,$$

which implies $|\,z\,| \le 3$. Since $z \in A$ was arbitrary, letting $M = 3$ in Definition 1.1 yields the result. Note that the points in the set A are those on the ellipse centered at $z = 1/2$, with foci at $z = 0$ and $z = 1$, in the complex plane. One could also argue as follows. If $|z| + |1-z| = 5$, then clearly $|z| \le 5$. Hence for all $z \in A$, $|z| \le 5$. This also shows that A is bounded, although the bound M we get is larger than before. ◀

► **2.1** Determine whether the given set is bounded or unbounded.

a) $S = \{z \in \mathbb{C} : |z| + |z-1| < |z-2|\,\}$
b) $S = \{z \in \mathbb{C} : |z| + |z-1| > |z-2|\,\}$
c) $S = \left\{\mathbf{x}_n \in \mathbb{R}^3 : \mathbf{x}_n = \left(n^2 2^{-n}, \frac{n^2+1}{2n^3+1}, \frac{\cos n}{\sqrt{n}}\right)\right\}$

► **2.2** Consider a set $S \subset \mathbb{X}$.

a) If S is bounded, show that the complement of S, denoted by S^C, is unbounded.
b) If S is unbounded, show that S^C is not necessarily bounded.

► **2.3** Suppose $\{S_\alpha\} \subset \mathbb{X}$ is a collection of bounded sets.

a) Is $\bigcup S_\alpha$ bounded? b) Is $\bigcap S_\alpha$ bounded? (Answers: a) Depends, b) Yes. For a), the answer is always yes if the collection is finite.)

In the following subsection, we consider the important case of bounded sets in $\mathbb{R}$. The real numbers possess special order properties that give rise to notions of boundedness not found in $\mathbb{R}^k$ or $\mathbb{C}$.

1.2 Bounded Sets in $\mathbb{R}$

Because $\mathbb{R}$ possesses a "natural" ordering that plays a significant role in characterizing the real numbers, there are several concepts associated with boundedness in this case that require special attention.

Upper Bounds and Lower Bounds

We begin with a definition.

Definition 1.6 A set of real numbers S is said to be **bounded above** if there exists a real number M such that $s \leq M$ for all $s \in S$. Such a number M is called an **upper bound** on S.

Sets of real numbers can have more than one upper bound, or no upper bound, as the following example illustrates.

Example 1.7

a) *Consider the set $\mathbb{R}^-$ of negative real numbers. Any nonnegative real number $M \geq 0$ serves as an upper bound for $\mathbb{R}^-$.*

b) *Consider the set of all real numbers $\mathbb{R}$. This set is not bounded above. To see this, assume M is an upper bound on $\mathbb{R}$. Then, since $M + 1$ is an element of $\mathbb{R}$ greater than M, we have a contradiction. Hence, no such upper bound exists.*

c) *Suppose we generate a set of real numbers A according to the following algorithm. Let $x_1 = 1$ and $x_{n+1} = 2 + \sqrt{x_n}$ for $n \geq 1$. That is,*

$$A = \left\{1, 2 + \sqrt{1}, 2 + \sqrt{2 + \sqrt{1}}, \dots\right\}.$$

We will show that A is bounded above. To do this, we will establish that $x_n \leq 4$ for all $n \geq 1$ by induction. This is obviously true for $n = 1$. Now assume that $x_N \leq 4$. We have then,

$$x_{N+1} = 2 + \sqrt{x_N} \leq 2 + \sqrt{4} = 4,$$

and the claim is proved. ◀

A *lower bound* for a set of real numbers is defined in a manner similar to that of an upper bound.

Definition 1.8 A set of real numbers S is said to be **bounded below** if there exists a real number m such that $m \leq s$ for all $s \in S$. Such a number m is called a **lower bound** on S.

Example 1.9

a) *The set of positive real numbers $\mathbb{R}^+$ is bounded below. Any nonpositive real number $m \leq 0$ serves as a lower bound on $\mathbb{R}^+$.*

b) *The set of real numbers* $\mathbb{R}$ *is not bounded below. For any assumed lower bound* m*, consider the real number* $m - 1$.

c) *Let* $A = \left\{ \frac{(-1)^n n}{n+1} : n \in \mathbb{N} \right\}$. *To show that* A *is bounded below, we note that if* n *is even, then* $\frac{(-1)^n n}{n+1} = \frac{n}{n+1} \geq 0$*, and if* n *is odd, then* $\frac{(-1)^n n}{n+1} = -\frac{n}{n+1} \geq -1$*. Therefore,* $\frac{(-1)^n n}{n+1} \geq -1$ *for every* $n \in \mathbb{N}$. ◀

▶ **2.4** *Prove that a set of real numbers* S *is both bounded above and bounded below if and only if it is bounded.*

▶ **2.5** *Show that the set* A *described in each part c) of the previous two examples is bounded.*

The Least Upper Bound and the Greatest Lower Bound

Suppose a set of real numbers is bounded above. If we consider the collection of all upper bounds of this set, it is reasonable to ask whether there is a *least upper bound* from this collection of upper bounds.

Definition 1.10 For a nonempty set of real numbers $S \subset \mathbb{R}$, the real number M_S is called the **least upper bound,** or the **supremum,** or the **sup** of the set S, if the following both hold:

(i) M_S is an upper bound on S,
(ii) $M_S \leq M$ for all upper bounds M of S.

In this case, we write $M_S = \sup S$.

It follows at once from the definition that $\sup S$, if it exists, is unique. In fact, suppose both M_S and M_S' satisfied the above properties. Then both are upper bounds for S, and so we get $M_S \leq M_S'$ and $M_S' \leq M_S$ by the second axiom. Hence $M_S = M_S'$.

Likewise, if a set S is bounded below, it is reasonable to ask whether there is a *greatest lower bound* for S.

Definition 1.11 For a nonempty set of real numbers $S \subset \mathbb{R}$, the real number m_S is called the **greatest lower bound,** or the **infimum,** or the **inf** of the set S, if the following both hold:

(i) m_S is a lower bound on S
(ii) $m \leq m_S$ for all lower bounds m of S.

In this case, we write $m_S = \inf S$.

Similarly, $\inf S$ is unique if it exists.

Of course, merely defining the properties of a new mathematical object does not, in itself, ensure its existence. Also, it is important to note that if either the sup or the inf of a set S exists, it may or may not be a member of the set S. "Having a sup" does not necessarily mean that $\sup S$ is a member of the set S, but only that the sup *exists*. Likewise for the inf. The following important theorem tells us when we are guaranteed the existence of a sup or an inf. As the proof will show, their existence is a direct consequence of the completeness of $\mathbb{R}$.

Theorem 1.12 *If S is a nonempty set of real numbers that is bounded above, then* $\sup S$ *exists. If S is a nonempty set of real numbers that is bounded below, then* $\inf S$ *exists.*

PROOF We prove the case where S is nonempty and bounded above, leaving the bounded below case as an exercise. Let B be the set of all upper bounds on S, and let $A = B^C$. Then $\mathbb{R} = A \cup B$, and $a < b$ for all $a \in A$ and for all $b \in B$. (Why?) Since A and B satisfy our Dedekind completeness property criteria, either A has a maximal element or B has a minimal element. Suppose A has a maximal element, α. Then $\alpha \notin B$, so α is not an upper bound for S, i.e., there exists $s \in S$ such that $\alpha < s$. Since this is a strict inequality, there must exist a real number between α and s. In fact, $c \equiv (\alpha + s)/2$ is such a real number, and so $\alpha < c < s$. But $c < s$ implies $c \notin B$, i.e., $c \in A$. Since $\alpha < c$, we have a contradiction, since $\alpha \in A$ was assumed to be the largest element of A. Therefore, since our completeness criteria are satisfied, and A does not have a maximal element, it must be true that B has a minimal element, call it β. Clearly β is the least upper bound or $\sup S$. ♦

► **2.6** *Complete the proof of the above theorem. That is, suppose S is a nonempty set of real numbers that is bounded below, and show from this that* $\inf S$ *exists.*

Example 1.13 Let $S = \left\{\frac{(-1)^n n}{n+1} : n \in \mathbb{N}\right\} = \left\{-\frac{1}{2}, \frac{2}{3}, -\frac{3}{4}, \frac{4}{5}, \ldots\right\}$. *We will show that $M_S = \sup S = 1$. To do this, we begin by noting that*

$$\frac{(-1)^n n}{n+1} \le \frac{n}{n+1} \le 1 \text{ for all } n.$$

Therefore, 1 is an upper bound for S. To show that 1 is the least upper bound for S, we assume there is a smaller upper bound M such that

$$\frac{(-1)^n n}{n+1} \le M < 1 \text{ for all } n.$$

To get a contradiction, it suffices to find an even integer n such that $M < \frac{n}{n+1} < 1$. This is equivalent to finding an even integer $n > \frac{M}{1-M}$, which is certainly possible according to the Archimedean property. Therefore, $M_S = \sup S = 1$. ◄

► **2.7** *For the set S of the previous example, show that $m_S = \inf S = -1$.*

▶ **2.8** *Recall that a* maximal element *of a set S is an element $x \in S$ such that $s \leq x$ for all $s \in S$. Likewise, a* minimal element *of a set S is an element $y \in S$ such that $y \leq s$ for all $s \in S$. When they exist, we may denote a maximal element of a set S by* $\max S$, *and a minimal element of a set S by* $\min S$. *Show that if S has a maximal element, then* $\max S = \sup S$. *Likewise, show that if S has a minimal element, then* $\min S = \inf S$.

The following proposition involving properties of sups and infs will be of much use to us in our work. The proof follows readily from the definitions of supremum and infimum and is left to the reader.

Proposition 1.14 *Let $M_S = \sup S$ and let $m_S = \inf S$ for some set S of real numbers.*

a) *For every $u < M_S$ there exists $s \in S$ such that $u < s \leq M_S$.*

b) *For every $v > m_S$ there exists $s \in S$ such that $m_S \leq s < v$.*

▶ **2.9** *Prove the above proposition.*

▶ **2.10** *Use Proposition 1.14 to show that if* $\sup S$ *exists for $S \subset \mathbb{R}$, then* $\sup S$ *is unique. Do likewise for* $\inf S$.

1.3 Spheres, Balls, and Neighborhoods

Among the simplest types of bounded set, and among the most useful in analysis, are *spheres, balls,* and *neighborhoods.*

Spheres

Consider a point x_0 in $\mathbb{X}$. Now consider all other points $x \in \mathbb{X}$ that lie a distance r from x_0. We call this collection of points the *sphere of radius r centered at x_0*. More formally, we state the following definition.

Definition 1.15 For x_0 in $\mathbb{X}$, and for any $r > 0$, the collection of points $x \in \mathbb{X}$ satisfying $|\,x - x_0\,| = r$ is called the **sphere** of radius r centered at x_0.

In most cases in which we will work, the intuition provided by thinking of a sphere as the familiar object from three-dimensional geometry is a valuable aid in understanding and in problem solving. Nonetheless, spheres need not be round, even in our familiar spaces $\mathbb{R}$, $\mathbb{R}^k$, and $\mathbb{C}$. (For $r > 0$ fixed, what, after all, is the sphere of radius r in $\mathbb{R}$?)[2]

[2]In fact, since the definition of sphere depends directly on the norm being used to measure distance, and since there are many possible norms for any normed space, the intuitive notion of "sphere" can become somewhat distorted. We will not have need of such counterintuitive cases in our development of analysis, but refer to Example 2.6 on page 17 of Chapter 1 for a norm that induces a distance function with nonround spheres.

Example 1.16 *As already seen in Chapter 1, polar notation and the usual norm give us an especially easy way to describe spheres in $\mathbb{C}$ (although in Chapter 1 they were referred to as circles). Recall that we considered the equation of such a sphere of radius $R > 0$ centered at the origin in $\mathbb{C}$ by the simple equation $|\,z - 0\,| = |\,z\,| = R$. We saw that the sphere of radius R centered at $z_0 = x_0 + iy_0$ is given by points $z = x + iy$ in the plane $\mathbb{C}$ satisfying the equation*

$$|\,z - z_0\,| = R.$$

Note that any point $z \in \mathbb{C}$ that satisfies this equation also satisfies $z - z_0 = R\,e^{i\theta}$ for some angle θ (after all, the complex number $w \equiv z - z_0$ is such that its magnitude is R, and its argument is as yet unspecified), and so the points z on this sphere are those points satisfying

$$z = z_0 + R\,e^{i\theta}, \quad \text{for } 0 \le \theta < 2\pi.$$

◀

Balls

We now give the definitions of an *open ball* and a *closed ball* in $\mathbb{X}$.

Definition 1.17 Consider a point x_0 in $\mathbb{X}$, and let $r > 0$ be fixed.

1. The collection of points $x \in \mathbb{X}$ satisfying $|\,x - x_0\,| < r$ is called the **open ball** of radius r centered at x_0.
2. The collection of points $x \in \mathbb{X}$ satisfying $|\,x - x_0\,| \le r$ is called the **closed ball** of radius r centered at x_0.

Like the sphere, the intuitive notion of "ball" is directly dependent on the norm used in the space $\mathbb{X}$.

Example 1.18 *Consider the set of real numbers described by $A = \{x \in \mathbb{R} : |\,x - 1\,| < 2\}$ and $B = \{x \in \mathbb{R} : |\,x + 1\,| \le 3\}$. The set A is an open ball of radius 2 centered at $x = 1$, and the set B is a closed ball of radius 3 centered at $x = -1$.* ◀

Of course, as the previous example illustrates, open and closed balls in $\mathbb{R}$ with the usual distance are just open and closed intervals, respectively. The situation in $\mathbb{R}^k$ is similarly straightforward. In fact, using the usual distance on $\mathbb{R}^k$, the open ball of radius r centered at $\boldsymbol{\xi} = (\xi_1, \ldots, \xi_k) \in \mathbb{R}^k$ is given by those $\mathbf{x} = (x_1, \ldots, x_k) \in \mathbb{R}^k$ satisfying $|\mathbf{x} - \boldsymbol{\xi}| < r$, or, equivalently,

$$(x_1 - \xi_1)^2 + \cdots + (x_k - \xi_k)^2 < r^2,$$

and the closed ball of radius r centered at $\boldsymbol{\xi} \in \mathbb{R}^k$ is given by those $\mathbf{x} \in \mathbb{R}^k$ satisfying $|\mathbf{x} - \boldsymbol{\xi}| \le r$, or, equivalently,

$$(x_1 - \xi_1)^2 + \cdots + (x_k - \xi_k)^2 \le r^2.$$

Finally, open and closed balls in the complex plane $\mathbb{C}$ are easy to describe. The open ball of radius r centered at $z_0 \in \mathbb{C}$ is given by the set of points $z \in \mathbb{C}$ satisfying

$$|z - z_0| < r,$$

and the closed ball of radius r centered at z_0 is given by the set of points z satisfying

$$|z - z_0| \leq r.$$

Neighborhoods

For a fixed point x_0 in $\mathbb{X}$, open balls provide a precise and intuitive way to describe the part of $\mathbb{X}$ that "surrounds" the point x_0. In fact, two different types of open ball will be of interest to us in our work, and we define them now. Their use in referring to that portion of $\mathbb{X}$ that surrounds a specified point x_0 is reflected in their names—they are called *neighborhoods* of x_0.

Definition 1.19

1. A **neighborhood** of $x_0 \in \mathbb{X}$ is an open ball of radius $r > 0$ in $\mathbb{X}$ that is centered at x_0. Such a neighborhood is denoted by $N_r(x_0)$.
2. A **deleted neighborhood** of $x_0 \in \mathbb{X}$ is an open ball of radius $r > 0$ in $\mathbb{X}$ that is centered at x_0 and that excludes the point x_0 itself. Such a deleted neighborhood is denoted by $N_r'(x_0)$.

To illustrate the above definition, consider the following sets in $\mathbb{R}$:

$$N_{\frac{1}{2}}(3) = \left\{x \in \mathbb{R} : |x - 3| < \tfrac{1}{2}\right\},$$

and

$$N_{\frac{1}{2}}'(3) = \left\{x \in \mathbb{R} : 0 < |x - 3| < \tfrac{1}{2}\right\}.$$

The first set, the neighborhood of radius $\frac{1}{2}$ centered at $x = 3$, is just the open interval $\left(\frac{5}{2}, \frac{7}{2}\right)$ centered at $x = 3$. The second set, the deleted neighborhood of radius $\frac{1}{2}$ centered at $x = 3$, consists of the union $\left(\frac{5}{2}, 3\right) \cup \left(3, \frac{7}{2}\right)$. For an example in $\mathbb{R}^2$, consider the following sets relative to the point $\mathbf{x_0} = (3, 1)$:

$$\begin{aligned} N_1(\mathbf{x_0}) &= \left\{\mathbf{x} \in \mathbb{R}^2 : |\mathbf{x} - \mathbf{x_0}| < 1\right\} \\ &= \left\{(x_1, x_2) \in \mathbb{R}^2 : \sqrt{(x_1 - 3)^2 + (x_2 - 1)^2} < 1\right\}, \\ N_1'(\mathbf{x_0}) &= \left\{\mathbf{x} \in \mathbb{R}^2 : 0 < |\mathbf{x} - \mathbf{x_0}| < 1\right\} \\ &= \left\{(x_1, x_2) \in \mathbb{R}^2 : 0 < \sqrt{(x_1 - 3)^2 + (x_2 - 1)^2} < 1\right\}. \end{aligned}$$

The first set, the neighborhood of radius 1 centered at $\mathbf{x_0} = (3, 1)$, is the interior of the disk of radius 1 centered at $\mathbf{x_0}$. The second set, the deleted neighborhood of radius 1 centered at $\mathbf{x}_0 = (3, 1)$, is the "punctured" interior of the disk of radius 1 centered at $\mathbf{x}_0$ that excludes $\mathbf{x}_0$ itself.

► **2.11** *Recall that any interval of real numbers contains infinitely many rational and infinitely many irrational numbers. Therefore, if x_0 is a real number, then every neighborhood $N_r(x_0)$ centered at x_0 contains infinitely many rational and infinitely many irrational numbers. Let A and B be subsets of $\mathbb{R}^2$ defined as follows:*

$$A = \left\{(x, y) \in \mathbb{R}^2 : x, y \in \mathbb{Q}\right\} \text{ and } B = \left\{(x, y) \in \mathbb{R}^2 : x, y \in \mathbb{I}\right\}.$$

That is, A consists of those vectors from $\mathbb{R}^2$ having only rational components, and B consists of those vectors of $\mathbb{R}^2$ having only irrational components. If $\mathbf{x}_0$ is an arbitrary vector in $\mathbb{R}^2$, show that every neighborhood $N_r(\mathbf{x}_0)$ centered at $\mathbf{x}_0$ contains infinitely many elements of A and infinitely many elements of B. Investigate the comparable situation in $\mathbb{R}^k$ and in $\mathbb{C}$.

2 Classification of Points

2.1 Interior, Exterior, and Boundary Points

The concept of *neighborhood* allows us to characterize any point x_0 in $\mathbb{X}$ in a very convenient way. In particular, consider an arbitrary set $A \subset \mathbb{X}$. Then any point in $\mathbb{X}$ can be unambiguously classified according to its relation to the set A.

Definition 2.1 For a given set $A \subset \mathbb{X}$, a point $x \in \mathbb{X}$ is called

1. an **interior point** of A if there exists a neighborhood of x that is entirely contained in A.
2. an **exterior point** of A if there exists a neighborhood of x that is entirely contained in A^C.
3. a **boundary point** of A if every neighborhood of x contains a point in A and a point in A^C.

For a given set $A \subset \mathbb{X}$, each point $x \in \mathbb{X}$ must satisfy exactly one of the properties described in the previous definition; each point $x \in \mathbb{X}$ is an interior point of A, an exterior point of A, or a boundary point of A, the designation being mutually exclusive of the other two possibilities.

► **2.12** *Prove the above claim in the following two steps: First, consider an arbitrary point in $\mathbb{X}$ and show that it must satisfy at least one of the above three characterizations. Then, prove that if a point in $\mathbb{X}$ satisfies one of the above three characterizations, it cannot satisfy the other two.*

Definition 2.2 A given set $A \subset \mathbb{X}$ implicitly defines the following three subsets of $\mathbb{X}$:

1. the **boundary** of A, denoted by ∂A, and consisting of all the boundary points of A,
2. the **interior** of A, denoted by $\text{Int}(A)$, and consisting of all the interior points of A,
3. the **exterior** of A, denoted by $\text{Ext}(A)$, and consisting of all the exterior points of A.

Example 2.3 *Consider the closed ball in $\mathbb{R}^2$ centered at $\mathbf{x}_0 = (1,2)$ and given by $A = \{\mathbf{x} \in \mathbb{R}^2 : |\mathbf{x} - \mathbf{x}_0| \leq 5\}$. We will show that*

a) the point $\mathbf{x}_1 = (1,3)$ is an interior point of A.
b) the point $\mathbf{x}_2 = (7,2)$ is an exterior point of A.
c) the point $\mathbf{x}_3 = (1,7)$ is a boundary point of A.

These classifications are easily determined by considering Figure 2.1 below. However, pictures are not sufficient, and in many cases not easily drawn or interpreted, as a means for properly classifying points. In what follows, for each point, we apply the relevant definition to establish the proper characterization.

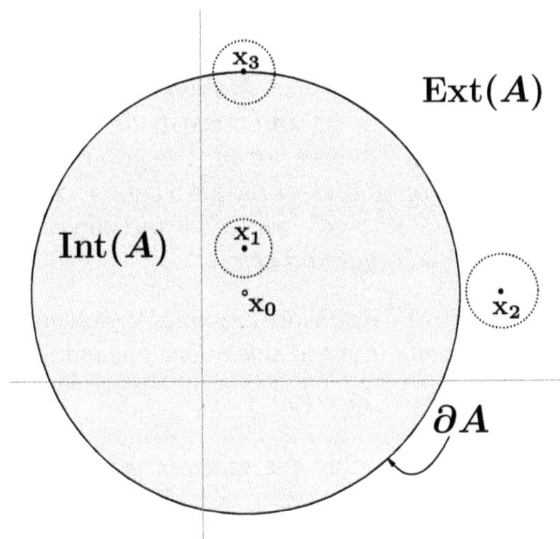

Figure 2.1 *A set with interior, exterior, and boundary points.*

To establish a), consider $N_{\frac{1}{2}}(\mathbf{x}_1)$. We will show that $N_{\frac{1}{2}}(\mathbf{x}_1) \subset A$. To see this, *let* $\mathbf{x} \in N_{\frac{1}{2}}(\mathbf{x}_1)$. *Then* $|\mathbf{x} - \mathbf{x}_1| < \frac{1}{2}$. *Now it follows that*

$$|\mathbf{x} - \mathbf{x}_0| \le |\mathbf{x} - \mathbf{x}_1| + |\mathbf{x}_1 - \mathbf{x}_0| < \tfrac{1}{2} + 1 < 5,$$

and therefore $\mathbf{x} \in A$. *To establish b), consider* $N_{\frac{1}{4}}(\mathbf{x}_2)$. *We will show that* $N_{\frac{1}{4}}(\mathbf{x}_2) \subset A^C$. *To see this, let* $\mathbf{x} \in N_{\frac{1}{4}}(\mathbf{x}_2)$. *Then* $|\mathbf{x} - \mathbf{x}_2| < \frac{1}{4}$. *From this we have*

$$\begin{aligned} |\mathbf{x} - \mathbf{x}_0| &= \left|\mathbf{x} - \mathbf{x}_2 + \mathbf{x}_2 - \mathbf{x}_0\right| \\ &= \left|\mathbf{x} - \mathbf{x}_2 + (6,0)\right| \\ &\ge 6 - |\mathbf{x} - \mathbf{x}_2| \\ &> 6 - \tfrac{1}{4} > 5. \end{aligned}$$

Therefore, $\mathbf{x} \in A^C$. *To establish c), we will show that every neighborhood centered at* $\mathbf{x}_3$ *contains at least one point in* A *and at least one point in* A^C. *To do this, we consider the case of neighborhoods having radius* $0 < r < 1$ *first. Let* $N_r(\mathbf{x}_3)$ *be an arbitrary such neighborhood centered at* $\mathbf{x}_3$. *Note that*

$$\mathbf{x} = \mathbf{x}_3 - \left(0, \tfrac{r}{2}\right) \in A \cap N_r(\mathbf{x}_3),$$

and that

$$\mathbf{x} = \mathbf{x}_3 + \left(0, \tfrac{r}{2}\right) \in A^C \cap N_r(\mathbf{x}_3).$$

Since the neighborhood $N_r(\mathbf{x}_3)$ *was an arbitrary neighborhood centered at* $\mathbf{x}_3$ *having radius* $0 < r < 1$, *all such neighborhoods must also contain at least one point in* A *and at least one point in* A^C. *We leave as an exercise the task of proving that any neighborhood centered at* $\mathbf{x}_3$ *having radius* $r \ge 1$ *also contains at least one point in* A *and at least one point in* A^C, *and that therefore* $\mathbf{x}_3$ *is a boundary point of* A. *Note that the boundary of* A *is given by* $\partial A = \{\mathbf{x} \in \mathbb{R}^2 : |\mathbf{x} - \mathbf{x}_0| = 5\}$ *and consists of those points on the circle of radius 5 centered at* $\mathbf{x}_0$. *The interior of* A *is given by* $\mathrm{Int}(A) = \{\mathbf{x} \in \mathbb{R}^2 : |\mathbf{x} - \mathbf{x}_0| < 5\}$, *and is the open disk of radius 5 centered at* $\mathbf{x}_0$. *The exterior of* A *is the set* $\mathrm{Ext}(A) = \{\mathbf{x} \in \mathbb{R}^2 : |\mathbf{x} - \mathbf{x}_0| > 5\}$ *and consists of all points of* $\mathbb{R}^2$ *outside the disk of radius 5 centered at* $\mathbf{x}_0$. ◀

► **2.13** *Complete part (iii) of the previous example by showing that* $N_r(\mathbf{x}_3)$ *for* $r \ge 1$ *must contain at least one point in* A *and at least one point in* A^C.

► **2.14** *For* $A \subset \mathbb{X}$, *show that* $\partial A = \partial\left(A^C\right)$.

It is an interesting fact that within the space of real numbers $\mathbb{R}$, every point of $\mathbb{Q}$ is a boundary point of $\mathbb{Q}$, i.e., $\mathbb{Q} \subset \partial\mathbb{Q}$. To see this, consider any point $q_0 \in \mathbb{Q}$. Then, for any real number $r > 0$ the neighborhood $N_r(q_0)$ must contain both rational and irrational numbers. Hence, q_0 is a boundary point of $\mathbb{Q}$. Perhaps even more interesting is the fact that the boundary of $\mathbb{Q}$ is all of $\mathbb{R}$.

► **2.15** *Show that* $\partial\mathbb{Q} = \mathbb{R}$. *Show the same for* $\partial\mathbb{I}$.

2.2 Limit Points and Isolated Points

Limit Points

For a subset A of $\mathbb{X}$, we have seen that any point $x \in \mathbb{X}$ is exactly one of the following three types: an *interior point* of A, an *exterior point* of A, or a *boundary point* of A. There is another classification of interest to us. For any subset A of $\mathbb{X}$, a point $x \in \mathbb{X}$ might also be what is referred to as a *limit point*[3] of A. We formalize this important concept in the following definition.

Definition 2.4 Let $A \subset \mathbb{X}$. Then $x_0 \in \mathbb{X}$ is called a **limit point** of A if $N'_r(x_0) \cap A \neq \varnothing$ for all $r > 0$, i.e., if every deleted neighborhood of x_0 contains at least one point of A.

► **2.16** *Show that, in fact, if x_0 is a limit point of A, then every deleted neighborhood of x_0 contains infinitely many points of A.*

► **2.17** *Show that if $x_0 \in \mathbb{X}$ is not a limit point of $A \subset \mathbb{X}$, then x_0 is a limit point of A^C.*

For any set $A \subset \mathbb{X}$, we have seen that there are two ways of classifying $x_0 \in \mathbb{X}$ with respect to A. First, x_0 must be an interior point, an exterior point, or a boundary point of A. Second, x_0 can be a limit point of A, or not a limit point of A. One might wonder if there is a relationship between these two classifications. The following examples give a partial answer to this question.

Example 2.5 *Consider $A \subset \mathbb{X}$ and suppose x_0 is an interior point of A. We will show that x_0 is also a limit point of A. To do this, note that since x_0 is an interior point of A, there exists a neighborhood $N_r(x_0) \subset A$ for some $r > 0$. Let $N'_\rho(x_0)$ be any deleted neighborhood of x_0. There are two cases to consider: $\rho \leq r$ and $\rho > r$. The first leads to $N_\rho(x_0) \subset N_r(x_0) \subset A$, while the second leads to $N_r(x_0) \subset N_\rho(x_0)$. In both situations there are clearly infinitely many points of A inside of $N'_\rho(x_0)$, and so x_0 is a limit point of A.* ◄

Example 2.6 *Let $A = \left\{\frac{1}{n} : n \in \mathbb{N}\right\} \subset \mathbb{R}$. We will show that 0, a boundary point of A that is not a member of A, is a limit point of A. To this end, consider $N_r(0)$ for any $r > 0$, and choose an integer $M > \frac{1}{r}$ (How do we know we can do this?). Then $0 < \frac{1}{M} < r$, and so $\frac{1}{M} \in N'_r(0)$. But $\frac{1}{M} \in A$, and since $r > 0$ was arbitrary, this shows that there is at least one element of A in every deleted neighborhood of 0. Hence 0 is a limit point of A.* ◄

Example 2.7 *Let $A = \left\{\mathbf{x} \in \mathbb{R}^2 : |\mathbf{x}| \leq 1\right\}$. We will show that $\mathbf{x_0} = (0,1)$, a boundary point that is an element of A, is a limit point of A. To this end, we*

[3]Some texts also use the term *accumulation point* or *cluster point* in addition to *limit point*. In fact, some authors use two of these three terms to distinguish between two situations: limit points of sequences of numbers and limit points of sets of points. We choose to use the single term *limit point* for both cases. The context is typically sufficient to clarify the interpretation.

first consider an arbitrary neighborhood $N_r'(\mathbf{x_0})$ for any $0 < r < 1$ (the $r \geq 1$ case will be handled separately). Choose x_2 such that $1 - r < x_2 < 1$. Then $\mathbf{x} = (0, x_2) \in N_r'(\mathbf{x_0}) \cap A$. Now consider an arbitrary neighborhood $N_r(\mathbf{x_0})$ with $r \geq 1$. In this case, any $\mathbf{x} = (0, x_2)$ with $0 < x_2 < 1$, is in $N_r'(\mathbf{x_0}) \cap A$. Since every deleted neighborhood of $\mathbf{x_0}$ has been shown to contain at least one point of A, $\mathbf{x_0}$ is a limit point of A. ◄

► **2.18** *It is not necessarily true that if x is a boundary point of a subset A of $\mathbb{X}$, then x must be a limit point of A. Give a counterexample.*

► **2.19** *Suppose x_0 is an exterior point of $A \subset \mathbb{X}$. Then show that x_0 can never be a limit point of A. But show that x_0 is necessarily a limit point of A^C.*

While it is useful to be able to identify individual points as limit points for a given set, it will also be useful to consider the collection of all the limit points associated with a given set. We formally define this collection of points now.

Definition 2.8 The collection of limit points of $A \subset \mathbb{X}$ is denoted by A' and is called **the derived set** of A.

Example 2.9 *We will show that $\mathbb{Z}' = \varnothing$. To this end, suppose $x_0 \in \mathbb{Z}'$. We will obtain a contradiction. Note that the point x_0 cannot be an integer, since if it were, $N_{\frac{1}{2}}'(x_0) \cap \mathbb{Z} = \varnothing$, contradicting that x_0 is a limit point. Therefore x_0 must be in $\mathbb{Z}^C$. But for $x_0 \in \mathbb{Z}^C$, there exists $M \in \mathbb{Z}$ such that $M < x_0 < M+1$. Taking $r < \min\{\,|\,x_0 - M\,|,\ |\,x_0 - (M+1)\,|\,\}$ obtains a neighborhood $N_r'(x_0)$ such that $N_r'(x_0) \cap \mathbb{Z} = \varnothing$. Since this contradicts our assumption that x_0 is a limit point of $\mathbb{Z}$, we must have that $\mathbb{Z}' = \varnothing$.* ◄

Example 2.10 *We will show that $\mathbb{Q}' = \mathbb{R}$. To this end, let x_0 be an arbitrary real number, and consider any deleted neighborhood $N_r'(x_0)$. Since there must be a rational number between x_0 and $x_0 + r$ (Why?), it follows that $N_r'(x_0) \cap \mathbb{Q} \neq \varnothing$, and so x_0 is a limit point of $\mathbb{Q}$. Since x_0 was arbitrary, $\mathbb{Q}' = \mathbb{R}$.* ◄

► **2.20** *If $A \subset B \subset \mathbb{X}$, show that $A' \subset B'$.*

Isolated Points

We now define the concept of an isolated point.

Definition 2.11 For $A \subset \mathbb{X}$, a point $x \in A$ is called an **isolated point** of A if there exists $r > 0$ such that $N_r'(x) \cap A = \varnothing$.

That is, $x \in A$ is an isolated point of A if and only if there exists a deleted neighborhood of x that contains no points of A. The following examples illustrate the idea.

Example 2.12 *Consider $\mathbb{Z} \subset \mathbb{R}$. Every $n \in \mathbb{Z}$ is an isolated point of $\mathbb{Z}$ because $N'_{\frac{1}{2}}(n) \cap \mathbb{Z} = \varnothing$.* ◀

Example 2.13 *Consider $S \subset \mathbb{R}$ where $S = \left\{1, \frac{1}{2}, \frac{1}{3}, \dots\right\}$. We will show that every point of S is isolated. To do this, let s be an arbitrary element of S. Then $s = \frac{1}{n}$ for some $n \in \mathbb{N}$. Choose $r = \frac{1}{n} - \frac{1}{n+1} = \frac{1}{n(n+1)}$ and consider $N'_r(s)$. Clearly $N'_r(s) \cap S = \varnothing$, and so $s = \frac{1}{n}$ is an isolated point of S. Since s was an arbitrary point of S, every point of S is isolated.* ◀

▶ **2.21** *In the last example, find the limit points of S.*

▶ **2.22** *Prove or disprove: The point $x \in A \subset \mathbb{X}$ is not an isolated point of A if and only if x is a limit point of A.*

▶ **2.23** *Suppose $S \subset \mathbb{R}$ is bounded and infinite. Let $M_S = \sup S$ and $m_S = \inf S$. Are M_S and m_S always in S? Are M_S and m_S always limit points of S? (Answer: In both cases, no. Give counterexamples.)*

3 Open and Closed Sets

3.1 Open Sets

With the fundamental notions of spheres, open and closed balls, neighborhoods, and classification of points established, we are ready to define the important topological concept of an *open* set.

Definition 3.1 A subset G of $\mathbb{X}$ is called **open** if for each $x \in G$ there is a neighborhood of x that is contained in G.

With a little thought, one realizes that the above definition is equivalent to the statement that a set $G \subset \mathbb{X}$ is *open* exactly when all of the points of G are interior points. We state this as a proposition.

Proposition 3.2 *The set $G \subset \mathbb{X}$ is open if and only if all of its points are interior points.*

▶ **2.24** *Prove the above proposition.*

Example 3.3 *In this example, we show that for any $r > 0$ and any $x_0 \in \mathbb{X}$, the neighborhood $N_r(x_0)$ is an open set. To establish this, let x be an arbitrary point in $N_r(x_0)$. We must find a neighborhood $N_\rho(x)$ that is contained in $N_r(x_0)$. (See Figure 2.2.) To this end, choose $\rho = r - |x - x_0|$, and consider an arbitrary point y in $N_\rho(x)$. We will show that y must be in $N_r(x_0)$. Since $y \in N_\rho(x)$, we have that $|y - x| < \rho$. From this, it follows that*

$$|y - x_0| \leq |y - x| + |x - x_0| < \rho + |x - x_0| = r.$$

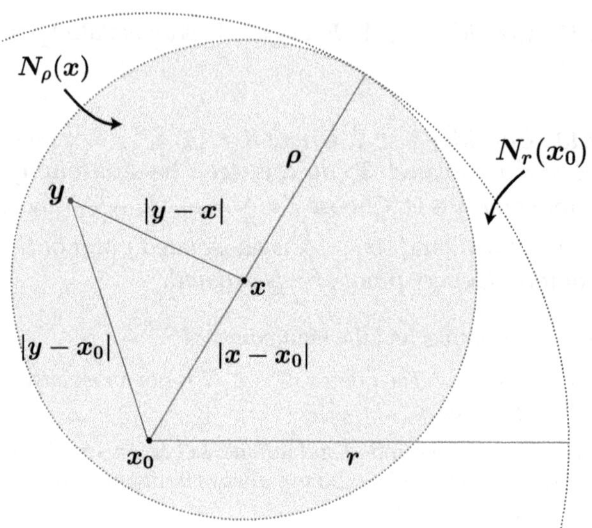

Figure 2.2 *The proof that $N_r(x_0)$ is an open set.*

Therefore $y \in N_r(x_0)$, and so $N_\rho(x) \subset N_r(x_0)$. Since $x \in N_r(x_0)$ was arbitrary, $N_r(x_0)$ is open. ◀

► **2.25** *For any real $r > 0$ and any $x_0 \in \mathbb{X}$, show that $N'_r(x_0)$ is an open set.*

► **2.26** *Using Definition 3.1, prove that every interval of the form $(a, b) \subset \mathbb{R}$ where $a < b$ is an open set in $\mathbb{R}$. Handle as well the cases where a is replaced by $-\infty$ or b is replaced by ∞.*

► **2.27** *Using Definition 3.1, prove that every open interval of the form $(a, b) \times (c, d) \subset \mathbb{R}^2$ where $a < b$ and $c < d$ is an open set in $\mathbb{R}^2$. Extend this result to open intervals $(a_1, b_1) \times (a_2, b_2) \times \cdots \times (a_k, b_k)$ in $\mathbb{R}^k$. Consider the infinite intervals of $\mathbb{R}^k$ as well.*

Open sets are at the root of many of the most important and fundamental ideas in analysis. It is worth our while, then, to investigate the properties of open sets.

Proposition 3.4

a) *The entire space $\mathbb{X}$ and the empty set $\varnothing$ are open sets in $\mathbb{X}$.*

b) *The intersection of any two open sets in $\mathbb{X}$ is an open set in $\mathbb{X}$.*

c) *The union of any collection of open sets in $\mathbb{X}$ is an open set in $\mathbb{X}$.*

PROOF a) Clearly $\mathbb{X}$ is open. The set $\varnothing$ has no elements, so the result holds vacuously in this case.

b) Let A and B be two open sets in $\mathbb{X}$, and let $x \in A \cap B$. Then $x \in A$

and $x \in B$, and since A and B are open there exist open neighborhoods $N_{r_1}(x) \subset A$ and $N_{r_2}(x) \subset B$. Therefore $x \in N_{r_1}(x) \cap N_{r_2}(x) \subset A \cap B$. All that is left to do is to find r small enough so that $x \in N_r(x)$ where $N_r(x) \subset \left(N_{r_1}(x) \cap N_{r_2}(x)\right)$. We leave this as an exercise for the reader.

c) Let $\{A_\alpha\}$ be a collection of open sets in $\mathbb{X}$, and let $x \in \bigcup_\alpha A_\alpha$. Then $x \in A_{\alpha'}$ for some α', so there exists a neighborhood $N_r(x)$ such that $x \in N_r(x) \subset A_{\alpha'} \subset \bigcup A_\alpha$. Therefore $\bigcup A_\alpha$ is open. ♦

► **2.28** *Complete the proof of part b) of the above proposition.*

► **2.29** *Generalize part b) of the above proposition as follows. Let $G_1, G_2, \ldots, G_n$ be open sets in $\mathbb{X}$. Prove that $\cap_{j=1}^n G_j$ is also open in $\mathbb{X}$. The generalization does not hold for arbitrary collections $\{G_\alpha\}$ of open sets. That is, it is not true that $\bigcap G_\alpha$ is necessarily open if each G_α is open. Find a counterexample to establish the claim.*

Example 3.5 *Let $G = \left\{\mathbf{x} \in \mathbb{R}^2 : |x_1| + |x_2| < 1\right\}$ be a subset of $\mathbb{R}^2$. The reader is urged to draw the set G in the plane. We will show that G is open in $\mathbb{R}^2$. To this end, fix an arbitrary $\mathbf{x_0} = (\xi, \eta) \in G$ and let $r = \frac{1}{10}(1 - |\xi| - |\eta|) > 0$. We will show that $N_r(\mathbf{x_0}) \subset G$. Let $\mathbf{x} = (x_1, x_2) \in N_r(\mathbf{x_0})$, and compute*

$$\begin{aligned} |x_1| + |x_2| &= \left|(x_1 - \xi) + \xi\right| + \left|(x_2 - \eta) + \eta\right| \\ &\le |x_1 - \xi| + |\xi| + |x_2 - \eta| + |\eta| \\ &\le 2\,|\mathbf{x} - \mathbf{x_0}| + |\xi| + |\eta| && \textit{(Why?)} \\ &< 2\left(\tfrac{1}{10}\right)(1 - |\xi| - |\eta|) + |\xi| + |\eta| \\ &= \tfrac{1}{5} + \tfrac{4}{5}\left(|\xi| + |\eta|\right) < 1. && \textit{(Why?)} \end{aligned}$$

Therefore, $\mathbf{x} \in G$. Since $\mathbf{x}$ was an arbitrary element of $N_r(\mathbf{x_0})$, we have shown that $N_r(\mathbf{x_0}) \subset G$. But $\mathbf{x_0}$ was an arbitrary element of G, and so G must be open. ◄

The following proposition allows us to conveniently characterize any open set in $\mathbb{X}$ in terms of open intervals. This is useful since intervals are often the easiest sets with which to work.

Proposition 3.6 *Let G be an open subset of $\mathbb{X}$. Then $G = \bigcup I_\alpha$ where $\{I_\alpha\}$ is a collection of open intervals in $\mathbb{X}$.*

PROOF Consider an arbitrary $x \in G$. Since G is open, there exists a neighborhood $N_r(x)$ such that $N_r(x) \subset G$. It is left as an exercise to verify that one can find an open interval I_x containing x such that $I_x \subset N_r(x) \subset G$. From this it is easy to see that

$$G = \bigcup_{x \in G} I_x.$$

♦

► **2.30** *Show in each of the cases $\mathbb{X} = \mathbb{R}$, $\mathbb{C}$, and $\mathbb{R}^k$ that one can find an open interval I_x such that $I_x \subset N_r(x) \subset G$ as claimed in the above proposition. Also, give the detailed argument justifying the final claim that $G = \bigcup_{x \in G} I_x$.*

3.2 Closed Sets

We now introduce what amounts to the complementary idea of an open set, namely, a *closed* set.

Definition 3.7 A subset F of $\mathbb{X}$ is called **closed** if its complement F^C is open.

Applying Definition 3.7 in order to verify that a set is closed is fairly straightforward, as the following examples show.

Example 3.8 *We will show that* $A = (-\infty, 1] \cup [3, \infty) \subset \mathbb{R}$ *is closed. Using the above definition, accomplishing this amounts to showing that* $A^C = (1, 3)$ *is open, a result the reader was asked to establish (for any open interval) in a previous exercise.* ◀

Example 3.9 *We will show that* $A = (0, 1]$ *is neither open nor closed in* $\mathbb{R}$*. First, assume that* A *is open. Then there exists a neighborhood centered at* $x = 1$ *for some* $r > 0$ *such that* $N_r(1) \subset A$*. But this implies that* $x = 1 + \frac{1}{2}r \in A$*, which is false, and so* A *cannot be open. Assume now that* A *is closed. This implies that* $A^C = (-\infty, 0] \cup (1, \infty)$ *is open. Therefore, there exists some neighborhood centered at* $x = 0$ *for some* $r > 0$ *such that* $N_r(0) \subset A^C$*. But this too yields a contradiction (Why?), and so* A *cannot be closed.* ◀

The counterpart to Proposition 3.4 on open sets is the following proposition highlighting some important properties of closed sets. The proof is left to the reader, since it follows directly from the definition of closed set and Proposition 3.4.

Proposition 3.10

a) *The entire space* $\mathbb{X}$ *and the empty set* $\varnothing$ *are closed sets in* $\mathbb{X}$*.*

b) *The union of any two closed sets in* $\mathbb{X}$ *is a closed set in* $\mathbb{X}$*.*

c) *The intersection of any collection of closed sets in* $\mathbb{X}$ *is a closed set in* $\mathbb{X}$*.*

► **2.31** *Prove the above proposition.*

► **2.32** *Part b) of the above proposition can be generalized in the same way as the corresponding part in Proposition 3.4 as described in a previous exercise. That is, let* $F_1, F_2, \ldots, F_n$ *be closed sets in* $\mathbb{X}$*. Prove that* $\bigcup_{j=1}^{n} F_j$ *is also closed in* $\mathbb{X}$*. Does the generalization hold for arbitrary collections* $\{F_\alpha\}$ *of closed sets? That is, is it true that* $\bigcup F_\alpha$ *is necessarily closed if each* F_α *is closed? (Answer: See the next exercise.)*

► **2.33** *Find an infinite collection of closed sets* $F_1, F_2, F_3, \ldots$ *where* $\bigcup_{j=1}^{\infty} F_j$ *is not closed.*

As the previous examples and theorems indicate, sets can be open, closed, both open and closed, or neither open nor closed. We now conveniently characterize closed sets in terms of their boundary points.

Proposition 3.11 *The set $F \subset \mathbb{X}$ is closed if and only if F contains all of its boundary points.*

PROOF Suppose the set F is closed, and let x be a boundary point of F. We will show by contradiction that $x \in F$. To this end, suppose $x \in F^C$. Then since F^C is open, there exists a neighborhood $N_r(x) \subset F^C$. That is, there are no points of F in $N_r(x)$. But this contradicts x being a boundary point of F. Therefore, the boundary point x must *not* be an element of F^C. This shows that if x is a boundary point of the closed set F, then it must be in F. Conversely, suppose the set F contains all of its boundary points. To show that F must be closed, consider an arbitrary $x \in F^C$. This x cannot be a boundary point of F, and so x must be an exterior point of F. (Why?) Therefore, there exists a neighborhood $N_r(x) \subset F^C$. Since the point x was an arbitrary point from F^C, we have shown that F^C is open, and therefore F is closed. ♦

► **2.34** *Suppose F is a finite set in $\mathbb{X}$. Show that F is closed.*

► **2.35** *If $F \subset \mathbb{X}$ is closed, and $x_0 \in \mathbb{X}$ is a limit point of F, prove that $x_0 \in F$.*

It is convenient at this point to define what is meant by *the closure* of a set. There are actually several equivalent notions of closure, and we leave a couple of them for the reader to explore in the exercises.

Definition 3.12 For any subset A of $\mathbb{X}$, the **closure** of A is denoted by $\overline{A}$ and is given by

$$\overline{A} = A \cup A'.$$

The closure of a set is the set itself along with all of its limit points.

► **2.36** *Show that for $A \subset \mathbb{X}$, if x is a boundary point of A then $x \in \overline{A}$. Also, if $\sup A$ exists then $\sup A \in \overline{A}$, and if $\inf A$ exists then $\inf A \in \overline{A}$.*

► **2.37** *Show that for any set $A \subset \mathbb{X}$, the point x is in $\overline{A}$ if and only if every neighborhood of x contains a point of A.*

► **2.38** *If $A \subset B \subset \mathbb{X}$, show that $\overline{A} \subset \overline{B}$.*

Example 3.13 *We will show that $\overline{\mathbb{Z}} = \mathbb{Z}$. According to a previous example, $\mathbb{Z}' = \varnothing$, and so we have $\overline{\mathbb{Z}} = \mathbb{Z} \cup \mathbb{Z}' = \mathbb{Z} \cup \varnothing = \mathbb{Z}$.* ◄

Example 3.14 *In this example we establish that $\overline{\mathbb{Q}} = \mathbb{R}$. According to a previous example, $\mathbb{Q}' = \mathbb{R}$, and so $\overline{\mathbb{Q}} = \mathbb{Q} \cup \mathbb{Q}' = \mathbb{Q} \cup \mathbb{R} = \mathbb{R}$.* ◄

Example 3.15 *Consider the set* $A = \left\{ \left(\frac{m}{n}, \frac{m}{n}\right) \in \mathbb{R}^2 : m, n \in \mathbb{N} \right\}$. *We will determine* $\overline{A}$. *Note that each point of* A *is of the form* $\left(\frac{m}{n}, \frac{m}{n}\right) = \frac{m}{n}(1,1) = q(1,1)$ *for* $q \in \mathbb{Q}^+$, *and therefore lies on the ray* $\mathcal{R}$ *that is the portion of the line* $y = x$ *within the first quadrant of the* xy*-plane. That is,* $A \subset \mathcal{R}$. *Consider the origin* $\mathbf{0}$, *and note that* $\mathbf{0}$ *is not in* A. *However, for any* $r > 0$, *there exists an integer* $M > 0$ *large enough so that the neighborhood* $N_r(\mathbf{0})$ *contains the points* $\left(\frac{1}{n}, \frac{1}{n}\right)$ *for all* $n > M$. *(Why?) Therefore,* $\mathbf{0}$ *is a limit point of* A *and is in* $\overline{A}$. *Now consider an arbitrary* $x > 0$ *so that* $\mathbf{x} \equiv x(1,1)$ *is an arbitrary point lying on* $\mathcal{R}$. *For any* $r > 0$, *consider the neighborhood* $N_r(\mathbf{x})$. *We will show that there are infinitely many points from* A *that lie within this neighborhood. To establish this, recall that within the interval* $(x - \frac{r}{\sqrt{2}}, x + \frac{r}{\sqrt{2}}) \subset \mathbb{R}$ *there are infinitely many rational numbers. Among these, there are infinitely many positive rational numbers. (Why?) Each such positive rational number* $q \in (x - \frac{r}{\sqrt{2}}, x + \frac{r}{\sqrt{2}})$, *corresponds to a point* $(q, q) \subset A$ *that lies within* $N_r(\mathbf{x})$. *(Why?) Hence,* $\mathbf{x}$ *is a limit point of* A. *That is, every point of the ray* $\mathcal{R}$ *is a limit point of* A, *and so* $\mathcal{R} \subset \overline{A}$. *We leave it to the reader to show that any point of* $\mathbb{R}^2$ *not on the ray* $\mathcal{R}$ *is not a limit point of* A, *and therefore* $\overline{A} = \mathcal{R}$. ◀

► **2.39** *Give a detailed answer for each (Why?) question from the previous example. Also, show that any point of* $\mathbb{R}^2$ *not on the ray* $\mathcal{R}$ *is not a limit point of* A, *and therefore* $\overline{A} = \mathcal{R}$.

The following proposition gives two useful criteria for the closure of a set.

Proposition 3.16 *Let* $A \subset \mathbb{X}$. *Then:*

a) $\overline{A}$ *is the smallest closed set containing* A.

b) We have $x_0 \in \overline{A}$ *if and only if* $N_r(x_0) \cap A \neq \varnothing$ *for all* $r > 0$.

PROOF We start with the second claim. Suppose $N_r(x_0) \cap A \neq \varnothing$ for all $r > 0$. We will show that $x_0 \in \overline{A}$. There are two possibilities: either $x_0 \in A$, or $x_0 \notin A$. In the former case, $x_0 \in \overline{A}$ clearly. In the latter, we see that $N_r(x_0) \cap A = N_r'(x_0) \cap A$ for all r, hence $N_r'(x_0) \cap A \neq \varnothing$ for all r, and $x_0 \in A' \subset \overline{A}$. The rest of the proof of *b)* is left to the reader. We now prove part *a)*. Let x_1 be an arbitrary element of $\partial\overline{A}$. We will show that $x_1 \in \overline{A}$, and therefore that $\overline{A}$ is closed. Each neighborhood $N_r(x_1)$ satisfies $N_r(x_1) \cap \overline{A} \neq \varnothing$. Thus there exists $a_1 \in N_r(x_1) \cap \overline{A}$, and we can choose s small and positive so that $N_s(a_1) \subset N_r(x_1)$. Now since $a_1 \in \overline{A}$, we see that $N_s(a_1) \cap A \neq \varnothing$ by part b), whence $N_r(x_1) \cap A \neq \varnothing$, and $x_1 \in \overline{A}$ by part *b)* w again. Thus $\overline{A}$ is closed. Moreover, any closed set containing A must contain $\overline{A}$. Indeed, it must contain A' by Exercise 2.35, hence must contain $\overline{A} = A \cup A'$. ♦

► **2.40** *Complete the proof of part b) of the above proposition.*

► **2.41** *For a set $S \subset \mathbb{X}$, prove that $\overline{S} = S \cup \partial S$. It is, however, not necessarily true that $S' = \partial S$. Give examples of this phenomenon.*

► **2.42** *For $A, B \subset \mathbb{X}$, prove the following: a) $\overline{A \cup B} = \overline{A} \cup \overline{B}$ b) $\bigcup \overline{A_\alpha} \subset \overline{\bigcup A_\alpha}$*

► **2.43** *For $\{A_\alpha\}$ with each $A_\alpha \subset \mathbb{X}$, find an example where $\bigcup \overline{A_\alpha} \neq \overline{\bigcup A_\alpha}$.*

► **2.44** *Presuming all sets are from $\mathbb{X}$, how do the following sets compare?*

a) $\overline{A \cap B}$, $\overline{A} \cap \overline{B}$ *b)* $\bigcap \overline{A_\alpha}$, $\overline{\bigcap A_\alpha}$
c) $(A \cup B)'$, $A' \cup B'$ *d)* $(A \cap B)'$, $A' \cap B'$

► **2.45** *Show that $\overline{(a, b)} = \overline{(a, b]} = \overline{[a, b)} = \overline{[a, b]} = [a, b]$.*

► **2.46** *Let I_1 and I_2 be intervals in $\mathbb{R}$. Then $\mathcal{I} \equiv I_1 \times I_2$ is an interval in $\mathbb{R}^2$. Show that $\overline{\mathcal{I}} = \overline{I_1} \times \overline{I_2}$, i.e., that $\overline{I_1 \times I_2} = \overline{I_1} \times \overline{I_2}$. Generalize this result to intervals in $\mathbb{R}^k$.*

3.3 Relatively Open and Closed Sets

It is sometimes useful to work within a subset S of a space $\mathbb{X}$. In such cases, we would like to characterize when a set $U \subset S$ is open or closed *relative to* S. The following definition and examples make this idea more clear.

Definition 3.17 Consider $U \subset S \subset \mathbb{X}$.

1. The set U is called **open relative to** S or **open in** S if there exists an open set $V \subset \mathbb{X}$ such that $U = S \cap V$.
2. The set U is called **closed relative to** S or **closed in** S if $S \setminus U$ is open in S.

Example 3.18 *Consider the subsets $U_1 = (\frac{1}{2}, 1)$ and $U_2 = (\frac{1}{2}, 1]$, each contained in $S = [0, 1] \subset \mathbb{R}$. We will show that both U_1 and U_2 are, in fact, open in S according to Definition 3.17. We begin with $U_1 = (\frac{1}{2}, 1)$. Since U_1 is open in $\mathbb{R}$, this might not be unexpected, yet we must properly apply Definition 3.17 to make the conclusion. To see this, choose the open set $V_1 \subset \mathbb{R}$ as $V_1 = U_1 = (\frac{1}{2}, 1)$. Then $U_1 = S \cap V_1$, and therefore U_1 is open in S. In the case of $U_2 = (\frac{1}{2}, 1]$, which is not open in $\mathbb{R}$, we choose the open set $V_2 \subset \mathbb{R}$ as $V_2 = (\frac{1}{2}, 2)$. Then $U_2 = S \cap V_2$, and therefore U_2 is open in S.* ◄

► **2.47** *Is it true in general that if U is open in $\mathbb{X}$, then for any subset $S \subset \mathbb{X}$ with $U \subset S$ the set U will be open in S? (Answer: Yes.)*

► **2.48** *Suppose $U \subset S \subset \mathbb{X}$ where S is closed. Is it possible for U to be open in S if U is not open in $\mathbb{X}$? (Yes. Give examples.)*

► **2.49** *Suppose $U \subset S \subset \mathbb{X}$ where U is open in S, and S is open. Prove that U is necessarily open in $\mathbb{X}$.*

▶ **2.50** *Suppose U_1 and U_2 are subsets of $S \subset \mathbb{X}$ and are relatively open in S. Prove the following:*

a) $U_1 \cup U_2$ is relatively open in S. *b) $U_1 \cap U_2$ is relatively open in S.*
c) Generalize a) to infinite unions. *d) Why does b) fail for infinite intersections?*

▶ **2.51** *Suppose $S \subset \mathbb{X}$. Show that $\varnothing$ and S are both relatively open in S, and that $\varnothing$ and S are both relatively closed in S.*

3.4 Density

For two sets A and B satisfying $A \subset B \subset \mathbb{X}$, it is of interest to know how the elements of A are distributed within B. Are they merely "sprinkled" in, or are they packed in very tightly throughout B? If they are packed in tightly enough throughout B, we say that A is *dense* in B. We make this idea more precise in the following definition.

Definition 3.19 Let A and B be subsets of $\mathbb{X}$, with $A \subset B$. We say that A is **dense** in B if

$$\overline{A} \supset B,$$

i.e., if every point of B is either in A or is a limit point of A.

The density of one set within another is a very useful concept that will be exploited later in our work. Its value is especially notable in the theory of approximation. The following proposition establishes another way of thinking about the idea of density that is less set-theoretic than the above definition, and more geometrically intuitive. Basically, it says that a set A is dense in B if, no matter where you go within the set B, you'll be arbitrarily close to elements of A. To be dense in B, then, the set A must be distributed throughout B and packed in very tightly indeed!

Proposition 3.20 *Let A and B be subsets of $\mathbb{X}$, with $A \subset B$. Then the following are equivalent.*

a) The set A is dense in B.

b) For each $b \in B$, every neighborhood of b contains elements of A.

c) Each $b \in B$ is in A or is a boundary point of A.

▶ **2.52** *Prove the above proposition.*

Example 3.21 *The set of integers $\mathbb{Z}$ is not dense in $\mathbb{R}$. This is so since we can find a real number with a deleted neighborhood around it containing no integers. An example is $x = \frac{1}{2}$, and $N_{\frac{1}{10}}(\frac{1}{2})$. In fact, for any $x \in \mathbb{R}$, there exists a deleted neighborhood centered at x that contains no integers.* ◀

Example 3.22 *The set of rationals $\mathbb{Q}$ is dense in $\mathbb{R}$. To show this, we will use Proposition 3.20. Consider an arbitrary $x \in \mathbb{R}$ and let $N_r(x)$ be a neighborhood around x of radius $r > 0$. According to the argument on page 6, there exists a rational number q satisfying $x < q < x + r$, and hence, $q \in N_r(x)$. Since $N_r(x)$ was an arbitrary neighborhood around x, and x was an arbitrary element of $\mathbb{R}$, this shows that $\mathbb{Q}$ is dense in $\mathbb{R}$.* ◀

We will find in our work that the density of $\mathbb{Q}$ in $\mathbb{R}$ is an important property of the real line.

4 NESTED INTERVALS AND THE BOLZANO-WEIERSTRASS THEOREM

4.1 Nested Intervals

More elaborate sets of real numbers can be constructed from the basic "intervals building blocks" by making use of unions, intersections, and complements. Recall that unions and intersections of sets need not be *finite* unions or intersections. Examples such as the following from $\mathbb{R}$ are common.

$$\mathbb{R} = (-\infty, \infty) = \bigcup_{n=1}^{\infty} [-n, n],$$

and

$$[\,0, 1\,] = \bigcap_{n=1}^{\infty} \left(-\tfrac{1}{n},\ 1 + \tfrac{1}{n}\right).$$

In both of these examples the sets on the right-hand side of the equality are *nested*. That is, each subsequent set in the sequence comprising the union (or intersection) contains (or is contained in) the previous set in the union (or intersection). In the first example, we have $[-1, 1] \subset [-2, 2] \subset \cdots \subset [-n, n] \subset \cdots$, and so the nested sets are said to be *increasing*, since each set in the union includes all previous sets in the union. In the second example the sets are nested and *decreasing*, since each set in the intersection is included in all previous sets in the intersection. To the newly initiated it might be surprising to learn that a nested sequence of decreasing sets may have either an empty or nonempty intersection. The second example above is a case of the latter, the intersection being the unit interval. (For an example of a nested sequence of decreasing sets whose intersection is empty, see the example that follows.)

While unions and intersections such as those in the two examples above might seem difficult to interpret at first, it helps to remember the corresponding interpretations of union and intersection. The first set in the above examples consists of all elements that are members of *at least one* of the intervals $[-n, n]$. The second consists of only those elements that are members of *all* the intervals $(-\frac{1}{n}, 1 + \frac{1}{n})$.

Example 4.1 *Consider the set expressed by $\bigcap_{n=1}^{\infty} [n, \infty)$. We claim this set* *is empty, and will establish this fact by contradiction. Assume there exists $x \in \bigcap_{n=1}^{\infty}[n,\infty)$. Then $x \in [n,\infty)$ for all $n \in \mathbb{N}$. By the Archimedean property, there exists $M \in \mathbb{N}$ such that $x < M$. This implies that $x \notin [M,\infty)$, a contradiction.* ◀

Example 4.2 *Let $I_n = \{z \in \mathbb{C} : 0 \le \text{Re}(z) < \frac{1}{n},\ 0 < \text{Im}(z) < \frac{1}{2n}\}$. We will show that $\bigcap_{n=1}^{\infty} I_n = \varnothing$. To see this, suppose $\bigcap_{n=1}^{\infty} I_n \neq \varnothing$ and that $z_0 \in \bigcap_{n=1}^{\infty} I_n$, where $z_0 = x_0 + i\,y_0$. Then $z_0 \in I_n$ for all $n \ge 1$, and hence $0 \le x_0 < \frac{1}{n}$ for all $n \ge 1$ and $0 < y_0 < \frac{1}{2n}$ for all $n \ge 1$. It is easy to see that $x_0 = 0$. To satisfy $y_0 < \frac{1}{2n}$ for all $n \ge 1$, we must have $y_0 = 0$. But this contradicts $y_0 > 0$. Therefore, no such z_0 exists, and $\bigcap_{n=1}^{\infty} I_n = \varnothing$.* ◀

Example 4.3 *Consider $\mathcal{I}_n \subset \mathbb{R}^2$ given by $\mathcal{I}_n = \left[\frac{1}{2}, 1+\frac{1}{n}\right) \times \left[\frac{n}{n+1}, 1\right]$. We will find $\bigcap_{n=1}^{\infty} \mathcal{I}_n$ and show that the sequence of intervals $\mathcal{I}_n$ is decreasing. To determine the intersection, suppose $(x,y) \in \mathcal{I}_n$ for all $n \ge 1$. Then*

$$\tfrac{1}{2} \le x < 1+\tfrac{1}{n} \quad \text{and} \quad \tfrac{n}{n+1} \le y \le 1 \quad \text{for all} \quad n \ge 1.$$

This implies $x \in \left[\frac{1}{2}, 1\right]$ and $y = 1$, and so

$$\bigcap_{n=1}^{\infty} \mathcal{I}_n = \left\{(x,1) \in \mathbb{R}^2 : \tfrac{1}{2} \le x \le 1\right\}.$$

To see that $\{\mathcal{I}_n\}$ is decreasing, let (x,y) be an arbitrary element of $\mathcal{I}_{n+1}$. Then

$$\tfrac{1}{2} \le x < 1+\tfrac{1}{n+1} < 1+\tfrac{1}{n} \quad \text{and} \quad \tfrac{n}{n+1} \le \tfrac{n+1}{n+2} \le y < 1,$$

and so $(x,y) \in \mathcal{I}_n$. ◀

It is a particularly important property of the real numbers $\mathbb{R}$, following from the completeness property, that a nested sequence of decreasing, closed, and bounded intervals is not empty. This property extends naturally to $\mathbb{R}^k$ and $\mathbb{C}$ as stated in the following theorem.

Theorem 4.4 (The Nested Closed Bounded Intervals Theorem in $\mathbb{X}$)
Suppose $\{\mathcal{I}_n\}_{n=1}^{\infty}$ is a nested sequence of decreasing, closed, bounded intervals in $\mathbb{X}$. That is, $\mathcal{I}_1 \supseteq \mathcal{I}_2 \supseteq \ldots \supseteq \mathcal{I}_n \supseteq \ldots$. Then $\bigcap_{n=1}^{\infty} \mathcal{I}_n \neq \varnothing$. That is, there exists at least one element common to all the intervals.

PROOF We begin by establishing the result for $\mathbb{R}$. To this end, suppose $I_n = [a_n, b_n]$ for $n = 1, 2, \ldots$, and that $\{I_n\}$ is a nested sequence of decreasing, closed, bounded intervals in $\mathbb{R}$, i.e., $I_1 \supseteq I_2 \supseteq \ldots \supseteq I_n \supseteq \ldots$. Our first step will be to show that every a_n is less than every b_m. Since $I_{n+1} = [a_{n+1}, b_{n+1}] \subset [a_n, b_n] = I_n$ for all n, it follows that

$$a_1 \le a_2 \le a_3 \le \cdots,$$

i.e., the $\{a_i\}$ are nondecreasing. Similarly,

$$b_1 \geq b_2 \geq b_3 \geq \cdots,$$

and the $\{b_i\}$ are nonincreasing. If we now fix n and m and let $k = \max(m, n)$, it is not hard to see that

$$a_n \leq a_k < b_k \leq b_m.$$

Therefore, *every* a_n is less than *every* b_m. Now fix the value of n. By the last expression above, we see that the set $B = \{b_1, b_2, b_3, \ldots\}$ is bounded below by any a_n. Therefore, $b \equiv \inf B$ exists, and $a_n \leq t$ for all n. This last inequality implies that the set $A = \{a_1, a_2, a_3, \ldots\}$ is bounded *above* by b, and so $a \equiv \sup A$ exists. Also, $a \leq b$. (Why?) Altogether we can write

$$a_n \leq a \leq b \leq b_n \text{ for all } n.$$

Therefore, $[a, b] \subset [a_n, b_n] = I_n$ for all n. (If $a = b$, we temporarily abuse notation and write $[a, b]$ for the set $\{a\}$.) That is, $[a, b] \subset \bigcap_n I_n$.

We now prove the case for $\mathbb{R}^2$, leaving the general case of $\mathbb{R}^k$ to the reader (clearly, the case for $\mathbb{C}$ will be equivalent to that of $\mathbb{R}^2$). Our proof for the $\mathbb{R}^2$ case will build on the result already established for $\mathbb{R}$. In particular, let $\{\mathcal{I}_n\}$ be a decreasing sequence of closed bounded intervals in $\mathbb{R}^2$, i.e.,

$$\mathcal{I}_1 \supset \mathcal{I}_2 \supset \cdots \supset \mathcal{I}_n \supset \cdots.$$

Then for each n, we have $\mathcal{I}_n = I_{1n} \times I_{2n}$, where I_{1n} and I_{2n} are closed bounded intervals in $\mathbb{R}$, and the sequences $\{I_{1n}\}$ and $\{I_{2n}\}$ are each decreasing. (Why?) From the case of the theorem for $\mathbb{R}$ proved above, there exists $x_0 \in \bigcap_{n=1}^{\infty} I_{1n}$ and $y_0 \in \bigcap_{n=1}^{\infty} I_{2n}$, which implies that

$$\begin{aligned}(x_0, y_0) \in \left(\bigcap_{n=1}^{\infty} I_{1n}\right) \times \left(\bigcap_{n=1}^{\infty} I_{2n}\right) &= \bigcap_{n=1}^{\infty} (I_{1n} \times I_{2n}) \quad \text{(Why?)} \\ &= \bigcap_{n=1}^{\infty} \mathcal{I}_n.\end{aligned}$$

Therefore $\bigcap_{n=1}^{\infty} \mathcal{I}_n$ is nonempty, and the case for $\mathbb{R}^2$ is proved. ♦

► **2.53** *Give the detailed answers to the three (Why?) questions in the above proof.*

Example 4.5 Let $I_n = (0, \frac{1}{n}]$ *for* $n \geq 1$ *where* $I_{n+1} \subset I_n$ *for all* n. *We will show that in this case,* $\bigcap_{n=1}^{\infty} I_n = \varnothing$. *To show this, suppose there exists* $x_0 \in \bigcap_{n=1}^{\infty} I_n$. *Then* $x_0 \in I_n$ *for all* n, *and so* $0 < x_0 \leq \frac{1}{n}$ *for all* n. *But since* $x_0 > 0$, *we can find a positive integer* $N > \frac{1}{x_0}$. *(Why?) This is a contradiction, since* x_0 *must satisfy* $x_0 \leq \frac{1}{N}$. *Therefore, no such* x_0 *exists, and* $\bigcap_{n=1}^{\infty} I_n$ *must be empty.* ◄

► **2.54** *What is the significance of the last example? How about Examples 4.1 and 4.2?*

We now give a useful corollary to Theorem 4.4. It gives sufficient conditions which, when satisfied, imply that the common intersection of the nested closed intervals is exactly one point.

Corollary 4.6 (to the Nested Closed Bounded Intervals Theorem in $\mathbb{X}$) *Suppose the hypotheses of Theorem 4.4 hold. In addition, suppose that there exists $r \in \mathbb{R}$ such that $0 < r < 1$ and*

$$\operatorname{diam}(\mathcal{I}_n) = r^{n-1} \operatorname{diam}(\mathcal{I}_1) \quad \textit{for } n \geq 1.$$

Then there exists exactly one point $x_0 \in \bigcap_{n=1}^{\infty} \mathcal{I}_n$.

PROOF Suppose two such points exist, call them x_0 and y_0. Then $x_0, y_0 \in \mathcal{I}_n$ for all $n \geq 1$. Therefore,

$$|x_0 - y_0| \leq \operatorname{diam}(\mathcal{I}_n) = r^{n-1} \operatorname{diam}(\mathcal{I}_1) \quad \text{for all } n \geq 1.$$

We may now choose n so large that $r^{n-1} \operatorname{diam}(\mathcal{I}_1) < |x_0 - y_0|$. (Why?) From this we obtain

$$|x_0 - y_0| \leq r^{n-1} \operatorname{diam}(\mathcal{I}_1) < |x_0 - y_0|,$$

which is clearly a contradiction. Therefore, we conclude that there can only be one point in $\bigcap_{n=1}^{\infty} \mathcal{I}_n$. ◆

► **2.55** *Answer the (Why?) question in the above proof.*

Note that in the statement of the above corollary we assume that the diameters of the nested intervals decrease to 0 *geometrically.* Later we will see that, in order to draw the same conclusion, all we really need is for the diameters to decrease to 0. However, this more general result will have to wait until we have defined the concept of *convergence.*

4.2 The Bolzano-Weierstrass Theorem

The Bolzano-Weierstrass theorem makes a significant statement about the structure of the spaces represented by $\mathbb{X}$ in which we have such a particular interest.[4] Its proof makes clever use of nested intervals, although the statement of the theorem makes no reference to such geometry. In fact, the result pertains to limit points, not intervals. Recall from our discussion of limit points in subsection 2.2 that some sets have no limit points, some sets have a finite number of limit points, and some sets have infinitely many. The Bolzano-Weierstrass theorem establishes that if $S \subset \mathbb{X}$ is a set of a particular type, then it must have at least one limit point.

[4]The analogous statement to the Bolzano-Weierstrass theorem with $\mathbb{X}$ replaced by a more general space is not true.

Theorem 4.7 (The Bolzano-Weierstrass Theorem)
Every bounded infinite set $S \subset \mathbb{X}$ has at least one limit point.

PROOF We begin with the case $S \subset \mathbb{R}$. Since S is bounded, there exists a closed interval I_1 such that $S \subset I_1$. Bisect I_1 to form two closed subintervals I_1', I_1''. Since S is an infinite set, either I_1' or I_1'' (or both) must contain an infinite number of points of S. Denote this subinterval by I_2. Note that the length of I_2 is half the length of I_1, i.e., $\ell(I_2) = \frac{1}{2}\ell(I_1)$. Repeat this process by bisecting I_2 into two closed subintervals I_2', I_2''. Similarly, at least one of these subintervals must contain infinitely many points of S. Denote it by I_3. Note that $\ell(I_3) = \frac{1}{2^2}\ell(I_1)$. Continuing in this way, we generate a sequence of closed intervals $\{I_n\}$, where

(*i*) $I_1 \supset I_2 \supset I_3 \supset \cdots$.
(*ii*) Each interval contains infinitely many points of S.
(*iii*) $\ell(I_n) = \frac{1}{2^{n-1}}\ell(I_1)$.

According to Corollary 4.6, there exists exactly one point in $\bigcap_{n=1}^{\infty} I_n$. Let x_0 be the unique point in $\bigcap_{n=1}^{\infty} I_n$. We will show that x_0 is a limit point of S. To see this, let $N_r'(x_0)$ be a deleted neighborhood of x_0. Since $x_0 \in I_n$ for all n, we may choose $K \in \mathbb{N}$ so large that $\ell(I_K) < r$, and therefore guarantee that $x_0 \in I_K \subset N_r(x_0)$. Since I_K contains infinitely many points of S, there must be a point y of S other than x_0 that lies in $N_r(x_0)$, i.e., $y \in N_r'(x_0)$. This establishes the result for $S \subset \mathbb{R}$.

Now let $S \subset \mathbb{R}^2$ be an infinite, bounded set. In a completely similar way, we construct a sequence of closed intervals $\{\mathcal{I}_n\}$ in $\mathbb{R}^2$ such that

(*i*) $\mathcal{I}_1 \supset \mathcal{I}_2 \supset \mathcal{I}_3 \supset \cdots$.
(*ii*) Each interval contains infinitely many points of S.
(*iii*) $\operatorname{diam}(\mathcal{I}_n) = \frac{1}{2^{n-1}} \operatorname{diam}(\mathcal{I}_1)$.

Then $\mathbf{x_0} \in \bigcap_{n=1}^{\infty} \mathcal{I}_n$ has the required properties to make it a limit point of S. The case for $S \subset \mathbb{C}$ is equivalent to that of $\mathbb{R}^2$, and the case for $S \subset \mathbb{R}^k$ is handled similarly. ♦

► **2.56** *Write the details of the proof of the above theorem for $\mathbb{R}^2$. Write the analogous proofs for $\mathbb{R}^k$ and for $\mathbb{C}$.*

In an intuitive, geometric sense, the Bolzano-Weierstrass theorem says that when you put an infinite number of points in a bounded subset of $\mathbb{X}$, the points must "accumulate" in at least one place within that bounded subset. Such a point of accumulation is, in fact, a limit point. We illustrate this important theorem with an example.

Example 4.8 *Let* $S = \left\{\frac{(-1)^n n}{n+1} : n \geq 1\right\} = \left\{-\frac{1}{2}, \frac{2}{3}, -\frac{3}{4}, \frac{4}{5}, \dots\right\} \subset \mathbb{R}$. *The set* S

is bounded since

$$\left|\frac{(-1)^n n}{n+1}\right| = \frac{n}{n+1} \le 1.$$

S is clearly also infinite, and so by the Bolzano-Weierstrass theorem, S must have at least one limit point. ◀

▶ **2.57** *In fact, both 1 and -1 are limit points for the set S of the previous example. Prove this.*

We will now use the Bolzano-Weierstrass theorem to generalize our nested closed bounded intervals theorem to nested, decreasing sets of a more arbitrary kind.

Theorem 4.9 (The Nested Closed Bounded Sets Theorem in $\mathbb{X}$)
Let $\mathcal{F}_1 \supset \mathcal{F}_2 \supset \mathcal{F}_3 \supset \cdots$ be a sequence of decreasing, closed, bounded, nonempty sets all contained in $\mathbb{X}$. Then $\bigcap_{j=1}^{\infty} \mathcal{F}_j \neq \varnothing$.

PROOF Let $x_j \in \mathcal{F}_j$ for each j, and let $S = \{x_1, x_2, x_3, \ldots\}$. There are two cases to consider: S possessing finitely many distinct values, and S possessing infinitely many distinct values. Suppose S possesses finitely many distinct values. Then one of the elements, say, x_L, must be repeated infinitely often in S. (Why?) This element must be in *every* set $\mathcal{F}_j$. (Why?) Therefore, $x_L \in \bigcap_{j=1}^{\infty} \mathcal{F}_j$. Now suppose S possesses infinitely many distinct values. Since S is also bounded, it must have a limit point, call it x. If we can show that x is in fact a limit point for *every* set $\mathcal{F}_j$, then $x \in \mathcal{F}_j$ for all j (Why?) and we are done. Consider an arbitrary $\mathcal{F}_j$ from the sequence, and let $N_r'(x)$ be any deleted neighborhood of x. This neighborhood must contain infinitely many points of S, since x is a limit point of S. That is, there are infinitely many x_ks in $N_r'(x)$, and in particular, at least one with index m_r greater than j. (Why?) Since the sequence $\{\mathcal{F}_j\}$ is decreasing, it is easy to see that x_{m_r} must be in $\mathcal{F}_j \cap N_r'(x)$. Since the deleted neighborhood $N_r'(x)$ of x was arbitrary, such a point x_{m_r} exists for every deleted neighborhood $N_r'(x)$ centered at x, and so x is a limit point of $\mathcal{F}_j$. Since $\mathcal{F}_j$ was arbitrary, x must be a limit point of every $\mathcal{F}_j$. ◆

Previously, we had defined what is meant by the diameter of an interval in $\mathbb{X}$. We now extend this idea to more general bounded sets in $\mathbb{X}$. Specifically, for any bounded set $S \subset \mathbb{X}$, we define the *diameter* of S by

$$\operatorname{diam}(S) \equiv \sup_{x,y \in S} \Big\{ |x - y| \Big\}.$$

That is, the diameter of S is the supremum of the set of all distances between pairs of points in S. With this definition in hand, we can state the following corollary to Theorem 4.9. This corollary is a natural generalization of Corollary 4.6.

Corollary 4.10 (to the Nested Closed Bounded Sets Theorem in $\mathbb{X}$) *Suppose the hypotheses of Theorem 4.9 hold. In addition, suppose that there exists $r \in \mathbb{R}$ such that $0 < r < 1$ and*

$$\operatorname{diam}(\mathcal{F}_n) = r^{n-1}\operatorname{diam}(\mathcal{F}_1) \quad \text{for } n \geq 1.$$

Then there exists exactly one point $x_0 \in \bigcap_{n=1}^{\infty} \mathcal{F}_n$.

► **2.58** *Prove the above corollary.*

5 Compactness and Connectedness

5.1 Compact Sets

We begin with a definition.

Definition 5.1 Let $A \subset \mathbb{X}$, and suppose S is a collection of elements. We call $\{\mathcal{O}_\alpha\}_{\alpha\in S}$ an **open covering** of A if the following two properties hold:

1. $\mathcal{O}_\alpha$ is open in $\mathbb{X}$ for all $\alpha \in S$.
2. $A \subset \bigcup_{\alpha\in S} \mathcal{O}_\alpha$.

As the above definition indicates, the number of open sets in an open covering need not be finite, or even countable. A somewhat general example gets us started.

Example 5.2 *Let $A \subset \mathbb{X}$ be any set. Then since $\mathbb{X}$ itself is open, it serves as an open covering of A. To obtain another covering, fix $r > 0$, and consider $N_r(x)$ for each $x \in A$. Then*

$$A \subset \bigcup_{x\in A} N_r(x)$$

is also an open covering of A. ◄

In dealing with open coverings of a set A, one might be troubled by the prospect of working with possibly uncountable index sets S as allowed in Definition 5.1. But one needn't worry. If $\{\mathcal{O}_\alpha\}_{\alpha\in S}$ is an open covering for a set $A \subset \mathbb{X}$, and if $A \subset \{\mathcal{O}_{\alpha'}\}_{\alpha'\in S'}$ where $S' \subset S$, we refer to $\{\mathcal{O}_{\alpha'}\}_{\alpha'\in S'}$ as an open *subcover* of A contained in the original open covering $\{\mathcal{O}_\alpha\}_{\alpha\in S}$. We will now exploit the density of the rationals in $\mathbb{R}$ to establish a rather convenient truth. For any set $A \subset \mathbb{X}$, if $\{\mathcal{O}_\alpha\}_{\alpha\in S}$ is an open covering of A, then we can always extract a *countable* open subcover $\{\mathcal{O}_j\}_{j=1}^{\infty} \subset \{\mathcal{O}_\alpha\}$ such that $A \subset \{\mathcal{O}_j\}_{j=1}^{\infty}$. To prove this, suppose the index set S on the sets $\{\mathcal{O}_\alpha\}$ is uncountable. For $p \in A$, there exists $\alpha_p \in S$ such that $p \in \mathcal{O}_{\alpha_p}$. Since $\mathcal{O}_{\alpha_p}$

is open, there exists a neighborhood of p such that $N_r(p) \subset \mathcal{O}_{\alpha_p}$. There also exists a point $p' \in N_r(p)$ having coordinates in $\mathbb{Q}$ that is the center of a neighborhood $N_\beta(p')$ of radius $\beta \in \mathbb{Q}$ such that $p \in N_\beta(p') \subset N_r(p) \subset \mathcal{O}_{\alpha_p}$. (Why?) Since $p \in A$ was arbitrary, this shows that each point of A is contained in such a neighborhood $N_\beta(p')$ for some rational β and some $p' \in \mathbb{X}$ having rational coordinates. We then have

$$A \subset \bigcup_{\beta, p'} N_\beta(p').$$

Therefore, the set A is covered by a collection of "rational" neighborhoods, a collection that *must* be countable. (Why?) From now on we may therefore presume that an open covering of a set $A \subset \mathbb{X}$ is a countable open covering.

► **2.59** *In the above, we argued that for arbitrary $p \in A$, there exists a neighborhood of p such that $N_r(p) \subset \mathcal{O}_{\alpha_p}$, and that there also exists a point $p' \in \mathbb{X}$ having coordinates in $\mathbb{Q}$ that is the center of a neighborhood $N_\beta(p')$ of radius $\beta \in \mathbb{Q}$ such that $p \in N_\beta(p') \subset N_r(p) \subset \mathcal{O}_{\alpha_p}$. Give the detailed argument for the existence of p' and β with the claimed properties.*

As convenient as it is to be able to presume that an open covering of any set $A \in \mathbb{X}$ is *countable*, it would be too greedy on our part to presume even more. In fact, it is a special class of sets that can always be covered by only a *finite* number of open sets. This special class of sets is defined next.

Definition 5.3 The set $A \subset \mathbb{X}$ is called **compact** if every open covering of A has a finite subcover.

Although not obvious at this stage of our development, the concept of compactness will be extremely important in our work. In fact, we have already made use of it in this chapter without explicitly referring to it. It happens to be a fact that within the spaces of interest to us, a set is compact if and only if it is closed and bounded. That is, within $\mathbb{X}$ the compactness of a set is equivalent to the set being closed and bounded.[5] As we will see, many significant results are related to compactness, primarily involving convergence (the topic of the next chapter), as well as several key properties of continuous functions (which we discuss in Chapter 4).[6] Such results might seem far removed from the above definition involving open coverings of a set. The following example illustrates the use of Definition 5.3 to establish that a set is compact.

Example 5.4 *Consider the set $A = \{0, 1, \frac{1}{2}, \frac{1}{3}, \dots\} \subset \mathbb{R}$. We will show that A is compact. To do this, let $\{\mathcal{O}_n\}$ be any open covering of A. Then $0 \in \mathcal{O}_M$*

[5]This equivalence is not true in more general spaces.

[6]Probably the best-known example to students of calculus is that a continuous function defined on a compact set assumes a maximum value and a minimum value somewhere within the set.

for some $M \geq 1$. Since $\mathcal{O}_M$ is open, there exists $N_r(0) \subset \mathcal{O}_M$. Now, choose positive integer $N > \frac{1}{r}$. Then, if $j > N$, we see that $\frac{1}{j} \in \mathcal{O}_M$. (Why?) Finally, for every integer $1 \leq j \leq N$ we have $\frac{1}{j} \in A$ and so $\frac{1}{j} \in \mathcal{O}_{m_j}$ for some $\mathcal{O}_{m_j} \in \{\mathcal{O}_n\}$. Therefore, $A \subset (\mathcal{O}_M \cup \mathcal{O}_{m_1} \cup \mathcal{O}_{m_2} \cup \cdots \cup \mathcal{O}_{m_N})$, a finite subcover for A. ◀

► **2.60** *Is the set given by $\left\{1, \frac{1}{2}, \frac{1}{3}, \dots\right\} \subset \mathbb{R}$ compact? (Answer: No. Prove it.)*

► **2.61** *Let $A = \left\{(0,0), (\frac{1}{m}, \frac{1}{n}) \in \mathbb{R}^2 : m, n \in \mathbb{N}\right\}$. Is this set compact? (Answer: No. Why is that?)*

5.2 The Heine-Borel Theorem

The Heine-Borel theorem[7] establishes the equivalence of compact sets and closed and bounded sets within $\mathbb{X}$. This theorem is of great practical importance since it allows for a simpler characterization of compactness when working in the spaces $\mathbb{R}$, $\mathbb{R}^k$, or $\mathbb{C}$.

Theorem 5.5 (The Heine-Borel Theorem)
The set $A \subset \mathbb{X}$ is compact if and only if it is closed and bounded.

PROOF[8] We begin by showing that if A is closed and bounded in $\mathbb{X}$, then A is compact. Suppose otherwise, namely, that there exists an open covering $\{\mathcal{O}_i\}_{i=1}^{\infty}$ having no finite subcover for A. Then, for every union $\bigcup_{i=1}^{n} \mathcal{O}_i$, there must be at least one point of A excluded from the union, i.e.,

$$E_n = A \cap \left[\bigcup_{i=1}^{n} \mathcal{O}_i\right]^C \neq \varnothing \text{ for } n = 1, 2, 3 \dots .$$

Since the sets E_n are closed, bounded, and decreasing (Why?), we may infer that there exists a point $x_0 \in \bigcap_{n=1}^{\infty} E_n$. Since $x_0 \in E_n = (A \cap \mathcal{O}_1^C \cap \mathcal{O}_2^C \cap \mathcal{O}_3^C \cap \cdots \cap \mathcal{O}_n^C)$ for all n, we see that $x_0 \in A$, and $x_0 \notin \mathcal{O}_i$ for all i. This contradicts the fact that $A \subset \bigcup_{i=1}^{\infty} \mathcal{O}_i$. Therefore, a finite subcover of A must, in fact, exist. To show that compactness of A implies that A is closed, assume A is compact. We will show that A^C is open. Consider an arbitrary $x \in A^C$, and let

$$\mathcal{O}_j(x) = \left\{y \in \mathbb{X} : |x - y| > \tfrac{1}{j}\right\}.$$

Then $\{\mathcal{O}_j(x)\}$ is a nested, increasing sequence of open sets in $\mathbb{X}$. Also, we have $\bigcup_{j=1}^{\infty} \mathcal{O}_j(x) = \mathbb{X} \setminus \{x\}$, and so $A \subset \bigcup_{j=1}^{\infty} \mathcal{O}_j(x)$. This means that the union

[7] It is interesting to note, although we shall not prove it here, that the Heine-Borel theorem is equivalent to the Bolzano-Weierstrass theorem.

[8] We borrow the proof from [Bar76].

$\bigcup_{j=1}^{\infty} \mathcal{O}_j(x)$ is an open cover for A, and so there exists a positive integer m such that

$$A \subset \bigcup_{j=1}^{m} \mathcal{O}_j(x) = \mathcal{O}_m(x).$$

Consider the neighborhood of radius $\frac{1}{m}$ centered at x, i.e., $N_{\frac{1}{m}}(x)$. Since $N_{\frac{1}{m}}(x) = \{y \in \mathbb{X} : |x - y| < \frac{1}{m}\}$, the established fact that $A \subset \mathcal{O}_m(x)$ implies that $N_{\frac{1}{m}}(x)$ contains no points of A. We have shown that there exists a neighborhood of x entirely contained in A^C. Since $x \in A^C$ was arbitrary, this shows that A^C is open and hence A is closed. To show that compactness of A implies boundedness of A, simply consider the open balls of radius j centered at the origin, denoted by $N_j(0)$. It is certainly true that $\mathbb{X} = \bigcup_{j=1}^{\infty} N_j(0)$, and so $A \subset \bigcup_{j=1}^{\infty} N_j(0)$. Since A is compact, there exists some positive integer m such that $A \subset \bigcup_{j=1}^{m} N_j(0) = N_m(0)$, and so A is bounded. ♦

An immediate consequence of the Heine-Borel theorem is that closed, bounded intervals in $\mathbb{X}$ are compact. Of course, such theorems as the nested closed bounded intervals theorem can now be more succinctly renamed.

► **2.62** *Give more concise names for Theorem 4.4 and Theorem 4.9.*

5.3 Connected Sets

Connectedness is one of those mathematical properties that seems easy to understand at first. In fact, most students probably feel that they already know what it means for a set to be connected. For them, this concept can only be complicated by mathematicians. But the notion is more subtle than novices realize. In making the determination of whether a set in $\mathbb{X}$ is *connected* or *disconnected* mathematically precise, mathematicians must keep in mind not just the obvious cases (although the definitions should certainly yield the intuitive results for them), but also the cases that are not so obvious. The value of mathematics is in its precision and its consistency, even if the price required is a bit of patience and the determined development of some technical skill. In fact, mathematicians find use for various kinds of connectedness, although we will limit ourselves to only those that are absolutely necessary for our development of analysis. We introduce what we consider our primary notion of connectedness in this section, leaving other types of connectedness to be defined later in our development as the need arises.

Connected and Disconnected Sets

We begin with our definition for a disconnected set.

Definition 5.6 A set $S \subset \mathbb{X}$ is called **disconnected** if there exist nonempty sets $A, B \subset \mathbb{X}$ such that the following are true:

1. $S = A \cup B$.
2. $A \cap \overline{B} = \overline{A} \cap B = \varnothing$.

Any set $S \subset \mathbb{X}$ that is not disconnected is called **connected.**

While this definition might initially appear overly complicated, a little thought reveals its straightforward logic. Any disconnected set must consist of at least two separated subsets, which motivates condition 1. The sense in which these subsets are separated from each other is characterized by condition 2, namely, no point of the subset A can be a member of the subset B or its boundary, and no point of the subset B can be a member of the subset A or its boundary. A few examples will illustrate the idea.

Example 5.7 *Let $S = \{x \in \mathbb{R} : 0 < |x - 3| < 1\} = N_1'(3)$. We will show that S is disconnected. To do this, make the rather obvious choices of letting $A = (2, 3)$ and $B = (3, 4)$. Then clearly $A \cup B = S$, while $A \cap \overline{B} = (2, 3) \cap [3, 4] = \varnothing$ and $\overline{A} \cap B = [2, 3] \cap (3, 4) = \varnothing$.* ◀

Example 5.8 *The set $\mathbb{Q}$ is disconnected. To see this, let $A = \mathbb{Q} \cap (-\infty, \sqrt{2})$ and let $B = \mathbb{Q} \cap (\sqrt{2}, \infty)$. We leave it to the reader to verify that*

$$\overline{A} = \overline{\mathbb{Q}} \cap (-\infty, \sqrt{2}] = \mathbb{R} \cap (-\infty, \sqrt{2}] = (-\infty, \sqrt{2}], \quad \text{and} \quad \overline{B} = [\sqrt{2}, \infty). \qquad (2.1)$$

From this we see that

$$A \cap \overline{B} = \mathbb{Q} \cap (-\infty, \sqrt{2}) \cap [\sqrt{2}, \infty) = \varnothing,$$

and

$$\overline{A} \cap B = (-\infty, \sqrt{2}] \cap \mathbb{Q} \cap (\sqrt{2}, \infty) = \varnothing.$$ ◀

► **2.63** *In general, for two sets $E, F \subset \mathbb{X}$, we have that $\overline{E \cap F} \subset \overline{E} \cap \overline{F}$. Find a pair of sets in $\mathbb{R}$ such that the closure of their intersection is a proper subset of the intersection of their closures. Now note that in (2.1) of the above example we have equalities, that is, in this case we have that the closure of the intersection equals the intersection of the closures. Verify (2.1), and determine why equalities hold here.*

We continue our discussion of connectedness with two useful propositions.

Proposition 5.9 *Suppose $S \subset \mathbb{X}$ is disconnected, i.e., there exist nonempty sets $A, B \subset \mathbb{X}$ such that*

(i) $S = A \cup B$,

(ii) $A \cap \overline{B} = \overline{A} \cap B = \varnothing$.

If C is a connected subset of S, then either $C \subset A$ or $C \subset B$.

PROOF Suppose $C \subset S$ is connected. Then since C is a subset of S, in combination with property (i) in the statement of the proposition, we have

$$C = C \cap S = C \cap (A \cup B) = (C \cap A) \cup (C \cap B).$$

Now, by property *(ii)* in the statement of the proposition, we have

$$(C \cap A) \cap \overline{(C \cap B)} = \overline{(C \cap A)} \cap (C \cap B) = \varnothing. \quad \text{(Why?)}$$

Since C is connected, it follows that $C \cap A = \varnothing$ or $C \cap B = \varnothing$ (Why?), which implies $C \subset B$, or $C \subset A$, respectively. ♦

► **2.64** *Answer the two (Why?) questions in the above proof.*

Proposition 5.10 *Suppose $\{A_\alpha\}$ is a collection of connected subsets of $\mathbb{X}$ and $\bigcap A_\alpha \neq \varnothing$. Then $\bigcup A_\alpha$ is connected.*

PROOF Suppose $\bigcup A_\alpha$ is disconnected. That is, suppose C and D are both nonempty subsets of $\mathbb{X}$, where

(*i*) $\bigcup A_\alpha = C \cup D$,

(*ii*) $C \cap \overline{D} = \overline{C} \cap D = \varnothing$.

We will show that this leads to a contradiction. Since each A_α is connected, it follows from the previous proposition that for each A_α, we have $A_\alpha \subset C$, or $A_\alpha \subset D$, but not both. Since $\bigcap A_\alpha \neq \varnothing$, there exists a point $p \in \bigcap A_\alpha$. This point p must also be in $\bigcup A_\alpha$, and so p belongs to either C or D, but not both. From this, it must be true that either $\bigcup A_\alpha \subset C$, in which case D must be empty, or $\bigcup A_\alpha \subset D$, in which case C must be empty. But this contradicts the assumption that C and D are both nonempty. Therefore, we must have that $\bigcup A_\alpha$ is connected. ♦

Probably the most obvious class of sets that are connected are the intervals contained in $\mathbb{X}$. We state this broad claim in the form of a proposition.

Proposition 5.11 *Let $\mathcal{I} \subset \mathbb{X}$ be any interval. Then $\mathcal{I}$ is connected.*

PROOF We prove the case for $\mathbb{R}$, leaving the other cases to the exercises. To begin, we will assume that an arbitrary interval $I \subset \mathbb{R}$ is disconnected, and derive a contradiction. Since I is assumed to be disconnected, there exists a pair of nonempty sets $A, B \subset \mathbb{R}$ such that

(*i*) $I = A \cup B$,

(*ii*) $A \cap \overline{B} = \overline{A} \cap B = \varnothing$.

Without loss of generality, choose $a \in A$, and $b \in B$ such that $a < b$. Since $a, b \in I$, it follows that $[a, b] \subset I$. Now let $c = \sup(A \cap [a, b])$. Then clearly $c \in \overline{A \cap [a, b]} \subset \overline{A} \cap [a, b] \subset I$ (see Exercise 2.63). By *(ii)* it follows that $c \notin B$. By *(i)* we then have that $c \in A$, and so again by *(ii)* we may conclude that $c \notin \overline{B}$. But, since $c = \sup(A \cap [a, b])$, we must have that $(c, b] \subset B$ and so $c \in \overline{B}$, a contradiction. ♦

► **2.65** *Complete the proof of the above proposition for the cases $\mathbb{R}^k$ and $\mathbb{C}$. To get started, consider the case of $\mathbb{R}^2$ as follows. Suppose $E, F \subset \mathbb{R}$ are connected sets. Prove that $E \times F$ is a connected set in $\mathbb{R}^2$. (Hint: Fix a point $(a, b) \in E \times F$, and show that $\{a\} \subset E$ and $\{b\} \subset F$ are connected. Then show that $(\{a\} \times F) \cup (E \times \{b\})$ is connected.) Extend this result to $\mathbb{R}^k$.*

The following important property of $\mathbb{X}$ is an immediate corollary to the previous proposition.

Corollary 5.12 *$\mathbb{X}$ is connected.*

► **2.66** *Prove the above corollary.*

6 Supplementary Exercises

1. *More than one "distance" might exist on a given space, and they need not all be induced by norms. As an example, consider the "discrete distance" defined as follows. For any $x, y \in \mathbb{X}$ denote the discrete distance between x and y by $d(x, y)$ and define it by $d(x,y) = \begin{cases} 0 & \text{if } x = y \\ 1 & \text{if } x \neq y \end{cases}$. First, show that the four properties of distance given on page 42 all hold for $d(\cdot,\cdot)$. Then, fix a point $\mathbf{x}_0$ in $\mathbb{R}^2$. Show that, with the discrete distance, all spheres around $\mathbf{x}_0 \in \mathbb{R}^2$ having radius $0 < r < 1$ are empty.*

2. *Consider the discrete distance function described in the previous exercise, and fix y at 0. Is the resulting function $d(\cdot, 0)$ a norm on $\mathbb{X}$? That is, is the function $d(\cdot, 0) : \mathbb{X} \to \mathbb{R}$ where $d(x,0) = \begin{cases} 0 & \text{if } x = 0 \\ 1 & \text{if } x \neq 0 \end{cases}$ a norm on $\mathbb{X}$? (Answer: No.)*

3. *Why do we specify in Definitions 1.10 and 1.11 that the set S be nonempty?*

4. *Let $S = \left\{\frac{(-1)^n}{n+1} : n \geq 1\right\} = \left\{-\frac{1}{2}, \frac{1}{3}, -\frac{1}{4}, \frac{1}{5}, \dots\right\}$. What are the inf and sup of S? Show that neither belongs to S. What are the max and min of S?*

5. *Suppose S is a nonempty, bounded set of real numbers, and let $M_S = \sup S$ and $m_S = \inf S$. For $c \in \mathbb{R}$ fixed, let cS be the set defined by $cS = \{cs : s \in S\}$. Prove that cS is bounded. Also, find the relationships among $M_S, m_S, M_{cS} \equiv \sup(cS)$, and $m_{cS} \equiv \inf(cS)$.*

6. *Suppose S and T are nonempty, bounded sets of real numbers, and let $M_S = \sup S, m_S = \inf S, M_T = \sup T$, and $m_T = \inf T$. Define the set $S + T$ by $S + T \equiv \{s + t : s \in S, t \in T\}$. Prove that $S + T$ is bounded and find the relationships among M_S, m_S, M_T, m_T, and $M_{S+T} \equiv \sup(S + T)$ and $m_{S+T} \equiv \inf(S + T)$.*

7. *Suppose S and T are nonempty, bounded sets of real numbers, and let $M_S = \sup S, m_S = \inf S, M_T = \sup T$, and $m_T = \inf T$. Define the set ST by $ST \equiv \{st : s \in S, t \in T\}$. Investigate the set ST as in the previous two exercises.*

8. *In the previous chapter, we stated the Archimedean property. Give a proof of this result that uses the method of contradiction and Theorem 1.12.*

9. *Consider $\mathbf{x_0} \in \mathbb{R}^2$ and the discrete distance. Describe the sphere of radius $r = 1$. Describe the spheres of radius $r > 1$.*

10. *Extend the norm $|\cdot|_1$ on $\mathbb{R}^2$ as defined in Example 2.6 on page 17 to the space $\mathbb{R}^3$ in the natural way, and verify that the result is, in fact, a norm on $\mathbb{R}^3$. Using the distance associated with this norm, describe the sphere of radius r centered at $\mathbf{x_0} = (\xi, \eta, \zeta) \in \mathbb{R}^3$.*

11. *Let $A = \left\{\frac{1}{n} : n \in \mathbb{N}\right\} \subset \mathbb{R}$. Show that 0 is a boundary point of A that is not an element of A.*

12. *Let $A = \left\{\frac{(-1)^n n}{n+1} : n \in \mathbb{N}\right\} \subset \mathbb{R}$. Show that ± 1 are boundary points of A that are not elements of A.*

13. *Let $A = \left\{\mathbf{x} = (x_1, x_2, x_3) \in \mathbb{R}^3 : |x_1| + |x_2| + |x_3| < 1\right\}$. Let $\mathbf{x_1} = \left(\frac{1}{2}, 0, 0\right)$, $\mathbf{x_2} = (5, 1, -2)$, and $\mathbf{x_3} = (0, 1, 0)$. Show that $\mathbf{x_1}$ is an interior point, $\mathbf{x_2}$ is an exterior point, and $\mathbf{x_3}$ is a boundary point of A.*

14. *Consider the distance on $\mathbb{R}^3$ defined by $d_1(\mathbf{x}, \mathbf{y}) \equiv |y_1 - x_1| + |y_2 - x_2| + |y_3 - x_3|$. Using this distance, determine whether each of the points $\mathbf{x_1}, \mathbf{x_2}$, and $\mathbf{x_3}$ as specified in the previous exercise is an interior, exterior, or boundary point relative to the set $A = \left\{\mathbf{x} = (x_1, x_2, x_3) \in \mathbb{R}^3 : |x_1| + |x_2| + |x_3| < 1\right\}$.*

15. *Let $A \subset \mathbb{R}^2$ consist of the pairs of points $(x_1, x_2) \in \mathbb{R}^2$ such that x_1 and x_2 are irrational and positive. Find the interior, exterior, and boundary points of A.*

16. *Let $A = \left\{\frac{(-1)^n n}{n+1} : n \in \mathbb{N}\right\} \subset \mathbb{R}$. Show that 1 is a boundary point of A that is not an element of A. Also show that 1 is a limit point of A.*

17. *Consider $\mathbb{Q}^2 = \{(q_1, q_2) : q_1, q_2 \in \mathbb{Q}\} \subset \mathbb{R}^2$. Find $\left(\mathbb{Q}^2\right)'$.*

18. *Consider $A \subset \mathbb{R}^2$ where $A = \left\{ \left(\frac{1}{n}, \frac{1}{m}\right) : m, n \in \mathbb{N} \right\}$. Find A'.*

19. *Consider $\mathbb{Z}^2 \subset \mathbb{R}^2$ where $\mathbb{Z}^2 = \{(m, n) : m, n \in \mathbb{Z}\}$. Find $(\mathbb{Z}^2)'$. What are the isolated points and the limit points of $\mathbb{Z}^2$?*

20. *Consider $\mathbb{Q}^2 \subset \mathbb{R}^2$. Find the isolated points of $\mathbb{Q}^2$.*

21. *Suppose $S \subset \mathbb{X}$ is a finite set. Find the isolated points and the limit points of S.*

22. *Consider $S \subset \mathbb{C}$ where $S = \left\{\frac{1}{m} + i\frac{1}{n} : m, n \in \mathbb{N}\right\}$. What are the isolated points and limit points of S?*

23. *Consider $S \subset \mathbb{R}^2$ where $S = \left\{\left(\frac{m}{m+1}, \frac{n}{n+1}\right) : m, n \in \mathbb{N}\right\}$. What are the isolated points and limit points of S?*

24. *Show that if $x \in A \subset \mathbb{X}$ is an interior point of A, then x is not an isolated point of A.*

25. *Show that if $x \in \mathbb{X}$ is an exterior point of A, then x can't be an isolated point of A^C.*

26. *Suppose $x \in A \subset \mathbb{X}$ is a boundary point of A. Then x can be an isolated point of A. Give an example.*

27. *Suppose $G \subset \mathbb{X}$ is open, and $x_0 \in \mathbb{X}$ is not a limit point of G. Show that x_0 must be an exterior point of G.*

28. *Show that the union in the statement of Proposition 3.6 on page 57 can be replaced by a countable union.*

29. *Consider $A = \{(-\infty, 1] \cup [3, \infty)\} \times \{0\} \subset \mathbb{R}^2$. Show that A is closed. How about $A = \{(-\infty, 1] \cup [3, \infty)\} \times \{0\} \times \{0\} \subset \mathbb{R}^3$?*

30. *Is the real axis in $\mathbb{C}$ open or closed? (Answer: closed.)*

31. *Find $\overline{A}$ for each of the following.*

a) $A = \left\{\left(\frac{m}{n}, \frac{n}{m}\right) \in \mathbb{R}^2 : m, n \in \mathbb{N}\right\}$ b) $A = \left\{\left(\frac{n}{n+1}, \frac{m}{n}\right) \in \mathbb{R}^2 : m, n \in \mathbb{N}\right\}$
c) $A = \left\{\left(\frac{m}{n}, \frac{k}{n}\right) \in \mathbb{R}^2 : k, m, n \in \mathbb{N}\right\}$ d) $A = \left\{\frac{1}{n} + i\,\frac{1}{m} \in \mathbb{C} : m, n \in \mathbb{N}\right\}$
e) $A = \left\{\left(\cos(\frac{m}{n}\pi), \sin(\frac{m}{n}\pi)\right) \in \mathbb{R}^2 : m, n \in \mathbb{N}\right\}$

32. *Find $\overline{\mathbb{X}}$ and $\overline{\varnothing}$.*

33. *Find a set in $\mathbb{R}^2$ having no interior points, yet whose closure is $\mathbb{R}^2$.*

34. *Consider $S \subset \mathbb{X}$, and let $\{F_\alpha\}$ be the collection of all closed sets that contain S. Prove that $\overline{S} = \bigcap_\alpha F_\alpha$.*

35. *Suppose A_1 and A_2 are arbitrary sets of real numbers. Then $A_1 \times A_2$ is a subset of $\mathbb{R}^2$. Show that $\overline{A_1 \times A_2} = \overline{A_1} \times \overline{A_2}$. More generally, if A_j is an arbitrary set of real numbers for $j = 1, \ldots, k$, then $\mathcal{A} = A_1 \times \cdots \times A_k$ is a subset of $\mathbb{R}^k$. Show that $\overline{\mathcal{A}} = \overline{A_1} \times \cdots \times \overline{A_k}$.*

36. *In Proposition 3.4 on on page 56 the claim was made that the entire space $\mathbb{X}$ is open. Of course, this refers to the spaces $\mathbb{R}$, $\mathbb{R}^k$, and $\mathbb{C}$. What if the space is $\mathbb{Z}$?*

37. *Let $A = \left\{\mathbf{x} = (x_1, x_2) \in \mathbb{R}^2 : |\,x_1\,| + |\,x_2\,| = 1\right\}$. Show that A is closed.*

38. *Consider the countably infinite set of natural numbers $\mathbb{N} \subset \mathbb{R}$. Is $\mathbb{N}$ open or closed? (Answer: closed.) What does your answer tell you about the integers $\mathbb{Z} \subset \mathbb{R}$?*

39. *Consider the set* $A = \{1, \frac{1}{2}, \frac{1}{3}, \frac{1}{4}, \dots\}$. *Is* A *open or closed? (Hint: Consider* A^C *and write it as a union of sets.)*

40. *Consider the countably infinite set of rationals* $\mathbb{Q} \subset \mathbb{R}$. *Is* $\mathbb{Q}$ *open or closed?*

41. *Let* $A = \left\{\frac{(-1)^n n}{n+1} : n \in \mathbb{N}\right\}$. *Find* $\overline{A}$.

42. *Consider the set* $\mathbb{Q}^2 = \{(q_1, q_2) : q_1, q_2 \in \mathbb{Q}\} \subset \mathbb{R}^2$. *Show that* $\overline{\mathbb{Q}^2} = \mathbb{R}^2$.

43. *Consider* $A, B \subset \mathbb{X}$.

a) *Show that it is not always true that* $\overline{A - B} = \overline{A} - \overline{B}$.

b) *Show that* $\partial A = \overline{A} \cap \overline{A^C}$.

44. *Show that the set* $U = \{z \in \mathbb{C} : \text{Im } z > 0\}$ *is open relative to* $S = \{z \in \mathbb{C} : \text{Im } \geq 0\}$.

45. *Prove that there exists no proper subset of* $[0,1]$ *that is closed in* $[0,1]$ *but not closed in* $\mathbb{R}$.

46. *Consider the set* $\mathbb{Q} \subset \mathbb{R}$. *Find a nontrivial subset of* $\mathbb{Q}$ *(i.e., not* $\varnothing$ *or* $\mathbb{Q}$ *itself) that is open in* $\mathbb{Q}$. *Find one that is closed in* $\mathbb{Q}$.

47. *Redo the previous exercise by replacing* $\mathbb{Q}$ *with* $\mathbb{I}$, *the set of irrational numbers. Redo it again by replacing* $\mathbb{Q}$ *with* $\mathbb{Z}$.

48. *In this exercise, you will derive another proof that* $\mathbb{Z}$ *is not dense in* $\mathbb{R}$. *Prove that* $\overline{\mathbb{Z}} = \mathbb{Z}$, *and apply the definition for denseness. In fact, the elements of* $\mathbb{Z}$ *are isolated since for each element in* $\mathbb{Z}$ *you can find a neighborhood within* $\mathbb{R}$ *containing no other elements of* $\mathbb{Z}$. $\mathbb{Z}$ *is an example of a* nowhere dense *set, the antithesis of a dense set.*

49. *An alternative characterization of a set* $A \subset \mathbb{X}$ *being nowhere dense in* $\mathbb{X}$ *is to check that the interior of the closure of* A *is empty. Show that* $\mathbb{Z}$ *is nowhere dense in* $\mathbb{R}$ *according to this definition. Is the interior of the closure of a set the same thing as the closure of the interior of a set? Determine the answer to this question by considering* $\mathbb{Z}$ *and* $\mathbb{Q}$ *in* $\mathbb{R}$.

50. *A set that is nowhere dense in* $\mathbb{X}$ *need not consist only of isolated points. Give an example.*

51. *Prove that* $\mathbb{Q} \times \mathbb{Q}$ *is dense in* $\mathbb{R}^2$.

52. *In part b) of Proposition 3.20 on page 62, if "neighborhood" is replaced with "deleted neighborhood," is the resulting statement still equivalent to parts a) and c)? Why or why not?*

53. *Consider the collection of intervals given by* $\{[n, n+1]\}$ *for* $n \in \mathbb{N}$. *Verify that* $\bigcap_{n=1}^{\infty}[n, n+1]$ *is empty. Does this contradict Theorem 4.4 on page 64?*

54. *Consider the collection of intervals given by* $\{[n-\frac{1}{n}, n+\frac{1}{n}]\}$ *for* $n \in \mathbb{N}$. *Verify that* $\bigcap_{n=1}^{\infty}[n-\frac{1}{n}, n+\frac{1}{n}]$ *is empty. Does this contradict Theorem 4.4?*

55. *Write a different, "component-wise" proof of the Bolzano-Weierstrass theorem for the case* $\mathbb{R}^2$ *that relies on the* $\mathbb{R}$ *case of the theorem.*

56. *Let* $\mathbb{Y}$ *be the set of all rational numbers between 0 and 3, and consider the usual distance as given by* $|x-y|$ *for all* $x, y \in \mathbb{Y}$. *Let* s_1 *be any element of* $\mathbb{Y} \cap (\sqrt{2}-1, \sqrt{2}+1)$. *Let* s_2 *be any element from* $\mathbb{Y} \cap (\sqrt{2}-\frac{1}{2}, \sqrt{2}+\frac{1}{2})$ *different from* s_1. *Similarly, let* s_n *be any element of* $\mathbb{Y} \cap (\sqrt{2}-\frac{1}{n}, \sqrt{2}+\frac{1}{n})$ *such that* s_n *is different from every previous* s_m *for* $m < n$. *This defines a set* $S = \{s_1, s_2, \ldots\} \subset \mathbb{Y}$ *that is bounded and infinite. Show that* S *has no limit points in* $\mathbb{Y}$. *Does this contradict the Bolzano-Weierstrass theorem? Why or why not? What property does* $\mathbb{Y}$ *lack that seems to be critical to the theorem according to this example?*

57. *Let* F_1 *be the points on and interior to the triangle in* $\mathbb{C}$ *having vertices* $0, 2$, *and* i. *Connect the midpoints of each side of* F_1 *to form the triangle* F_2. *Continue in this way to connect the midpoint of each side of* F_2 *to form the triangle* F_3, *and so on. Show that there is a unique point* $z_0 \in \bigcap_{n=1}^{\infty} F_n$.

58. *Suppose* $F \subset \mathbb{X}$ *is closed,* $K \subset \mathbb{X}$ *is compact, and* $F \subset K$. *Show that* F *is compact.*

59. *Suppose* K_j *for* $j \geq 1$ *are compact subsets of* $\mathbb{X}$. *Show that* $K_1 \cap K_2$ *is compact. Show the same for* $\bigcap_{j=1}^{m} K_j$. *This is no longer necessarily true for* $\bigcap_{j=1}^{\infty} K_j$. *Give a counterexample. Suppose* $\{K_\alpha\}$ *is a collection of compact subsets of* $\mathbb{X}$. *Is* $\bigcap K_\alpha$ *compact?*

60. *For two subsets* A, B *of* $\mathbb{X}$, *define* $A + B \equiv \{a + b \in \mathbb{X} : a \in A, b \in B\}$. *Assume* A *is open and* B *is arbitrary. Prove that* $A + B$ *is open.*

61. *Suppose* A *and* B *are compact subsets of* $\mathbb{R}^n$. *Prove that* $A + B$ *is compact by first showing that* $A + B$ *is bounded, and then showing that* $A + B$ *is closed.*

62. *Exhibit an infinite disconnected set whose only connected subsets contain one point.*

63. *Suppose* A *is a subset of* $\mathbb{X}$.

a) *Show that* $A' = \left(\overline{A}\right)'$.

b) *Is it necessarily true that* $\left(\overline{A}\right)' = \overline{(A')}$? *If so, prove it. If not, provide a counterexample.*

c) *Show that* A' *is closed.*

64. *Consider* $\mathbb{X}$ *with the discrete distance* $d(x,y) = \begin{cases} 1 & \text{if } x \neq y \\ 0 & \text{if } x = y \end{cases}$. *With this distance function, which subsets of* $\mathbb{X}$ *are open? Which are closed? Which (if any) are both open and closed?*

65. *Consider $\mathbb{R}$ with $d(x,y) = \frac{|x-y|}{1+|x-y|}$. Does d have the distance properties specified on page 42?*

66. *Show that every open set in $\mathbb{R}$ is the union of disjoint open intervals. Can the same be said of every open set in $\mathbb{R}^k$?*

67. *Is $\mathbb{I}$ connected? How about $\mathbb{Z}$?*

68. *Can you show that for any $x \in \mathbb{X}$, and any $r > 0$, the set $N_r(x)$ is connected? What about $N_r'(x)$?*

69. *Let $A \subset \mathbb{X}$ be connected and $A \subset S \subset \overline{A}$. Show that S is also connected.*

70. *Is $\varnothing$ connected?*

71. *Prove that intervals are the only connected subsets of $\mathbb{R}$. How about in $\mathbb{R}^2$?*

72. *Suppose for integers $n \geq 1$, the sets $\{A_n\} \subset \mathbb{X}$ are such that every A_n is connected and $A_n \cap A_{n+1} \neq \varnothing$. Show that $\bigcup A_n$ is connected.*

73. *Show that the only subsets of $\mathbb{X}$ that are both open and closed are $\varnothing$ and $\mathbb{X}$ itself.*

74. *Suppose $S \subset \mathbb{X}$ is such that S and S^C are each nonempty. Show from this that $\partial S \neq \varnothing$.*

75. The Cantor set
Let $C_0 = [0,1]$ be the closed unit interval in $\mathbb{R}$. Remove the middle third $G_1 = \left(\frac{1}{3}, \frac{2}{3}\right)$ to form the set $C_1 = \left[0, \frac{1}{3}\right] \cup \left[\frac{2}{3}, 1\right]$. Remove the middle third of each interval comprising C_1, that is, $G_2 = \left(\frac{1}{3^2}, \frac{2}{3^2}\right) \cup \left(\frac{7}{3^2}, \frac{8}{3^2}\right)$, to form C_2. Continue in this manner of removing middle thirds of each interval comprising C_n to form a sequence of nested sets,

$$C_0 \supset C_1 \supset C_2 \supset \cdots.$$

Defining $C \equiv \bigcap_{n=1}^{\infty} C_n$ to be the Cantor set, *establish the following:*

a) $C \neq \varnothing$.

b) C is closed.

c) C is compact.

d) C_n is the union of 2^n disjoint intervals, each having length $\left(\frac{1}{3}\right)^n$. What is the total length of C_n? What is the total "length" of C? (Hint: Can you determine the length of G_n and hence of $\bigcup G_j$?)

e) No interval of the form $\left(\frac{3k+1}{3^m}, \frac{3k+2}{3^m}\right)$ where $k, m > 0$ contains any elements of C.

f) C contains no intervals (α, β) for any $\alpha < \beta \in \mathbb{R}$.

g) C contains at least a countably infinite subset.

h) What are the boundary points of C?

i) What are the limit points of C?

j) Does C have any isolated points? If so, what are they?

k) Does C have any interior points? If so, what are they?

l) Is C connected? Is there any proper subset of C that is connected? (Answer: No.)

m) Each element of $\mathbb{Z} \subset \mathbb{R}$ is an isolated point of $\mathbb{Z}$, since one can find a deleted neighborhood of any integer that contains no integers. This is not so of the Cantor set, however. Can you show that any deleted neighborhood of any point $c \in C$ must contain elements of C?

n) A set $A \subset \mathbb{X}$ is called nowhere dense *if the set's closure has empty interior. The set of integers $\mathbb{Z} \subset \mathbb{R}$ is clearly nowhere dense. Can you show that the Cantor set $C \subset \mathbb{R}$, is, like $\mathbb{Z}$, nowhere dense? That is, can you show that $\overline{C}$ has empty interior?*

3

Limits and Convergence

The infinite! No other question has ever moved so profoundly the spirit of man.

David Hilbert

One could argue that the concept of *limit* is the most fundamental one in analysis. Two of the most important operations in a first-year calculus course, the derivative and the integral, are in fact defined in terms of limits, even though many first-year calculus students willingly forget this is so. We begin this chapter by considering limits of sequences. The limit of a sequence of numbers, while the simplest kind of limit, is really just a special case of the limit of a sequence of vectors. In fact, real numbers can be considered geometrically as points in one-dimensional space, while vectors with k real components are just points in k-dimensional space. The special case of $k = 2$ corresponds to limits of sequences of points in $\mathbb{R}^2$ and to limits of sequences of points in $\mathbb{C}$. Whether a sequence of real numbers, a sequence of real vectors, or a sequence of complex numbers has a well-defined limit is just a matter of determining whether the sequence of points is *converging* in some sense to a unique point in the associated space. This notion of *convergence* is one of the distance-related concepts referred to in the previous chapter, and it is common to all the spaces of interest to us. For this reason, we will again use the symbol $\mathbb{X}$ to denote any of the spaces $\mathbb{R}$, $\mathbb{R}^k$, or $\mathbb{C}$ in those cases where the results apply to all of them. After establishing the ideas underlying convergence of sequences in $\mathbb{X}$, we develop the related notion of a *series*, whereby the terms of a sequence are added together. As we will see, whether a series converges to a well-defined sum depends on the behavior of its associated sequence of *partial sums*. While this definition of convergence for a series is both efficient and theoretically valuable, we will also develop tests for convergence that in many cases are easier to apply.

1 Definitions and First Properties

1.1 Definitions and Examples

We begin by defining the basic idea of a sequence of elements in $\mathbb{X}$. In what follows, unless specifically noted otherwise, we emphasize that references to elements $x \in \mathbb{X}$ include the cases of $x \in \mathbb{R}$, $\mathbf{x} \in \mathbb{R}^k$, and $z \in \mathbb{C}$. In statements of results that rely on the associated algebra of the space (in particular, whether or not the space possesses the field properties), the definition or theorem will mention this fact, and refer to $x \in \mathbb{R}$ and $z \in \mathbb{C}$ more explicitly and exclusively.

Definition 1.1 Consider the infinite, ordered list of elements from $\mathbb{X}$ given by $x_1, x_2, x_3, \ldots$. We refer to such an ordered list as a **sequence**, and denote it more compactly by $\{x_n\}$ for $n \in \mathbb{N}$.

It is important to note that while we define a sequence as starting with $n = 1$ for convenience, in fact, a sequence can start with n at any integer value. For this reason, if the context is clear, the specification of the index set for n is often omitted. Several examples of sequences follow.

Example 1.2 *We look at several examples of sequences.*

a) *Consider* $x_n = \frac{1}{n} \in \mathbb{R}$ *for* $n \geq 1$. *This is called* the harmonic sequence,

$$1, \tfrac{1}{2}, \tfrac{1}{3}, \tfrac{1}{4}, \ldots.$$

b) *Consider* $x_n = (-1)^n n \in \mathbb{R}$ *for* $n \geq 1$. *This is the sequence*

$$-1, 2, -3, 4, -5, 6, \ldots.$$

c) *Let* $\mathbf{x}_n = \left(n, \frac{1}{n}\right) \in \mathbb{R}^2$ *for* $n \geq 1$. *This is the sequence*

$$(1, 1), \left(2, \tfrac{1}{2}\right), \left(3, \tfrac{1}{3}\right), \left(4, \tfrac{1}{4}\right), \ldots.$$

d) *Let* $z_n = \left(\frac{1+i}{4}\right)^n \in \mathbb{C}$ *for* $n \geq 0$. *This is the sequence*

$$1, \tfrac{1+i}{4}, \left(\tfrac{1+i}{4}\right)^2, \left(\tfrac{1+i}{4}\right)^3, \left(\tfrac{1+i}{4}\right)^4, \ldots.$$

◀

As can be seen from the above examples, along with a bit of intuition, some sequences seem to "converge" to a fixed, particular member of the corresponding space, while others clearly do not. Even more interesting are those sequences for which this "convergence" behavior is not so obvious either way. To clarify this determination, and to give it precision, we present the following definition.

Definition 1.3 Let $\{x_n\}$ be a sequence of elements from $\mathbb{X}$. We say that the sequence **converges** to $x \in \mathbb{X}$, and we write

$$\lim_{n\to\infty} x_n = x,$$

if for every $\epsilon > 0$, there exists an $N \in \mathbb{N}$ such that

$$n > N \Rightarrow |x_n - x| < \epsilon.$$

Otherwise, we say that the sequence **diverges**.

A more concise notation for $\lim_{n\to\infty} x_n$ when no possibility of ambiguity exists is $\lim x_n$. This alternate notation will also be used in what follows.

Loosely speaking, the above definition says that the sequence $\{x_n\}$ converges to x if we can get (and remain) as close as we like to x by going out far enough in the sequence. A couple of things are worth noting:

1. In general N will depend on ϵ.
2. N is not unique, since any $N' \geq N$ will work as well.

A few examples will show how to effectively use Definition 1.3 to establish limits of sequences. The goal is always to find an $N \in \mathbb{N}$ that accomplishes the task, given some value of ϵ with which to work. In these examples it might seem strange at first that we begin with the conclusion—that is, we start with the expression $|x_n - x|$, and set it less than ϵ. We then determine from this inequality how large N needs to be. Ultimately, once we've found the N that works, we've done the hard part. To be completely formal, we should then rewrite the overall statement in the order presented in Definition 1.3.

Example 1.4 *Consider the sequence $\{x_n\}$ where $x_n = \frac{n+1}{n} \in \mathbb{R}$ for $n \geq 1$. We will show, according to Definition 1.3, that the limit of this sequence is 1. To begin, suppose we have an arbitrary $\epsilon > 0$ with which to work. The distance from the nth term in the sequence to the proposed limit value of 1 is given by*

$$\left|\frac{n+1}{n} - 1\right| = \left|\frac{1}{n}\right| = \frac{1}{n}.$$

This distance will be less than ϵ if $\frac{1}{n} < \epsilon$, or $n > \frac{1}{\epsilon}$. That is,

$$n > \frac{1}{\epsilon} \Rightarrow \left|\frac{n+1}{n} - 1\right| = \left|\frac{1}{n}\right| = \frac{1}{n} < \epsilon.$$

From this, we see that choosing $N \in \mathbb{N}$ such that $N > \frac{1}{\epsilon}$ does the trick. That is, for $N > \frac{1}{\epsilon}$, we have that $n > N \Rightarrow |\frac{n+1}{n} - 1| < \epsilon$, which establishes the result. We have shown that for any given value of ϵ, there is an index value $N \in \mathbb{N}$ such that all the terms of the sequence beyond the Nth one are within ϵ of the limit. ◀

Note from this example that a smaller value of ϵ would require a larger value of N. This is usually the case. Also note that $x_n > 1$ for all x_n in the sequence, and yet $\lim x_n = 1$. From this we see that "taking the limit on each side" of the inequality $x_n > 1$ does not yield a correct result. This example points to an important fact, namely, there is a subtlety to taking limits that students might easily overlook. All students of calculus (or even precalculus) know that when faced with an equality or an inequality, there are several mathematical operations that can be applied to each side of the expression that leave the equality or inequality unchanged; for example, adding a constant to both sides, or multiplying both sides by a positive constant. Students might naturally feel that the same principle applies when taking limits on both sides of such an expression. While this is true in many situations, such as those cases involving =, $\leq$, and $\geq$, an exception to this "rule" is the case where the expression involves a strict inequality. For example, each term of the sequence given by $\{\frac{1}{n}\}$ for $n \geq 1$ is a positive number, and so $\frac{1}{n} > 0$ for all $n \geq 1$. However, when taking the limit as $n \to \infty$ on each side of the expression $\frac{1}{n} > 0$, one must weaken the strict inequality and change it to $\geq$, yielding $\lim_{n\to\infty} \frac{1}{n} \geq 0$. The reason is that the limit of a sequence is a special type of point called a *limit point* that is often not a member of the sequence itself, and, as in this case, may not share in all of the attributes of the members of that set. We will learn more about limit points of sequences later in this chapter. For now, this new "rule" for handling limits involving strict inequalities should be remembered!

► **3.1** *Show that* $\lim \left(\frac{1+i}{4}\right)^n = 0$.

The techniques illustrated in the following examples can be extended to other, more complicated limit problems.

Example 1.5 *Consider the sequence $\{x_n\}$ where $x_n = \frac{n^2-n-1}{2n^3+n^2+7} \in \mathbb{R}$ for $n \geq 1$. We will show, according to Definition 1.3, that the limit of this sequence is 0. This suspected value of the limit should be clear. After all, for very large n, each x_n is very close to $\frac{n^2}{2n^3} = \frac{1}{2n}$, which clearly goes to 0 as n increases. To begin, suppose we have an arbitrary $\epsilon > 0$. Then,*

$$\left|\frac{n^2-n-1}{2n^3+n^2+7} - 0\right| = \frac{|n^2-n-1|}{2n^3+n^2+7}.$$

To remove the absolute value in the numerator, note that $x^2 - x - 1 \geq 0$ if $x > 2$ (Why?), and so

$$\left|\frac{n^2-n-1}{2n^3+n^2+7}\right| = \frac{n^2-n-1}{2n^3+n^2+7} \quad \text{for } n > 2.$$

We would like to find an upper bound to $\frac{n^2-n-1}{2n^3+n^2+7}$ that is an even simpler function of n, and that applies when n is large. In particular, we seek simple polynomials $p(n)$ and $q(n)$ such that for some $M \in \mathbb{N}$, if $n > M$, then $n^2 - n -$

$1 \le p(n)$ and $2n^3 + n^2 + 7 \ge q(n)$. We already require n to be greater than 2. Note that for large n values, the largest term in $n^2 - n - 1$ is n^2. Likewise, for large n values, the largest term in $2n^3 + n^2 + 7$ is $2n^3$. This leads us to choose as a candidate for $p(n)$ a constant multiple of n^2, and as a candidate for $q(n)$ a constant multiple of n^3. In fact, $p(n) = n^2$ and $q(n) = n^3$ satisfy all of our conditions for all $n \ge 1$, and so we have

$$n > 2 \;\Rightarrow\; \left|\frac{n^2 - n - 1}{2n^3 + n^2 + 7}\right| = \frac{n^2 - n - 1}{2n^3 + n^2 + 7} \le \frac{n^2}{n^3} = \frac{1}{n}.$$

The right-hand side of the above inequality will be less than ϵ if $n > \frac{1}{\epsilon}$ and $n > 2$. That is, overall,

$$n > \max\left(2, \frac{1}{\epsilon}\right) \;\Rightarrow\; \left|\frac{n^2 - n - 1}{2n^3 + n^2 + 7}\right| = \frac{n^2 - n - 1}{2n^3 + n^2 + 7} \le \frac{n^2}{n^3} = \frac{1}{n} < \epsilon.$$

From this, we see that choosing $N \in \mathbb{N}$ such that $N > \max\left(2, \frac{1}{\epsilon}\right)$ will satisfy Definition 1.3. That is, for $N > \max\left(2, \frac{1}{\epsilon}\right)$ we have that $n > N \Rightarrow \left|\frac{n^2-n-1}{2n^3+n^2+7} - 0\right| < \epsilon$. ◀

▶ **3.2** *Why can't we take $N = \max\left(2, \frac{1}{\epsilon}\right)$ in the above example?*

Example 1.6 *We will show that the limit of the sequence $\{x_n\}$ where $x_n = \frac{n^2+n-10}{2n^3-5n^2+1}$ is 0. Again suppose that $\epsilon > 0$ is given, and consider*

$$\left|\frac{n^2 + n - 10}{2n^3 - 5n^2 + 1} - 0\right| = \frac{|n^2 + n - 10|}{|2n^3 - 5n^2 + 1|}.$$

To remove the absolute value in the numerator, we note that $x^2 + x - 10 \ge 0$ if $x > 3$ (Why?). To remove the absolute value in the denominator, we note that $2x^3 - 5x^2 + 1 > 0$ if $x > 3$ (Why?). This yields

$$\left|\frac{n^2 + n - 10}{2n^3 - 5n^2 + 1} - 0\right| = \frac{n^2 + n - 10}{2n^3 - 5n^2 + 1} \quad \text{if } n > 3.$$

It would be nice if we could say that $n^2+n-10 \le n^2$ and $2n^3-5n^2+1 \ge n^3$ for n large enough, but the first of these inequalities is not true. (The reader can verify that the second inequality is true for $n > 5$.) To work on the numerator a bit more, we note that $x^2 + x - 10 \le 2x^2$ if $x > 0$. Therefore,

$$\frac{n^2 + n - 10}{2n^3 - 5n^2 + 1} \le \frac{2n^2}{n^3} = \frac{2}{n} \quad \text{if } n > 5.$$

Overall then, we obtain

$$\left|\frac{n^2 + n - 10}{2n^3 - 5n^2 + 1} - 0\right| = \frac{n^2 + n - 10}{2n^3 - 5n^2 + 1} \le \frac{2}{n} < \epsilon \quad \text{if } n > \max\left(5, \frac{2}{\epsilon}\right).$$

That is, if we choose $N \in \mathbb{N}$ such that $N > \max\left(5, \frac{2}{\epsilon}\right)$, we have that

$$n > N \Rightarrow \left|\frac{n^2+n-10}{2n^3-5n^2+1} - 0\right| < \epsilon.$$

◀

Example 1.7 *Let $\{\mathbf{x}_n\}$ be the sequence in $\mathbb{R}^2$ with $\mathbf{x}_n = \left(\frac{1}{n}, \frac{2n}{n+12}\right)$. We will show that $\lim \mathbf{x}_n = (0,2)$. To do this, let $\epsilon > 0$ be given. Then*

$$\left|\left(\frac{1}{n}, \frac{2n}{n+12}\right) - (0,2)\right| = \left|\left(\frac{1}{n}, \frac{-24}{n+12}\right)\right| = \sqrt{\left(\frac{1}{n}\right)^2 + \left(\frac{-24}{n+12}\right)^2}.$$

If we can force $\left(\frac{1}{n}\right)^2 < \frac{\epsilon^2}{2}$ and $\left(\frac{-24}{n+12}\right)^2 < \frac{\epsilon^2}{2}$ then we are done. To establish the first inequality, we just need to take $n > \frac{\sqrt{2}}{\epsilon}$. For the second inequality, we need $\frac{24}{n+12} < \frac{\epsilon}{\sqrt{2}}$, or $n > \frac{24\sqrt{2}}{\epsilon} - 12$. Choosing $N \in \mathbb{N}$ such that $N > \max\left(\frac{\sqrt{2}}{\epsilon}, \frac{24\sqrt{2}}{\epsilon} - 12\right)$ yields

$$n > N \Rightarrow \left|\left(\frac{1}{n}, \frac{2n}{n+12}\right) - (0,2)\right| < \epsilon,$$

i.e., $\lim\left(\frac{1}{n}, \frac{2n}{n+12}\right) = (0,2)$.

◀

▶ **3.3** *Show that $\lim\left(\frac{n}{n^2-4n+1}, \frac{n^2+5}{n^2+1}\right) = (0,1)$.*

Suppose you suspect a given sequence is converging to the limit L when in fact it converges to H. The following example illustrates how applying Definition 1.3 will show you your error.

Example 1.8 *Consider the sequence $\{z_n\}$ in $\mathbb{C}$ where $z_n = \left(\frac{1}{1+i}\right)^n$. We will show that $\lim z_n \neq 1$. To do this, assume the limit is 1. We will derive a contradiction. In particular, we will exhibit a value of ϵ for which Definition 1.3 doesn't apply. According to the definition, for $\epsilon = \frac{1}{2}$, we have*

$$n > N \Rightarrow \left|\left(\frac{1}{1+i}\right)^n - 1\right| < \frac{1}{2}. \tag{3.1}$$

But the reverse triangle inequality yields

$$1 - \left|\frac{1}{1+i}\right|^n \le \left|\left(\frac{1}{1+i}\right)^n - 1\right|,$$

and so (3.1) becomes

$$n > N \Rightarrow 1 - \left|\frac{1}{1+i}\right|^n < \frac{1}{2},$$

which after algebra is just

$$n > N \;\Rightarrow\; \frac{1}{2} < \left(\frac{1}{\sqrt{2}}\right)^n.$$

This is clearly a contradiction. Therefore, $\lim z_n \neq 1$. ◄

► **3.4** *Was there anything special about our choosing* $\epsilon = \frac{1}{2}$ *in the above example? What if you chose* $\epsilon = \frac{1}{4}$ *instead? How about* $\epsilon = 2$*? You should find that large* ϵ *choices do not necessarily work.*

► **3.5** *Show that* $\lim i^n$ *does not exist. (Hint: Assume it does.)*

1.2 First Properties of Sequences

When a sequence converges, it is reassuring to know that the limit is unique.

Proposition 1.9 *Suppose* $\{x_n\} \in \mathbb{X}$ *is a convergent sequence. Then the limit of the sequence is unique.*

PROOF Suppose x and $\widetilde{x}$ are both limits of the convergent sequence $\{x_n\}$. We will use Definition 1.3 to establish that $x = \widetilde{x}$. In particular, consider any $\epsilon > 0$, and divide it in two. According to Definition 1.3 there exists $N_1 \in \mathbb{N}$ such that $|\,x_n - x\,| < \frac{\epsilon}{2}$ for $n > N_1$. Likewise, there exists $N_2 \in \mathbb{N}$ such that $|\,x_n - \widetilde{x}\,| < \frac{\epsilon}{2}$ for $n > N_2$. Consider the number $|x - \widetilde{x}|$, and note that

$$|\,x - \widetilde{x}\,| = \left|\,x - x_n + x_n - \widetilde{x}\right| \leq |x - x_n\,| + |\,x_n - \widetilde{x}\,| \text{ for any } n \in \mathbb{N}. \quad (3.2)$$

Choose $n > \max(N_1, N_2)$, and (3.2) becomes

$$|\,x - \widetilde{x}\,| = \left|\,x - x_n + x_n - \widetilde{x}\right| \leq |x - x_n\,| + |\,x_n - \widetilde{x}\,| < \frac{\epsilon}{2} + \frac{\epsilon}{2} = \epsilon.$$

That is, $|\,x - \widetilde{x}\,| < \epsilon$. Since ϵ was arbitrary, it follows that $x = \widetilde{x}$. ♦

The following proposition allows us to handle sequences of vectors from $\mathbb{R}^k$ or $\mathbb{C}$ as a finite number of sequences of real numbers. That is, we can choose to handle a sequence of vectors or complex numbers one component at a time. The practical advantages of this should be clear. Stated informally, the proposition establishes that a sequence of vectors from $\mathbb{R}^k$ or $\mathbb{C}$ converges to a vector limit if and only if each component sequence converges to the corresponding component of the limit.

Proposition 1.10

a) Let $\{\mathbf{x}_n\}$ *be a sequence in* $\mathbb{R}^k$ *with* $\mathbf{x}_n = (x_n^{(1)}, x_n^{(2)}, \ldots, x_n^{(k)})$*, and suppose* $\mathbf{x} = (x_1, x_2, \ldots, x_k) \in \mathbb{R}^k$*. Then for* $j = 1, \ldots, k$ *we have*

$$\lim_{n\to\infty} x_n^{(j)} = x_j \text{ if and only if } \lim_{n\to\infty} \mathbf{x}_n = \mathbf{x}.$$

b) Let $\{z_n\}$ be a sequence in $\mathbb{C}$ with $z_n = x_n + i\,y_n$ and let $z = x + i\,y \in \mathbb{C}$. Then,
$\lim_{n\to\infty} z_n = z$ *if and only if* $\lim_{n\to\infty} x_n = x$ *and* $\lim_{n\to\infty} y_n = y$.

PROOF We will prove the first result and leave the proof of the second to the reader. Assume $\lim_{n\to\infty} \mathbf{x}_n = \mathbf{x}$ and let $\epsilon > 0$ be given. Then there exists $N \in \mathbb{N}$ such that $n > N \Rightarrow |\mathbf{x}_n - \mathbf{x}| < \epsilon$. Since $|x_n^{(j)} - x_j| \le |\mathbf{x}_n - \mathbf{x}|$ for each $j = 1, 2, \ldots, k$, we have that

$$n > N \Rightarrow |x_n^{(j)} - x_j| \le |\mathbf{x}_n - \mathbf{x}| < \epsilon \text{ for } j = 1, 2, \ldots, k,$$

i.e., $\lim_{n\to\infty} x_n^{(j)} = x_j$ for $j = 1, 2, \ldots, k$. Now assume that $\lim_{n\to\infty} x_n^{(j)} = x_j$ for $j = 1, 2, \ldots, k$. Then for any given $\epsilon > 0$, there exists for each j an $N_j \in \mathbb{N}$ such that $n > N_j \Rightarrow |x_n^{(j)} - x_j| < \frac{\epsilon}{k}$. From this we have

$$n > \max_j(N_j) \Rightarrow |\mathbf{x}_n - \mathbf{x}| \le \sum_{j=1}^{k} |x_n^{(j)} - x_j| < \epsilon,$$

i.e., $\lim_{n\to\infty} \mathbf{x}_n = \mathbf{x}$. ◆

▶ **3.6** *Prove part* b) *of the above proposition.*

▶ **3.7** *Reconsider the limits* $\lim\left(\frac{1}{n}, \frac{2n}{n+12}\right)$ *and* $\lim\left(\frac{1+i}{4}\right)^n$ *with this proposition in mind.*

2 CONVERGENCE RESULTS FOR SEQUENCES

In this section, we present some results that allow one to determine the convergence behavior of a sequence without having to resort to the definition given in Definition 1.3. This can often be a great convenience.

2.1 General Results for Sequences in $\mathbb{X}$

To begin, we define what it means for a sequence to be *bounded.*

Definition 2.1 A sequence $\{x_n\} \in \mathbb{X}$ is called **bounded** if there exists a real number M such that $|x_n| \le M$ for all n.

In the case of a sequence of real numbers, boundedness implies that the elements of the sequence are restricted to a finite interval of the real line. In the case of a sequence of vectors in $\mathbb{R}^2$ or points in $\mathbb{C}$, boundedness implies that the elements of the sequence are restricted to a disk of finite radius in the plane. More generally, boundedness of a sequence in $\mathbb{R}^k$ implies that the vectors of the sequence are restricted to a ball of finite radius in k-dimensional Euclidean space. Clearly, the sequence elements considered as a set $S \subset \mathbb{X}$ is a bounded set as defined in Chapter 2. That a sequence is bounded is often a very useful piece of information to determine.

▶ **3.8** *Consider the sequence* $\{\mathbf{x}_n\}$ *in* $\mathbb{R}^k$, *where* $\mathbf{x}_n = (x_{1n}, x_{2n}, \ldots, x_{kn})$. *Show that* $\{\mathbf{x}_n\}$ *is bounded in* $\mathbb{R}^k$ *if and only if each sequence* $\{x_{jn}\}$ *is bounded in* $\mathbb{R}$ *for* $j = 1, 2, \ldots, k$. *State and prove the analogous situation for a sequence* $\{z_n\} \subset \mathbb{C}$.

▶ **3.9** *Consider the case where* $\{x_n\}$ *is a sequence of real numbers. Such a sequence is called* bounded below *if there exists a real number* M_1 *such that* $M_1 \le x_n$ *for all* n. *Similarly, the sequence is called* bounded above *if there exists a real number* M_2 *such that* $x_n \le M_2$ *for all* n. *Show that a sequence of real numbers is bounded if and only if it is bounded below and bounded above. We will not be using the notions of "bounded below" and "bounded above" for sequences in higher dimensions. Why is that?*

Proposition 2.2 *Let* $\{x_n\} \in \mathbb{X}$ *be a convergent sequence. Then* $\{x_n\}$ *is bounded.*

PROOF Suppose $\lim x_n = x$ exists. Then there exists $N \in \mathbb{N}$ such that $n > N \Rightarrow |x_n - x| < 1$. From this we have that, for all $n > N$,

$$|x_n| = \big|x_n - x + x\big| \le |x_n - x| + |x| < 1 + |x|.$$

Now it is easy to see that

$$|x_n| \le \max\big\{|x_1|, |x_2|, \ldots, |x_N|, 1 + |x|\big\} \text{ for all } n.$$ ♦

As the proof of the above proposition implies, any finite subset of terms from a sequence is bounded. Determining that a sequence is bounded, therefore, says something important about the "tail" of the sequence, that is, those terms x_n in the sequence for all $n > N$ for some finite $N \in \mathbb{N}$. We may actually disregard the first N terms, if it is convenient to do so, to determine boundedness. We will see that the same idea applies to the notion of convergence of a sequence—it too is a "tail property" of the sequence.

The following proposition summarizes several convenient results that allow for easier manipulation of limits involving sequences in $\mathbb{X}$. Most of them should be familiar to any student of calculus.

Proposition 2.3 *Let* $\{x_n\}$ *and* $\{y_n\} \in \mathbb{X}$ *be two sequences such that* $\lim x_n = x$ *and* $\lim y_n = y$ *for some* x *and* $y \in \mathbb{X}$. *Then*

a) $\lim(x_n \pm x_0) = x \pm x_0$ *for any* $x_0 \in \mathbb{X}$.

b) $\lim(x_n \pm y_n) = x \pm y$.

c) $\lim(c\,x_n) = c\,x$ *for any* $c \in \mathbb{R}$ *or* $\mathbb{C}$.

d) $\lim(x_n \cdot y_n) = x \cdot y$.

e) $\lim|x_n| = |x|$.

Note that in the statement of part *d)* in the above proposition, the product should be understood as the usual product in $\mathbb{R}$ and in $\mathbb{C}$, and as the dot product in $\mathbb{R}^k$.

PROOF We prove *b)* and *d)*, and leave the rest for the reader as exercises.

To establish *b)*, note that for any $\epsilon > 0$ there exists $N_1 \in \mathbb{N}$ such that $n > N_1 \Rightarrow |x_n - x| < \frac{\epsilon}{2}$, and there exists $N_2 \in \mathbb{N}$ such that $n > N_2 \Rightarrow |y_n - y| < \frac{\epsilon}{2}$. Choose $N = \max(N_1, N_2)$. Then for $n > N$, we have

$$|(x_n \pm y_n) - (x \pm y)| = |(x_n - x) \pm (y_n - y)| \leq |x_n - x| + |y_n - y| < \epsilon.$$

To establish *d)*, note that since $\{x_n\}$ is convergent there exists an $M \in \mathbb{N}$ such that $|x_n| \leq M$ for all n. It is also true that for any $\epsilon > 0$ there exists an $N_1 \in \mathbb{N}$ such that

$$n > N_1 \Rightarrow |y_n - y| < \frac{\epsilon}{2(M+1)},$$

and there exists an $N_2 \in \mathbb{N}$ such that

$$n > N_2 \Rightarrow |x_n - x| < \frac{\epsilon}{2(|y|+1)}.$$

From all this, we obtain that whenever $n > \max(N_1, N_2)$,

$$\begin{aligned} |x_n \cdot y_n - x \cdot y| &= |x_n \cdot (y_n - y) + (x_n - x) \cdot y| \\ &\leq |x_n||y_n - y| + |y||x_n - x| \qquad (3.3) \\ &\leq M\left[\frac{\epsilon}{2(M+1)}\right] + |y|\left[\frac{\epsilon}{2(|y|+1)}\right] \\ &< \frac{\epsilon}{2} + \frac{\epsilon}{2} = \epsilon, \end{aligned}$$

and the result is proved. Note that in expression (3.3) we have applied the triangle inequality, and in the case of $\mathbb{X} = \mathbb{R}^k$, the Cauchy-Schwarz inequality. ♦

► **3.10** *Complete the proof of the above proposition.*

► **3.11** *Show that the converse of property e) is true only when* $x = 0$. *That is, show the following: If* $\lim |x_n| = 0$, *then* $\lim x_n = 0$. *If* $\lim |x_n| = x \neq 0$, *then it is not necessarily true that* $\lim x_n = x$.

► **3.12** *In the previous proposition, show that property d) is still true when* $\{x_n\}$ *is a sequence of real numbers while* $\{y_n\}$ *is a sequence from* $\mathbb{R}^k$ *or* $\mathbb{C}$.

2.2 Special Results for Sequences in $\mathbb{R}$ and $\mathbb{C}$

Certain results are particular to $\mathbb{R}$ and $\mathbb{C}$ due to their possessing the field properties, or to $\mathbb{R}$ alone due to its possessing special order properties. We state these results next.

A Field Property Result for Sequences in $\mathbb{R}$ and $\mathbb{C}$

We begin with a proposition that relies on the field properties, and therefore applies only to sequences in $\mathbb{R}$ and $\mathbb{C}$.

Proposition 2.4 *Let $\{z_n\}$ and $\{w_n\}$ be sequences in $\mathbb{R}$ or in $\mathbb{C}$ such that $w_n \neq 0$ for all n. If $\lim z_n = z$ and $\lim w_n = w \neq 0$ for some real or complex z and w, then*

$$\lim \left(\frac{z_n}{w_n}\right) = \frac{z}{w}.$$

PROOF We first prove the result for the case $z_n = 1$ for all $n \geq 1$. Note that $\left|\frac{1}{w_n} - \frac{1}{w}\right| = \frac{|w_n - w|}{|w||w_n|}$. Now, since $w \neq 0$, we can choose $\epsilon = \frac{1}{2}|w|$. In particular, there exists $N_1 \in \mathbb{N}$ such that $n > N_1 \Rightarrow |w_n - w| < \frac{1}{2}|w|$. Then from the triangle inequality, we have

$$n > N_1 \Rightarrow |w| - |w_n| \leq |w - w_n| < \tfrac{1}{2}|w|,$$

which in turn implies that $|w_n| > \frac{1}{2}|w|$ if $n > N_1$. From this, we have that

$$n > N_1 \quad \Rightarrow \quad \left|\frac{1}{w_n} - \frac{1}{w}\right| = \frac{|w_n - w|}{|w||w_n|} \leq \frac{2|w_n - w|}{|w|^2}. \tag{3.4}$$

Now, for $\epsilon > 0$ there exists $N_2 \in \mathbb{N}$ such that $n > N_2 \Rightarrow |w_n - w| < \frac{\epsilon|w|^2}{2}$. This combined with (3.4) yields, for $N \equiv \max(N_1, N_2)$,

$$n > N \quad \Rightarrow \quad \left|\frac{1}{w_n} - \frac{1}{w}\right| \leq \frac{2|w_n - w|}{|w|^2} < \epsilon.$$

We leave the remainder of the proof as an exercise. ♦

► **3.13** *Complete the proof of the above proposition by handling the case where $z_n \neq 1$ for at least one $n \in \mathbb{N}$.*

Order Property Results for Sequences in $\mathbb{R}$

We now develop certain results particular to sequences in $\mathbb{R}$. These results involve the special order properties possessed by $\mathbb{R}$ that are so familiar to students of calculus. We begin with a proposition.

Proposition 2.5 *Suppose $\{x_n\} \in \mathbb{R}$ is a sequence of real numbers with $x_n \geq 0$ for all n, and for which $\lim x_n = x$. Then $x \geq 0$.*

PROOF We employ the method of proof by contradiction. Assume the limit x of the convergent sequence is negative. Then there exists an $N \in \mathbb{N}$ such that $n > N \Rightarrow |x_n - x| < -\frac{1}{2}x$. This in turn implies

$$x_n = (x_n - x) + x \leq |x_n - x| + x < \tfrac{1}{2}x < 0 \text{ for } n > N.$$

This contradicts what was given about each x_n, namely, that $x_n \geq 0$ for all n. ♦

Corollary 2.6 *Suppose $\{x_n\}$ and $\{y_n\} \in \mathbb{R}$ are sequences of real numbers with $x_n \leq y_n$ for all n, and for which $\lim x_n = x$ and $\lim y_n = y$. Then $x \leq y$.*

▶ **3.14** *Prove the above corollary.*

We apply the result of Proposition 2.5 in the following example.

Example 2.7 *Fix a real convergent sequence $\{x_n\}$ where $x_n \geq 0$ for all n and $\lim x_n = x$. We will show that $\lim \sqrt{x_n} = \sqrt{x}$. According to Proposition 2.5, we know that $x \geq 0$. If $x > 0$, then*

$$\left|\sqrt{x_n} - \sqrt{x}\right| = \frac{|x_n - x|}{\sqrt{x_n} + \sqrt{x}} \leq \frac{|x_n - x|}{\sqrt{x}}.$$

For a given $\epsilon > 0$, there exists an $N \in \mathbb{N}$ such that

$$n > N \Rightarrow |x_n - x| < \epsilon \sqrt{x}.$$

Therefore,

$$n > N \Rightarrow \left|\sqrt{x_n} - \sqrt{x}\right| \leq \frac{|x_n - x|}{\sqrt{x}} < \epsilon,$$

i.e., $\lim \sqrt{x_n} = \sqrt{x}$. What if $x = 0$? Then $|\sqrt{x_n} - 0| = \sqrt{x_n}$. For a given $\epsilon > 0$, there exists an $N \in \mathbb{N}$ such that

$$n > N \Rightarrow |x_n| < \epsilon^2 \tag{3.5}$$

But $|x_n| = (\sqrt{x_n})^2$, and so (3.5) is equivalent to $n > N \Rightarrow \sqrt{x_n} < \epsilon$. Therefore, $\lim \sqrt{x_n} = 0 = \sqrt{x}$. ◀

▶ **3.15** *Suppose $\{x_n\} \in \mathbb{R}$ is a sequence of real numbers with $x_n \leq 0$ for all n, and for which $\lim x_n = x$. Prove that $x \leq 0$.*

We now define *the supremum* and *the infimum* of a sequence. In particular, suppose $\{x_n\}$ is a sequence of real numbers whose elements comprise the set $S \subset \mathbb{R}$. If the sequence is bounded above, then $S \subset \mathbb{R}$ is bounded above, and by Theorem 1.12 in Chapter 2 the supremum $\sup S$ exists. Likewise, if $\{x_n\}$ is bounded below, then $S \subset \mathbb{R}$ is bounded below, and by Theorem 1.12 in Chapter 2 the infimum $\inf S$ exists. We use these facts as the basis for the following definition.

Definition 2.8 Suppose $\{x_n\}$ is a sequence of real numbers whose elements comprise the set $S \subset \mathbb{R}$.

1. If the sequence $\{x_n\}$ is bounded above, then we define the **supremum** of the sequence $\{x_n\}$ to be $\sup S$, and we denote it by $\sup x_n$.
2. If the sequence $\{x_n\}$ is bounded below, then we define the **infimum** of the sequence $\{x_n\}$ to be $\inf S$, and we denote it by $\inf x_n$.

► **3.16** *Suppose* $\{x_n\}$ *and* $\{y_n\}$ *are sequences of real numbers that are bounded above. If* $x_n \leq y_n$ *for all* $n \geq 1$*, show the following:*

a) $\sup x_n \leq \sup y_n$.
b) $\sup(c\,x_n) = c \sup x_n$*, if* $c \geq 0$.
c) If $c < 0$ *in part b), what is the conclusion? (You should find the appearance of the inf.)*

► **3.17** *Suppose* $\{x_n\}$ *and* $\{y_n\}$ *are sequences of real numbers that are bounded below. If* $x_n \leq y_n$ *for all* $n \geq 1$*, show the following:*

a) $\inf x_n \leq \inf y_n$.
b) $\inf(c\,x_n) = c \inf x_n$*, if* $c \geq 0$.
c) If $c < 0$ *in part b), what is the conclusion? (You should find a similar reversal as in the previous exercise.)*

We continue our discussion of special sequence results relating to the order properties of the real numbers with the following proposition.

Proposition 2.9 (The Squeeze Theorem)

Suppose $\{w_n\}$, $\{x_n\}$*, and* $\{y_n\}$ *are sequences of real numbers. If* $x_n \leq w_n \leq y_n$ *for all* n*, and if* $\lim x_n = \lim y_n = L$*, then* $\lim w_n = L$.

PROOF Since $x_n \leq w_n \leq y_n$ for any n, it is also true that for any n we have $x_n - L \leq w_n - L \leq y_n - L$. Let $\epsilon > 0$ be given. Then there exists $N \in \mathbb{N}$ such that whenever $n > N$, we have both $|\,x_n - L\,| < \epsilon$ and $|\,y_n - L\,| < \epsilon$. Combining these results we see that

$$n > N \;\Rightarrow\; -\epsilon < x_n - L \leq w_n - L \leq y_n - L < \epsilon.$$

That is,

$$n > N \;\Rightarrow\; -\epsilon < w_n - L < \epsilon,$$

or

$$n > N \;\Rightarrow\; |\,w_n - L\,| < \epsilon,$$

and the result is proved. ♦

Example 2.10 *Suppose* $x_n = 1 + \frac{\cos n}{n}$ *for* $n \geq 1$*. We will show that* $\lim x_n = 1$ *by using the squeeze theorem. To begin, note that*

$$1 - \tfrac{1}{n} \leq x_n \leq 1 + \tfrac{1}{n}.$$

Since $\lim \left(1 - \frac{1}{n}\right) = \lim \left(1 + \frac{1}{n}\right) = 1$ *(Why?), we have via the squeeze theorem that* $\lim x_n = 1$. ◄

► **3.18** *Show that* $\lim \left(1 - \frac{1}{n}\right) = \lim \left(1 + \frac{1}{n}\right) = 1$.

Finally, we define what it means for a real sequence to be *monotone*.

Definition 2.11 Let $\{x_n\}$ be a sequence in $\mathbb{R}$. The sequence is called **monotone** if any of the following hold true:

1. $x_n \leq x_{n+1}$ for all n. In this case, the monotone sequence is called **nondecreasing.**
2. $x_n < x_{n+1}$ for all n. In this case, the monotone sequence is called **increasing.**
3. $x_n \geq x_{n+1}$ for all n. In this case, the monotone sequence is called **nonincreasing.**
4. $x_n > x_{n+1}$ for all n. In this case, the monotone sequence is called **decreasing.**

Theorem 2.12 (The Monotone Sequence Theorem)

Suppose $\{x_n\}$ is a monotone sequence of real numbers.

a) If $\{x_n\}$ is a nondecreasing or an increasing sequence that is bounded above, then $\{x_n\}$ converges to $x = \sup x_n$.

b) If $\{x_n\}$ is a nonincreasing or a decreasing sequence that is bounded below, then $\{x_n\}$ converges to $x = \inf x_n$.

PROOF We prove part *a*) and leave part *b*) to the reader. Let $x = \sup x_n$. Then for each $\epsilon > 0$ there exists an $N \in \mathbb{N}$ such that $x - \epsilon < x_N$. For $n > N$, we have

$$x - \epsilon < x_N \leq x_n \leq x < x + \epsilon.$$

Subtracting x from each part of this multiple inequality leads to the result we seek. That is, $n > N \Rightarrow |x_n - x| < \epsilon$. ♦

► **3.19** *Prove part b) of the above theorem.*

Example 2.13 *Suppose the real sequence $\{x_n\}$ is defined recursively as follows: $x_1 = 1$, $x_2 = 2 + \sqrt{x_1}$, $x_3 = 2 + \sqrt{x_2}, \dots, x_{n+1} = 2 + \sqrt{x_n}$. We will show that $\lim x_n$ exists. Recall that we showed in Example 1.7 of Chapter 2 that $x_n \leq 4$ for all $n \geq 1$. Therefore, the sequence is bounded above. We now use induction to show that $x_n \leq x_{n+1}$ for all $n \geq 1$. Note that $x_2 = 2 + \sqrt{x_1} = 2 + \sqrt{1} = 3 > x_1$. Also, if $x_N \leq x_{N+1}$ for some $N \in \mathbb{N}$, then*

$$x_{N+2} = 2 + \sqrt{x_{N+1}} \geq 2 + \sqrt{x_N} = x_{N+1}.$$

This proves that the sequence $\{x_n\}$ is nondecreasing. Since we've already established that the sequence is bounded above, by the monotone sequence theorem $\lim x_n = \sup x_n$ exists. ◄

For some divergent real sequences, we can be more specific about *how* they diverge. The following definition describes what it means for a real sequence to diverge to ∞ or to $-\infty$.

Definition 2.14 Let $\{x_n\}$ be a sequence in $\mathbb{R}$.

1. If for every real number M there exists an $N \in \mathbb{N}$ such that $n > N$ implies $x_n > M$, then we say that the sequence **diverges to ∞**, and we write $\lim x_n = \infty$.
2. If for every real number M there exists an $N \in \mathbb{N}$ such that $n > N$ implies $x_n < M$, then we say that the sequence **diverges to $-\infty$**, and we write $\lim x_n = -\infty$.

Example 2.15 *We will show that the real sequence $\{x_n\}$ with $x_n = n^2-3n-2$ diverges to ∞ as n increases to ∞. To show this, note that if $n > 6$ then $n^2 - 3n - 2 > \frac{1}{2}n^2$ (Why?). Now let M be any real number. There are two cases to consider:*

(i) If $M < 0$, then $n^2 - 3n - 2 > \frac{1}{2}n^2 > M$ whenever $n > 6$.

(ii) If $M \geq 0$, then $n^2 - 3n - 2 > \frac{1}{2}n^2 > M$ whenever $n > 6$ and $n > \sqrt{2M}$.

If we choose $N > \max\left(6, \sqrt{2|M|}\right)$, then in either case we have

$$n > N \Rightarrow n^2 - 3n - 2 > M,$$

and so $\lim x_n = \infty$. ◀

3 Topological Results for Sequences

3.1 Subsequences in $\mathbb{X}$

The notion of a subsequence of a given sequence is a useful one in analysis. It has both conceptual and practical use, as we will soon see.

Definition 3.1 Let $\{x_n\}$ be a sequence in $\mathbb{X}$, and let $n_1, n_2, n_3, \ldots$ be a set of positive integers such that $1 \leq n_1 < n_2 < n_3 < \cdots$. Then $\{x_{n_m}\}$ for $m \geq 1$ is called a **subsequence** of $\{x_n\}$.

A subsequence, loosely speaking, is a sampling from the original sequence. If one were to progress through the terms of a sequence and write down only those that correspond to an even index value, retaining their relative order as presented in the original sequence, one obtains a subsequence from the original sequence. The odd-indexed terms also comprise a subsequence, as do the prime-indexed terms, and so forth. There are in fact an infinite number of possible subsequences one could obtain from a given sequence.

Example 3.2 *Suppose the sequence* $\{x_n\} \in \mathbb{R}$ *is given by* $x_n = \frac{1}{n} + \cos\left(\frac{n\pi}{2}\right)$ *for* $n \geq 1$. *We will exhibit the terms of the subsequences* $\{x_{2m}\}$ *and* $\{x_{4m}\}$ *for* $m \in \mathbb{N}$. *Note that* $\cos(m\pi) = (-1)^m$, *and* $\cos(2\pi m) = 1$ *for all* m, *and so*

$$x_{2m} = \frac{1}{2m} + \cos(m\,\pi) = \frac{1}{2\,m} + (-1)^m \text{ for all } m \in \mathbb{N},$$

while

$$x_{4m} = \frac{1}{4m} + \cos(2m\,\pi) = \frac{1}{4\,m} + 1 \text{ for all } m \in \mathbb{N}.$$

◀

While there are an infinite number of different subsequences one could construct from a given sequence, it is helpful to know that when the original sequence converges, all subsequences of that sequence converge to the same limit. We state this result in the following proposition.

Proposition 3.3 *Let* $\{x_n\}$ *be a sequence in* $\mathbb{X}$, *and suppose* $\lim x_n = x$. *Then* $\lim_{m\to\infty} x_{n_m} = x$ *for every subsequence* $\{x_{n_m}\}$ *of the sequence.*

PROOF Let $\epsilon > 0$ be given. Then there exists an $N \in \mathbb{N}$ such that

$$n > N \Rightarrow |x_n - x| < \epsilon.$$

Therefore,

$$n_m > N \Rightarrow |x_{n_m} - x| < \epsilon.$$

Now, since $n_m \geq m$, it follows that

$$m > N \Rightarrow n_m > N \Rightarrow |x_{n_m} - x| < \epsilon,$$

and the proposition is proved. ◆

Definition 3.4 Let $\{x_n\}$ be a sequence in $\mathbb{X}$. The point x is called a **limit point** of the sequence if there exists a subsequence $\{x_{n_m}\}$ of $\{x_n\}$ such that $\lim_{m\to\infty} x_{n_m} = x$.

Limit points of a sequence are those points having infinitely many elements of the sequence arbitrarily close to them. They are analogous to limit points of sets as described in Chapter 2.

► **3.20** *Show that if* x *is a limit point of the sequence* $\{x_n\}$ *in* $\mathbb{X}$, *then for any* $\epsilon > 0$, *the neighborhood* $N_\epsilon(x)$ *contains infinitely many elements of* $\{x_n\}$.

Example 3.5 *Let* $\{\mathbf{x}_n\}$ *be a sequence in* $\mathbb{R}^2$ *with* $\mathbf{x}_n = \left(\frac{n}{n+1}, (-1)^n\right)$. *Note that the subsequence of even indexed terms* $\mathbf{x}_{2m} = \left(\frac{2m}{2m+1}, 1\right)$ *has the limit given by* $\lim_{m\to\infty} \mathbf{x}_{2m} = (1, 1)$. *The subsequence of odd indexed terms* $\mathbf{x}_{2m+1} = \left(\frac{2m+1}{2m+2}, -1\right)$ *has the limit given by* $\lim_{m\to\infty} \mathbf{x}_{2m+1} = (1, -1)$. *Therefore* $(1, 1)$ *and* $(1, -1)$ *are limit points of the original sequence. It is also worth noting that according to Proposition 3.3,* $\lim \mathbf{x}_n$ *does not exist.* ◀

▶ **3.21** *Prove that if a sequence is convergent, the limit of the sequence is the only limit point of that sequence.*

▶ **3.22** *Suppose $\{x_n\}$ is a bounded sequence in $\mathbb{X}$. Define the sequence $\{x'_n\}$ such that $x'_n = x_{n+L}$ for some fixed $L \in \mathbb{N}$. Prove that x is a limit point of $\{x_n\}$ if and only if x is a limit point of $\{x'_n\}$. This result establishes that deleting any finite number of terms from a sequence does not affect the set of limit points of that sequence. What if you delete an infinite number of terms from the original sequence? Show that this is no longer necessarily true by giving a counterexample.*

The previous examples and exercises bring to light several facts about limit points of sequences. In particular, the limit points of a given sequence need not be members of the sequence. Also, not every sequence has a limit point, while some sequences have more than one. Despite the apparently unbridled freedom the relationship between limit points and their associated sequences seems to possess, there are some things that can be counted on. We start by showing that, under certain circumstances, a sequence is guaranteed to have at least one limit point. The theorem that establishes this fact can be thought of as the "sequence version" of the Bolzano-Weierstrass theorem, discussed earlier in the context of limit points of sets.

Theorem 3.6 (The Bolzano-Weierstrass Theorem for Sequences)
Every bounded sequence in $\mathbb{X}$ has at least one limit point.

It is worth noting that this theorem is often alternatively stated as follows: "Every bounded sequence in $\mathbb{X}$ has a convergent subsequence." Clearly the two statements are equivalent.

PROOF Let $\{x_n\}$ be a bounded sequence whose elements comprise the set $S \subset \mathbb{X}$. If S has only finitely many distinct elements, then at least one element of S, call it x_L, must occur infinitely many times in $\{x_n\}$. That is, $x_L = x_{n_1} = x_{n_2} = x_{n_3} = \cdots$, where $n_1 < n_2 < n_3 < \ldots$. In this case, it is easy to see that $\lim_{m\to\infty} x_{n_m} = x_L$, and so x_L is a limit point of the sequence $\{x_n\}$. Suppose now that there are infinitely many distinct elements in the sequence $\{x_n\}$ and that no element occurs infinitely many times. Then if we consider S as a set of points in $\mathbb{X}$, according to Theorem 4.7 in Chapter 2 the set S has at least one limit point, x. We will show that this "set-sense" limit point is in fact a "sequence-sense" limit point as well. In particular, we will show that there exists a subsequence of $\{x_n\}$ that converges to x. Now, according to the exercise immediately following Definition 2.4 on page 53 in Chapter 2, every neighborhood of x contains infinitely many points of S. In particular, in the neighborhood of radius 1 centered at x there exists $x_{n_1} \in S$. That is, there exists $x_{n_1} \in S$ satisfying $|\,x_{n_1} - x\,| < 1$. Now consider the neighborhood of radius $\frac{1}{2}$ centered at x. It too contains infinitely many elements of S, and one of these must have index n_2 greater than n_1 (Why?). That is, there exists $x_{n_2} \in S$ such that $|\,x_{n_2} - x\,| < \frac{1}{2}$ with $n_2 > n_1$. Continuing in this fashion, we can find $x_{n_m} \in S$ such that $|\, x_{n_m} - x \,| < \frac{1}{m}$ with $n_m > n_{m-1} > \cdots > n_2 > n_1$. Note from this that for any given $\epsilon > 0$, we can specify an $N \in \mathbb{N}$ such that $N > \frac{1}{\epsilon}$ (that is, $\frac{1}{N} < \epsilon$) and $m > N \Rightarrow |\, x_{n_m} - x \,| < \epsilon$. But this means that the subsequence $\{x_{n_m}\}$ so generated has the property that $\lim_{m\to\infty} x_{n_m} = x$. That is, x is a (sequence-sense) limit point of S. ♦

If you reconsider the previous examples and exercises, you will see that some sequences have limit points that are members of the sequence itself, and some have limit points that are not members of the sequence. How a limit point is related to the set of points S that constitutes the elements of the sequence is not completely arbitrary, however, as the following proposition shows.

Proposition 3.7 *Let $\{x_n\} \subset S \subset \mathbb{X}$ be a sequence. Then the limit points of $\{x_n\}$ are in the closure of S. That is, for each limit point x, we have $x \in \overline{S}$.*

PROOF Suppose the limit point x is in $(\overline{S})^C$. Then, since $(\overline{S})^C$ is open, there exists a neighborhood $N_r(x) \subset (\overline{S})^C$. Also, since x is a limit point of $\{x_n\}$, there exists a subsequence $\{x_{n_m}\}$ convergent to x. This implies there exists $M \in \mathbb{N}$ such that $m > M \Rightarrow |\, x_{n_m} - x \,| < r$, and so $x_{n_m} \in N_r(x) \subset (\overline{S})^C$ for $m > M$. But then $x_{n_m} \notin S$ for $m > M$. This is a contradiction, and so x must be in $\overline{S}$. ♦

▶ **3.23** *Consider a set of points $S \subset \mathbb{X}$, and let S' be the set of limit points of S. Show that if $x \in S'$, there exists a sequence $\{x_n\} \subset S$ for $n \ge 1$ such that $\lim x_n = x$.*

▶ **3.24** *Suppose $\{x_n\} \subset S \subset \mathbb{X}$ is a sequence convergent to x. Show that $x \in \overline{S}$, but it is not necessarily true that $x \in S'$.*

▶ **3.25** *Suppose $A \subset \mathbb{X}$ is compact and $B \subset \mathbb{X}$ is closed such that $A \cap B = \varnothing$. Then* $\text{dist}(A, B) \equiv \inf\limits_{\substack{a\in A \\ b\in B}} |a - b| > 0$. *To show this, assume* $\text{dist}(A, B) = 0$. *Then there exist sequences $\{a_n\} \subset A$ and $\{b_n\} \subset B$ such that $\lim |a_n - b_n| = 0$. Exploit the compactness of A to derive a contradiction.*

3.2 The Limit Superior and Limit Inferior

The results of this section apply to bounded sequences of real numbers. Of course, as was seen in a previous exercise, $\{x_n\} \subset \mathbb{X}$ is a bounded sequence in $\mathbb{X}$ if and only if each component of $\{x_n\}$ is a bounded sequence in $\mathbb{R}$. With this fact in mind, we can apply the results of this subsection more generally in a component-wise way. They are therefore not as limited as they might otherwise first appear.

Definitions of the lim sup *and* lim inf

As we have seen, a sequence can have more than one limit point. In the case of a bounded sequence of real numbers $\{x_n\}$ with $|x_n| \le M$ for all n, the collection of limit points $\mathcal{L}$ is nonempty. It is not too hard to show in this case that the set $\mathcal{L}$ is also bounded. In fact, for arbitrary $L \in \mathcal{L}$, there exists a subsequence $\{x_{n_m}\}$ of the original sequence that converges to L. That is, for any $\epsilon > 0$ there exists an $N \in \mathbb{N}$ such that $m > N \Rightarrow |x_{n_m} - L| < \epsilon$. In particular, for $\epsilon = 1$ there is an $N \in \mathbb{N}$ such that $m > N \Rightarrow |x_{n_m} - L| < 1$. From this, and the triangle inequality, we have that $m > N \Rightarrow |L| \leq |L - x_{n_m}| + |x_{n_m}| < 1 + M$. Since the element $L \in \mathcal{L}$ was arbitrary, we have shown that $\mathcal{L}$ is bounded.

► **3.26** *Show that the M referred to above is a sharper bound on $\mathcal{L}$. That is, show that if $L \in \mathcal{L}$, then $|L| \leq M$.*

Since the set $\mathcal{L}$ of limit points of a bounded sequence $\{x_n\}$ in $\mathbb{R}$ is itself just a bounded set of real numbers, we can consider $\sup \mathcal{L}$ and $\inf \mathcal{L}$, which are also real numbers. These values are of particular importance since they relate to the sequence $\{x_n\}$, and we define them now.

Definition 3.8 Let $\{x_n\}$ be a bounded sequence in $\mathbb{R}$, and let $\mathcal{L}$ be its associated set of limit points. We define the **limit superior** of $\{x_n\}$ to be the supremum of the set $\mathcal{L}$, and we denote this new quantity by $\limsup x_n$. Similarly, the **limit inferior** of $\{x_n\}$ is defined to be the infimum of the set $\mathcal{L}$, and is denoted by $\liminf x_n$. That is,

$$\limsup x_n \equiv \sup \mathcal{L} \quad \text{and} \quad \liminf x_n \equiv \inf \mathcal{L}.$$

In the following discussion, we will often use the symbol Λ to represent the real number that is the $\limsup x_n$, and likewise, the symbol λ will be used to represent the real number that is the $\liminf x_n$.

Example 3.9 *Consider the sequence $\{x_n\}$ in $\mathbb{R}$ with terms given by $x_n = 1 + \sin\left(\frac{n\pi}{2}\right)$ for $n \in \mathbb{N}$. We will find Λ and λ for this sequence. To this end, note that since $|x_n| \leq 2$, it follows that Λ and λ each exist. Also, since $\sin\left(\frac{n\pi}{2}\right)$ only takes values from the set $\{-1, 0, 1\}$, the set of limit points for the sequence is given by $\mathcal{L} = \{0, 1, 2\}$. From this we see that $\Lambda = 2$ and $\lambda = 0$.* ◄

Properties of the lim sup *and* lim inf

We now establish some results relating to Λ and λ associated with a bounded sequence $\{x_n\}$ in $\mathbb{R}$.

Proposition 3.10 *Λ and λ are elements of $\mathcal{L}$, and therefore*

$$\Lambda = \max \mathcal{L} \geq \min \mathcal{L} = \lambda.$$

PROOF The relation $\Lambda \geq \lambda$ is a direct consequence of the definitions of lim sup and lim inf. We prove that $\Lambda = \max \mathcal{L}$, and leave the proof for λ as an exercise. Since $\Lambda = \sup \mathcal{L}$, any $L \in \mathcal{L}$ must satisfy $L \leq \Lambda$. Also, there exists $L_1 \in \mathcal{L}$ such that $\Lambda - 1 < L_1$. Together, these two facts require that L_1 satisfies

$$\Lambda - 1 < L_1 < \Lambda + 1.$$

That is, L_1 is a point in the interior of the interval $(\Lambda - 1, \Lambda + 1)$. We also know that since L_1 is a limit point, there exists a subsequence of $\{x_n\}$ that converges to L_1. Therefore, there exists an x_{n_1} within the interval containing L_1, that is,

$$\Lambda - 1 < x_{n_1} < \Lambda + 1.$$

By a similar argument, we can find an $L_2 \in \mathcal{L}$ that lies within the interval $\left(\Lambda - \frac{1}{2}, \Lambda + \frac{1}{2}\right)$, and an associated x_{n_2} with $n_2 > n_1$ (Why?) satisfying

$$\Lambda - \tfrac{1}{2} < x_{n_2} < \Lambda + \tfrac{1}{2}.$$

Continuing in this way, we generate a subsequence $\{x_{n_m}\}$ such that

$$\Lambda - \tfrac{1}{m} < x_{n_m} < \Lambda + \tfrac{1}{m}.$$

This implies that

$$\lim_{m\to\infty} x_{n_m} = \Lambda,$$

and so $\Lambda \in \mathcal{L}$. ♦

► **3.27** *Answer the (Why?) in the above proof, and then prove that $\lambda \in \mathcal{L}$.*

Since Λ is the supremum of all the limit points associated with the bounded sequence $\{x_n\}$, there are no limit points of $\{x_n\}$ greater than Λ. Recalling the idea behind the Bolzano-Weierstrass theorem, we realize that for any $\epsilon > 0$ there should be at most a finite number of sequence terms greater than $\Lambda + \epsilon$. Similarly, for any $\epsilon > 0$ there should be at most a finite number of sequence terms less than $\lambda - \epsilon$. We summarize these conclusions in the following proposition.

Proposition 3.11 *For any $\epsilon > 0$, there exist $N_1, N_2 \in \mathbb{N}$ such that*

a) $n > N_1 \Rightarrow x_n < \Lambda + \epsilon$,
b) $n > N_2 \Rightarrow \lambda - \epsilon < x_n$.

PROOF We prove *a)* and leave the proof of *b)* as an exercise. Let $\epsilon > 0$ be given, and suppose that infinitely many terms of the bounded sequence $\{x_n\}$ are greater than or equal to $\Lambda + \epsilon$, i.e., that there exists a subsequence $\{x_{n_m}\}$ such that $\Lambda + \epsilon \le x_{n_m}$ for $m \ge 1$. We will show that this leads to a contradiction. Since the sequence $\{x_n\}$ is bounded, there exists some real number M such that $|x_n| \le M$ for all $n \ge 1$, and so certainly the same bound applies to the terms of the subsequence, namely, $|x_{n_m}| \le M$ for all $m \ge 1$. Therefore we have that $x_{n_m} \in [\Lambda + \epsilon, M]$ for all $m \ge 1$. It follows from the Bolzano-Weierstrass theorem for sequences that the subsequence $\{x_{n_m}\}$ must have a limit point, L. Clearly, $L \in [\Lambda + \epsilon, M]$. But this means that $L > \Lambda$ is a limit point of the original sequence $\{x_n\}$ as well, which contradicts the fact that Λ

is the supremum of all the limit points of the sequence $\{x_n\}$. Therefore our initial assumption, that infinitely many terms of the bounded sequence $\{x_n\}$ are greater than $\Lambda + \epsilon$, must be false. This proves *a)*. ♦

► **3.28** *Prove part b) of the above proposition.*

The following proposition characterizes the relationship between λ and Λ in the case where the bounded sequence $\{x_n\}$ is convergent.

Proposition 3.12 *$\Lambda = \lambda$ if and only if the bounded sequence $\{x_n\}$ converges to $\Lambda = \lambda$.*

PROOF Suppose $\Lambda = \lambda$. Then by the previous proposition, for any $\epsilon > 0$ there exists a positive integer N_1 such that $n > N_1 \Rightarrow x_n < \Lambda + \epsilon$. For this same $\epsilon > 0$ there exists a positive integer N_2 such that $n > N_2 \Rightarrow \Lambda - \epsilon < x_n$. Letting $N = \max(N_1, N_2)$, we have that

$$n > N \Rightarrow \Lambda - \epsilon < x_n < \Lambda + \epsilon,$$

which in turn is equivalent to

$$n > N \Rightarrow |x_n - \Lambda| < \epsilon.$$

This proves that $\Lambda = \lambda \Rightarrow \lim x_n = \Lambda$. Now suppose the bounded sequence $\{x_n\}$ converges to L. Since *every* subsequence of $\{x_n\}$ must converge to L we see that $\mathcal{L} = \{L\}$, and so $\Lambda = \lambda = L$. ♦

Example 3.13 *In this example we will show that $\lim_{n\to\infty} \sqrt[n]{n} = 1$. Since $\sqrt[n]{n} \geq 1$ for $n \geq 1$, we have that*

$$1 \leq \liminf \sqrt[n]{n} \leq \limsup \sqrt[n]{n}.$$

If we can show that $\limsup \sqrt[n]{n} = 1$ we are done. We will argue by contradiction. To this end, assume $\Lambda = \limsup \sqrt[n]{n} > 1$, so that $\beta \equiv \Lambda - 1 > 0$, and $1 < \left(1 + \frac{\beta}{2}\right) < \Lambda$. Then there exists a subsequence $\{\sqrt[n_j]{n_j}\}$ for $j \geq 1$ that converges to Λ, and therefore for some large enough $N \in \mathbb{N}$ we have that

$$j > N \Rightarrow \sqrt[n_j]{n_j} \geq \left(1 + \tfrac{\beta}{2}\right),$$

i.e.,

$$j > N \Rightarrow n_j \geq \left(1 + \tfrac{\beta}{2}\right)^{n_j} \geq 1 + n_j\left(\tfrac{\beta}{2}\right) + \frac{n_j(n_j - 1)}{2}\left(\tfrac{\beta}{2}\right)^2. \quad \textit{(Why?)}$$

But this is a contradiction since

$$\begin{aligned} 1 + n_j\left(\tfrac{\beta}{2}\right) + \frac{n_j(n_j - 1)}{2}\left(\tfrac{\beta}{2}\right)^2 &> \frac{n_j(n_j - 1)}{2}\left(\tfrac{\beta}{2}\right)^2 \\ &> n_j \quad \text{for } j \text{ large, } \textit{(Why?)} \end{aligned}$$

and therefore $\Lambda = \limsup \sqrt[n]{n} = 1$. ◄

► **3.29** *Answer the two (Why?) questions in the above example.*

▶ **3.30** *Let $\{x_n\}$ be a bounded sequence of real numbers. Show that for any $c \geq 0$,*
a) $\limsup(c\,x_n) = c \limsup x_n$ *b)* $\liminf(c\,x_n) = c \liminf x_n$

▶ **3.31** *Show that the equalities in parts a) and b) of the last exercise do not necessarily hold if $c < 0$.*

▶ **3.32** *Let $\{x_n\}$ and $\{y_n\}$ be bounded sequences of real numbers such that $x_n \leq y_n$ for all $n = 1, 2, \ldots$. Show that a)* $\limsup x_n \leq \limsup y_n$ *b)* $\liminf x_n \leq \liminf y_n$

▶ **3.33** *Suppose $\{x_n\}$ is a bounded sequence of real numbers where $x_n \geq 0$ for $n = 1, 2, \ldots$. Suppose also that $\limsup x_n = 0$. Show in this case that $\lim x_n = 0$.*

3.3 Cauchy Sequences and Completeness

Classifying a sequence as convergent or divergent can be a technically challenging problem. The usefulness of theorems such as the monotone sequence theorem is in their accomplishing this task through the verification of conditions other than those in the definition of convergence given in Definition 1.3. Often a theorem's conditions are easier to verify or are intuitively clearer than those in the definition. Unfortunately, however, applicability of such theorems is typically limited. A drawback of the monotone sequence theorem, for example, is its explicit dependence on the order property of the real number system, and hence its limited applicability to sequences in $\mathbb{R}$. The classification we introduce in this subsection, that of *Cauchy sequences*, possesses no such limitations, and in fact will be seen to be equivalent to convergence as described in Definition 1.3 for sequences in $\mathbb{X}$. We begin by defining what it means for a sequence to be a *Cauchy sequence*. The significant and convenient equivalence of this new class of sequences and the class of *convergent* sequences will be expressed in the form of a theorem. This equivalence is so important, that we will also characterize completeness of the real numbers $\mathbb{R}$, and hence also that of $\mathbb{R}^k$ and $\mathbb{C}$, in terms of it.

Definition 3.14 Let $\{x_n\}$ be a sequence in $\mathbb{X}$. Then $\{x_n\}$ is called a **Cauchy sequence** if for every $\epsilon > 0$, there exists an $N \in \mathbb{N}$ such that

$$n, m > N \Rightarrow |x_n - x_m| < \epsilon.$$

Suppose someone could write down all the elements of a sequence (this person must have an infinite amount of ink, paper, and time). For convenience, label the elements of the sequence 1st, 2nd, 3rd,.... A Cauchy sequence is a sequence with the following property: given an arbitrarily small distance $\epsilon > 0$, one can determine an $N \in \mathbb{N}$ such that all the elements in the sequence beyond the Nth element remain within a distance ϵ from each other. Often, a sequence that satisfies Definition 3.14 will be said to be "Cauchy."

Example 3.15 *We will show that the sequence $\{x_n\} \subset \mathbb{R}$ given by $x_n = \frac{n}{n+1}$*

is a Cauchy sequence. To this end, let $\epsilon > 0$ be any positive number. Then for $n, m \in \mathbb{N}$ we have

$$\begin{aligned}|x_n - x_m| = \left|\frac{n}{n+1} - \frac{m}{m+1}\right| &= \left|\frac{nm+n-(mn+m)}{(n+1)(m+1)}\right| \\ &= \frac{|n-m|}{(n+1)(m+1)} \\ &\le \frac{n}{(n+1)(m+1)} + \frac{m}{(n+1)(m+1)} \\ &< \frac{1}{m+1} + \frac{1}{n+1} \\ &< \epsilon \quad \text{if } n, m > \left(\frac{2}{\epsilon} - 1\right).\end{aligned}$$

Choosing $N > \left(\frac{2}{\epsilon} - 1\right)$ gives the result. ◀

Example 3.16 *Consider the sequence of complex numbers $\{z_n\}$ given by $z_n = \frac{n+1}{n} + \frac{i}{n^2}$. We will show that this sequence is Cauchy. To this end, consider*

$$\begin{aligned}|z_n - z_m| &= \left|\left(\frac{n+1}{n} - \frac{m+1}{m}\right) + i\left(\frac{1}{n^2} - \frac{1}{m^2}\right)\right| \\ &= \left|\left(\frac{1}{n} - \frac{1}{m}\right) + i\left(\frac{1}{n^2} - \frac{1}{m^2}\right)\right| \\ &\le \frac{1}{n} + \frac{1}{m} + \frac{1}{n^2} + \frac{1}{m^2}.\end{aligned}$$

If we can make each term on the right-hand side of the above inequality less than $\frac{\epsilon}{4}$, we'll be done. This is accomplished by taking both n and m to be greater than $\max\left(\frac{4}{\epsilon}, \sqrt{\frac{4}{\epsilon}}\right)$. That is, choosing $N > \max\left(\frac{4}{\epsilon}, \sqrt{\frac{4}{\epsilon}}\right)$, we obtain

$$n, m > N \;\Rightarrow\; |z_n - z_m| < \epsilon,$$

and so the sequence $\{z_n\}$ is Cauchy. ◀

The following theorem is the key result of this subsection.

Theorem 3.17 *Let $\{x_n\}$ be a sequence in $\mathbb{X}$. Then $\{x_n\}$ is convergent if and only if it is a Cauchy sequence.*

PROOF Assume $\{x_n\}$ in $\mathbb{X}$ is convergent, and denote the limit of the sequence by x. We will show that the sequence is Cauchy. For $\epsilon > 0$, there exists an $N \in \mathbb{N}$ such that $n > N \Rightarrow |x_n - x| < \frac{\epsilon}{2}$. Similarly, $m > N \Rightarrow |x_m - x| < \frac{\epsilon}{2}$. Combining these inequalities, we have that

$$n, m > N \;\Rightarrow\; |x_n - x_m| = \big|(x_n - x) + (x - x_m)\big| \le |x_n - x| + |x - x_m| < \epsilon.$$

This proves that the convergent sequence is a Cauchy sequence. Assume now that the sequence $\{x_n\}$ is a Cauchy sequence. We will show that $\{x_n\}$ must be convergent. Since the sequence is Cauchy, there exists an $N \in \mathbb{N}$ such that $m, n > N \Rightarrow |x_n - x_m| < 1$. Therefore, choosing m to be $N+1$, and making use of the triangle inequality, we obtain $|x_n| < 1 + |x_{N+1}|$ for $n > N$. From this it follows that

$$|x_n| \le \max\Big(|x_1|, |x_2|, \ldots, |x_N|, 1 + |x_{N+1}|\Big) \text{ for all } n,$$

i.e., the sequence $\{x_n\}$ is bounded. By Theorem 3.6, we know that the sequence must have at least one limit point. Let x be a limit point for this sequence. We will show that the sequence must converge to x. First, note that since x is a limit point there exists a subsequence $\{x_{n_j}\}$ convergent to x. That is, for $\epsilon > 0$ there exists an $N_1 \in \mathbb{N}$ such that $j > N_1 \Rightarrow |x_{n_j} - x| < \frac{\epsilon}{2}$. Also, since $\{x_n\}$ is a Cauchy sequence, for each $\epsilon > 0$ there exists an $N_2 \in \mathbb{N}$ such that $n, m > N_2 \Rightarrow |x_n - x_m| < \frac{\epsilon}{2}$. Therefore, $n, n_j > N_2 \Rightarrow |x_n - x_{n_j}| < \frac{\epsilon}{2}$. But $n_j \ge j$, so it follows that $n, j > N_2 \Rightarrow |x_n - x_{n_j}| < \frac{\epsilon}{2}$. Now, fix $j > \max(N_1, N_2)$. Then, if $n > N_2$ we have

$$|x_n - x| = \big|x_n - x_{n_j} + x_{n_j} - x\big| \le |x_n - x_{n_j}| + |x_{n_j} - x| < \epsilon.$$

This shows that the Cauchy sequence $\{x_n\}$ in $\mathbb{X}$ is convergent. ◆

► **3.34** *Show that* $\{\mathbf{x}_n\}$ *where* $\mathbf{x}_n = (x_n, y_n)$ *is a Cauchy sequence in* $\mathbb{R}^2$ *if and only if* $\{x_n\}$ *and* $\{y_n\}$ *are Cauchy sequences in* $\mathbb{R}$. *What does this imply about sequences* $\{z_n\} \in \mathbb{C}$, *where* $z_n = x_n + i\, y_n$? *(Think about convergence.) Generalize this result to sequences* $\{\mathbf{x}_n\} = (x_{1n}, x_{2n}, \ldots, x_{kn}) \in \mathbb{R}^k$.

► **3.35** *Suppose* $\{x_n\}$ *is a sequence in* $\mathbb{X}$ *such that* $|x_{n+1} - x_n| \le c\, r^n$ *for some positive* $c \in \mathbb{R}$ *and* $0 < r < 1$. *Prove that* $\{x_n\}$ *is a Cauchy sequence, and therefore* $\lim x_n = x$ *exists.*

The significance of Theorem 3.17 cannot be overemphasized. The property that all Cauchy sequences converge is referred to as *Cauchy completeness,* and can be taken as an axiomatic property of the real numbers. In fact, it can replace Dedekind completeness as a description of completeness in $\mathbb{R}$. Of course, both notions of completeness must be equivalent in this case. This means that if we assume all the axioms on $\mathbb{R}$ including the Dedekind completeness property, then we can prove $\mathbb{R}$ satisfies the Cauchy completeness property. Moreover, if we take all the axioms on $\mathbb{R}$ *except the Dedekind completeness property* and *replace it* with the Cauchy completeness property, then we can deduce the Dedekind completeness property. We prove this now.

Theorem 3.18 *$\mathbb{R}$ is Cauchy complete if and only if $\mathbb{R}$ possesses the Dedekind completeness property.*

PROOF Note that since we have presumed throughout our development that $\mathbb{R}$ possesses the Dedekind completeness property, the proof of Theorem 3.17 establishes that this implies that $\mathbb{R}$ must also be Cauchy complete.

We now prove the converse. To begin, suppose $\mathbb{R}$ is Cauchy complete, and that $\mathbb{R} = A \cup B$ where A and B are nonempty subsets of $\mathbb{R}$ satisfying the following:

(i) $\mathbb{R} = A \cup B$,

(ii) $a < b$ for every $a \in A$ and every $b \in B$.

We will show that either A has a maximal element or B has a minimal element, and hence that the Dedekind completeness property holds. To see this, take any $a_1 \in A$ and $b_1 \in B$. Then $a_1 < b_1$ and for $I_1 = [a_1, b_1]$, the length of I_1 is given by $\ell(I_1) = b_1 - a_1$. Bisect I_1 to form two intervals I_1' and I_1''. One of these intervals must be of the form $I_2 = [a_2, b_2]$ where $a_2 \in A$, $b_2 \in B$, and $a_2 < b_2$. (Why?) Bisect I_2 similarly and continue in this way to construct a sequence of nested intervals

$$I_1 \supset I_2 \supset I_3 \supset \cdots$$

where

(i) $I_n = [a_n, b_n]$ with $a_n \in A$, $b_n \in B$,

(ii) $\ell(I_n) = \frac{1}{2^{n-1}}\, \ell(I_1)$.

From this we have that

$$\lim(b_n - a_n) = \lim \ell(I_n) = 0. \tag{3.6}$$

We will now show that both sequences $\{a_n\}$ and $\{b_n\}$ are Cauchy sequences (and hence both converge by assumption). To show this, we examine $a_{n+1} - a_n$, which is either 0 or $\frac{1}{2}\, \ell(I_n)$, and so

$$a_{n+1} - a_n \le \frac{1}{2}\, \ell(I_n) = \frac{1}{2^{n-1}}\, \ell(I_1).$$

It follows from Exercise 3.35 that $\{a_n\}$ is a Cauchy sequence, and therefore $\lim a_n = a$ exists. Similarly, $\lim b_n = b$ exists. Since, according to (3.6), $\lim(b_n - a_n) = 0$, we have

$$b - a = \lim b_n - \lim a_n = \lim(b_n - a_n) = 0,$$

and so $a = b$. All that remains is to show that either A has a maximal element, or B has a minimal element. We handle this in two cases: $a = b \in A$, and $a = b \in B$. For the case $a = b \in A$, there can be no element of A that is greater than a. If there were, say, $a^* > a$ such that $a^* \in A$, then because $\lim b_n = a$, there exists $N \in \mathbb{N}$ large enough that $a < b_N < a^*$, a contradiction. Therefore, $a = \max A$. The set B cannot simultaneously have a minimal value $b_{\min}$, since if it did, then $a < b_{\min}$, which contradicts $a = \lim b_n$. (Why?) The case $a = b \in B$ is handled similarly, and the result is proved. ♦

► **3.36** *Answer the two (Why?) questions, and write the details of the $a = b \in B$ case to complete the above proof.*

Having established the equivalence of Cauchy completeness and Dedekind completeness in the case of $\mathbb{R}$, we may now adopt Cauchy completeness as our fundamental notion of completeness for all of the spaces of interest to us. That each of the spaces represented by $\mathbb{X}$ is complete in this sense is of great significance in analysis.

4 Properties of Infinite Series

In this section, we consider what happens when the terms of a sequence $\{x_n\}$ in $\mathbb{X}$ are added together to form what is called an *infinite series*. Such an infinite series is denoted by $\sum_{j=1}^{\infty} x_j = x_1 + x_2 + \cdots$. Does the result of such a sum even make sense? We will find that under the right circumstances, it does. When it does, we say the series *converges* to the sum. We will develop results for determining whether or not a series converges, and in some cases, in what manner the series converges. For many convergent series, no means for determining the exact value of the sum is known. While methods for approximating the sum of a convergent series exist, we will not present such techniques in our development of analysis, techniques more suited to a course in numerical methods or applied mathematics.

4.1 Definition and Examples of Series in $\mathbb{X}$

We begin with a definition.

Definition 4.1 For a sequence $\{x_j\} \in \mathbb{X}$, we define the ***n*th partial sum** by $s_n = \sum_{j=1}^{n} x_j = x_1 + x_2 + \cdots + x_n$. Considering the sequence of partial sums given by $\{s_n\}$, if $\lim s_n = s$ exists, then we say that the infinite series $\sum_{j=1}^{\infty} x_j$ **converges** to s, and we write

$$\sum_{j=1}^{\infty} x_j = s.$$

Otherwise, we say that the infinite series $\sum_{j=1}^{\infty} x_j$ **diverges**.

Note that while this definition introduces a new concept, that of an infinite series, it provides a way to determine whether an infinite series converges or diverges in terms of an already familiar concept, that of a sequence. This is a common strategy in mathematics—seeing new problems in the light of old ones already solved. Note too that the term "infinite" in infinite series refers to the fact that we are adding an infinite number of terms. It does not imply that the sum itself is necessarily infinite. Also, since each x_j in the infinite series $\sum_{j=1}^{\infty} x_j$ is finite, omitting any finite number of terms from the series will not alter its behavior as far as convergence or divergence is concerned. We say that whether an infinite series converges or diverges depends only on the "tail" of the series, the terms beyond the Nth term for any finite choice of $N \in \mathbb{N}$. This means that for a given sequence $\{x_j\}$, the associated infinite series given by $\sum_{j=1}^{\infty} x_j$ and $\sum_{j=m}^{\infty} x_j$ for any $m \in \mathbb{N}$ will both converge or both diverge. Of course, in the case where both series converge, they will not in general converge to the same sum. Finally, it should be pointed out that while Definition 4.1 describes infinite series with the summation index starting at the value 1, this is not necessary. We should, however, clarify our interpretation of the associated partial sums for such a series. If the series starts at $j = j_0$ rather than $j = 1$, then the associated sequence of partial sums $\{s_n\}$ is defined for $n \geq j_0$ by $s_n = \sum_{j=j_0}^{n} x_j$. Note that this definition of the nth partial sum means that we are summing the terms of the sequence up to the nth term, not that we are summing the first n terms of the sequence.

Example 4.2 *Consider the series $\sum_{j=1}^{\infty} x_j$ where $x_j = \frac{1}{j(j+1)} \in \mathbb{R}$ for each $j \in \mathbb{N}$. We will show that this series converges, and we will exhibit the sum. This is significant since finding the sum to a convergent series is not usually possible. Note that the nth partial sum is given by*

$$\begin{aligned} s_n = \sum_{j=1}^{n} \frac{1}{j(j+1)} &= \sum_{j=1}^{n} \left(\tfrac{1}{j} - \tfrac{1}{j+1}\right) \quad \textit{by partial fractions,} \\ &= \left(1 - \tfrac{1}{2}\right) + \left(\tfrac{1}{2} - \tfrac{1}{3}\right) + \left(\tfrac{1}{3} - \tfrac{1}{4}\right) + \cdots + \left(\tfrac{1}{n} - \tfrac{1}{n+1}\right) \\ &= 1 - \tfrac{1}{n+1} \quad \textit{after cancelation.} \end{aligned}$$

From this we see that $\lim s_n = 1$, and so the series converges to 1. Such a series where the partial sum expression "collapses" in this way is called a **telescoping series.** ◀

Example 4.3 *Consider the series $\sum_{j=0}^{\infty} \frac{1}{2^j}$. The nth partial sum is given by*

$$s_n = 1 + \frac{1}{2} + \frac{1}{2^2} + \frac{1}{2^3} + \cdots + \frac{1}{2^n},$$

so

$$\frac{1}{2} s_n = \frac{1}{2} + \frac{1}{2^2} + \frac{1}{2^3} + \frac{1}{2^4} \cdots + \frac{1}{2^{n+1}},$$

and we find

$$s_n - \frac{1}{2} s_n = 1 - \frac{1}{2^{n+1}}.$$

This in turn implies that $s_n = 2 - \frac{1}{2^n}$, and so $\lim s_n = 2$. Therefore, $\sum_{j=0}^{\infty} \frac{1}{2^j} = 2$, i.e., the series is convergent to 2. This is an example of what is called a **geometric series**. ◀

Note that, as in the case of the convergent telescoping series in Example 4.2, we were able to determine the sum for the convergent geometric series in Example 4.3. This is not always possible for convergent series more generally. The classes of series represented by these two examples are special in this regard. We will look at other special classes of series of real numbers later in this section.

Example 4.4 *Consider the series $\sum_{j=1}^{\infty} \frac{1}{j}$, also known as* the harmonic se*ries. We will show that this series is divergent. To this end, consider the subsequence of partial sums given by $\{s_{2^n}\} = \{s_2, s_4, s_8, s_{16}, \ldots\}$. If the* *harmonic series converges, then* $\lim s_n = s$ *exists, and according to Proposition 3.3, the above-mentioned subsequence must have the same limit, that is,* $\lim s_{2^n} = \lim s_n = s$. *But*

$$\begin{aligned} s_2 &= 1 + \tfrac{1}{2} \ \geq \ \tfrac{1}{2} + \tfrac{1}{2} = 2\left(\tfrac{1}{2}\right), \\ s_4 &= \left(1 + \tfrac{1}{2}\right) + \left(\tfrac{1}{3} + \tfrac{1}{4}\right) \ \geq \ \tfrac{1}{2} + \tfrac{1}{2} + \tfrac{1}{4} + \tfrac{1}{4} = 3\left(\tfrac{1}{2}\right), \\ s_8 &= s_4 + \tfrac{1}{5} + \tfrac{1}{6} + \tfrac{1}{7} + \tfrac{1}{8} \ \geq \ 3\left(\tfrac{1}{2}\right) + \tfrac{1}{8} + \tfrac{1}{8} + \tfrac{1}{8} + \tfrac{1}{8} = 4\left(\tfrac{1}{2}\right), \end{aligned}$$

and in general,

$$s_{2^n} \geq (n+1)\left(\tfrac{1}{2}\right). \quad \textit{(Why?)}$$

Therefore, $\lim s_{2^n} = \infty \neq s$, *and so the series must diverge.* ◀

► **3.37** *Answer the (Why?) in the above example with an appropriate induction proof.*

4.2 Basic Results for Series in $\mathbb{X}$

In the examples of the previous subsection, we saw some special techniques that can be useful in working with certain kinds of infinite series. In this subsection, we formalize some general rules for manipulating series in $\mathbb{X}$. We also prove some general results that enable one to determine whether a given series converges or diverges. This is an important first step in dealing with any infinite series. It is important to note that often all one can do is distinguish whether a given series converges or diverges. While in some cases one can go further and describe *how* the series converges or diverges, in many of the convergent cases determining the exact sum is not possible. We begin with a proposition, the proof of which we leave to the exercises.

Proposition 4.5 *Suppose* $\sum_{j=1}^{\infty} x_j$ *and* $\sum_{j=1}^{\infty} y_j$ *are each convergent series in* $\mathbb{X}$. *Then, for any* $c \in \mathbb{R}$,

a) $\sum_{j=1}^{\infty} (x_j \pm y_j) = \sum_{j=1}^{\infty} x_j \pm \sum_{j=1}^{\infty} y_j$,

b) $\sum_{j=1}^{\infty} c\, x_j = c \sum_{j=1}^{\infty} x_j$.

► **3.38** *Prove the above proposition.*

► **3.39** *Consider part b) of the above proposition. What if* $c \in \mathbb{C}$ *and* $\{x_j\} \subset \mathbb{R}$? *How about if* $c \in \mathbb{C}$ *and* $\{x_j\} \subset \mathbb{C}$? *Prove that analogous results hold.*

Example 4.6 Consider the infinite series given by $\sum_{j=1}^{\infty} \left(\frac{1}{2^j} + \frac{5}{j(j+1)}\right)$. According to the previous proposition, since $\sum_{j=1}^{\infty} \frac{1}{2^j}$ and $\sum_{j=1}^{\infty} \frac{1}{j(j+1)}$ are each convergent infinite series, we know that $\sum_{j=1}^{\infty} \left(\frac{1}{2^j} + \frac{5}{j(j+1)}\right)$ is convergent. In fact,

$$\sum_{j=1}^{\infty}\left(\frac{1}{2^j}+\frac{5}{j(j+1)}\right)=\sum_{j=1}^{\infty}\frac{1}{2^j}+5\sum_{j=1}^{\infty}\frac{1}{j(j+1)}=1+5(1)=6.$$

◀

Proposition 4.7 *Let* $\sum_{j=1}^{\infty} x_j$ *be a series in* $\mathbb{X}$. *Then*

a) $\sum_{j=1}^{\infty} x_j$ *converges if and only if* $\sum_{j=m}^{\infty} x_j$ *converges for all* $m \in \mathbb{N}$, *and* $\sum_{j=1}^{\infty} x_j = s$ *if and only if* $\sum_{j=m}^{\infty} x_j = s - (x_1 + x_2 + \cdots + x_{m-1})$.

b) $\sum_{j=1}^{\infty} x_j$ *converges if and only if for any* $\epsilon > 0$, *there exists a positive integer* N *such that* $n, m > N \Rightarrow \left|\sum_{j=m+1}^{n} x_j\right| < \epsilon$.

PROOF The first result follows directly from the definition of convergence for infinite series. The second result is simply the statement that an infinite series converges if and only if its associated sequence of partial sums is a Cauchy sequence. To establish this, apply Theorem 3.17 to the sequence of partial sums $\{s_n\}$. ◆

We now apply this proposition in the following example.

Example 4.8 *We will show that the series* $\sum_{j=1}^{\infty}\left(\frac{1}{2^j}, \frac{1}{3^j}\right)$ *converges in* $\mathbb{R}^2$. *Note that for* $m, n \in \mathbb{N}$,

$$\begin{aligned}\left|\sum_{j=m+1}^{n}\left(\frac{1}{2^j}, \frac{1}{3^j}\right)\right| &\le \sum_{j=m+1}^{n}\left|\left(\frac{1}{2^j}, \frac{1}{3^j}\right)\right| \\ &= \sum_{j=m+1}^{n}\sqrt{\frac{1}{2^{2j}}+\frac{1}{3^{2j}}} \\ &\le \sum_{j=m+1}^{n}\left(\frac{1}{2^j}+\frac{1}{3^j}\right) \\ &\le 2\sum_{j=m+1}^{n}\left(\tfrac{1}{2}\right)^j \\ &< 2\left(\tfrac{1}{2^m}\right).\end{aligned}$$

This last will be less than any given $\epsilon > 0$ *if* $m, n > \ln\left(\frac{2}{\epsilon}\right)/\ln 2$. *Therefore, choosing* $N \in \mathbb{N}$ *such that* $N > \ln\left(\frac{2}{\epsilon}\right)/\ln 2$, *we have that*

$$m, n > N \Rightarrow \left|\sum_{j=m+1}^{n}\left(\frac{1}{2^j}, \frac{1}{3^j}\right)\right| < \epsilon,$$

i.e., $\sum_{j=1}^{\infty}\left(\frac{1}{2^j}, \frac{1}{3^j}\right)$ *converges.* ◀

The following result is sometimes referred to as the "component-wise convergence theorem" for series.

Proposition 4.9

a) *Let* $\{\mathbf{x}_j\}$ *be a sequence in* $\mathbb{R}^k$ *with* $\mathbf{x}_j = (x_j^{(1)}, x_j^{(2)}, \ldots, x_j^{(k)})$. *Then* $\sum_{j=1}^\infty \mathbf{x}_j$ *converges if and only if* $\sum_{j=1}^\infty x_j^{(i)}$ *converges for* $1 \le i \le k$. *Moreover, if* $\sum_{j=1}^\infty \mathbf{x}_j$ *converges, then*

$$\sum_{j=1}^\infty \mathbf{x}_j = \left(\sum_{j=1}^\infty x_j^{(1)}, \sum_{j=1}^\infty x_j^{(2)}, \ldots, \sum_{j=1}^\infty x_j^{(k)}\right).$$

b) *Let* $\{z_j\}$ *be a sequence in* $\mathbb{C}$ *with* $z_j = x_j + i\,y_j$. *Then* $\sum_{j=1}^\infty z_j$ *converges if and only if* $\sum_{j=1}^\infty x_j$ *and* $\sum_{j=1}^\infty y_j$ *converge. Moreover, if* $\sum_{j=1}^\infty z_j$ *converges, then* $\sum_{j=1}^\infty z_j = \sum_{j=1}^\infty x_j + i\sum_{j=1}^\infty y_j$.

PROOF We prove part *a*) and leave part *b*) to the exercises. Define the partial sums $\mathbf{s}_n$ and s_{m_n} as follows:

$$\mathbf{s}_n = \sum_{j=1}^n \mathbf{x}_j, \text{ and } s_n^{(m)} = \sum_{j=1}^n x_j^{(m)} \text{ for } m = 1, 2, \ldots, k.$$

Then,

$$\mathbf{s}_n = \left(s_n^{(1)}, s_n^{(2)}, \ldots, s_n^{(k)}\right),$$

and by Proposition 1.10 on page 89 we know that $\lim \mathbf{s}_n$ exists if and only if $\lim s_n^{(m)}$ exists for each m. Moreover, when $\lim \mathbf{s}_n$ exists we have

$$\lim \mathbf{s}_n = \left(\lim s_n^{(1)}, \lim s_n^{(2)}, \ldots, \lim s_n^{(k)}\right),$$

proving the result. ♦

► **3.40** *Prove part b) of the above proposition.*

Showing that a divergent series diverges is often easier than showing that a convergent series converges. This is due to a very important fact that the examples and exercises so far may have already revealed. In order for an infinite series $\sum x_j$ to converge to a finite sum, the terms being added must be decreasing to 0 in magnitude as $j \to \infty$. This is a *necessary* condition for convergence of the series, but not a sufficient condition. This is illustrated most simply by considering the series $\sum_{j=1}^\infty \frac{1}{j}$ and $\sum_{j=1}^\infty \frac{1}{j^2}$. In fact, a series converges only if the terms decrease to 0 *fast enough* as $j \to \infty$. The following proposition, often referred to as the test for divergence, gives us a very practical test.

Proposition 4.10 (Test for Divergence)
For $\{x_j\} \in \mathbb{X}$, if $\lim x_j \neq 0$, then $\sum_{j=1}^{\infty} x_j$ is divergent.

A statement that is equivalent to the above proposition is, "*If $\sum_{j=1}^{\infty} x_j$ converges, then* $\lim x_j = 0$."

PROOF Assume that $\sum_{j=1}^{\infty} x_j$ converges. For $\epsilon > 0$, there exists a positive integer N such that

$$m, n > N \Rightarrow \left| \sum_{j=m+1}^{n} x_j \right| < \epsilon.$$

Now, let $m = n + 1$. Then for $n > N$, we have

$$\left| \sum_{k=n+1}^{n+1} x_j \right| < \epsilon,$$

i.e.,

$$n + 1 > N \Rightarrow |x_{n+1}| < \epsilon.$$

This proves that $\lim x_{n+1} = 0$, and so $\lim x_n = 0$. ♦

Example 4.11 *Consider $\sum_{j=1}^{\infty} \mathbf{x}_j$ where $\mathbf{x}_j = \left(\frac{1}{j}, \frac{j+1}{j+2}\right) \in \mathbb{R}^2$. Then, since $\lim \frac{j+1}{j+2} = 1 \neq 0$, we have that $\lim \mathbf{x}_j \neq \mathbf{0}$, and so the series must diverge.* ◄

► **3.41** *Suppose $|x_j| \geq a\,r^j$, where $a > 0$ and $r > 1$. Show that $\sum_{j=1}^{\infty} x_j$ diverges.*

When a series converges, one can describe more specifically *how* it converges. The following definition distinguishes between those series that converge *absolutely* and those that converge only *conditionally*.

Definition 4.12 For a convergent series $\sum_{j=1}^{\infty} x_j$ in $\mathbb{X}$, if the series $\sum_{j=1}^{\infty} |x_j|$ converges, we say that the series $\sum_{j=1}^{\infty} x_j$ is **absolutely convergent.** If the series $\sum_{j=1}^{\infty} |x_j|$ diverges, we say that $\sum_{j=1}^{\infty} x_j$ is **conditionally convergent.**

Example 4.13 *We will show that the series $\sum_{j=1}^{\infty} \frac{(-1)^j 4(j+1)}{(2j+1)(2j+3)}$ converges conditionally. To this end, let $s_n = \sum_{j=1}^{n} \frac{(-1)^j 4(j+1)}{(2j+1)(2j+3)}$. Using partial fractions, it is easy to see that*

$$s_n = \sum_{j=1}^{n} (-1)^j \left(\frac{1}{2j+1} + \frac{1}{2j+3}\right) = -\frac{1}{3} + \frac{(-1)^n}{2n+3},$$

and that $\lim s_n = -\frac{1}{3}$ exists. Therefore, the series converges. To show that the series is not absolutely convergent, consider $\sum_{j=1}^{\infty} \frac{4(j+1)}{(2j+1)(2j+3)}$ and its partial sums given by $\widetilde{s_n} = \sum_{j=1}^{n} \frac{4(j+1)}{(2j+1)(2j+3)}$. We estimate the jth term of this partial *sum as follows. Note that when j is large, the numerator "looks like" $4j$ while the denominator "looks like" $(2j)(2j) = 4j^2$, resulting in an overall ratio that "looks like" $\frac{1}{j}$. Therefore the "tail" of the series "looks like" the tail of the divergent harmonic series, and so we suspect we can bound $\frac{4(j+1)}{(2j+1)(2j+3)}$ below by a harmonic-like term. To this end, we note that $4(j+1) > 4j$ for all $j \geq 1$, and $(2j+1) \leq (2j+j)$ for all $j \geq 1$. Finally, $(2j+3) \leq (2j+3j)$ for all $j \geq 1$, and so,*

$$\frac{4(j+1)}{(2j+1)(2j+3)} > \frac{4j}{(2j+j)(2j+3j)} = \frac{4}{15\,j} \quad \text{for } j \geq 1.$$

From this we see that

$$\widetilde{s_n} = \sum_{j=1}^{n} \frac{4(j+1)}{(2j+1)(2j+3)} > \frac{4}{15} \sum_{j=1}^{n} \frac{1}{j}.$$

Since the sum on the right-hand side diverges as $n \to \infty$, so too $\widetilde{s_n} \to \infty$ as $n \to \infty$, and the convergence of $\sum_{j=1}^{\infty} \frac{(-1)^j 4(j+1)}{(2j+1)(2j+3)}$ is only conditional. ◀

As the above definition and example indicate, a series that does not converge absolutely may still converge conditionally. The following proposition tells us that absolute convergence of a series implies conditional convergence.

Proposition 4.14 *For $\{x_j\} \in \mathbb{X}$, if $\sum_{j=1}^{\infty} |\,x_j\,|$ converges, then $\sum_{j=1}^{\infty} x_j$ converges.*

PROOF Let $\epsilon > 0$ be given. Then, from result *b)* from Proposition 4.7 on page 111, there exists an $N \in \mathbb{N}$ such that

$$m, n > N \Rightarrow \left| \sum_{j=m+1}^{n} |\,x_j\,| \right| < \epsilon.$$

But,

$$\left| \sum_{j=m+1}^{n} x_j \right| \leq \sum_{j=m+1}^{n} |\,x_j\,| = \left| \sum_{j=m+1}^{n} |\,x_j\,| \right| < \epsilon \text{ if } m, n > N.$$

Therefore $\sum_{j=1}^{\infty} x_j$ converges, and the result is proved. ♦

► **3.42** *For $\{x_j\} \subset \mathbb{X}$ such that $\sum_{j=1}^{\infty} |x_j|$ converges, show that*

$$\left| |x_1| - \left| \sum_{j=2}^{\infty} x_j \right| \right| \leq \left| \sum_{j=1}^{\infty} x_j \right| \leq \sum_{j=1}^{\infty} |x_j|.$$

This is the natural extension of the triangle and reverse triangle inequalities to infinite series.

Many results for series, and in particular, many tests for convergence of series, apply only to series in $\mathbb{R}$. We will explore these results and tests in the next subsection. The significance of the above proposition is that it allows us to use many of these "$\mathbb{R}$-specific" results and tests even when working with a series in $\mathbb{R}^k$ or $\mathbb{C}$. In particular, we can use them to test any series in $\mathbb{X}$ for absolute convergence, since for any $x_j \in \mathbb{X}$, the quantity $|x_j|$ is a real number.

4.3 Special Series

We now describe several special classes of series, each of which occurs frequently in practice and has its own convergence and divergence criteria. We start with one of the most useful classes of infinite series in analysis and in applications.

Geometric Series

Definition 4.15 An infinite series of the form $\sum_{j=0}^{\infty} r^j = 1 + r + r^2 + \cdots$ for constant $r \in \mathbb{R}$ or $\mathbb{C}$ is called a **geometric series**.

Note from the above that the first summand of a geometric series is 1 by definition. Example 4.3 on page 109 is an example of a geometric series. This is possibly the most important class of series in analysis and will be critical to our development in later chapters. It is also important because of its appearance in many applications, as well as the fact that when a geometric series converges, we can determine its sum. The following exercises allow you to characterize the convergence properties of geometric series.

▶ **3.43** *Show that, if $|r| < 1$, the corresponding geometric series converges to the sum $\frac{1}{1-r}$. To get started, show that $s_n = \frac{1-r^{n+1}}{1-r}$.*

▶ **3.44** *Show that, if $|r| \geq 1$, the corresponding geometric series diverges.*

Alternating Series

An important class of series within $\mathbb{R}$ that contains examples that are conditionally convergent is the class of *alternating series.*

Definition 4.16 A series $\sum_{j=1}^{\infty}(-1)^{j+1}x_j$ with $0 < x_j \in \mathbb{R}$ for all j is called an **alternating series.**

There is an especially convenient test for convergence for alternating series. We establish this test in the following proposition.

Proposition 4.17 (Alternating Series Test)

Suppose $\sum_{j=1}^{\infty}(-1)^{j+1}x_j$ is an alternating series. If $x_j \geq x_{j+1}$ for all j, and if $\lim x_j = 0$, then the series converges.

PROOF Let $s_n = \sum_{j=1}^{n}(-1)^{j+1}x_j$. We will show that the odd and even subsequences, $\{s_{2k+1}\}$ and $\{s_{2k}\}$, both converge to the same limit. To see this, note that

$$s_{2k+1} = s_{2k-1} - x_{2k} + x_{2k+1} \le s_{2k-1},$$

and so $\{s_{2k+1}\}$ is nonincreasing. It is also easy to see that $s_{2k+1} > 0$, and so $\{s_{2k+1}\}$ is bounded below. Therefore, $\lim s_{2k+1} = s_{\mathrm{odd}}$ exists. Similarly, we see that

$$s_{2k} = s_{2k-2} + x_{2k-1} - x_{2k} \ge s_{2k-2},$$

and $s_{2k} \le x_1$ for all k (Why?), and so $\{s_{2k}\}$ is nondecreasing and bounded above. Therefore, it follows that $\lim s_{2k} = s_{\mathrm{even}}$ exists. We will now show that $s_{\mathrm{odd}} = s_{\mathrm{even}}$. To establish this, note that $s_{2k+1} = s_{2k} + x_{2k+1}$. Taking the limit as k goes to ∞ on both sides of this equality yields the result that $s_{\mathrm{odd}} = s_{\mathrm{even}}$. The completion of the proof is left to the reader. ♦

► **3.45** *Complete the proof of the above proposition by showing that for every $\epsilon > 0$, there exists $N \in \mathbb{N}$ such that $n > N \Rightarrow |s_n - s| < \epsilon$. (Hint: The value of n must be either even or odd.)*

Example 4.18 *We will show that the alternating series $\sum_{j=1}^{\infty}(-1)^j \frac{1}{j}$ is conditionally convergent. (We already know that it is not absolutely convergent.) First, note that $\sum_{j=1}^{\infty}(-1)^j \frac{1}{j} = (-1)\sum_{j=1}^{\infty}(-1)^{j+1}\frac{1}{j}$, provided the series both converge. The series on the right converges by the alternating series test, since $\frac{1}{j} > 0$ for all $j \ge 1$, and $\left\{\frac{1}{j}\right\}$ is a decreasing sequence (Why?) with $\lim \frac{1}{j} = 0$. Therefore, the series $\sum_{j=1}^{\infty}(-1)^j \frac{1}{j}$ converges conditionally.* ◄

Series with Nonnegative Terms

The following important results pertain to series whose terms are presumed to be nonnegative real numbers. This is not as restrictive a category as one might think. It includes situations where one tests a series for *absolute* convergence, even when the series has negative terms, and even when the series is a sum of vectors.

Proposition 4.19 *Suppose $x_j \ge 0$ for all $j = 1, 2, \ldots$. Then $\sum_{j=1}^{\infty} x_j$ converges if and only if the sequence of associated partial sums $\{s_n\}$ is bounded above.*

PROOF Since $x_j \ge 0$, it follows that the sequence of partial sums $\{s_n\}$ is increasing. If $\{s_n\}$ is also bounded above, then $\lim s_n$ exists, and $\sum_{j=1}^{\infty} x_j$ converges. Conversely, if $\sum_{j=1}^{\infty} x_j$ converges, then $\lim s_n = s$ exists, and so the sequence $\{s_n\}$ must be bounded. ♦

Example 4.20 *We illustrate the use of the above proposition by establishing the convergence of the series $\sum_{j=0}^{\infty}\left(\frac{1}{2}\right)^{j^2}$. To see this, simply note that for any integer $n \ge 0$,*

$$s_n = \sum_{j=0}^{n} \left(\tfrac{1}{2}\right)^{j^2} \le \sum_{j=0}^{n} \left(\tfrac{1}{2}\right)^{j} = 2\left(1 - \left(\tfrac{1}{2}\right)^{n+1}\right) < 2.$$

Therefore, according to Proposition 4.19, the series $\sum_{j=0}^{\infty} \left(\frac{1}{2}\right)^{j^2}$ converges. ◂

▸ **3.46** *Give an example of a series from* $\mathbb{R}^2$ *that converges absolutely.*

The following test for convergence is of great use in proving theorems. Of course, it can also be used as a practical tool to show that a series converges.

Proposition 4.21 (The Comparison Test)

Suppose $0 \le x_j \le y_j$ *for* $j = 1, 2, \ldots$. *Then the following are true:*

a) *If* $\sum_{j=1}^{\infty} y_j$ *converges, then* $\sum_{j=1}^{\infty} x_j$ *converges, and* $\sum_{j=1}^{\infty} x_j \le \sum_{j=1}^{\infty} y_j$.

b) *If* $\sum_{j=1}^{\infty} x_j$ *diverges, then* $\sum_{j=1}^{\infty} y_j$ *diverges.*

PROOF To prove *a*), let $s_n = \sum_{j=1}^{n} x_j$, and $t_n = \sum_{j=1}^{n} y_j$ be the nth partial sums of the two series, and assume that $\sum_{j=1}^{\infty} y_j$ converges. By Proposition 2.2, the sequence of partial sums $\{t_n\}$ is bounded above. Clearly, we have $s_n \le t_n$ for any n, and so the sequence $\{s_n\}$ is also bounded above. But since $x_j \ge 0$ for all j, it follows that the sequence $\{s_n\}$ is increasing, and so it must be convergent according to the monotone sequence theorem. It easily follows that $\sum_{j=1}^{\infty} x_j \le \sum_{j=1}^{\infty} y_j$. The proof of *b*) is a direct consequence of part *a*), since, according to *a*) if $\sum_{j=1}^{\infty} y_j$ converges, then $\sum_{j=1}^{\infty} x_j$ converges. This means that the divergence of $\sum_{j=1}^{\infty} x_j$ implies the divergence of $\sum_{j=1}^{\infty} y_j$. ♦

▸ **3.47** *Show that the condition "*$0 \le x_j \le y_j$ *for* $j = 1, 2, \ldots$*" in the comparison theorem can be weakened to "*$0 \le x_j \le y_j$ *for* $j \ge N$ *for some* $N \in \mathbb{N}$*" and the same conclusion still holds.*

Example 4.22 *Consider the series given by* $\sum_{j=1}^{\infty} \left(\frac{j}{2j+1}\right)^j$. *Clearly for all* $j \ge 1$ *we have* $0 \le \left(\frac{j}{2j+1}\right)^j \le \left(\frac{1}{2}\right)^j$, *where* $\left(\frac{1}{2}\right)^j$ *are the terms of a convergent geometric series. Therefore, by the comparison test,* $\sum_{j=1}^{\infty} \left(\frac{j}{2j+1}\right)^j$ *converges.* ◂

We now state a particularly potent test for convergence. Its proof relies on the comparison test.

Theorem 4.23 (The Cauchy Condensation Test)

Suppose $\{x_j\}$ *is a nonincreasing sequence, and* $x_j \ge 0$ *for all* j. *Then the series* $\sum_{j=1}^{\infty} x_j$ *and* $\sum_{j=0}^{\infty} 2^j\, x_{2^j}$ *both converge, or both diverge.*

PROOF Consider $s_n = x_1+x_2+\cdots+x_n$, and $t_n = x_1+2\,x_2+4\,x_4+\cdots+2^{n-1}\,x_{2^{n-1}}$ for $n \geq 1$. Then note that

$$\begin{aligned}
s_1 &= x_1 = t_0,\\
s_2 &= x_1 + x_2 \leq t_1,\\
s_3 &= x_1 + x_2 + x_3 \leq x_1 + 2x_2 = t_1,\\
s_4 &= x_1 + x_2 + x_3 + x_4 \leq x_1 + 2x_2 + x_4 \leq t_2,\\
s_5 &\leq t_2,\\
s_6 &\leq t_2,\\
s_7 &\leq t_2,\\
s_8 &= x_1 + x_2 + \cdots + x_8 \leq x_1 + 2x_2 + 4x_4 + x_8 \leq t_3.
\end{aligned}$$

From this it can be shown that

$$s_M \leq t_n \quad \text{for } 2^n \leq M < 2^{n+1}. \tag{3.7}$$

Now suppose $\sum_{j=0}^{\infty} 2^j\, x_{2^j}$ converges. Then the sequence of partial sums $\{t_n\}$ is bounded. Inequality (3.7) then implies that the sequence of partial sums $\{s_n\}$ is bounded, and hence that $\sum_{j=1}^{\infty} x_j$ converges. Now consider

$$\begin{aligned}
2s_1 &= 2x_1 \geq t_0,\\
2s_2 &= 2x_1 + 2x_2 = x_1 + (x_1 + 2x_2) \geq t_1,\\
2s_3 &= 2x_1 + 2x_2 + 2x_3 \geq x_1 + (x_1 + 2x_2) + 2x_3 \geq t_1,\\
2s_4 &= 2x_1 + 2x_2 + 2x_3 + 2x_4 \geq x_1 + (x_1 + 2x_2 + 4x_4) \geq t_2,\\
2s_5 &\geq t_2,\\
2s_6 &\geq t_2,\\
2s_7 &\geq t_2,\\
2s_8 &= 2x_1 + \cdots + 2x_8 \geq x_1 + (x_1 + 2x_2) + 4x_3 + 8x_8 \geq t_3.
\end{aligned}$$

From this it can be shown that

$$t_n \leq 2s_{2^n} \quad \text{for } n \geq 1. \tag{3.8}$$

If $\sum_{j=1}^{\infty} x_j$ converges, then the sequence of partial sums $\{s_n\}$ is bounded. Inequality (3.8) then implies that the sequence of partial sums $\{t_n\}$ is bounded, and hence that $\sum_{j=0}^{\infty} 2^j\, x_{2^j}$ converges. ♦

► **3.48** *Prove that $s_M \leq t_n$ for $2^n \leq M < 2^{n+1}$, and that $t_n \leq 2s_{2^n}$ for $n \geq 1$.*

Example 4.24 *We use the Cauchy condensation test to show that the harmonic series, $\sum_{j=1}^{\infty} \frac{1}{j}$, diverges. Note that $\sum_{j=1}^{\infty} 2^j \left(\frac{1}{2^j}\right) = \sum_{j=1}^{\infty} 1$ clearly diverges, and so the harmonic series diverges.* ◄

Example 4.25 *We now use the Cauchy condensation test to show that the series, $\sum_{j=1}^{\infty} \frac{1}{j^2}$, converges. Note that $\sum_{j=1}^{\infty} 2^j \left(\frac{1}{2^j}\right)^2 = \frac{1}{2} \sum_{j=1}^{\infty} \frac{1}{2^{j-1}}$ is a con-*

stant multiple of a convergent geometric series, and so the series $\sum_{j=1}^{\infty} \frac{1}{j^2}$ *converges.* ◀

Each of the two previous examples is a member of the more general category of series known as *p-series*. Such series have the form $\sum_{j=1}^{\infty} \frac{1}{j^p}$ for some constant real number p. We formally define the p-series below.

Definition 4.26 A series of the form $\sum_{j=1}^{\infty} \frac{1}{j^p}$ for real constant p is called a **p-series**.

► **3.49** *Show that a given p-series is convergent if and only if* $p > 1$.

While not as useful in proving theorems, the following test is usually more easily applied than the comparison test. Its proof, like that of the Cauchy condensation test, relies on the comparison test.

Theorem 4.27 (The Limit Comparison Test)

Suppose $\sum_{j=1}^{\infty} x_j$ *and* $\sum_{j=1}^{\infty} y_j$ *are series of positive terms such that*

$$\lim \left(\frac{x_j}{y_j}\right) = L > 0 \text{ exists.}$$

Then $\sum_{j=1}^{\infty} x_j$ *converges if and only if* $\sum_{j=1}^{\infty} y_j$ *converges.*

PROOF There exists an $N \in \mathbb{N}$ such that

$$n > N \Rightarrow \left|\frac{x_j}{y_j} - L\right| < \frac{L}{2}.$$

That is,

$$n > N \Rightarrow \frac{1}{2} L\, y_j < x_j < \frac{3}{2} L\, y_j.$$

The comparison test now proves the result. ♦

Example 4.28 *We will show that the series* $\sum_{j=1}^{\infty} \frac{1}{j^2-5j+1}$ *converges by use of the limit comparison test. To find the appropriate candidate for* y_j *to compare to our* $x_j = \frac{1}{j^2-5j+1}$*, we note that to determine the convergence behavior of a series requires only that we determine the convergence behavior of its tail. That is, how do the terms of the series behave when* j *is very large? For very large* j*, we see that* $x_j = \frac{1}{j^2-5j+1}$ *is approximately* $\frac{1}{j^2}$*, and so the natural choice for* y_j *is* $y_j = \frac{1}{j^2}$*. Applying the limit comparison test we find*

$$\lim \left(\frac{x_j}{y_j}\right) = \lim \left(\frac{j^2}{j^2 - 5j + 1}\right) = \lim \left(\frac{1}{1 - \frac{5}{j} + \frac{1}{j^2}}\right) = 1.$$

Since $\sum_{j=1}^{\infty} y_j = \sum_{j=1}^{\infty} \frac{1}{j^2}$ is a convergent p-series, it follows by the limit comparison test that $\sum_{j=1}^{\infty} \frac{1}{j^2-5\,j+1}$ converges. ◄

Note that the limit comparison test allows us to be a bit less "fine" in estimating the terms of the given series. In the above example, we determined that when j is very large, x_j "looks like" $\frac{1}{j^2}$, and we can stop there. In applying the comparison test, we had to find an appropriate scalar multiple of $\frac{1}{j^2}$ with which to compare x_j, a step that the limit comparison test does not require. As the previous examples and exercises involving the comparison test probably revealed to you, the actual value of the scalar multiple is not really important. What matters as far as convergence is concerned is that such a scalar exists. In determining whether a series converges, it is after all just the "tail" behavior we are interested in.

► **3.50** *Suppose $\sum_{j=1}^{\infty} \mathbf{x}_j$ and $\sum_{j=1}^{\infty} \mathbf{y}_j$ are each series of nonzero terms in $\mathbb{R}^k$ such that*

$$\lim \frac{|\,\mathbf{x}_j\,|}{|\,\mathbf{y}_j\,|} = L > 0.$$

What conclusion can you make? A similar conclusion to the limit comparison test should hold with absolute convergence. What if $\sum_{j=1}^{\infty} z_j$ and $\sum_{j=1}^{\infty} w_j$ are each series of nonzero terms in $\mathbb{C}$ such that

$$\lim \frac{|\,z_j\,|}{|\,w_j\,|} = L > 0.$$

What conclusion can you make?

4.4 Testing for Absolute Convergence in $\mathbb{X}$

The following test for convergence will be especially useful to us in our discussion of *power series* in Chapter 9.

Theorem 4.29 (The Ratio Test)

Let $\{x_j\}$ be a sequence of nonzero elements of $\mathbb{X}$, and suppose

$$\lim \frac{|\,x_{j+1}\,|}{|\,x_j\,|} = r \quad \textit{exists.}$$

Then,

a) $r < 1 \;\Rightarrow\; \sum_{j=1}^{\infty} |\,x_j\,|$ *converges, and hence,* $\sum_{j=1}^{\infty} x_j$ *converges.*

b) $r > 1 \;\Rightarrow\; \sum_{j=1}^{\infty} x_j$ *diverges.*

c) $r = 1 \;\Rightarrow\;$ *The test is inconclusive.*

PROOF The proof of *c)* is easiest, so we begin with that. Note that $\sum_{j=1}^{\infty} \frac{1}{j}$ diverges and has $r = 1$, while $\sum_{j=1}^{\infty} \frac{1}{j^2}$ converges and has $r = 1$. To prove part

a), note that there exists an $N \in \mathbb{N}$ such that

$$j > N \Rightarrow \left| \frac{|\, x_{j+1}\,|}{|\, x_j\,|} - r \right| < \frac{1-r}{2}.$$

From this we obtain

$$j > N \Rightarrow |\, x_{j+1}\,| < \left(\frac{1+r}{2}\right) |\, x_j\,|.$$

Letting $t \equiv \frac{1+r}{2} < 1$, this gives

$$j > N \Rightarrow |\, x_{j+1}\,| < t\,|\, x_j\,|,$$

which in turn yields

$$|\, x_j\,| < t^{j-N-1}\,|\, x_{N+1}\,| \text{ for } j \geq N+2.$$

That is,

$$|\, x_j\,| < \left(t^{-N-1}\,|\, x_{N+1}\,|\right) t^j, \text{ for } j \geq N+2,$$

and the result follows by use of the comparison test (and Exercise 3.47) where we note that the right-hand side of the above inequality represents the jth term of a convergent geometric series. To prove part b), note that there exists a positive integer N such that for $j > N$,

$$\left| \frac{|\, x_{j+1}\,|}{|\, x_j\,|} - r \right| < \frac{r-1}{2}.$$

This in turn implies that for $j > N$,

$$|\, x_{j+1}\,| > \left(\frac{1+r}{2}\right) |\, x_j\,|,$$

or equivalently, for $t \equiv \frac{1+r}{2} > 1$,

$$|\, x_j\,| > t^{j-N-1}\,|\, x_{N+1}\,| \text{ for } j \geq N+2.$$

From this it follows that $\lim x_j \neq 0$. Therefore, by the test for divergence, the series $\sum_{j=1}^{\infty} x_j$ diverges. ♦

Example 4.30 *Fix $z_0 \in \mathbb{C}$. We will show that the series $\sum_{n=0}^{\infty} \frac{z_0^n}{n!}$ converges absolutely by use of the ratio test. Simply consider the limit*

$$\lim \left| \frac{z_0^{n+1}/(n+1)!}{z_0^n/n!} \right| = \lim \left| \frac{z_0}{n+1} \right| = 0.$$

Note that since $z_0 \in \mathbb{C}$ was arbitrary, this result holds for any $z_0 \in \mathbb{C}$. ◄

The idea behind the truth of parts a) and b) of the ratio test is that, in each of these two cases, the tails of the given series can be seen to "look like" that of a convergent or a divergent geometric series, respectively. Since the convergence behavior of a series depends only on the behavior of its tail, we obtain a useful test for convergence in these two cases. Case *c*) is a case that the ratio test is not fine enough to resolve. Other methods need to be tried in this case to properly determine the series" behavior. Similar reasoning gives rise to another useful test that is actually more general than the ratio test, although not always as easy to apply. That test is the root test. Our version of the root test defines ρ as a lim sup rather than as a limit, which gives the test more general applicability since the lim sup always exists (if we include the possible value of ∞), whereas the limit might not. Of course, when the limit *does* exist, it equals the lim sup.

Theorem 4.31 (The Root Test)

Let $\{x_j\}$ be a sequence of nonzero elements of $\mathbb{X}$, and suppose

$$\limsup \sqrt[j]{|\,x_j\,|} = \rho \quad \textit{exists.}$$

Then,

a) $\rho < 1 \Rightarrow \sum_{j=1}^{\infty} |\,x_j\,|$ *converges, and hence,* $\sum_{j=1}^{\infty} x_j$ *converges.*

b) $\rho > 1 \Rightarrow \sum_{j=1}^{\infty} x_j$ *diverges.*

c) $\rho = 1 \Rightarrow$ *The test is inconclusive.*

PROOF To prove part *c*), consider $\sum_{j=1}^{\infty} \frac{1}{j}$, which diverges, and $\sum_{j=1}^{\infty} \frac{1}{j^2}$, which converges. Both have $\rho = \lim\limits_{j\to\infty} \sqrt[j]{|a_j|} = 1$. Now suppose $\rho < 1$. Then there exists $N \in \mathbb{N}$ such that

$$n > N \quad \Rightarrow \quad \sqrt[j]{|a_j|} < \frac{1+\rho}{2} < 1.$$

From this we have

$$n > N \quad \Rightarrow \quad |a_j| < \left(\frac{1+\rho}{2}\right)^j,$$

and so $\sum_{j=0}^{\infty} |a_j|$ is bounded above by a geometric series and therefore converges by the comparison test and Exercise 3.47. Now suppose $\rho > 1$. If $\sum_{j=0}^{\infty} a_j$ converges then $\lim_{j\to\infty} a_j = 0$. Since $\rho = \limsup \sqrt[j]{|a_j|} > 1$, there must be a subsequence of $\{\sqrt[j]{|a_j|}\}$ that converges to ρ. But this contradicts the fact that every subsequence of $\{a_j\}$ must converge to 0, and so $\sum_{j=0}^{\infty} a_j$ must diverge. ♦

Example 4.32 *Consider the series $\sum_{j=0}^{\infty} a_j$, where a_j is given by*

$$a_j = \begin{cases} \left(\frac{1}{2}\right)^j & \text{for odd } j \geq 1, \\ \left(\frac{1}{3}\right)^j & \text{for even } j \geq 2. \end{cases}$$

In this case, even though $\lim\limits_{j\to\infty} \sqrt[j]{|a_j|}$ does not exist, we do have that

$$\limsup \sqrt[j]{|a_j|} = \tfrac{1}{2} < 1,$$

and so by the root test, the series converges. ◀

Example 4.33 *In this example we will see that the root test can work where the ratio test fails. Consider the series given by*

$$\sum_{j=1}^{\infty} e^{-j+\sin(j\pi/2)}.$$

Then $a_j = e^{-j+\sin(j\pi/2)}$ *and so*

$$\left|\frac{a_{j+1}}{a_j}\right| = e^{-1+\sin((j+1)\pi/2)-\sin(j\pi/2)},$$

which does not have a limit as $j \to \infty$. *Yet,*

$$\sqrt[j]{|a_j|} = e^{-1+(\sin(j\pi/2))/j},$$

which converges to e^{-1} *as* $j \to \infty$. *Therefore, the series converges by the root test.* ◀

5 MANIPULATIONS OF SERIES IN $\mathbb{R}$

Several interesting results pertain specifically to the case of series of real numbers. We discuss these next.

5.1 Rearrangements of Series

We begin with a definition.

Definition 5.1 Let $\sum_{j=1}^{\infty} a_j$ be an infinite series of real numbers corresponding to the sequence $\{a_n\}$. If $\{a'_n\}$ is a permutation[1] of $\{a_n\}$, the infinite series $\sum_{j=1}^{\infty} a'_j$ is called a **rearrangement** of $\sum_{j=1}^{\infty} a_j$.

It is an interesting fact that if $\sum_{j=1}^{\infty} a_j$ is conditionally convergent, one can find a rearrangement that converges to any specified real number α. However, if $\sum_{j=1}^{\infty} a_j$ is absolutely convergent we will show that every rearrangement converges to the same value. In order to establish these results we will make use of the following convenient notation. In particular, we define

$$a_j^+ \equiv \frac{|a_j| + a_j}{2} = \begin{cases} a_j & \text{if } a_j \geq 0 \\ 0 & \text{if } a_j < 0 \end{cases}, \tag{3.9}$$

and

$$a_j^- \equiv \frac{|a_j| - a_j}{2} = \begin{cases} 0 & \text{if } a_j > 0 \\ -a_j & \text{if } a_j \leq 0 \end{cases}. \tag{3.10}$$

[1]A *permutation* of a_n means that the terms are placed in a different order.

Note that $a_j^+ \geq 0$ and $a_j^- \geq 0$. Also, it is easy to see that $a_j = a_j^+ - a_j^-$, and $|a_j| = a_j^+ + a_j^-$.

Proposition 5.2 *Consider the series $\sum_{j=1}^{\infty} a_j$ and the associated series $\sum_{j=1}^{\infty} a_j^+$ and $\sum_{j=1}^{\infty} a_j^-$ with a_j^+ and a_j^- as defined above.*

a) If $\sum_{j=1}^{\infty} a_j$ converges conditionally, then both of the series $\sum_{j=1}^{\infty} a_j^+$ and $\sum_{j=1}^{\infty} a_j^-$ diverge.

b) If $\sum_{j=1}^{\infty} a_j$ converges absolutely, then both of the series $\sum_{j=1}^{\infty} a_j^+$ and $\sum_{j=1}^{\infty} a_j^-$ converge absolutely.

PROOF We prove part *a*), and leave part *b*) to the reader. Suppose $\sum_{j=1}^{\infty} a_j$ converges conditionally and that $\sum_{j=1}^{\infty} a_j^+$ converges. Then we have that $\sum_{j=1}^{\infty} a_j^- = \sum_{j=1}^{\infty} \left(a_j^+ - a_j\right)$ converges, and so $\sum_{j=1}^{\infty} |a_j| = \sum_{j=1}^{\infty} \left(a_j^+ + a_j^-\right)$ converges, giving a contradiction. Presuming $\sum_{j=1}^{\infty} a_j^-$ converges yields a similar contradiction. ♦

► **3.51** *Prove part b) of the above proposition. To get started, consider that $a_j^+ - a_j^- = a_j$, and $a_j^+ + a_j^- = |a_j|$.*

The proof of the next result is left to the reader.

Proposition 5.3 *Suppose $\sum_{j=1}^{\infty} a_j$ converges conditionally. If S_n and T_n are the partial sums of $\sum_{j=1}^{\infty} a_j^+$ and $\sum_{j=1}^{\infty} a_j^-$, respectively, then* $\lim S_n = \lim T_n = \infty$.

► **3.52** *Prove the above proposition.*

The following result is often surprising at first. Of course, as with many theorems, the proof should make things clearer.

Theorem 5.4 *If $\sum_{j=1}^{\infty} a_j$ converges conditionally, then there exists a rearrangement of $\sum_{j=1}^{\infty} a_j$ that converges to any real number α.*

PROOF Let α be an arbitrary real number, and let $S_n = \sum_{j=1}^{n} a_j^+$ and $T_n = \sum_{j=1}^{n} a_j^-$. Here, a_j^+ and a_j^- are as defined in equations (3.9) and (3.10) on page 123. The strategy is to add together just enough of the a_j^+ terms to exceed α, then to subtract just enough of the a_j^- terms to make the sum no greater than α, then add more a_j^+ terms to just exceed α again, and so on, thereby "zeroing in" on α. To this end, we recall that $\lim S_n = \infty$, so there exists $n_1 \geq 1$ such that

$$S_{n_1} > \alpha \quad \text{while} \quad S_\ell \leq \alpha \quad \text{for } \ell < n_1.$$

Similarly, since $\lim T_n = \infty$, we can find $k_1 \geq 1$ such that

$$S_{n_1} - T_{k_1} < \alpha \quad \text{while} \quad S_{n_1} - T_\ell \geq \alpha \quad \text{for} \quad \ell < k_1.$$

Continuing in this way, we can find $n_2 \geq 1$ such that

$$S_{n_1+n_2} - T_{k_1} > \alpha \quad \text{while} \quad S_{n_1+n_2-1} - T_{k_1} \leq \alpha,$$

and $k_2 \geq 1$ such that

$$S_{n_1+n_2} - T_{k_1+k_2} < \alpha \quad \text{while} \quad S_{n_1+n_2} - T_{k_1+k_2-1} \geq \alpha.$$

After such an n_j and k_j have been found, we have that

$$S_{n_1+\cdots+n_j} - T_{k_1+\cdots+k_j} < \alpha \quad \text{while} \quad S_{n_1+\cdots+n_j} - T_{k_1+\cdots+k_j-1} \geq \alpha,$$

and therefore,

$$\alpha \leq S_{n_1+\cdots+n_j} - T_{k_1+\cdots+k_j-1} = S_{n_1+\cdots+n_j} - T_{k_1+\cdots+k_j} + a^-_{k_j} < \alpha + a^-_{k_j}.$$

That is,

$$\left| S_{n_1+\cdots+n_j} - T_{k_1+\cdots+k_j-1} - \alpha \right| < a^-_{k_j},$$

and since $\lim_{j\to\infty} a^-_{k_j} = 0$ (Why?), it follows that

$$\lim_{j\to\infty} \left(S_{n_1+\cdots+n_j} - T_{k_1+\cdots+k_j-1} \right) = \alpha.$$

♦

► **3.53** *Answer the (Why?) question in the above proof.*

The following result is often convenient.

Theorem 5.5 *If $\sum_{j=1}^\infty a_j$ converges absolutely, then every rearrangement of $\sum_{j=1}^\infty a_j$ converges to the same value.*

PROOF Let $\sum_{j=1}^\infty a'_j$ be a rearrangement of $\sum_{j=1}^\infty a_j$. We will first assume that $a_j \geq 0$ for all j. Then, defining $S_n \equiv a_1 + \cdots + a_n$, we have that $\lim S_n = S$ for some real number S. Define $T_n \equiv a'_1 + \cdots + a'_n$. Then $T_n \leq S$ (Why?), and since T_n is bounded above, it follows that $\lim T_n = T$ exists and $T \leq S$. Similarly, $S_n \leq T$ (Why?), and so $S \leq T$. Therefore, we must have $T = S$. The case where $a_j < 0$ for at least one j is left to the reader. ♦

► **3.54** *Answer the two (Why?) questions in the above proof, and prove the above theorem in the case where $a_j < 0$ for at least one j.*

5.2 Multiplication of Series

If $\sum_{j=1}^\infty a_j$ and $\sum_{j=1}^\infty b_j$ are convergent infinite series, it might seem natural to consider their product $\left(\sum_{j=1}^\infty a_j\right)\left(\sum_{j=1}^\infty b_j\right)$. How might one define such a product? Suppose we try to formally multiply the terms as follows,

$$\begin{aligned}\Big(\sum_{j=1}^{\infty} a_j\Big)\Big(\sum_{j=1}^{\infty} b_j\Big) &= (a_1 + a_2 + a_3 + \cdots)(b_1 + b_2 + b_3 + \cdots) \\ &= (a_1 b_1) + (a_1 b_2 + a_2 b_1) + (a_1 b_3 + a_2 b_2 + a_3 b_1) + \cdots \\ &\qquad \text{(Why?)} \\ &= \sum_{j=1}^{\infty} c_j, \quad \text{where } c_j = \sum_{k=1}^{j} a_k b_{j-k+1}.\end{aligned}$$

We wish to investigate when $\sum_{j=1}^{\infty} c_j$ converges, and if it converges to the product $\big(\sum_{j=1}^{\infty} a_j\big)\big(\sum_{j=1}^{\infty} b_j\big)$. In particular, if $\sum_{j=1}^{\infty} a_j$ and $\sum_{j=1}^{\infty} b_j$ each converge to A and B, respectively, it is not unreasonable to expect that $\sum_{j=1}^{\infty} c_j$ converges to the product $\big(\sum_{j=1}^{\infty} a_j\big)\big(\sum_{j=1}^{\infty} b_j\big) = AB$.

To set the notation for what follows,[2] we define

$$\begin{aligned} A_n &\equiv a_1 + \cdots + a_n, \quad A_n' \equiv a_{n+1} + a_{n+2} + \cdots, \\ B_n &\equiv b_1 + \cdots + b_n, \quad B_n' \equiv b_{n+1} + b_{n+2} + \cdots. \end{aligned}$$

We also define

$$A \equiv \lim A_n = \sum_{j=1}^{\infty} a_j \quad \text{and} \quad B \equiv \lim B_n = \sum_{j=1}^{\infty} b_j.$$

Note that $A = A_n + A_n'$ and $B = B_n + B_n'$. Also, it is easy to see that for $C_n \equiv c_1 + \cdots + c_n$, we have

$$\begin{aligned} C_n &= a_1 b_1 + (a_1 b_2 + a_2 b_1) + \cdots + (a_1 b_n + a_2 b_{n-1} + \cdots + a_n b_1) \\ &= a_1(b_1 + \cdots + b_n) + a_2(b_1 + \cdots + b_{n-1}) + \cdots + a_n b_1 \\ &= a_1\left(B - B_n'\right) + a_2\left(B - B_{n-1}'\right) + \cdots + a_n\left(B - B_1'\right) \\ &= A_n B - a_1 B_n' - a_2 B_{n-1}' - \cdots - a_n B_1'. \end{aligned}$$

The following theorem is the result we seek.

Theorem 5.6 *If $\sum_{j=1}^{\infty} a_j$ converges absolutely and $\sum_{j=1}^{\infty} b_j$ converges, then the series $\sum_{j=1}^{\infty} c_j$ with $c_j = \sum_{k=1}^{j} a_k b_{j-k+1}$ converges to $\left(\sum_{j=1}^{\infty} a_j\right)\left(\sum_{j=1}^{\infty} b_j\right)$.*

PROOF Since $C_n = A_n B - a_1 B_n' - a_2 B_{n-1}' - \cdots - a_n B_1'$, all we need show is that

$$\lim\left(a_1 B_n' + a_2 B_{n-1}' + \cdots + a_n B_1'\right) = 0, \tag{3.11}$$

since then $\sum_{j=1}^{\infty} c_j = \lim C_n = AB = \left(\sum_{j=1}^{\infty} a_j\right)\left(\sum_{j=1}^{\infty} b_j\right)$. To establish the

[2]We borrow the notation from [Rud76].

limit (3.11), note that $\lim B_n' = 0$ (Why?), and so for any $\epsilon > 0$ there exists an $N \in \mathbb{N}$ such that $n > N \Rightarrow |B_n'| < \epsilon$. That is, there exists an $N \in \mathbb{N}$ such that $m > 0 \Rightarrow |B_{N+m}'| < \epsilon$. We now consider the following estimate:

$$\begin{aligned}
|a_1 B_n' + a_2 B_{n-1}' + \cdots + a_n B_1'| &\le |a_1||B_n'| + |a_2||B_{n-1}'| + \cdots + |a_n||B_1'| \\
&= \Big(|a_n||B_1'| + \cdots + |a_{n-(N-1)}||B_N'|\Big) \\
&\quad + \Big(|a_{n-N}||B_{N+1}'| + \cdots + |a_1|\,|B_n'|\Big) \\
&< \Big(|a_n||B_1'| + \cdots + |a_{n-(N-1)}||B_N'|\Big) \\
&\quad + \epsilon \sum_{j=1}^{\infty} |a_j|. \qquad \text{(Why?)}
\end{aligned}$$

Since $\lim a_n = 0$, we have for all $\epsilon > 0$,

$$\begin{aligned}
\limsup |a_1 B_n' + a_2 B_{n-1}' + \cdots + a_n B_1'| &\le \limsup \Bigg[\Big(|a_n||B_1'| + \cdots \\
&\quad + |a_{n-(N-1)}||B_N'|\Big) + \epsilon \sum_{j=1}^{\infty} |a_j|\Bigg] \\
&= \epsilon \sum_{j=1}^{\infty} |a_j|. \qquad \text{(Why?)}
\end{aligned}$$

Since this is true for all $\epsilon > 0$, we have that

$$\limsup |a_1 B_n' + a_2 B_{n-1}' + \cdots + a_n B_1'| = 0.$$

Clearly,

$$\liminf |a_1 B_n' + a_2 B_{n-1}' + \cdots + a_n B_1'| = 0, \quad \text{(Why?)}$$

and therefore

$$\lim |a_1 B_n' + \cdots + a_n B_1'| = 0.$$

This establishes the limit (3.11), and the theorem is proved. ♦

► **3.55** *Answer the four (Why?) questions from the proof of the previous theorem.*

Example 5.7 *Recall from Example 4.30 that $\sum_{j=0}^{\infty} \frac{z_0^j}{j!}$ converges absolutely for any $z_0 \in \mathbb{C}$. Fix z_0 and w_0 in $\mathbb{C}$. We will show that*

$$\left(\sum_{j=0}^{\infty} \frac{z_0^j}{j!}\right)\left(\sum_{j=0}^{\infty} \frac{w_0^j}{j!}\right) = \sum_{j=0}^{\infty} \frac{(z_0 + w_0)^j}{j!}.$$

To see this, note that by Theorem 5.6 we may write

$$\left(\sum_{j=0}^{\infty} \frac{z_0^j}{j!}\right)\left(\sum_{j=0}^{\infty} \frac{w_0^j}{j!}\right) = \sum_{j=0}^{\infty} c_j,$$

where

$$c_j = \sum_{m=0}^{j} \frac{z_0^m}{m!} \frac{w_0^{j-m}}{(j-m)!}.$$

Simple algebra and use of the binomial theorem gives

$$c_j = \frac{1}{j!} \sum_{m=0}^{j} \binom{j}{m} z_0^m \, w_0^{j-m} = \frac{1}{j!} (z_0 + w_0)^j.$$

◀

5.3 Definition of e^x for $x \in \mathbb{R}$

In this subsection, we formally define the real number denoted by e^x for any fixed $x \in \mathbb{R}$. We do this in terms of a convergent series.

Definition 5.8 For any fixed $x \in \mathbb{R}$, the real number denoted by e^x is defined as

$$e^x \equiv \sum_{j=0}^{\infty} \frac{x^j}{j!}.$$

To establish that this series converges absolutely for every $x \in \mathbb{R}$, refer to Example 4.30 on page 121. Several familiar properties of e^x are a direct consequence of the above definition. In particular, note that the number e is now defined to be $e = \sum_{j=0}^{\infty} \frac{1}{j!} = 1 + 1 + \frac{1}{2!} + \frac{1}{3!} + \cdots$, a convergent sum. Also, $e^0 = 1$. While we will assume the reader is familiar with the other properties associated with e^x, several of them are available to the reader to establish in the supplementary exercises at the end of the chapter.

6 Supplementary Exercises

1. *Suppose the sequence $\{x_n\}$ converges to $x \in \mathbb{X}$ with the usual norm $|\cdot|$. We will show that convergence in $\mathbb{X}$ is "norm-independent." To see this, consider another norm $|\cdot|_1$ on $\mathbb{X}$ and recall as stated in an earlier exercise that all norms on $\mathbb{X}$ are equivalent. That is, there exists $\beta > 0$ such that*

$$\frac{1}{\beta} |x|_1 \le |x| \le \beta \, |x|_1$$

for every $x \in \mathbb{X}$. From this, show that for any $\epsilon > 0$, there exists an integer $N > 0$ such that $n > N \Rightarrow |x - x_n|_1 < \epsilon$.

2. *Show that* $\lim \frac{n^3 - 2n + 12}{2n^4 - 12} = 0$.

3. *Does the sequence* $\left\{ \left(\frac{n^2+n+1}{n^2-n+1} \cos \frac{n\pi}{3}, \frac{n^2+n+1}{n^2-n+1} \sin \frac{n\pi}{3} \right) \right\}$ *in $\mathbb{R}^2$ converge? (It does not. Prove this.)*

4. *If it exists, determine the limit of the sequence* $\{z_n\} \subset \mathbb{C}$ *given by* $z_n = \frac{n}{2n+1} + i(-1)^n$. *If it doesn't exist, show why not.*

5. *Suppose* $\{x_n\}$ *and* $\{y_n\}$ *are sequences in* $\mathbb{X}$ *such that* $\lim(x_n \pm y_n) = L$ *exists.*

a) *If* $\lim x_n = x$ *exists, what can you say about* $\lim y_n$*? Show that the limit exists, and evaluate it.*

b) *If* $\lim x_n$ *does not exist, what can you say about* $\lim y_n$*? Try finding an example where* $\lim x_n$ *and* $\lim y_n$ *do not exist but* $\lim(x_n + y_n)$ *does.*

6. *Come up with another example of a convergent sequence where the strict inequality* $<$ *must be replaced by* $\leq$ *when taking the limit.*

7. *Find* $\lim \frac{4n^2+1}{n^3+n}$ *by using the squeeze theorem.*

8. *Use the result of Example 2.7 to determine the limit in Example 2.13.*

9. *Show that* $\lim \frac{n^3+n-5}{3n^2+1} = \infty$ *and that* $\lim \frac{1-n^2}{1+3n} = -\infty$.

10. *Show that* $\lim x_n = -\infty$ *if and only if* $\lim(-x_n) = \infty$.

11. *Suppose* $\{x_n\}$ *and* $\{y_n\}$ *are sequences in* $\mathbb{X}$ *such that* $\lim(x_n \pm y_n) = L$ *exists. If* $\lim x_n = \infty$, *what can you say about* $\lim y_n$*? (You should find that the limit is either* $\pm\infty$, *depending on the various cases.)*

12. *Suppose the sequence* $\{\mathbf{x}_n\} \in \mathbb{R}^2$ *is given by* $\mathbf{x}_n = \left(\frac{(-1)^n}{n^2}, 4 + \sin\left(\frac{n\pi}{2}\right)\right)$. *Find expressions for the terms of the subsequences* $\{\mathbf{x}_{2m}\}$ *and* $\{\mathbf{x}_{2m+1}\}$.

13. *In the previous exercise, determine whether* $\lim \mathbf{x}_{2m}$ *or* $\lim \mathbf{x}_{4m}$ *exists. Based on your conclusions, what do you think is true of* $\lim \mathbf{x}_n$*? (You should conclude* $\{\mathbf{x}_n\}$ *does not converge.)*

14. *Suppose* $\{x_n\}$ *is a decreasing sequence of positive real numbers. If* $\lim x_{2^m} = 1$, *find* $\lim x_n$.

15. *Does the sequence in Example 3.5 on page 98 have any other limit points? (Yes.)*

16. *Consider the sequence* $\{z_n\} \in \mathbb{C}$ *with* $z_n = \frac{1}{n} + i\cos\left(\frac{n\pi}{4}\right)$. *Find the limit points of* $\{z_n\}$.

17. *If* $\{x_n\}$ *in* $\mathbb{X}$ *has exactly one limit point, must* $\{x_n\}$ *converge? (No. Find a counterexample.)*

18. *Is it true that if the sequence* $\{x_n\} \subset S \subset \mathbb{X}$ *converges to the limit* x, *then* x *must be a limit point of* S*? (Yes. Find a quick proof.)*

19. *Give an example of a sequence* $\{z_n\} \subset \mathbb{C}$ *that does not have a limit point.*

20. *Let* $\{x_n\} \subset S \subset \mathbb{X}$ *be a sequence where* S *is closed. Prove that every limit point of* $\{x_n\}$ *lies in* S.

21. *Find Λ and λ for the real sequence $\{x_n\}$ given by*

a) $x_n = \frac{(-1)^n\, n}{n+1}$ *b)* $x_n = \frac{1}{e^n} + \cos\left(\frac{n\,\pi}{4}\right)$ *c)* $x_n = \frac{1}{n} + \sin\left(\frac{n\,\pi}{2}\right)$

22. *Suppose $\{x_n\}$ and $\{y_n\}$ are bounded sequences of nonnegative real numbers. Show that* $\limsup(x_n y_n) \le (\limsup x_n)\,(\limsup y_n)$.

23. *Suppose $\{y_n\}$ is a bounded sequence of real numbers such that* $\limsup y_n = 0$. *Also suppose that for all $\epsilon > 0$, there exists $N \in \mathbb{N}$ such that $n > N \Rightarrow |x_n| \le y_n + \epsilon$.*

a) Does $\lim y_n$ *exist? (Yes.) b) Does* $\lim x_n$ *exist? (Yes.)*

24. *In the previous exercise, if the condition on* $\limsup y_n$ *is changed to* $\limsup y_n = \Lambda \neq 0$, *what happens to your conclusions in parts a) and b)? (Neither is true any more.)*

25. *Show that the sequence $\{\mathbf{x}_n\} \in \mathbb{R}^2$ given by $\mathbf{x}_n = \left(\left(\frac{1}{2}\right)^n, \frac{n}{n+1}\right)$ is Cauchy.*

26. *Find the sum of the series given by* $\sum_{j=1}^{\infty} \frac{1}{j\,(j+2)}$.

27. *Show that the series $\sum_{j=1}^{\infty} \frac{1}{j^2}$ converges. (Hint: First show by induction that $s_n \le \frac{2n}{n+1}$ for all $n \ge 1$.)*

28. *Use the comparison test to show that the real series given by $\sum_{j=1}^{\infty} \frac{1}{j^2 - 5\,j + 1}$ converges.*

29. *Suppose $x_j = (-1)^j$ and $y_j = (-1)^{j+1}$. What can you say about $\sum_{j=1}^{\infty}(x_j + y_j)$? What can you say about $\sum_{j=1}^{\infty} x_j + \sum_{j=1}^{\infty} y_j$? Prove that one series converges, but the other does not.*

30. *Suppose the sequences $\{x_j\}$ and $\{y_j\}$ in $\mathbb{X}$ are such that $\sum_{j=1}^{\infty}(x_j + y_j)$ converges. What can you say about $\sum_{j=1}^{\infty} x_j$ and $\sum_{j=1}^{\infty} y_j$? What if $\sum_{j=1}^{\infty}(x_j - y_j)$ also converges? In the final case, show that $\sum_{j=1}^{\infty} x_j$ and $\sum_{j=1}^{\infty} y_j$ converge as well; this is not necessarily true without that assumption.*

31. *Show that the series $\sum_{j=1}^{\infty} \left(\frac{\pi}{6} + \frac{i}{4}\right)^j$ converges in $\mathbb{C}$.*

32. *In Example 4.8 on page 111 we confirmed the convergence of a series in $\mathbb{R}^2$. Determine the sum.*

33. *Show that the series $\sum_{j=5}^{\infty} \frac{1}{2^j}$ converges. To what number does this series converge?*

34. *Determine the limit as n goes to ∞ of the product $\sqrt{\pi}\,\sqrt[4]{\pi}\,\sqrt[8]{\pi}\,\cdots\,\sqrt[2n]{\pi}$.*

35. *Consider the series given by $\sum_{j=0}^{\infty} z_j$, where $z_j = \left(\frac{1}{2^j} + \frac{i}{3^j}\right) \in \mathbb{C}$ for $j = 1, 2, \ldots$. Show that the series converges and find the sum.*

36. *Consider the series in* $\mathbb{C}$ *given by* $\sum_{j=0}^{\infty}\left(\frac{1+i}{2}\right)^j$. *Determine whether it converges, and if so, the sum.*

37. *Show that the real series given by* $\sum_{j=1}^{\infty}\frac{j}{e^j}$ *converges by using the comparison test.*

38. *Consider the series given by* $\sum_{j=2}^{\infty}\frac{1}{j\ln j}$. *How does this series compare to a p-series? Use the Cauchy condensation test to determine whether it converges. (It does not converge.)*

39. *Show that the real series given by* $\sum_{j=1}^{\infty}\frac{j}{e^j}$ *and by* $\sum_{j=1}^{\infty}\frac{1}{j(j+1)}$ *are convergent by using the limit comparison test.*

40. *Using the ratio test, show that* $r=1$ *for the series* $\sum_{j=1}^{\infty}\frac{1}{j}$ *and* $\sum_{j=1}^{\infty}\frac{1}{j^2}$.

41. *In the application of the ratio test, suppose the limit condition is replaced by the following:*

$$r=\limsup\frac{|x_{j+1}|}{|x_j|}\quad\text{exists.}$$

Show that if $r<1$ *the series converges, while if* $\frac{|x_{j+1}|}{|x_j|}\geq 1$ *for* $j\geq N$ *for some* $N\in\mathbb{N}$, *the series diverges.*

42. *Using the root test, show that* $\rho=1$ *for the series* $\sum_{j=1}^{\infty}\frac{1}{j}$ *and* $\sum_{j=1}^{\infty}\frac{1}{j^2}$.

43. *Abel's identity. Suppose* $\{a_n\}$ *is a sequence in* $\mathbb{R}$ *or* $\mathbb{C}$, *and* $\{b_n\}$ *is a sequence in* $\mathbb{X}$. *Let the sequence* $\{T_n\}$ *in* $\mathbb{X}$ *be such that* $T_n-T_{n-1}=b_n$. *Show that for* $n\geq m+2$,

$$\sum_{j=m+1}^{n}a_j\,b_j=a_n\,T_n-a_{m+1}\,T_m-\sum_{j=m+1}^{n-1}(a_{j+1}-a_j)\,T_j.$$

Does this result remind you of integration by parts?

44. *Abel's test. Suppose* $\{a_n\}$ *is a sequence in* $\mathbb{R}$ *or* $\mathbb{C}$, *and* $\{b_n\}$ *is a sequence in* $\mathbb{X}$ *such that*

(*i*) $\sum_{j=1}^{\infty}b_j$ *converges,*

(*ii*) $\sum_{j=2}^{\infty}|a_j-a_{j-1}|$ *converges.*

Then, $\sum_{j=1}^{\infty}a_j\,b_j$ *converges. To show this, define the following four quantities:*

$$K\equiv\max\left(\sum_{j=2}^{\infty}|a_j-a_{j-1}|,\,|a_1|\right),\quad B\equiv\sum_{j=1}^{\infty}b_j,\ \ B_n\equiv\sum_{j=1}^{n}b_j,\ \ T_n\equiv B_n-B.$$

For now, assume that $K>0$. *Clearly* $T_{n+1}-T_n=b_n$. *Since* $\lim T_n=0$, *we know that for any* $\epsilon>0$ *there exists* $N\in\mathbb{N}$ *such that*

$$n>N\ \Rightarrow\ |T_n|<\frac{\epsilon}{6K}.$$

Note that

$$a_n=a_1+\sum_{j=2}^{n}(a_j-a_{j-1}),$$

and so it follows that

$$|a_n| \le |a_1| + \sum_{j=2}^{n} |a_j - a_{j-1}| \le K + K = 2K \quad \text{for all } n.$$

To determine the convergence of $\sum_{j=1}^{\infty} a_j\, b_j$ *we will show that the partial sums are Cauchy. To this end, let* $m < n$*, and consider that*

$$\begin{aligned} \left| \sum_{m+1}^{n} a_j\, b_j \right| &= \left| a_n T_n - a_{m+1} T_m - \sum_{m+2}^{n-1} (a_{j+1} - a_j)\, T_j \right| \\ &\le |a_n|\,|T_n| + |a_{m+1}|\,|T_m| + \sum_{m+2}^{n-1} |a_{j+1} - a_j|\,|T_j| \\ &\le 2K\left(\frac{\epsilon}{6K}\right) + 2K\left(\frac{\epsilon}{6K}\right) + K\left(\frac{\epsilon}{6K}\right) \\ &< \epsilon \quad \text{whenever } m, n > N. \end{aligned}$$

a) *What if* $K = 0$ *in the above argument? Show that it is a trivial case.*

b) *What if* $\{a_n\}$ *is a sequence in* $\mathbb{R}^k$*? Does Abel's test still apply in some form? (It does, if* $\{b_n\}$ *is a sequence in* $\mathbb{R}$*.)*

45. *Prove the following corollary to Abel's test. Suppose* $\{a_n\}$ *is a sequence in* $\mathbb{R}$ *or* $\mathbb{C}$*, and* $\{b_n\}$ *is a sequence in* $\mathbb{X}$ *such that*

(*i*) $\sum_{j=1}^{\infty} b_j$ *converges,*

(*ii*) $\{a_n\}$ *is monotone and converges.*

Then, $\sum_{j=1}^{\infty} a_j\, b_j$ *converges.*

46. *Dirichlet's test. Suppose* $\{b_n\}$ *is a sequence in* $\mathbb{X}$ *such that the sequence of partial sums* $\{B_n\}$ *given by* $B_n = \sum_{j=1}^{n} b_j$ *is bounded. Show that if* $\{a_n\}$ *is a sequence in* $\mathbb{R}$ *or* $\mathbb{C}$ *such that* $\lim a_n = 0$ *and* $\sum_{j=2}^{\infty} |a_j - a_{j-1}|$ *converges, then* $\sum_{j=1}^{\infty} a_j\, b_j$ *converges.*

47. *Show that the alternating series test is a special case of Dirichlet's test.*

48. *For a sequence* $\{x_n\} \in \mathbb{R}$*, suppose* $\lim x_n = x$ *exists. Prove that* $\lim \sqrt[3]{x_n} = \sqrt[3]{x}$*. (Hint: Use the fact that* $a^3 - b^3 = (a-b)(a^2 + ab + b^2)$ *for any* $a, b \in \mathbb{R}$*, and consider the cases* $x > 0$*,* $x < 0$*, and* $x = 0$ *separately.) Can you generalize this result to show that* $\lim \sqrt[m]{x_n} = \sqrt[m]{x}$ *for any* $m \in \mathbb{N}$*? How about generalizing the result to show that* $\lim (x_n)^q = x^q$ *for any* $q \in \mathbb{Q}$*?*

49. *Suppose* $0 < x_1 < y_1$*, and that* $x_{n+1} = \sqrt{x_n y_n}$ *and* $y_{n+1} = \frac{1}{2}(x_n + y_n)$ *for* $n \in \mathbb{N}$*.*
a) *Show that* $0 < x_n < y_n$ *for all* n*.* b) *Show that* $x_n < x_{n+1}$ *for all* n*.*
c) *Show that* $y_{n+1} < y_n$ *for all* n*.* d) *Show that* $\lim x_n = \lim y_n = L$*.*

50. *Is it possible that* $\lim a_n = 0$ *and* a) $\sum_{j=1}^{\infty} \frac{a_j}{j}$ *diverges? (Yes. Find an example.)* b) $\sum_{j=1}^{\infty} \frac{a_j}{j^2}$ *diverges? (No. Prove this.)*

51. *Is it possible that* $\sum_{j=1}^{\infty} a_j$ *converges and* $\sum_{j=1}^{\infty} a_j^2$ *diverges? (No. Why?)*

52. *Determine whether the series* $\sum_{j=1}^{\infty} \frac{(-1)^j \sin j}{\sqrt{j}}$ *diverges, converges conditionally, or converges absolutely.*

53. *Suppose* $\{y_j\}$ *is a nonnegative sequence, and that* $\sum_{j=1}^{\infty} y_j$ *has bounded partial sums. Also, suppose the sequence of real numbers* $\{x_n\}$ *is such that* $x_n > 0$ *for all* n*, and* $\lim x_n = 0$*. Show that* $\sum_{j=1}^{\infty} x_j y_j$ *converges.*

54. *Let* a_n*, for* $n \geq 0$ *be a sequence of real numbers such that* $a = \lim a_n$ *exists. Define* $\sigma_n \equiv \frac{a_0+a_1+\cdots+a_n}{n+1}$*. Prove that* $\lim \sigma_n = a$*. To do so, fix* $\epsilon > 0$*, and split in two the sum in the definition of* σ_n *at* ϵn*. Then take* n *very large.*

55. *Suppose* $\{a_n\}, \{b_n\}$ *are real sequences for* $n \geq 1$*. Suppose further that* $\sum_{n=1}^{\infty} a_n{}^2$ *and* $\sum_{n=1}^{\infty} b_n{}^2$ *converge. Then it can be shown that* $\sum_{n=1}^{\infty} a_n b_n$ *converges absolutely and satisfies*

$$\left|\sum_{n=1}^{\infty} a_n b_n\right| \leq \sqrt{\sum_{n=1}^{\infty} a_n{}^2}\sqrt{\sum_{n=1}^{\infty} b_n{}^2}.$$

a) *Does this look familiar for finite sums? You may wish to look back at Chapter 1 here where we discussed inner product spaces. Assume the* a_n *and* b_n *are nonnegative to start with.*

b) *Prove that the inequality for finite sums can be extended to infinite sums.*

56. *Prove the following facts relating to* e^x*:*

a) $e^x e^y = e^{x+y}$ for all $x, y \in \mathbb{R}$.   b) $e^{-x} = \frac{1}{e^x}$ for all $x \in \mathbb{R}$.
c) $e^x > 0$ for all $x \in \mathbb{R}$.   d) $e^x > x$ for all $x \in \mathbb{R}$.
e) $x > 0 \Rightarrow e^x > 1$.   f) $x < 0 \Rightarrow e^x < 1$.
g) $e^x = 1 \Leftrightarrow x = 0$.   h) $x < y \Rightarrow e^x < e^y$ for all $x, y \in \mathbb{R}$.
i) $(e^x)^n = e^{nx}$ for all $n \in \mathbb{Z}$.   j) $(e^x)^r = e^{rx}$ for all $r \in \mathbb{Q}$.

To establish j), begin by showing that $(e^x)^{1/n} = e^{x/n}$ *for all nonzero* $n \in \mathbb{Z}$*.*

57. The Cantor Set
Recall the construction of the Cantor set, C*, from Exercise 75 on page 80. We will characterize points from this set according to their ternary expansion, a concept analagous to a decimal expansion. In fact, for any* $x \in [0,1]$ *we can write* x *in* decimal expansion *form as*

$$x = \frac{a_1}{10} + \frac{a_2}{10^2} + \frac{a_3}{10^3} + \cdots,$$

where each $a_j \in \{0,1,2,\ldots,9\}$ *for* $j = 1,2,\ldots$*. This expansion is unique for some* $x \in [0,1]$*. For others, there might be more than one decimal expansion representation.*

a) *Show that for* $x = \frac{1}{3}$*, we can write*

$$\frac{1}{3} = \frac{3}{10} + \frac{3}{10^2} + \frac{3}{10^3} + \cdots,$$

and that this representation is unique.

b) *Show that for* $x = \frac{1}{2}$*, we can write* x *in a decimal expansion in two ways. (In fact, any* x *with a decimal expansion that terminates in an infinite string of* $a_j s$ *equal to 0 can also be written to terminate with an infinite string of 9s as well, and vice* *versa.)*

c) *Show that for $x \in [0,1]$, we can write x in a* ternary *expansion so that*

$$x = \frac{c_1}{3} + \frac{c_2}{3^2} + \frac{c_3}{3^3} + \cdots,$$

where each $c_j \in \{0,1,2\}$ for $j = 1,2,\ldots$. Show that the number $x = \frac{1}{3}$ has two ternary expansions. (In fact, any x with a ternary expansion that terminates in an infinite string of $c_j s$ equal to 0 can also be written to terminate with an infinite string of 2s as well, and vice-versa.)

d) *Recall from Exercise 75 on page 80 that the construction of the Cantor set $C = \bigcap_{j=0}^{\infty} C_j$ involves generating a sequence $\{C_j\}$ of sets starting with the interval $C_0 = [0,1]$ in $\mathbb{R}$ and removing the open middle-third segment $G_1 = \left(\frac{1}{3}, \frac{2}{3}\right)$ to form C_1, removing the two open middle-third segments comprising $G_2 = \left(\frac{1}{3^2}, \frac{2}{3^2}\right) \cup \left(\frac{7}{3^2}, \frac{8}{3^2}\right)$ from C_1 to form C_2, and so on. The set G_n in this process is the union of "open middle-thirds" that is removed from C_{n-1} to form C_n. Show that if $x \in G_1$ then $c_1 = 1$, if $x \in G_2$ then $c_2 = 1$, and more generally, if $x \in G_n$ then $c_n = 1$. From this, conclude that $x \notin C$ if and only if $c_n = 1$ for some n. That is, $x \in C$ if and only if x has a ternary expansion such that $c_j \neq 1$ for all j. From this, can you determine the cardinality of the Cantor set (i.e., how many points are in the Cantor set)?*

e) *From the results of parts c) and d), you should conclude that $x = \frac{1}{3}$ is in C. In fact, $x = \frac{1}{3}$ is a boundary point of G_1, the first removed middle-third from the Cantor set construction algorithm. Define $K_1 \equiv \bigcup \partial G_j$. Can you show that any $x \in K_1$ is a Cantor set point? How many such points are there?*

f) *In fact, $C = K_1 \cup K_2$ where $K_2 \subset [0,1]$, $K_1 \cap K_2 = \varnothing$, and $K_2 \neq \varnothing$. To prove this last fact, show that $x = \frac{1}{4}$ is in C, but $\frac{1}{4} \notin K_1$. How many points must exist in the set K_2?*

g) *Are there any irrational numbers in C? (Yes.)*

h) *Can you show that K_1 is dense in C? That is, can you show that $\overline{K_1} = C$?*

i) *From the result in part h), show that for any Cantor set point x, there exists a sequence of Cantor set points $\{x_j\} \subset K_1$ such that $\lim x_j = x$.*

j) *Show that for any Cantor set point x, there exists an increasing sequence of Cantor set points $\{\overline{x_j}\} \subset K_1$ such that $\lim \overline{x_j} = x$. Similarly, show that there exists a decreasing sequence of Cantor set points $\{\underline{x_j}\} \subset K_1$ such that $\lim \underline{x_j} = x$.*

4

FUNCTIONS: DEFINITIONS AND LIMITS

Mathematics may be defined as the subject in which we never know what we are talking about, nor whether what we are saying is true.

Bertrand Russell

Much of the material of this chapter should be familiar to students who have completed a "transition course" to higher mathematics, and for this reason many of the proofs are omitted. We define the general concept of a function and we assume the reader is well acquainted with real-valued functions of a single real variable, and even somewhat familiar with real-valued and vector-valued functions of a vector. We introduce the important class of complex functions in a bit more detail, spending extra effort to investigate the properties of some fundamental examples from this special class. While some of these examples are seemingly natural extensions of their real-function counterparts, we will see that there are some significant differences between the real and complex versions. We also consider functions as mappings and review some basic terminology for discussing how sets of points in the function's domain are mapped to their corresponding image sets in the codomain. We summarize the rules for combining functions to make new functions, and we note the special circumstances under which a given function has an associated inverse function. Finally, we investigate the important process of taking the limit of a function in our various cases of interest, a process that will be used throughout our remaining development of analysis.

1 DEFINITIONS

Although readers are probably familiar with a variety of functions from previous experience, we take the time now to formally discuss what constitutes a function in the many contexts we will be exploring. Several examples will familiarize the reader with each case. We begin with a summary of our adopted notation and the definition for function in the most general case.

1.1 Notation and Definitions

We will again use the symbol $\mathbb{X}$ in all that follows to represent any of the spaces $\mathbb{R}$, $\mathbb{C}$, or $\mathbb{R}^k$ in those situations where the result may apply to any of those spaces. Related to this, we use the symbol $\mathbb{D}$ to denote the generic domain of a function, where $\mathbb{D}$ is to be understood as a nonempty subset of $\mathbb{X}$. In specific examples or results, we will use the more specific notation D^1 when the domain lies in $\mathbb{R}$, D when the domain lies in $\mathbb{C}$, and the notation D^k when the domain lies in $\mathbb{R}^k$. In a similar fashion to our use of the symbols $\mathbb{X}$ and $\mathbb{D}$, we use the notation $\mathbb{Y}$ to denote the codomain of a function, where $\mathbb{Y}$ should be interpreted as a subset (possibly a *proper* subset!) of any of the spaces $\mathbb{R}$, $\mathbb{C}$, or $\mathbb{R}^p$ for integer values $p \geq 2$. Note that we use the letter p to denote the dimension of $\mathbb{R}^p$ associated with the symbol $\mathbb{Y}$, since the use of the letter k is reserved for the space $\mathbb{R}^k$ associated with the symbol $\mathbb{X}$. Also, recall that the norm notation $|\cdot|$ will be used for all of the spaces of interest. Finally, as usual, when results do not apply to *all* the spaces $\mathbb{R}$, $\mathbb{C}$, $\mathbb{R}^k$, and $\mathbb{R}^p$, we will refer to the relevant spaces explicitly.

We now provide some definitions that will be used throughout our discussion of functions.

Definition 1.1 Suppose for some nonempty $\mathbb{D} \subset \mathbb{X}$ each element in $\mathbb{D}$ can be associated with a unique element in $\mathbb{Y}$. We make the following definitions.

1. The association, denoted by f, is called a **function** from $\mathbb{D}$ into $\mathbb{Y}$, and we write $f : \mathbb{D} \to \mathbb{Y}$.
2. The set $\mathbb{D}$ is called the **domain** of f, and the set $\mathbb{Y}$ is called the **codomain** of f.
3. For a particular associated $x \in \mathbb{D}$ and $y \in \mathbb{Y}$ we write $y = f(x)$. In this case y is referred to as the **image** of x under f.
4. The **range** of f consists of all those elements of $\mathbb{Y}$ that are images of some $x \in \mathbb{D}$ under f. We denote the range of f by

$$\mathcal{R}_f = \{y \in \mathbb{Y} : y = f(x) \text{ for some } x \in \mathbb{D}\}.$$

Note that a function f is actually a set of ordered pairs (x, y) with $x \in \mathbb{D}$ and $y = f(x) \in \mathbb{Y}$. In fact, the function f can be understood to be a subset of the Cartesian product $\mathbb{D} \times \mathbb{Y}$. Note too that the domain $\mathbb{D}$ does not have to be the entire space $\mathbb{X}$. We can think of $\mathbb{X}$ as the ambient space in which the domain $\mathbb{D}$ resides. Finally, note that in general, the range of a function may be a proper subset of its codomain. We can think of $\mathbb{Y}$ as the ambient space in which the range $\mathcal{R}_f$ resides. The range of $f : \mathbb{D} \to \mathbb{Y}$ is also sometimes denoted by $f(\mathbb{D})$.

In referring to a function, we will often allude to the specific spaces $\mathbb{D}$ and $\mathbb{Y}$ on which it is defined. For example, in the special case where $\mathbb{D}$ and $\mathbb{Y}$ are specified as subsets of $\mathbb{R}$ we will simply write $f : D^1 \to \mathbb{R}$, or even $f : I \to \mathbb{R}$

if the domain is an interval I of real numbers. This case corresponds to the familiar situation from calculus, that of a *real-valued function of a single real variable*. When $\mathbb{D}$ is specified as a subset of $\mathbb{R}^k$ and $\mathbb{Y}$ as $\mathbb{R}$ we will write $f : D^k \to \mathbb{R}$. This corresponds to the case where f is a *real-valued function of several real variables*, or a *real-valued function of a vector*. Such real-valued functions will often be denoted more explicitly in either of the following ways,

$$f(\mathbf{x}) \quad \text{or} \quad f(x_1, \dots, x_k).$$

Here, $\mathbf{x}$ denotes the vector (usually considered as a row vector) having components $x_1, \dots, x_k$. In the case where $\mathbb{D}$ is specified as a subset of $\mathbb{R}^k$ and $\mathbb{Y}$ is specified as $\mathbb{R}^p$, we will write $\boldsymbol{f} : D^k \to \mathbb{R}^p$, and we are dealing with a *vector-valued function of several real variables*, or a *vector-valued function of a vector*. Such functions will be more explicitly denoted by either

$$\boldsymbol{f}(\mathbf{x}) \quad \text{or} \quad \boldsymbol{f}(x_1, \dots, x_k).$$

Here, the bold $\boldsymbol{f}$ indicates its vector value. We will also consider functions from a subset of $\mathbb{C}$ to $\mathbb{C}$. For such functions we write $f : D \to \mathbb{C}$ and refer to them as *complex functions*. This class of functions is associated with some of the most interesting and beautiful results in analysis. Complex functions will often be denoted as $f(z)$ where $z = x + i\,y \in \mathbb{C}$. Finally, in those cases where the domain of a function has not been specified, but only its rule or association given, the reader should interpret the domain to be the largest set of allowable values within the appropriate space.

1.2 Complex Functions

The special case of complex-valued functions of a complex variable, hereafter referred to as *complex functions*, is of great importance in analysis. Not only is complex function theory regarded as one of the most beautiful subjects in all of mathematics, it is also one of the most useful, being routinely employed by mathematicians, physicists, and engineers to solve practical problems. To many readers the category of complex functions is entirely new, and so we will develop it a bit more carefully than the more general description of functions already presented.

Definition 1.2 A **complex function** $f : D \to \mathbb{C}$ with $D \subset \mathbb{C}$ is an association that pairs each complex number $z = x + i\,y \in D$ with a complex number $f(z) = u + i\,v \in \mathbb{C}$. Here, u and v are real-valued functions of x and y. That is,

$$f(z) = u(x, y) + i\,v(x, y).$$

We refer to $u(x, y)$ as the **real part** of $f(z)$, and to $v(x, y)$ as the **imaginary part** of $f(z)$.

As we have already seen, $\mathbb{C}$ is different from both $\mathbb{R}$ and $\mathbb{R}^2$ in significant ways. Yet we can rely on our knowledge of these more familiar Euclidean spaces to guide our exploration of $\mathbb{C}$. In fact, as shown in Chapter 1, the space $\mathbb{C}$ is geometrically similar to $\mathbb{R}^2$. Also, $\mathbb{C}$ is a field as $\mathbb{R}$ is, and therefore shares certain algebraic similarities with $\mathbb{R}$. Recognizing this, we can view complex functions in two different ways. From a geometric point of view, since the complex variable $z = x + i\,y$ can be thought of as a vector having components x and y, a function $f : D \to \mathbb{C}$ can be thought of as a special case of $\boldsymbol{f} : D^2 \to \mathbb{R}^2$, where $D \subset \mathbb{C}$ is like $D^2 \subset \mathbb{R}^2$. That is, complex functions are functions from the plane to itself. From an algebraic point of view, since $\mathbb{C}$ is a field, a complex function can be thought of as a function that takes an element of the field $\mathbb{C}$ to another element of the field $\mathbb{C}$. That is, just like real-valued functions of a single real variable, complex functions are functions from a field to itself. The notation we adopt for complex functions follows from this algebraic perspective. Since we are discussing complex-valued functions of a single complex variable, it is common practice to use the same style font as for a real-valued function of a single real variable, namely, $f(z)$ (rather than $\boldsymbol{f}(\mathbf{z})$). In the end, we hope to reconcile both points of view as they pertain to the world of complex functions.

Example 1.3 *Consider the complex function given by the rule $f(z) = z^2$. This can be written as*

$$f(z) = z^2 = (x + i\,y)^2 = \left(x^2 - y^2\right) + i\,2\,xy,$$

where $u(x,y) = x^2 - y^2$ and $v(x,y) = 2\,xy$. The largest domain one can associate with f can be determined by scrutinizing the functions of x and y given by $u(x,y)$ and $v(x,y)$. Note that they are both well defined for all $(x,y) \in \mathbb{R}^2$, and so f is well defined for all $z = x + i\,y \in \mathbb{C}$. Note too that if $z = 1 + i$ then $f(z) = 2i$, and so the codomain of f should be taken as $\mathbb{C}$. The range of f is that portion of the complex plane in which the ordered pairs (u,v) lie. Consider an arbitrary $w = u + iv \in \mathbb{C}$. Does there exist a pair (x,y) such that $u = x^2 - y^2$ and $v = 2\,xy$? These are two equations in the two unknowns x and y. If $v \neq 0$, then $x \neq 0$, and the condition $v = 2\,xy$ leads algebraically to $y = \frac{v}{2\,x}$. This in turn converts the condition $u = x^2 - y^2$ into $u = x^2 - \left(\frac{v}{2\,x}\right)^2$, or $u = x^2 - \frac{v^2}{4\,x^2}$. Solving this for x^2 yields $x^2 = \frac{1}{2}\left(u + \sqrt{u^2 + v^2}\right)$, and so $x = \pm\sqrt{\frac{1}{2}\left(u + \sqrt{u^2 + v^2}\right)}$. Subbing these values into $y = \frac{v}{2\,x}$ yields the corresponding values of y. Hence, any ordered pair (u,v) in the complex plane with $v \neq 0$ lies in the range of f. We leave it to the reader to verify that points $(u,0) \in \mathbb{C}$, i.e., points on the real axis, also lie in the range of f. ◀

▶ **4.1** *Show that the real axis lies in the range of f in the previous example, and therefore the range of f is all of $\mathbb{C}$.*

Example 1.4 *Consider the complex function given by the rule $f(z) = \frac{1}{z}$. This function can be written as*

$$w = f(z) = \frac{1}{z} = \frac{1}{x + iy} = \frac{1}{x + iy}\,\frac{(x - iy)}{(x - iy)} = \frac{x}{x^2 + y^2} - i\,\frac{y}{x^2 + y^2},$$

where $u(x,y) = \frac{x}{x^2+y^2}$ *and* $v(x,y) = -\frac{y}{x^2+y^2}$. *The largest domain one can associate with* f *can be determined by scrutinizing the functions of* x *and* y *given by* $u(x,y)$ *and* $v(x,y)$*, and noting that they are both well defined for all* $(x,y) \in \mathbb{R}^2$*, except* $(x,y) = (0,0)$*, and so* f *is well defined for all* $z \in D = \mathbb{C}\setminus\{0\}$. *As in the previous example, the codomain can be readily seen to be* $\mathbb{C}$. *To determine the range, first note that for the set of* $z = (x + i\,y) \in D$*, the functions* $u(x,y) = \frac{x}{x^2+y^2}$ *and* $v(x,y) = -\frac{y}{x^2+y^2}$ *can each take any real value, but* x *and* y *can never be simultaneously equal to* 0. *From this we conclude that the range must exclude* $w = 0$. ◀

▶ **4.2** *Show that the range of the function in the previous example is, in fact,* $\mathbb{C} \setminus \{0\}$ *by considering an arbitrary* $\xi + i\,\eta \in \mathbb{C}\setminus\{0\}$ *and showing that there exists an* $x + i\,y \in \mathbb{C}$ *that corresponds to it.*

▶ **4.3** *Accomplish the same end as the previous exercise by letting* $w = f(z)$ *and considering* $z = \frac{1}{w}$. *For which complex numbers* w *does this equation have complex solutions? Will this technique always work, rather than the lengthier method of the previous exercise?*

2 Functions as Mappings

Sometimes functions are referred to as *mappings*. The use of the term *mapping* instead of function is common when we are discussing sets of points rather than an individual point or its image. This point of view stresses the geometric nature of a function, viewing it as a transformation that takes a set in the domain to an image set in the codomain. Within this context, the range of a function $f : \mathbb{D} \to \mathbb{Y}$ is often denoted by $f(\mathbb{D})$, in which case it can be interpreted as *the image of* $\mathbb{D}$ *under the mapping* f. Eventually we will explore whether certain topological properties of the domain set are retained after mapping it to the codomain by a particular function. For now, we merely lay out the basic ideas associated with thinking of functions as mappings.

2.1 Images and Preimages

Definition 2.1 Consider $f : \mathbb{D} \to \mathbb{Y}$ and $A \subset \mathbb{D}$. Then the **image of** A under f is denoted by $f(A)$ and is defined as

$$f(A) = \{y \in \mathbb{Y} : y = f(x) \text{ for some } x \in A\}.$$

For any set $B \subset \mathbb{Y}$, the **preimage of** B under f is denoted by $f^{-1}(B)$ and is defined as

$$f^{-1}(B) = \{x \in \mathbb{D} : f(x) \in B\}.$$

The preimage of B under f is sometimes also referred to as *the inverse image of* B *under* f. Note that the notation f^{-1} here is not meant to denote the inverse function associated with f. Indeed, such an inverse function may not even exist. We will review the concept of inverse function a bit later in this chapter. Here, $f^{-1}(B)$ denotes a subset of the domain of f.

We now consider some higher-dimensional examples.

Example 2.2 *Suppose $\boldsymbol{f} : \mathbb{R}^2 \to \mathbb{R}^2$ is given by $\boldsymbol{f}(\mathbf{x}) = \mathbf{x} + \mathbf{a}$ where $\mathbf{a} \in \mathbb{R}^2$ is a constant vector. Then $\boldsymbol{f}$ is said to "translate" points of $\mathbb{R}^2$ a distance $|\mathbf{a}|$ in the direction of $\mathbf{a}$. Let's apply this function to the set $A = \{\mathbf{x} \in \mathbb{R}^2 : |\mathbf{x} - \mathbf{c}| = 4$, where $\mathbf{c} = (1,0)\} \subset \mathbb{R}^2$. Note that A is just the circle of radius 4 centered at the point $\mathbf{c} = (1,0)$ in $\mathbb{R}^2$. From this we have that*

$$\begin{aligned} \boldsymbol{f}(A) &= \{\mathbf{y} \in \mathbb{R}^2 : \mathbf{y} = \boldsymbol{f}(\mathbf{x}) \ \textit{for some} \ \mathbf{x} \in A\} \\ &= \{\mathbf{y} \in \mathbb{R}^2 : \mathbf{y} = \mathbf{x} + \mathbf{a} \ \textit{for} \ \mathbf{x} \ \textit{satisfying} \ |\mathbf{x} - \mathbf{c}| = 4\} \\ &= \{\mathbf{y} \in \mathbb{R}^2 : |(\mathbf{y} - \mathbf{a}) - \mathbf{c}| = 4\} \\ &= \{\mathbf{y} \in \mathbb{R}^2 : |\mathbf{y} - (\mathbf{a} + \mathbf{c})| = 4\}. \end{aligned}$$

This is just the circle of radius 4 centered at $\mathbf{a} + \mathbf{c}$ in $\mathbb{R}^2$. ◄

Example 2.3 *Consider $f : \mathbb{C} \setminus \{0\} \to \mathbb{C}$ given by $f(z) = \frac{1}{z}$. We will show that f maps circles in $\mathbb{C} \setminus \{0\}$ into circles. Note that, as seen previously,*

$$f(z) = \frac{1}{z} = \frac{x - i\,y}{x^2 + y^2} = \left(\frac{x}{x^2 + y^2}\right) - i\left(\frac{x}{x^2 + y^2}\right).$$

Letting $w = u + i\,v = f(z)$, we see that

$$u = \frac{x}{x^2 + y^2} \ \text{ and } \ v = -\frac{y}{x^2 + y^2},$$

from which we obtain that

$$x = \frac{u}{u^2 + v^2} \ \text{ and } \ y = -\frac{v}{u^2 + v^2}.$$

Now, for $z = x + i\,y$ and $z_0 = x_0 + i\,y_0$, consider the circle of radius $r > 0$ centered at z_0 and described by the equation $|z - z_0| = r$. Notice that

$$|z - z_0| = r \quad \Rightarrow \quad (x - x_0)^2 + (y - y_0)^2 = r^2.$$

Subbing our formulas for x and y in terms of u and v into this last equality obtains

$$\left(\frac{u}{u^2 + v^2} - x_0\right)^2 + \left(-\frac{v}{u^2 + v^2} - y_0\right)^2 = r^2.$$

This, in turn, is equivalent to

$$(x_0^2 + y_0^2 - r^2)\,(u^2 + v^2) + 2\,(y_0\,v - x_0\,u) + 1 = 0.$$

Now, since our original circle of radius r does not pass through the origin, we have that $x_0^2 + y_0^2 - r^2 \neq 0$, and the above equation can be seen to be the equation of a circle in the variables u and v. ◄

Example 2.4 *Let $f : \mathbb{C} \setminus \{-1\} \to \mathbb{C}$ be given by $f(z) = \frac{z-1}{z+1}$. We will find the image of the points on the circle of radius 1 centered at the origin given by the set $A = \{z \in \mathbb{C} \setminus \{-1\} : |z| = 1\}$. Note first that the set A excludes the point $z = -1$, and so is not the whole circle of radius 1 centered at the origin. Then note that*

$$f(z) = \frac{z-1}{z+1} = \frac{(x+iy)-1}{(x+iy)+1} = \frac{(x^2+y^2-1)+i2y}{(x+1)^2+y^2}. \tag{4.1}$$

Also, since $|z| = 1$ for each $z \in A$, we have $x^2 + y^2 = 1$, and so (4.1) becomes

$$f(z) = i\left(\frac{y}{x+1}\right) \quad \text{for } z \in A. \tag{4.2}$$

Hence, f has pure imaginary values on the set A. We leave it to the reader to show that $f(A)$ is the whole imaginary axis. ◀

▶ **4.4** *Consider the function f and the set A from the previous example. Determine that $f(A)$ is the whole imaginary axis by using equation (4.2) and considering an arbitrary pure imaginary number $i\eta$. Show that there exists an x and a y such that $\eta = \frac{y}{x+1}$ and $x + iy \in A$.*

We now summarize several useful properties regarding images and preimages of functions. These results are usually proved in a transition course to higher mathematics, and so we omit the proof for brevity.

Proposition 2.5 *For $f : \mathbb{D} \to \mathbb{Y}$, suppose $\{A_\alpha\} \subset \mathbb{D}$, and $\{B_\beta\} \subset \mathbb{Y}$. Then the following all hold.*

a) $f\left(\cup A_\alpha\right) = \bigcup f(A_\alpha)$

b) $f^{-1}\left(\cup B_\beta\right) = \bigcup f^{-1}(B_\beta)$

c) $f\left(\cap A_\alpha\right) \subset \bigcap f(A_\alpha)$

d) $f^{-1}\left(\cap B_\beta\right) = \bigcap f^{-1}(B_\beta)$

Suppose $A \subset \mathbb{D}$ and $B \subset \mathbb{Y}$. Then the following hold.

e) $f^{-1}\left(f(A)\right) \supseteq A$

f) $f\left(f^{-1}(B)\right) \subset B$

2.2 Bounded Functions

The following definition specifies a special case where the range is a bounded subset of the codomain.

Definition 2.6 A function $f : \mathbb{D} \to \mathbb{Y}$ is **bounded** on $\mathbb{D}$ if there exists an $M \in \mathbb{R}$ such that

$$|f(x)| \le M \text{ for all } x \in \mathbb{D}.$$

That is, a function $f : \mathbb{D} \to \mathbb{Y}$ is bounded on $\mathbb{D}$ if the range of f is a bounded subset of $\mathbb{Y}$.

Example 2.7 *Let $f : \mathbb{R}^k \to \mathbb{R}$ be given by $f(\mathbf{x}) = e^{-|\mathbf{x}|^2}$. Then f is bounded on its domain. To see this, note that since $|\mathbf{x}| \ge 0$, it follows that $-|\mathbf{x}|^2 \le 0$, and therefore*

$$e^{-|\mathbf{x}|^2} \le e^0 = 1 \quad \textit{for all } \mathbf{x} \in \mathbb{R}^k.$$ ◀

Example 2.8 *Let $f : D \to \mathbb{C}$ be given by $f(z) = \frac{1}{(z+1)^2}$ where $D = \{z \in \mathbb{C} : \mathrm{Re}(z) \ge 0\}$. Then, since $\mathrm{Re}(z) \ge 0$, it follows that $|z + 1| \ge 1$ (Why?), and therefore*

$$|f(z)| = \frac{1}{|z+1|^2} \le 1 \quad \textit{on } D.$$

Therefore f is bounded on D. ◀

▶ **4.5** *Answer the (Why?) question in the previous example.*

2.3 Combining Functions

The following rules for combining functions should be familiar to the reader. We list them here for completeness. For simplicity in expressing and properly interpreting the statements that follow, the codomain $\mathbb{Y}$ should be understood to be the whole space $\mathbb{R}$, $\mathbb{C}$, or $\mathbb{R}^p$, and not a proper subset of any of them.

Sums, Differences, and Scalar Multiples of Functions

Suppose f and g are functions from the same domain $\mathbb{D}$ to codomain $\mathbb{Y}$. Then,

1. $f \pm g : \mathbb{D} \to \mathbb{Y}$ are functions, and $(f \pm g)(x) = f(x) \pm g(x)$ for all $x \in \mathbb{D}$.
2. $c\,f : \mathbb{D} \to \mathbb{Y}$ for $c \in \mathbb{R}$ is a function, and $(c\,f)(x) = c\,f(x)$ for all $x \in \mathbb{D}$.

Note that the constant c in statement 2 above can also be a complex number in the case of a complex function f. While the same is true even when f is real valued, we won't have need to work with such products in our development of analysis.[1]

[1]One could also define a complex multiple of a vector-valued function, but the resulting function is complex vector valued, a case we don't consider in our development of analysis.

Products and Quotients of Functions

Suppose $f : \mathbb{D} \to \mathbb{Y}_f$ and $g : \mathbb{D} \to \mathbb{Y}_g$ are functions from the same domain $\mathbb{D}$. Then,

1. $(f \cdot g) : \mathbb{D} \to \mathbb{Y}_{f \cdot g}$ where $(f \cdot g)(x) = f(x) \cdot g(x)$ for all $x \in \mathbb{D}$ is a function in the following cases:

   a) $\mathbb{Y}_f = \mathbb{R}$, $\mathbb{Y}_g = \mathbb{Y}_{f \cdot g} = \mathbb{Y}$,

   b) $\mathbb{Y}_f = \mathbb{Y}_g = \mathbb{R}^p$, $\mathbb{Y}_{f \cdot g} = \mathbb{R}$,

   c) $\mathbb{Y}_f = \mathbb{Y}_g = \mathbb{Y}_{f \cdot g} = \mathbb{C}$,

2. $(f/g) : \mathbb{D} \to \mathbb{Y}_{f/g}$ where $(f/g)(x) = f(x)/g(x)$ for all $x \in \mathbb{D}$, provided $g(x) \neq 0$ is a function in the following cases:

   a) $\mathbb{Y}_f = \mathbb{Y}$, $\mathbb{Y}_g = \mathbb{R}$, and $\mathbb{Y}_{f/g} = \mathbb{Y}$

   b) $\mathbb{Y}_f = \mathbb{Y}_g = \mathbb{Y}_{f/g} = \mathbb{C}$.

Note that in 1 the product between f and g should be interpreted as the dot product when $\mathbb{Y}_f$ and $\mathbb{Y}_g$ are both $\mathbb{R}^p$, and as the usual product when $\mathbb{Y}_f$ and $\mathbb{Y}_g$ are both $\mathbb{R}$ or $\mathbb{C}$.

Composition of Functions

We now formally define the composition of functions. In what follows, each of $\mathbb{Y}_f$ and $\mathbb{Y}_g$ should be understood to be any of the spaces $\mathbb{R}$, $\mathbb{R}^p$, or $\mathbb{C}$.

Definition 2.9 Consider $f : \mathbb{D} \to \mathbb{Y}_f$ with $\mathcal{R}_f = f(\mathbb{D})$, and $g : \mathcal{R}_f \to \mathbb{Y}_g$. Define the function $g \circ f : \mathbb{D} \to \mathbb{Y}_g$ by

$$(g \circ f)(x) = g\left(f(x)\right) \quad \text{for all } x \in \mathbb{D}.$$

The function $g \circ f$ is called the **composition** of g with f.

Example 2.10 *Consider $f : \mathbb{R} \to \mathbb{R}^2$ and $g : \mathcal{R}_f \to \mathbb{C}$, where $f(x) = (x^2, 3 - x)$, and $g(r, s) = r^2\, e^{i\, s}$. Then, $g \circ f : \mathbb{R} \to \mathbb{C}$, and*

$$(g \circ f)(x) = g\left(f(x)\right) = g(x^2, 3 - x) = x^4\, e^{i\,(3-x)}.$$

Note that the domain of g is the range of f given by $\mathcal{R}_f = \{(x, y) \in \mathbb{R}^2 : x \geq 0\}$. Also note that $f \circ g$ does not exist in this case since $f\left(g(r, s)\right) = f\left(r^2\, e^{i\, s}\right)$ is not defined. ◀

Example 2.11 *Consider $f : \mathbb{C} \setminus \{-1\} \to \mathbb{C}$ given by $f(z) = \frac{z+1}{z-1}$. We will write f as a composition of simpler functions. To this end, note that*

$$f(z) = \frac{z+1}{z-1} = \frac{(z-1)+2}{z-1} = 1 + \frac{2}{z-1}. \tag{4.3}$$

We now define the following:

$$f_1 : \mathbb{C} \setminus \{1\} \to \mathbb{C} \setminus \{0\}, \text{ given by } f_1(z) = z - 1,$$
$$f_2 : \mathbb{C} \setminus \{0\} \to \mathbb{C}, \text{ given by } f_2(z) = \tfrac{2}{z},$$
$$f_3 : \mathbb{C} \to \mathbb{C}, \text{ given by } f_3(z) = 1 + z.$$

Note from this that

$$f(z) = 1 + \frac{2}{z-1} = 1 + \frac{2}{f_1(z)} = 1 + f_2\left(f_1(z)\right) = f_3\big(f_2\left(f_1(z)\right)\big),$$

and therefore, $f = f_3 \circ f_2 \circ f_1$. ◀

2.4 One-to-One Functions and Onto Functions

There are properties a function might possess that indicate in significant ways what kind of function it is, what operations it will allow, or to what, if any, other functions it might be related. In this subsection, we review two such important properties. We begin with a definition.

Definition 2.12 Suppose $f : \mathbb{D} \to \mathbb{Y}$ is a function such that for all pairs $x, \widetilde{x} \in \mathbb{D}$, the following property holds,

$$f(x) = f(\widetilde{x}) \Rightarrow x = \widetilde{x}.$$

Then we say that f is a **one-to-one** function, or that f is 1-1.

A statement that is equivalent to the above definition is that a function $f : \mathbb{D} \to \mathbb{Y}$ is one-to-one if and only if for $x, \widetilde{x} \in \mathbb{D}$, $x \neq \widetilde{x} \Rightarrow f(x) \neq f(\widetilde{x})$. One-to-one functions are those functions that map no more than one element of their domains into a given element of their codomains. Another term for one-to-one that is often used in analysis is *injective*. Such a function is referred to as an *injection*. Our first example establishes that linear functions are one-to-one.

Example 2.13 *Consider* $f : \mathbb{C} \to \mathbb{C}$ *given by* $f(z) = az + b$, *where* $a, b \in \mathbb{C}$, *and* $a \neq 0$. *We will show that* f *is one-to-one. To see this, let* $z, \widetilde{z} \in \mathbb{C}$ *be such that* $f(z) = f(\widetilde{z})$. *Simple algebra obtains* $z = \widetilde{z}$, *and therefore* f *is one-to-one.* ◀

Example 2.14 *Consider* $f : \mathbb{R}^k \to \mathbb{R}$ *given by* $f(\mathbf{x}) = \frac{1}{|\mathbf{x}|+1}$. *Let* $\mathbf{e_j}$ *for* $j = 1, \ldots, k$ *denote the unit vectors in* $\mathbb{R}^k$ *whose components are all zeros except for the* j*th component, which has the value* 1. *Clearly* f *is not one-to-one, since* $f(\mathbf{e_j}) = f(\mathbf{e_m})$ *for* $j \neq m$ *such that* $1 \leq j, m \leq k$. ◀

The first result we will establish about one-to-one functions involves a particular type of real-valued function of a single real variable, namely, *strictly monotone functions*. We define *monotone functions* and *strictly monotone functions* next.

Definition 2.15 A real-valued function $f : D^1 \to \mathbb{R}$ is called **monotone** if any of the following hold true. For all $x_1, x_2 \in D^1$,

1. $x_1 < x_2 \Rightarrow f(x_1) \leq f(x_2)$. In this case, the monotone function is called **nondecreasing.**
2. $x_1 < x_2 \Rightarrow f(x_1) < f(x_2)$. In this case, the monotone function is called **increasing.**
3. $x_1 < x_2 \Rightarrow f(x_1) \geq f(x_2)$. In this case, the monotone function is called **nonincreasing.**
4. $x_1 < x_2 \Rightarrow f(x_1) > f(x_2)$. In this case, the monotone function is called **decreasing.**

A function f is called **strictly monotone** if it is either increasing or decreasing.

Since the definition relies on the order property exclusively possessed by the real numbers, it does not make sense to refer to a monotone function that is not real-valued or whose domain is not a subset of $\mathbb{R}$. Monotone functions are convenient functions with which to work, as we will soon see. For example, strictly monotone functions are necessarily one-to-one.[2]

Proposition 2.16 *Suppose $f : D^1 \to \mathbb{R}$ is a strictly monotone function. Then f is one-to-one on D^1.*

PROOF Suppose $f(x_1) = f(x_2)$ and that f is decreasing (the increasing case is handled similarly). If $x_1 < x_2$ then $f(x_1) > f(x_2)$, a contradiction. If $x_1 > x_2$ then $f(x_1) < f(x_2)$, also a contradiction. Therefore we must have $x_1 = x_2$, and so f is one-to-one. ♦

The above proposition allows us a convenient means, in some cases anyway, for establishing that a function is one-to-one. If we can show that the function is strictly monotone, then it must be one-to-one as well.

► **4.6** *Consider $D^1 \subset \mathbb{R}$ and consider $f_j : D^1 \to \mathbb{R}$ for $j = 1, 2, \ldots, p$. Is it true that if f_j is a one-to-one function for $j = 1, 2, \ldots, p$, then the function $\boldsymbol{f} : D^1 \to \mathbb{R}^p$ given by $\boldsymbol{f} = (f_1, f_2, \ldots, f_p)$ is one-to-one? (Yes.) How about the converse? (No.)*

► **4.7** *Consider $D^k \subset \mathbb{R}^k$ and consider $f_j : D^k \to \mathbb{R}$ for $j = 1, 2, \ldots, p$. Is it true that if f_j is a one-to-one function for $j = 1, 2, \ldots, p$, then the function $\boldsymbol{f} : D^k \to \mathbb{R}^p$ given by $\boldsymbol{f}(\mathbf{x}) = (f_1(\mathbf{x}), f_2(\mathbf{x}), \ldots, f_p(\mathbf{x}))$ is one-to-one? (Yes.) How about the converse? (No.)*

[2]Note that some texts use the terms *increasing* and *strictly increasing* in place of *nondecreasing* and *increasing*, respectively. Likewise, such texts also use the terms *decreasing* and *strictly decreasing* in place of *nonincreasing* and *decreasing*.

► **4.8** *Show that the real exponential function,* $\exp : \mathbb{R} \to \mathbb{R}$ *given by* $\exp(x) = e^x$ *is increasing, and therefore one-to-one.*

We now define another property of interest in this subsection.

Definition 2.17 The function $f : \mathbb{D} \to \mathbb{Y}$ is called **onto** if the range of f is $\mathbb{Y}$, i.e., if $\mathcal{R}_f = f(\mathbb{D}) = \mathbb{Y}$.

Onto functions are functions that map their domain onto every element of the codomain. That is, each element of $\mathbb{Y}$ is the image of at least one $x \in \mathbb{D}$. Functions that are onto are sometimes also referred to as *surjective*. A surjective function is then also referred to as a *surjection*.

It is important to note that the properties one-to-one and onto are not mutually exclusive. Some functions are one-to-one and onto, some are just one-to-one, some are just onto, and some are neither. The following example should make this clear.

Example 2.18 *We consider three functions in this example, each given by the same rule but with a different domain or codomain. First, consider the function* $f : \mathbb{R} \to \mathbb{R}$ *given by* $f(x) = x^2$*. This function is not one-to-one since* $f(-1) = f(1) = 1$*, and so* $x = -1$ *and* $x = 1$ *both get mapped to 1. The function is not onto, since every element of the domain* $\mathbb{R}$ *gets mapped to a nonnegative number. More explicitly, any* $y \in \mathbb{R}$ *satisfying* $y < 0$ *has no preimage in* $\mathbb{R}$*. For example, there is no* $x \in \mathbb{R}$ *such that* $f(x) = -3$*. Next, consider* $f : \mathbb{R} \to [0, \infty)$ *given by* $f(x) = x^2$*. Note that this function has the same rule as the one we just considered above, but the codomain has been restricted from* $\mathbb{R}$ *to* $[0, \infty) \subset \mathbb{R}$*. This new function is still not one-to-one, but it* **is** *onto. Each element of the newly defined codomain has at least one preimage in* $\mathbb{R}$*. Finally, consider the rule given by* $f(x) = x^2$ *once more, but this time restrict both the domain and codomain to the interval* $[0, \infty) \subset \mathbb{R}$*. This new function* $f : [0, \infty) \to [0, \infty)$ *is both one-to-one and onto. It is one-to-one since, if* $f(x) = f(\widetilde{x})$*, we have that* $x^2 = (\widetilde{x})^2$*, which is true if and only if* $|x| = |\widetilde{x}|$*, which in turn means* $x = \widetilde{x}$*. (Why?) The function is onto since if* y *is any element in the codomain,* $\sqrt{y}$ *exists and is in the domain. That is, for any* $y \in [0, \infty)$ *we have* $y = f(\sqrt{y})$ *with* $\sqrt{y} \in [0, \infty)$*.* ◄

The moral of the above example is clear. By changing the domain and/or the codomain associated with a given function, we can make a new function with the same rule as the original that is one-to-one and/or onto.

► **4.9** *Show that* $f : [0, \infty) \to \mathbb{R}$ *given by* $f(x) = x^2$ *is one-to-one, but not onto.*

► **4.10** *Suppose* $f : \mathbb{D} \to \mathbb{Y}$ *is not onto. What set is the natural choice to replace* $\mathbb{Y}$ *that obtains a new function with the same rule as the original, the same domain as the original, but that is now onto?*

Of course, as one might expect, functions that are both one-to-one and onto are special.

Definition 2.19 A function that is both one-to-one and onto is called a **one-to-one correspondence.**

One-to-one correspondences are important in many areas of mathematics. Such a function may also be referred to as a *bijection*. The function, possessing both the one-to-one and the onto properties, is then said to be *bijective*. The following example illustrates that linear complex functions are one-to-one correspondences.

Example 2.20 *Let $f : \mathbb{C} \to \mathbb{C}$ be given by $f(z) = az + b$ where $a, b \in \mathbb{C}$ and $a \neq 0$. We will show that f is onto. To this end, consider an arbitrary $w \in \mathbb{C}$. If we can find a $z \in \mathbb{C}$ such that $f(z) = w$, i.e., such that $az + b = w$, we will be done. But clearly $z = \frac{w-b}{a}$ has the desired property, and so f is onto.* ◀

As learned in typical transition courses to higher mathematics, it is exactly one-to-one correspondences that have inverses. We review this idea next.

2.5 Inverse Functions

We begin by formally defining the inverse function for a function.

Definition 2.21 Consider the function $f : \mathbb{D} \to \mathcal{R}_f$, where $\mathcal{R}_f = f(\mathbb{D})$. If there exists a function $g : \mathcal{R}_f \to \mathbb{D}$ such that

$$(i) \quad f(g(y)) = y \text{ for all } y \in \mathcal{R}_f.$$

$$(ii) \quad g(f(x)) = x \text{ for all } x \in \mathbb{D},$$

then g is called **the inverse function** of f.

A few remarks relating to this definition are in order.

1. Note that if g is the inverse of f, then f is the inverse of g. The pair of functions, f and g, are called *inverse functions*, and the inverse of f is often denoted by f^{-1} instead of g.
2. When f^{-1} exists, $f\left(f^{-1}(y)\right) = y$ for all $y \in \mathcal{R}_f$. That is, $f \circ f^{-1} = I_{\mathcal{R}_f}$, the *identity function on* $\mathcal{R}_f$. Similarly, $f^{-1}(f(x)) = x$ for all $x \in \mathbb{D}$, i.e., $f^{-1} \circ f = I_{\mathbb{D}}$, the *identity function on* $\mathbb{D}$.
3. The two identity functions described in the last remark are in general not the same, since $\mathbb{D}$ and $\mathcal{R}_f$ are in general not the same.

Example 2.22 *Let $f : \mathbb{C} \setminus \{1\} \to \mathbb{C} \setminus \{1\}$ be given by $f(z) = \frac{z+1}{z-1}$. The reader is asked to verify that this function is, in fact, a one-to-one correspondence. We* *will determine its inverse function. To do so, we seek a function* $g : \mathbb{C} \setminus \{1\} \to \mathbb{C} \setminus \{1\}$ *such that*

$$f(g(w)) = \frac{g(w)+1}{g(w)-1} = w \quad \text{for all } w \neq 1.$$

Solving the above equation for $g(w)$ *in terms of* w *yields that* $g(w) = \frac{w+1}{w-1}$, *and so*

$$f^{-1}(w) = \frac{w+1}{w-1} = f(w).$$

Note that this is an exceptional case where $f^{-1} = f$. ◀

Again, the following result is included for completeness only, and we omit the proof.

Proposition 2.23 *Consider* $f : \mathbb{D} \to \mathbb{Y}$. *Then* f *has an inverse function defined on all of* $\mathbb{Y}$ *if and only if* f *is a one-to-one correspondence.*

3 Some Elementary Complex Functions

Recall that a complex function $f : D \to \mathbb{C}$ has a real and an imaginary part, namely, $f(z) = u(x, y) + i\, v(x, y)$. We take the time now to carefully consider some elementary complex functions that will play a larger role in our study of analysis. These functions can be viewed as complex extensions of some of the more familiar real-valued functions from calculus. Beginning with complex polynomials and rational functions, we ultimately summarize the main properties of the complex square root, exponential, logarithm, and trigonometric functions as well. We presume the reader is already familiar with the basic properties of the real versions of these functions. Our purpose here is to develop the corresponding properties of the complex versions and to point out how they differ from their real counterparts.

3.1 Complex Polynomials and Rational Functions

Among the simplest complex functions are complex polynomials $p : \mathbb{C} \to \mathbb{C}$, which, for some integer $n \geq 0$, referred to as the *degree* of the polynomial, and for some set of constants $a_0, a_1, a_2, \ldots, a_n \in \mathbb{C}$ with $a_n \neq 0$, can be written as

$$p(z) = a_0 + a_1 z + a_2 z^2 + \cdots + a_n z^n.$$

The importance of polynomials of a single real variable should already be appreciated by students of calculus. This class of functions is equally valuable in developing the theory of functions of a complex variable.

Related to polynomials, and comprising another class of functions traditionally handled in a first course in calculus, are the so-called rational functions. Rational functions are simply ratios of polynomials. In particular, in the complex case, if $p, q : \mathbb{C} \to \mathbb{C}$ are complex polynomials, then the function $r : D \to \mathbb{C}$ given by $r(z) = \frac{p(z)}{q(z)}$ is a complex rational function, where $D \subset \mathbb{C}$ is the set of complex numbers that excludes the zeros of the polynomial q. An example of such a rational function is given by $r(z) = \frac{z^3+z+7}{z^4-z^2}$ defined on $D \subset \mathbb{C}$ with $D = \mathbb{C} \setminus \{-1, 0, 1\}$.

3.2 The Complex Square Root Function

Recall from Chapter 1 that every $z \in \mathbb{C} \setminus \{0\}$ can be written in polar form as $z = r\, e^{i\,\theta}$, where $r = |z|$ and θ is an angle measured counterclockwise from the positive real axis to the ray in the complex plane emanating from the origin through z. We called θ an *argument* of z, denoted it by $\arg z$, and noted that θ is not unique since

$$z = r\, e^{i\,\theta} = r\, e^{i\,(\theta + 2\,\pi\,k)} \text{ for } k \in \mathbb{Z}.$$

This means there are infinitely many ways to represent any particular $z \in \mathbb{C} \setminus \{0\}$ in polar form. Restricting $\theta = \arg z$ to the interval $(-\pi, \pi\,]$ distinguished a unique choice of argument for z and defined a function of z that we called *the principal branch of the argument* and denoted by $\mathrm{Arg}(z)$ or Θ. The key point here is that $\mathrm{Arg}(z)$ is a function while the unrestricted $\arg z$ is not. In fact, $\mathrm{Arg}(z)$ is a function from $\mathbb{C} \setminus \{0\}$ to $\mathbb{R}$. Recall also that $f(x) = \sqrt{x}$ was made into a function by requiring the symbol $\sqrt{x}$ to represent only the nonnegative square root of x. Can something similar be done in $\mathbb{C}$ where x is replaced by $z \in \mathbb{C}$? Yes! We will use the Arg function to define a square root function in $\mathbb{C}$. To this end, consider $g : \mathbb{C} \setminus \{0\} \to \mathbb{C}$ given by

$$g(z) = z^{\frac{1}{2}} = \left(r\, e^{i\Theta}\right)^{1/2} = r^{\frac{1}{2}} e^{i\frac{1}{2}\Theta}.$$

Note that 0 has been excluded from the domain of g because the argument of $z = 0$ is not well defined. We can, however, "patch" our function g at $z = 0$. We do this in the following definition.

Definition 3.1 Let $z = r\, e^{i\Theta} \in \mathbb{C}$, where $r = |z|$ and $\Theta = \mathrm{Arg}(z)$. Then **the complex square root function** $f : \mathbb{C} \to \mathbb{C}$ is defined as

$$f(z) = \begin{cases} r^{\frac{1}{2}}\, e^{i\,\frac{1}{2}\Theta} & \text{for } z \neq 0 \\ 0 & \text{for } z = 0 \end{cases}.$$

Note that $\left(f(z)\right)^2 = \left(r^{\frac{1}{2}}\, e^{i\,\frac{1}{2}\Theta}\right)^2 = r\, e^{i\Theta} = z$ for $z \neq 0$, and $\left(f(0)\right)^2 = 0^2 = 0$, which implies $(f(z))^2 = z$ for all $z \in \mathbb{C}$. Therefore the function $h : \mathbb{C} \to \mathbb{C}$ given by $h(z) = z^2$ is a function that takes our defined square root of z back to z.

▶ **4.11** *Show that the square root function $f : \mathbb{C} \to \mathbb{C}$ given in Definition 3.1 is one-*

to-one. Is it onto? (No.) What can you conclude about the function $h : \mathbb{C} \to \mathbb{C}$ given by $h(z) = z^2$? Is it f^{-1}? (No.)

Two well-known properties of the real square root function $f : [0, \infty) \to \mathbb{R}$ given by $f(x) = \sqrt{x}$ are

$$(i) \quad f(x_1 x_2) = f(x_1) f(x_2),$$
$$(ii) \quad f\left(\frac{x_1}{x_2}\right) = \frac{f(x_1)}{f(x_2)}, \text{ provided } x_2 \neq 0.$$

However, these familiar properties do not hold for the complex square root function given in Definition 3.1. We establish this for (i) and leave (ii) to the reader. Let $z_1 = -1 - i$ and $z_2 = -i$. Then

$$f(z_1) = 2^{1/4} e^{i\,5\pi/8}, \quad \text{and} \quad f(z_2) = e^{i\,3\pi/4}.$$

But $z_1 z_2 = -1 + i$, and so

$$f(z_1 z_2) = 2^{1/4} e^{i\,3\pi/8}, \quad \text{while} \quad f(z_1) f(z_2) = 2^{1/4} e^{i\,11\pi/8}.$$

► **4.12** *Find a pair of complex numbers z_1, z_2 that violates property (ii) above. Why exactly do properties (i) and (ii) fail to hold for the complex square root function?*

► **4.13** *Note that for the real square root function $f : [0, \infty) \to \mathbb{R}$ given by $f(x) = \sqrt{x}$, we know that $f(x^2) = |x| = x$. Now consider the complex square root function given by Definition 3.1. It is still true that $f(z^2) = z$, but $f(z^2) \neq |z|$. Why is that?*

3.3 The Complex Exponential Function

Consider the complex function given by the rule $f(z) = e^z$. Recall that thus far we have defined $e^{i\theta} \equiv \cos\theta + i \sin\theta$ for $\theta \in \mathbb{R}$. This formula defines exponentials for pure imaginary numbers $i\,\theta$. We would now like to extend this definition to handle exponentials of more general complex numbers $z = x + i\,y$.

Definition 3.2 Let $z = x + i\,y \in \mathbb{C}$. Then the function $\exp : \mathbb{C} \to \mathbb{C}$ given by $\exp(z)$, or more commonly denoted by e^z, is defined by

$$\exp(z) = e^z \equiv e^x e^{i\,y}.$$

The motivation for this definition should be clear. We are merely extending the traditional rules of manipulating exponents for real numbers as learned in calculus. Note that this definition is consistent with our notion of exponentiation of pure imaginary numbers (corresponding to $x = 0$), as it must be. Note too, however, that the complex exponential function differs from the real exponential function in significant ways. For instance, the complex exponential function is not one-to-one, since

$$\exp(z + 2\pi i) = e^{z+2\pi i} = e^z e^{2\pi i} = e^z = \exp(z)$$

for every $z \in \mathbb{C}$, and so exp is periodic with period $2\pi i$. Also, $\exp(i\pi) = \cos\pi + i\sin\pi = -1$, and so the complex exponential function takes negative real values.

We now consider some important algebraic properties of e^z. We state them in the form of a proposition.

Proposition 3.3 *Let z, z_1, and z_2 be in $\mathbb{C}$. Then*

a) $e^{z_1+z_2} = e^{z_1}e^{z_2}$,

b) $e^{-z} = (e^z)^{-1}$,

c) $|e^z| = e^{\text{Re}(z)}$,

d) $e^{z_1} = e^{z_2}$ *if and only if* $z_2 = z_1 + i\,2\pi\,k$ *for some* $k \in \mathbb{Z}$,

e) $e^z \neq 0$.

PROOF Properties *a)* and *b)* are simple consequences of the definition. These and property *e)* are left to the exercises. To prove *c)*, we take norms to obtain

$$|e^z| = \left|e^x\,e^{i\,y}\right| = |e^x|\,\left|e^{i\,y}\right| = |e^x| = e^x.$$

To establish property *d)*, we equate (after application of the definition)

$$e^{x_1}\,e^{i\,y_1} = e^{x_2}\,e^{i\,y_2},$$

which implies that $e^{x_1} = e^{x_2}$. (Why?) From this it follows that $x_1 = x_2$. We also have that $e^{i\,y_1} = e^{i\,y_2}$, which implies that $y_2 = y_1 + 2\pi\,k$ for some integer k. Overall this gives

$$z_2 = x_2 + i\,y_2 = x_1 + i\,(y_1 + 2\pi\,k) = z_1 + 2\pi\,i\,k \quad \text{for some } k \in \mathbb{Z}.$$

We have shown that $e^{z_1} = e^{z_2} \Rightarrow z_2 = z_1 + i\,2\pi\,k$ for some integer k. The converse follows readily from Definition 3.2. ♦

► **4.14** *Prove properties a), b), and e) of the above proposition.*

► **4.15** *Show that the range of the complex exponential function is $\mathbb{C}\setminus\{0\}$.*

► **4.16** *Show that the following properties hold for the complex exponential function: a) $e^{z_1-z_2} = \frac{e^{z_1}}{e^{z_2}}$ b) $e^0 = 1$ c) $(e^z)^n = e^{nz}$ for all $n \in \mathbb{Z}$.*
More general powers of complex numbers will be postponed until we define the complex logarithm function.

3.4 The Complex Logarithm

In what follows we will denote the natural logarithm of a general complex number by "log" and reserve the calculus notation "ln" for those cases where we are taking the natural logarithm of a positive real number.[3] In defining a complex logarithm function, we would like to preserve the inverse relationship between the logarithm and the exponential as learned in calculus, and so in our definition of the log of a complex number z we require that

$$e^{\log z} = z.$$

Certainly this will preserve the relevant inverse relationship in the special case when z is real and positive. Also note that since $e^w \neq 0$ for any $w \in \mathbb{C}$, we immediately see that the logarithm of 0 will remain undefined, even in the complex case. With these ideas in mind, consider $z = r\,e^{i\,\theta} \neq 0$ and let $w = \log z = \phi + i\,\psi$, where here ϕ and ψ are real numbers. It then follows that $e^w = e^{\phi + i\,\psi}$. But $e^w = e^{\log z} = z = r\,e^{i\,\theta}$ as well, and so we have that $e^{\phi+i\,\psi} = r\,e^{i\,\theta}$. From this we obtain

$$e^{\phi+i\,\psi} = r\,e^{i\,\theta} = e^{\ln r}\,e^{i\,\theta} = e^{\ln r + i\,\theta}.$$

(Note our use of ln in this last expression, since it is known that r is a positive real value.) From this last equality we may conclude that

$$\phi + i\,\psi = \ln r + i\,\theta + i\,2\pi\,k \quad \text{for some } k \in \mathbb{Z}.$$

Equating the real and imaginary parts obtains

$$\phi = \ln r \quad \text{and} \quad \psi = \theta + 2\pi\,k \quad \text{for } k \in \mathbb{Z}.$$

Overall, we see that a nonzero complex number $z = r\,e^{i\theta}$ has infinitely many logarithms given by

$$\log z = \ln r + i\,(\theta + 2\pi\,k) \quad \text{for } k \in \mathbb{Z}.$$

Again we point out that the $\ln r$ appearing on the right side of the above equality is the "ordinary," single-valued natural logarithm of a positive real number as encountered in calculus, while $\log z$ is the newly defined, multiple-valued natural logarithm of a complex number.

We now consider a few examples to illustrate how to compute logarithms in the context of complex numbers. To do so, it is often easiest to first convert the given z to its most general polar form.

Example 3.4 *We will find the natural logarithm of the following:*

a) $z = i$ *b)* $z = 1 + i$ *c)* $z = -1$ *d)* $z = 1$

In all that follows, k should be understood to be an arbitrary integer value.

a) Here, $z = i = e^{i\,(\frac{\pi}{2} + 2\pi\,k)}$. *Taking the logarithm gives*

$$\log i = \ln 1 + i\,\left(\tfrac{\pi}{2} + 2\pi\,k\right) = i\,\left(\tfrac{\pi}{2} + 2\pi\,k\right).$$

[3]This is not necessarily standard notation. The notation log is typically used for either context described here, at least in higher mathematics. For most mathematics students, in fact, the notation ln is rarely encountered after a first-year calculus course, and rarely is any base other than e ever used in analysis. For these reasons, there is never really any confusion as to which base is implied by either notation. Our reasons for distinguishing between the two situations will become clear shortly.

b) $z = 1 + i = \sqrt{2}\, e^{i\,(\frac{\pi}{4}+2\pi\, k)}$. *Taking the logarithm gives*

$$\log(1+i) = \ln\sqrt{2} + i\,\left(\tfrac{\pi}{4} + 2\pi\, k\right).$$

c) $z = -1 = e^{i(\pi+2\pi\, k)}$. *Taking the logarithm gives*

$$\log(-1) = \ln 1 + i\,(\pi + 2\pi\, k) = i\,(\pi + 2\pi\, k).$$

d) $z = 1 = e^{i\,(0+2\pi\, k)}$. *Taking the logarithm gives*

$$\log 1 = \ln 1 + i\,(0 + 2\pi\, k) = i\,2\pi\, k.$$ ◀

Note from this example that our new, generalized definition of natural logarithm allows us to take the logarithm of negative real numbers. This wasn't possible with the old logarithm from calculus. Note also that with the newly defined logarithm, any nonzero real number (in fact, any nonzero complex number too) has *infinitely many* logarithms instead of just one. This last fact means that our complex logarithm, as defined so far, is not yet a function.

We now state the relationship between the complex logarithm and the complex exponential in the form of a proposition.

Proposition 3.5

a) For nonzero $z \in \mathbb{C}$, $\;e^{\log z} = z$.

b) For any $z \in \mathbb{C}$, $\;\log e^{z} = z + i\,2\pi k\;$ *for* $\;k \in \mathbb{Z}$.

PROOF To prove *a)*, note that $\log z = \ln|z| + i\,(\theta + 2\pi\, k)$, so exponentiating gives

$$e^{\log z} = e^{\ln|z| + i\,(\theta+2\pi\, k)} = e^{\ln|z|}e^{i\,(\theta+2\pi\, k)} = |z|\, e^{i\,\theta} = z.$$

To establish *b)*, note that $e^{z} = e^{x}\, e^{i\,y}$, and so taking logarithms gives

$$\log e^{z} = \log\left(e^{x}\, e^{i\,y}\right) = \ln e^{x} + i\,(y + 2\pi\, k) = (x + i\,y) + i\,2\pi\, k = z + i\,2\pi\, k.$$ ♦

As explained prior to the statement of the above proposition, just as for complex square roots, if $z \neq 0 \in \mathbb{C}$ the complex quantity $\log z$ has *infinitely many possible values*. We will once again use the principal branch of the argument, Arg, to define a single-valued function from our complex logarithm.

Definition 3.6 Let $z = r\, e^{i\theta} \neq 0$ be a complex number. Then if $\Theta = \text{Arg}(z)$ is the principal branch of the argument of z, we define **the principal branch of the complex logarithm function** $\text{Log} : \mathbb{C} \setminus \{0\} \to \mathbb{C}$ by

$$\text{Log}(z) \equiv \ln r + i\;\text{Arg}(z) = \ln r + i\,\Theta.$$

Note that other "branches" of the complex logarithm can be defined as functions as well. This is done by restricting $\arg z$ to intervals of length $2\,\pi$ other than $(-\pi, \pi]$. The *principal* branch is distinguished from other possible definitions of single-valued complex logarithm functions by its designation $\mathrm{Log}(z)$ rather than $\log(z)$.

► **4.17** *Determine the range of the complex logarithm function,* $\mathrm{Log} : \mathbb{C} \setminus \{0\} \to \mathbb{C}$. *(Hint: Consider its real and imaginary parts.)*

► **4.18** *Show that the complex logarithm function* $\mathrm{Log} : \mathbb{C} \setminus \{0\} \to \mathbb{C}$ *is one-to-one.*

► **4.19** *Proposition 3.5 established that* $e^{\log z} = z$ *and that* $\log e^z = z + i\,2\,\pi\,k$ *for* $k \in \mathbb{Z}$ *for all nonzero* $z \in \mathbb{C}$. *What can you say about* $e^{\mathrm{Log}(z)}$ *and* $\mathrm{Log}\,(e^z)$*? (The latter need not equal* z.*)*

3.5 Complex Trigonometric Functions

We have already seen in Chapter 1 that

$$\sin\theta = \frac{e^{i\,\theta} - e^{-i\,\theta}}{2\,i}, \quad \text{and} \quad \cos\theta = \frac{e^{i\,\theta} + e^{-i\,\theta}}{2}.$$

We would like any generalization of these formulas with the real value θ replaced by the more general complex number z to reduce to the above in the special case where z is real. To this end, we make the following definition.

Definition 3.7 For $z \in \mathbb{C}$ define the complex functions $\sin : \mathbb{C} \to \mathbb{C}$ and $\cos : \mathbb{C} \to \mathbb{C}$ by

$$\sin z \equiv \frac{e^{i\,z} - e^{-i\,z}}{2\,i}, \quad \text{and} \quad \cos z \equiv \frac{e^{i\,z} + e^{-i\,z}}{2}.$$

► **4.20** *Find* $\mathrm{Re}(\sin z)$, $\mathrm{Im}(\sin z)$, $\mathrm{Re}(\cos z)$, *and* $\mathrm{Im}(\cos z)$.

We illustrate the use of Definition 3.7 to calculate some complex trigonometric values in the following examples.

Example 3.8 *In this example, we will determine* $\cos(i)$ *and* $\sin(1+i)$. *Applying the definitions, we have*

$$\cos(i) = \frac{e^{i\,i} + e^{-i\,i}}{2} = \frac{e^{-1} + e}{2}, \quad \textit{a real number.}$$

Similarly, we find that

$$\sin(1+i) = \frac{e^{i\,(1+i)} - e^{-i\,(1+i)}}{2\,i} = \frac{e^{i}e^{-1} - e^{-i}e}{2\,i}.$$

But $e^{i} = \cos(1) + i\,\sin(1)$, *and* $e^{-i} = \cos(1) - i\,\sin(1)$. *Overall then, we have*

$$\sin(1+i) = \left(\frac{e^{-1} + e}{2}\right)\sin(1) - i\left(\frac{e^{-1} - e}{2}\right)\cos(1).$$

◄

We now consider the properties of the complex trigonometric functions to determine how they might differ from their real function counterparts. To this end, consider the function $\sin : \mathbb{C} \to \mathbb{C}$ as defined in Definition 3.7. The range is the set $\mathcal{R} = \left\{w \in \mathbb{C} : w = (e^{i\,z} - e^{-i\,z})/(2\,i) \text{ for some } z \in \mathbb{C}\right\}$. We will show that $\mathcal{R} = \mathbb{C}$. To prove this, we must solve the equation $w = (e^{i\,z} - e^{-i\,z})/(2\,i)$ for z. Multiplying this equation by $2\,i$, by $e^{i\,z}$, and rearranging, yields the quadratic equation in $e^{i\,z}$ given by

$$\left(e^{i\,z}\right)^2 - 2\,i\,w\left(e^{i\,z}\right) - 1 = 0.$$

Solving for $e^{i\,z}$ via the quadratic formula obtains $e^{i\,z} = i\,w \pm \sqrt{1-w^2}$. From this we get

$$z = \frac{1}{i}\log\left(i\,w \pm \sqrt{1-w^2}\right).$$

For any choice of $w \in \mathbb{C}$ this formula provides a $z \in \mathbb{C}$ (in fact, infinitely many) as a preimage of w, and so $\mathcal{R} = \mathbb{C}$. Therefore, $\sin z$ is unbounded, a very different situation from the case of real-valued trigonometric functions. This unboundedness property can also be established as follows:

$$\begin{aligned} \left|\sin z\right| &= \left|\frac{e^{i\,z} - e^{-i\,z}}{2\,i}\right| \\ &= \frac{1}{2}\left|e^{i\,(x+i\,y)} - e^{-i\,(x+i\,y)}\right| \\ &= \frac{1}{2}\left|e^{i\,x}\,e^{-y} - e^{-i\,x}\,e^{y}\right| \\ &\geq \frac{1}{2}\left(e^{y} - e^{-y}\right), \end{aligned}$$

from which it follows that $\sin z$ is unbounded on $\mathbb{C}$. (Why?)

▶ **4.21** *Answer the (Why?) question above. Also, show that the complex function* $\cos z$ *has range* $\mathbb{C}$*, and is therefore unbounded on* $\mathbb{C}$*.*

▶ **4.22** *Establish the following properties of the complex sine and cosine functions. Note that each one is a complex analog to a well-known property of the real sine and cosine functions.*

a) $\sin(-z) = -\sin z$ *b)* $\cos(-z) = \cos z$ *c)* $\sin^2 z + \cos^2 z = 1$

d) $\sin(2\,z) = 2\,\sin z\,\cos z$ *e)* $\sin(z + 2\,\pi) = \sin z$ *f)* $\cos(z + 2\,\pi) = \cos z$

g) $\sin(z + \pi) = -\sin z$ *h)* $\cos(z + \pi) = -\cos z$ *i)* $\cos(2\,z) = \cos^2 z - \sin^2 z$

j) $\sin\left(z + \frac{\pi}{2}\right) = \cos z$

▶ **4.23** *Show that* $\sin(z_1 + z_2) = \sin z_1\,\cos z_2 + \cos z_1\,\sin z_2$ *and that* $\cos(z_1 + z_2) = \cos z_1\,\cos z_2 - \sin z_1\,\sin z_2$.

▶ **4.24** *Show that* $e^{i\,z} = \cos z + i\,\sin z$ *for any* $z \in \mathbb{C}$.

4 LIMITS OF FUNCTIONS

Consider a function $f : \mathbb{D} \to \mathbb{Y}$, and suppose $x_0 \in \mathbb{D}'$, the derived set of $\mathbb{D}$. Since there exist points $x \in \mathbb{D}$ arbitrarily close to x_0, we can consider the values $f(x)$ as x gets closer and closer to x_0. In particular, if we let x get arbitrarily close to x_0, does $f(x)$ get arbitrarily close to an element in $\mathbb{Y}$? In order to consider such a limit as well defined, we require that $f(x)$ get arbitrarily close to the *same* element in $\mathbb{Y}$ no matter how we let x approach x_0. We formalize these ideas in defining the limit of a function, a concept that is crucial to the development of analysis.

4.1 Definition and Examples

We begin with a definition.

Definition 4.1 Suppose $f : \mathbb{D} \to \mathbb{Y}$ is a function and $x_0 \in \mathbb{D}'$. We say the **limit** of $f(x)$ as $x \in \mathbb{D}$ approaches x_0 exists and equals L, and we write

$$\lim_{x \to x_0} f(x) = L,$$

if for each $\epsilon > 0$ there exists a $\delta > 0$ such that

$$x \in \mathbb{D} \text{ and } 0 < |x - x_0| < \delta \ \Rightarrow \ |f(x) - L| < \epsilon.$$

Several remarks are worth noting about this definition.

1. The point x_0 must be in $\mathbb{D}'$, since it is only such points that can be approached arbitrarily closely by points $x \in \mathbb{D}$.
2. The limit $\lim_{x \to x_0} f(x)$ may exist even if f is not defined at x_0. That is, the limit point x_0 might not be in the domain of f.
3. In general, δ will depend on ϵ and x_0.
4. The value of δ is not unique, since any $\delta' < \delta$ will also work.
5. Even in cases where both $f(x_0)$ and $\lim_{x \to x_0} f(x)$ exist, they may not be equal to each other.
6. In those cases where $\mathbb{Y}$ is a proper subset of $\mathbb{R}$, $\mathbb{R}^p$, or $\mathbb{C}$, it is possible for the limit L to be in $\overline{\mathbb{Y}}$ but not in $\mathbb{Y}$.
7. An alternative, equivalent notation for the condition "$x \in \mathbb{D}$ and $0 < |x - x_0| < \delta$" that is sometimes used is "$x \in \mathbb{D} \cap N_\delta'(x_0)$."
8. The limit process described here is not the same as that associated with a limit of a sequence as defined in the last chapter. Here, x is not constrained to approach x_0 via a particular sequence of values $\{x_n\}$ within the domain of f. The limit process described here is one that implicitly applies for *any* sequence $\{x_n\}$ converging to x_0, and for which $x_n \neq x_0$ for each x_n. Of

course, in those cases where one knows that the limit according to Definition 4.1 exists, one can take the limit along a particular sequence. That is, suppose $f : \mathbb{D} \to \mathbb{Y}$ is a function where x_0 is a limit point of $\mathbb{D}$. If $\lim_{x\to x_0} f(x) = L$, and if $\{x_n\}$ is a sequence in $\mathbb{D}$ such that $x_n \neq x_0$ for all n and $\lim_{n\to\infty} x_n = x_0$, then $\lim_{n\to\infty} f(x_n) = L$.

▶ **4.25** *Prove the statement at the end of remark 8 above. That is, for a function $f : \mathbb{D} \to \mathbb{Y}$ and x_0 a limit point of $\mathbb{D}$, suppose $\lim_{x\to x_0} f(x) = L$. If $\{x_n\}$ is a sequence in $\mathbb{D}$ such that $x_n \neq x_0$ for all n and $\lim_{n\to\infty} x_n = x_0$, then $\lim_{n\to\infty} f(x_n) = L$. What if the sequence does not satisfy the condition that $x_n \neq x_0$ for all n? (In that case, the result fails.)*

▶ **4.26** *Consider $f : \mathbb{R}^2 \setminus \{\mathbf{0}\} \to \mathbb{R}$ given by*

$$f(x, y) = \frac{x\,y}{x^2 + y^2} \quad \text{for any } (x, y) \in \mathbb{R}^2 \setminus \{\mathbf{0}\}.$$

Show that the limit $\lim_{(x,y)\to(0,0)} f(x, y)$ does not exist. (Hint: Argue by contradiction, using the result of remark 8 above.)

We now provide several examples to illustrate the use of Definition 4.1 in establishing the limit of a function.

Example 4.2 *For $f : \mathbb{R} \to \mathbb{R}$ given by $f(x) = x^2$, we will show that $\lim_{x\to 3} f(x) = 9$. Note that*

$$|x^2 - 9| = |x - 3|\,|x + 3| = |x - 3|\,|(x - 3) + 6| \le |x - 3|^2 + 6\,|x - 3|.$$

For arbitrary $\epsilon > 0$, if we force $|x - 3|^2 < \frac{\epsilon}{2}$, i.e., $|x - 3| < \sqrt{\frac{\epsilon}{2}}$, and also $6\,|x - 3| < \frac{\epsilon}{2}$, i.e., $|x - 3| < \frac{\epsilon}{12}$, then we obtain $|x^2 - 9| < \epsilon$ as desired. Taking $\delta = \min\left(\sqrt{\frac{\epsilon}{2}}, \frac{\epsilon}{12}\right)$, we have that x satisfying $0 < |x - 3| < \delta \Rightarrow |x^2 - 9| < \epsilon$.◀

Example 4.3 *For $f : \mathbb{C} \setminus \{-1, 1\} \to \mathbb{C}$ given by $f(z) = \frac{z^2+1}{z^2-1}$, we will show that $\lim_{z\to 0} f(z) = -1$. Let $\epsilon > 0$ be given, and note that*

$$\left|\frac{z^2 + 1}{z^2 - 1} + 1\right| = \frac{2\,|z|^2}{|z^2 - 1|}.$$

Suppose $0 < |z| < \frac{1}{2}$. Then $|z^2 - 1| \ge 1 - |z|^2 > \frac{3}{4}$, so that

$$\left|\frac{z^2 + 1}{z^2 - 1} + 1\right| = \frac{2\,|z|^2}{|z^2 - 1|} < \frac{8}{3}\,|z|^2.$$

This last expression will be less than ϵ as long as $|z| < \sqrt{\frac{3\epsilon}{8}}$. We therefore choose $\delta \equiv \min\left(\frac{1}{2}, \sqrt{\frac{3\epsilon}{8}}\right)$, and we have

$$0 < |z| < \delta \Rightarrow \left|\frac{z^2 + 1}{z^2 - 1} + 1\right| < \epsilon.$$

◀

Note that in the two previous examples, $\lim_{x\to x_0} f(x) = f(x_0)$. This is not always the case.

Example 4.4 *Consider $f : \mathbb{R} \to \mathbb{R}$ given by $f(x) = \begin{cases} 3x+1 & \text{for } x \neq 2 \\ 1 & \text{for } x = 2 \end{cases}$. We will show that $\lim_{x\to 2} f(x) = 7$. Note that for $x \neq 2$, i.e., for $0 < |x-2|$, we have that $|f(x) - 7| = |(3x+1) - 7| = 3|x-2|$. This last expression will be less than any arbitrary $\epsilon > 0$ if $|x-2| < \frac{\epsilon}{3}$. That is, letting $\delta = \frac{\epsilon}{3}$ obtains*

$$0 < |x-2| < \delta \Rightarrow |f(x) - 7| = 3|x-2| < \epsilon.$$ ◀

Note in the last example that $f(2)$ is defined, and $\lim_{x\to 2} f(x)$ exists, but $\lim_{x\to 2} f(x) \neq f(2)$.

Example 4.5 *Consider $f : \mathbb{R} \to \mathbb{R}$ given by $f(x) = \sin x$. We will establish that $\lim_{x\to 0} \sin x = 0$. To see this, let $0 < \epsilon < 1$ and choose $\delta < \sin^{-1} \epsilon$. Then,*

$$0 < |x| < \delta \Rightarrow |\sin x - 0| = |\sin x| < \sin \delta < \epsilon.$$ ◀

▶ **4.27** *Show that $\lim_{\theta\to 0} \cos\theta = 1$. To do this, draw the unit circle and consider the ray with angle θ having small magnitude as measured from the positive x-axis. Recall that $\cos\theta$ corresponds to the x coordinate of the point of intersection of this ray with the unit circle. Use the definition of limit along with the geometry of the problem to derive the conclusion. That is, for $\epsilon > 0$ find $\delta > 0$ such that $0 < |\theta| < \delta \Rightarrow |\cos\theta - 1| < \epsilon$.*

In the following example we show how to confirm that a proposed limiting value for a function is *not* the correct limit.

Example 4.6 *For $f : \mathbb{R} \to \mathbb{R}$ given by $f(x) = x^2 - 2$ we will show that $\lim_{x\to 0} f(x) \neq -1$. To this end, assume that $\lim_{x\to 0} f(x) = -1$. We will obtain a contradiction. Assuming the limit, let $\epsilon = \frac{1}{2}$. Then there exists a $\delta > 0$ such that*

$$0 < |x| < \delta \Rightarrow \left|x^2 - 2 + 1\right| < \tfrac{1}{2}.$$

This in turn yields that

$$0 < |x| < \delta \Rightarrow \left|x^2 - 1\right| < \tfrac{1}{2}.$$

But by the triangle inequality we have that $1 - x^2 \leq |x^2 - 1| < \frac{1}{2}$, i.e.,

$$0 < |x| < \delta \Rightarrow |x| > \tfrac{1}{\sqrt{2}}.$$

This is clearly a contradiction, since whenever $|x|$ is close enough to 0 it cannot simultaneously be greater than $\frac{1}{\sqrt{2}}$. Therefore $\lim_{x\to 0} f(x) \neq -1$. ◀

Finally, in the following examples, we illustrate the case where the argument of the function approaches a limit point that is not in the function's domain.

Example 4.7 *Consider $f : \mathbb{R} \setminus \{0\} \to \mathbb{R}$ given by $f(x) = \frac{x^3 - 2x}{x}$. Note that $x = 0$ is not in the domain of f, and hence, $f(0)$ is not defined. We will show that $\lim_{x\to 0} f(x) = -2$. To this end, let $\epsilon > 0$ be arbitrary and consider*

$$|f(x) - (-2)| = \left|\frac{x^3 - 2x}{x} + 2\right| = \left|\frac{x^3}{x}\right| = x^2 \text{ if } x \neq 0.$$

Since $x^2 < \epsilon$ as long as $|x| < \sqrt{\epsilon}$, it follows that taking $\delta = \sqrt{\epsilon}$ yields

$$0 < |x| < \delta \Rightarrow |f(x) - (-2)| < \epsilon.$$

◀

▶ **4.28** *Show that the real function $f : \mathbb{R} \setminus \{0\} \to \mathbb{R}$ given by $f(x) = \exp\left(-1/x^2\right)$ has a limit of 0 as $x \to 0$, but that the complex function $f : \mathbb{C} \setminus \{0\} \to \mathbb{C}$ given by $f(z) = \exp\left(-1/z^2\right)$ does not have a limit as $z \to 0$.*

We now consider a higher-dimensional example.

Example 4.8 *Consider $\boldsymbol{f} : \mathbb{R}^2 \setminus \{\mathbf{0}\} \to \mathbb{R}^2$ given by*

$$\boldsymbol{f}(\mathbf{x}) = \left(\frac{x_1^2}{|\mathbf{x}|}, \frac{x_2^2}{|\mathbf{x}|}\right) \quad \text{for any } \mathbf{x} = (x_1, x_2) \in \mathbb{R}^2.$$

We will show that $\lim_{\mathbf{x}\to\mathbf{0}} \boldsymbol{f}(\mathbf{x}) = \mathbf{0}$. As in the previous example, note that the point $\mathbf{x} = \mathbf{0}$ is not in the domain of $\boldsymbol{f}$, and hence $\boldsymbol{f}(\mathbf{0})$ is not defined. However, since $\mathbf{x} = \mathbf{0}$ is a limit point of the domain of $\boldsymbol{f}$ we may consider the limit $\lim_{\mathbf{x}\to\mathbf{0}} \boldsymbol{f}(\mathbf{x})$. To this end, note that for $\mathbf{x} \neq \mathbf{0}$,

$$\begin{aligned} |\boldsymbol{f}(\mathbf{x}) - \mathbf{0}| = \left|\left(\frac{x_1^2}{|\mathbf{x}|}, \frac{x_2^2}{|\mathbf{x}|}\right)\right| &= \frac{1}{|\mathbf{x}|}\sqrt{x_1^4 + x_2^4} \\ &\leq \frac{x_1^2 + x_2^2}{|\mathbf{x}|} \\ &= |\mathbf{x}|. \end{aligned}$$

Therefore, taking $\delta = \epsilon$ yields

$$0 < |\mathbf{x} - \mathbf{0}| < \delta \Rightarrow |\boldsymbol{f}(\mathbf{x}) - \mathbf{0}| < \epsilon,$$

and the limit is established. ◀

Example 4.9 *Consider $f : \mathbb{R}^2 \setminus \{\mathbf{0}\} \to \mathbb{R}$ given by*

$$f(x, y) = \frac{x\,y}{x^2 + y^2},$$

and consider the limit $\lim_{(x,y)\to(0,0)} f(x, y)$. In order for the limit to exist, the same limiting value must be obtained regardless of the path of approach (x, y) takes toward $(0, 0)$. We will show that the limit does not exist. To do so, we will investigate the limit along straight lines passing through the origin. Let $y = m\,x$ for some fixed $m \in \mathbb{R}$ and note that along this line,

$$\lim_{(x,y)\to(0,0)} f(x, y) = \lim_{x\to 0} \frac{m\,x^2}{x^2 + m^2 x^2} = \lim_{x\to 0} \frac{m}{1 + m^2},$$

a value that clearly depends on the slope of the line of approach of (x, y) toward $(0, 0)$. Therefore, the limit does not exist. ◀

The last example illustrates a significant difference between establishing a limit for a function of more than one real variable and establishing a limit for a function of a single real variable. In the latter case, the limit exists as long as the same value is obtained as x approaches its limit point from either the left or the right (these two limits are referred to as *left-hand* and *right-hand* limits, respectively, and will be discussed later in this chapter). When the function depends on two or more variables, there are an infinite number of ways to approach the limit point. Since one can't exhaust all possible paths (even checking all possible straight line paths is not sufficient), the definition is often the only way to establish the limit for such functions, when it exists.

4.2 Properties of Limits of Functions

In this subsection, we discuss some properties of limits of functions that either allow for easier computation of the limits, or that imply certain nice characteristics about the functions in question if the limits exist. We'll begin with a result that proves the uniqueness of the limit, when it exists.

Proposition 4.10 *For $f : \mathbb{D} \to \mathbb{Y}$ and $x_0 \in \mathbb{D}'$, suppose $\lim_{x\to x_0} f(x)$ exists. Then the limit is unique.*

PROOF Suppose $\lim_{x\to x_0} f(x) = A$ and $\lim_{x\to x_0} f(x) = B$. Then, for any $\epsilon > 0$ there exist $\delta_1 > 0$ and $\delta_2 > 0$ such that

$$x \in \mathbb{D} \text{ and } 0 < |x - x_0| < \delta_1 \Rightarrow |f(x) - A| < \frac{\epsilon}{2},$$

and

$$x \in \mathbb{D} \text{ and } 0 < |x - x_0| < \delta_2 \Rightarrow |f(x) - B| < \frac{\epsilon}{2}.$$

Choosing $\delta < \min(\delta_1, \delta_2)$, if $x \in \mathbb{D}$ is such that $0 < |x - x_0| < \delta$ we have

$$|A - B| = |A - f(x) + f(x) - B| \leq |A - f(x)| + |f(x) - B| < \epsilon.$$

Since this is true for arbitrary $\epsilon > 0$, it follows that $A = B$. Therefore the limit is unique. ♦

When we learn that the limit of a function $\lim_{x\to x_0} f(x)$ exists we have actually learned more about the function than merely the value of the limit itself. When such a limit exists, the function cannot behave too wildly near that limit point. That is, the behavior of the function is constrained to a certain degree in the immediate vicinity of the limit point x_0. In fact, the next result establishes that such a function must be bounded near that limit point.

Proposition 4.11 *Consider $f : \mathbb{D} \to \mathbb{Y}$ and suppose $x_0 \in \mathbb{D}'$. If $\lim_{x\to x_0} f(x) = A$ exists, then there exists a deleted neighborhood $N'_\delta(x_0)$ centered at x_0 such that f is bounded on $N'_\delta(x_0) \cap \mathbb{D}$.*

PROOF According to Definition 4.1 there exists a $\delta > 0$ such that for $x \in \mathbb{D}$ and $0 < |x - x_0| < \delta$, i.e., for $x \in N'_\delta(x_0) \cap \mathbb{D}$, we have $|f(x) - A| < 1$. Applying the triangle inequality, we easily obtain for all such x that

$$|f(x)| = |f(x) - A + A| \leq |f(x) - A| + |A| < 1 + |A|.$$

This proves the result. ♦

Example 4.12 *Recall that in Example 4.7 on page 158 we showed the the function $f : \mathbb{R} \setminus \{0\} \to \mathbb{R}$ given by $f(x) = \frac{x^3 - 2x}{x}$ has the limit $\lim_{x \to 0} f(x) = -2$. We now will show that f is bounded on $N'_1(0)$. To see this, consider an arbitrary $x \in N'_1(0)$. Then $0 < |x| < 1$, and so we have*

$$|f(x)| = \left| \frac{x^3 - 2x}{x} \right| = \left| x^2 - 2 \right| \leq |x|^2 + 2 < 3.$$ ◀

The following useful result might be called "the component-wise limit theorem." To make its usefulness even more plain, note that for any $\boldsymbol{f} : D^k \to \mathbb{R}^p$ we may write $\boldsymbol{f}(\mathbf{x})$ in terms of its component functions $f_j : D^k \to \mathbb{R}$ for $1 \leq j \leq p$. That is, $\boldsymbol{f}(\mathbf{x}) = \big(f_1(\mathbf{x}), f_2(\mathbf{x}), \ldots, f_p(\mathbf{x})\big)$. Similarly, for any complex function $f : D \to \mathbb{C}$ where $z = x + i\,y \in D$, we have $f(z) = u(x, y) + i\,v(x, y)$ with $u(x, y)$ and $v(x, y)$ considered as "component functions" of $f(z)$. The following result summarizes the idea.

Proposition 4.13

a) Consider $\boldsymbol{f} : D^k \to \mathbb{R}^p$ where $\boldsymbol{f}(\mathbf{x}) = \big(f_1(\mathbf{x}), f_2(\mathbf{x}), \ldots, f_p(\mathbf{x})\big)$ for each $\mathbf{x} \in \mathbb{D}^k$, and let $\mathbf{x_0}$ be a limit point of D^k. Then,

$$\lim_{\mathbf{x} \to \mathbf{x_0}} \boldsymbol{f}(\mathbf{x}) = \mathbf{A} \ \ \textit{with} \ \ \mathbf{A} = \big(A_1, A_2, \ldots, A_p\big) \in \mathbb{R}^p$$

if and only if $\lim_{\mathbf{x} \to \mathbf{x_0}} f_j(\mathbf{x}) = A_j$ for each $1 \leq j \leq p$.

b) Consider the complex function $f : D \to \mathbb{C}$ where $f(z) = u(x, y) + i\,v(x, y)$ for each $z = x + i\,y \in D$, and let $z_0 = x_0 + i\,y_0$ be a limit point of D. Then,

$$\lim_{z \to z_0} f(z) = u_0 + i\,v_0 \in \mathbb{C}$$

if and only if $\lim_{(x,y) \to (x_0, y_0)} u(x, y) = u_0$ and $\lim_{(x,y) \to (x_0, y_0)} v(x, y) = v_0$.

PROOF We prove *a)* and leave the proof of *b)* to the reader. Suppose $\lim_{\mathbf{x} \to \mathbf{x_0}} \boldsymbol{f}(\mathbf{x}) = \mathbf{A}$. Then, for each $\epsilon > 0$ there exists a $\delta > 0$ such that

$$0 < |\mathbf{x} - \mathbf{x_0}| < \delta \ \Rightarrow \ |\boldsymbol{f}(\mathbf{x}) - \mathbf{A}| < \epsilon. \tag{4.4}$$

But

$$\left| f_j(\mathbf{x}) - A_j \right| \leq |\boldsymbol{f}(\mathbf{x}) - \mathbf{A}| \ \ \text{for each} \ \ 1 \leq j \leq p,$$

and so expression (4.4) implies

$$0 < |\mathbf{x} - \mathbf{x_0}| < \delta \ \Rightarrow \ \left| f_j(\mathbf{x}) - A_j \right| < \epsilon \ \ \text{for each} \ \ 1 \leq j \leq p.$$

This shows that $\lim_{\mathbf{x}\to\mathbf{x}_0} f_j(\mathbf{x}) = A_j$ for each $j = 1, 2, \ldots, p$. Conversely, suppose $\lim_{\mathbf{x}\to\mathbf{x}_0} f_j(\mathbf{x}) = A_j$ for each $1 \le j \le p$. Then, for each $\epsilon > 0$ there exists a $\delta > 0$ such that

$$0 < |\mathbf{x} - \mathbf{x}_0| < \delta \;\Rightarrow\; \left|f_j(\mathbf{x}) - A_j\right| < \tfrac{\epsilon}{p}.$$

Note that in the last expression, we can use the same δ that works for $1 \le j \le p$. (Why?) From this we see that

$$0 < |\mathbf{x} - \mathbf{x}_0| < \delta \;\Rightarrow\; |\boldsymbol{f}(\mathbf{x}) - \mathbf{A}| \le \sum_{j=1}^{p} |f_j(\mathbf{x}) - A_j| < \epsilon.$$

◆

► **4.29** *Answer the (Why?) question in the above proof, and also prove part b).*

Example 4.14 *Suppose $\boldsymbol{f} : \mathbb{R}^2 \to \mathbb{R}^3$ is given by*

$$\boldsymbol{f}(\mathbf{x}) = \big(f_1(\mathbf{x}),\, f_2(\mathbf{x}),\, f_3(\mathbf{x})\big) = \left(x_1 + x_2,\, x_1\, x_2,\, \tfrac{x_1}{1+x_2^2}\right),$$

and let $\mathbf{x}_0 = (0, 1)$. To determine the limit $\lim_{\mathbf{x}\to\mathbf{x}_0} \boldsymbol{f}(\mathbf{x})$, we consider the limit componentwise. That is,

$$\lim_{\mathbf{x}\to\mathbf{x}_0} \boldsymbol{f}(\mathbf{x}) = \left(\lim_{\mathbf{x}\to\mathbf{x}_0} f_1(\mathbf{x}),\, \lim_{\mathbf{x}\to\mathbf{x}_0} f_2(\mathbf{x}),\, \lim_{\mathbf{x}\to\mathbf{x}_0} f_3(\mathbf{x})\right),$$

and since $\lim_{\mathbf{x}\to\mathbf{x}_0} f_1(\mathbf{x}) = 1$, $\lim_{\mathbf{x}\to\mathbf{x}_0} f_2(\mathbf{x}) = 0$, and $\lim_{\mathbf{x}\to\mathbf{x}_0} f_3(\mathbf{x}) = 0$, we have that $\lim_{\mathbf{x}\to\mathbf{x}_0} \boldsymbol{f}(\mathbf{x}) = (1, 0, 0)$. ◄

A Note on Limits of Rational Functions

The following discussion is particularly relevant in those cases where the codomain $\mathbb{Y}$ possesses the field properties, i.e., when the range of the functions of interest are subsets of either $\mathbb{R}$ or $\mathbb{C}$. The practical question of interest is "How does one "guess" at the correct value of $\lim_{x\to x_0} f(x)$ before attempting to verify that the hunch is correct, via Definition 4.1, for example?" Consider the case where $f(x)$ is a real-valued rational function of x, i.e., $f(x) = \frac{p(x)}{q(x)}$ for polynomials $p(x)$ and $q(x)$. All students are taught the "trick" in a first-year calculus course of factoring $p(x)$ and $q(x)$ and cancelling any common factors that appear. This technique is especially useful when the common factor involves the limit point x_0. For example, consider $\lim_{x\to 1} \frac{x^2-1}{x-1}$. The function given by $\frac{x^2-1}{x-1}$ is not even defined at $x = 1$ due to the denominator's vanishing there. Yet, most students have no trouble justifying the following steps:

$$\lim_{x\to 1} \frac{x^2 - 1}{x - 1} = \lim_{x\to 1} \frac{(x - 1)(x + 1)}{x - 1} = \lim_{x\to 1} (x + 1) = 2.$$

Technically, each step in this argument is presumed to occur under the assumption that x *approaches* 1, but that x *never equals* 1. This is what allows us to cancel the common factor of $(x - 1)$ from the numerator and denominator in cases such as this. The fact that the original function is not defined at

$x = 1$ does not deter us from readily determining the function's limit as x *approaches* 1. We have really replaced the original function $f(x) = \frac{x^2-1}{x-1}$ with another function $g(x) = (x+1)$ in this limit calculation. The reason the limit result for g equals that of the function f is that these functions give the same values everywhere in a neighborhood of the limit point x_0, *except at the limit point itself.* We formalize this idea in the following result.

Proposition 4.15 *Suppose $\mathbb{D}$ is a subset of $\mathbb{X}$, and that $x_0 \in \mathbb{D}' \cap \mathbb{D}^C$. Consider $f : \mathbb{D} \to \mathbb{Y}$ and $g : \mathbb{D}_0 \to \mathbb{Y}$, where $\mathbb{D}_0 = \mathbb{D} \cup \{x_0\}$. If $f = g$ on $\mathbb{D}$ and $\lim_{x\to x_0} g(x) = A$, then $\lim_{x\to x_0} f(x) = A$ as well.*

PROOF For any $\epsilon > 0$ there exists a $\delta > 0$ such that

$$x \in \mathbb{D}_0 \text{ and } 0 < |x - x_0| < \delta \quad \Rightarrow \quad |g(x) - A| < \epsilon.$$

Since $f(x) = g(x)$ for all $x \in \mathbb{D}$, we have that

$$x \in \mathbb{D} \text{ and } 0 < |x - x_0| < \delta \quad \Rightarrow \quad |f(x) - A| < \epsilon,$$

i.e.,

$$\lim_{x\to x_0} f(x) = A.$$

♦

► **4.30** *If the condition "$f = g$ on $\mathbb{D}$" in the statement of the above result is replaced by "$f = g$ on $N_r'(x_0) \cap \mathbb{D}$," does the conclusion of the proposition still hold? (Yes.)*

Example 4.16 *Consider $f : \mathbb{R} \setminus \{0\} \to \mathbb{R}$ given by*

$$f(x) = \frac{x^3 - 2x}{x}.$$

Since $\frac{x^3-2x}{x} = x^2 - 2$ if $x \neq 0$, let $g : \mathbb{R} \to \mathbb{R}$ be given by $g(x) = x^2 - 2$. Since $\lim_{x\to 0} g(x) = -2$ and $f = g$ on $\mathbb{R} \setminus \{0\}$, the previous result allows us to conclude that

$$\lim_{x\to 0} \frac{x^3 - 2x}{x} = \lim_{x\to 0} (x^2 - 2) = -2.$$

◄

4.3 Algebraic Results for Limits of Functions

General Results

We now present some limit rules that allow for easy manipulation of limits of functions. Most of these rules are familiar to students of calculus, and many of the proofs are left as exercises for the reader. The following proposition is analogous to Proposition 2.3 of the last chapter, which established a similar set of rules for manipulating limits of sequences.

Proposition 4.17 *Consider the functions f and g, both with domain $\mathbb{D}$ and codomain $\mathbb{Y}$, where $\mathbb{Y}$ is here understood to be the whole space $\mathbb{R}$, $\mathbb{R}^p$, or $\mathbb{C}$. Also suppose x_0 is a limit point of $\mathbb{D}$. If $\lim_{x\to x_0} f(x) = A$ and $\lim_{x\to x_0} g(x) = B$, then* *a)* $\lim_{x\to x_0}(f(x)\pm y_0) = A \pm y_0$ *for* $y_0 \in \mathbb{Y}$,

b) $\lim_{x\to x_0}(f\pm g)(x) = A \pm B$,

c) $\lim_{x\to x_0}(c\,f)(x) = c\,A$ *for* $c \in \mathbb{R}$ *or* $\mathbb{C}$,

d) $\lim_{x\to x_0}(f\cdot g)(x) = A \cdot B$,

e) $\lim_{x\to x_0}|f(x)| = |A|$.

Note that the product in part *d)* is to be interpreted as the dot product in $\mathbb{R}^p$, and the usual product in $\mathbb{R}$ or $\mathbb{C}$.

PROOF We prove *b)* and *d)* and leave the other properties to the reader as exercises. To prove *b)*, note that for a given $\epsilon > 0$, there exists a $\delta_1 > 0$ such that

$$x \in \mathbb{D} \text{ and } 0 < |x - x_0| < \delta_1 \ \Rightarrow\ |f(x) - A| < \frac{\epsilon}{2}, \tag{4.5}$$

and also a $\delta_2 > 0$ such that

$$x \in \mathbb{D} \text{ and } 0 < |x - x_0| < \delta_2 \ \Rightarrow\ |g(x) - B| < \frac{\epsilon}{2}. \tag{4.6}$$

For $x \in \mathbb{D}$ satisfying $0 < |x - x_0| < \delta = \min(\delta_1, \delta_2)$, expressions (4.5) and (4.6) yield

$$\begin{aligned} |(f\pm g)(x) - (A \pm B)| &= |(f(x) - A) \pm (g(x) - B)| \\ &\le |f(x) - A| + |g(x) - B| \\ &< \epsilon. \end{aligned}$$

The proof of *d)* is similar to the proof of the corresponding property from Proposition 2.3. In particular, note that

$$\begin{aligned} |f(x)\cdot g(x) - A\cdot B| &= |f(x)\cdot(g(x) - B) + B\cdot(f(x) - A)| \\ &\le |f(x)|\,|g(x) - B| + |B|\,|f(x) - A|. \end{aligned} \tag{4.7}$$

The idea is to find an upper bound for each x-dependent term on the right-hand side of inequality (4.7). To this end, note that since $\lim_{x\to x_0} f(x)$ exists there exists a deleted neighborhood $N_r'(x_0)$ such that f is bounded on $N_r'(x_0) \cap \mathbb{D}$. That is, $|f(x)| \le M$ on $N_r'(x_0) \cap \mathbb{D}$ for some number M. Now consider that for $\epsilon > 0$ there exists $\delta_1 > 0$ such that $x \in \mathbb{D}$ satisfying $0 < |x - x_0| < \delta_1$ yields

$$|g(x) - B| < \frac{\epsilon}{2(M+1)}.$$

Finally, for $\epsilon > 0$ there exists $\delta_2 > 0$ such that $x \in \mathbb{D}$ satisfying $0 < |x - x_0| < \delta_2$ yields

$$|f(x) - A| < \frac{\epsilon}{2(|B|+1)}.$$

Considering only those $x \in \mathbb{D}$ satifying $0 < |x - x_0| < \delta = \min(\delta_1, \delta_2, r)$, we can substitute these upper bounds into the original inequality (4.7), yielding

$$\begin{aligned} |f(x)\cdot g(x) - A\cdot B| &\le |f(x)|\,|g(x)-B| + |B|\,|f(x)-A| \\ &< M\,\frac{\epsilon}{2\,(M+1)} + |B|\,\frac{\epsilon}{2\,(|B|+1)} \\ &= \frac{\epsilon}{2}\left(\frac{M}{M+1}\right) + \frac{\epsilon}{2}\left(\frac{|B|}{|B|+1}\right) \\ &< \frac{\epsilon}{2} + \frac{\epsilon}{2} = \epsilon. \end{aligned}$$

Since ϵ was arbitrary, the result is proved. ♦

► **4.31** *Prove the remaining results of the above proposition.*

► **4.32** *Suppose the functions f and g are such that $f : \mathbb{D} \to \mathbb{Y}_1$ and $g : \mathbb{D} \to \mathbb{Y}_2$ where $\mathbb{Y}_1$ is either $\mathbb{R}$ or $\mathbb{C}$, and $\mathbb{Y}_2$ is one of the spaces $\mathbb{R}$, $\mathbb{C}$, or $\mathbb{R}^p$. What can you say about property d) from the above proposition? (Similar results should hold.)*

A Field Property Result for Limits of Real or Complex Valued Functions

As described in previous chapters, certain results rely explicitly on the properties of a *field*. As we did in Chapter 3 when discussing sequences of numbers, we now prove one such result dealing with limits of functions where one of the functions takes its values in either the real or complex field, $\mathbb{R}$ or $\mathbb{C}$. The reader is urged to compare the following proposition with Proposition 2.4 on page 93 of the above referenced chapter.

Proposition 4.18 *Consider two functions f and g, both having domain $\mathbb{D}$. Assume f has codomain $\mathbb{Y}$ and that g has codomain $\mathbb{R}$ or $\mathbb{C}$. (If its codomain is $\mathbb{C}$, then we assume that of f is $\mathbb{C}$ too to avoid complex vectors.) Suppose also that $x_0 \in \mathbb{D}'$. Then, if $\lim_{x\to x_0} g(x) = B \neq 0$, and if $\lim_{x\to x_0} f(x) = A$, we have that*

$$\lim_{x\to x_0} \left(\frac{f}{g}\right)(x) = \frac{A}{B}.$$

Note that if $\mathbb{Y} = \mathbb{R}^k$ and g has codomain $\mathbb{R}$, then the division in the statement of the theorem is legitimate: one is dividing a vector by a scalar, or multiplying the former by the inverse of the latter. Also, that scalar is nonzero in a small deleted neighborhood of x_0 since $B \neq 0$, as we shall see in the proof.

PROOF Consider the case $f \equiv 1$ on $\mathbb{D}$. To prove the result, note that

$$\left|\frac{1}{g(x)} - \frac{1}{B}\right| = \frac{|g(x) - B|}{|B|\,|g(x)|}.$$

Since $B \neq 0$ there exists a $\delta_1 > 0$ such that $x \in \mathbb{D}$ satisfying $0 < |x - x_0| < \delta_1$ implies $|g(x) - B| < \frac{|B|}{2}$. This in turn implies, for $x \in \mathbb{D}$ satisfying $0 < |x - x_0| < \delta_1$, that $|g(x)| > \frac{|B|}{2}$. (Why?) In particular, $g(x) \neq 0$ for such x. Finally, for $\epsilon > 0$ there exists a $\delta_2 > 0$ such that $x \in \mathbb{D}$ satisfying $0 < |x - x_0| < \delta_2$ yields $|g(x) - B| < \frac{\epsilon\,|B|^2}{2}$. From all this, we see that for $\delta = \min(\delta_1, \delta_2)$ and $x \in \mathbb{D} \cap N'_\delta(x_0)$, we obtain

$$\left|\frac{1}{g(x)} - \frac{1}{B}\right| = \frac{|g(x) - B|}{|B|\,|g(x)|} < \epsilon.$$

We leave the general case to the reader. ♦

▶ **4.33** *Complete the proof of the proposition by handling the case where $f \not\equiv 1$.*

▶ **4.34** *If $p(z) = c_0 + c_1\,z + c_2\,z^2 + \cdots + c_n\,z^n$ with $c_j \in \mathbb{R}$ or $\mathbb{C}$ for $1 \le j \le n$, show that $\lim_{z\to z_0} p(z) = p(z_0)$.*

▶ **4.35** *Consider $f : D \to \mathbb{C}$ where $f(z) = \frac{p(z)}{q(z)}$, and $p(z)$ and $q(z)$ are polynomials in z. Let $D = \mathbb{C} \setminus Z_q$ where Z_q is the set of roots of the polynomial q. Show that $\lim_{z\to z_0} f(z) = \frac{p(z_0)}{q(z_0)}$ as long as $z_0 \in D$.*

Order Property Results for Real-Valued Functions

Finally, as we did for sequences of real numbers in Chapter 3, we now consider those special rules relating to the order property of $\mathbb{R}$. In particular, for real-valued functions f the following proposition is analogous to Proposition 2.5 on page 93 and its corollary from section 2.2.

Proposition 4.19 *Consider f and g where both are functions from $\mathbb{D}$ to codomain $\mathbb{R}$. Suppose x_0 is a limit point of $\mathbb{D}$, and that $\lim_{x\to x_0} f(x) = A$ and $\lim_{x\to x_0} g(x) = B$.*

a) If $f(x) \ge 0$ on $N'_\delta(x_0) \cap \mathbb{D}$ for some $\delta > 0$, then $A \ge 0$.

b) If $f(x) \ge g(x)$ on $N'_\delta(x_0) \cap \mathbb{D}$ for some $\delta > 0$, then $A \ge B$.

PROOF To prove *a*), suppose $A < 0$. Then there exists a $\delta_1 > 0$ such that

$$x \in \mathbb{D} \text{ and } 0 < |x - x_0| < \delta_1 \;\Rightarrow\; |f(x) - A| < -\tfrac{1}{2}\,A.$$

Application of the triangle inequality yields that

$$x \in \mathbb{D} \cap N'_{\delta_1}(x_0) \;\Rightarrow\; f(x) < \tfrac{1}{2}\,A < 0.$$

But this is a contradiction since $f(x) \ge 0$ on $N'_\delta(x_0) \cap \mathbb{D}$ for some $\delta > 0$. We leave the proof of *b*) to the reader. ♦

▶ **4.36** *Prove part b) of the above proposition.*

Example 4.20 *In this example, we will show that if $f : \mathbb{D} \to \mathbb{R}$ is any nonnegative real-valued function such that $\lim_{x\to x_0} f(x) = A$ exists, then*

$$\lim_{x\to x_0} \sqrt{f(x)} = \sqrt{A}.$$

Note that this effectively means that $\lim_{x\to x_0}\sqrt{f(x)} = \sqrt{\lim_{x\to x_0} f(x)}$, *that is, the limit can be passed "inside" the square root function. Functions that allow the limit to be passed inside to their arguments in this way are special and will be discussed more thoroughly in the next chapter. According to the previous proposition we know that* $A \geq 0$. *If* $A > 0$, *for each* $\epsilon > 0$ *there exists a* $\delta > 0$ *such that*

$$x \in \mathbb{D} \text{ and } 0 < |x - x_0| < \delta \;\Rightarrow\; |f(x) - A| < \epsilon\sqrt{A}.$$

Now, for such $x \in \mathbb{D}$ *satisfying* $0 < |x - x_0| < \delta$, *we have that*

$$\begin{aligned} \left|\sqrt{f(x)} - \sqrt{A}\right| &= \left|\frac{\sqrt{f(x)} - \sqrt{A}}{\sqrt{f(x)} + \sqrt{A}}\left(\sqrt{f(x)} + \sqrt{A}\right)\right| \\ &= \frac{|f(x) - A|}{\sqrt{f(x)} + \sqrt{A}} \leq \frac{|f(x) - A|}{\sqrt{A}} < \epsilon. \end{aligned}$$

The $A = 0$ *case is left to the reader as an exercise.* ◀

A special case of the result from this example is worth mentioning, namely, if $f : D \to \mathbb{R}$ is given by $f(x) = x$ with $D = [0, \infty)$ and $x_0 \in D$, then $\lim_{x\to x_0}\sqrt{x} = \sqrt{x_0}$.

► **4.37** *Complete the case* $A = 0$ *from the previous example.*

The following proposition is called the squeeze theorem for functions. It is analogous to Proposition 2.9 on page 95 in Chapter 3, and its proof is similar. We therefore leave the proof to the exercises.

Proposition 4.21 (Squeeze Theorem for Functions)

Consider the functions f, g, *and* h, *all with domain* $\mathbb{D}$ *and codomain* $\mathbb{R}$, *and suppose* x_0 *is a limit point of* $\mathbb{D}$. *Also suppose the following:*

(i) There exists $\delta > 0$ *such that* $f(x) \leq g(x) \leq h(x)$ *for all* $x \in N'_\delta(x_0) \cap \mathbb{D}$,

(ii) $\lim_{x\to x_0} f(x) = \lim_{x\to x_0} h(x) = L$.

Then $\lim_{x\to x_0} g(x) = L$.

► **4.38** *Prove the above proposition.*

Example 4.22 *We will apply the squeeze theorem for functions to establish that* $\lim_{\theta\to 0}\frac{\sin\theta}{\theta} = 1$. *To begin, note from the comparison of areas in Figure 4.1 that if* $0 < |\theta| < \frac{\pi}{4}$ *we have*

$$\tfrac{1}{2}\sin\theta\cos\theta < \tfrac{1}{2}\,\theta\,(1)^2 < \tfrac{1}{2}(1)\tan\theta.$$

This is equivalent to

$$\cos\theta < \frac{\sin\theta}{\theta} < \frac{1}{\cos\theta}. \tag{4.8}$$

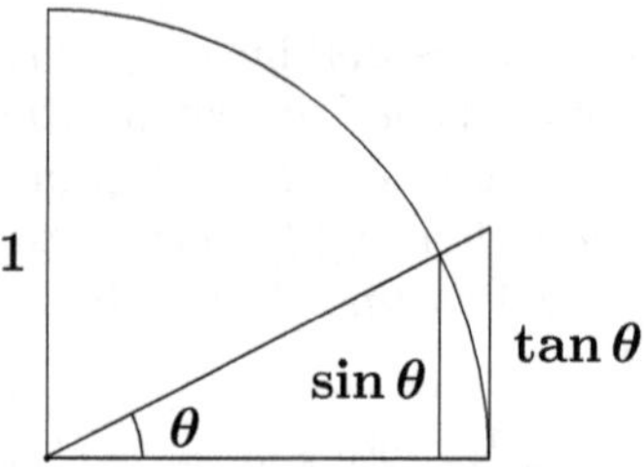

Figure 4.1 *Demonstration that* $\lim_{\theta \to 0} \frac{\sin\theta}{\theta} = 1$.

We saw in a previous exercise that $\lim_{\theta \to 0} \cos\theta = 1$, *and so application of the squeeze theorem for functions to the double inequality* (4.8) *yields the result.* ◀

Our final order property results pertaining to limits of real-valued functions have to do with one-sided limits. In particular, for $f : (a, b) \to \mathbb{R}$ and $x_0 \in [a, b)$, we wish to investigate the behavior of $f(x)$ as $x \to x_0$ from the right, i.e., through values of $x > x_0$. Similarly, for $x_0 \in (a, b]$, we wish to investigate the behavior of $f(x)$ as $x \to x_0$ from the left, i.e., through values of $x < x_0$. A careful look at Definition 4.1 on page 156 will show that such one-sided limits are already well defined, since that definition handles the case where x_0 is a limit point. Of course, this includes limits such as $\lim_{x \to a} f(x)$ or $\lim_{x \to b} f(x)$ for a function $f : (a, b) \to \mathbb{R}$. However, we make this idea more explicit in the case of real-valued functions through the following special definition.

Definition 4.23 Consider $f : \mathbb{D}^1 \to \mathbb{R}$, and x_0 such that for any $r > 0$, $\mathbb{D}^1 \cap (x_0, x_0 + r) \neq \varnothing$. We write

$$\lim_{x \to x_0^+} f(x) = A$$

if for any $\epsilon > 0$ there exists a $\delta > 0$ such that

$$x \in \mathbb{D}^1 \cap (x_0, x_0 + \delta) \quad \Rightarrow \quad |f(x) - A| < \epsilon.$$

When it exists, the limit A is called the **right-hand limit** of f as x approaches x_0, or *the limit of f as x approaches x_0 from the right*. Similarly, for x_0 such that for any $r > 0$, $\mathbb{D}^1 \cap (x_0 - r, x_0) \neq \varnothing$, we write

$$\lim_{x \to x_0^-} f(x) = B$$

if for any $\epsilon > 0$ there exists a $\delta > 0$ such that

$$x \in \mathbb{D}^1 \cap (x_0 - \delta, x_0) \quad \Rightarrow \quad |f(x) - B| < \epsilon.$$

When it exists, the limit B is called the **left-hand limit** of f as x approaches x_0, or *the limit of f as x approaches x_0 from the left*.

Note from the above definition that the right-hand limit of f might exist at x_0 even though the function f might not be defined at x_0. Similarly for the left-hand limit of f at x_0. Also, while we will not formalize these ideas in a separate definition, note in the definition of right-hand limit that the condition on x_0 that for any $r > 0$, $D^1 \cap (x_0, x_0 + \delta) \neq \varnothing$ ensures that x_0 is a "right-hand limit point" of D^1. Similarly, in the case of the left-hand limit the corresponding condition on x_0 that for any $r > 0$, $D^1 \cap (x_0 - \delta, x_0) \neq \varnothing$ ensures that x_0 is a "left-hand limit point" of D^1. A limit point x_0 of D^1 is either a right-hand limit point of D^1, a left-hand limit point of D^1, or both a right- and left-hand limit point of D^1.

As one might expect, a natural relationship exists between the limit of f and the right- and left-hand limits of f as $x \to x_0$.

Proposition 4.24 *Consider $f : D^1 \to \mathbb{R}$.*

a) If $\lim_{x \to x_0^+} f(x)$ and $\lim_{x \to x_0^-} f(x)$ both exist and equal $L \in \mathbb{R}$, then $\lim_{x \to x_0} f(x)$ exists and equals L.

b) Suppose x_0 is such that for any $r > 0$, $D^1 \cap (x_0 - r, x_0) \neq \varnothing$, and $D^1 \cap (x_0, x_0 + r) \neq \varnothing$. If $\lim_{x \to x_0} f(x) = L$ exists, then $\lim_{x \to x_0^+} f(x)$ and $\lim_{x \to x_0^-} f(x)$ exist and equal L.

▶ **4.39** *Prove the previous proposition.*

▶ **4.40** *Suppose $f : D^1 \to \mathbb{R}$ and x_0 are such that $\lim_{x \to x_0} f(x) = L$ exists. Must f have both left- and right-handed limits? (Yes, if the set has appropriate nonempty intersections.)*

Example 4.25 *Suppose $f : (0, 5) \to \mathbb{R}$ is given by*

$$f(x) = \begin{cases} 2x - 1 & \text{for } 0 < x \leq 3 \\ 5 - x & \text{for } 3 < x < 5. \end{cases}$$

We will investigate the left- and right-hand limits of f as x approaches 3. Considering the left-hand limit first, note that for $0 < x \leq 3$,

$$|f(x) - 5| = |2x - 1 - 5| = |2x - 6| = 2|x - 3|.$$

This last expression will be less than any given $\epsilon > 0$ if $|x - 3| < \delta = \frac{\epsilon}{2}$. That is, for $\epsilon > 0$ take $\delta = \frac{\epsilon}{2}$ to obtain

$$x < 3 \text{ and } 0 < |x - 3| < \delta \quad \Rightarrow \quad |f(x) - 5| < \epsilon.$$

Hence, the left-hand limit is 5. Considering the right-hand limit now, note that for $3 < x < 5$,

$$|f(x) - 2| = |5 - x - 2| = |3 - x| = |x - 3|.$$

This last will be less than any given $\epsilon > 0$ *if* $|x-3| < \delta = \epsilon$. *That is, for* $\epsilon > 0$ *take* $\delta = \epsilon$ *to obtain*

$$x > 3 \text{ and } 0 < |x-3| < \delta \quad \Rightarrow \quad |f(x) - 2| < \epsilon.$$

Hence, the right-hand limit is 2. Clearly the left- and right-hand limits are not equal, and so according to Proposition 4.24 the limit $\lim_{x\to 3} f(x)$ *does not exist.* ◀

► **4.41** *Using the function from the previous example, show that the left- and right-hand limits of* $f(x)$ *as* x *approaches 2 are the same, and hence the limit* $\lim_{x\to 2} f(x)$ *exists.*

For a function $f : [a, \infty) \to \mathbb{R}$ we may be interested in determining whether the function approaches a single value, called a *horizontal asymptote,* as x approaches $+\infty$. Likewise, for a function $g : (-\infty, b] \to \mathbb{R}$ we may investigate whether g approaches a single value as x approaches $-\infty$.

Definition 4.26 Consider $f : [a, \infty) \to \mathbb{R}$. We say **the limit of f as x approaches ∞ is L,** and we write

$$\lim_{x\to\infty} f(x) = L,$$

if for any $\epsilon > 0$ there exists $M \in \mathbb{R}$ such that

$$x > M \quad \Rightarrow \quad |f(x) - L| < \epsilon.$$

Similarly, for $g : (-\infty, b] \to \mathbb{R}$, we say **the limit of g as x approaches $-\infty$ is L,** and we write

$$\lim_{x\to -\infty} g(x) = L,$$

if for any $\epsilon > 0$ there exists $M \in \mathbb{R}$ such that

$$x < M \quad \Rightarrow \quad |g(x) - L| < \epsilon.$$

► **4.42** *Apply the above definition to establish the following limits.*

a) $\lim_{x\to\infty} \frac{1}{x} = 0$ *b)* $\lim_{x\to-\infty} \frac{1}{x} = 0$ *c)* $\lim_{x\to\infty} \frac{x}{x+1} = 1$ *d)* $\lim_{x\to-\infty} \frac{x}{x+1} = 1$

► **4.43** *It is sometimes necessary to investigate functions that are unbounded as their argument approaches* $\pm\infty$ *or* x_0. *In this exercise we define what it means for a function to have such an infinite limit. Suppose* $f : (a, \infty) \to \mathbb{R}$ *for some* $a \in \mathbb{R}$ *and* $g : (-\infty, b) \to \mathbb{R}$ *for some* $b \in \mathbb{R}$. *Then we say that* f ***diverges to ∞ as x goes to ∞*** *and we write*

$$\lim_{x\to\infty} f(x) = \infty,$$

if for any $K > 0$ *there exists* $M \in \mathbb{R}$ *such that*

$$x > M \quad \Rightarrow \quad f(x) > K.$$

Likewise, we say that g ***diverges to ∞ as x goes to $-\infty$*** *and we write*

$$\lim_{x\to-\infty} g(x) = \infty,$$

if for any $K > 0$ there exists $M \in \mathbb{R}$ such that

$$x < M \Rightarrow g(x) > K.$$

Similar definitions exist for functions that approach $-\infty$ in the limit as x approaches either ∞ or $-\infty$. Write them. It is also possible for functions to grow without bound toward ∞ or $-\infty$ as x approaches a finite limit x_0. We say in such cases that the function has a vertical asymptote at x_0. Write these definitions as well. Finally, while it is also possible for a function $f : \mathbb{R}^k \to \mathbb{R}$ to have an infinite limit such as $\lim_{\mathbf{x}\to\mathbf{x}_0} f(\mathbf{x}) = \infty$, or $\lim_{\mathbf{x}\to\mathbf{x}_0} f(\mathbf{x}) = -\infty$, we will not try to formalize the notion of a vector-valued function having an infinite limit. Why is that?

▶ **4.44** *Evaluate or verify the following limits.*

a) $\lim\limits_{x\to\infty} e^x = \infty$ *b)* $\lim\limits_{x\to-\infty} e^x = 0$ *c)* $\lim\limits_{x\to\infty} \frac{e^x - e^{-x}}{e^x + e^{-x}}$ *d)* $\lim\limits_{x\to-\infty} \frac{e^x - e^{-x}}{e^x + e^{-x}}$

e) $\lim\limits_{x\to 0} \frac{e^x - 1}{x} = 1$

5 Supplementary Exercises

1. *Specify the domain, codomain, and range for the function $\boldsymbol{f} : \mathbb{R}^3 \to \mathbb{R}^2$, given by the rule $\boldsymbol{f}(x, y, z) = (x + y + z, xyz)$.*

2. *Consider $f(z) = \overline{z}$. Find $u(x, y)$ and $v(x, y)$, and determine the domain, codomain, and range of f.*

3. *Consider the complex function given by the rule $f(z) = |z|$. Find $u(x, y)$ and $v(x, y)$, and determine the domain, codomain, and range of f.*

4. *Consider the function $f : \mathbb{R} \setminus \{0\} \to \mathbb{R}$ given by $f(x) = x + \frac{1}{x}$. What is $f(A)$ for the following sets?* *a)* $A = (0, 1)$ *b)* $A = [1, 2]$ *c)* $A = (1, \infty)$

5. *If $f : \mathbb{R} \setminus \{0\} \to \mathbb{R}$ is given by $f(x) = \frac{1}{x}$, what is $f^{-1}(B)$ if $B = [0, 2)$?*

6. *In Example 2.2 on page 140, let $B = \{\mathbf{x} \in \mathbb{R}^2 : |\mathbf{x}| = 1\}$. Find $f^{-1}(B)$.*

7. *Consider the same function as in Example 2.4 on page 141, namely, $f : \mathbb{C} \setminus \{-1\} \to \mathbb{C}$ given by $f(z) = \frac{z-1}{z+1}$. Find the image of the points on the circle of radius a centered at the origin for the two cases $0 < a < 1$ and $1 < a$.*

8. *Suppose $f : \mathbb{X} \to \mathbb{X}$ is given by $f(x) = ax + b$ where $a \in \mathbb{R}$ such that $a \neq 0$, and $b \in \mathbb{X}$. Show that f is one-to-one and onto.*

9. *Suppose $f : \mathbb{X} \to \mathbb{R}$ is given by $f(x) = a \cdot x + b$ where $a \in \mathbb{X}$ such that $a \neq 0$, and $b \in \mathbb{R}$. Is f one-to-one? Is it onto? (Yes in both cases.)*

10. *For the function $f : \mathbb{C} \to \mathbb{C}$ given by $f(z) = e^z$,*

a) What happens to horizontal lines under this mapping?

b) What happens to vertical lines under this mapping?

c) What happens to the strip $S = \{z \in \mathbb{C} : 1 \le \text{Re}(z) \le 4\}$?

11. *Consider the function* $f : \mathbb{C} \setminus \{-1\} \to \mathbb{C}$ *given by* $f(z) = \frac{z-1}{z+1}$ *once more. Find* $f(A)$ *where* A *is the set given by the following (in part c),* $a \in \mathbb{R}$ *is fixed):*

a) $A = \{z \in \mathbb{C} : |z| = 2\}$ b) $A = \{z \in \mathbb{C} : |z| = \frac{1}{2}\}$ c) $A = \{z \in \mathbb{C} : |z| \ge a\}$

12. *Let* $\boldsymbol{f} : \mathbb{R} \to \mathbb{R}^2$ *be given by* $\boldsymbol{f}(x) = (x, x^2)$. *Find* $\boldsymbol{f}(A)$ *for the following:*

a) $A = [0, 1]$ b) $A = (-\infty, \infty)$

13. *Suppose* $f : \mathbb{C} \to \mathbb{R}$ *is given by* $f(z) = |z|$. *Find* $f(A)$ *for the following:*

a) $A = \{z \in \mathbb{C} : |z| = 5\}$ b) $A = \{z \in \mathbb{C} : \text{Re}(z) = 2\}$ c) $A = \{z \in \mathbb{C} : \text{Im}(z) = 2\}$

14. *For the complex function* $f : D \to \mathbb{C}$, *suppose you know that* $\overline{f(z)} = f(\overline{z})$ *for all* $z \in D$. *What does this tell you about the function* f? *(Hint: Think of* f *as* $f = u + i\,v$.*)*

15. *Consider* $f : \mathbb{C} \setminus \{0\} \to \mathbb{C}$ *given by* $f(z) = z + \frac{1}{z}$. *If* A *is the unit circle in the complex plane centered at the origin, find* $f(A)$. *(Hint: Consider the points on the circle in polar form.)*

16. *Consider* $f : \mathbb{R} \to \mathbb{R}$ *given by* $f(x) = x^2$. *Is* $f^{-1}(\mathbb{Q}) = \mathbb{Q}$? *(No.) Is* $f^{-1}(\mathbb{Q}) = \mathbb{R}$? *(No.) Is* $f^{-1}(\mathbb{I}) = \mathbb{I}$? *(Yes.) Is* $f^{-1}(\mathbb{I}) = \mathbb{R}$? *(No.)*

17. *Consider* $f : \mathbb{R}^2 \to \mathbb{R}$ *given by* $f(x, y) = x^2 + y$, *and suppose* $B = (0, 1)$. *Find* $f^{-1}(B)$.

18. *Consider a function* $f : \mathbb{D} \to \mathbb{Y}$. *Find the following:*

a) $f(\varnothing)$ b) $f^{-1}(\varnothing)$ c) $f^{-1}(\mathbb{Y})$

19. *Find an example where equality in part 6 of Proposition 2.5 on page 141 does not hold.*

20. *For each of the following pairs of sets, determine whether one is a subset of the other, or if they are equal.*

a) $f(A \times B),\ f(A) \times f(B)$ b) $\overline{f^{-1}(B)},\ f^{-1}(\overline{B})$ c) $\left[f^{-1}(B)\right]',\ f^{-1}(B')$

d) $f(\overline{A}),\ \overline{f(A)}$

21. *Show that* $f : \mathbb{C} \to \mathbb{R}$ *given by* $f(z) = e^{-|z|^2}$ *is bounded on* $\mathbb{C}$.

22. *Suppose* $g : \mathbb{R}^2 \to \mathbb{C}$ *and* $f : \mathbb{C} \to \mathbb{R}$ *are given by* $g(x, y) = x^2 - i\,y$ *and* $f(z) = |z|$, *respectively. Find* $f \circ g$ *and identify its domain and codomain.*

23. *Determine whether each of the following functions is one-to-one.*

a) $f : \mathbb{R} \setminus \{-\frac{c}{b}\} \to \mathbb{R}$ *given by* $f(x) = \frac{a}{bx+c}$ *where* $a, b, c \in \mathbb{R}$ *and* $b \neq 0$.

b) $f : \mathbb{C} \setminus \{1\} \to \mathbb{C}$ *given by* $f(z) = \frac{z}{z-1}$.

24. *Consider the function $f : \mathbb{R} \to \mathbb{R}$ given by $f(x) = 2^x$. Show that f is increasing, and therefore one-to-one. Determine the range of f. Is f onto? (No.)*

25. *Let $\boldsymbol{f} : \mathbb{R}^2 \to \mathbb{R}^3$ be given by $\boldsymbol{f}(x_1, x_2) = (x_1,\, x_1 + x_2,\, 2^{x_2})$. Is $\boldsymbol{f}$ one-to-one?*

26. *Give an example of a function that is one-to-one but not strictly monotone.*

27. *Determine whether the function $\boldsymbol{f} : \mathbb{R}^3 \to \mathbb{R}^2$ given by $\boldsymbol{f}(x, y, z) = (x + y + z,\, xyz)$ is one-to-one and/or onto.*

28. *Consider $\boldsymbol{f} : \mathbb{R}^2 \to \mathbb{R}^2$ given by $\boldsymbol{f}(\mathbf{x}) = (x_1 + x_2, x_1 - x_2)$ where $\mathbf{x} = (x_1, x_2)$. Is $\boldsymbol{f}$ onto? (Yes.)*

29. *Consider $\boldsymbol{f} : \mathbb{R}^2 \to \mathbb{Y} \subset \mathbb{R}^2$ given by $\boldsymbol{f}(\mathbf{x}) = (x_1 + x_2, x_1 x_2)$ where $\mathbf{x} = (x_1, x_2)$. Find a space $\mathbb{Y} \subset \mathbb{R}^2$ that makes $\boldsymbol{f}$ onto.*

30. *We define the complex tangent function by $\tan z \equiv \frac{\sin z}{\cos z}$. The other complex trigonometric functions are defined in terms of the sine and cosine in a similar manner. Find the domain and range of the following: $\tan z$, $\cot z$, $\sec z$, and $\csc z$. Also, show that $1 + \tan^2 z = \sec^2 z$ and that $1 + \cot^2 z = \csc^2 z$.*

31. *Define $\sin^{-1} z$ in the natural way by letting $w = \sin^{-1} z$ if and only if $z = \sin w = \frac{e^{iw} - e^{-iw}}{2i}$ and solving for w. Show that $\sin\left(\sin^{-1} z\right) = z$. However, $\sin^{-1}(\sin z)$ is not just equal to z. Why?*

32. *Define $\cos^{-1} z$, $\tan^{-1} z$, $\cot^{-1} z$, $\sec^{-1} z$, and $\csc^{-1} z$ in the natural way.*

33. *Define the complex hyperbolic sine and cosine functions $\sinh : \mathbb{C} \to \mathbb{C}$ and $\cosh : \mathbb{C} \to \mathbb{C}$ according to*

$$\sinh z \equiv \frac{e^z - e^{-z}}{2}, \quad \text{and} \quad \cosh z \equiv \frac{e^z + e^{-z}}{2}.$$

Show that $\cosh^2 z - \sinh^2 z = 1$ for all $z \in \mathbb{C}$.

34. *Define $\sinh^{-1} z$ and $\cosh^{-1} z$ in the natural way.*

35. *Show that the function $f : \mathbb{C} \setminus \{1\} \to \mathbb{C} \setminus \{1\}$ given by $f(z) = \frac{z+1}{z-1}$ in Example 2.22 on page 147 is, in fact, a one-to-one correspondence.*

36. *In Example 2.22 on page 147, we saw that the function in the previous exercise is its own inverse. Can you think of another function that is its own inverse?*

37. *Is the function $f : \mathbb{C} \to \mathbb{C}$ given by $f(z) = e^{-z^2}$ bounded on $\mathbb{C}$? (No.)*

38. *Let $\boldsymbol{f} : \mathbb{R}^2 \setminus \{\mathbf{0}\} \to \mathbb{R}^2$ be given by $\boldsymbol{f}(\mathbf{x}) = \left(\frac{x_1^2}{|\mathbf{x}|}, \frac{x_2^2}{|\mathbf{x}|}\right)$ for any $\mathbf{x} = (x_1, x_2) \in \mathbb{R}^2$, and let $\mathbf{x_n} = \left(\frac{1}{n^2+1}, \frac{1}{n}\right)$ for $n \geq 1$. Show that $\lim_{n\to\infty} \boldsymbol{f}(\mathbf{x_n}) = \mathbf{0}$.*

39. *Suppose $f : \mathbb{R} \setminus \{-1\} \to \mathbb{R}$ is given by $f(x) = \frac{x^3+1}{x+1}$. Prove that $\lim_{x\to-1} \frac{x^3+1}{x+1} = 3$.*

40. *Suppose $\boldsymbol{f} : \mathbb{R} \to \mathbb{R}^2$ is given by $\boldsymbol{f}(x) = \left(x^2, \frac{x+1}{x-1}\right)$. Prove that $\lim_{x\to 2} \boldsymbol{f}(x) = (4, 3)$.*

41. *Suppose $\boldsymbol{f} : \mathbb{R} \setminus \{0, -1\} \to \mathbb{R}^3$ is given by $\boldsymbol{f}(x) = \left(1, \frac{1}{x}, \frac{x^2+1}{x+1}\right)$. Prove that $\lim_{x\to -1} \boldsymbol{f}(x) = (1, -1, -2)$.*

42. *Rework the last two exercises using Proposition 4.13 on page 161.*

43. *A function $f : \mathbb{R} \to \mathbb{R}$ is called* even *if $f(-x) = f(x)$ for all x. What is the relation between the left-hand and right-hand limits of an even function at 0? A function $f : \mathbb{R} \to \mathbb{R}$ is called* odd *if $f(-x) = -f(x)$ for all x. Answer the same question for odd functions.*

44. *Prove that any function $f : \mathbb{R} \to \mathbb{R}$ is the sum of an even and an odd function. (See the previous exercise for the definitions.)*

45. *Suppose $P : \mathbb{C} \to \mathbb{C}$ is a polynomial and $P(\sin z) \equiv 0$. Prove that P is the zero polynomial. Prove a similar result for $\cos z$ and e^z.*

46. *Let $\{a_n\}$ for $n \geq 0$ be a sequence in $\mathbb{X}$. Define $f : [0, \infty) \to \mathbb{X}$ by the formula: $f(x) \equiv a_n$, if $n \leq x < n+1$. What is $\lim_{x\to\infty} f(x)$ if it exists? When does it exist?*

47. *Define $f : \mathbb{R} \to \mathbb{R}$ such that $f(x) = 1$ if $x \in \mathbb{Q}$, and $f(x) = 0$ otherwise. Describe the limits of f.*

48. *Recall that $f : \mathbb{C} \setminus \{-1, 1\} \to \mathbb{C}$ given by $f(z) = \frac{z^2+1}{z^2-1}$ has the limit $\lim_{z\to 0} f(z) = -1$. Find a deleted neighborhood $N_r'(0)$ such that f is bounded on $N_r'(0) \cap D$ where $D \equiv \mathbb{C} \setminus \{-1, 1\}$.*

49. *In Example 4.8 on page 159 we showed that $\lim_{\mathbf{x}\to\mathbf{0}} \boldsymbol{f}(\mathbf{x}) = \mathbf{0}$. Find a deleted neighborhood of $\mathbf{0}$ where $\boldsymbol{f}$ is bounded.*

50. *Suppose $\boldsymbol{f} : \mathbb{R}\setminus\{0\} \to \mathbb{R}^2$ is given by $\boldsymbol{f}(x) = \left(\frac{x^3-2x}{x}, x^2\right)$. Prove that $\lim_{x\to 0} \boldsymbol{f}(x) = (-2, 0)$.*

51. *Suppose $\boldsymbol{f} : \mathbb{R} \setminus \{-1, 0\} \to \mathbb{R}^3$ is given by $\boldsymbol{f}(x) = \left(\frac{x^2-1}{x+1}, \frac{x^2+x}{x+1}, \frac{1}{x}\right)$. Prove that $\lim_{x\to -1} \boldsymbol{f}(x) = (-2, -1, -1)$.*

52. *Consider $f : \mathbb{C}\setminus\{-1\} \to \mathbb{C}$ given by $f(z) = \frac{z^2-1}{z+1}$. Use Proposition 4.15 on page 163 to find the limit $\lim_{z\to -1} f(z)$.*

53. *Use Proposition 4.15 to show that, for $n \in \mathbb{N}$, $\lim_{x\to 1} \frac{x^n-1}{x-1} = n$.*

54. *What are the domains and codomains of the functions in a) through e) of Proposition 4.17 on page 163?*

55. *Consider the function given by $f(z) = \frac{z^2+1}{z^2-1}$ on $\mathbb{D} = \mathbb{C} \setminus \{-1, 1\}$ from Example 4.3 on page 157. Determine the limit $\lim_{z\to 0} f(z)$ by applying Proposition 4.18 on page 165, and compare it to the example.*

56. *Determine the following limits if they exist.*

a) $\lim_{x\to 0} \frac{1-\cos x}{x}$ *b)* $\lim_{z\to 0} \frac{1-\cos z}{z}$ *c)* $\lim_{z\to 0} \frac{\sin z}{z}$

57. *Suppose* $f : \mathbb{R}\setminus\{1\} \to \mathbb{R}$ *is given by* $f(x) = \begin{cases} 3\,x & -\infty < x < 1 \\ 2\,x-4 & 1 \le x < \infty \end{cases}$. *Show that* $\lim_{x\to 1^-} f(x) = 3$ *and* $\lim_{x\to 1^+} f(x) = -2$.

58. *Consider* $f : \mathbb{R}\setminus\{0\} \to \mathbb{R}$ *be given by* $f(x) = x\sin\left(\frac{1}{x}\right)$. *Find*

a) $\lim_{x\to 0^-} f(x)$ *b)* $\lim_{x\to 0^+} f(x)$

59. *Let* $f : \mathbb{R} \to \mathbb{R}$ *be given by* $f(x) = \begin{cases} \frac{1}{2} & \text{if } x \in (-\infty, 0), \\ 0 & \text{if } x \in [0,\infty)\cap\mathbb{Q}, \\ 1 & \text{if } x \in (0,\infty)\cap\mathbb{I}. \end{cases}$

Find: a) $\lim_{x\to 0^-} f(x)$ *b)* $\lim_{x\to 0^+} f(x)$

60. *Evaluate or verify the following limits.*

a) $\lim\limits_{x\to\infty} \frac{\sin x}{x}$ *b)* $\lim\limits_{x\to -\infty} \frac{\sin x}{x}$ *c)* $\lim\limits_{x\to\infty} x\,\sin x$

d) $\lim\limits_{x\to\infty} \frac{x\,\cos x}{x+1}$ *e)* $\lim\limits_{x\to\infty} \left(x^2+x+1\right) = \infty$ *f)* $\lim\limits_{x\to -\infty} \left(x - x^2\right) = -\infty$.

61. *Using the real exponential function and its inverse, the natural logarithm function, define* y^x *for any pair of real numbers,* $x \in \mathbb{R}$ *and* $y > 0$. *Extend this definition appropriately for the case* $y < 0$.

62. *Definition of* z^w.

Recall that for $x, y \in \mathbb{R}$ *with* $x > 0$, *we have that* $x^y = e^{\ln(x^y)} = e^{y\ln x}$. *With our new definitions of* e^z *and* $\log z$, *we may extend this idea as follows. Suppose* $w \in \mathbb{C}$ *is fixed and that* $z \in \mathbb{C}$ *is nonzero. Then we define*

$$z^w \equiv e^{w\log z}. \tag{4.9}$$

Note that since $\log z$ *is multiple valued,* z^w *is also multiple valued. To illustrate this fact, we determine the values of* i^i. *Using (4.9), we have*

$$i^i = e^{i\log i} = e^{i\left[\ln 1 + i\left(\frac{\pi}{2}+2\pi k\right)\right]} = e^{-\left(\frac{\pi}{2}+2\pi k\right)}, \quad \text{where } k \in \mathbb{Z}.$$

For parts a) through g) below, determine the values of z^w:

a) i^{-i} *b)* $i^{1/4}$ *c)* $(-i)^{1/3}$ *d)* 3^{1+i} *e)* $(\sqrt{2})^{2-i}$ *f)* $(\sqrt{2}-i)^{\frac{2}{3}+2i}$ *g)* $5^{\sqrt{2}-i}$

h) Recall that we already have a single-valued definition for z^n *where* $n \in \mathbb{Z}$ ($z \ne 0$ *if* $n < 0$). *Does the new definition given by (4.9) above agree with the old one when* $n \in \mathbb{Z}$ *even though the new definition is multiple valued?*

i) How many values does $z^{p/q}$ *have where* $z \ne 0$ *and* p/q *is a rational number in lowest terms?*

j) How many values does z^π *have where* $z \ne 0$?

k) Relate your findings above with our discussion of the n*th roots of unity given in Chapter 1.*

l) Consider our definition of e^z given earlier in this chapter. Verify that it is consistent with the definition of z^w given here (that is, though we didn't define it this way, e^z can be interpreted as a complex number raised to another complex number, just as z^w is). Also verify that e^z is single-valued via (4.9).

m) Just as in the case of the complex logarithm, we may define the principal value of z^w as

$$z^w \equiv e^{w\,\mathrm{Log}z}. \tag{4.10}$$

According to (4.10), the principal value of i^i is $e^{-\pi/2}$. Assuming $z \neq 0$, does the principal value of z^w obey the traditional rules of exponents given by $z^{w_1}z^{w_2} = z^{w_1+w_2}$ and $\frac{z^{w_1}}{z^{w_2}} = z^{w_1-w_2}$?

n) Define the function $F : \mathbb{C} \to \mathbb{C}$ by

$$F(z) = \begin{cases} e^{\frac{1}{2}\mathrm{Log}z} & \text{if } z \neq 0, \\ 0 & \text{if } z = 0. \end{cases}$$

Does F agree with our definition of the complex square root function on page 149?

o) Define the complex nth root function in the obvious way.

5

FUNCTIONS: CONTINUITY AND CONVERGENCE

The continuous function is the only workable and usable function. It alone is subject to law and the laws of calculation. It is a loyal subject of the mathematical kingdom. Other so-called or miscalled functions are freaks, anarchists, disturbers of the peace, malformed curiosities which one and all are of no use to anyone, least of all to the loyal and burden-bearing subjects who by keeping the laws maintain the kingdom and make its advance possible.

E.D. Roe, Jr.

In this chapter, we formally define what it means for a function to be *continuous* at a point or on a set. As we will see, continuous functions have many nice properties. We will also consider a special type of continuity called *uniform continuity*. Functions that are uniformly continuous are even nicer in certain ways than functions that are only continuous. We will also apply the limit concept to a sequence of functions, and in this way effectively generalize the notion of sequence, and that of convergence, to "points" that are functions in a function space. Within this context, we take special note of a type of convergence called *uniform convergence*, and determine the particular consequences to which this type of convergence can lead. Finally, we will consider series of functions. While the idea of a series of functions is analogous to that of a series of numbers or a series of vectors, there are some subtleties to explore. Later, series of functions will play a more prominent role in our development of analysis.

1 CONTINUITY

1.1 Definitions

We begin by defining what it means for a function to be continuous at a point.

Definition 1.1 Consider the function $f : \mathbb{D} \to \mathbb{Y}$.

1. We say that f is **continuous at the point** $x_0 \in \mathbb{D}$ if for any $\epsilon > 0$, there exists a $\delta > 0$ such that
$$x \in \mathbb{D} \text{ and } |x - x_0| < \delta \;\Rightarrow\; |f(x) - f(x_0)| < \epsilon.$$
2. If f is continuous at every point $x_0 \in S \subset \mathbb{D}$, we say that f is **continuous on** S.

We follow Definition 1.1 with a several important remarks.

1. The δ that makes the definition "work" depends, in general, on both ϵ and on x_0. A different choice of either ϵ or $x_0 \in \mathbb{D}$ might require a different δ. This is especially significant as it relates to part 2 of Definition 1.1, which involves the continuity of f on a set of points, not just at one particular point. We will explore this subtlety in our discussion of the concept of *uniform continuity* later in this section.
2. If $x_0 \in \mathbb{D}$ is also a limit point of $\mathbb{D}$, then our definition of continuity at x_0 reduces to $\lim_{x\to x_0} f(x) = f(x_0)$. In this case, the intuitive idea behind Definition 1.1 is fairly clear. It states that f is continuous at x_0 if when values of $x \in \mathbb{D}$ get arbitrarily close to x_0 the associated values of $f(x)$ get arbitrarily close to $f(x_0)$. In this case, one can interpret Definition 1.1 as $\lim_{x\to x_0} f(x) = f(\lim_{x\to x_0} x)$, i.e., the limit process can be taken "inside" the function f to its argument.
3. If $x_0 \in \mathbb{D}$ is an isolated point of $\mathbb{D}$, then our definition of continuity at x_0 holds *vacuously*. In this case, even the relevance of Definition 1.1 may not seem clear. In fact, however, when $x_0 \in \mathbb{D}$ is an isolated point of $\mathbb{D}$ there exists a deleted neighborhood $N'_\delta(x_0)$ such that $N'_\delta(x_0) \cap \mathbb{D} = \varnothing$, and so there are no points $x \in \mathbb{D}$ available to contradict the definition of continuity. For this reason, if x_0 is an isolated point of its domain, f is continuous there. This seemingly strange conclusion may take some getting used to.
4. If $f : \mathbb{D} \to \mathbb{Y}$ is continuous throughout a set $S \subset \mathbb{D}$, as described in part 2 of the definition, it is *not* true in general that f maps convergent sequences in S to convergent sequences in $\mathbb{Y}$. That is, f being continuous on S is not enough to imply that f preserves convergence on S. We refer the reader to the exercises that follow for evidence of this. We will discover that this convergence-preserving property is only possessed by functions that are *uniformly continuous*.

► **5.1** *Prove remark 2 above.*

► **5.2** *As a special case of remark 2, consider the following situation. Suppose $f : \mathbb{D} \to \mathbb{Y}$ is continuous at $x_0 \in \mathbb{D}$ where x_0 is a limit point of $\mathbb{D}$. If $\{x_n\} \subset \mathbb{D}$ is a sequence that converges to x_0, then show that $\lim_{n\to\infty} f(x_n) = f(\lim_{n\to\infty} x_n) = f(x_0)$.*

► **5.3** *What is wrong with the following statement: "Loosely speaking, Definition 1.1 says that the function f is continuous at the point x_0 if, the closer x is to x_0, the closer $f(x)$ is to $f(x_0)$." In light of this statement, consider the function $f : \mathbb{R} \to \mathbb{R}$ given by*

$$f(x) = \begin{cases} |x| & \textit{if } x \neq 0 \\ -1 & \textit{if } x = 0 \end{cases}$$

near $x = 0$. Does the statement apply at $x_0 = 0$? Is the function continuous at $x_0 = 0$?

1.2 Examples of Continuity

We now consider some examples, the first of which establishes that linear real-valued functions of a single real variable are continuous at any point $x \in \mathbb{R}$. In fact, later in this section we will easily establish that all polynomials $p : \mathbb{R} \to \mathbb{R}$ are continuous on $\mathbb{R}$, as are all complex polynomials $p : \mathbb{C} \to \mathbb{C}$ on $\mathbb{C}$, and all real-valued polynomials of several real variables $p : \mathbb{R}^k \to \mathbb{R}$ on $\mathbb{R}^k$.

Example 1.2 *Let $f : \mathbb{R} \to \mathbb{R}$ be given by $f(x) = ax + b$ where $a, b \in \mathbb{R}$. Then, for any $x_0 \in \mathbb{R}$,*

$$|f(x) - f(x_0)| = |(ax + b) - (ax_0 + b)| = |a(x - x_0)| = |a|\,|x - x_0|.$$

If $a \neq 0$, the above yields that for any $\epsilon > 0$ and $\delta = \frac{\epsilon}{|a|}$,

$$|x - x_0| < \delta \Rightarrow |f(x) - f(x_0)| < \epsilon.$$

Since $x_0 \in \mathbb{R}$ was arbitrary, we have shown that f is continuous on $\mathbb{R}$ in the case $a \neq 0$. ◄

► **5.4** *Show that f in the above example is continuous in the case $a = 0$.*

► **5.5** *Let $f : \mathbb{C} \to \mathbb{C}$ be given by $f(z) = az + b$ where $a, b \in \mathbb{C}$. Show that f is continuous on $\mathbb{C}$.*

► **5.6** *Let $\boldsymbol{f} : \mathbb{R}^k \to \mathbb{R}^k$ be given by $\boldsymbol{f}(\mathbf{x}) = a\mathbf{x} + \mathbf{b}$ where $a \in \mathbb{R}$, and $\mathbf{x} \in \mathbb{R}^k$. Show that $\boldsymbol{f}$ is continuous on $\mathbb{R}^k$.*

The next two examples are examples of *rational functions*, that is, ratios of polynomials, $\frac{p(x)}{q(x)}$. Such functions are continuous everywhere except at zeros of the polynomial q (in fact, these zeros are not even in the rational function's domain). We will see in a later section of this chapter that in some cases we can "extend" such functions to be continuously defined at such points.

Example 1.3 *Let $f : (0, \infty) \to \mathbb{R}$ be given by $f(x) = \frac{1}{x}$. We will show that f is continuous on its whole domain. To this end, consider any $x_0 \in (0, \infty)$, and note that for $x \in (0, \infty)$,*

$$|f(x) - f(x_0)| = \left|\frac{1}{x} - \frac{1}{x_0}\right| = \frac{|x - x_0|}{|x_0|\,|x|}.$$

If we assume that $|x - x_0| < \frac{1}{2}\,x_0$, *then* $|x| \geq \frac{1}{2}\,x_0$, *and therefore*

$$|f(x) - f(x_0)| \leq \tfrac{2}{x_0^2}\,|x - x_0| < \epsilon,$$

as long as $|x - x_0| < \frac{\epsilon\, x_0^2}{2}$. *That is, for* $\delta = \min\left(\frac{1}{2}x_0, \frac{\epsilon\, x_0^2}{2}\right)$, *we have that for* $x \in (0, \infty)$,

$$|x - x_0| < \delta \;\Rightarrow\; |f(x) - f(x_0)| < \epsilon.$$

◀

▶ **5.7** *Show in the above example that it is impossible to choose* δ *to be independent of* x_0. *(Hint: For given* $\epsilon > 0$, *consider* $x_0 = \frac{1}{n\,\epsilon}$ *and* $x = \frac{1}{(n+1)\,\epsilon}$ *for sufficiently large* $n \in \mathbb{N}$.)

▶ **5.8** *Consider* $f : (0, \infty) \to \mathbb{R}$ *given by* $f(x) = \frac{1}{x}$. *Use this example to show that, as claimed in remark 4 after Definition 1.1, not all continuous functions preserve convergence.*

In the following example, we apply Definition 1.1 to a function of more than one variable.

Example 1.4 *Let* $D^2 = \{\mathbf{x} = (x, y) \in \mathbb{R}^2 : x > 0,\, y > 0\}$ *and suppose* $f : D^2 \to \mathbb{R}$ *is given by*

$$f(\mathbf{x}) = \frac{x}{x + y}.$$

We will show that f *is continuous at the point* $(2, 1)$, *an interior point (and therefore a limit point) of its domain. To this end, let* $\mathbf{x} = (x, y) \in D^2$. *Then*

$$\begin{aligned}
|f(\mathbf{x}) - f(2,1)| &= \left|\frac{x}{x+y} - \frac{2}{3}\right| \\
&= \frac{|(x-2) + 2\,(1-y)|}{3\,(x+y)} \\
&\leq \frac{|x-2|}{3\,(x+y)} + \frac{2\,|y-1|}{3\,(x+y)} \\
&\leq \frac{|\mathbf{x} - (2,1)|}{3\,(x+y)} + \frac{2\,|\mathbf{x} - (2,1)|}{3\,(x+y)} \\
&= \frac{|\mathbf{x} - (2,1)|}{(x+y)}.
\end{aligned}$$

If we suppose for the moment that $|\mathbf{x} - (2,1)| < \frac{1}{2}$, *then* $|x - 2| < \frac{1}{2}$ *and* $|y - 1| < \frac{1}{2}$, *and so* $x + y > 2$ *(Why?), and* $\frac{1}{x+y} < \frac{1}{2}$. *This yields*

$$|f(\mathbf{x}) - f(2,1)| < \tfrac{1}{2}\,|\mathbf{x} - (2,1)|,$$

and the right side of the above inequality will be less than any given $\epsilon > 0$ *as long as* $|\mathbf{x} - (2,1)| < \min\left(2\,\epsilon, \frac{1}{2}\right) \equiv \delta$. *That is, overall we have shown that for any given* $\epsilon > 0$ *and* $\mathbf{x} \in D^2$,

$$|\mathbf{x} - (2,1)| < \delta \;\Rightarrow\; |f(\mathbf{x}) - f(2,1)| < \epsilon.$$

Therefore, f *is continuous at* $(2, 1)$. ◀

► **5.9** *Show that the function $f(\mathbf{x}) = \frac{x}{x+y}$ in the last example is continuous at every point in its domain.*

Example 1.5 *Consider $f : \mathbb{C} \setminus \{1\} \to \mathbb{C}$ given by $f(z) = z + \frac{1}{z-1}$, and let $z_n = \left(\frac{1+i}{3}\right)^n$ for $n \in \mathbb{N}$. We will use the result of Exercise 5.2 to show that $\lim_{n\to\infty} f(z_n) = -1$. To do this, we will first show that f is continuous at every point in its domain. Consider an arbitrary $z_0 \in \mathbb{C} \setminus \{1\}$ and note that for any other $z \in \mathbb{C} \setminus \{1\}$,*

$$\begin{aligned} |f(z) - f(z_0)| &= \left| z - z_0 + \frac{1}{z-1} - \frac{1}{z_0 - 1} \right| \\ &= \left| z - z_0 + \frac{z_0 - z}{(z-1)(z_0-1)} \right| \\ &\le |z - z_0| + \frac{|z - z_0|}{|z-1|\,|z_0-1|}. \end{aligned} \tag{5.1}$$

Now, consider those points $z \in \mathbb{C}$ that satisfy $|z - z_0| < \frac{1}{2}|z_0 - 1|$. For such z,

$$|z-1| = |z - z_0 + z_0 - 1| \ge |z_0 - 1| - |z - z_0| > \tfrac{1}{2}|z_0 - 1|,$$

and therefore (5.1) becomes

$$\begin{aligned} |f(z) - f(z_0)| &\le |z - z_0| + \frac{|z - z_0|}{|z-1|\,|z_0-1|} \\ &< |z - z_0| + \frac{2\,|z - z_0|}{|z_0 - 1|^2} \\ &= \left(1 + \frac{2}{|z_0 - 1|^2}\right) |z - z_0|. \end{aligned}$$

This last expression will be less than ϵ provided $z \in D$ and $|z - z_0| < \delta$, where

$$\delta \equiv \min\left(\tfrac{1}{2}|z_0 - 1|,\ \epsilon\left(1 + \tfrac{2}{|z_0-1|^2}\right)^{-1}\right).$$

This establishes that f is continuous at every $z_0 \in \mathbb{C} \setminus \{1\}$. We now will show that $\lim_{n\to\infty} z_n = 0$. To see this, note that

$$|z_n - 0| = \left|\left(\tfrac{1+i}{3}\right)^n\right| = \left|\tfrac{1+i}{3}\right|^n = \left(\tfrac{\sqrt{2}}{3}\right)^n,$$

and the above clearly goes to zero as n goes to ∞. Finally, by remark 2 following Definition 1.1, we have

$$\lim_{n\to\infty} f(z_n) = f\left(\lim_{n\to\infty} z_n\right) = f(0) = -1.$$ ◄

Example 1.6 *Let $f : [0, \infty) \to \mathbb{R}$ be given by $f(x) = \sqrt{x}$. Then for $x_0 \in (0, \infty)$ and $x \in [0, \infty)$, we have that*

$$|f(x) - f(x_0)| = \left|\sqrt{x} - \sqrt{x_0}\right| = \left|\left(\sqrt{x} - \sqrt{x_0}\right) \frac{\sqrt{x} + \sqrt{x_0}}{\sqrt{x} + \sqrt{x_0}}\right|$$
$$= \frac{|x - x_0|}{\sqrt{x} + \sqrt{x_0}}$$
$$< \frac{|x - x_0|}{\sqrt{x_0}}.$$

The above yields that for any $\epsilon > 0$, taking $\delta = \epsilon\sqrt{x_0}$ obtains

$$|x - x_0| < \delta \Rightarrow \left|\sqrt{x} - \sqrt{x_0}\right| < \epsilon.$$

From this we see that f is continuous on $(0, \infty)$. ◀

▶ **5.10** *Show that the function $f : [0, \infty) \to \mathbb{R}$ given by $f(x) = \sqrt{x}$ is also continuous at $x_0 = 0$, and hence is continuous on its whole domain.*

▶ **5.11** *Show that if $f : \mathbb{D} \to \mathbb{R}$ is continuous at x_0 and if $f \geq 0$ on $\mathbb{D}$, then $\sqrt{f}$ is continuous at x_0.*

Example 1.7 *Let $f : \mathbb{R} \to \mathbb{R}$ be given by $f(x) = \sin x$. We will show that f is continuous on $\mathbb{R}$. To this end, fix $x_0 \in \mathbb{R}$ and recall from an exercise in the last chapter that $\lim_{x\to 0} \sin x = 0$, and so for $\epsilon > 0$ there exists $\delta > 0$ such that for $x \in \mathbb{R}$,*

$$\left|\tfrac{x-x_0}{2}\right| < \tfrac{\delta}{2} \Rightarrow \left|\sin\left(\tfrac{x-x_0}{2}\right)\right| < \tfrac{\epsilon}{2}. \tag{5.2}$$

We now make use of a trigonometric identity, namely,

$$\sin x - \sin x_0 = 2\sin\left(\tfrac{x-x_0}{2}\right)\cos\left(\tfrac{x+x_0}{2}\right), \tag{5.3}$$

to obtain

$$|\sin x - \sin x_0| = \left|2\sin\left(\tfrac{x-x_0}{2}\right)\cos\left(\tfrac{x+x_0}{2}\right)\right| \leq 2\left|\sin\left(\tfrac{x-x_0}{2}\right)\right|.$$

Applying (5.2) to the above, we have that

$$\left|\tfrac{x-x_0}{2}\right| < \tfrac{\delta}{2}, \text{ or } |x - x_0| < \delta \Rightarrow |\sin x - \sin x_0| \leq 2\left|\sin\left(\tfrac{x-x_0}{2}\right)\right| < \epsilon,$$

and so $\sin x$ is continuous on $\mathbb{R}$. ◀

▶ **5.12** *Establish the trigonometric identity given in (5.3).*

▶ **5.13** *Let $f : \mathbb{R} \to \mathbb{R}$ be given by $f(x) = \cos x$. Show that f is continuous on $\mathbb{R}$.*

Example 1.8 *Let $f : \mathbb{C}\setminus\{0\} \to \mathbb{R}$ be given by $f(z) = \mathrm{Arg}(z)$. We will establish that f is continuous on $\mathbb{C} \setminus K$ where $K = \{z \in \mathbb{C} : \mathrm{Re}(z) \leq 0, \mathrm{Im}(z) = 0\}$. In fact, the function $\mathrm{Arg}(z)$ is discontinuous at each $z \in K$. The set $K \subset \mathbb{C}$ forms what is known as a "branch cut" associated with the Arg function. We will begin by showing that f is continuous for any $z_0 \in \mathbb{C} \setminus K$. To this end, suppose $z_0 \in \mathbb{C} \setminus K$. Let $\epsilon > 0$ and define the sector $S_{z_0} \subset \mathbb{C}$ by*

$$S_{z_0} \equiv \left\{z \in \mathbb{C} : |\mathrm{Arg}(z) - \mathrm{Arg}(z_0)| < \min\left(\pi - \mathrm{Arg}(z_0), \pi + \mathrm{Arg}(z_0), \epsilon\right)\right\}.$$

Note that this ensures that $S_{z_0} \subset \mathbb{C}\backslash K$. Figure 5.1 illustrates the case where z_0 lies in the second quadrant of the complex plane. The reader is encouraged to sketch the cases corresponding to z_0 lying in the other quadrants to see the resulting geometry associated with S_{z_0}. Now choose $\delta > 0$ such that

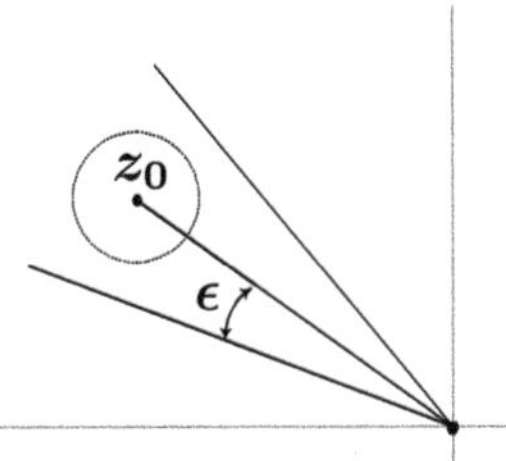

Figure 5.1 *Demonstration that* Arg(z) *is continuous.*

$N_\delta(z_0) \subset S_{z_0}$. It is easy to see that

$$z \in N_\delta(z_0) \ \Rightarrow \ z \in S_{z_0}, \ \text{i.e.,} \ |\operatorname{Arg}(z) - \operatorname{Arg}(z_0)| < \epsilon,$$

which in turn implies that $\operatorname{Arg} : \mathbb{C} \setminus K \to \mathbb{C}$ *is continuous on its domain. To see that $f(z) = \operatorname{Arg}(z)$ is not continuous at $z_0 = -1$ (for example), we assume the opposite and argue by contradiction. If f is continuous at $z_0 = -1$, then there exists $\delta > 0$ such that*

$$|z+1| < \delta \ \Rightarrow \ |\operatorname{Arg}(z) - \operatorname{Arg}(-1)| = |\operatorname{Arg}(z) - \pi| < \tfrac{\pi}{4}.$$

We may choose $N \in \mathbb{N}$ large enough so that $z_N \equiv e^{i(-\pi + \frac{1}{N})}$ is within δ of $z_0 = -1$, that is $|z_N + 1| < \delta$. (Why?) This yields

$$|\operatorname{Arg}(z_N) - \pi| = |(-\pi + \tfrac{1}{N}) - \pi| = |2\pi - \tfrac{1}{N}| < \tfrac{\pi}{4},$$

a contradiction. The proof that Arg(z) *is discontinuous on the rest of K is left to the reader.* ◀

▶ **5.14** *Complete the proof of the claim in the above example, and show that* Arg(z) *is discontinuous on all of K.*

▶ **5.15** *Why does the proof given in the above example for the continuity of* Arg(z) *on $\mathbb{C} \setminus K$ not work for $z_0 \in K$?*

▶ **5.16** *Let $f : \mathbb{C} \to \mathbb{C}$ be given by $f(z) = \sqrt{z}$ as given in Definition 3.1 on page 149. Show that f is not continuous on all of $\mathbb{C}$. Where is this function continuous?*

▶ **5.17** *Let $f : \mathbb{C} \setminus \{0\} \to \mathbb{C}$ be given by $f(z) = \operatorname{Log}(z)$ as given in Definition 3.6 on page 153. Show that f is not continuous on all of $\mathbb{C} \setminus \{0\}$. Where is this function continuous?*

▶ **5.18** *Consider the function $\boldsymbol{f} : \mathbb{R} \setminus \{0\} \to \mathbb{R}^2$ given by $\boldsymbol{f}(x) = \left(\frac{1}{x}, 2\,x^2\right)$. Show that $\boldsymbol{f}$ is continuous at every point in $\mathbb{R} \setminus \{0\}$.*

▶ **5.19** *Consider the function from the previous exercise, and let* $x_n = \frac{2n^2}{n^2+1}$ *for* $n \in \mathbb{N}$. *Show that* $\lim_{n\to\infty} f(x_n) = \left(\frac{1}{2}, 8\right)$.

▶ **5.20** *Show that the complex function* $f : \mathbb{C} \to \mathbb{C}$ *given by* $f(z) = a\,z^2 + b\,z + c$ *for* $a, b, c \in \mathbb{C}$ *is continuous on* $\mathbb{C}$.

▶ **5.21** *Prove that the real exponential function* $\exp : \mathbb{R} \to \mathbb{R}$ *given by* $\exp(x) = e^x$ *is continuous on its domain.*

1.3 Algebraic Properties of Continuous Functions

In this subsection we describe the many convenient properties possessed by continuous functions in various settings. We begin with the algebraic properties that allow for simple constructions of more elaborate continuous functions from more basic ones.

Proposition 1.9 *Consider* f *and* g *where both are functions from* $\mathbb{D}$ *to codomain* $\mathbb{Y}$, *and both* f *and* g *are continuous at* $x_0 \in \mathbb{D}$. *Then*

a) $f \pm y_0$ *is continuous at* x_0 *for any* $y_0 \in \mathbb{Y}$.

b) $f \pm g$ *is continuous at* x_0.

c) $c\,f$ *is continuous at* x_0 *for* $c \in \mathbb{R}$ *or* $\mathbb{C}$.

d) $f \cdot g$ *is continuous at* x_0.

e) f/g *is continuous at* x_0 *if* f *and* g *are real or complex valued and* g *is nonzero at* x_0.

f) $|f|$ *is continuous at* x_0.

Note that in property *c)*, the constant c should be understood as complex only in the case where f is complex valued. Note also that in property *d)*, the multiplication indicated is the dot product in the case $\mathbb{Y} = \mathbb{R}^p$, and the usual multiplication in $\mathbb{R}$ or $\mathbb{C}$ otherwise.

▶ **5.22** *Prove the previous proposition.*

▶ **5.23** *Does property d) in the above proposition still hold if* f *has codomain* $\mathbb{R}$ *while* g *has codomain* $\mathbb{R}^p$?

▶ **5.24** *In property e), is it necessary that* f *be real or complex valued? Explain.*

▶ **5.25** *Prove that the following functions* $f : \mathbb{R} \to \mathbb{R}$ *are continuous on* $\mathbb{R}$.

a) $f(x) = c$ *for* $c \in \mathbb{R}$ *b)* $f(x) = x$ *c)* $f(x) = x^n$ *for* $n \in \mathbb{Z}^+$

d) $f(x) = c_0 + c_1\,x + c_2\,x^2 + \cdots + c_n\,x^n$ *for* $c_j \in \mathbb{R}$ *and* $n \in \mathbb{N}$ *with* $c_n \neq 0$.

This establishes that all real polynomials are continuous on $\mathbb{R}$.

▶ **5.26** *Generalize the conclusion of the previous two exercises to the complex function case. That is, show that complex polynomials are continuous on* $\mathbb{C}$ *and complex rational functions are continuous on their domains.*

► **5.27** *Prove that the following functions $f : \mathbb{R}^2 \to \mathbb{R}$ are continuous on $\mathbb{R}^2$.*

a) $f(x,y) = c$ for $c \in \mathbb{R}$ *b)* $f(x,y) = x$ *c)* $f(x,y) = y$

d) $f(x,y) = x^n$ for $n \in \mathbb{Z}^+$ *e)* $f(x,y) = y^n$ for $n \in \mathbb{Z}^+$

f) $f(x,y) = c_{ij}\, x^i\, y^j$ for $i,j \in \mathbb{Z}^+$ and $c_{ij} \in \mathbb{R}$

g) $f(x,y) = \sum_{1 \le i,j \le m} c_{ij}\, x^i\, y^j$

This establishes that real-valued polynomials in two real variables are continuous on all of $\mathbb{R}^2$.

► **5.28** *Generalize the previous exercise to the case of $f : \mathbb{R}^k \to \mathbb{R}$.*

► **5.29** *Let $f : D^2 \to \mathbb{R}$ be given by $f(x,y) = \frac{p(x,y)}{q(x,y)}$ where $p : \mathbb{R}^2 \to \mathbb{R}$ and $q : \mathbb{R}^2 \to \mathbb{R}$ are of the form described in part g) of the previous exercise, and $D^2 = \{(x,y) \in \mathbb{R}^2 : q(x,y) \neq 0\}$. Show that f is continuous on D^2. This establishes that real-valued rational functions of two real variables are continuous on all of $\mathbb{R}^2 \setminus Z_q$ where Z_q is the set of zeros of $q(x,y)$.*

► **5.30** *Generalize the previous exercise to the case of the rational function $f : D^k \to \mathbb{R}$ given by $f(x_1, \dots, x_k) = \frac{p(x_1,\dots,x_k)}{q(x_1,\dots,x_k)}$ defined on D^k given by $D^k = \{(x_1, \dots, x_k) \in \mathbb{R}^k : q(x_1, \dots, x_k) \neq 0\}$.*

► **5.31** *Consider $\boldsymbol{f} : \mathbb{D} \to \mathbb{R}^p$ where $\boldsymbol{f}(x) = (f_1(x), f_2(x), \dots, f_p(x))$ for each $x \in \mathbb{D}$, and $f_j : \mathbb{D} \to \mathbb{R}$ for $1 \le j \le p$. Show that $\boldsymbol{f}$ is continuous at $x_0 \in \mathbb{D}$ if and only if f_j is continuous at $x_0 \in \mathbb{D}$ for each $1 \le j \le p$. What does this say about the complex function case $f : D \to \mathbb{C}$ with $z_0 \in D$?*

Example 1.10 *Let $f : \mathbb{C} \to \mathbb{C}$ be given by $f(z) = \sin z$. We will establish that f is continuous on $\mathbb{C}$. To this end, note that*

$$f(z) = \sin z = \left(\tfrac{e^y + e^{-y}}{2}\right) \sin x + i \left(\tfrac{e^{-y} - e^y}{2}\right) \cos x. \quad \textit{(Why?)}$$

Since $\mathrm{Re}\,(f(z))$ *and* $\mathrm{Im}\,(f(z))$ *are continuous functions of (x,y) (Why?), it follows that $f(z)$ is continuous in z.* ◄

► **5.32** *Answer the two (Why?) questions in the previous example.*

► **5.33** *Consider $f : \mathbb{C} \to \mathbb{C}$ given by $f(z) = \cos z$. Show that f is continuous on $\mathbb{C}$.*

As described in an earlier chapter, each of $\mathbb{Y}_1$ and $\mathbb{Y}_2$ in the statement of the following proposition should be understood to be any of the spaces $\mathbb{R}$, $\mathbb{R}^p$, or $\mathbb{C}$. The proposition basically states that a continuous function of a continuous function is a continuous function.

Proposition 1.11 *Let f and g be functions such that $f : \mathbb{D} \to \mathbb{Y}_1$ has range given by $\mathcal{R}_f = f(\mathbb{D})$, and $g : \mathcal{R}_f \to \mathbb{Y}_2$. Suppose also that f is continuous at $x_0 \in \mathbb{D}$, and g is continuous at $y_0 = f(x_0) \in \mathcal{R}_f$. Then the composite function $g \circ f : \mathbb{D} \to \mathbb{Y}_2$ is continuous at $x_0 \in \mathbb{D}$.*

PROOF Suppose $\epsilon > 0$ is given. Then there exists $\eta > 0$ such that

$$y \in \mathcal{R}_f \text{ and } |y - y_0| < \eta \Rightarrow |g(y) - g(y_0)| < \epsilon.$$

Similarly, for this same $\eta > 0$ there exists $\delta > 0$ such that

$$x \in \mathbb{D} \text{ and } |x - x_0| < \delta \Rightarrow |f(x) - f(x_0)| < \eta.$$

For $y = f(x)$ we have $|g(y) - g(y_0)| = |g(f(x)) - g(f(x_0))|$ and $|f(x) - f(x_0)| = |y - y_0|$. From this we see that for any given $\epsilon > 0$ there exists a $\delta > 0$ such that

$$x \in \mathbb{D} \text{ and } |x - x_0| < \delta \Rightarrow |g(f(x)) - g(f(x_0))| < \epsilon,$$

and the proposition is proved. ♦

Example 1.12 *Consider the function $h : \mathbb{R}^k \to \mathbb{R}$ given by $h(\mathbf{x}) = e^{-|\mathbf{x}|^2}$. Let $f : \mathbb{R}^k \to \mathbb{R}$ be given by $f(\mathbf{x}) = -|\mathbf{x}|^2$, and let $g : \mathcal{R}_f \to \mathbb{R}$ be given by $g(x) = e^x$. By a previous exercise we know that g is continuous on $\mathbb{R}$, and therefore is continuous on $\mathcal{R}_f$. The reader should verify that f is continuous on $\mathbb{R}^k$. By Proposition 1.11, it follows that $h = g \circ f$ is continuous on $\mathbb{R}^k$.* ◄

► **5.34** *Show that the function $f : \mathbb{R}^k \to \mathbb{R}$ given by $f(\mathbf{x}) = -|\mathbf{x}|^2$ is continuous on $\mathbb{R}^k$.*

The following proposition is an unsurprising consequence of the continuity of $f : D^k \to \mathbb{Y}$, a function of a vector. It basically states that if f is a continuous function of the vector $\mathbf{x}$, then f is continuous in each component of $\mathbf{x}$. One should be careful, however, in noting that the converse is not true in general.

Proposition 1.13 *Suppose $f : D^k \to \mathbb{Y}$ is continuous at $\boldsymbol{a} = (a_1, \ldots, a_k) \in D^k$. Let $D_j \equiv \left\{x_j \in \mathbb{R} : (a_1, \ldots, a_{j-1}, x_j, a_{j+1}, \ldots, a_k) \in D^k\right\}$, and define the function $f^{(j)} : D_j \to \mathbb{Y}$ by*

$$f^{(j)}(x_j) \equiv f(a_1, \ldots, a_{j-1}, x_j, a_{j+1}, \ldots, a_k) \quad \text{for } 1 \le j \le k.$$

Then $f^{(j)}$ is continuous at $x_j = a_j$.

PROOF Since f is continuous at $\boldsymbol{a}$, given any $\epsilon > 0$ there exists $\delta > 0$ such that

$$\mathbf{x} \in D^k \text{ and } |\mathbf{x} - \boldsymbol{a}| < \delta \Rightarrow |f(\mathbf{x}) - f(\boldsymbol{a})| < \epsilon.$$

Note that if $x_j \in D_j$ then $\mathbf{x_j} \equiv (a_1, \ldots, a_{j-1}, x_j, a_{j+1}, \ldots, a_k) \in D^k$, and that $f^{(j)}(x_j) = f(\mathbf{x_j})$. All of this yields

$$x_j \in D_j \text{ and } |x_j - a_j| < \delta \Rightarrow \mathbf{x_j} \in D^k \text{ and } |\mathbf{x_j} - \boldsymbol{a}| < \delta.$$

Therefore,

$$x_j \in D_j \text{ and } |x_j - a_j| < \delta \Rightarrow |f(\mathbf{x_j}) - f(\boldsymbol{a})| = |f^{(j)}(x_j) - f^{(j)}(a_j)| < \epsilon,$$

that is, $f^{(j)}$ is continuous at $x_j = a_j$. ♦

To illustrate that the converse of Proposition 1.13 is not true in general, we give the following example.

Example 1.14 *Consider* $f : \mathbb{R}^2 \to \mathbb{R}$ *given by*

$$f(x,y) = \begin{cases} \frac{2\,x\,y}{x^2+y^2} & \text{if } (x,y) \neq (0,0) \\ 0 & \text{if } (x,y) = (0,0) \end{cases}.$$

This function is not continuous at $(0,0)$*, since*

$$f(t,t) = \frac{2\,t^2}{t^2+t^2} = 1 \quad \text{if } t \neq 0,$$

and $f(0,0) = 0$*. However, if* $x = c$ *is fixed, then*

$$f^{(2)}(y) \equiv f(c,y) = \frac{2\,c\,y}{c^2+y^2}$$

is a continuous function of y *for each value of* c*. Similarly, if* $y = k$ *is fixed, then the corresponding function*

$$f^{(1)}(x) \equiv f(x,k) = \frac{2\,k\,x}{x^2+k^2}$$

is a continuous function of x *for each value of* k*.* ◀

The above example illustrates that a function of a vector may be continuous in each component of the vector separately, with the other components held fixed, but it may *not* be continuous as a function of the vector according to Definition 1.1. We sometimes refer to such a function as being *separately continuous in each variable*, but not *jointly continuous*.

1.4 Topological Properties and Characterizations

We now consider how the continuity of a function $f : \mathbb{D} \to \mathbb{Y}$ is related to the topology of the sets $\mathbb{D}$ and $\mathbb{Y}$. In particular, bounded sets, open sets, compact sets, and connected sets have special significance as they relate to continuous functions. Before we state our first result, recall from Chapter 2 that for $\mathbb{D} \subset \mathbb{X}$, a subset $U \subset \mathbb{D}$ is *open in* $\mathbb{D}$ if $U = \mathbb{D} \cap V$ for some open set $V \subset \mathbb{X}$. Also, U is said to be *closed in* $\mathbb{D}$ if $\mathbb{D} \setminus U$ is open in $\mathbb{D}$.

We begin by showing that if a function is continuous at a point x_0, then it must be bounded on a neighborhood of that point.

Proposition 1.15 *Suppose* $f : \mathbb{D} \to \mathbb{Y}$ *is continuous at* $x_0 \in \mathbb{D}$*. Then there exists* $\delta > 0$ *such that* f *is bounded on* $N_\delta(x_0) \cap \mathbb{D}$*.*

PROOF Since f is continuous at x_0, there exists a $\delta > 0$ such that if $x \in \mathbb{D}$ and $|x - x_0| < \delta$, i.e., if $x \in N_\delta(x_0) \cap \mathbb{D}$, then $|f(x) - f(x_0)| < 1$. Application of the triangle inequality then yields that

$$x \in N_\delta(x_0) \cap \mathbb{D} \;\Rightarrow\; |f(x)| < 1 + |f(x_0)|,$$

and the proposition is proved. ♦

► **5.35** *Does Proposition 1.15 simply follow from Proposition 4.11 on page 160?*

► **5.36** *Prove the following corollary to Proposition 1.15:* Suppose $f : \mathbb{D} \to \mathbb{Y}$ is continuous at $x_0 \in \mathbb{D}$ where x_0 is an interior point of $\mathbb{D}$. Then there exists a $\delta > 0$ such that f is bounded on $N_\delta(x_0)$.

Our next result states that a function f is continuous on its domain if and only if inverse images of open sets in its codomain are open in its domain. Stated more concisely, "f is continuous if and only if inverse images of open sets are open sets."

Theorem 1.16 *The function $f : \mathbb{D} \to \mathbb{Y}$ is continuous on $\mathbb{D}$ if and only if for any set U open in $\mathbb{Y}$ the set $f^{-1}(U)$ is open in $\mathbb{D}$.*

PROOF We prove the case where $\mathbb{Y}$ is all of $\mathbb{R}$, $\mathbb{R}^p$, or $\mathbb{C}$, and leave the case where $\mathbb{Y}$ is a proper subset of one of these spaces to the reader. Suppose $f : \mathbb{D} \to \mathbb{Y}$ is continuous on $\mathbb{D}$, and let U be an open subset of $\mathbb{Y}$. We must show that $f^{-1}(U)$ is open in $\mathbb{D}$, i.e., that $f^{-1}(U) = \mathbb{D} \cap V$ for some open set V in $\mathbb{X}$. To see this, consider an arbitrary $x_0 \in f^{-1}(U)$. Then $f(x_0) \in U$, and since U is open in $\mathbb{Y}$, there exists a neighborhood $N_r(f(x_0)) \subset U$. Since f is continuous at x_0, there exists a $\delta > 0$ such that

$$x \in \mathbb{D} \text{ and } |x - x_0| < \delta \Rightarrow |f(x) - f(x_0)| < r.$$

That is,

$$x \in \mathbb{D} \cap N_\delta(x_0) \Rightarrow f(x) \in N_r(f(x_0)) \subset U.$$

This implies that $x \in f^{-1}(U)$, and so

$$\mathbb{D} \cap N_\delta(x_0) \subset f^{-1}(U). \tag{5.4}$$

Note that every $x_0 \in f^{-1}(U)$ has a neighborhood $N_\delta(x_0)$ such that (5.4) is true, and note also that δ will in general depend on x_0. Finally, define V by $V \equiv \bigcup_{x_0 \in f^{-1}(U)} N_\delta(x_0)$. It then follows from (5.4) that $\mathbb{D} \cap V \subset f^{-1}(U)$. Since it is also true that any $x \in f^{-1}(U)$ must also be in $\mathbb{D} \cap V$ (Why?), we have that $f^{-1}(U) = \mathbb{D} \cap V$, and hence, $f^{-1}(U)$ is open in $\mathbb{D}$. For the converse, suppose that $f : \mathbb{D} \to \mathbb{Y}$ is such that $f^{-1}(U)$ is open in $\mathbb{D}$ for every open subset $U \subset \mathbb{Y}$. Fix $x_0 \in \mathbb{D}$, and consider an arbitrary $\epsilon > 0$. We must find $\delta > 0$ such that

$$x \in \mathbb{D} \text{ and } |x - x_0| < \delta \Rightarrow |f(x) - f(x_0)| < \epsilon,$$

that is,

$$x \in \mathbb{D} \cap N_\delta(x_0) \Rightarrow f(x) \in N_\epsilon(f(x_0)).$$

Note that $N_\epsilon(f(x_0))$ is open in $\mathbb{Y}$, and therefore $f^{-1}[N_\epsilon(f(x_0))] = \mathbb{D} \cap V$, where V is open in $\mathbb{X}$. Since $x_0 \in f^{-1}[N_\epsilon(f(x_0))]$, it follows that $x_0 \in \mathbb{D} \cap V$, and since V is open in $\mathbb{X}$ there exists a $\delta > 0$ such that $N_\delta(x_0) \subset V$. Finally, if $x \in \mathbb{D} \cap N_\delta(x_0)$, we have that $x \in \mathbb{D} \cap V$, and, since $\mathbb{D} \cap V = f^{-1}[N_\epsilon(f(x_0))]$, that $f(x) \in N_\epsilon(f(x_0))$. ♦

▶ **5.37** *Answer the (Why?) question in the above proof, and then prove the case of the above theorem where $\mathbb{Y}$ is a proper subset of $\mathbb{R}$, $\mathbb{R}^p$, or $\mathbb{C}$.*

▶ **5.38** *Note that the condition on U that it be "open in $\mathbb{Y}$" cannot be changed to "open and a subset of $\mathbb{Y}$." Trace through the example of the function $f : \mathbb{R} \to \mathbb{Y}$ given by $f(x) = \begin{cases} -1 & \text{if } x < 0 \\ 1 & \text{if } x \geq 0 \end{cases}$, with $\mathbb{Y} = \{-1, 1\}$ to see why. In this example, the conditions of the altered theorem would be satisfied, but the conclusion obviously doesn't hold for f.*

Example 1.17 *Suppose $f : \mathbb{R} \to \mathbb{R}$ has the property that $f^{-1}(I)$ is open for every open interval $I \subset \mathbb{R}$. We will show that f must be continuous on $\mathbb{R}$. To this end, let $B \subset \mathbb{R}$ be an open set. From Proposition 3.6 on page 57 we know that $B = \bigcup I_\alpha$ where $\{I_\alpha\}$ is a collection of open intervals in $\mathbb{R}$. Since $f^{-1}(B) = f^{-1}\left(\bigcup I_\alpha\right) = \bigcup f^{-1}(I_\alpha)$, and since $f^{-1}(I_\alpha)$ is open for each α, it follows that $f^{-1}(B)$ is also open. Hence, f is continuous.* ◀

▶ **5.39** *Does Example 1.17 generalize to functions $f : \mathbb{R}^k \to \mathbb{R}$? How about to functions $\boldsymbol{f} : \mathbb{R}^k \to \mathbb{R}^p$? How about to complex functions $f : \mathbb{C} \to \mathbb{C}$?*

▶ **5.40** *Show that $f : \mathbb{D} \to \mathbb{Y}$ is continuous on $\mathbb{D}$ if and only if $f^{-1}(B)$ is closed in $\mathbb{D}$ for every B closed in $\mathbb{Y}$.*

Yet another equivalence for the continuity of f on $\mathbb{D}$ is given by the following result. While its meaning may be unclear initially, it will be useful in proving the next result whose meaning is clearer.

Proposition 1.18 *The function $f : \mathbb{D} \to \mathbb{Y}$ is continuous on $\mathbb{D}$ if and only if $f\left(\overline{A}\right) \subset \overline{f(A)}$ for every subset $A \subset \mathbb{D}$.*

PROOF We consider the case where $\mathbb{D} = \mathbb{X}$ and $\mathbb{Y}$ is all of $\mathbb{R}$, $\mathbb{R}^p$, or $\mathbb{C}$, leaving the more general case to the reader. Suppose $f : \mathbb{D} \to \mathbb{Y}$ is such that $f\left(\overline{A}\right) \subset \overline{f(A)}$ for every subset $A \subset \mathbb{D}$. We will show that $f^{-1}(B)$ is closed in $\mathbb{D}$ for every B closed in $\mathbb{Y}$. By Exercise 5.40, this will establish the continuity of f on $\mathbb{D}$. To this end, suppose B is a closed subset of $\mathbb{Y}$ and let $A = f^{-1}(B)$. To see that A is closed in $\mathbb{D}$, let $x \in \overline{A}$. Then

$$f(x) \in f\left(\overline{A}\right) \subset \overline{f(A)} \subset \overline{B} = B,$$

and therefore $x \in f^{-1}(B) = A$. We have shown that $\overline{A} \subset A$, which implies that A is closed. Now assume $f : \mathbb{D} \to \mathbb{Y}$ is continuous on $\mathbb{D}$. Let $A \subset \mathbb{D}$ and $y_0 \in f(\overline{A})$. Then $y_0 = f(x_0)$ for $x_0 \in \overline{A}$, and since there exists a sequence $\{x_n\} \subset A$ such that $\lim x_n = x_0$, it follows that the sequence $\{f(x_n)\} \subset f(A)$ converges to $f(x_0)$, that is, $y_0 = f(x_0) \in \overline{f(A)}$. This completes the proof in the special case. The general case is left to the reader. ♦

► **5.41** *Generalize the argument given in the proof above to account for the cases where* $\mathbb{D} \subset \mathbb{X}$ *and* $\mathbb{Y} \subset \mathbb{R}$, $\mathbb{R}^p$, *or* $\mathbb{C}$.

► **5.42** *Can you find an example of a discontinuous function* $f : \mathbb{D} \to \mathbb{Y}$ *such that for some* $A \subset \mathbb{D}$, $f(\overline{A}) \supset \overline{f(A)}$?

The following proposition establishes a convenient sufficient condition for continuity of a function on its domain. In a certain sense, it can be seen as a refinement of remark 2 immediately following the definition of continuity on page 178. Note the difference between that remark and the conditions stated in the proposition.

Proposition 1.19 *Suppose* $f : \mathbb{D} \to \mathbb{Y}$ *is such that for each* $x \in \mathbb{D}$, $\lim f(x_n) = f(x)$ *for every sequence* $\{x_n\} \subset \mathbb{D}$ *that converges to* $x \in \mathbb{D}$. *Then* f *is continuous on* $\mathbb{D}$.

PROOF Let A be any subset of $\mathbb{D}$. We will show that $f\left(\overline{A}\right) \subset \overline{f(A)}$. Recall that according to Proposition 1.18, this is equivalent to f being continuous on $\mathbb{D}$. To establish the result, let $x_0 \in \overline{A}$. Then there exists $\{x_n\} \subset A$ such that $\lim x_n = x_0$, and therefore $\lim f(x_n) = f(x_0)$. Since $f(x_n) \in f(A)$, it follows that $f(x_0) \in \overline{f(A)}$, which is equivalent to $f(\overline{A}) \subset \overline{f(A)}$. ♦

The next result effectively states that "continuous functions take compact sets to compact sets," that is, "continuous functions preserve compactness."

Theorem 1.20 *Let* $K \subset \mathbb{X}$ *be a compact set and* $f : K \to \mathbb{Y}$ *a continuous function on* K. *Then* $f(K)$ *is compact.*

PROOF Suppose $f(K) \subset \bigcup_\alpha W_\alpha$, where $\{W_\alpha\}$ is a collection of open sets in $\mathbb{Y}$. That is, $\bigcup_\alpha W_\alpha$ is an open cover for $f(K)$. Then $K \subset f^{-1}\left(\bigcup_\alpha W_\alpha\right) = \bigcup_\alpha f^{-1}(W_\alpha)$, and since $f^{-1}(W_\alpha)$ is open in K we may write $f^{-1}(W_\alpha) = K \cap V_\alpha$ where V_α is open in $\mathbb{X}$. From this, we may conclude that $K \subset \bigcup_\alpha V_\alpha$. But since K is compact and $K \subset \bigcup_\alpha V_\alpha$, we can extract a finite subcover for K, namely, $K \subset V_{\alpha_1} \cup V_{\alpha_2} \cup \cdots \cup V_{\alpha_r}$. From this we can conclude that

$$f(K) \subset W_{\alpha_1} \cup W_{\alpha_2} \cup \cdots \cup W_{\alpha_r}.$$

Therefore, $f(K)$ is compact. ♦

► **5.43** *As claimed in the above proof, verify that* $K \subset \bigcup_\alpha V_\alpha$ *and that* $f(K) \subset \bigcup_{i=1}^r W_{\alpha_i}$.

Example 1.21 Let $a, b > 0$ and define $K \equiv \{(x, y) \in \mathbb{R}^2 : \frac{x^2}{a^2} + \frac{y^2}{b^2} \leq 1\}$, that is, K is the interior and boundary of an ellipse centered at the origin in the plane. Consider the continuous function $f : \mathbb{R}^2 \to \mathbb{R}$ given by $f(x, y) = \frac{1}{|xy|+1}$. *By the previous theorem, we know that the image $f(K)$ must be a compact subset of $\mathbb{R}$, even if we can't visualize it.* ◄

► **5.44** *In the previous example, show that f is continuous on $\mathbb{R}^2$.*

► **5.45** *Let $f : (0,1) \to \mathbb{R}$ be given by $f(x) = \frac{1}{x}$. Note that f is continuous on $(0,1)$ and $f((0,1)) = (1,\infty)$ is not compact. Does this contradict Theorem 1.20?*

► **5.46** *Let $f : (0,1) \to \mathbb{R}$ be given by $f(x) = \frac{1}{x}\sin\frac{1}{x}$. Show that f is continuous on $(0,1)$. Is $f((0,1))$ compact?*

Finally, our last result in this discussion of topological properties and continuous functions states that "continuous functions take connected sets to connected sets," or "continuous functions preserve connectedness."

Theorem 1.22 *Suppose $f : \mathbb{D} \to \mathbb{Y}$ is continuous on $\mathbb{D}$. If $\mathbb{D}$ is connected, then $f(\mathbb{D}) \subset \mathbb{Y}$ is connected.*

PROOF Suppose $f(\mathbb{D})$ is disconnected. Then $f(\mathbb{D}) = V \cup W$ where V and W are nonempty and $\overline{V} \cap W = V \cap \overline{W} = \varnothing$. From this, it follows that

$$\mathbb{D} = f^{-1}(V) \cup f^{-1}(W).$$

We leave it as an exercise to show that

$$\begin{aligned} &a)\quad f^{-1}(V) \neq \varnothing \text{ and } f^{-1}(W) \neq \varnothing \\ &b)\quad f^{-1}(V) \cap \overline{f^{-1}(W)} = \overline{f^{-1}(V)} \cap f^{-1}(W) = \varnothing. \end{aligned}$$

This contradiction (that $\mathbb{D}$ is disconnected) shows that $f(\mathbb{D})$ must be connected. ♦

► **5.47** *Finish the proof of the above theorem.*

Example 1.23 *Let $g : (0,1) \to \mathbb{R}$ be given by $g(x) = \frac{1}{x}\sin\frac{1}{x}$. Since g is continuous and $(0,1)$ is connected, it follows that $g((0,1)) = \{y \in \mathbb{R} : y = \frac{1}{x}\sin\frac{1}{x}$ for $x \in (0,1)\}$ is also connected.* ◄

1.5 Real Continuous Functions

Here we present some results that are relevant only to real-valued functions. Each depends on the special order properties that, among our Euclidean spaces of interest, are only possessed by the real number system $\mathbb{R}$.

Proposition 1.24 *Suppose $f : \mathbb{D} \to \mathbb{R}$ is continuous at $x_0 \in \mathbb{D}$, and suppose $f(x_0) > 0$. Then there exists a neighborhood $N_\delta(x_0)$ centered at x_0 such that $f(x) > 0$ for $x \in N_\delta(x_0) \cap \mathbb{D}$.*

PROOF By the continuity of f at x_0, there exists a $\delta > 0$ such that

$$x \in \mathbb{D} \text{ and } |x - x_0| < \delta \quad\Rightarrow\quad |f(x) - f(x_0)| < \tfrac{1}{2}\,f(x_0).$$

Since
$$f(x_0) - f(x) \le |f(x) - f(x_0)| < \tfrac{1}{2} f(x_0),$$
we have that
$$x \in \mathbb{D} \cap N_\delta(x_0) \Rightarrow f(x) > \tfrac{1}{2} f(x_0) > 0,$$
and the proposition is proved. ♦

Note that in the special case where x_0 is an interior point of $\mathbb{D}$, the above result can be stated more simply. Namely, if f is continuous at the interior point $x_0 \in \mathbb{D}$ and if $f(x_0) > 0$, then f must be positive throughout some neighborhood of x_0.

► **5.48** *If $f : \mathbb{D} \to \mathbb{R}$ is continuous at $x_0 \in \mathbb{D}$ and if $f(x_0) < 0$, show that there exists a neighborhood $N_\delta(x_0)$ centered at x_0 such that $f(x) < 0$ for all $x \in \mathbb{D} \cap N_\delta(x_0)$.*

The next result has great practical significance, as every student of calculus knows.

Theorem 1.25 (The Max-Min Theorem)
Let $K \subset \mathbb{X}$ be a compact set, and $f : K \to \mathbb{R}$ a continuous, real-valued function on K. Then f attains a maximum value and a minimum value on K, i.e., there exist x_1 and x_2 in K such that $f(x_1) \le f(x) \le f(x_2)$ for all $x \in K$.

PROOF The set $f(K)$ is compact, and therefore closed and bounded. Let $M = \sup f(K)$ and $m = \inf f(K)$. Since $f(K)$ is closed, both M and m are in $f(K)$. That is, there exist x_1 and x_2 in K such that $f(x_1) = m$ and $f(x_2) = M$. ♦

Example 1.26 *Consider again $D^2 = \{\mathbf{x} = (x,y) \in \mathbb{R}^2 : x > 0, y > 0\}$, and $f : D^2 \to \mathbb{R}$ given by $f(\mathbf{x}) = \frac{x}{x+y}$. As already seen in a previous example and a previous exercise, f is continuous at every point in D^2. If we take $K = [1,2] \times [1,4]$ then $K \subset D^2$ and f must attain a maximum and a minimum somewhere on K. Note that Theorem 1.25 does not provide a means for finding these extreme values.* ◄

► **5.49** *Even though Theorem 1.25 is not useful in this regard, can you find the maximum and minimum of the function f from the previous example?*

► **5.50** *Let $K \subset \mathbb{X}$ be a compact set, and $f : K \to \mathbb{Y}$ a continuous function on K. Prove that $|f| : K \to \mathbb{R}$ attains a maximum value and a minimum value on K.*

In some ways the following significant theorem sheds more light on Proposition 1.24 on page 191, although it is not as generally applicable as Proposition 1.24.

Theorem 1.27 (The Intermediate Value Theorem)
Suppose $f : [a,b] \to \mathbb{R}$ is continuous on $[a, b] \subset \mathbb{R}$ and that $f(a) \neq f(b)$. Then for every y between $f(a)$ and $f(b)$, there exists $c \in (a, b)$ such that $f(c) = y$.

PROOF We prove a special case and leave the general proof to the reader. Suppose $f(a) < 0$ and $f(b) > 0$. We will show that there exists $c \in (a,b)$ such that $f(c) = 0$. To this end, let $S = \{x \in [a, b] : f(x) < 0\}$. Then $a \in S$, and since f is continuous at a there exists a $\delta_1 > 0$ such that $[a, a + \delta_1) \subset S$. (Why?) Similarly, $b \in S^C$, and there exists a $\delta_2 > 0$ such that $(b - \delta_2, b] \subset S^C$. Now let $c = \sup S$. It follows that $a + \frac{\delta_1}{2} \le c \le b - \frac{\delta_2}{2}$. (Why?) This shows that $c \in (a, b)$. We will now show that $f(c) = 0$ with a proof by contradiction. Suppose $f(c) < 0$. Then there exists a $\delta_3 > 0$ such that $(c - \delta_3, c + \delta_3) \subset [a, b]$, where $f(x) < 0$ for all $x \in (c - \delta_3, c + \delta_3)$. But then $f\left(c + \frac{\delta_3}{2}\right) < 0$, which implies that $c + \frac{\delta_3}{2} \in S$, a contradiction. Therefore $f(c) \ge 0$. But if $f(c) > 0$, then there exists a $\delta_4 > 0$ such that $(c - \delta_4, c + \delta_4) \subset [a, b]$, where $f(x) > 0$ for all $x \in (c - \delta_4, c + \delta_4)$. By the definition of c there must exist $\xi \in (c - \delta_4, c]$ such that $\xi \in S$, i.e., $f(\xi) < 0$. This contradicts the fact that $f(x) > 0$ on $(c - \delta_4, c + \delta_4)$, and so we must have $f(c) = 0$. ♦

► **5.51** *Prove the general case of the intermediate value theorem. Why is it not as generally applicable as Proposition 1.24?*

► **5.52** *Is the value c referenced in the intermediate Value theorem unique?*

The following examples illustrate an interesting consequences of the intermediate value theorem.

Example 1.28 *Suppose $f : [0,1] \to [0,1]$ is continuous. We will show that there exists at least one point $x_0 \in [0,1]$ such that $f(x_0) = x_0$. Such a point is called a fixed point for f. To show this, first note that if $f(0) = 0$ or if $f(1) = 1$ then we are done. So assume that $f(0) > 0$ and that $f(1) < 1$. Let $g(x) = f(x) - x$, and note that g is continuous on $[0,1]$. Since $g(0) > 0$ and $g(1) < 0$, according to the intermediate value theorem there must be a point $x_0 \in (0,1)$ such that $g(x_0) = 0$, i.e., $f(x_0) = x_0$.* ◄

Example 1.29 *Suppose $f : \mathbb{R} \to \mathbb{R}$ is the continuous function given by $f(x) = x^3 - x + 1$. Are there any real roots to the equation $f(x) = 0$? If so, can they be identified? While the intermediate value theorem can yield an answer to the first question, we often can't get a precise answer to the second. Note that $f(-2) = -5$, while $f(-1) = 1$. Therefore, there exists $c \in (-2, -1)$ such that $f(c) = 0$.* ◄

► **5.53** *Can you think of a procedure for specifying the numerical value of c in the previous example to any desired degree of accuracy?*

Our next order-related continuity property is that of left- and right-handed continuity of real-valued functions. Just as with our description of left- and right-handed limits in the previous section, the concept of left- and right-handed continuity is already implicitly well defined. Our definition of continuity given on page 178 includes it as a special case. However, because it is a traditional topic in the development of the theory of functions of a single real variable, and because the reader is likely to encounter the terms *left continuous* and *right continuous* in later work, we make the idea more explicit in the following special definition.

Definition 1.30 Consider $f : \mathcal{D}^1 \to \mathbb{R}$, and $x_0 \in \mathcal{D}^1$.

1. We say that f is **right continuous** at x_0 if for any $\epsilon > 0$ there exists $\delta > 0$ such that
$$x \in \mathcal{D}^1 \cap [x_0, x_0 + \delta) \;\Rightarrow\; |f(x) - f(x_0)| < \epsilon.$$
If f is right continuous at x for all $x \in \mathcal{D}^1$, we say that f is **right continuous on $\mathcal{D}^1$**.

2. We say that f is **left continuous** at x_0 if for any $\epsilon > 0$ there exists $\delta > 0$ such that
$$x \in \mathcal{D}^1 \cap (x_0 - \delta, x_0] \;\Rightarrow\; |f(x) - f(x_0)| < \epsilon.$$
If f is left continuous at x for all $x \in \mathcal{D}^1$, we say that f is **left continuous on $\mathcal{D}^1$**.

Example 1.31 *Let $f : [0,2] \to \mathbb{R}$ be given by $f(x) = \begin{cases} 3x-1 & \text{if } 0 \le x < 1 \\ -x+7 & \text{if } 1 \le x \le 2 \end{cases}$. We will show that f is right continuous at $x = 1$. To this end, let $\epsilon > 0$ be given, and for $1 \le x \le 2$ consider that*

$$|f(x) - f(1)| = |-x+7-6| = |-x+1| = |x-1|.$$

This last will be less than the given ϵ as long as $|x-1| < \delta = \epsilon$. That is, for any given $\epsilon > 0$, we have shown that

$$x \in [1,2] \text{ and } |x-1| < \delta = \epsilon \;\Rightarrow\; |f(x) - 6| < \epsilon.$$

Hence f is right continuous at $x = 1$. ◀

▶ **5.54** *Show that the function f in the previous example is not left continuous at $x = 1$.*

▶ **5.55** *What kind of continuity does f possess at $x_0 = 0$? Is it left, right, or just plain continuous there? How about at $x_0 = 2$?*

As might have been expected, continuity of a function at a point is directly related to left and right continuity at that point. The following proposition makes the relationship clear.

Proposition 1.32 *Consider $f : \mathcal{D}^1 \to \mathbb{R}$ and $x_0 \in \mathcal{D}^1$. Then f is continuous at x_0 if and only if f is right continuous and left continuous at x_0.*

▶ **5.56** *Prove the above proposition.*

Finally, we consider the case where a real-valued function $f : D^1 \to \mathbb{R}$ has an inverse function $f^{-1} : \mathcal{R}_f \to D^1$, where here $\mathcal{R}_f$ is the range of f. If f is continuous on D^1, it is natural to wonder whether f^{-1} is continuous on $\mathcal{R}_f$. To establish the relevant result requires some work, including the following exercise.

► **5.57** *Let $I = [a, b]$ be an interval of real numbers.*

a) *Show that any subset $J \subset I$ of the form $J_1 = [a, b]$, $J_2 = (c, b]$, $J_3 = [a, c)$, or $J_4 = (c, d)$ is open in I.*

b) *Suppose $f : I \to \mathbb{R}$ is one-to-one and continuous on I. Let $m \equiv \min\{f(a), f(b)\}$, and let $M \equiv \max\{f(a), f(b)\}$. Show that $f(I) = [m, M]$, and that f must be either increasing or decreasing on I.*

c) *Suppose $f : I \to [m, M]$ is one-to-one, onto, and continuous on I. For each of the J intervals described in part a), prove that $f(J)$ is open in $[m, M]$.*

We are now ready to state and prove the relevant theorem.

Theorem 1.33 *Suppose $f : [a, b] \to [m, M]$ is a continuous one-to-one correspondence. Then $g \equiv f^{-1} : [m, M] \to [a, b]$ is also continuous.*

PROOF To establish that g is continuous, we will show that $g^{-1}(A) = f(A)$ is open in $[m, M]$ for every open subset $A \subset [a, b]$. To this end, note that for any open subset $A \subset [a, b]$ there exists an open set $V \subset \mathbb{R}$ such that $A = [a, b] \cap V$. By Proposition 3.6 on page 57 we may write $V = \bigcup_\alpha I_\alpha$ where $\{I_\alpha\}$ is a collection of open intervals in $\mathbb{R}$. From this, we then have that

$$A = [a, b] \cap \Big(\bigcup_\alpha I_\alpha\Big) = \bigcup_\alpha \big([a, b] \cap I_\alpha\big) \equiv \bigcup_\alpha J_\alpha,$$

where $J_\alpha \equiv [a, b] \cap I_\alpha$ is of the form of J_1, J_2, J_3, or J_4 in the exercise preceding the theorem. Since

$$f(A) = f\Big(\bigcup_\alpha J_\alpha\Big) = \bigcup_\alpha f(J_\alpha)$$

is the union of open subsets of $[m, M]$, it follows from the previous exercise that $f(A)$ is open in $[m, M]$, and the theorem is proved. ♦

A familiar example that illustrates the use of the above theorem is the following. Consider $f : [-a, a] \to \mathbb{R}$ given by $f(x) = \tan x$, where $a < \frac{\pi}{2}$. It is not hard to show that f is a continuous one-to-one correspondence, and in fact is increasing on $[-a, a]$. Therefore, according to the above theorem, the inverse function $f^{-1} : \big[\tan(-a), \tan(a)\big] \to [-a, a]$ given by $f^{-1}(x) = \tan^{-1} x$ is continuous on $\big[\tan(-a), \tan(a)\big]$.

► **5.58** *Show that $f : [-a, a] \to \mathbb{R}$ given by $f(x) = \tan x$ is a continuous one-to-one correspondence, and is increasing on $[-a, a]$ for $a < \frac{\pi}{2}$.*

► **5.59** *Within the context of the above discussion where f is a continuous one-to-one correspondence, prove that if f is increasing so is f^{-1}. Likewise, if f is decreasing so is f^{-1}.*

▶ **5.60** *Suppose $f : [a, b) \to \mathcal{R}_f$ is continuous and one-to-one where $\mathcal{R}_f$ is the range of f. Can you use a similar argument as in the above proof to show that $f^{-1} : \mathcal{R}_f \to [a, b)$ is continuous?*

▶ **5.61** *Suppose $f : [a, b] \to [c, d]$ is a continuous one-to-one correspondence. If $A \subset [a, b]$ is open in $[a, b]$ then there exists an open $V \subset \mathbb{R}$ such that $A = [a, b] \cap V$. It is also true in this case that $f(A) = f([a, b]) \cap f(V)$. (Note that, in general for sets B and C, we may conclude only that $f(B \cap C) \subset f(B) \cap f(C)$.) What allows for the equality in this case?*

Classification of Discontinuities

Suppose $f : \mathbb{D} \to \mathbb{Y}$ is a function and x_0 is a point in $\overline{\mathbb{D}}$ at which f is either not defined or not continuous. In such cases, we often say that f has a *discontinuity* at x_0. Clearly not all points of discontinuity are the same. After all, according to this convention, some might be points that are in the domain of the function, and some might lie outside the domain of the function. There are still other differences as well. We will provide a definition describing three kinds of discontinuities for a real-valued function of a real variable. To do so, we first define two different ways that a one-sided limit of a function f can fail to exist as x approaches x_0.

Definition 1.34 Consider $f : D^1 \to \mathbb{R}$ and suppose x_0 is a limit point of D^1.

1. Suppose $\lim_{x \to x_0^+} f(x)$ fails to exist and that, in addition, x_0 is such that there exists $r > 0$ where $D^1 \cap (x_0, x_0 + r) = \varnothing$. In this case, we will say that the right-hand limit of f at x_0 fails to exist **in the weak sense.** Similarly, if $\lim_{x \to x_0^-} f(x)$ fails to exist where x_0 is such that there exists $r > 0$ where $D^1 \cap (x_0 - r, x_0) = \varnothing$, we will say that the left-hand limit of f at x_0 fails to exist **in the weak sense.**
2. Suppose $\lim_{x \to x_0^+} f(x)$ fails to exist and that x_0 is such that for all $r > 0$, $D^1 \cap (x_0, x_0 + r) \neq \varnothing$. In this case, we will say that the right-hand limit of f at x_0 fails to exist **in the strict sense.** Similarly, if $\lim_{x \to x_0^-} f(x)$ fails to exist where x_0 is such that for all $r > 0$, $D^1 \cap (x_0 - r, x_0) \neq \varnothing$, we will say that the left-hand limit of f at x_0 fails to exist **in the strict sense.**

Note that the above definition relates to our informal discussion of "left-hand limit points" and "right-hand limit points" in Chapter 4. In particular, a function $f : D^1 \to \mathbb{R}$ fails in the weak sense to have a left-hand limit as x approaches x_0 precisely when x_0 fails to be a "left-hand limit point" of D^1. In this case, we cannot even approach x_0 from the left, and that is why the left-hand limit fails to exist there. Failure in the strict sense implies that we can approach x_0 from the relevant side, but the one-sided limit fails to exist anyway. A similar explanation for the case of right-hand limits also applies.

Definition 1.35 Consider $f : \mathbb{D}^1 \to \mathbb{R}$, and suppose $x_0 \in \overline{\mathbb{D}^1}$ is a point where f is not continuous or not defined. Such a point x_0 is sometimes referred to as a **point of discontinuity** or a **discontinuity point** of f. We consider three kinds of discontinuity.

1. The point x_0 is called a **removable discontinuity** of f if either of the following is true:
   a) The point x_0 is an interior point of D^1 and both $\lim_{x\to x_0^+} f(x)$ and $\lim_{x\to x_0^-} f(x)$ exist and equal L.
   b) The point x_0 is a boundary point of D^1 and both $\lim_{x\to x_0^+} f(x)$ and $\lim_{x\to x_0^-} f(x)$ exist and equal L, or one of either $\lim_{x\to x_0^+} f(x)$ or $\lim_{x\to x_0^-} f(x)$ exists while the other fails to exist in the weak sense.
2. The point x_0 is called a **jump discontinuity** of f if $\lim_{x\to x_0^+} f(x)$ and $\lim_{x\to x_0^-} f(x)$ exist but are not equal.
3. The point x_0 is called an **essential discontinuity** of f if either of the following is true:
   a) The point x_0 is an interior point of D^1 and $\lim_{x\to x_0^+} f(x)$ or $\lim_{x\to x_0^-} f(x)$ fails to exist.
   b) The point x_0 is a boundary point of D^1 and either $\lim_{x\to x_0^+} f(x)$ or $\lim_{x\to x_0^-} f(x)$ fails to exist in the strict sense.

Example 1.31 on page 194 illustrates a function with a jump discontinuity at $x_0 = 1$. The function $f : \mathbb{R} \setminus \{0\} \to \mathbb{R}$ given by $f(x) = \frac{1}{x}$ has an essential discontinuity at $x = 0$. In the case of a removable discontinuity, we may assign f the common limit value at x_0 and hence "make" f continuous at x_0. We illustrate the removable discontinuity case in the following example.

Example 1.36 *Let $f : \mathbb{R} \setminus \{0\} \to \mathbb{R}$ be given by $f(x) = \frac{\sin x}{x}$. Then f is clearly continuous on its domain. Near 0 we have*

$$\lim_{x\to 0} \left(\frac{\sin x}{x}\right) = \lim_{x\to 0^+} \left(\frac{\sin x}{x}\right) = \lim_{x\to 0^-} \left(\frac{\sin x}{x}\right) = 1.$$

According to Definition 1.35, the function f has a removable discontinuity at $x = 0$. Therefore the function $g : \mathbb{R} \to \mathbb{R}$ defined by

$$g(x) = \begin{cases} \frac{\sin x}{x} & \text{if } x \neq 0 \\ 1 & \text{if } x = 0 \end{cases},$$

is continuous on $\mathbb{R}$. ◀

► **5.62** *If x_0 is an essential discontinuity of $f : D^1 \to \mathbb{R}$, is it necessarily true that f becomes unbounded as x approaches x_0?*

2 Uniform Continuity

We are now ready to introduce the special property of uniform continuity. We will see that functions that are uniformly continuous are somewhat more "tame" than functions that are merely continuous. Also, there are some convenient consequences to uniform continuity that do not hold for merely continuous functions. We begin by motivating the definition.

2.1 Definition and Examples

Suppose $f : \mathbb{D} \to \mathbb{Y}$ is continuous on $\mathbb{D}$. Then, fixing $x_0 \in \mathbb{D}$, continuity implies that for any $\epsilon > 0$ there exists a $\delta > 0$ such that for $x \in \mathbb{D}$,

$$|x - x_0| < \delta \Rightarrow |f(x) - f(x_0)| < \epsilon.$$

As noted in the remarks following our definition of continuity in Definition 1.1 on page 178, for any given $\epsilon > 0$ the δ that makes the definition "work" generally depends on ϵ and on x_0. Specifying a smaller ϵ usually requires a smaller δ. Considering the function f near a different point than x_0, say $x_1 \in \mathbb{D}$, might also require a smaller δ. However, there are functions for which this last possibility is not a concern, in that a positive δ can be found for which the continuity definition "works" independently of the point within $\mathbb{D}$ at which one investigates f. For such a function, determining the continuity of f on $\mathbb{D}$ depends on the nearness of pairs of points in $\mathbb{D}$, but not on the location within $\mathbb{D}$ where this nearness is measured. Such functions are called *uniformly continuous on* $\mathbb{D}$.

Definition 2.1 A function $f : \mathbb{D} \to \mathbb{Y}$ is **uniformly continuous on** $A \subset \mathbb{D}$ if for any $\epsilon > 0$ there exists a $\delta > 0$ such that for any $\xi, \eta \in A$,

$$|\xi - \eta| < \delta \Rightarrow |f(\xi) - f(\eta)| < \epsilon.$$

It is important to note that when speaking of a function as being uniformly continuous, one must always specify on what set this uniform continuity applies, whether it be the whole domain or only a subset of it. It is also useful to understand what it means when a function $f : \mathbb{D} \to \mathbb{Y}$ is *not* uniformly continuous on $\mathbb{D}$. Negating the above definition gives the following useful statement:

> *A function $f : \mathbb{D} \to \mathbb{Y}$ is* not *uniformly continuous on $A \subset \mathbb{D}$ if and only if there exists an $\epsilon > 0$ such that for every $\delta > 0$ there are points $\xi, \eta \in A$ such that $|\xi - \eta| < \delta$ and yet $|f(\xi) - f(\eta)| \geq \epsilon$.*

We will now use Definition 2.1 to establish uniform continuity in several examples. In the first example, we show that linear complex functions are uniformly continuous on their whole domains. The argument is exactly the same for establishing the analogous fact about real-valued linear functions of a single real variable.

Example 2.2 *Consider $f : \mathbb{C} \to \mathbb{C}$ given by $f(z) = a\,z + b$ for $a, b \in \mathbb{C}$. We will show that f is uniformly continuous on $\mathbb{C}$. For the case $a \neq 0$, merely consider that for any given $\epsilon > 0$ and $z, w \in \mathbb{C}$, we have*

$$|f(z) - f(w)| = |a|\,|z-w| < \epsilon \quad \text{as long as } |z-w| < \tfrac{\epsilon}{|a|}.$$

That is, for $\delta = \frac{\epsilon}{|a|}$, we have $|z-w| < \delta \Rightarrow |f(z) - f(w)| < \epsilon$. Since the δ we obtained to establish the continuity of f does not depend on z or w, the function f is uniformly continuous on its whole domain $\mathbb{C}$. The case where $a = 0$ is left to the reader. ◄

► **5.63** *Establish the uniform continuity of the function f given in the previous example for the case where $a = 0$.*

Example 2.3 *Consider $f : [-2, 5] \to \mathbb{R}$ given by $f(x) = x^2 + 2x - 7$. We will show that f is uniformly continuous on $[-2, 5]$ by applying Definition 2.1. Consider an arbitrary pair of points ξ and η from $[-2, 5]$. Then,*

$$\begin{aligned} |f(\xi) - f(\eta)| &= \left|\left(\xi^2 + 2\xi - 7\right) - \left(\eta^2 + 2\eta - 7\right)\right| \\ &\leq |\xi^2 - \eta^2| + 2\,|\xi - \eta| \\ &= |\xi - \eta|\,|\xi + \eta| + 2\,|\xi - \eta| \\ &\leq 10\,|\xi - \eta| + 2\,|\xi - \eta| \\ &= 12\,|\xi - \eta| \\ &< \epsilon \ \text{ if } \ |\xi - \eta| < \tfrac{\epsilon}{12}. \end{aligned}$$

Choosing $\delta = \frac{\epsilon}{12}$ yields the result. Note that the δ we use is independent of both ξ and η, and so f is uniformly continuous on $[-2, 5]$. ◄

► **5.64** *What if the domain of the function in the previous example is changed to $(-2, 5)$? Does the domain matter at all?*

► **5.65** *Consider $f : [a, b] \to \mathbb{R}$ given by $f(x) = \alpha x^2 + \beta x + \gamma$ where $\alpha, \beta, \gamma \in \mathbb{R}$. Is f uniformly continuous on $[a, b]$?*

► **5.66** *Consider $\boldsymbol{f} : [a, b] \times [c, d] \to \mathbb{R}^2$ given by $\boldsymbol{f}(\mathbf{x}) = \alpha\mathbf{x} + \boldsymbol{\beta}$ where $\alpha \in \mathbb{R}$ and $\boldsymbol{\beta} \in \mathbb{R}^2$. Is $\boldsymbol{f}$ uniformly continuous on $[a, b] \times [c, d]$?*

Example 2.4 *Consider $f : [0, \infty) \to \mathbb{R}$ given by $f(x) = \frac{x}{x+1}$. We will show that f is uniformly continuous on $[0, \infty)$. Consider for $\xi, \eta \in [0, \infty)$ that*

$$\begin{aligned} |f(\xi) - f(\eta)| = \left|\frac{\xi}{\xi+1} - \frac{\eta}{\eta+1}\right| = \left|\frac{\xi - \eta}{(\xi+1)(\eta+1)}\right| &= \frac{|\xi - \eta|}{(\xi+1)(\eta+1)} \\ &< |\xi - \eta|, \end{aligned}$$

which is less than any $\epsilon > 0$ if $|\xi - \eta| < \delta = \epsilon$. Therefore, f is uniformly continuous on $[0, \infty)$. ◄

▶ **5.67** *What if the domain in the previous example is changed to $(-1, \infty)$? Is f uniformly continuous on $(-1, \infty)$?*

And how does one show that a given continuous function is, in fact, *not* uniformly continuous on a given set? We illustrate this in the next example.

Example 2.5 *Consider $f : (0, 1) \to \mathbb{R}$ given by $f(x) = \frac{1}{x}$. As seen in Example 1.3 on page 179, this function is continuous on $(0, \infty)$. However, as indicated in that example, the derived δ was dependent upon x. We will show, in fact, that f is not uniformly continuous on $(0, 1)$. We will use the method of proof by contradiction. To this end, assume f is uniformly continuous on $(0, 1)$. Then for $\epsilon = \frac{1}{2}$ there exists $\delta > 0$ such that*

$$\xi, \eta \in (0,1) \text{ and } |\xi - \eta| < \delta \quad \Rightarrow \quad \left|\frac{1}{\xi} - \frac{1}{\eta}\right| = \frac{|\xi - \eta|}{|\xi|\,|\eta|} < \tfrac{1}{2}.$$

With no loss in generality, we may assume that $\delta < 1$. (Why?) Note that the numerator of $\frac{|\xi-\eta|}{|\xi|\,|\eta|}$ is restricted in magnitude to be no larger than δ, yet the denominator can be made arbitrarily small (and hence, the overall ratio arbitrarily large) by taking ξ and η as close as we like to 0. That is, no matter how close together ξ and η are within $(0, 1)$, we can find such a ξ and η as close to 0 as we need in order to derive the contradiction we seek. In particular, suppose $|\xi - \eta| = \frac{\delta}{2} < \delta$, and consider such a ξ and η that satisfy $0 < \xi, \eta < \sqrt{\delta}$. This yields

$$\frac{|\xi - \eta|}{|\xi|\,|\eta|} = \frac{\delta/2}{\xi\,\eta} > \frac{\delta/2}{\delta} = \tfrac{1}{2},$$

a contradiction. Therefore, our initial assumption that f is uniformly continuous on $(0, 1)$ cannot be true. ◀

▶ **5.68** *Answer the (Why?) question in the above example.*

▶ **5.69** *Can you change the domain of the function in the last example in such a way that the resulting function is uniformly continuous on the new domain?*

Example 2.6 *In this example we will show that $f : \mathbb{R} \to \mathbb{R}$ is not uniformly continuous on $[a, \infty)$ for any fixed $a \in \mathbb{R}$. To do so, assume the contrary. Then there exists $\delta > 0$ such that for any $\xi, \eta \in [a, \infty)$ satisfying $|\xi - \eta| < \delta$ we will have $|e^\xi - e^\eta| < 1$. If we choose $\xi = \eta + \frac{\delta}{2} > \eta > 0$, we have that $e^{\eta+\delta/2} - e^\eta < 1$ for all $\eta > 0$. From this we may write*

$$e^{\eta+\delta/2} - e^\eta = e^\eta(e^{\delta/2} - 1) = e^\eta \sum_{j=1}^{\infty} \frac{(\delta/2)^j}{j!} \geq e^\eta\left(\tfrac{\delta}{2}\right),$$

which implies $e^\eta\left(\frac{\delta}{2}\right) < 1$ for all $\eta > 0$, i.e., $e^\eta < \frac{2}{\delta}$ for all $\eta > 0$. This last is a contradiction. ◀

▶ **5.70** *Suppose $[a, b] \subset \mathbb{R}$. Show that the real exponential function $\exp : \mathbb{R} \to \mathbb{R}$ given by $\exp(x) = e^x$ is uniformly continuous on $[a, b]$. Show that it is also uniformly continuous on $(-\infty, a]$ for any $a \in \mathbb{R}$.*

Example 2.7 *Consider $f : D^2 \to \mathbb{R}$ given by $f(x,y) = \frac{1}{x+y}$, where $D^2 = \{(x,y) \in \mathbb{R}^2 : x+y \geq 1\}$. We will show that f is uniformly continuous on its domain. To this end, let $\epsilon > 0$ be given, and consider $\boldsymbol{\xi} = (\xi_1, \xi_2)$ and $\boldsymbol{\eta} = (\eta_1, \eta_2)$ from D^2. Note that*

$$\begin{aligned} |f(\boldsymbol{\xi}) - f(\boldsymbol{\eta})| = \left| \frac{1}{\xi_1 + \xi_2} - \frac{1}{\eta_1 + \eta_2} \right| &= \frac{|(\eta_1 - \xi_1) + (\eta_2 - \xi_2)|}{(\xi_1 + \xi_2)(\eta_1 + \eta_2)} \\ &\leq |\eta_1 - \xi_1| + |\eta_2 - \xi_2| \\ &\leq |\boldsymbol{\eta} - \boldsymbol{\xi}| + |\boldsymbol{\eta} - \boldsymbol{\xi}| \\ &= 2\,|\boldsymbol{\eta} - \boldsymbol{\xi}| \\ &< \epsilon \quad \text{if} \quad |\boldsymbol{\eta} - \boldsymbol{\xi}| < \tfrac{\epsilon}{2}. \end{aligned}$$

Therefore, choosing $\delta \equiv \frac{\epsilon}{2}$ establishes the result. ◀

2.2 Topological Properties and Consequences

Interestingly enough, what can often determine the *uniform* continuity of a continuous function on a certain set are the topological characteristics of the set, rather than any particular qualities the function possesses. The examples and exercises up to this point might have already led the reader to suspect this very idea. The following theorem points out how compactness of the set in question plays the key role in this issue.

Theorem 2.8 *Suppose $f : \mathbb{D} \to \mathbb{Y}$ is continuous on a compact set $K \subset \mathbb{D}$. Then f is uniformly continuous on K.*

PROOF Suppose $\epsilon > 0$ is given. For each $x_0 \in K$ there exists a $\delta_{x_0} > 0$ such that

$$x \in K \text{ and } |x - x_0| < \delta_{x_0} \;\Rightarrow\; |f(x) - f(x_0)| < \tfrac{\epsilon}{2},$$

i.e., we have that

$$x \in \left(K \cap N_{\delta_{x_0}}(x_0)\right) \;\Rightarrow\; |f(x) - f(x_0)| < \tfrac{\epsilon}{2}.$$

Here, δ_{x_0} emphasizes the potential dependence of δ on x_0. Of course, we will show in this case that the dependence does not, in fact, exist. It is certainly true that for each $x_0 \in K$ we have $x_0 \in N_{\frac{1}{2}\delta_{x_0}}(x_0)$, and so it follows that

$$K \subset \bigcup_{x \in K} N_{\frac{1}{2}\delta_x}(x).$$

But since K is compact, there exists a finite collection $x_1, x_2, \ldots, x_m \in K$ such that

$$K \subset \bigcup_{j=1}^{m} N_{\frac{1}{2}\delta_{x_j}}(x_j).$$

Now choose

$$\delta \equiv \min\left(\tfrac{1}{2}\delta_{x_1}, \tfrac{1}{2}\delta_{x_2} \ldots, \tfrac{1}{2}\delta_{x_m}\right), \tag{5.5}$$

and consider any pair of points $\xi, \eta \in K$ such that $|\xi - \eta| < \delta$. Since $\xi \in$

$N_{\frac{1}{2}\delta_{x_{j*}}}(x_{j*})$ for some $j* \in \{1,2,\ldots,m\}$, we know that $|\xi - x_{j*}| < \frac{1}{2}\delta_{x_{j*}}$. But we also know that

$$|\eta - x_{j*}| \le |\eta - \xi| + |\xi - x_{j*}| < \delta + \tfrac{1}{2}\delta_{x_{j*}} < \delta_{x_{j*}},$$

and so ξ and η both belong to $N_{\delta_{x_{j*}}}(x_{j*})$. This means that

$$|f(\xi) - f(x_{j*})| < \tfrac{\epsilon}{2} \text{ and } |f(\eta) - f(x_{j*})| < \tfrac{\epsilon}{2}.$$

From this we obtain

$$|f(\xi) - f(\eta)| \le |f(\xi) - f(x_{j*})| + |f(x_{j*}) - f(\eta)| < \epsilon.$$

That is, for $\delta > 0$ chosen according to (5.5), we have that

$$\xi, \eta \in K \text{ and } |\xi - \eta| < \delta \Rightarrow |f(\xi) - f(\eta)| < \epsilon,$$

and the theorem is proved. ♦

Example 2.9 *Consider $f : [0,1] \to \mathbb{R}$ given by*

$$f(x) = \begin{cases} x \sin\frac{1}{x} & \text{for } 0 < x \le 1 \\ 0 & \text{for } x = 0 \end{cases}.$$

We will show that f is uniformly continuous on $[0,1]$. Since $[0,1]$ is compact, by Theorem 2.8, we only need to show that f is continuous on $[0,1]$. The function f is clearly continuous on $(0,1]$, and so we need only show that it is also continuous at $x_0 = 0$. To see this, consider

$$\left|x \sin\tfrac{1}{x} - 0\right| = |x| \left|\sin\tfrac{1}{x}\right| \le |x| < \epsilon \text{ if } |x| < \delta = \epsilon.$$

Hence, f is continuous at $x_0 = 0$, and therefore on all of $[0,1]$. By Theorem 2.8, f is uniformly continuous on $[0,1]$. ◀

▶ **5.71** *Fix $\alpha \in \mathbb{R}$ and consider the function $f : [0,1] \to \mathbb{R}$ given by*

$$f(x) = \begin{cases} x^\alpha \sin\frac{1}{x} & \text{for } 0 < x \le 1 \\ 0 & \text{for } x = 0 \end{cases}.$$

For what values of α is the function uniformly continuous on $[0,1]$?

Consequences of Uniform Continuity

There are some significant consequences of a function being uniformly continuous on its domain. Our first result establishes that uniform continuity preserves the Cauchy property.

Proposition 2.10 *Suppose $f : \mathbb{D} \to \mathbb{Y}$ is uniformly continuous on $\mathbb{D}$. If $\{x_n\}$ is a Cauchy sequence in $\mathbb{D}$, then $\{f(x_n)\}$ is a Cauchy sequence in $\mathbb{Y}$.*

PROOF Assume f is uniformly continuous on $\mathbb{D}$. Then given $\epsilon > 0$, there exists $\delta > 0$ such that

$$\xi, \eta \in \mathbb{D} \text{ and } |\xi - \eta| < \delta \;\Rightarrow\; |f(\xi) - f(\eta)| < \epsilon. \tag{5.6}$$

Since $\{x_n\}$ is Cauchy, there exists $N \in \mathbb{N}$ such that

$$n, m > N \;\Rightarrow\; |x_n - x_m| < \delta. \tag{5.7}$$

Combining expressions (5.6) and (5.7) yields

$$n, m > N \;\Rightarrow\; |f(x_n) - f(x_m)| < \epsilon,$$

and the result is proved. ♦

► **5.72** *What happens if in the statement of the above proposition, you replace uniform continuity with mere continuity?*

The following proposition establishes the fact that a uniformly continuous function "preserves boundedness."

Proposition 2.11 *If $\mathbb{D} \subset \mathbb{X}$ is bounded and $f : \mathbb{D} \to \mathbb{Y}$ is uniformly continuous on $\mathbb{D}$, then $f(\mathbb{D})$ is bounded.*

PROOF Since f is uniformly continuous on $\mathbb{D}$ there exists $\delta > 0$ such that

$$\xi, \eta \in \mathbb{D} \text{ and } |\xi - \eta| < \delta \;\Rightarrow\; |f(\xi) - f(\eta)| < 1.$$

Assume that f is not bounded on $\mathbb{D}$. Then there exists a sequence $\{x_n\} \subset \mathbb{D}$ such that $|f(x_n)| \geq n$ for all integers $n \geq 1$. Since the sequence $\{x_n\}$ is bounded, there exists a convergent subsequence $\{x_{n_m}\}$ with the property that $\lim_{m\to\infty} x_{n_m} = x_0$ for some $x_0 \in \overline{\mathbb{D}}$. Therefore the deleted neighborhood $N'_{\frac{\delta}{2}}(x_0)$ contains x_j for infinitely many values of j. Fix one of these points, say, x_J. Then it follows that $|x_j - x_J| < \delta$ for infinitely many values of j, which in turn implies that $|f(x_j) - f(x_J)| < 1$, and hence, $|f(x_j)| < 1 + |f(x_J)|$ for infinitely many values of j. But this is a contradiction, since $|f(x_j)| \geq j$ for all j. Therefore, f must be bounded on $\mathbb{D}$. ♦

Recall that the function $f : (0,1) \to \mathbb{R}$ given by $f(x) = \frac{1}{x}$ maps its bounded domain to $(1, \infty)$, an unbounded range. Therefore by the above proposition, we may conclude that f cannot be uniformly continuous on its domain.

► **5.73** *Suppose $f : \mathbb{D} \to \mathbb{R}$ is uniformly continuous on $\mathbb{D}$. If $\mathbb{D}$ is not bounded, what can you say about $f(\mathbb{D})$? What if $\mathbb{D}$ is bounded, but f is merely continuous?*

2.3 Continuous Extensions

Recall that in the case of a removable discontinuity at $x_0 \in D^1 \subset \mathbb{R}$, a function $f : D^1 \to \mathbb{R}$ that is continuous on $D^1 \setminus \{x_0\}$ can be "redefined" at x_0 so as to be continuous on all of D^1. We now consider a slightly different problem. Stated generally, suppose $f : \mathbb{D} \to \mathbb{Y}$ possesses certain properties throughout its domain $\mathbb{D}$, but one would like to "extend" the function so as to possess them at points beyond $\mathbb{D}$. Such extensions are not always possible in general. In this subsection, we will consider the more specific problem of how one might extend a continuous function $f : \mathbb{D} \to \mathbb{Y}$ to a continuous function $\widetilde{f} : \mathbb{B} \supset \mathbb{D} \to \mathbb{Y}$, such that $\widetilde{f} = f$ on $\mathbb{D}$. We refer to $\widetilde{f}$ as a *continuous extension of* f, and we say that f has been *continuously extended to* $\mathbb{B}$. As we will see, this particular type of extension is not always possible either, unless the domain $\mathbb{D}$ is closed. We begin with a simple example.

Example 2.12 *Consider $f : (0,1) \to \mathbb{R}$ given by $f(x) = x^2$. Clearly, f is continuous on its specified domain. Perhaps one would like to extend f to a larger set, say $(-2,2)$. One might also wish, along with the extension equalling f on $(0,1)$, that it be continuous throughout $(-2,2)$. One obvious way to accomplish this is with the function $\widetilde{f}_1 : (-2,2) \to \mathbb{R}$ given by $\widetilde{f}_1(x) = x^2$. If we also require that the extension be bounded by the same bounds that applied to the original function f, where $0 \le |f(x)| \le 1$ for all $x \in (0,1)$, then this particular extension won't do. However, a continuous extension that satisfies this additional requirement is given by the function $\widetilde{f}_2 : (-2,2) \to \mathbb{R}$ defined as*

$$\widetilde{f}_2(x) = \begin{cases} 0 & \text{if } -2 < x \le 0, \\ x^2 & \text{if } 0 < x < 1, \\ 1 & \text{if } 1 \le x < 2. \end{cases}$$

The reader can verify that $\widetilde{f}_2 = f$ on $(0,1)$, that $\widetilde{f}_2$ is continuous throughout $(-2,2)$, and that $0 \le \left|\widetilde{f}_2(x)\right| \le 1$ for all $x \in (-2,2)$. ◀

While the above example illustrates that what we wish to do may be possible, it makes no claim as to the uniqueness of the extensions presented. There may be many ways to extend a given function to larger domains while retaining certain specified properties of the original function. While such questions of uniqueness can ultimately be of interest, it is first useful to know when such continuous extensions exist. The following theorem, called the pasting theorem, gives sufficient conditions for the existence of such an extension.

Theorem 2.13 (The Pasting Theorem)
Suppose $\mathbb{D} = A \cup B$ where A and B are closed subsets of $\mathbb{D}$. Let $f : A \to \mathbb{Y}$ and $g : B \to \mathbb{Y}$ be continuous functions. If $f(x) = g(x)$ for all $x \in A \cap B$, then f and g extend to the continuous function $h : \mathbb{D} = A \cup B \to \mathbb{R}$ given by

$$h(x) = \begin{cases} f(x) & \text{if } x \in A, \\ g(x) & \text{if } x \in B. \end{cases}$$

PROOF [1] First, note that the function h is well defined. Let V be closed in $\mathbb{Y}$. We will show that $h^{-1}(V)$ is closed in $\mathbb{D}$. According to a previous exercise, this will imply that h is continuous on $\mathbb{D}$. We leave it to the reader to verify that

$$h^{-1}(V) = f^{-1}(V) \cup g^{-1}(V).$$

Also, since f and g are continuous on their domains, $f^{-1}(V)$ is closed in A and $g^{-1}(V)$ is closed in B. It follows that both $f^{-1}(V)$ and $g^{-1}(V)$ are closed in $\mathbb{D}$ (Why?), and therefore $h^{-1}(V)$ is closed in $\mathbb{D}$. Hence, h is continuous on $\mathbb{D}$. ◆

► **5.74** *Show that $h^{-1}(V) = f^{-1}(V) \cup g^{-1}(V)$ as claimed in the above proof. Also, answer the (Why?) question posed there.*

► **5.75** *In the statement of the pasting theorem, suppose A and B are presumed to be open sets rather than closed. Prove that the result of the pasting theorem still holds.*

► **5.76** *Suppose $A = (-\infty, 1]$, $B = (1, \infty)$, and that $h : \mathbb{R} = A \cup B \to \mathbb{R}$ is given by $h(x) = \begin{cases} x^2 & \text{if } x \in A \\ 2 & \text{if } x \in B \end{cases}$. Is h continuous at $x = 1$? What does this exercise illustrate?*

Example 2.14 *Let $f : [0, \infty) \to \mathbb{R}$ be given by $f(x) = \frac{1}{(x+1)^2}$, and $g : (-\infty, 0] \to \mathbb{R}$ be given by $g(x) = \frac{1}{(x-1)^2}$. Then clearly f and g are continuous on their domains, and since $f(0) = g(0)$, by the pasting theorem we have that $h : \mathbb{R} \to \mathbb{R}$ given by*

$$h(x) = \begin{cases} g(x) & \text{if } -\infty < x \le 0 \\ f(x) & \text{if } 0 < x < \infty \end{cases}$$

is continuous on all of $\mathbb{R}$. Note in this case that f cannot be extended continuously to all of $\mathbb{R}$ with the same rule that defines it on $[0, \infty)$, since this rule is not well defined at $x = -1$. Similarly for g at $x = 1$. Yet, through g we obtain a continuous extension of f, namely, h. Likewise, through f we obtain a continuous extension of g. ◄

Example 2.15 *The previous example can be easily "extended" to a higher-dimensional case. Let $A = \{(x, y) \in \mathbb{R}^2 : x \ge 0\}$ and let $B = \{(x, y) \in \mathbb{R}^2 : x \le 0\}$. Consider the functions $f : A \to \mathbb{R}$ and $g : B \to \mathbb{R}$ given by*

$$f(x, y) = \frac{1}{(x+1)^2 + y^2} \quad \text{and} \quad g(x, y) = \frac{1}{(x-1)^2 + y^2}.$$

Then the pasting theorem yields that $h : \mathbb{R}^2 \to \mathbb{R}$ given by

$$h(x, y) = \begin{cases} g(x, y) & \text{if } x \le 0 \\ f(x, y) & \text{if } x > 0 \end{cases},$$

is continuous on all of $\mathbb{R}^2$. ◄

[1] We follow [Mun00] in the proof.

Uniformly continuous functions can always be extended continuously to the boundaries of their domains, as the following theorem establishes.

Theorem 2.16 *Let $f : \mathbb{D} \to \mathbb{Y}$ be uniformly continuous on $\mathbb{D}$. Then f can be extended to a continuous function $\widetilde{f} : \overline{\mathbb{D}} \to \overline{\mathbb{Y}}$. Moreover, the extension $\widetilde{f}$ is unique and uniformly continuous on $\overline{\mathbb{D}}$.*

PROOF We wish to define $\widetilde{f}$ on $\overline{\mathbb{D}}$ so that $\widetilde{f} = f$ on $\mathbb{D}$. To this end, we define $\widetilde{f}(x) \equiv f(x)$ for all $x \in \mathbb{D}$. We now set about defining the values of $\widetilde{f}(x)$ for $x \in \overline{\mathbb{D}} \backslash \mathbb{D}$. Consider such a point, $x_0 \in \overline{\mathbb{D}} \backslash \mathbb{D}$. There exists a sequence $\{x_n\} \subset \mathbb{D}$ such that $\lim x_n = x_0$ (Why?), and since the sequence $\{x_n\}$ converges, it must be a Cauchy sequence in $\mathbb{D}$. Therefore, $\{\widetilde{f}(x_n)\}$ must be a Cauchy sequence in $\mathbb{Y}$ (Why?), and so must converge to some point in $\overline{\mathbb{Y}}$. We define $\widetilde{f}(x_0) \equiv \lim f(x_n) \in \overline{\mathbb{Y}}$. The function $\widetilde{f}$ is now defined for all $x \in \overline{\mathbb{D}}$ so that $\widetilde{f} = f$ on $\mathbb{D}$, and $\widetilde{f}\left(\overline{\mathbb{D}}\right) \subset \overline{\mathbb{Y}}$. We now show that $\widetilde{f}$ is uniformly continuous on $\overline{\mathbb{D}}$. Note that since f is uniformly continuous on $\mathbb{D}$, so is $\widetilde{f}$, and therefore for any $\epsilon > 0$ there exists a $\delta_1 > 0$ such that

$$\xi, \eta \in \mathbb{D} \text{ and } |\xi - \eta| < \delta_1 \quad \Rightarrow \quad \left|\widetilde{f}(\xi) - \widetilde{f}(\eta)\right| < \tfrac{\epsilon}{3}. \tag{5.8}$$

Now suppose $\xi, \eta \in \overline{\mathbb{D}}$ and that $|\xi - \eta| < \frac{1}{3}\delta_1 \equiv \delta$. There exists a sequence $\{x_n\} \subset \mathbb{D}$ such that $\lim x_n = \xi$, and a sequence $\{x_n'\} \subset \mathbb{D}$ such that $\lim x_n' = \eta$, and therefore $\widetilde{f}(\xi) = \lim \widetilde{f}(x_n)$ and $\widetilde{f}(\eta) = \lim \widetilde{f}(x_n')$. Finally, there exists a positive integer N large enough so that the following all hold (Why?):

$$|x_N - \xi| < \tfrac{\delta_1}{3}, \;\; |x_N' - \eta| < \tfrac{\delta_1}{3}, \;\; |\widetilde{f}(x_N) - \widetilde{f}(\xi)| < \tfrac{\epsilon}{3}, \;\; |\widetilde{f}(x_N') - \widetilde{f}(\eta)| < \tfrac{\epsilon}{3}. \tag{5.9}$$

From this we have that

$$|x_N - x_N'| \le |x_N - \xi| + |\xi - \eta| + |\eta - x_N'| < \tfrac{\delta_1}{3} + \tfrac{\delta_1}{3} + \tfrac{\delta_1}{3} = \delta_1,$$

which in turn implies $|\widetilde{f}(x_N) - \widetilde{f}(x_N')| < \frac{\epsilon}{3}$, via (5.8). This last, along with (5.9), yields

$$\left|\widetilde{f}(\xi) - \widetilde{f}(\eta)\right| \le \left|\widetilde{f}(\xi) - \widetilde{f}(x_N)\right| + \left|\widetilde{f}(x_N) - \widetilde{f}(x_N')\right| + \left|\widetilde{f}(x_N') - \widetilde{f}(\eta)\right| < \epsilon.$$

We have shown that for $\epsilon > 0$ there exists $\delta = \frac{\delta_1}{3} > 0$ such that

$$\xi, \eta \in \overline{\mathbb{D}} \text{ and } |\xi - \eta| < \delta \quad \Rightarrow \quad \left|\widetilde{f}(\xi) - \widetilde{f}(\eta)\right| < \epsilon,$$

i.e., $\widetilde{f}$ is uniformly continuous on $\overline{\mathbb{D}}$. We leave the proof of the uniqueness of $\widetilde{f}$ to the reader. ♦

► **5.77** *Answer the three (Why?) questions in the proof of the above theorem. Also, note that there were three places where we claimed a sequence existed convergent to x_0, ξ, and η, respectively. Show that the result of the theorem does not depend on the choice of convergent sequence chosen in each case. Finally, establish that the $\widetilde{f}$ constructed in the above proof is unique.*

Note that the function $f:(0,1)\to\mathbb{R}$ given by $f(x)=\frac{1}{x}$ is continuous on $(0,1)$. However, there is no way to extend f continuously to $x_0=0$ since $\lim_{x\to 0^+} f(x)=\infty$. In fact, the sequence $\{x_n\}\subset(0,1)$ with $x_n=\frac{1}{n}$ for $n\in\mathbb{N}$ is such that $\lim x_n=0$, but $\lim f(x_n)$ does not exist. This example shows that not all continuous functions can be extended continuously to the boundary of their domains. The key in this case is that the function f is not *uniformly* continuous on $(0,1)$. This example points to the following result, the proof of which is left to the reader.

Proposition 2.17 *Suppose $f:\mathbb{D}\to\mathbb{Y}$ is continuous on the bounded domain $\mathbb{D}$. Then f can be extended continuously to the closure $\overline{\mathbb{D}}$ if and only if f is uniformly continuous on $\mathbb{D}$.*

► **5.78** *Prove the above proposition.*

Recall that the function $\mathrm{Arg}(z)$ was defined on $\mathbb{C}\setminus\{0\}$ and that it is continuous only on $\mathbb{C}\setminus K$ where $K=\{z\in\mathbb{C}:\mathrm{Re}(z)\le 0,\ \mathrm{Im}(z)=0\}$. In the following example, we establish that the restriction to $\mathbb{C}\setminus K$ of the function $\mathrm{Arg}(z)$ cannot be extended continuously to all of $\mathbb{C}$, or even to $\mathbb{C}\setminus\{0\}$.

Example 2.18 *Consider the function $f:\mathbb{C}\setminus K\to\mathbb{R}$ given by $f(z)=\mathrm{Arg}(z)$. We will show that $f(z)$ cannot be continuously extended to all of $\mathbb{C}\setminus\{0\}$. To this end, suppose there exists such an extension, $\tilde{f}:\mathbb{C}\setminus\{0\}\to\mathbb{C}$, continuous on $\mathbb{C}\setminus\{0\}$ such that $\tilde{f}(z)=\mathrm{Arg}(z)$ for all $z\in\mathbb{C}\setminus K$. We will derive a contradiction. Consider the sequence of points $\{z_n\}$ with $z_n=e^{i\left(\pi-\frac{1}{n}\right)}$ for $n\in\mathbb{N}$. These points approach $z=-1$ from above the real axis along the the arc of the unit circle centered at the origin. Since $\lim z_n=-1$, continuity of $\tilde{f}$ on K implies*

$$\tilde{f}(-1)=\tilde{f}(\lim z_n)=\lim\tilde{f}(z_n)=\lim\mathrm{Arg}(z_n)=\lim(\pi-\tfrac{1}{n})=\pi. \qquad (5.10)$$

If we now consider the points $\{w_n\}$ with $w_n=e^{i\left(-\pi+\frac{1}{n}\right)}$ for $n\in\mathbb{N}$, a sequence approaching $z=-1$ from below the real axis along the arc of the unit circle centered at the origin, then clearly $\lim w_n=-1$. However, by the continuity of $\tilde{f}$ on K we now conclude that

$$\tilde{f}(-1)=\tilde{f}(\lim w_n)=\lim\tilde{f}(w_n)=\lim\mathrm{Arg}(w_n)=\lim\left(-\pi+\tfrac{1}{n}\right)=-\pi.$$

This contradicts (5.10). ◄

► **5.79** *Is $f:\mathbb{C}\setminus\{0\}\to\mathbb{R}$ given by $f(z)=\mathrm{Arg}(z)$ uniformly continuous on $\mathbb{C}\setminus K$?*

► **5.80** *What are the implications of the previous example as it relates to the functions $\mathrm{Log}(z)$ and $z^{1/2}$?*

Another significant continuous extension result is the Tietze extension theorem. It states that if $\mathbb{D}$ is closed, and if $f:\mathbb{D}\to\mathbb{Y}$ is continuous and bounded on $\mathbb{D}$, i.e., $|f| \leq M$, then f can be extended to a continuous function on the whole space $\mathbb{X} \supset \mathbb{D}$ such that the extension is also bounded by M. To establish this result requires the use of sequences and series of functions, topics we develop next.

3 Sequences and Series of Functions

In this section we broaden our concept of sequences to include sequences of functions. This is a more sophisticated notion than a mere sequence of numbers, or even of vectors. One can develop the same intuitive understanding that the geometry of convergence of points allows, and some texts do so. In this view, one considers functions as points in a function space, and a sequence of functions is considered to converge if these points are determined to be getting closer to a well-defined limit function in that space. We choose not to pursue this point of view in our treatment of analysis, as it requires development of the more abstract notion of a *metric space*. In choosing a less sophisticated approach, we remind the reader that functions *are* more complicated mathematical objects than mere numbers or vectors.

3.1 Definitions and Examples

Definition 3.1 Consider $f_n : \mathbb{D} \to \mathbb{Y}$ for $n = 1, 2, \ldots$. We call the collection $\{f_n\}$ a **sequence of functions.** We say the sequence is **convergent** and we write $\lim_{n\to\infty} f_n(x) = f(x)$, where f is called **the limit function** of the sequence, if both of the following are true:

1. The limit function has the same domain and codomain as the functions in the sequence. That is, $f : \mathbb{D} \to \mathbb{Y}$.
2. For each $\epsilon > 0$ and each $x_0 \in \mathbb{D}$, there exists an $N \in \mathbb{N}$ such that $n > N \Rightarrow |f_n(x_0) - f(x_0)| < \epsilon$.

It is worth noting that the N referred to in Definition 3.1 will generally depend on both ϵ and x_0. In the special case where N does *not* depend on x_0, the convergence is rather special. Also, the definition of convergence given above is sometimes more specifically referred to as "pointwise convergence" to distinguish it from other, more subtle modes of convergence that might be defined for sequences of functions. The term "pointwise" refers to the fact that the sequence of functions must converge to the limit function at every point in the domain. While interesting, and of great use in more advanced topics in analysis, we won't have need of any of the other types of convergence in what follows. For this reason, our use of the unadorned term "convergence" will always mean "pointwise convergence."

An alternative version of condition 2 in the above definition of convergence is the following equivalent statement:

2′. For each $x_0 \in \mathbb{D}$, the sequence of points $\{f_n(x_0)\} \in \mathbb{Y}$ converges to the point $f(x_0) \in \mathbb{Y}$. That is,
$$x_0 \in \mathbb{D} \ \Rightarrow\ \lim_{n\to\infty} f_n(x_0) = f(x_0).$$

Example 3.2 *Consider the functions $f_n : (-1,1) \to \mathbb{R}$ given by $f_n(x) = x^n$ for $n \in \mathbb{N}$, and let $f : (-1,1) \to \mathbb{R}$ be given by $f(x) \equiv 0$. We will show that $\lim_{n\to\infty} f_n(x) = f(x) = 0$ on $(-1,1)$. To this end, fix an arbitrary $x_0 \in (-1,1)$. Then,*
$$|f_n(x_0) - f(x_0)| = |x_0^n| = |x_0|^n.$$
Certainly the above expression will be less than any given $\epsilon > 0$ for $x_0 = 0$. In the case $x_0 \neq 0$ the expression will be less than any given $\epsilon > 0$ as long as $n > \frac{\ln \epsilon}{\ln |x_0|}$. That is, for any $\epsilon > 0$ and any specified $x_0 \in (-1,1)$, we choose $N \in \mathbb{N}$ such that $N > \frac{\ln \epsilon}{\ln |x_0|}$. We then have that $n > N \ \Rightarrow\ |f_n(x_0) - f(x_0)| < \epsilon$, i.e., $\lim_{n\to\infty} f_n(x_0) = f(x_0)$. ◀

Note in the previous example that the value of N that fulfills our definition of convergence, Definition 3.1, depends on x_0.

► **5.81** *Consider the sequence of complex functions $f_n : N_1(0) \to \mathbb{C}$ described by $f_n(z) = z^n$ for $n \in \mathbb{N}$, and the complex function $f : N_1(0) \to \mathbb{C}$ given by $f(z) \equiv 0$ on $N_1(0) \subset \mathbb{C}$. What can you say about $\lim_{n\to\infty} f_n(z)$ on $N_1(0)$?*

We consider another example.

Example 3.3 *Let $D^2 = \{\mathbf{x} = (x,y) \in \mathbb{R}^2 : |x| < 1\}$, and consider the functions $\boldsymbol{f}_n : D^2 \to \mathbb{R}^2$ given by*
$$\boldsymbol{f}_n(\mathbf{x}) = \left(x^n,\ \frac{n\,y}{n+1}\right).$$
We outline the argument for showing that $\lim_{n\to\infty} \boldsymbol{f}_n(\mathbf{x}) = (0,\,y) \equiv \boldsymbol{f}(\mathbf{x})$, and leave the details to the reader. We denote $\boldsymbol{f}_n(\mathbf{x})$ component-wise by $\boldsymbol{f}_n(\mathbf{x}) = (f_{1n}(x,y),\ f_{2n}(x,y))$, and we proceed to examine the component functions $f_{1n}(x,y) = x^n$ and $f_{2n}(x,y) = \frac{n\,y}{n+1}$ for fixed values of x and y. One only needs to show that for all $(x,y) \in D^2$, $\lim_{n\to\infty} f_{1n}(x,y) = 0$, and $\lim_{n\to\infty} f_{2n}(x,y) = y$ to determine the limit $\lim_{n\to\infty} \boldsymbol{f}_n(\mathbf{x})$. ◀

► **5.82** *Complete the argument of the last example. Does the N required to establish convergence depend on x and/or y, and hence on $(x,y) = \mathbf{x}$?*

► **5.83** *Let $f_n : [0,1] \to \mathbb{R}$ be given by $f_n(x) = x^n$ for $n \in \mathbb{N}$. Use the Definition 3.1 to show that $\lim\limits_{n\to\infty} f_n(x) = \begin{cases} 0 & \text{for } 0 \le x < 1 \\ 1 & \text{for } x = 1 \end{cases}$. Does the N required depend on x? Note that each f_n is continuous on $[0,1]$, but the limit function f is not.*

► **5.84** *Consider the sequence of functions $f_n : \mathbb{R} \to \mathbb{R}$ given by $f_n(x) = \frac{\sin(nx)}{\sqrt{n}}$ for $n \in \mathbb{N}$. Find $\lim_{n\to\infty} f_n(x)$.*

► **5.85** *For $D = \{z \in \mathbb{C} : \mathrm{Re}(z) \neq 0\}$, consider the sequence of functions $f_n : D \to \mathbb{C}$ given by $f_n(z) = e^{-nz}$ for $n \in \mathbb{N}$. Find $\lim_{n\to\infty} f_n(z)$.*

3.2 Uniform Convergence

We now define a special form of convergence for a sequence of functions, a form that carries with it some rather important implications. Recall that uniform continuity implied more about a function's behavior than regular continuity. Likewise, we will find that *uniform convergence* implies more about a sequence's behavior than regular convergence. We begin with a definition.

Definition 3.4 Let $\{f_n\}$ be a sequence of functions, each with domain $\mathbb{D}$ and codomain $\mathbb{Y}$, that converges to the limit function f according to Definition 3.1 on page 208. We will say that the convergence is **uniform on** $\mathbb{D}$, or that the sequence converges **uniformly on** $\mathbb{D}$ if the following is true:

For each $\epsilon > 0$, there exists an $N \in \mathbb{N}$ such that for all $x_0 \in \mathbb{D}$,
$n > N \Rightarrow |f_n(x_0) - f(x_0)| < \epsilon$.

It is worth emphasizing how uniform convergence differs from ordinary convergence. In comparing Definition 3.1 on page 208 with the definition of uniform convergence given above, one should note that uniform convergence has the added condition that the value of N required to satisfy the definition must be *independent of* $x_0 \in \mathbb{D}$. This difference will be seen to have significant implications.

Example 3.5 *Recall our sequence of functions from Example 3.2, namely,* $f_n : (-1, 1) \to \mathbb{R}$ *given by* $f_n(x) = x^n$ *for* $n \geq 1$. *In that example, we showed that* $\lim f_n(x) = 0$ *on* $(-1, 1)$. *While we alluded to the convergence being nonuniform, we now show this to be the case unequivocally. To show this, let's assume otherwise. That is, assume there exists an* $N \in \mathbb{N}$ *such that for any* $x \in (-1, 1)$,

$$n > N \Rightarrow |x^n - 0| < \tfrac{1}{2}.$$

Note that we have chosen the ϵ *value to be* $\frac{1}{2}$ *here for convenience. The definition for convergence must be true for any positive* ϵ, *so certainly it must also be true for this particular choice. From the above mentioned assumption, we may conclude that*

$$\left|x^{N+1}\right| < \tfrac{1}{2} \quad \text{for all } x \in (-1, 1).$$

Taking the left-handed limit as x *approaches* 1 *on each side of the above inequality yields*

$$\lim_{x \to 1^-} \left|x^{N+1}\right| = 1 \leq \tfrac{1}{2},$$

a contradiction. Hence, our initial assumption that the convergence is uniform on $(-1, 1)$ *is incorrect.* ◀

Example 3.6 *Fix* $0 < a < 1$ *and consider the functions* $f_n : [-a, a] \to \mathbb{R}$ *given by* $f_n(x) = x^n$ *for* $n \geq 1$. *We will show that in this case, with the domain of definition for each* f_n *altered from the previous example, the convergence of* $\lim_{n \to \infty} f_n(x) = f(x) \equiv 0$ *is uniform on* $[-a, a]$. *To establish this, note* *that for any $x \in [-a, a]$ we have that $|x^n - 0| = |x|^n$. Clearly, for any given $\epsilon > 0$ we have $|x|^n < \epsilon$ for any n when $x = 0$. Also, for any $x \in [-a, a]$ we have $|x|^n \le a^n$, and this will be less than any given ϵ as long as $n > \frac{\ln \epsilon}{\ln a}$. Clearly neither case requires an N that depends on x, and so the convergence is uniform as claimed.* ◀

Example 3.7 *Consider the functions $f_n : D \to \mathbb{C}$ for $n \in \mathbb{N}$ and $D = \{z \in \mathbb{C} : |z| \le 1\}$ given by $f_n(z) = \frac{n\,z^2}{n+1+z}$. We will show that the sequence $\{f_n\}$ converges uniformly to some limit function $f : D \to \mathbb{C}$. In fact, it is easy to see that*

$$\lim f_n(z) = \lim \left(\frac{z^2}{1 + \frac{1}{n} + \frac{z}{n}} \right) = z^2 = f(z) \text{ on } D.$$

To show that the convergence is uniform, note that

$$|f_n(z) - f(z)| = \left| \frac{n\,z^2}{n+1+z} - z^2 \right| = \frac{|z^2 + z^3|}{|n+1+z|} \le \frac{|z|^2 + |z|^3}{n+1-|z|} \le \frac{2}{n+1-1} = \frac{2}{n}.$$

Clearly, given any $\epsilon > 0$ the above will be less than ϵ if $n > \frac{2}{\epsilon}$. Therefore, the convergence is uniform on D. ◀

There are many interesting consequences of uniform convergence that we will explore in later chapters. For now, we satisfy ourselves with one of the simpler ones, namely, the preservation of continuity.

Theorem 3.8 *Suppose $\{f_n\}$ is a sequence of continuous functions on $\mathbb{D}$ where $f_n : \mathbb{D} \to \mathbb{Y}$ for each $n = 1, 2, \ldots$. If $\{f_n\}$ converges uniformly to f on $\mathbb{D}$, then f is also continuous on $\mathbb{D}$.*

PROOF For any $\epsilon > 0$ there exists $N \in \mathbb{N}$ such that for all $x \in \mathbb{D}$,

$$n > N \Rightarrow |f_n(x) - f(x)| < \tfrac{\epsilon}{3}. \tag{5.11}$$

Fix $M > N$. Then, $|f_M(x) - f(x)| < \frac{\epsilon}{3}$ for all $x \in \mathbb{D}$. Now fix $x_0 \in \mathbb{D}$. Since f_M is continuous at x_0, there exists a $\delta > 0$ such that

$$x \in \mathbb{D} \text{ and } |x - x_0| < \delta \Rightarrow |f_M(x) - f_M(x_0)| < \tfrac{\epsilon}{3}. \tag{5.12}$$

Combining inequalities (5.11) and (5.12), we obtain that $x \in \mathbb{D}$ and $|x - x_0| < \delta$ together imply

$$\begin{aligned} |f(x) - f(x_0)| &= |f(x) - f_M(x) + f_M(x) - f_M(x_0) + f_M(x_0) - f(x_0)| \\ &\le |f(x) - f_M(x)| + |f_M(x) - f_M(x_0)| + |f_M(x_0) - f(x_0)| \\ &< \tfrac{\epsilon}{3} + \tfrac{\epsilon}{3} + \tfrac{\epsilon}{3} = \epsilon, \end{aligned}$$

and so f is continuous at x_0. ♦

► **5.86** *In a previous exercise, we saw that $f_n : [0, 1] \to \mathbb{R}$ given by $f_n(x) = x^n$ for $n \in \mathbb{N}$ converged to $f(x) = \begin{cases} 0 & \text{for } 0 \le x < 1 \\ 1 & \text{for } x = 1 \end{cases}$. Is the convergence uniform?*

► **5.87** *Let $f_n : (0,1] \to \mathbb{R}$ be given by $f_n(x) = \begin{cases} 1 & \text{for } 0 < x \le \frac{1}{n} \\ 0 & \text{for } \frac{1}{n} < x \le 1. \end{cases}$ Note that the functions $\{f_n\}$ are not continuous on $(0,1]$. Show that the limit function $f(x) \equiv \lim f_n(x)$ is continuous on $(0,1]$, and the convergence is not uniform. What does this say about Theorem 3.8?*

There is an important consequence following from the above theorem. Suppose $\{f_n(x)\}$ is a sequence of continuous functions on $\mathbb{D}$ that converges uniformly to $f(x)$ on $\mathbb{D}$. If x_0 is a limit point of $\mathbb{D}$, it can be shown that

$$\lim_{x \to x_0} \lim_{n \to \infty} f_n(x) = \lim_{n \to \infty} \lim_{x \to x_0} f_n(x), \tag{5.13}$$

i.e., the two limits may be interchanged. Such limit interchanges can be convenient.

► **5.88** *Prove the equality (5.13) under the conditions described above.*

Criteria for Uniform Convergence

The notion of a Cauchy sequence is relevant for sequences of functions just as it is for sequences of vectors. In particular, we define what it means for a sequence of functions $\{f_n\}$ to be *uniformly Cauchy* on their common domain.

Definition 3.9 Suppose $\{f_n\}$ is a sequence of functions where $f_n : \mathbb{D} \to \mathbb{Y}$. We say the sequence is **uniformly Cauchy** on $\mathbb{D}$ if for each $\epsilon > 0$ there exists an $N \in \mathbb{N}$ such that

$$m, n > N \Rightarrow |f_n(x) - f_m(x)| < \epsilon \text{ for all } x \in \mathbb{D}.$$

Note that the N that corresponds to a given ϵ in the above definition is independent of $x \in \mathbb{D}$. This is the "uniform" part of the definition of *uniformly Cauchy.*

The following proposition establishes the important fact that, under certain conditions, the sequence $\{f_n\}$ converges uniformly to a function $f : \mathbb{D} \to \mathbb{Y}$ if and only if the convergence is uniformly Cauchy on $\mathbb{D}$.

Proposition 3.10 *Suppose $\{f_n\}$ is a sequence of functions where $f_n : \mathbb{D} \to \mathbb{Y}$ and $\mathbb{Y}$ is $\mathbb{R}$, $\mathbb{R}^p$, or $\mathbb{C}$ for all of $n = 1, 2, \ldots$. Then $\{f_n\}$ converges uniformly to a function $f : \mathbb{D} \to \mathbb{Y}$ if and only if the convergence is uniformly Cauchy on $\mathbb{D}$.*

The key point here is that $\mathbb{Y}$ be complete, as the proof will show.

PROOF Suppose $\{f_n\}$ converges uniformly to f on $\mathbb{D}$, and let $\epsilon > 0$ be given. Then there exists an $N \in \mathbb{N}$ such that

$$n > N \text{ and } x \in \mathbb{D} \Rightarrow |f_n(x) - f(x)| < \tfrac{\epsilon}{2}.$$

Similarly,

$$m > N \text{ and } x \in \mathbb{D} \quad \Rightarrow \quad |f_m(x) - f(x)| < \tfrac{\epsilon}{2}.$$

Therefore, $m, n > N$ and $x \in \mathbb{D}$ implies that, for all $x \in \mathbb{D}$,

$$\begin{aligned} |f_n(x) - f_m(x)| &= |f_n(x) - f(x) + f(x) - f_m(x)| \\ &\leq |f_n(x) - f(x)| + |f(x) - f_m(x)| \\ &< \tfrac{\epsilon}{2} + \tfrac{\epsilon}{2} = \epsilon. \end{aligned}$$

Hence, the convergence is uniformly Cauchy on $\mathbb{D}$. Suppose now that $\{f_n\}$ is a sequence of functions from $\mathbb{D}$ to $\mathbb{Y}$ with the property that for each $\epsilon > 0$ there exists an $N \in \mathbb{N}$ such that

$$m, n > N \text{ and } x \in \mathbb{D} \quad \Rightarrow \quad |f_n(x) - f_m(x)| < \epsilon. \tag{5.14}$$

We will show that $\lim_{n\to\infty} f_n(x) = f(x)$ exists, and the convergence is uniform. To this end, note that according to (5.14), for any fixed $x \in \mathbb{D}$ the sequence $\{f_n(x)\}$ is a Cauchy sequence of elements in $\mathbb{Y}$. Therefore, since $\mathbb{Y}$ is complete, $\lim_{n\to\infty} f_n(x) = f(x)$ exists. To prove that the convergence is uniform, let $\epsilon > 0$ be given. Then there exists an $N \in \mathbb{N}$ such that

$$m, n > N \text{ and } x \in \mathbb{D} \quad \Rightarrow \quad |f_n(x) - f_m(x)| < \tfrac{\epsilon}{2}.$$

Now fix $n > N$ and $x \in \mathbb{D}$, and take the limit as $m \to \infty$ on each side of the last inequality. This yields

$$n > N \text{ and } x \in \mathbb{D} \quad \Rightarrow \quad |f_n(x) - f(x)| \leq \tfrac{\epsilon}{2} < \epsilon. \tag{5.15}$$

Since the N in expression (5.15) does not depend on x, the proposition is proved. ♦

► **5.89** *Suppose in the statement of the previous proposition that $\mathbb{Y}$ is presumed to be a subset of $\mathbb{R}$, $\mathbb{R}^p$, or $\mathbb{C}$. In this case, what else must be presumed about $\mathbb{Y}$ in order for the argument given in the above proof to remain valid?*

► **5.90** *Consider the sequence of functions $f_n : [-a, a] \to \mathbb{R}$ given by $f_n(x) = x^n$ where $0 \leq a < 1$. Establish that the sequence $\{f_n\}$ is uniformly Cauchy on $[-a, a]$, first by applying Proposition 3.10, and then by applying Definition 3.9. Which method is easier?*

Proposition 3.10 is a nice result in that, when $\mathbb{Y}$ is complete it characterizes uniformly convergent sequences of functions as those that have the uniformly Cauchy property. This has great value, particularly in proving theorems where it is not always necessary to know, or even possible to determine, the exact form of the limit function. Just confirming that the sequence converges uniformly to *some* limit function is often enough. For example, if each f_n is continuous and $\{f_n\}$ is a uniformly Cauchy sequence, Proposition 3.10 implies that f_n converges to some function f uniformly. Theorem 3.8 then implies that the limit function f is continuous, even if it's not possible to characterize its explicit form.

Example 3.11 *Consider a sequence of complex functions $f_n : D \to \mathbb{C}$ such that for all $n \in \mathbb{N}$ and for all $z \in D$,*

$$|f_{n+1}(z) - f_n(z)| < c\,r^n, \quad \text{where } c \in \mathbb{R} \text{ and } 0 < r < 1.$$

We will show that $\{f_n\}$ is uniformly Cauchy on D, and hence converges uniformly to some function $f : D \to \mathbb{C}$. To this end, if $n > m$ we have

$$\begin{aligned} |f_n(z) - f_m(z)| &\le |f_n(z) - f_{n-1}(z)| + |f_{n-1}(z) - f_{n-2}(z)| + \cdots \\ &\qquad + |f_{m+1}(z) - f_m(z)| \\ &< c\,r^{n-1} + c\,r^{n-2} + \cdots + c\,r^m \\ &< c\,r^m \left(1 + r + r^2 + \cdots\right) \\ &= \frac{c\,r^m}{1-r}. \end{aligned}$$

In general, we then have that

$$|f_n(z) - f_m(z)| \le \frac{c\,r^{\min(m,n)}}{1-r},$$

and the last expression above will be less than ϵ if $m, n > N$ for some $N \in \mathbb{N}$. (Why?) Therefore $\{f_n\}$ is uniformly Cauchy on D and hence converges uniformly to some limit function $f : D \to \mathbb{C}$. ◀

▶ **5.91** *Answer the (Why?) question in the above example. Also, come up with an explicit example of a sequence of complex functions $f_n : D \to \mathbb{C}$ behaving as described there. Do the functions f_n have to be complex functions?*

In fact, when dealing with sequences of functions it is often not possible to characterize the form of the limit function, even in cases where the convergence is uniform. If one *does* know the limiting function, a situation that is relatively rare, the following result provides a convenient test for determining whether the convergence is uniform. The test is easy to use because it relates the convergence of functions to the convergence of real numbers.

Theorem 3.12 (The M-Test)
Suppose $\{f_n\}$ is a sequence of functions where $f_n : \mathbb{D} \to \mathbb{Y}$ for each $n = 1, 2, \ldots,$ that $f : \mathbb{D} \to \mathbb{Y}$, and that $\lim_{n\to\infty} f_n(x) = f(x)$ for all $x \in \mathbb{D}$. Suppose too that $M_n \equiv \sup_{x\in\mathbb{D}} |f_n(x) - f(x)|$ exists for all n. Then $\{f_n\}$ converges uniformly to f on $\mathbb{D}$ if and only if $\lim M_n = 0$.

PROOF Suppose $\lim M_n = 0$, and let $\epsilon > 0$ be given. Then there exists an $N \in \mathbb{N}$ such that $n > N \Rightarrow M_n < \epsilon$. Note that this N is independent of any $x \in \mathbb{D}$. With this same N we also have that

$$n > N \Rightarrow |f_n(x) - f(x)| \le M_n < \epsilon,$$

and so the sequence $\{f_n\}$ converges to f uniformly. Conversely, suppose

$\lim_{n\to\infty} f_n(x) = f(x)$ uniformly on $\mathbb{D}$, and let $\epsilon > 0$ be given. Then there exists an $N \in \mathbb{N}$ such that

$$n > N \text{ and } x \in \mathbb{D} \Rightarrow |f_n(x) - f(x)| < \tfrac{\epsilon}{2}. \tag{5.16}$$

Since N in expression (5.16) is independent of $x \in \mathbb{D}$, we clearly have

$$n > N \Rightarrow M_n = \sup_{x\in\mathbb{D}} |f_n(x) - f(x)| \le \tfrac{\epsilon}{2} < \epsilon.$$

Hence, $\lim M_n = 0$ and the theorem is proved. ♦

The M-test says, roughly, that for a sequence of functions $\{f_n\}$ converging to f on $\mathbb{D}$, the convergence will be uniform on $\mathbb{D}$ exactly when the biggest difference in values between $f_n(x)$ and $f(x)$ over $\mathbb{D}$ goes to 0 as n goes to ∞.

Example 3.13 *Let $f_n : [0,1] \to \mathbb{R}$ be given by $f_n(x) = x^n$ for $n \in \mathbb{N}$. Note that*

$$\lim_{n\to\infty} f_n(x) = f(x) = \begin{cases} 0 & \text{for } 0 \le x < 1 \\ 1 & \text{for } x = 1 \end{cases},$$

and that $M_n = \sup_{x\in[0,1]} |f_n(x) - f(x)| = 1$. Since $\lim M_n \neq 0$, it follows from the M-test that the convergence is not uniform on $[0,1]$. ◀

Example 3.14 *Let $f_n : D \to \mathbb{C}$ be given by $f_n(z) = z^n$ for $n \in \mathbb{N}$, where $D = N_a(0)$ for $0 < a < 1$. Then $\lim_{n\to\infty} f_n(z) = 0$ (Why?), and*

$$M_n = \sup_{z\in D} |z^n| = \sup_{z\in D} |z|^n = a^n.$$

Since $\lim M_n = 0$, it follows by the M-test that the convergence is uniform on D. ◀

▶ **5.92** *Answer the (Why?) question in the last example.*

Example 3.15 *Let $f_n : [1,2] \to \mathbb{R}$ be given by $f_n(x) = \frac{n^2 x}{1+n^3x^2}$ for $n \in \mathbb{N}$. Then $\lim_{n\to\infty} f_n(x) = 0$ on $[1,2]$, and the convergence is uniform on $[1,2]$. To see this, note that*

$$M_n = \sup_{[1,2]} \left|\tfrac{n^2 x}{1+n^3x^2}\right| = \max_{[1,2]} \left(\tfrac{n^2 x}{1+n^3x^2}\right).$$

To evaluate M_n, it is not difficult to show that for n fixed, $f_n(x) = \frac{n^2 x}{1+n^3x^2}$ is decreasing on $[1,2]$, and therefore $M_n = \frac{n^2}{1+n^3}$. Clearly then, $\lim M_n = 0$. ◀

▶ **5.93** *Show that $f_n(x) = \frac{n^2 x}{1+n^3x^2}$ is decreasing on $[1,2]$.*

▶ **5.94** *Let $f_n : D \to \mathbb{C}$ be given by $f_n(z) = \frac{n^2 z}{1+n^3z^2}$ for $n = 2, 3, \ldots$, where $D = \{z \in \mathbb{C} : 1 \le |z| \le 2\}$. Find $f(z) = \lim f_n(z)$. Is the convergence uniform on D?*

Example 3.16 *Let f_n be given by $f_n(x) = \frac{n^2 x}{1+n^3x^2}$ for $n \in \mathbb{N}$ as in the previous example, but change the domain to $[0,1]$. We will show that although*

$\lim_{n\to\infty} f_n(x) = 0$ *as before, the convergence is not uniform on* $[0,1]$. *To see this, note that*

$$M_n = \sup_{[0,1]} \left| \tfrac{n^2 x}{1+n^3x^2} \right| = \max_{[0,1]} \left(\tfrac{n^2 x}{1+n^3x^2} \right),$$

but

$$M_n \geq \frac{n^2 \left(\frac{1}{n^{3/2}}\right)}{1 + n^3 \left(\frac{1}{n^{3/2}}\right)^2} = \frac{\sqrt{n}}{2}.$$

Therefore, $\lim M_n \neq 0$, *and so the convergence is not uniform on* $[0,1]$. ◀

3.3 Series of Functions

Just as we generalized the notion of sequences to sequences of functions, we can also generalize the notion of series to series of functions. We begin with a definition.

Definition 3.17 Let $f_j : \mathbb{D} \to \mathbb{Y}$ for $j \in \mathbb{N}$ be a sequence of functions.

1. The series $\sum_{j=1}^{\infty} f_j(x)$ is referred to as the **series of functions** associated with the sequence $\{f_j\}$.
2. The function $s_n : \mathbb{D} \to \mathbb{Y}$ for each $n \in \mathbb{N}$, defined by $s_n(x) \equiv \sum_{j=1}^{n} f_j(x)$, is called the n**th partial sum** of the series, and the sequence of functions $\{s_n\}$ is called the **sequence of partial sums** associated with the series $\sum_{j=1}^{\infty} f_j(x)$.

Of course, we are primarily concerned with those series of functions that converge in some sense. We define this concept next.

Definition 3.18 Suppose $\sum_{j=1}^{\infty} f_j(x)$ is a series of functions $f_j : \mathbb{D} \to \mathbb{Y}$ for $j \in \mathbb{N}$, with the associated sequence of partial sums $\{s_n\}$. If the sequence $\{s_n\}$ converges on $\mathbb{D}$ to some function $s : \mathbb{D} \to \mathbb{Y}$, then we say that the series $\sum_{j=1}^{\infty} f_j(x)$ **converges to the function** $s : \mathbb{D} \to \mathbb{Y}$, and we write

$$s(x) = \sum_{j=1}^{\infty} f_j(x) \quad \text{for each } x \in \mathbb{D}.$$

Example 3.19 *Let* $f_j : (-1,1) \to \mathbb{R}$ *be given by* $f_j(x) = x^j$ *for* $j \geq 0$. *Then,* $\sum_{j=0}^{\infty} f_j(x)$ *is a geometric series with ratio* x, *and so*

$$\sum_{j=0}^{\infty} f_j(x) = \sum_{j=0}^{\infty} x^j = \frac{1}{1-x} = s(x) \quad \text{for } |x| < 1, \textit{i.e.,} \text{ on } (-1,1).$$

◀

Example 3.20 *Consider the series of complex functions given by* $\sum_{j=0}^{\infty} \left(\frac{4}{z}\right)^j$. *Since this is a geometric series with ratio* $\frac{4}{z}$, *we may conclude that the series* *converges to* $s(z) \equiv \frac{1}{1-\frac{4}{z}} = \frac{z}{z-4}$ *if* $\left|\frac{4}{z}\right| < 1$. *That is, the series converges to the function* $s(z) = \frac{z}{z-4}$ on $|z| > 4$. ◀

▶ **5.95** *What if the series in the previous example is replaced by* $\sum_{j=0}^{\infty} \left(\frac{z_0}{z}\right)^j$, *for some* $z_0 \in \mathbb{C}$? *On what part of the complex plane is the series convergent now?*

Just as with sequences of functions, series of functions can converge *uniformly*. We specify this idea next.

Definition 3.21 Let $f_j : \mathbb{D} \to \mathbb{Y}$ for $j \in \mathbb{N}$ be a sequence of functions, and suppose the series $\sum_{j=1}^{\infty} f_j(x)$ converges to the function $s : \mathbb{D} \to \mathbb{Y}$. We say that the series **converges uniformly on** $\mathbb{D}$ if the sequence of associated partial sums $\{s_n\}$ converges uniformly to s on $\mathbb{D}$.

Example 3.22 *Consider again the functions* $f_j : (-1,1) \to \mathbb{R}$ *given by* $f_j(x) = x^j$ *for* $j \geq 0$. *We learned in Example 3.19 above that*

$$\sum_{j=0}^{\infty} f_j(x) = \sum_{j=0}^{\infty} x^j = \frac{1}{1-x} = s(x) \text{ on } (-1,1).$$

So certainly the series also converges to $s(x)$ *on* $[-a,a] \subset (-1,1)$. *In fact, the convergence on such an interval* $[-a,a]$ *is uniform. To see this, consider*

$$s_n(x) = \sum_{j=1}^{n} x^j = \frac{1-x^{n+1}}{1-x},$$

and compute

$$M_n \equiv \sup_{x\in[-a,a]} \left|s_n(x) - \frac{1}{1-x}\right| = \max_{x\in[-a,a]} \left(\frac{|x|^{n+1}}{1-x}\right) \leq \frac{a^{n+1}}{1-a}.$$

Clearly, $\lim M_n = 0$, *and the result follows by the M-test.* ◀

▶ **5.96** *In Example 3.20 on page 216, is the convergence uniform on* $D = \{z \in \mathbb{C} : |z| > 4\}$? *If not, can you modify* D *so that the convergence is uniform on* D? *What if* $f_j(z) = \left(\frac{4}{z}\right)^j$ *is replaced by* $\left(\frac{z_0}{z}\right)^j$ *for some* $z_0 \in \mathbb{C}$? *What is the largest subset* $D \subset \mathbb{C}$ *where the convergence is uniform now?*

Another test for the uniform convergence of a series of functions is the Weierstrass M-test.

Theorem 3.23 (The Weierstrass M-test)
Suppose $f_j : \mathbb{D} \to \mathbb{Y}$ *where* $\mathbb{Y}$ *is the whole space* $\mathbb{R}$, $\mathbb{R}^p$, *or* $\mathbb{C}$ *for* $j \in \mathbb{N}$ *is a sequence of functions. Suppose also that* $|f_j(x)| \leq M_j$ *for all* $x \in \mathbb{D}$, *where* $\sum_{j=1}^{\infty} M_j$ *converges. Then* $\sum_{j=1}^{\infty} f_j(x)$ *converges uniformly to* $s(x)$ *on* $\mathbb{D}$ *for some function* $s : \mathbb{D} \to \mathbb{Y}$.

PROOF Let $\epsilon > 0$ be given. Then there exists an $N \in \mathbb{N}$ such that

$$m < n \text{ and } m, n > N \Rightarrow \left| \sum_{j=m+1}^{n} M_j \right| < \epsilon.$$

But $m, n > N$ implies that

$$|s_n(x) - s_m(x)| = \left| \sum_{j=m+1}^{n} f_j(x) \right| \le \sum_{j=m+1}^{n} M_j < \epsilon,$$

i.e., the sequence of functions $\{s_j(x)\}$ is uniformly Cauchy on $\mathbb{D}$. Therefore, since $\mathbb{Y}$ is complete, the sequence converges uniformly to some function $s(x)$ on $\mathbb{D}$. ♦

► **5.97** *Suppose in the statement of the previous theorem that $\mathbb{Y}$ is presumed to be a subset of $\mathbb{R}$, $\mathbb{R}^p$, or $\mathbb{C}$. In this case, what else must be presumed about $\mathbb{Y}$ in order for the argument given in the above proof to remain valid?*

The benefits of the Weierstrass M-test are that it is fairly easy to apply and it does not require us to know the sum s in order to apply it. The drawback is that the limit function s is not always evident, and the Weierstrass M-test does not provide a means for determining it, even when convergence is confirmed.

Example 3.24 *We can show that $\sum_{j=0}^{\infty} x^j$ converges uniformly to $\frac{1}{1-x}$ on $[-a, a]$ for $0 < a < 1$. This follows easily from the Weierstrass M-test since*

$$|x^j| = |x|^j \le a^j \equiv M_j \text{ for all } x \in [-a, a],$$

and $\sum_{j=1}^{\infty} a^j$ is a convergent geometric series. ◄

Example 3.25 *Consider the series of functions given by $\sum_{j=1}^{\infty} \frac{\sin(jx)}{j^5}$. Note that*

$$\left| \frac{\sin(jx)}{j^5} \right| \le \frac{1}{j^5} = M_j,$$

and since $\sum_{j=1}^{\infty} M_j$ converges, it follows that $\sum_{j=1}^{\infty} \frac{\sin(jx)}{j^5}$ converges uniformly on $\mathbb{R}$. ◄

The following result is yet another important consequence of uniform convergence.

Theorem 3.26 *Suppose $f_j : \mathbb{D} \to \mathbb{Y}$ for $j \in \mathbb{N}$ is a sequence of continuous functions on $\mathbb{D}$. Suppose also that $\sum_{j=1}^{\infty} f_j(x)$ converges uniformly to $s(x)$ on $\mathbb{D}$ for some function $s : \mathbb{D} \to \mathbb{Y}$. Then the function s is continuous.*

► **5.98** *Prove the above theorem.*

Example 3.27 *Fix $0 < a < 1$, and let $x \in [-a, a]$. Then $\sum_{j=0}^{\infty} x^j = \frac{1}{1-x}$. Since the convergence is uniform on $[-a, a] \subset (-1,1)$, and $f_j(x) = x^j$ is continuous on $[-a, a] \subset (-1,1)$ for each $j \geq 1$, we see that the sum $\frac{1}{1-x}$ must also be continuous on $[-a, a] \subset (-1,1)$. Since this is true for all closed intervals $[-a, a] \subset (-1,1)$, the function given by $f(x) = \frac{1}{1-x}$ is continuous on $(-1,1)$. To see this, consider any $x_0 \in (-1,1)$. Then there exists $0 < a < 1$ such that $x_0 \in [-a, a] \subset (-1,1)$, and the result follows from the above.* ◀

▶ **5.99** *Is the convergence of $\sum_{j=0}^{\infty} x^j$ to $\frac{1}{1-x}$ uniform on $(-1,1)$?*

Uniform convergence can be used to confirm the continuity of the exponential function $\exp : \mathbb{R} \to \mathbb{R}$ given by $\exp(x) = e^x$.

Example 3.28 *Consider the real exponential function $\exp : \mathbb{R} \to \mathbb{R}$ denoted by $\exp(x) = e^x$ and defined by $\sum_{j=0}^{\infty} \frac{x^j}{j!}$. We will (again) establish that this function is continuous on $\mathbb{R}$. To this end, consider $N_r(0)$ with $r > 0$. Then, for $x \in N_r(0)$ we have*

$$\left|\frac{x^j}{j!}\right| \leq \frac{r^j}{j!},$$

and since $\sum_{j=0}^{\infty} \frac{r^j}{j!}$ converges, it follows by the Weierstrass M-test that $\sum_{j=0}^{\infty} \frac{x^j}{j!}$ converges uniformly to some function, in fact, $\exp(x) = e^x$, on $N_r(0)$. By Theorem 3.8 on page 211, since $\exp(x) = \sum_{j=0}^{\infty} \frac{x^j}{j!} = \sum_{j=0}^{\infty} f_j(x)$ converges uniformly on $N_r(0)$ and each $f_j(x)$ is continuous on $N_r(0)$, the function $\exp(x)$ is also continuous on $N_r(0)$. Since $r > 0$ was arbitrary, $\exp(x)$ is continuous on all of $\mathbb{R}$. ◀

Note in the above example that while we are able to conclude that $\exp(x)$ is continuous on all of $\mathbb{R}$, we have not concluded that the sum $\exp(x) = \sum_{j=0}^{\infty} \frac{x^j}{j!} = \sum_{j=0}^{\infty} f_j(x)$ converges uniformly on all of $\mathbb{R}$ (it doesn't!).

▶ **5.100** *Consider $\sum_{j=0}^{\infty} \frac{z^j}{j!}$ where $z \in \mathbb{C}$.*

a) For what values of z, if any, does the series converge?

b) Is the convergence uniform on $\mathbb{C}$?

c) What if z is restricted to $\{z \in \mathbb{C} : |z| \leq r\}$, for some $r > 0$?

3.4 The Tietze Extension Theorem

Finally, we conclude this section with an important result called the Tietze extension theorem.[2] It states that if $\mathbb{D}$ is closed, and if $f : \mathbb{D} \to \mathbb{Y}$ is continuous and bounded on $\mathbb{D}$, i.e., $|f| \leq M$, then f can be extended to a continuous function $\widetilde{f}$ on the whole space $\mathbb{X} \supset \mathbb{D}$ such that $\widetilde{f}$ is bounded by M on $\mathbb{X}$. While this result would more naturally find its place in subsection 2.3 of this

[2]We follow [Bar76] in the proof.

chapter, its proof requires Theorem 3.26, and so it is presented here. The theorem is often a great convenience in that, when the conditions are satisfied, one can effectively consider a given function $f : \mathbb{D} \to \mathbb{Y}$ as if it were defined on all of $\mathbb{X}$, retaining the often crucial properties of being continuous and bounded. In order to prove the theorem, we require a few technical results. We begin by defining the "distance function" from a point to a set. More specifically, consider any subset $\mathbb{D} \subset \mathbb{X}$, and define the distance function $d_{\mathbb{D}} : \mathbb{X} \to \mathbb{R}$ by

$$d_{\mathbb{D}}(x) = \inf_{a \in \mathbb{D}} |x - a|.$$

Then $d_{\mathbb{D}}$ can be interpreted as measuring the distance from any point in $\mathbb{X}$ to the set $\mathbb{D}$.

► **5.101** *Establish the following properties of $d_{\mathbb{D}}$ defined above.*

a. *Show that $|d_{\mathbb{D}}(x) - d_{\mathbb{D}}(y)| \leq |x - y|$ for all $x, y \in \mathbb{X}$, and therefore $d_{\mathbb{D}}$ is uniformly continuous on $\mathbb{X}$.*

b. *Show that $d_{\mathbb{D}}(x) \geq 0$ for all $x \in \mathbb{X}$, and that $d_{\mathbb{D}}(x) = 0$ if and only if $x \in \overline{\mathbb{D}}$.*

The proof of the theorem will require the following technical lemma.

Lemma 3.29 *Suppose $\mathbb{D} \subset \mathbb{X}$ is closed, and suppose $g : \mathbb{D} \to \mathbb{R}$ is continuous on $\mathbb{D}$ and bounded, i.e., $|g(x)| \leq M$ on $\mathbb{D}$ for some $M > 0$. Then there exists a continuous function $G : \mathbb{X} \to \mathbb{R}$ such that $|G(x)| \leq \frac{M}{3}$ on $\mathbb{X}$, and $|g(x) - G(x)| \leq \frac{2M}{3}$ for all $x \in \mathbb{D}$.*

The lemma states that there exists a bounded, continuous function G defined on all of $\mathbb{X}$ whose function values are "never too far from" those of g on $\mathbb{D}$.

PROOF We begin by defining the following subsets of $\mathbb{D}$:

$$\mathbb{D}^- \equiv \left\{x \in \mathbb{D} : -M \leq g(x) \leq -\tfrac{M}{3}\right\},$$

and

$$\mathbb{D}^+ \equiv \left\{x \in \mathbb{D} : \tfrac{M}{3} \leq g(x) \leq M\right\}.$$

We leave it to the reader to show that $\mathbb{D}^+$ and $\mathbb{D}^-$ are closed in $\mathbb{D}$, and therefore closed in $\mathbb{X}$. We now define $G : \mathbb{X} \to \mathbb{R}$ by

$$G(x) = \frac{M}{3}\left[\frac{d_{\mathbb{D}^-}(x) - d_{\mathbb{D}^+}(x)}{d_{\mathbb{D}^-}(x) + d_{\mathbb{D}^+}(x)}\right],$$

and leave it to the reader to show that G has the following properties:

(i) $G(x) = -\frac{M}{3}$ on $\mathbb{D}^-$.

(ii) $G(x) = \frac{M}{3}$ on $\mathbb{D}^+$.

(iii) $|G(x)| \leq \frac{M}{3}$ on $\mathbb{X}$.

We will now show that $|g(x) - G(x)| \leq \frac{2M}{3}$ if $x \in \mathbb{D}$. To see this, suppose first that $x \in \mathbb{D}^-$. Then,

$$|g(x) - G(x)| = \left|g(x) + \tfrac{M}{3}\right| \le \tfrac{2M}{3}.$$

A similar argument shows that for $x \in \mathbb{D}^+$,

$$|g(x) - G(x)| = \left|g(x) - \tfrac{M}{3}\right| \le \tfrac{2M}{3}.$$

Finally, if $x \in \mathbb{D} \setminus \left(\mathbb{D}^- \cup \mathbb{D}^+\right)$, we have $|g(x)| \le \frac{M}{3}$, and so,

$$|g(x) - G(x)| \le |g(x)| + |G(x)| \le \tfrac{2M}{3}. \qquad \blacklozenge$$

► **5.102** *Show that $\mathbb{D}^+$ and $\mathbb{D}^-$ are closed in $\mathbb{D}$, and that $\mathbb{D}^- \cap \mathbb{D}^+ = \varnothing$. This is sufficient to show that $d_{\mathbb{D}^-}(x) + d_{\mathbb{D}^+}(x)$ is never 0, and therefore G is well defined.*

► **5.103** *Prove properties (i), (ii), and (iii) listed in the above proof.*

We are now ready to state and prove the Tietze extension theorem.

Theorem 3.30 (The Tietze Extension Theorem)
Suppose $\mathbb{D}$ is closed and that $f : \mathbb{D} \to \mathbb{Y}$ is continuous and bounded on $\mathbb{D}$, i.e., $|f| \le M$ for some $M > 0$. Then f can be extended to a continuous function $\widetilde{f}$ on the whole space $\mathbb{X} \supset \mathbb{D}$ such that $\left|\widetilde{f}\right| \le M$.

PROOF Suppose $\mathbb{D} \subset \mathbb{X}$ is closed, and let $f : \mathbb{D} \to \mathbb{R}$ be continuous on $\mathbb{D}$. Suppose also that there exists $M > 0$ such that $|f(x)| \le M$ for all $x \in \mathbb{D}$. Then by Lemma 3.29 there exists a continuous function $F_1 : \mathbb{X} \to \mathbb{R}$ such that

$$|f(x) - F_1(x)| \le \tfrac{2}{3} M \quad \text{on } \mathbb{D}, \quad \text{and} \quad |F_1(x)| \le \tfrac{1}{3} M \quad \text{on } \mathbb{X}.$$

Now consider the function $f - F_1 : \mathbb{D} \to \mathbb{R}$, and apply Lemma 3.29 to obtain another continuous function $F_2 : \mathbb{X} \to \mathbb{R}$ such that

$$|f - F_1 - F_2| \le \left(\tfrac{2}{3}\right)^2 M \quad \text{on } \mathbb{D}, \quad \text{and} \quad |F_2| \le \tfrac{1}{3}\left(\tfrac{2}{3}\right) M \quad \text{on } \mathbb{X}.$$

Continuing in this way, we generate a sequence of continuous functions $F_j : \mathbb{X} \to \mathbb{R}$ such that for any $n \in \mathbb{N}$ and $1 \le j \le n$,

$$\left|f - \sum_{j=1}^{n} F_j\right| \le \left(\tfrac{2}{3}\right)^n M \quad \text{on } \mathbb{D}, \quad \text{and} \quad |F_j| \le \tfrac{1}{3}\left(\tfrac{2}{3}\right)^{j-1} M \quad \text{on } \mathbb{X}.$$

By the Weierstrass M-test and Theorem 3.26, it is easy to see that $\sum_{j=1}^{\infty} F_j(x)$ converges uniformly to a continuous function $F : \mathbb{X} \to \mathbb{R}$. The reader should verify that $F(x) = f(x)$ for all $x \in \mathbb{D}$, and that $|F(x)| \le M$ on $\mathbb{D}$.[3] ♦

► **5.104** *Verify that $F(x) = f(x)$ for all $x \in \mathbb{D}$, and that $|F(x)| \le M$ on $\mathbb{D}$.*

[3]However, even if we did not know that $|F(x)| \le M$ for all $x \in \mathbb{X}$, we could make that happen by defining $F_1(x) \equiv \max(M, \min(-M, F(x)))$. Then F_1 still agrees with f on $\mathbb{D}$, F_1 is continuous, and F_1 is bounded by M.

The Tietze extension theorem is an existence theorem. Its statement makes a claim about the existence of a function with certain properties but makes no claims about how to find this function. It might seem that the proof of the theorem provides a practical algorithm to do just that, but exploration of an example or two will unfortunately convince the reader that this is not the case. Despite this, the theorem is of great theoretical value, and of practical use too, but the practicality is of a slightly different variety than most would initially presume. In very simple cases, the theorem might not even seem necessary. For example, consider the function $f : [-1,1] \to \mathbb{R}$ given by $f(x) = x^3$. This function is continuous on its domain and is bounded by 1 there. Clearly we can extend f continuously to all of $\mathbb{R}$ so as to maintain the bound by simply "capping off" the function at the value -1 for $x < -1$ and at the value 1 for $x > 1$. The continuous extension of f that we obtain in this way is $\widetilde{f} : \mathbb{R} \to \mathbb{R}$ given by

$$\widetilde{f}(x) = \begin{cases} -1 & \text{if } x < -1 \\ x^3 & \text{if } -1 \leq x < 1 \, . \\ 1 & \text{if } x > 1 \end{cases}$$

Similarly, the function $g : D^2 \to \mathbb{R}$ given by $g(x,y) = x^2 + y^2$ where $D^2 = \{(x,y) \in \mathbb{R}^2 : x^2+y^2 \leq 1\}$ is readily continuously extended to the whole plane while maintaining the bound of 1 by "capping off" the function outside the unit disk. The continuous extension of g obtained in this way is $\widetilde{g} : \mathbb{R}^2 \to \mathbb{R}$ given by

$$g(x,y) = \begin{cases} x^2 + y^2 & \text{if } (x,y) \in D^2 \\ 1 & \text{if } (x,y) \in \mathbb{R}^2 \setminus D^2 \end{cases} .$$

It is not clear in either of the simple examples described above whether the algorithm described in the proof of the Tietze extension theorem would give rise to the same extensions, even if the algorithm were easy to apply. The theorem, after all, makes no claims of uniqueness. For an example whose extension is not so obvious, consider a function such as $h : D^2 \to \mathbb{R}$ given by $h(x,y) = xy$ where $D^2 = \{(x,y) \in \mathbb{R}^2 : x^2 + y^2 \leq 1\}$. Initially, it might not even seem possible to continuously extend such a function to the whole plane so as to maintain the bound on h for all values in D^2 (consider the differing values of h along the boundary of D^2). Yet, the theorem guarantees that just such an extension exists.

4 SUPPLEMENTARY EXERCISES

1. *Prove the inequality $|\sin x| \leq |x|$ by a geometric argument (look at a circle). Use that to prove the continuity of $\sin x$ in a similar fashion as in the example.*

2. *Determine whether the given function is continuous anywhere in its domain.*

a) $f : \mathbb{D} \to \mathbb{R}$ *with* $f(x) = x$ *for* $\mathbb{D} = \{0,1\}$.

b) $f : \mathbb{D} \to \mathbb{R}$ *with* $f(x) = x$ *for* $\mathbb{D} = \{\frac{1}{n} : n \in \mathbb{N}\}$.

c) $f : \mathbb{D} \to \mathbb{R}$ *with* $f(x) = x$ *for* $\mathbb{D} = \{\frac{1}{n} : n \in \mathbb{N}\} \cup \{0\}$.

d) $f : \mathbb{D} \to \mathbb{R}$ with $f(x) = x$ for $\mathbb{D} = \mathbb{Q} \cap [0,1]$.
e) $f : \mathbb{D} \to \mathbb{R}$ with $f(x) = x^{-1}$ for $\mathbb{D} = \{\frac{1}{n} : n \in \mathbb{N}\}$.
f) $f : \mathbb{D} \to \mathbb{R}$ with $f(x) = x^{-1}$ for $\mathbb{D} = \mathbb{Q} \cap [0,1]$.
g) $f : \mathbb{D} \to \mathbb{R}$ with $f(x) = \begin{cases} 0 & \text{for } x \in \mathbb{Q} \text{ such that } x < \sqrt{2} \\ 1 & \text{for } x \in \mathbb{Q} \text{ such that } x > \sqrt{2} \end{cases}$.

3. *Suppose $f : \mathbb{D} \to \mathbb{Y}$ is such that $\mathbb{D}$ is a discrete set in $\mathbb{X}$. Can f be continuous anywhere in its domain?*

4. *Consider the domain $D^2 = \{\mathbf{x} = (x,y) \in \mathbb{R}^2 : x > 0, y > 0\}$ and the function $f : D^2 \to \mathbb{R}$ given by $f(\mathbf{x}) = \frac{x}{x+y}$ as in Example 1.4 on page 180. Let $\mathbf{x}_n = \left(\frac{1}{n}, 1 + \frac{1}{n^2}\right)$ and use remark 2 following Definition 1.1 to show that $\lim_{n\to\infty} f(\mathbf{x}_n) = 0$.*

5. *Consider the function $f : \mathbb{R} \to \mathbb{R}$ given by $f(x) = \begin{cases} 0 & \text{if } x \in \mathbb{Q} \\ 1 & \text{if } x \in \mathbb{I} \end{cases}$. Is f continuous at any point in $\mathbb{R}$?*

6. *This exercise establishes the continuity of the dot product on $\mathbb{R}^k$. Let the function $f : \mathbb{R}^k \times \mathbb{R}^k \to \mathbb{R}$ be given by $f(\mathbf{x}, \mathbf{y}) = \mathbf{x} \cdot \mathbf{y}$.*
a) *Fix $\mathbf{y} = \mathbf{y}_0$. Show that the function defined by $f_1(\mathbf{x}) \equiv f(\mathbf{x}, \mathbf{y}_0)$ is continuous, i.e., the dot product is continuous in the first variable.*
b) *Show that the dot product is continuous in the second variable.*
c) *Consider the function $f : \mathbb{R}^k \times \mathbb{R}^k \to \mathbb{R}$ described in this problem as the function $F : \mathbb{R}^{2k} \to \mathbb{R}$ where $F(x_1, x_2, \ldots, x_k, y_1, y_2, \ldots, y_k) = \sum_{j=1}^k x_j\, y_j$. Show that this function is continuous on $\mathbb{R}^{2k}$.*

7. *Suppose $f : \mathbb{D} \to \mathbb{R}$ is continuous on $\mathbb{D}$. Show that for any $a \in \mathbb{R}$, the set $A = \{x \in \mathbb{D} : f(x) = a\}$ is closed in $\mathbb{D}$. Also show that the set $A_- = \{x \in \mathbb{D} : f(x) < a\}$ is open in $\mathbb{D}$. Use this to prove the following, more general statement of the intermediate value theorem: Suppose $f : \mathbb{D} \to \mathbb{R}$ is continuous on the connected set $\mathbb{D}$, and that $a < b$ are any two distinct values of f on $\mathbb{D}$. Then for each $y \in \mathbb{R}$ such that $a < y < b$, there exists an $x \in \mathbb{D}$ such that $f(x) = y$.*

8. *Consider $f : \mathbb{Q} \cap [0,2] \to \mathbb{R}$ given by $f(x) = \begin{cases} 0 & \text{if } 0 \le x < 1 \\ 1 & \text{if } 1 \le x \le 2 \end{cases}$. Is f continuous on $\mathbb{Q} \cap [0,2]$?*

9. *Consider $f : \mathbb{Q} \cap [0,2] \to \mathbb{R}$ given by $f(x) = \begin{cases} 0 & \text{if } 0 \le x < \sqrt{2} \\ 1 & \text{if } \sqrt{2} \le x \le 2 \end{cases}$. Is f continuous on $\mathbb{Q} \cap [0,2]$?*

10. *What does the intermediate value theorem have to say about the function in the previous exercise?*

11. *Suppose $f : [0,1] \to \mathbb{R}$ is continuous on $[0,1]$. Is f continuous on $\mathbb{Q} \cap [0,1]$?*

12. *Suppose $f : [a,b] \to \mathbb{C}$ is continuous on $[\,a,\,b\,] \subset \mathbb{R}$ and that $f(a)$ and $f(b)$ are real numbers such that $f(a) < f(b)$. Is it still true that for every real number y such that $f(a) < y < f(b)$ there exists $c \in (a,\,b)$ such that $f(c) = y$?*

13. *Suppose $f : [a,b] \to \mathbb{R}$ is continuous on $[a,b]$. Will the graph of the function always include at least one point (x,y) with both x and y in $\mathbb{Q}$? What if f is also presumed to be strictly increasing?*

14. *Let $f : \mathbb{R}^2 \to \mathbb{R}$ be given by $f(x,y) = \begin{cases} \frac{xy}{x^2+y^2} & \text{if } (x,y) \neq (0,0) \\ 0 & \text{if } (x,y) = (0,0) \end{cases}$. Show that f is separately continuous in each variable, but is not jointly continuous.*

15. *Let $f : \mathbb{R} \to \mathbb{R}$ have the property that $\lim_{h\to 0}\,[f(x+h) - f(x-h)] = 0$ for all x. Is f continuous on $\mathbb{R}$?*

16. *Suppose $f : \mathbb{D} \to \mathbb{Y}$ is such that f takes open sets to open sets. That is, if U is open in $\mathbb{D}$, then $f(U)$ is open in $\mathbb{Y}$. Is f necessarily continuous?*

17. *Let $f : \mathbb{X} \to \mathbb{R}$ be continuous on $\mathbb{X}$. Define the set of zeros of the function f as $Z(f) \equiv \{x \in \mathbb{X} : f(x) = 0\}$. Show that $Z(f)$ is a closed subset of $\mathbb{X}$. Must $Z(f)$ be connected? Must $Z(f)$ be discrete?*

18. *Consider the function $f : K \to \mathbb{R}$ where $K \subset \mathbb{R}$ is a compact set, and let $G = \{(x, f(x)) \in \mathbb{R}^2 : x \in K\}$ be the graph of f. Show that f is continuous on K if and only if G is compact in $\mathbb{R}^2$.*

19. *Suppose $f : \mathbb{D} \to \mathbb{Y}$ takes every compact $K \subset \mathbb{D}$ to a compact set $f(K) \subset \mathbb{Y}$. Is f necessarily continuous on $\mathbb{D}$? (Hint: Consider a function whose range is finite.)*

20. *Suppose $f : \mathbb{D} \to \mathbb{Y}$ is such that $f(A) \subset \mathbb{Y}$ is connected for every connected subset $A \subset \mathbb{D}$. Is f necessarily continuous on $\mathbb{D}$? (Hint: Consider $\sin \frac{1}{x}$).*

21. *Suppose $f : \mathbb{R} \to \mathbb{Q}$ is a continuous function on $\mathbb{R}$. What can you conclude about f?*

22. *Can you find a function that is one-to-one but not continuous on its whole domain? Can you find a function that is a one-to-one correspondence but is not continuous on its whole domain?*

23. *Consider the function $f : \mathbb{R} \setminus \{\frac{\pi}{2}\} \to \mathbb{R}$ given by $f(x) = \frac{\cos x}{\frac{\pi}{2} - x}$. Can f be assigned a value at $x = \frac{\pi}{2}$ so as to be continuous there?*

24. *Recall the function from Example 4.9 on page 159, given by $f : \mathbb{R}^2 \setminus (0,0) \to \mathbb{R}$ where $f(x,y) = \frac{x\,y}{x^2+y^2}$. If this function is assigned the value 0 at $(0,0)$, will the resulting "extended" function be continuous at $(0,0)$?*

25. *Consider the function* $f : \mathbb{R}^2 \to \mathbb{R}$ *given by* $f(x,y) = \begin{cases} \frac{x^2 y}{x^4+y^2} & \text{if } (x,y) \neq (0,0) \\ 0 & \text{if } (x,y) = (0,0) \end{cases}$. *Where is* f *continuous?*

26. *Does the result of Example 1.28 on page 193 still hold if the codomain is changed to* $[0, \frac{1}{2}]$*? What if the codomain is changed to any interval* $I \subset [0,1]$*? What if the codomain is changed to any closed subset* $F \subset [0,1]$*?*

27. *Suppose* $f : [a,b] \to [a,b]$ *is a continuous function.*
a) *Must there exist a point* $x_0 \in [a,b]$ *such that* $f(x_0) = x_0$*?*
b) *What if the codomain is changed to any interval* $I \subset [a,b]$*?*
c) *What if the codomain is changed to any closed subset* $F \subset [a,b]$*?*

28. *Let* $I_1 = [0,1] \subset \mathbb{R}$*, and suppose* $f : I_1 \times I_1 \to I_1 \times I_1$ *is continuous.*
a) *Does there exist at least one point* $(x_0, y_0) \in I_1 \times I_1$ *such that* $f(x_0, y_0) = (x_0, y_0)$*?*
b) *What if the codomain is changed to* $I_{ab} \times I_{ab} \subset I_1 \times I_1$*, where* $I_{ab} = [a,b] \subset [0,1]$*?*
c) *What if the codomain is changed to any closed subset* $F \subset I_1 \times I_1$*?*

29. *Suppose* $f : \mathbb{R} \to \mathbb{R}$ *is given by* $f(x) = 3\sin x - \cos x$*. Are there any roots to the equation* $f(x) = 1$*? If so, can they be identified?*

30. *Let* $f : \mathbb{R} \to \mathbb{R}$ *be given by* $f(x) = \begin{cases} x+1 & \text{if } -\infty < x < 1 \\ 2x-3 & \text{if } 1 \leq x < 3 \\ 4 & \text{if } x = 3 \\ 2x & \text{if } 3 < x < \infty \end{cases}$. *Determine whether* f *is left or right continuous at* $x = 1$ *and* $x = 3$.

31. *Consider the function* $h : [0,1) \to \mathbb{R}$ *given by* $h(x) = \begin{cases} x^\alpha \sin \frac{1}{x} & \text{on } (0,1) \\ 0 & \text{if } x = 0 \end{cases}$. *For each of the values* $\alpha \in \{-1, 0, 1, 2\}$*, determine*

(i) *if* h *is right continuous at* $x = 0$*, and*
(ii) *if* $h([0,1))$ *is connected.*

32. *Determine whether the function* $f : [0,\infty) \to \mathbb{R}$ *given by* $f(x) = \sqrt{x}$ *is right continuous at* 0.

33. *Consider the functions and the specified discontinuity points given below. Determine the type of discontinuity in each case.*

a) $f : \mathbb{R} \setminus \{0\} \to \mathbb{R}$ *given by* $f(x) = \sin \frac{1}{x}$*;* $x = 0$.
b) $f : \mathbb{R} \setminus \{0\} \to \mathbb{R}$ *given by* $f(x) = \cos \frac{1}{x}$*;* $x = 0$.
c) $f : \mathbb{R} \setminus \{0\} \to \mathbb{R}$ *given by* $f(x) = e^{\frac{1}{x}}$*;* $x = 0$.

34. *Suppose* $f : (-1,1) \to \mathbb{R}$ *is given by* $f(x) = \begin{cases} x & \text{if } -1 < x \leq 0 \\ \sin \frac{1}{x} & \text{if } 0 < x \leq 1 \end{cases}$. *What type of discontinuity does* f *have at* $x = 0$*?*

35. *Consider the function* $f : \mathbb{R} \to \mathbb{R}$ *given by* $f(x) = \begin{cases} 1 & \text{if } x \in \mathbb{Q} \\ 0 & \text{if } x \in \mathbb{I} \end{cases}$. *What kind(s) of discontinuities does this function have? Note that this function is bounded for all* $x \in \mathbb{R}$.

36. *How many nonessential discontinuities can a function have? If it is infinitely many, can it be uncountably many?*

37. *Let* $f : \mathbb{R} \to \mathbb{Y}$ *be continuous and periodic on* $\mathbb{R}$. *Show that* f *is uniformly continuous on* $\mathbb{R}$. *Must* f *be bounded as well?*

38. *Show that if* $f : \mathbb{D} \to \mathbb{Y}$ *is uniformly continuous on* $\mathbb{D}$, *and if* $\mathbb{D}$ *is not bounded, then* $f(\mathbb{D})$ *is not necessarily bounded.*

39. *Show that if* $f : \mathbb{D} \to \mathbb{Y}$ *is merely continuous on* $\mathbb{D}$, *and if* $\mathbb{D}$ *is bounded, then* $f(\mathbb{D})$ *is not necessarily bounded.*

40. *Consider* $f : \mathbb{R}^2 \to \mathbb{R}$ *given by* $f(\mathbf{x}) = 3\,|\mathbf{x}|$. *Show that* f *is uniformly continuous on* $\mathbb{R}^2$.

41. *Consider* $D = \{z \in \mathbb{C} : |z| \le 1\}$ *and let* $f : D \to \mathbb{C}$ *be given by* $f(z) = \frac{z}{z+2}$. *Show that* f *is uniformly continuous on* D. *If* D *is changed to* $D = \{z \in \mathbb{C} : |z| < 2\}$, *is* f *still uniformly continuous on* D?

42. *Here is the idea for another proof of Proposition 2.11 on page 203, the details of which are left to the reader. The theorem states that if* $\mathbb{D} \subset \mathbb{X}$ *is bounded and if* $f : \mathbb{D} \to \mathbb{Y}$ *is uniformly continuous on* $\mathbb{D}$, *then* $f(\mathbb{D})$ *must be bounded. To prove this, one can show that there exists a finite number of points* $S = \{x_1, \ldots, x_N\} \subset \mathbb{D}$ *distributed within* $\mathbb{D}$ *so that every point in* $\mathbb{D}$ *is within distance* $\delta > 0$ *of some* $x_j \in S$. *Here,* δ *should be chosen via uniform continuity for* $\epsilon = 1$ *for simplicity. One can then show that* f *is bounded by* $1 + \max\big(f(x_1), \ldots, f(x_N)\big)$.

43. *Prove that* $f(x) = \sin \frac{1}{x}$ *is not uniformly continuous on* $(0, 1)$. *What can you say about* $f(x) = x^{\alpha} \sin \frac{1}{x}$ *for different real powers* α?

44. *Suppose* $f : \mathbb{R} \to \mathbb{R}$ *is such that there exists a positive real number* M *where*

$$|f(x) - f(y)| \le M\,|x - y| \text{ for all } x, y \in \mathbb{R}.$$

Prove that f *is continuous and uniformly continuous on* $\mathbb{R}$. *Functions that obey this property are called* **Lipschitz continuous.** *The constant* M *is referred to as the* **Lipschitz constant** *associated with the function.*

45. *Show that the real exponential function* $\exp : \mathbb{R} \to \mathbb{R}$ *is* not *Lipschitz continuous on its whole domain.*

46. *Let* $f : \mathbb{R} \to \mathbb{R}$ *be given by* $f(x) = \begin{cases} x^2 & \text{if } x \le 1, \\ 1 & \text{if } x > 1. \end{cases}$ *Use the pasting theorem to prove that* f *is continuous on* $\mathbb{R}$.

47. *What if, in Example 2.15, the function $g : \{x \leq 0\} \subset \mathbb{R}^2 \to \mathbb{R}$ was given by $g(x, y) = f(0, y)$? What does this tell you about the uniqueness of the extension given in that example?*

48. *Suppose $f : \{\mathrm{Im}(z) \geq 0\} \to \mathbb{C}$ is given by $f(z) = z$. Can you find at least two different complex functions g_1 and g_2 defined on $\{\mathrm{Im}(z) \leq 0\} \subset \mathbb{C}$ which, via the pasting theorem, allow you to extend f to the whole complex plane $\mathbb{C}$? What if f were given instead by the rule $f(z) = \sin z$?*

49. *Suppose $f : \{\mathrm{Im}(z) \geq 0\} \to \mathbb{C}$ is given by $f(z) = \frac{1}{(z+i)^2}$. Find at least two complex functions g_1 and g_2 which, via the pasting theorem, allow you to extend f continuously to the whole complex plane $\mathbb{C}$. What if f were given instead by the rule $f(z) = \frac{1}{z+i}$?*

50. *Suppose $f : \left[-\frac{1}{2}, \infty\right) \to \mathbb{R}$ is given by $f(x) = \frac{1}{(x+1)^2}$. Can you find a function $g : \left(-\infty, \frac{1}{2}\right]$ such that $f = g$ on $\left[-\frac{1}{2}, \frac{1}{2}\right]$, and g extends f continuously to all of $\mathbb{R}$? This is not easy in general.*

51. *Let $|\cdot|_1$ be a norm on $\mathbb{R}^k$.*

a) *Show that there exists a constant M such that $|\mathbf{x}|_1 \leq K\,|\mathbf{x}|$. To do this, note that if $\mathbf{x} = (x_1, \ldots, x_n)$, then*

$$|\mathbf{x}|_1 = \left|\sum_{j=1}^{k} x_j \mathbf{e_j}\right|_1 \leq \left(\max_j |\mathbf{e_j}|_1\right) \sum_{j=1}^{k} |x_j| \leq k \left(\max_j |\mathbf{e_j}|_1\right) |\mathbf{x}|\,.$$

b) *Show that $|\cdot|_1$ is Lipschitz continuous with constant M. Use the reverse triangle inequality. Show that $|\cdot|$ is also continuous on $\mathbb{R}^k$.*

c) *Show that the function $g : \mathbb{R}^k \setminus \{\mathbf{0}\} \to \mathbb{R}$ given by $g(\mathbf{x}) \equiv \frac{|\mathbf{x}|_1}{|\mathbf{x}|}$ is continuous and nonzero on the unit sphere $\{\mathbf{x} : |\mathbf{x}| = 1\}$. Using a homogeneity argument, show that the norms $|\cdot|_1$ and $|\cdot|$ are equivalent in the sense of Supplementary Exercise 32 in Chapter 1.*

d) *Conclude that any two norms on $\mathbb{R}^k$ are equivalent.*

52. *Consider the sequence of functions $f_n : (-1, 1) \to \mathbb{R}$ given by $f_n(x) = 2\,x + x^n$ for $n \geq 1$. That is,*

$$\begin{aligned} f_1(x) &= 2\,x + x \\ f_2(x) &= 2\,x + x^2 \\ f_3(x) &= 2\,x + x^3 \\ &\vdots \end{aligned}$$

If $f : (-1, 1) \to \mathbb{R}$ is given by $f(x) = 2\,x$, show that $\lim_{n\to\infty} f_n(x) = f(x)$ on $(-1, 1)$. Does the N required to establish convergence depend on x?

53. *Define for $n \geq 1$ and $0 \leq k \leq 2^{n-1}$ the sequence of functions $\{f_n\}$ given by*

$$f_n(x) = \begin{cases} x - \frac{k}{2^{n-1}} & \text{for } x \in \left[\frac{2k}{2^n}, \frac{2k+1}{2^n}\right] \\[2ex] \frac{k}{2^{n-1}} - x & \text{for } x \in \left(\frac{2k+1}{2^n}, \frac{2k+1}{2^n}\right] \end{cases}.$$

a) *Show that $f_n(x)$ is continuous for each $n \geq 1$. Draw the first several functions in this sequence.*

b) *What is the value of $\lim_{n\to\infty} f_n\left(\frac{1}{2^k}\right)$ for $k \in \mathbb{Z}$?*

c) *What is $\lim_{n\to\infty} f_n\left(\frac{3}{5}\right)$? What is $\lim_{n\to\infty} f_n(\pi)$?*

d) *Find $f(x) \equiv \lim_{n\to\infty} f_n(x)$. Is f continuous?*

54. *Let $f_n : \mathbb{C} \to \mathbb{C}$ be given by $f_n(z) = \frac{n\,z}{n+1}$ for $n \in \mathbb{N}$. Show that $\lim_{n\to\infty} f_n(z) = z$ for all $z \in \mathbb{C}$. Is the convergence uniform on $\mathbb{C}$?*

55. *Modify the domain in the last exercise to be $D = \{z \in \mathbb{C} : |z - 1| < 12\}$. Is the convergence uniform on D? How about for $D = \{z \in \mathbb{C} : |z - 1| \leq 12\}$?*

56. *Let $\boldsymbol{f}_n : [4,5] \times [0, \frac{1}{2}] \to \mathbb{R}^2$ be given by $\boldsymbol{f}_n(\mathbf{x}) = \left(y^n, \frac{n\,x}{n+1}\right)$ for $n \in \mathbb{N}$. Find $\lim_{n\to\infty} \boldsymbol{f}_n(\mathbf{x})$ and determine whether the convergence is uniform.*

57. *Define $f : \mathbb{R} \to \mathbb{R}$ according to the following description. For each $x \in \mathbb{R}$, note that there exists a unique $k \in \mathbb{Z}$ such that $2k - 1 \leq x \leq 2k + 1$. Let $f(x) = |x - 2k|$ for this k.*

a) *Sketch the graph of f.*

b) *Show that $f(x + 2L) = f(x)$ for $L \in \mathbb{Z}$.*

c) *Show that for all $x_1, x_2 \in \mathbb{R}$, $|f(x_2) - f(x_1)| \leq |x_2 - x_1|$, and therefore f is uniformly continuous on $\mathbb{R}$.*

d) *Let $g : \mathbb{R} \to \mathbb{R}$ be given by $g(x) = \sum_{n=0}^{\infty} \left(\frac{1}{4}\right)^n f(4^n x)$. Show that the sum converges uniformly and that g is continuous on $\mathbb{R}$.*

58. *Suppose $f : \mathbb{X} \to \mathbb{Y}$ is continuous on $\mathbb{X}$. If A is open in $\mathbb{X}$, does it follow that $f(A)$ must be open in $\mathbb{Y}$? What if f has an inverse function? What if the inverse function is continuous?*

59. *Suppose $f : D^1 \to \mathbb{R}$ is continuous on $D^1 \subset \mathbb{R}$. Is $f(D^1)$ necessarily connected?*

60. *Suppose $f : D^1 \to \mathbb{R}$ is continuous on $D^1 \subset \mathbb{R}$ and D^1 is connected. Is $G_f \equiv \{(x, f(x)) : x \in D^1\}$ necessarily a connected subset of $\mathbb{R}^2$?*

61. *Using Theorem 1.22 on page 191, give another proof of the intermediate value theorem on page 192.*

62. *Let C be the Cantor set as described in the Supplementary Exercise section of Chapter 2, and let $f : C \to \mathbb{R}$ be given by $f(x) = 3x^2 - 5\sqrt{x}$. Show that $f(C) = \{y \in \mathbb{R} : y = 3x^2 - 5\sqrt{x}$ for $x \in C\}$ is compact.*

63. *Consider $f_n : (-1, 1) \to \mathbb{R}$ given by $f_n(x) = x^n$ for $n \in \mathbb{N}$. We already know this sequence of functions converges. Use the M-test to determine whether the convergence is uniform.*

64. *Use the M-test on the sequence of functions* $f_n : (-\frac{1}{2}, \frac{2}{3}) \to \mathbb{R}$ *given by* $f_n(x) = x^n$ *for* $n \in \mathbb{N}$ *to determine whether the convergence is uniform.*

65. *Use the M-test on the sequence of functions* $f_n : \mathbb{C} \to \mathbb{C}$ *given by* $f_n(z) = \frac{n\,z}{n+1}$ *for* $n \in \mathbb{N}$ *to determine whether the convergence is uniform.*

66. *Change the domain of definition for each* f_n *in the previous exercise to* $D = \{z \in \mathbb{C} : |z-1| < 12\}$ *and use the M-test to determine whether the convergence is uniform.*

67. *Change the domain of definition for each* f_n *in the previous exercise to* $D = \{z \in \mathbb{C} : |z-1| \leq 12\}$ *and use the M-test to determine whether the convergence is uniform.*

68. *Let* $f_n : [0,2] \to \mathbb{R}$ *be given by* $f_n(x) = \begin{cases} n\,x & \text{for } 0 \leq x < \frac{1}{n} \\ -\frac{n\,x}{2\,n-1} + \frac{2\,n}{2\,n-1} & \text{for } \frac{1}{n} \leq x \leq 2 \end{cases}$.

a) Sketch the first three functions in this sequence.

b) Find $\lim_{n\to\infty} f_n(x)$*.*

c) Is the convergence uniform on $[0,2]$*?*

69. *Let* $f_n : [0,1] \to \mathbb{R}$ *be given by* $f_n(x) = \begin{cases} 2\,n\,\sqrt{n}\,x & \text{for } 0 \leq x < \frac{1}{2n} \\ -2\,n\,\sqrt{n}\,x + 2\,\sqrt{n} & \text{for } \frac{1}{2n} \leq x < \frac{1}{n} \\ 0 & \text{for } \frac{1}{n} \leq x \leq 1 \end{cases}$.

a) Sketch the first three functions in this sequence.

b) Find $\lim_{n\to\infty} f_n(x)$*.*

c) Is the convergence uniform on $[0,1]$*?*

70. *Let* $f_n : [0,1] \to \mathbb{R}$ *be given by* $f_n(x) = \begin{cases} 2\,n\,x & \text{for } 0 \leq x < \frac{1}{2n} \\ -2\,n\,x + 2\,\sqrt{n} & \text{for } \frac{1}{2n} \leq x < \frac{1}{n} \\ 0 & \text{for } \frac{1}{n} \leq x \leq 1 \end{cases}$.

a) Sketch the first three functions in this sequence.

b) Find $\lim_{n\to\infty} f_n(x)$*.*

c) Is the convergence uniform on $[0,1]$*?*

d) What is the point of this exercise?

71. *Consider the functions* $f_n : [0,1] \to \mathbb{R}$ *given by* $f_n(x) = \frac{nx^2}{1+nx}$*.*

a) Find $\lim_{n\to\infty} f_n(x)$*.* *b) Show that the convergence is uniform on* $[0,1]$*.*

72. *The Cantor function*

Consider the function $f_C : [0,1] \to \mathbb{R}$ *defined as follows. Recall that the removed middle-third sets in the Cantor set construction (Supplementary Exercise 75 in Chap-*

ter 2) are of the form

$$G_1 = \left(\tfrac{1}{3}, \tfrac{2}{3}\right),$$
$$G_2 = \left(\tfrac{1}{3^2}, \tfrac{2}{3^2}\right) \cup \left(\tfrac{7}{3^2}, \tfrac{8}{3^2}\right),$$
$$G_3 = \left(\tfrac{1}{3^3}, \tfrac{2}{3^3}\right) \cup \left(\tfrac{7}{3^3}, \tfrac{8}{3^3}\right) \cup \left(\tfrac{19}{3^3}, \tfrac{20}{3^3}\right) \cup \left(\tfrac{25}{3^3}, \tfrac{26}{3^3}\right),$$

and more generally

$$G_n \equiv \bigcup_{j=1}^{2^{n-1}} I_j^{(n)}. \tag{5.17}$$

From this we see that each G_n is the union of 2^{n-1} disjoint open intervals $I_j^{(n)}$. Since $G_n \cap G_m = \varnothing$ for $n \neq m$, note that $I_{j_1}^{(n_1)} \cap I_{j_2}^{(n_2)} = \varnothing$ if either $j_1 \neq j_2$ or $n_1 \neq n_2$. The union $\bigcup_{n=1}^{\infty} G_n$ is the complement of the Cantor set in $[0,1]$. We will first define f_C for $x \in \bigcup_{n=1}^{\infty} G_n$. To this end, let

$$f_C(x) = \frac{2j-1}{2^n} \quad \text{if } x \in I_j^{(n)}. \tag{5.18}$$

Thus, f_C takes on a different constant value on each $I_j^{(n)}$. Next we will define f_C on the Cantor set C. Recall from Supplementary Exercise 57 in Chapter 3 that $C = K_1 \cup K_2$ where $K_1 \equiv \bigcup_{n=1}^{\infty} \partial G_n$ is the set of "Cantor set points of the first kind." From the same exercise, we saw that for any $x \in C$ there exists an increasing sequence of points $\{\overline{x_n}\} \subset K_1$ such that $\lim \overline{x_n} = x$, and likewise, there exists a decreasing sequence of points $\{\underline{x_n}\} \subset K_1$ such that $\lim \underline{x_n} = x$. With these facts in mind, we define f_C at $x \in C$ by

$$f_C(x) = \lim f_C(\overline{x_n}) = \lim f_C(\underline{x_n}) \quad \text{if } x \in C. \tag{5.19}$$

a) Show that expressions (5.18) and (5.19) yield a well-defined function f_C on all of $[0,1]$. Begin by considering what value f_C must take at $x \in \partial I_j^{(n)}$. What about $x \in K_2$?

b) Show that f_C is continuous on $[0,1]$.

c) Is f_C uniformly continuous on $[0,1]$?

73. Consider the sequence of functions $f_n : [0,1] \to \mathbb{R}$ given by $f_0(x) = x$, and

$$f_{n+1}(x) = \begin{cases} \frac{1}{2} f_n(3x) & \text{if } 0 \le x \le \frac{1}{3}, \\ \frac{1}{2} & \text{if } \frac{1}{3} < x \le \frac{2}{3}, \\ \frac{1}{2}\left(1 + f_n\left(3\left(x - \frac{2}{3}\right)\right)\right) & \text{if } \frac{2}{3} < x \le 1, \end{cases}$$

for integers $n \ge 0$.

a) Sketch the first few functions in this sequence.

b) Show that each f_n is continuous on $[0,1]$.

c) Show that $\lim_{n\to\infty} f_n(x) = f_C(x)$.

d) Is the convergence uniform?

e) What does your answer to part d) tell you about the function f_C?

74. *Let $D = \bigcup_{n=1}^{\infty} G_n \subset [0,1]$, where G_n is as described in (5.17) from the Cantor function construction given in a previous exercise, and consider the function $f : D \to \mathbb{R}$ given by $f = f_C$ on D. That is, $f(x) = \frac{2j-1}{2^n}$ if $x \in I_j^{(n)}$, where $I_j^{(n)} \subset G_n$ as described in (5.18). Show that f is uniformly continuous on D, and that the continuous extension of f to all of $[0,1]$ is the Cantor function f_C. Is f_C uniformly continuous on $[0,1]$?*

75. *We have seen that the function $\exp : \mathbb{R} \to \mathbb{R}$ given by $\exp(x) = e^x$ is not uniformly continuous on its domain. What about the function $f : \mathbb{R} \to \mathbb{R}$ given by $f(x) = \exp(-x) = e^{-x}$?*

76. *Suppose $f : \mathbb{R} \to \mathbb{R}$ and $g : \mathbb{R} \to \mathbb{R}$ are each uniformly continuous on $\mathbb{R}$. Determine whether the functions $f + g$ and fg must be uniformly continuous on $\mathbb{R}$. If not, give a counterexample.*

77. *Let $f_j : D_j \to \mathbb{C}$ be given by $f_j(z) = \frac{j\,z}{j^3\,(z^2+1)}$ for $j \in \mathbb{N}$.*

a) *Determine $D_j \subset \mathbb{C}$ for each f_j.*

b) *Determine the domain $D \subset \mathbb{C}$ where the series $\sum_{j=1}^{\infty} f_j$ converges.*

78. *Let $f_j : D_j \to \mathbb{C}$ be given by $f_j(z) = \frac{j\,z}{j^3\,z^2+1}$ for $j \in \mathbb{N}$.*

a) *Determine $D_j \subset \mathbb{C}$ for each f_j.*

b) *Determine the domain $D \subset \mathbb{C}$ where the series $\sum_{j=1}^{\infty} f_j$ converges.*
(Hint: Try to show the series converges absolutely.)

79. *Determine whether $\sum_{j=1}^{\infty} \frac{\cos(jx)}{j^{7/3}}$ converges. If so, does it converge uniformly on any subset of $\mathbb{R}$?*

80. *Determine whether the series $\sum_{j=1}^{\infty} \frac{j\,z}{j^3\,(z^2+1)}$ converges uniformly on the specified domain: a) $2 < |z| < 4$ b) $1 < |z| < 4$ c) $2 < |z|$*

81. *Determine whether the series $\sum_{j=1}^{\infty} \frac{j\,z}{j^3\,z^2+1}$ converges uniformly on the specified domain: a) $2 < |z| < 4$ b) $1 < |z| < 4$ c) $2 < |z|$*

82. *Consider the function $f : \mathbb{R} \to \mathbb{R}$ given by $f(x) = \sum_{j=0}^{\infty} \frac{|x-j|}{j!}$. Show that this function is well defined, i.e., converges, for any $x \in \mathbb{R}$. Show also that for $x \in [N, N+1]$ and $N \in \mathbb{Z}$, the convergence of $\sum_{j=0}^{\infty} \frac{|x-j|}{j!}$ is uniform, and hence f is continuous on $\mathbb{R}$.*

83. *As described at the end of the last section of this chapter, one can experience the difficulties in attempting to apply the algorithm described in the proof of the Tietze extension theorem. To do so, attempt to apply the algorithm to the simple case of $f : [-1, 1] \to \mathbb{R}$ given by $f(x) = x$.*

6

THE DERIVATIVE

I recoil with dismay and horror at this lamentable plague of functions which do not have derivatives.

Charles Hermite

In this chapter we abandon our more general notation involving $\mathbb{X}$, $\mathbb{D}$, and $\mathbb{Y}$ and work individually with the specific cases of interest in developing the derivative. We begin in Section 6.1 by considering functions $f : D^1 \to \mathbb{R}$. There, we remind the reader of the traditional difference quotient definition for the derivative of a function at a point, a formulation with very practical benefits. Most notable among these is that this version of the definition allows for the computation of the derivative at a specified point. This definition also provides conceptual clarity, directly exhibiting the derivative $f'(a)$ to be the rate of change of f with respect to changes in its argument at the point a. Yet despite these strengths, the difference quotient derivative definition has one significant shortcoming. While it extends to the class of complex functions $f : D \to \mathbb{C}$, it is not generalizable to other real classes of functions $f : D^k \to \mathbb{R}$ or $\boldsymbol{f} : D^k \to \mathbb{R}^p$ for $k, p > 1$. For this reason, we immediately develop two alternative versions of the derivative definition, "the ϵ, δ version" and "the linear approximation version." All three versions will be shown to be mathematically equivalent in the cases $f : D^1 \to \mathbb{R}$ and $f : D \to \mathbb{C}$, but the ϵ, δ version and the linear approximation version are both generalizable to the higher-dimensional real-function cases, while the difference quotient version is not. And yet these alternate versions are not perfect either. The ϵ, δ version is generalizable and practical in theoretical developments, but it lacks the conceptual clarity of the difference quotient version and cannot be used to actually compute the derivative of a function at a point. The linear approximation version is also generalizable and provides a conceptual clarity of its own, but it lacks the practical benefit that each of the other versions provides. Since each of the three derivative formulations is less than optimal in some way, we choose the linear approximation version as our "official" definition for its generalizability and its conceptual strengths. However, it is important to note that, depending on the circumstances, *all three versions of the definition are useful.* For this reason, it is important to develop a facility with *each* of them. The remaining part of Section 6.1 is devoted to establishing the many derivative related results for functions $f : D^1 \to \mathbb{R}$ that are so familiar to students of calculus. In Section 6.2, we extend many of these results to functions $f : D^k \to \mathbb{R}$ for $k > 1$, pointing out differences and subtleties associated with the higher-dimensional domain, including the concept of partial derivatives. In Section 6.3 we handle the class of real functions $\boldsymbol{f} : D^k \to \mathbb{R}^p$ for $k, p > 1$, where we find that some new tools are required to deal with the fact that the derivative at a point in this case can be represented as a matrix of constants. Finally, in Section 6.4 we consider the case of a complex-valued function of a complex variable, $f : D \to \mathbb{C}$. As we learned in a previous chapter, $\mathbb{C}$ is geometrically similar to $\mathbb{R}^2$ and algebraically similar to $\mathbb{R}$. In fact, we'll find that the class of complex functions can be compared to the two classes of real functions represented by $f : D^1 \to \mathbb{R}$ and $\boldsymbol{f} : D^2 \to \mathbb{R}^2$ and inherits the best of both worlds in many ways. Complex functions possess the extra algebraic structure that the field properties afford, as well as the richness of a two-dimensional geometric environment. Yet despite these similarities, we will discover that complex functions also possess many unique properties not found in any of the real counterparts to which we compare it. In the last section of the chapter, we state and prove the inverse and implicit function theorems. While they are relatively straightforward generalizations of what students already know about the existence of inverse and implicit functions from a first year calculus course, their proofs are subtle and involve much of the more advanced techniques of analysis developed thus far.

1 The Derivative for $f : D^1 \to \mathbb{R}$

1.1 Three Definitions Are Better Than One

Our goal in this subsection is to motivate what will ultimately be adopted as our formal definition for the derivative of a function $f : D^1 \to \mathbb{R}$ at a certain kind of point a in its domain. In particular, we require that the point a be *a limit point of* D^1 *that is also a member of* D^1. The set of such "included" limit points of a set D^1 will hereafter be denoted by $\mathcal{L}(D^1)$. Note that the set $\mathcal{L}(D^1)$ contains the interior points of D^1, so the derivative will be defined and may exist at any interior point of D^1. In the important case where D^1 is a closed interval $I \subset \mathbb{R}$, we have $\mathcal{L}(I) = I$, and so the derivative may exist at the endpoints as well as on the interior of the interval. We will see that there are three different versions of the derivative definition to choose from, and while only one of them will be adopted as our "official" definition (because it generalizes most naturally to the other cases we will be studying), we will show that all three formulations are mathematically equivalent. With a bit of experience, one finds that the initial ambiguity over the decision of which definition to apply gives way to a sense of appreciation for the three choices that are available. Each has its own strengths and limitations, as we will see.

The Difference Quotient Version

We begin by recalling the familiar derivative definition from calculus. For a real-valued function $f : D^1 \to \mathbb{R}$, and $a \in \mathcal{L}(D^1)$, consider the limit of the "difference quotient" given by

$$\lim_{x \to a} \frac{f(x) - f(a)}{x - a}. \tag{6.1}$$

Recall that in the case where a is an endpoint of an interval, this limit[1] is equivalent to the appropriate one-sided limit associated with a. In any case, if this limit exists, we call it *the derivative of f at a* and denote it by $f'(a)$ or $\frac{df}{dx}(a)$. We then say that f is *differentiable at a*. As mentioned in the introduction to this chapter, this formulation for the derivative has two particular merits. First, it is conceptually clear from the very form of (6.1) that, should the limit exist, it is the rate of change of the function f at the point a. Second, it is practically convenient as a tool for actually *computing* the derivative, a property unique to this, our first version of the derivative definition. And yet the difference quotient formulation is not perfect. Its failing is that it does not generalize to accommodate all the different kinds of functions we would like to study, such as $\boldsymbol{f} : D^k \to \mathbb{R}^p$ for $k, p > 1$. In fact, the difference quotient definition depends implicitly on the field property possessed by $\mathbb{R}$, which makes the quotient itself a well-defined quantity. The only other case where this will remain true is the complex function case, $f : D \to \mathbb{C}$.

Since students of calculus are already well acquainted with the difference quotient definition and its manipulation, we immediately move on to our next derivative formulation, using the difference quotient formulation as a starting point.

The ϵ, δ Version

As a first step toward our second formulation of the derivative definition, let us assume for a given real-valued function $f : D^1 \to \mathbb{R}$ that the derivative $f'(a)$ at $a \in \mathcal{L}(D^1)$ exists. In this case, the meaning of the expression (6.1) is that the quantity $\frac{f(x)-f(a)}{x-a}$ can be made to be as close to the quantity $f'(a)$ as we like by restricting x to within a certain distance of a. Stated more precisely, for any $\epsilon > 0$, there exists a number $\delta > 0$ small enough that

$$x \in D^1 \cap N'_\delta(a) \quad \Rightarrow \quad \left| \frac{f(x) - f(a)}{x - a} - f'(a) \right| < \epsilon. \tag{6.2}$$

With simple rearrangement, (6.2) becomes

$$x \in D^1 \cap N'_\delta(a) \quad \Rightarrow \quad \big| f(x) - f(a) - f'(a)\,(x - a) \big| < \epsilon\, |x - a|.$$

[1] Alternatively, the equivalent limit given by $\lim_{h \to 0} \frac{f(a+h)-f(a)}{h}$ serves just as well.

Including the possibility $x = a$, we obtain

$$x \in D^1 \cap N_\delta(a) \quad \Rightarrow \quad \big|f(x) - f(a) - f'(a)(x-a)\big| \le \epsilon\,|x-a|.$$

What we have just shown is that if $f'(a)$ exists according to expression (6.1), then for any $\epsilon > 0$, there exists $\delta > 0$ such that

$$x \in D^1 \cap N_\delta(a) \quad \Rightarrow \quad \big|f(x) - f(a) - f'(a)(x-a)\big| \le \epsilon\,|x-a|.$$

Now, without presuming the existence of the derivative of f at a, assume that there is a number $A \in \mathbb{R}$ such that for any $\epsilon > 0$, there exists $\delta > 0$ such that

$$x \in D^1 \cap N_\delta(a) \quad \Rightarrow \quad |f(x) - f(a) - A\,(x-a)| \le \epsilon\,|x-a|. \tag{6.3}$$

Then, replacing ϵ with $\frac{\epsilon}{2}$, we can certainly find $\delta > 0$ small enough so that

$$x \in D^1 \cap N_\delta(a) \quad \Rightarrow \quad |f(x) - f(a) - A\,(x-a)| \le \tfrac{\epsilon}{2}\,|x-a|.$$

If $x \ne a$, the above expression can be rearranged to obtain

$$x \in D^1 \cap N'_\delta(a) \quad \Rightarrow \quad \left|\frac{f(x)-f(a)}{x-a} - A\right| \le \tfrac{\epsilon}{2} < \epsilon,$$

or, in other words,

$$\lim_{x\to a} \frac{f(x)-f(a)}{x-a} = A.$$

This shows that if there exists an $A \in \mathbb{R}$ satisfying the ϵ, δ expression (6.3), then according to our difference quotient definition (6.1), the derivative of f exists at $x = a$ and is equal to A, i.e., $A = f'(a)$. Overall, we have just established the equivalence of the difference quotient expression (6.1) and the ϵ, δ expression (6.3) for the derivative of a function $f : D^1 \to \mathbb{R}$. While not as practical for *computing* derivatives, the ϵ, δ version provides an equivalent mathematical formulation for the differentiability of f at a that is often extremely useful in writing proofs. It also has the benefit of being generalizable (as we will see) to all the other cases we would like to explore, including $f : D^k \to \mathbb{R}^p$ for $k, p > 1$. However, despite these merits, the ϵ, δ version lacks a certain conceptual clarity. It is not as easy to determine, particularly for those new to the subject, just what it is saying about the function f at a. Our third version, the version we will ultimately adopt as our "official" definition, will regain this conceptual simplicity even while retaining the ϵ, δ version's generalizability. Yet we will still find it lacking in one respect. It is often not as practical a tool as either of the first two versions.

The Linear Approximation Version

Consider once again the ϵ, δ version for the definition of the derivative of $f : D^1 \to \mathbb{R}$ at $a \in \mathcal{L}(D^1)$. That is, there exists a number $A \in \mathbb{R}$ such that for any $\epsilon > 0$ there exists $\delta > 0$ such that

$$x \in D^1 \cap N_\delta(a) \quad \Rightarrow \quad |f(x) - f(a) - A\,(x-a)| \le \epsilon\,|x-a|.$$

Note that by defining the linear function $p : D^1 \to \mathbb{R}$ by $p(x) \equiv f(a)+A\,(x-a)$, we can express the ϵ, δ version of differentiability (6.3) as

$$x \in D^1 \cap N_\delta(a) \quad \Rightarrow \quad |f(x) - p(x)| \le \epsilon\,|x - a|.$$

By choosing $\delta > 0$ small enough to accommodate $\frac{\epsilon}{2}$ in place of ϵ, we obtain

$$x \in D^1 \cap N_\delta(a) \quad \Rightarrow \quad |f(x) - p(x)| \le \tfrac{\epsilon}{2}\,|x - a|,$$

which, for $x \neq a$, yields

$$x \in D^1 \cap N'_\delta(a) \quad \Rightarrow \quad \left|\frac{f(x) - p(x)}{x - a}\right| \le \tfrac{\epsilon}{2} < \epsilon.$$

This indicates a nice conceptual formulation for the derivative. A given $f : D^1 \to \mathbb{R}$ will have a derivative at $a \in \mathcal{L}(D^1)$ if and only if there exists a linear function $p : D^1 \to \mathbb{R}$ of the form $p(x) = f(a) + A\,(x - a)$ with $A \in \mathbb{R}$ such that

$$\lim_{x \to a} \frac{f(x) - p(x)}{x - a} = 0.$$

In this case, we refer to the number A as the derivative of f at a and denote it thereafter by $f'(a)$. The linear function p corresponding to this A has the same value as f at $x = a$, and, in a sense to be made clearer later, is the *best* such linear function to approximate the values of f when x is sufficiently near to $x = a$. We take this linear function approximation as the basis for our "official" definition of derivative, which we formally state below. Before doing so, however, we will replace $x - a$ in the denominator by $|x - a|$. This change does not alter the meaning of the result and is not even necessary in the case $f : D^1 \to \mathbb{R}$. We include it here merely to ensure consistency of form between this definition and the case $\boldsymbol{f} : D^k \to \mathbb{R}^p$ for $k, p > 1$ to follow.

Definition 1.1 (Derivative of $\boldsymbol{f} : D^1 \to \mathbb{R}$ at the point a)
For a function $f : D^1 \to \mathbb{R}$, and $a \in \mathcal{L}(D^1)$, we say that f is **differentiable** at a if there exists a real number A such that the linear function $p : D^1 \to \mathbb{R}$ given by

$$p(x) = f(a) + A\,(x - a) \quad \text{satisfies} \quad \lim_{x \to a} \frac{f(x) - p(x)}{|x - a|} = 0.$$

In this case, we refer to the number A as **the derivative** of f at a and denote it thereafter by $f'(a)$ or $\frac{df}{dx}(a)$.

In the case where a is an interior point of D^1, the geometric interpretation of the value of the derivative is even more evident in our adopted definition than in either of the previous two versions. Clearly, it is the slope term in the equation of the line given by the function $p : D^1 \to \mathbb{R}$. This linear function is, in fact, the equation of the tangent line to the graph of f at the point $(a, f(a))$ in the xy-plane. Finally, it is worth reiterating that while we choose this linear approximation version as our formal definition for the derivative of a function $f : D^1 \to \mathbb{R}$ at $a \in \mathcal{L}(D^1)$ for its conceptual elegance, we often will use the equivalent ϵ, δ version and the difference quotient version in our work because of their utility. The reader is therefore encouraged to become well acquainted with all three forms. To further emphasize this fact, we state the following theorem, the proof of which we have almost already established.

Theorem 1.2 *Consider $f : D^1 \to \mathbb{R}$ and $a \in \mathcal{L}(D^1)$. The following three statements are mathematically equivalent.*

a) $\lim_{x\to a} \frac{f(x)-f(a)}{x-a} = A$.

b) There exists $A \in \mathbb{R}$ such that for any $\epsilon > 0$ there exists $\delta > 0$ such that

$$x \in D^1 \cap N_\delta(a) \;\Rightarrow\; |f(x) - f(a) - A\,(x-a)| \le \epsilon\,|x-a|.$$

c) There exists $A \in \mathbb{R}$ such that the linear function $p : D^1 \to \mathbb{R}$ given by

$$p(x) = f(a) + A\,(x-a) \quad \textit{satisfies} \quad \lim_{x\to a} \frac{f(x)-p(x)}{|x-a|} = 0.$$

When any one (and therefore all three) of these statements is true, the function f is differentiable at $a \in \mathcal{L}(D^1)$ with derivative $f'(a) = A$.

► **6.1** *In the above discussion, we started with the ϵ, δ version of the definition and derived the linear approximation version from it. Beginning with the linear approximation version, derive the ϵ, δ version, thus establishing the equivalence of these two versions of the derivative definition, and hence completing the proof of Theorem 1.2.*

► **6.2** *Show that* $\lim_{x\to a} \frac{f(x)-p(x)}{x-a} = 0$ *if and only if* $\lim_{x\to a} \frac{f(x)-p(x)}{|x-a|} = 0$.

► **6.3** *Suppose* $\lim_{x\to a} \frac{f(x)-f(a)-A\,(x-a)}{|x-a|}$ *exists and equals L. Is it necessarily true that* $\lim_{x\to a} \frac{f(x)-f(a)-A\,(x-a)}{x-a}$ *exists as well? And if so, is it necessarily equal to L?*

► **6.4** *Now suppose that* $\lim_{x\to a} \frac{f(x)-f(a)-A\,(x-a)}{x-a}$ *exists and equals L. Is it necessarily true that* $\lim_{x\to a} \frac{f(x)-f(a)-A\,(x-a)}{|x-a|}$ *exists as well? And if so, is it necessarily equal to L?*

► **6.5** *Based on your results from the previous two problems, what is the relationship between the limits given by* $\lim_{x\to a} \frac{f(x)-f(a)-A\,(x-a)}{x-a}$ *and* $\lim_{x\to a} \frac{f(x)-f(a)-A\,(x-a)}{|x-a|}$? *How does the situation in Exercise 6.2 differ from that exhibited in Exercises 6.3 and 6.4?*

1.2 First Properties and Examples

We have shown three equivalent ways to determine whether $f : D^1 \to \mathbb{R}$ is differentiable at $a \in \mathcal{L}(D^1)$. But we have not yet shown that when such a derivative exists, it is unique. We do this next.

Proposition 1.3 *Consider $f : \mathbb{D}^1 \to \mathbb{R}$ and $a \in L(\mathbb{D}^1)$. If $f'(a)$ exists, then it is unique.*

PROOF Suppose there are two numbers A and B such that $p_A(x) = f(a) + A(x-a)$ satisfies $\lim_{x\to a} \frac{f(x)-p_A(x)}{|x-a|} = 0$, and $p_B(x) = f(a) + B(x-a)$ satisfies $\lim_{x\to a} \frac{f(x)-p_B(x)}{|x-a|} = 0$. From this it follows that

$$\begin{aligned} 0 &= \left(\lim_{x\to a} \frac{f(x)-p_A(x)}{|x-a|} - \lim_{x\to a} \frac{f(x)-p_B(x)}{|x-a|} \right) \\ &= \lim_{x\to a} \left(\frac{f(x)-p_A(x)}{|x-a|} - \frac{f(x)-p_B(x)}{|x-a|} \right) \\ &= \lim_{x\to a} \frac{(B-A)(x-a)}{|x-a|}. \end{aligned}$$

If we assume that $A \neq B$, dividing by $(B-A)$ yields

$$0 = \lim_{x\to a} \frac{(x-a)}{|x-a|}.$$

But this is a contradiction, since $\lim_{x\to a} \frac{x-a}{|x-a|}$ does not exist (Why?). Therefore $A = B$, and the result is proved. ♦

Example 1.4 *Consider the function $f : \mathbb{R} \to \mathbb{R}$ given by $f(x) = x^2$. We will show that f is differentiable at every real number a, and $f'(a) = 2a$. We will do so by first using the difference quotient formulation of the derivative, and then the ϵ, δ version. To this end, consider the limit*

$$\lim_{x\to a} \frac{f(x)-f(a)}{x-a} = \lim_{x\to a} \frac{x^2-a^2}{x-a} = \lim_{x\to a} (x+a) = 2a.$$

From this we see that $f'(a) = 2a$ for each $a \in \mathbb{R}$. Alternatively, by applying the ϵ, δ version of our definition with $2a$ as a candidate for $f'(a)$, we see that

$$|f(x) - f(a) - 2a(x-a)| = |x^2 - a^2 - 2a(x-a)| = |x-a|\,|x-a|.$$

This implies that as long as $x \in \mathbb{D}^1$ satisfies $|x-a| < \epsilon$, we will have

$$|f(x) - f(a) - 2a(x-a)| \le \epsilon\,|x-a|.$$

That is, given any $\epsilon > 0$, taking $\delta = \epsilon$ gives the result. ◄

Note in the above example that in applying the ϵ, δ version of our derivative definition we start with the left-hand side of the inequality

$$|f(x) - f(a) - A(x-a)| \le \epsilon\,|x-a|. \tag{6.4}$$

We then use it to determine what δ must be to imply that the inequality (6.4) is true. This is typically accomplished by showing that the term $|f(x) - f(a) -$

$A(x-a)|$ is less than or equal to a product of $|x-a|$ and another factor that is itself a function of $|x-a|$ (in this case the other factor just happened to be another copy of $|x-a|$), and then forcing this other factor to be less than ϵ in magnitude. The δ required for a given ϵ is then found, in general, to depend on ϵ. One more example should make the general idea clear.

Example 1.5 *Consider $f : \mathbb{R} \to \mathbb{R}$ defined by $f(x) = 2x^3 + x - 7$. We will use the ϵ, δ version of our derivative definition to show that $f'(a) = 6a^2 + 1$ for every $a \in \mathbb{R}$. First, suppose $a \neq 0$. Then we have*

$$\begin{aligned}
\big|f(x) - f(a) - f'(a)(x-a)\big| \\
&= \left|\left(2x^3 + x - 7\right) - \left(2a^3 + a - 7\right) - \left(6a^2 + 1\right)(x-a)\right| \\
&= \left|2\left(x^3 - a^3\right) - 6a^2(x-a)\right| \\
&= \left|2(x-a)\left(x^2 + ax + a^2\right) - 6a^2(x-a)\right| \\
&= 2|x-a|\left|\left(x^2 + ax + a^2\right) - 3a^2\right| \\
&= 2|x-a|\left|x^2 + ax - 2a^2\right| \\
&= 2|x-a|^2|x+2a| \\
&= 2|x-a|^2|x-a+3a| \\
&\leq 2|x-a|^3 + 6a|x-a|^2 \\
&= \left(2|x-a|^2 + 6a|x-a|\right)|x-a|.
\end{aligned}$$

Now, for any given $\epsilon > 0$, if $\left(2|x-a|^2 + 6a|x-a|\right) < \epsilon$ we will have the result. To obtain this, force each summand in $\left(2|x-a|^2 + 6a|x-a|\right)$ to be less than $\frac{\epsilon}{2}$. Then $|x-a| < \delta \equiv \min\left(\sqrt{\frac{\epsilon}{4}}, \frac{\epsilon}{12a}\right)$ yields

$$\big|f(x) - f(a) - f'(a)(x-a)\big| \leq \epsilon|x-a|.$$

The case of $a = 0$ is left to the reader. ◀

► **6.6** *Finish the previous example by using the ϵ, δ version of our derivative definition to show that $f'(0) = 1$.*

► **6.7** *For each $f(x)$ below, use the difference quotient formulation to verify that the given $f'(a)$ is as claimed. Then use the ϵ, δ version to prove that the derivative is the* *given* $f'(a)$.

a) $f(x) = c$ *for* $D^1 = \mathbb{R}$, *constant* $c \in \mathbb{R}$, $f'(a) = 0$ *for all* $a \in D^1$

b) $f(x) = x^3$ *for* $D^1 = \mathbb{R}$, $f'(a) = 3a^2$ *for all* $a \in D^1$

c) $f(x) = x^{-1}$ *for* $D^1 = \mathbb{R} \setminus \{0\}$, $f'(a) = -a^{-2}$ *for all* $a \in D^1$

d) $f(x) = x^n$ *for* $D^1 = \mathbb{R}$, $n \in \mathbb{Z}^+$, $f'(a) = na^{n-1}$ *for all* $a \in D^1$

e) $f(x) = x^n$ *for* $D^1 = \mathbb{R} \setminus \{0\}$, $n \in \mathbb{Z}^-$, $f'(a) = na^{n-1}$ *for all* $a \in D^1$

f) $f(x) = \sin x$ *for* $D^1 = \mathbb{R}$, $f'(a) = \cos a$, *for all* $a \in D^1$

g) $f(x) = \cos x$ *for* $D^1 = \mathbb{R}$, $f'(a) = -\sin a$, *for all* $a \in D^1$

A hint for part f) is to consider the trigonometric identity

$$\sin\theta - \sin\phi = 2\cos\left(\tfrac{\theta+\phi}{2}\right)\sin\left(\tfrac{\theta-\phi}{2}\right).$$

► **6.8** *In Example 1.4 on page 239 we established that the function* $f : \mathbb{R} \to \mathbb{R}$ *given by* $f(x) = x^2$ *was differentiable on* $\mathbb{R}$ *and had derivative* $f'(a) = 2\,a$ *at* $x = a$.

a) *Consider* $g : [1,3] \to \mathbb{R}$ *given by* $g(x) = x^2$. *Show that* $g'(3) = 6 = f'(3)$.

b) *Consider* $h : [1,\infty)$ *given by* $h(x) = \begin{cases} x^2 & \text{for } x \in [1,3] \\ 9 & \text{for } x \in (3,\infty) \end{cases}$. *Is* h *differentiable at* $a = 3$? *Explain.*

► **6.9** *Consider* $f : \mathbb{R} \to \mathbb{R}$ *given by* $f(x) = |x|$. *Show that* f *is not differentiable at* $a = 0$. *Let* $g : [0,\infty) \to \mathbb{R}$ *be given by* $g(x) = |x|$. *Show that* g *is differentiable at* $a = 0$.

In the following example, we establish the important fact that the real exponential function is its own derivative.

Example 1.6 *Consider the real exponential function* $\exp : \mathbb{R} \to \mathbb{R}$ *defined by* $\exp(x) = e^x$. *We will show that this function is its own derivative at every point in its domain. In fact, all we need to show is that the derivative exists at* $x = 0$ *and equals 1 there, since*

$$f'(a) = \lim_{x\to a} \frac{e^x - e^a}{x - a} = e^a \lim_{x\to a} \frac{e^{x-a} - 1}{x - a} = e^a \lim_{x\to 0} \frac{e^x - 1}{x} = e^a\, f'(0).$$

To show that $f'(0) = 1$, *note that*

$$\left|e^x - e^0 - x\right| = \left|\sum_{j=0}^{\infty} \frac{x^j}{j!} - 1 - x\right| = \left|\sum_{j=2}^{\infty} \frac{x^j}{j!}\right|.$$

Our goal, using the ϵ, δ *version of our definition of derivative, is to show that there exists* $\delta > 0$ *such that*

$$|x| < \delta \ \Rightarrow\ \left|f(x) - f(0) - f'(0)\,(x - 0)\right| = \left|e^x - e^0 - x\right| = \left|\sum_{j=2}^{\infty} \frac{x^j}{j!}\right| \le \epsilon\,|x|.$$

To obtain this, note that if $|x| < 1$, then

$$\left|\sum_{j=2}^{\infty} \frac{x^{j-2}}{j!}\right| \le \sum_{j=2}^{\infty} \frac{1}{j!} \le e,$$

and so $|x| < 1$ obtains

$$\left|\sum_{j=2}^{\infty} \frac{x^j}{j!}\right| = \left|\sum_{j=2}^{\infty} \frac{x^{j-2}}{j!}\right| |x|^2 \le e\,|x|^2.$$

Choosing $\delta \equiv \min\left(1, \frac{\epsilon}{e}\right)$ yields that

$$|x| < \delta \;\Rightarrow\; \left|\sum_{j=2}^{\infty} \frac{x^j}{j!}\right| \le e\,|x|^2 \le \epsilon\,|x|.$$

This establishes that $\exp(x) = e^x$ has derivative 1 at $x = 0$, and the result follows. ◀

As we will now see, if the derivative of a function $f : D^1 \to \mathbb{R}$ is posited to exist at $a \in \mathcal{L}(D^1)$, the ϵ, δ version of our derivative definition is still useful even when no explicit candidate for the derivative value is at hand. This is a common occurrence in proofs, such as the one associated with the following well-known result from calculus.

Proposition 1.7 *Let the function $f : D^1 \to \mathbb{R}$ be differentiable at $a \in \mathcal{L}(D^1)$. Then f is continuous at a as well.*

PROOF Using the ϵ, δ version of differentiability with $\epsilon = 1$, there exists $\delta_1 > 0$ such that

$$x \in D^1 \cap N_{\delta_1}(a) \;\Rightarrow\; \left|f(x) - f(a) - f'(a)(x-a)\right| \le |x-a|. \tag{6.5}$$

Also, by the reverse triangle inequality, we have

$$\left|f(x) - f(a) - f'(a)(x-a)\right| \ge |f(x) - f(a)| - \left|f'(a)\right|\,|x-a|. \tag{6.6}$$

Therefore, as long as $x \in D^1$ satisfies $|x-a| < \delta_1$, it follows from (6.5) and (6.6) that

$$\begin{aligned} |f(x) - f(a)| &\le |x-a| + |f'(a)|\,|x-a| \\ &= \left(1 + \left|f'(a)\right|\right)|x-a| \\ &< \epsilon \text{ whenever } |x-a| < \delta \equiv \min\left(\delta_1, \frac{\epsilon}{1 + |f'(a)|}\right), \end{aligned}$$

and continuity is proved. ♦

► **6.10** *Establish the result again using the difference quotient definition given by $\lim\limits_{x\to a} \frac{f(x)-f(a)}{x-a} = f'(a)$, and the fact that $\lim_{x\to a}(x-a) = 0$ with the product rule for limits.*

To some, the following example might exhibit one of the less intuitive consequences of our derivative definition.

Example 1.8 *Let $f : D^1 \to \mathbb{R}$ be given by $f(x) = x^3$ where $D^1 = \{0\} \cup \{\pm\frac{1}{n} : n \in \mathbb{N}\}$. We will show that $f'(0)$ exists and is zero. First, note that $0 \in \mathcal{L}(D^1)$. For $x \in D^1$ such that $x \neq 0$,*

$$\left|\frac{f(x) - f(0)}{x - 0} - 0\right| = |x^2| = x^2.$$

Therefore, for $\epsilon > 0$ we may choose $\delta = \sqrt{\epsilon}$ to obtain

$$x \in D^1 \cap N'_\delta(0) \quad \Rightarrow \quad \left|\frac{f(x) - f(0)}{x - 0} - 0\right| = x^2 < \epsilon.$$

◀

► **6.11** *Let $f : D^1 \to \mathbb{R}$ be given by $f(x) = \frac{1}{x}$ where $D^1 = \{1\} \cup \left\{\frac{n}{n+1} : n \in \mathbb{N}\right\}$. Does $f'(1)$ exist? If so, what is it?*

Higher-Order Derivatives

Of course, a function $f : D^1 \to \mathbb{R}$ might be differentiable at more than just a single point of its domain. In fact, we may consider all the points $D^1_{f'} \subset D^1$ at which f is differentiable as the domain of the function $f' : D^1_{f'} \to \mathbb{R}$. That is, the derivative of f is itself a real-valued function, and we say that *f is differentiable on $D^1_{f'}$*. If $a \in \mathcal{L}(D^1_{f'})$, we may consider whether f' has a derivative at a. According to Theorem 1.2 on page 238, we can determine this by checking whether the limit

$$\lim_{x \to a} \frac{f'(x) - f'(a)}{x - a} = B$$

for some real number B. If so, then B is called *the second derivative of f at a*, and we denote it by $f''(a)$ or $\frac{d^2 f}{dx^2}(a)$. In similar fashion one can consider whether the third derivative $f'''(a)$ exists at a, and more generally whether the nth derivative $f^{(n)}(a)$ exists at a. As this derivative-generating scheme indicates, it is clearly necessary for the $(n-1)^{st}$ derivative to exist before one can consider the existence of the nth derivative. It is also necessary for the point a to be an element of $\mathcal{L}(D^1_{f^{(n-1)}})$ to consider the existence of $f^{(n)}(a)$. Finally, this last fact implies that the domain of the nth derivative will be contained in the domain of the $(n-1)^{st}$. To illustrate these ideas, consider the function $f : \mathbb{R} \to \mathbb{R}$ given by $f(x) = x^5$. Since $f^{(n)}(x)$ exists for every x and every $n \geq 0$, it follows that $D^1_{f^{(n)}} = \mathcal{L}(D^1_{f^{(n)}}) = \mathbb{R}$ for all $n \geq 0$. For a simple example where the domain of f' is a proper subset of the domain of f, consider the function $f(x) : \mathbb{R} \to \mathbb{R}$ given by $f(x) = |x|$. The derivative does not exist at $x = 0$, and

$$f'(x) = \begin{cases} 1 & \text{if } x > 0 \\ -1 & \text{if } x < 0 \end{cases}.$$

It follows that $D^1_{f'} = \mathcal{L}(D^1_{f'}) = \mathbb{R} \setminus \{0\}$ is a proper subset of $D^1 = \mathcal{L}(D^1) = \mathbb{R}$.

We take the time now to develop a few more practical results that are of interest in the case $f : D^1 \to \mathbb{R}$. They should be familiar to the reader.

Algebraic Results

We begin with a theorem that allows for more efficient algebraic manipulation of derivatives. Many problems typically involve combinations of functions as sums, differences, products, and quotients, and so convenient means for finding the derivatives of such combinations is a valuable tool. Also of great importance is a convenient way to handle the computation of the derivative of a composition of functions. This last result, of course, is the familiar chain rule learned in calculus.

Theorem 1.9 (Algebraic Properties of the Derivative)
Suppose the functions $f : D^1 \to \mathbb{R}$ and $g : D^1 \to \mathbb{R}$ are differentiable at $a \in \mathcal{L}(D^1)$. Then

a) $f \pm g$ is differentiable at a, and $(f \pm g)'(a) = f'(a) \pm g'(a)$.

b) fg is differentiable at a, and $(fg)'(a) = f'(a)g(a) + f(a)g'(a)$.

c) cf is differentiable at a, and $(cf)'(a) = cf'(a)$ for $c \in \mathbb{R}$.

d) (f/g) is differentiable at a provided $g(a) \neq 0$. In this case,

$$(f/g)'(a) = \frac{g(a)f'(a) - f(a)g'(a)}{(g(a))^2}.$$

Note that property *b)* in Theorem 1.9 is often referred to as *the product rule*. Similarly, property *d)* is often called *the quotient rule.*

PROOF We prove *b)* and leave the others to the exercises. Consider the difference quotient associated with the function $fg : D^1 \to \mathbb{R}$ at $a \in \mathcal{L}(D^1)$, broken up cleverly as

$$\frac{f(x)g(x) - f(a)g(a)}{x - a} = f(x)\left[\frac{g(x) - g(a)}{x - a}\right] + g(a)\left[\frac{f(x) - f(a)}{x - a}\right].$$

Since f is differentiable at a, it follows from Proposition 1.7 on page 242 that f is continuous at a, and so $\lim_{x\to a} f(x) = f(a)$, yielding

$$\lim_{x\to a} \frac{f(x)g(x) - f(a)g(a)}{x - a} = f(a)g'(a) + g(a)f'(a),$$

and the theorem is proved. ♦

► **6.12** *Complete the above proof for parts a), c), and d). Also, determine the domains of the functions defined in a) through d).*

▶ **6.13** *Prove that the following functions $f: \mathbb{R} \to \mathbb{R}$ are differentiable on $\mathbb{R}$.*
a) $f(x) = x$ *b)* $f(x) = x^n$ *for* $n \in \mathbb{Z}^+$
c) $f(x) = c_0 + c_1 x + c_2 x^2 + \cdots + c_n x^n$ *for* $c_j \in \mathbb{R}$ *and* $n \in \mathbb{N}$ *with* $c_n \neq 0$.

▶ **6.14** *For each of the following, verify that the given $f'(a)$ is as claimed by using an induction proof.*
a) $f(x) = x^n$ *for* $D^1 = \mathbb{R}$, $n \in \mathbb{Z}^+$, $f'(a) = na^{n-1}$ *for all* $a \in D^1$
b) $f(x) = x^n$ *for* $D^1 = \mathbb{R} \setminus \{0\}$, $n \in \mathbb{Z}^-$, $f'(a) = na^{n-1}$ *for all* $a \in D^1$

▶ **6.15** *Suppose $p: \mathbb{R} \to \mathbb{R}$ is given by $p(x) = a_0 + a_1 x + a_2 x^2 + \cdots + a_n x^n$ for $n \in \mathbb{N}$, $a_j \in \mathbb{R}$ for all $0 \le j \le n$, and $a_n \neq 0$, and $q: \mathbb{R} \to \mathbb{R}$ is given by $q(x) = b_0 + b_1 x + b_2 x^2 + \cdots + b_m x^m$ for $m \in \mathbb{N}$, $b_j \in \mathbb{R}$ for all $0 \le j \le m$, and $b_m \neq 0$. Then, for $f: D^1 \to \mathbb{R}$ given by $f(x) = \frac{p(x)}{q(x)}$ where $D^1 = \{x \in \mathbb{R} : q(x) \neq 0\}$, show that f is differentiable on D^1.*

Suppose we are interested in finding the derivative of a more complicated function, such as $h(x) = \left(x^2 - 7x + 1\right)^{80}$. Here, $h: \mathbb{R} \to \mathbb{R}$ can be expressed as the composite function $h = g \circ f$ where $f: \mathbb{R} \to \mathbb{R}$ is given by $f(x) = x^2 - 7x + 1$, and $g: \mathbb{R} \to \mathbb{R}$ is given by $g(x) = x^{80}$. Must we use the difference quotient version of our definition directly to determine the derivative of h? Of course, thanks to the following frequently employed theorem known to all students of calculus, the answer is no. In fact, we can compute $h'(a)$ by knowing only $f'(a)$ and $g'(f(a))$. While of great practical value, this result will be important in our theoretical development as well.

Theorem 1.10 (The Chain Rule)
Suppose $f: D^1_f \to \mathbb{R}$ is differentiable at $a \in \mathcal{L}(D^1_f)$, and let $\mathcal{R}_f \equiv f(D^1_f)$. Suppose $\mathcal{R}_f \subset D^1_g$. If $g: D^1_g \to \mathbb{R}$ is differentiable at $b = f(a) \in \mathcal{L}(D^1_g)$, then $h \equiv g \circ f: D^1_f \to \mathbb{R}$ is differentiable at $a \in \mathcal{L}(D^1_f)$, and

$$h'(a) = g'(b) f'(a) = g'\left(f(a)\right) f'(a).$$

PROOF [2] Let the function $G: D^1_g \to \mathbb{R}$ be defined by

$$G(y) \equiv \begin{cases} \frac{g(y)-g(b)}{y-b} & \text{if } y \neq b \text{ and } y \in D^1_g \\ g'(b) & \text{if } y = b \end{cases}.$$

Note that since $\lim_{y \to b} G(y) = g'(b) = G(b)$, it follows that G is continuous at b. Since $g(y) - g(b) = G(y)(y - b)$ for all $y \in D^1_g$, we have for $y = f(x)$ and $b = f(a)$,

$$\frac{h(x) - h(a)}{x - a} = \frac{g\left(f(x)\right) - g\left(f(a)\right)}{x - a} = G\left(f(x)\right) \left(\frac{f(x) - f(a)}{x - a}\right). \qquad (6.7)$$

Since f is differentiable at a it follows that f is continuous at a, and so $G\left(f(x)\right)$

[2] We follow [Fit95].

is also continuous at a. Taking the limit as x approaches a in (6.7) above, we obtain

$$h'(a) = G(f(a))\, f'(a) = G(b) f'(a) = g'(b) f'(a),$$

and the theorem is proved. ♦

Example 1.11 *We now consider the example posed just before our statement of the chain rule. Suppose $f : \mathbb{R} \to \mathbb{R}$ is given by $f(x) = x^2 - 7\,x + 1$, and $g : \mathcal{R}_f \to \mathbb{R}$ by $g(x) = x^{80}$. Then $f'(a) = 2\,a - 7$ for all $a \in \mathbb{R}$, and $g'(a) = 80\,a^{79}$ for all $a \in \mathbb{R}$. If $h : \mathbb{R} \to \mathbb{R}$ is given by*

$$h(x) = g(f(x)) = \left(x^2 - 7\,x + 1\right)^{80},$$

then h is differentiable at every $a \in \mathbb{R}$ and

$$h'(a) = g'(f(a))\; f'(a) = 80\,\left(a^2 - 7\,a + 1\right)^{79}\,(2\,a - 7).$$ ◀

A significant result is that the natural logarithm is differentiable on its whole domain, having derivative $\frac{d}{dx}(\ln x) = \frac{1}{x}$. We establish this next.

Example 1.12 *Consider the natural logarithm function $\ln : \mathbb{R}^+ \to \mathbb{R}$. We will show that the derivative of this function is given by $\frac{d}{dx}(\ln x) = \frac{1}{x}$ for each $x \in \mathbb{R}^+$. First, we note that if $\ln x$ is presumed to be differentiable, its derivative must take the form $\frac{d}{dx}(\ln x) = \frac{1}{x}$. To see this, note that $e^{\ln x} = x$ for all $x \in \mathbb{R}^+$, and so application of the chain rule yields*

$$e^{\ln x}\,\frac{d}{dx}(\ln x) = 1,$$

which implies that $\frac{d}{dx}(\ln x) = \frac{1}{e^{\ln x}} = \frac{1}{x}$. We now show that the derivative of $\ln x$ exists at any $x_0 \in \mathbb{R}^+$. To this end, we wish to show that for any $\epsilon > 0$ there exists $\delta > 0$ small enough that $x \in \mathbb{R}^+$ satisfying $0 < |x - x_0| < \delta$ will imply

$$\left|\frac{\ln x - \ln x_0}{x - x_0} - \frac{1}{x_0}\right| < \epsilon.$$

Note that since the real exponential function is a one-to-one correspondence of $\mathbb{R}$ onto $\mathbb{R}^+$, and the natural logarithm function is its inverse, there are unique values $y, y_0 \in \mathbb{R}$ such that $\ln x = y$ and $\ln x_0 = y_0$. From this we obtain

$$\left|\frac{\ln x - \ln x_0}{x - x_0} - \frac{1}{x_0}\right| = \left|\frac{y - y_0}{e^y - e^{y_0}} - \frac{1}{e^{y_0}}\right|.$$

But

$$\lim_{y \to y_0}\left(\frac{e^y - e^{y_0}}{y - y_0}\right) = e^{y_0} \neq 0,$$

and so it follows that

$$\lim_{y \to y_0}\left(\frac{y - y_0}{e^y - e^{y_0}}\right) = \frac{1}{e^{y_0}}.$$

Therefore, for any $\epsilon > 0$ we can choose $\delta_1 > 0$ such that

$$0 < |y - y_0| < \delta_1 \;\Rightarrow\; \left|\frac{y - y_0}{e^y - e^{y_0}} - \frac{1}{e^{y_0}}\right| < \epsilon,$$

or, equivalently,

$$0 < |\ln x - \ln x_0| < \delta_1 \;\Rightarrow\; \left| \frac{\ln x - \ln x_0}{x - x_0} - \frac{1}{x_0} \right| < \epsilon.$$

Since the natural logarithm function is continuous and one-to-one, we can find $\delta > 0$ such that

$$0 < |x - x_0| < \delta \;\Rightarrow\; 0 < |\ln x - \ln x_0| < \delta_1,$$

and so overall we have

$$0 < |x - x_0| < \delta \;\Rightarrow\; \left| \frac{\ln x - \ln x_0}{x - x_0} - \frac{1}{x_0} \right| < \epsilon.$$

◀

▶ **6.16** *For $f : \mathbb{R}^+ \to \mathbb{R}$ given by $f(x) = x^{3/2}$, show that $f'(a) = \frac{3}{2}\, a^{1/2}$ for all $a \in \mathbb{R}^+$. (Hint: Consider $f(x) = e^{\frac{3}{2} \ln x}$.)*

▶ **6.17** *Consider the function $f : D^1 \to \mathbb{R}$ given by $f(x) = x^{p/q}$ with $p/q \in \mathbb{Q}$.*

a) Find the largest domain D^1 for this function. (Hint: The answer depends on p and q.)

b) For each distinct case determined in part a) above, find the values $a \in D^1$ where f is differentiable and find the derivative.

1.3 Local Extrema Results and the Mean Value Theorem

Local Extrema Results

Other derivative results familiar to students of calculus are those that deal with the identification of points in a function's domain that correspond to local maxima or minima (collectively referred to as the function's *local extrema*). We say that a function $f : D^1 \to \mathbb{R}$ has a *local maximum* at a point $a \in D^1$ if there exists a deleted neighborhood $N'(a) \subset D^1$ such that $f(x) < f(a)$ for all $x \in N'(a)$. Similarly, we say that a function $f : D^1 \to \mathbb{R}$ has a *local minimum* at a point $a \in D^1$ if there exists a deleted neighborhood $N'(a) \subset D^1$ such that $f(a) < f(x)$ for all $x \in N'(a)$. Students of calculus will recall that for a function $f : D^1 \to \mathbb{R}$ a *critical point* of f is any point $a \in \mathcal{L}(D^1)$ at which the derivative of f either fails to exist or is zero. Recall also that for a function $f : (c, d) \to \mathbb{R}$ differentiable on (c, d), local extrema can occur only at points $a \in (c, d)$ where $f'(a) = 0$. We prove this fact below.

Theorem 1.13 *Suppose $f : (c, d) \to \mathbb{R}$ is differentiable at $a \in (c, d)$. If f has a local maximum or a local minimum at a, then $f'(a) = 0$.*

PROOF We prove the case where f has a local maximum at a, leaving the local minimum case to the reader. Also, assuming that $f'(a) \neq 0$, we consider the case $f'(a) < 0$, and note that the $f'(a) > 0$ is handled similarly. With these assumptions there exists $\delta > 0$ such that for $0 < |x - a| < \delta$,

$$\left| \frac{f(x) - f(a)}{x - a} - f'(a) \right| < -\tfrac{1}{2} f'(a). \quad \text{(Why?)}$$

Since

$$\left|\frac{f(x)-f(a)}{x-a}-f'(a)\right| \ge \frac{f(x)-f(a)}{x-a}-f'(a),$$

it follows that for $0 < |x-a| < \delta$ we have that $\frac{f(x)-f(a)}{x-a} < 0$. (Why?) But this implies that $f(x) > f(a)$ for $x < a$ and $0 < |x-a| < \delta$, contradicting the assumption that $f(a)$ is a local maximum. The case $f'(a) > 0$ is handled similarly and is left to the reader. Therefore we must have $f'(a) = 0$. ♦

► **6.18** *Answer the two (Why?) questions in the proof given above, and complete the proof to the theorem by assuming $f(a)$ to be a local maximum with $f'(a) > 0$. Show that a contradiction results. Complete the similar argument for $f(a)$ presumed as a local minimum with $f'(a) \neq 0$.*

► **6.19** *Suppose $f : [c,d] \to \mathbb{R}$ is differentiable at $a \in [c,d]$. Does the conclusion of the above theorem necessarily hold?*

Note that this theorem in combination with Theorem 1.25 on page 192 from the previous chapter implies the following practical fact. We leave its proof to the reader.

Corollary 1.14 *If $f : [c,d] \to \mathbb{R}$ is continuous on $[c,d]$ and differentiable on (c,d), then f achieves a maximum value and a minimum value somewhere in $[c,d]$. If either extreme value is attained at an interior point, ξ, of the interval, then $f'(\xi) = 0$.*

► **6.20** *Prove the above result.*

Example 1.15 *Consider the function $f : [-\frac{1}{2}, \frac{1}{2}] \to \mathbb{R}$ given by $f(x) = x^4 - 2x^2 + 3$. We will find the local maxima and minima and the absolute maximum and minimum on the interval $\left[-\frac{1}{2}, \frac{1}{2}\right]$. As students of calculus will recall, we must check the value of the function on the boundary points of the interval separately. The only possible local extrema in $\left(-\frac{1}{2}, \frac{1}{2}\right)$ correspond to $x \in \left(-\frac{1}{2}, \frac{1}{2}\right)$ for which $f'(x) = 4x^3 - 4x = 0$, i.e., at $x = 0$. To determine whether $f(0) = 3$ is a local maximum or minimum, we examine the function at values of x near 0. Note that for $\epsilon > 0$, we have that $f(-\epsilon) = \epsilon^4 - 2\epsilon^2 + 3 = f(\epsilon)$. Since both $f(-\epsilon)$ and $f(\epsilon)$ are less than 3 for $0 < \epsilon < \sqrt{2}$, we conclude that the point $(0,3)$ is a local maximum. To find the absolute maximum and minimum of f on $\left[-\frac{1}{2}, \frac{1}{2}\right]$, we determine the function values at the boundary points of the interval and compare them to the function value at the critical point. In this case, $f\left(\pm\frac{1}{2}\right) = \frac{41}{16} < 3$, and so the function f achieves its absolute minimum value at each of the endpoints of the interval $\left[-\frac{1}{2}, \frac{1}{2}\right]$, while its absolute maximum is its local maximum value at the point $x = 0$.* ◄

Rolle's Theorem and the Mean Value Theorem

We now build on our extrema results to obtain a result known as Rolle's theorem. Used more as a tool for proving other, more significant theorems such as the mean value theorem and Taylor's theorem, which follow immediately after, one might wonder if "Rolle's lemma" would be a more appropriate name. However, as the exercises will reveal, Rolle's theorem is actually equivalent to the mean value theorem, and so the name is justified.

Theorem 1.16 (Rolle's Theorem)
Suppose the function $f : [c, d] \to \mathbb{R}$ is continuous on $[c, d]$ and differentiable on the open interval (c, d). If $f(c) = f(d) = 0$, then there exists $\xi \in (c, d)$ such that $f'(\xi) = 0$.

PROOF Let $M = f(x_M) = \max_{[c,d]} f$ and $m = f(x_m) = \min_{[c,d]} f$. If $M = m$, then $f(x) = M = m$ for all $x \in [c, d]$ and we have $f'(x) = 0$ for all $x \in (c, d)$, and the theorem is established. Therefore, assume $M > m$. It follows that at least one of the points x_M or x_m must lie in the interior of $[c, d]$. (Why?) Denote that x value by ξ. Since ξ is a point where f attains a maximum or minimum value, by the previous theorem we must have that $f'(\xi) = 0$, and the theorem is proved. ♦

► **6.21** *Answer the (Why?) question in the above proof.*

Theorem 1.17 (The Mean Value Theorem)
Suppose the function $f : [c, d] \to \mathbb{R}$ is continuous on $[c, d]$ and differentiable on the open interval (c, d). Then, there exists $\xi \in (c, d)$ such that

$$f(d) - f(c) = f'(\xi)(d - c).$$

PROOF Consider the points $(c, f(c))$ and $(d, f(d))$ connected by the line

$$y_L = f(c) + \frac{f(d) - f(c)}{d - c}(x - c).$$

Since the curve $y = f(x)$ and the line y_L contain these points, it must be true that the difference

$$F(x) \equiv y - y_L = f(x) - \left[f(c) + \frac{f(d) - f(c)}{d - c}(x - c) \right]$$

vanishes at the points $x = c$ and $x = d$. Applying Rolle's theorem to F we obtain

$$0 = F'(\xi) = f'(\xi) - \frac{f(d) - f(c)}{d - c} \quad \text{for some } \xi \in (c, d).$$

Simple rearrangement of the above gives the result. ♦

Example 1.18 *Consider the function $f : [0, 2\pi] \to \mathbb{R}$ given by $f(x) = x - \sin x$. By the mean value theorem, there must be a point $\xi \in (0, 2\pi)$ such that $f(2\pi) - f(0) = f'(\xi)\, 2\pi$, i.e., $2\pi = (1 - \cos\xi)2\pi$. From this it is easy to see that ξ must be $\pi/2$ or $3\pi/2$.* ◀

Example 1.19 *In this example we will use the mean value theorem to show that the function $f : \mathbb{R} \to \mathbb{R}$ given by $f(x) = e^x$ is uniformly continuous on $(-\infty, a]$ for any $a \in \mathbb{R}$, but is not uniformly continuous on $[a, \infty)$. To this end, let $\epsilon > 0$ be given. Then for $\xi, \eta \in (-\infty, a]$, the mean value theorem implies that there exists c between ξ and η such that*

$$\left|e^\xi - e^\eta\right| = \left|e^c\right| |\xi - \eta| \le e^a |\xi - \eta| < \epsilon \quad \text{if} \quad |\xi - \eta| < \delta \equiv \frac{\epsilon}{e^a}.$$

This establishes the uniform continuity on $(-\infty, a]$. To see that f is not uniformly continuous on $[a, \infty)$, assume the opposite. That is, assume that for any $\epsilon > 0$ there exists $\delta > 0$ such that

$$\left|e^\xi - e^\eta\right| < 1 \quad \text{if} \quad \xi, \eta \in [a, \infty) \;\text{ and }\; |\xi - \eta| < \delta.$$

If we let $\xi = \eta + \delta/2$, then we have

$$\left|e^{\eta + \delta/2} - e^\eta\right| < 1 \quad \text{for all} \quad \eta \in [a, \infty),$$

and by the mean value theorem we obtain for some c between η and $\eta + \delta/2$ that

$$\left|e^{\eta+\delta/2} - e^\eta\right| = \left|e^c\right| \tfrac{\delta}{2} \ge e^\eta \, \tfrac{\delta}{2},$$

which implies $e^\eta < \frac{2}{\delta}$ for all $\eta \in [a, \infty)$, a contradiction. ◀

Example 1.20 *Consider the function $f : [-1, 1] \to \mathbb{R}$ given by $f(x) = x^{2/3}$. Since f is not differentiable on the whole interval $(-1, 1)$, we should not expect the conclusion of the mean value theorem to necessarily hold in this case. In fact, it doesn't.* ◀

► **6.22** *Verify the claim in the previous example.*

► **6.23** *Prove the following result, known as the* Cauchy mean value theorem. *Let $f, g : [a, b] \to \mathbb{R}$ both be continuous on $[a, b]$ and differentiable on (a, b). Then there exists $c \in (a, b)$ such that $[f(b) - f(a)]\, g'(c) = [g(b) - g(a)]\, f'(c)$.*

► **6.24** *Suppose $f : (a, b) \to \mathbb{R}$ is differentiable on (a, b), and $f'(x) = 0$ for all $x \in (a, b)$. Use the mean value theorem to show that there exists a constant c such that $f(x) \equiv c$ for all $x \in (a, b)$.*

1.4 Taylor Polynomials

As our linear approximation definition of the derivative should have already implied, it is often of interest in analysis to be able to approximate complicated functions by simpler ones. The simplest and most useful functions for making such approximations are polynomials.

Definition 1.21 (Taylor Polynomial and Remainder)
Consider a function $f : D^1 \to \mathbb{R}$, and $a \in \mathcal{L}(D^1)$. If $f'(a), f''(a), \ldots, f^{(n)}(a)$ all exist, then the polynomial $P_n : D^1 \to \mathbb{R}$ defined by

$$P_n(x) \equiv f(a) + f'(a)(x-a) + \frac{f''(a)}{2!}(x-a)^2 + \cdots + \frac{f^{(n)}(a)}{n!}(x-a)^n$$

is called the nth-degree **Taylor polynomial** at a, and the difference $R_n(x)$ given by

$$R_n(x) \equiv f(x) - P_n(x)$$

is called the **remainder** associated with $P_n(x)$.

For a given function $f : D^1 \to \mathbb{R}$, a Taylor polynomial at $a \in \mathcal{L}(D^1)$ acts as a perfect approximation to the function f at $x = a$. In fact, $P_n(a) = f(a)$. But P_n may also be a good approximation to f for values of $x \neq a$ as long as x is sufficiently close to a. How good is *good*? And how close is *sufficiently close*? For a given Taylor polynomial P_n, *sufficiently close* is determined by how small we require the magnitude of the remainder $R_n(x) = f(x) - P_n(x)$ to be. Suppose one wishes to approximate f by P_n with an error no greater than E. Then one must restrict the domain of consideration to only those values of x near a that obtain $|R_n(x)| < E$. Depending on the function f, and on which point a of its domain the Taylor polynomial is centered, the set of x values that yield $|R_n(x)| < E$ might comprise a whole interval containing a, a half-interval with a as an endpoint, or perhaps just a itself. The Taylor polynomial P_n is said to be a *good* approximation if the error R_n goes to 0 *fast enough* as x approaches a. In particular, consider the first-degree Taylor polynomial approximation P_1 at a, which serves as a *linear* approximation to f at a. In this case, to be a *good* approximation, we require that R_1 go to 0 fast enough that even when divided by $|x-a|$, which itself is going to zero as x approaches a, the resulting ratio still goes to 0. That is,

$$\lim_{x \to a} \frac{R_1(x)}{|x-a|} = 0.$$

We say in this case that $R_1(x)$ is $o(x-a)$, read *"little-oh of $(x-a)$."* A *good* linear approximation, then, is one for which $R_1(x)$ is $o(x-a)$. Of course, the reader should recognize the linear approximation $P_1(x)$ as the $p(x)$ to which we refer in our definition of the derivative $f'(a)$ at a. This discussion then yields an insightful interpretation of our linear approximation derivative definition. Namely, *$f : D^1 \to \mathbb{R}$ has a derivative at the point $a \in \mathcal{L}(D^1)$, only if f has a good linear approximation via the first-degree Taylor polynomial at a.*

Formulas for the remainder associated with a Taylor polynomial are of particular interest in analysis. The following theorem expresses one form that is useful, although there are others as well.

Theorem 1.22 (Taylor's Theorem with Remainder)
Consider $f : [c,d] \to \mathbb{R}$. Suppose $f'(x), \ldots, f^{(n)}(x)$ exist and are continuous on $[c,d]$, and $f^{(n+1)}(x)$ exists on (c,d). Fix $a \in [c,d]$, and let P_n be the nth-degree Taylor polynomial at a. Then for every $x \in [c,d]$ such that $x \neq a$ there exists a real number ξ between a and x such that

$$f(x) = P_n(x) + R_n(x),$$

where $R_n(x)$ is given by

$$R_n(x) = \frac{f^{(n+1)}(\xi)}{(n+1)!}(x-a)^{n+1}.$$

PROOF [3] We prove the theorem for $a \in (c,d)$, leaving the cases $a = c$ and $a = d$ to the reader. Fix x and a in (c,d) such that $x > a$, and let $\theta \in [a,x]$ (the proof for $x < a$ is handled similarly). Define $F : [a,x] \to \mathbb{R}$ by

$$\begin{aligned} F(\theta) \equiv\ & f(x) - f(\theta) - (x-\theta)f'(\theta) \\ & - \frac{(x-\theta)^2}{2!}f''(\theta) - \cdots - \frac{(x-\theta)^n}{n!}f^{(n)}(\theta) - \frac{(x-\theta)^{n+1}}{(n+1)!}K, \end{aligned}$$

where K is a constant to be determined. We wish to apply Rolle's theorem to F, so we need to have $F(a) = F(x) = 0$. It is easy to see that $F(x) = 0$. To guarantee that $F(a) = 0$, we define K to make it so, i.e.,

$$K \equiv \frac{(n+1)!}{(x-a)^{n+1}}\left[f(x) - f(a) - (x-a)f'(a) - \cdots - \frac{(x-a)^n}{n!}f^{(n)}(a)\right]. \quad (6.8)$$

Recall in the above expression that x and a are fixed. From this, Rolle's theorem implies that there exists a point $\xi \in (a,x)$ such that $F'(\xi) = 0$. Differentiating F and evaluating at ξ gives

$$0 = F'(\xi) = -\frac{(x-\xi)^n}{n!}f^{(n+1)}(\xi) + \frac{(x-\xi)^n}{n!}K,$$

which when combined with the expression (6.8) for K yields

$$\begin{aligned} \frac{(n+1)!}{(x-a)^{n+1}}&\left[f(x) - f(a) - (x-a)f'(a) - \cdots - \frac{(x-a)^n}{n!}f^{(n)}(a)\right] \\ &= K = f^{(n+1)}(\xi). \end{aligned}$$

Now a simple rearrangement gives

$$f(x) = f(a) + f'(a)(x-a) + \cdots + \frac{f^{(n)}(a)}{n!}(x-a)^n + \frac{f^{n+1}(\xi)}{(n+1)!}(x-a)^{n+1},$$

where $\xi \in (a,x)$ and $\frac{f^{n+1}(\xi)}{(n+1)!}(x-a)^{n+1} = R_n(x)$. ♦

[3]Our proof is from Fulks [Ful78].

► **6.25** *Complete the above proof by considering the cases $a = c$ and $a = d$.*

Example 1.23 *We apply Taylor's theorem with remainder to the function $f : \mathbb{R} \to \mathbb{R}$ given by $f(x) = \sin x$, and $a = 0$. The utility of the theorem will be seen in our finding an explicit expression for $R_n(x)$ and an estimate for its magnitude. We have*

$$\sin x = x - \frac{x^3}{3!} + \frac{x^5}{5!} - \cdots + \frac{(-1)^n x^{2n+1}}{(2n+1)!} + R_{2n+1}(x), \quad \textit{where}$$

$$R_{2n+1}(x) = C \frac{x^{2n+2}}{(2n+2)!}, \quad \textit{and } C = \pm \sin \xi \ \textit{ for some } \xi \textit{ between } 0 \textit{ and } x.$$

For example, with $n = 2$ in the above expression for $\sin x$, we have

$$\sin x = x - \frac{x^3}{3!} + \frac{x^5}{5!} + R_5(x),$$

where $R_5(x)$ is given by

$$R_5(x) = -(\sin \xi) \frac{x^6}{6!} \quad \textit{for some } \xi \textit{ between } 0 \textit{ and } x.$$

As an explicit example of such an approximation, consider the angle $x = \frac{\pi}{90}$ radians (i.e., 2 degrees). The above Taylor polynomial gives us

$$\sin\left(\frac{\pi}{90}\right) = \frac{\pi}{90} - \frac{(\pi/90)^3}{3!} + \frac{(\pi/90)^5}{5!} + R_5(\pi/90),$$

where $|R_5(\pi/90)| \le \frac{(\pi/90)^6}{6!} < 2.52 \times 10^{-12}$. That is, estimating $\sin(\pi/90)$ by the sum $\frac{\pi}{90} - \frac{(\pi/90)^3}{3!} + \frac{(\pi/90)^5}{5!}$ is in error by at most $\frac{(\pi/90)^6}{6!}$, or less than 2.52×10^{-12}. This is a relative error of less than 10^{-8} percent! We note in this example that for any $x \in \mathbb{R}$,

$$|R_{2n+1}(x)| \le \frac{|x|^{2n+2}}{(2n+2)!} \to 0 \ \textit{ as } n \to \infty, \quad \textit{(Why?)}$$

and so $\lim_{n\to\infty} R_{2n+1}(x) = 0$ for all $x \in \mathbb{R}$. This means that for any specified degree of accuracy E, we can approximate the function $\sin x$ on any bounded interval by a polynomial of suitably high (yet finite) degree.[4] ◄

► **6.26** *Determine $P_6(x)$ and $R_6(x)$ in the above example. How does it compare to $P_5(x)$ and $R_5(x)$ obtained there?*

[4]Real functions that can be approximated by Taylor polynomials of arbitrarily high degree within a neighborhood of a point a, and whose remainders behave in this way are called *analytic* at a and are very special. In the case where $R_n(x) \to 0$ as $n \to \infty$, we refer to the associated "infinite degree Taylor polynomial" implied by $\lim_{n\to\infty} P_n(x)$ (actually called a *power series*) as *the Taylor series of f centered at a*. We will discuss Taylor series and analyticity more fully in Chapter 9.

► **6.27** *Consider the function $f(x) = \sin x$ and its fifth degree Taylor polynomial and remainder term at 0 as in the previous example. Evaluate the function, the polynomial, and the remainder at the value $x = \pi$. Is this as good an approximation to the function's value as at $x = \pi/90$? Is the remainder approximation of any use at all? What is the difference between this exercise and the example? Can the Taylor polynomial approximation at 0 evaluated at π be improved in any way?*

► **6.28** *In the above example, we showed that $\lim_{n\to\infty} R_{2n+1}(x) = 0$ for all $x \in \mathbb{R}$. Show that, in fact, $R_{2n+1}(x) \to 0$ uniformly on any bounded interval as $n \to \infty$.*

While useful in a first-year calculus course, we won't often use L'Hospital's rule in our work. Nonetheless, for completeness, we present it here.

Theorem 1.24 (L'Hospital's Rule)
Suppose f and g are differentiable on $(a-h, a+h) \subset \mathbb{R}$ for some number $h > 0$ such that the following all hold:

$$(i)\ \ f(a) = g(a) = 0,$$

$$(ii)\ \ g' \neq 0 \text{ and } g \neq 0 \text{on } (a-h, a+h),$$

$$(iii)\ \ \lim_{x\to a} \frac{f'(x)}{g'(x)} = L \text{ exists }.$$

Then, $\lim_{x\to a} \frac{f(x)}{g(x)} = L$.

PROOF Let $\epsilon > 0$ be given. Then there exists $\delta > 0$ such that

$$0 < |x-a| < \delta \ \Rightarrow\ \left|\frac{f'(x)}{g'(x)} - L\right| < \epsilon.$$

Fix $x > a$ such that $0 < |x-a| < \delta$. Then define $F : [a, x] \to \mathbb{R}$ by

$$F(\theta) = f(\theta)g(x) - g(\theta)f(x).$$

Note that $F(a) = F(x) = 0$. According to Rolle's theorem there must exist a point $\xi \in (a, x)$ such that $F'(\xi) = 0$, i.e.,

$$F'(\xi) = f'(\xi)g(x) - g'(\xi)f(x) = 0.$$

Rearranging this last expression yields

$$\frac{f'(\xi)}{g'(\xi)} = \frac{f(x)}{g(x)}.$$

Assuming $0 < |x-a| < \delta$, and noting that $\xi \in (a, x)$, we see that

$$\left|\frac{f'(\xi)}{g'(\xi)} - L\right| < \epsilon, \quad \text{which implies} \quad \left|\frac{f(x)}{g(x)} - L\right| < \epsilon.$$

We have shown that for a given $\epsilon > 0$ there exists a $\delta > 0$ such that

$$x > a \text{ and } 0 < |x-a| < \delta \ \Rightarrow\ \left|\frac{f(x)}{g(x)} - L\right| < \epsilon.$$

The case $x < a$ can be handled similarly, and so $\lim_{x\to a} \frac{f(x)}{g(x)} = L$. ♦

1.5 Differentiation of Sequences and Series of Functions

The following result is often of great practical use.

Theorem 1.25 *For $n \in \mathbb{N}$, consider the sequence of functions $f_n : (a, b) \to \mathbb{R}$ such that the following hold:*

(i) For some $x_0 \in (a, b)$, the limit $\lim f_n(x_0)$ exists.

(ii) For each $n \in \mathbb{N}$, $f_n'(x)$ exists on (a, b).

(iii) There exists a function $g : (a, b) \to \mathbb{R}$ such that $f_n'(x)$ converges uniformly to $g(x)$ on (a, b).

Then there exists a function $f : (a, b) \to \mathbb{R}$ such that for all $x \in (a, b)$,

$$\lim f_n(x) = f(x) \quad \textit{and} \quad f'(x) = g(x).$$

That is, for all $x \in (a, b)$, $\frac{d}{dx}\left(\lim f_n(x)\right) = \lim\left(f_n'(x)\right)$.

PROOF [5] We begin by showing that $f(x) = \lim f_n(x)$ exists on (a, b). To this end, consider $n, m \in \mathbb{N}$ and apply the mean value theorem to $f_n(x) - f_m(x)$ to obtain

$$f_n(x) - f_m(x) = f_n(x_0) - f_m(x_0) + (x - x_0)\left[f_n'(c) - f_m'(c)\right],$$

where c is a number between x and x_0. From this we see that

$$|f_n(x) - f_m(x)| \le |f_n(x_0) - f_m(x_0)| + (b - a)\,|f_n'(c) - f_m'(c)|.$$

Since $\{f_n(x_0)\}$ is a Cauchy sequence (Why?), we have for $\epsilon > 0$ that there exists $N_1 \in \mathbb{N}$ such that

$$n, m > N_1 \;\Rightarrow\; |f_n(x_0) - f_m(x_0)| < \tfrac{\epsilon}{2}, \tag{6.9}$$

and since $\{f_n'\}$ is uniformly Cauchy on (a, b) (Why?), there exists $N_2 \in \mathbb{N}$ such that for all $x \in (a, b)$,

$$n, m > N_2 \;\Rightarrow\; |f_n'(x) - f_m'(x)| < \frac{\epsilon}{2\,(b - a)}. \tag{6.10}$$

Combining inequalities (6.9) and (6.10) yields that, for all $x \in (a, b)$,

$$n, m > \max(N_1, N_2) \;\Rightarrow\; |f_n(x) - f_m(x)| < \epsilon,$$

i.e., $\{f_n\}$ is uniformly Cauchy on (a, b), and thus convergent to a limit function $f : (a, b) \to \mathbb{R}$ given by $f(x) = \lim f_n(x)$.

Now we will show that $f'(c) = g(c)$ for all $c \in (a, b)$. To see this, we again use

[5] We follow [Rud76].

the mean value theorem to obtain

$$f_n(x) - f_m(x) = f_n(c) - f_m(c) + (x - c)\left[f_n'(s) - f_m'(s)\right],$$

where s is a number between c and x (recall that c is between x and x_0, and therefore $x \neq c$). From this, it follows that

$$\left(\frac{f_n(x) - f_n(c)}{x - c}\right) - \left(\frac{f_m(x) - f_m(c)}{x - c}\right) = f_n'(s) - f_m'(s). \tag{6.11}$$

Again, since $\{f_n'\}$ is uniformly Cauchy on (a, b), we have for $\epsilon > 0$ that there exists $N_3 \in \mathbb{N}$ such that for all $x \in (a, b)$,

$$n, m > N_3 \Rightarrow |f_n'(x) - f_m'(x)| < \tfrac{\epsilon}{3}. \tag{6.12}$$

Combining inequality (6.12) with (6.11) obtains for all $x \in (a, b)$ and $c \neq x$ that

$$n, m > N_3 \Rightarrow \left|\left(\frac{f_n(x) - f_n(c)}{x - c}\right) - \left(\frac{f_m(x) - f_m(c)}{x - c}\right)\right| < \tfrac{\epsilon}{3}. \tag{6.13}$$

Now we let $n \to \infty$ in (6.13) to yield for all $x \in (a, b)$ and $c \neq x$ that

$$m > N_3 \Rightarrow \left|\left(\frac{f(x) - f(c)}{x - c}\right) - \left(\frac{f_m(x) - f_m(c)}{x - c}\right)\right| \le \tfrac{\epsilon}{3}. \tag{6.14}$$

Since $\lim f_m'(c) = g(c)$, there exists $N_4 \in \mathbb{N}$ such that

$$m > N_4 \Rightarrow |f_m'(c) - g(c)| < \tfrac{\epsilon}{3}. \tag{6.15}$$

Finally, for any m, there exists $\delta > 0$ such that

$$0 < |x - c| < \delta \Rightarrow \left|\frac{f_m(x) - f_m(c)}{x - c} - f_m'(c)\right| < \tfrac{\epsilon}{3}. \tag{6.16}$$

Combining inequalities (6.14), (6.15), and (6.16), and fixing $m > \max(N_3, N_4)$, we obtain that $0 < |x - c| < \delta \Rightarrow$

$$\begin{aligned}\left|\frac{f(x) - f(c)}{x - c} - g(c)\right| &\le \left|\left(\frac{f(x) - f(c)}{x - c}\right) - \left(\frac{f_m(x) - f_m(c)}{x - c}\right)\right| \\ &\quad + \left|\frac{f_m(x) - f_m(c)}{x - c} - f_m'(c)\right| + |f_m'(c) - g(c)| \\ &< \epsilon,\end{aligned}$$

i.e., $f'(c) = g(c)$. ◆

Corollary 1.26 *For $j \in \mathbb{N}$, consider the sequence of functions $f_j : (a, b) \to \mathbb{R}$ such that the following hold:*

(i) For some $x_0 \in (a, b)$, the sum $\sum_{j=1}^{\infty} f_j(x_0)$ converges.

(ii) For each $j \in \mathbb{N}$, $f_j'(x)$ exists for all $x \in (a, b)$.

(iii) There exists a function $g : (a, b) \to \mathbb{R}$ such that $\sum_{j=1}^{\infty} f_j'(x) = g(x)$ on (a, b), and the convergence is uniform on (a, b).

Then there exists a function $f : (a,b) \to \mathbb{R}$ such that for all $x \in (a,b)$,

$$\sum_{j=1}^{\infty} f_j(x) = f(x) \quad and \quad f'(x) = g(x).$$

That is, for all $x \in (a,b)$ we have that $\frac{d}{dx}\sum_{j=1}^{\infty} f_j(x) = \sum_{j=1}^{\infty} f_j'(x)$.

▶ **6.29** *Prove the above corollary.*

The great utility of the theorem and its corollary is in allowing for term-by-term differentiation of a series of functions. The following example illustrates this in an important case.

Example 1.27 *Recall that* $\exp(x) : \mathbb{R} \to \mathbb{R}$ *is given by* $e^x = \sum_{j=0}^{\infty} \frac{x^j}{j!}$. *We will show that* $\frac{d}{dx}e^x = e^x$ *via Corollary 1.26. To this end, let* $f_j : \mathbb{R} \to \mathbb{R}$ *be given by* $f_j(x) = \frac{x^j}{j!}$, *and note that*

(i) $\sum_{j=0}^{\infty} f_j(x) = \sum_{j=0}^{\infty} \frac{x^j}{j!}$ *converges for all* $x \in \mathbb{R}$,

(ii) *For each* $j \geq 1$, $f_j'(x) = \dfrac{x^{j-1}}{(j-1)!}$ *exists for all* $x \in \mathbb{R}$,

(iii) $\sum_{j=0}^{\infty} f_j'(x) = f_0'(x) + \sum_{j=1}^{\infty} f_j'(x) = \sum_{j=1}^{\infty} \frac{x^{j-1}}{(j-1)!} = \sum_{j=0}^{\infty} \frac{x^j}{j!} = e^x$ *for all* $x \in \mathbb{R}$.

Therefore, $\frac{d}{dx}e^x = \frac{d}{dx}\sum_{j=0}^{\infty} \frac{x^j}{j!} = \sum_{j=0}^{\infty} \frac{d}{dx}\frac{x^j}{j!} = \sum_{j=1}^{\infty} \frac{x^{j-1}}{(j-1)!} = e^x$ *on* $\mathbb{R}$. ◀

2 The Derivative for $f : D^k \to \mathbb{R}$

The functions we will consider next are those that take k-dimensional vectors to real numbers. Of course, the special case $k = 1$ corresponds to the case just considered in the last section, and our results here should reduce to those from before when setting $k = 1$. Recall that when $k > 1$ we refer to such functions as real-valued functions of several real variables, or real-valued functions of a vector. Such functions will be denoted by

$$f(\mathbf{x}) \quad \text{or} \quad f(x_1, \ldots, x_k).$$

Here, $\mathbf{x}$ denotes the vector having components $x_1, \ldots, x_k$. As described in a previous chapter, we will often dispense with transpose notation when referring to column vectors in a line of text. A vector $\mathbf{x}$ having components x_i for $i = 1, \ldots, k$ will also be denoted by $(x_1, \ldots, x_k)$ or $[x_i]$ where the context is clear regarding its dimension, as well as its being a column or row vector. Regardless of notation, when working with functions of several variables there are many derivatives to investigate. In fact, such a function has (potentially) k first-order derivatives, each taken with respect to one of the k independent variables x_i for $i = 1, \ldots, k$. We call such derivatives *partial derivatives*, and in a sense we will see that they form the building blocks of our more general definition of derivative.

In what follows, we denote the unit vector in the direction of the positive x_i-coordinate axis by $\mathbf{e}_i$. (In three-dimensional Euclidean space, $\mathbf{e}_1$ is the familiar $\hat{\imath}$, for example.)

2.1 Definition

We immediately present our formal definition for the derivative of a function $f : D^k \to \mathbb{R}$. We concern ourselves in this and other higher-dimensional cases with the derivative of such a function f only at interior points of its domain.

Definition 2.1 (The Derivative of $f : D^k \to \mathbb{R}$ at the Point a)
Consider $f : D^k \to \mathbb{R}$ and an interior point $\mathbf{a} \in D^k$. We say that f is **differentiable** at $\mathbf{a}$ if there exists a $1 \times k$ matrix $\boldsymbol{A}$ having real components, A_i, such that the mapping $p : D^k \to \mathbb{R}$ given by

$$p(\mathbf{x}) = f(\mathbf{a}) + \boldsymbol{A}\,(\mathbf{x} - \mathbf{a}) \quad \text{satisfies} \quad \lim_{\mathbf{x}\to\mathbf{a}} \frac{f(\mathbf{x}) - p(\mathbf{x})}{|\mathbf{x} - \mathbf{a}|} = 0.$$

In this case we refer to the matrix $\boldsymbol{A} = [A_1 \; A_2 \; \cdots \; A_k]$ as the **derivative** of f at $\mathbf{a}$ and denote it thereafter by $\boldsymbol{f'}(\mathbf{a})$.

Note that what was a product of the two real numbers A and $(x - a)$ in Theorem 1.2 on page 238 is replaced by a product between a matrix and a column vector in this case. Each quantity on the right side of the expression for $p(\mathbf{x})$ *must* be a real number (Why?). Since the scalar quantity $(x - a)$ from Definition 1.1 on page 237 is here changed to a vector quantity $(\mathbf{x} - \mathbf{a})$, the other factor and the binary operator between them *must* be such that the product is a real number.

▶ **6.30** *Answer the above (Why?) question. Then, suppose in our definition for the derivative of a function $f : D^k \to \mathbb{R}$ we replace $\boldsymbol{A}\,(\mathbf{x} - \mathbf{a})$ with $A\,|\mathbf{x} - \mathbf{a}|$ for some real number A. What is wrong with this proposal? (Hint: Does this definition reduce to the correct derivative value in the special case $k = 1$?)*

The above linear approximation definition is equivalent to the following ϵ, δ version:

The function $f : D^k \to \mathbb{R}$ is differentiable at interior point $\mathbf{a} \in D^k$ with the derivative given by the $1 \times k$ matrix $\boldsymbol{A}$ if and only if for any $\epsilon > 0$ there exists $\delta > 0$ such that

$$|\mathbf{x} - \mathbf{a}| < \delta \;\Rightarrow\; |f(\mathbf{x}) - f(\mathbf{a}) - \boldsymbol{A}\,(\mathbf{x} - \mathbf{a})| \le \epsilon\,|\mathbf{x} - \mathbf{a}|. \tag{6.17}$$

The reader should verify this. Note also that since we restrict our discussion here to interior points $\mathbf{a} \in D^k$, if we choose $\delta > 0$ small enough the condition

$|\mathbf{x} - \mathbf{a}| < \delta$ will ensure $\mathbf{x} \in D^k$ in expressions such as (6.17). Therefore the simpler condition $|\mathbf{x} - \mathbf{a}| < \delta$ will be used instead of $\mathbf{x} \in D^k \cap N_\delta(\mathbf{a})$.

► **6.31** *Verify that the above ϵ, δ version of Definition 2.1 is in fact equivalent to it.*

► **6.32** *Is it possible to formulate an equivalent difference quotient version to Definition 2.1? Refer to the exercises immediately following Exercise 6.1 on page 238 for a clue.*

Of course, just as in the $f : D^1 \to \mathbb{R}$ case, when the derivative of a function $f : D^k \to \mathbb{R}$ exists at an interior point $\mathbf{a} \in D^k$, it is unique. We formally state this result, leaving the proof to the reader.

Proposition 2.2 *Consider $f : D^k \to \mathbb{R}$ and an interior point $\mathbf{a} \in D^k$. If $\boldsymbol{f}'(\mathbf{a})$ exists, then it is unique.*

► **6.33** *Prove the above result.*

Example 2.3 *For the function $f : \mathbb{R}^2 \to \mathbb{R}$ given by $f(x, y) = x^2 y$, we will use the ϵ, δ version of the definition to show that $\boldsymbol{f}'(1, 2) = \boldsymbol{A} = [4 \;\; 1]$. To this end, let $\mathbf{x} = (x, y)$ and $\mathbf{a} = (1, 2)$. We must show that for any given $\epsilon > 0$ there exists $\delta > 0$ such that*

$$|\mathbf{x} - \mathbf{a}| < \delta \;\; \Rightarrow \;\; |f(\mathbf{x}) - f(\mathbf{a}) - \boldsymbol{A}\,(\mathbf{x} - \mathbf{a})| \le \epsilon\, |\mathbf{x} - \mathbf{a}|.$$

Once again, the idea is to begin with the ϵ inequality on the right-hand side of the implication, and to use it to determine the required value of δ in the inequality on the left-hand side. Substituting f and $\boldsymbol{A}$ into the ϵ inequality gives

$$\begin{aligned}
|f(\mathbf{x}) - f(\mathbf{a}) - \boldsymbol{A}\,(\mathbf{x} - \mathbf{a})| &= \left| x^2 y - 2 - 4x + 4 - y + 2 \right| \\
&= \left| (x^2 - 1)(y - 2) + 2(x - 1)^2 \right| \\
&\le |x - 1|\,|x + 1|\,|y - 2| + 2\,|x - 1|^2 \\
&= |x - 1|\,|(x - 1) + 2|\,|y - 2| + 2\,|x - 1|^2 \\
&\le |x - 1|^2\,|y - 2| + 2\,|x - 1|\,|y - 2| + 2\,|x - 1|^2 \\
&\le |\mathbf{x} - \mathbf{a}|^3 + 2\,|\mathbf{x} - \mathbf{a}|^2 + 2\,|\mathbf{x} - \mathbf{a}|^2 \\
&= \Big(|\mathbf{x} - \mathbf{a}|^2 + 4\,|\mathbf{x} - \mathbf{a}| \Big)\; |\mathbf{x} - \mathbf{a}|.
\end{aligned}$$

We would like the last expression above to be no greater than $\epsilon\,|\mathbf{x} - \mathbf{a}|$. This will be true if

$$\big(|\mathbf{x} - \mathbf{a}|^2 + 4\,|\mathbf{x} - \mathbf{a}| \big) < \epsilon. \tag{6.18}$$

But if $|\mathbf{x} - \mathbf{a}| < 1$, then $\big(|\mathbf{x} - \mathbf{a}|^2 + 4\,|\mathbf{x} - \mathbf{a}| \big) < 5\,|\mathbf{x} - \mathbf{a}| < \epsilon$ as long as $|\mathbf{x} - \mathbf{a}| < \frac{\epsilon}{5}$. Overall then, we have

$$|\mathbf{x} - \mathbf{a}| < \delta \equiv \min\left(\tfrac{\epsilon}{5}, 1\right) \;\; \Rightarrow \;\; |f(\mathbf{x}) - f(\mathbf{a}) - \boldsymbol{A}\,(\mathbf{x} - \mathbf{a})| \le \epsilon\,|\mathbf{x} - \mathbf{a}|,$$

and the result is established. ◄

► **6.34** *In Example 2.3, consider the inequality (6.18). Complete the proof differently by forcing each summand in the parentheses on the left-hand side to be less than* $\frac{\epsilon}{2}$.

We postpone the proof that differentiability of a function $f : D^k \to \mathbb{R}$ implies its continuity until we reach the more general case of functions $\boldsymbol{f} : D^k \to \mathbb{R}^p$ in the next section. The more general result will also apply to the special case of $p = 1$.

2.2 Partial Derivatives

Suppose $f : D^k \to \mathbb{R}$ is differentiable at the interior point $\mathbf{a} \in D^k$ according to Definition 2.1. What are the entries of the $1 \times k$ matrix $\boldsymbol{f}'(\mathbf{a}) = \boldsymbol{A} = [A_1\ A_2\ \cdots\ A_k]$? How do they depend on f and on $\mathbf{a}$? To answer these questions, we will refer to the equivalent ϵ, δ version of the definition. That is, we assume for any $\epsilon > 0$ there exists $\delta > 0$ such that

$$|\mathbf{x} - \mathbf{a}| < \delta \quad \Rightarrow \quad |f(\mathbf{x}) - f(\mathbf{a}) - \boldsymbol{A}\,(\mathbf{x} - \mathbf{a})| \le \epsilon\,|\mathbf{x} - \mathbf{a}|.$$

Denoting the ith components of the column vectors $\mathbf{x}$ and $\mathbf{a}$ by x_i and a_i, respectively, if we substitute $\mathbf{x} = \mathbf{a} + h\,\mathbf{e_i}$, i.e., we let $\mathbf{x}$ differ from $\mathbf{a}$ only in the ith coordinate. This obtains

$$|h| < \delta \quad \Rightarrow \quad |f(\mathbf{a} + h\,\mathbf{e_i}) - f(\mathbf{a}) - \boldsymbol{A}\,h\,\mathbf{e_i}| \le \epsilon\,|h|.$$

But since $\boldsymbol{A}\,h\,\mathbf{e_i} = A_i\,h$ and $h = x_i - a_i$ (Why?), this expression becomes

$$|x_i - a_i| < \delta \quad \Rightarrow$$

$$|f(a_1, \ldots, x_i, \ldots, a_k) - f(a_1, \ldots, a_i, \ldots, a_k) - A_i\,(x_i - a_i)| \le \epsilon\,|x_i - a_i|,$$

or, equivalently for $p_i(x_i) \equiv f(a_1, \ldots, a_i, \ldots, a_k) + A_i\,(x_i - a_i)$,

$$\lim_{x_i \to a_i} \frac{f(a_1, \ldots, x_i, \ldots, a_k) - p_i(x_i)}{|x_i - a_i|} = 0. \tag{6.19}$$

Note that the function $f(a_1, \ldots, x_i, \ldots, a_k)$ in the above limit expression is a function of only the variable x_i. Therefore, the limit expression (6.19) says that the real number A_i is just $\frac{d}{dx_i} f(a_1, \ldots, x_i, \ldots a_k)$ evaluated at $x_i = a_i$, according to Definition 1.1 on page 237. We have proved the following result.

Theorem 2.4 *Suppose $f : D^k \to \mathbb{R}$ is differentiable at interior point $\mathbf{a} \in D^k$, with the derivative given by the $1 \times k$ matrix $\boldsymbol{A} = \begin{bmatrix} A_1 & A_2 & \cdots & A_k \end{bmatrix}$ there. Then the entries A_i of the matrix $\boldsymbol{A}$ satisfy the following properties.*

a) For $p_i(x_i) \equiv f(a_1, \ldots, a_i, \ldots, a_k) + A_i\,(x_i - a_i)$,

$$\lim_{x_i \to a_i} \frac{f(a_1, \ldots, x_i, \ldots, a_k) - p_i(x_i)}{|x_i - a_i|} = 0.$$

b) A_i *is the real number given by* $\frac{d}{dx_i} f(a_1, \ldots, x_i, \ldots a_k)$ *evaluated at* $x_i = a_i$.

Note that in the above discussion the characterization of each A_i is a consequence of f being differentiable at $\mathbf{a}$, and the linear polynomial p_i represents the best linear approximation of f at $\mathbf{a}$ corresponding to changes in f in the x_i direction. That is, for $\mathbf{x}$ shifted from $\mathbf{a}$ by a small amount in the x_i direction, $p_i(x_i)$ is a good linear approximation to $f(\mathbf{x})$ in the sense of Definition 1.1. The associated A_i is a real number representing the rate of change of f at $\mathbf{a}$ with respect to a change in $\mathbf{a}$ *only* in the ith component. We will refer to this quantity as *the partial derivative of f with respect to x_i at* $\mathbf{a}$, and denote it hereafter by $\frac{\partial f}{\partial x_i}(\mathbf{a})$ or by $f_{x_i}(\mathbf{a})$. Overall then, when the *total* derivative $\boldsymbol{f}'(\mathbf{a})$ for a function $f : D^k \to \mathbb{R}$ exists, it is the $1 \times k$ matrix whose entries are the above described partial derivatives. In fact, however, we can define each partial derivative $A_i = f_{x_i}(\mathbf{a})$ according to either part *a)* or part *b)* of Theorem 2.4 *without assuming the existence of* $\boldsymbol{f}'(\mathbf{a})$. Of course, our definition should be formulated so that when the *total* derivative $\boldsymbol{f}'(\mathbf{a})$ *does* exist, the partial derivatives $f_{x_i}(\mathbf{a})$ given by the definition will correspond to the components of $\boldsymbol{f}'(\mathbf{a})$ as described above. We give this definition now.

Definition 2.5 (Partial Derivatives of $f : D^k \to \mathbb{R}$ at the Point a)
Consider $f : D^k \to \mathbb{R}$ and interior point $\mathbf{a} = (a_1, \ldots, a_k) \in D^k$. If there exists a real-valued function p_i of the form $p_i(x_i) = f(\mathbf{a}) + A_i\,(x_i - a_i)$ for some real number A_i such that

$$\lim_{x_i \to a_i} \frac{f(a_1, \ldots, x_i, \ldots, a_k) - p_i(x_i)}{|x_i - a_i|} = 0,$$

then we say that A_i is the **partial derivative of f with respect to x_i at a** and denote it by

$$A_i = \frac{\partial f}{\partial x_i}(\mathbf{a}) \quad \text{or} \quad A_i = f_{x_i}(\mathbf{a}).$$

The partial derivatives defined above are also referred to as *the first-order partial derivatives of f with respect to x_i at* $\mathbf{a}$. As noted in the discussion preceding Definition 2.5, *if* the total derivative $\boldsymbol{f}'$ exists at the interior point $\mathbf{a} \in D^k$, it is given by

$$\boldsymbol{f}'(\mathbf{a}) = \big[f_{x_1}(\mathbf{a}) \;\; f_{x_2}(\mathbf{a}) \;\; \cdots \;\; f_{x_k}(\mathbf{a})\big]. \tag{6.20}$$

► **6.35** *Formulate a difference quotient version and an ϵ, δ version for the definition of partial derivative $f_{x_i}(\mathbf{a})$. Show that they are equivalent to each other and to Definition 2.5.*

► **6.36** *Suppose $f : D^2 \to \mathbb{R}$ is differentiable on all of D^2, where $D^2 = (a, b) \times (c, d)$. Fix $\xi \in (a, b)$, and $\eta \in (c, d)$, and define the functions u and v by*

$$u : (a, b) \to \mathbb{R}, \quad \text{where } u(x) \equiv f(x, \eta)$$

$$v : (c, d) \to \mathbb{R}, \quad \text{where } v(y) \equiv f(\xi, y).$$

Are the functions u and v differentiable on their domains? If so, find their derivatives.

The following theorem is a higher-dimensional version of the result from Exercise 6.24 on page 250.

Theorem 2.6 *Suppose D^k is open and connected, and $f : D^k \to \mathbb{R}$ is such that $\mathbf{f}'(\mathbf{x}) = \mathbf{0}$ for all $\mathbf{x} \in D^k$. Then there exists $c \in \mathbb{R}$ such that $f(\mathbf{x}) = c$ for all $\mathbf{x} \in D^k$.*

PROOF We begin our proof by considering the special case where D^k is a neighborhood of a point $\mathbf{a} \in \mathbb{R}^k$. To this end, consider $N_r(\mathbf{a})$ for some $r > 0$, and suppose $[\mathbf{a}, \mathbf{b}]$ is a segment that lies in $N_r(\mathbf{a})$ where $\mathbf{b} = \mathbf{a} + h\,\mathbf{e_i}$, i.e., the segment $[\mathbf{a}, \mathbf{b}]$ is parallel to the ith coordinate axis. Then

$$f(\mathbf{b}) - f(\mathbf{a}) = f(\mathbf{a} + h\,\mathbf{e_i}) - f(\mathbf{a}).$$

Defining $F : \mathbb{R} \to \mathbb{R}$ by $F(x_i) \equiv f(a_1, \ldots, x_i, \ldots, a_k)$ we have that $f(\mathbf{b}) - f(\mathbf{a}) = F(a_i + h) - F(a_i)$. It follows from the mean value theorem that there exists c_i between a_i and $a_i + h$ such that

$$f(\mathbf{b}) - f(\mathbf{a}) = F(a_i + h) - F(a_i) = F'(c_i)\,h = \frac{\partial f}{\partial x_i}(a_1, \ldots, c_i, \ldots, a_k)\,h.$$

But $\frac{\partial f}{\partial x_i}(a_1, \ldots, c_i, \ldots, a_k) = 0$ (Why?), and so $f(\mathbf{b}) = f(\mathbf{a}) = c$ for some real number c. From this we may conclude that f is constant along segments parallel to the coordinate axes. *Every* pair of points $\mathbf{a}, \mathbf{b} \in N_r(\mathbf{a})$ can be connected by a polygonal path whose segments are parallel to the coordinate axes (Why?), and so for any point $\mathbf{x} \in N_r(\mathbf{a})$, $\mathbf{x}$ can be so connected to $\mathbf{b}$. From this we see that $f(\mathbf{x}) = c$ for all $\mathbf{x} \in N_r(\mathbf{a})$. Now for the more general D^k, fix $\mathbf{x_0} \in D^k$ and define $A \equiv \left\{\mathbf{x} \in D^k : f(\mathbf{x}) = f(\mathbf{x_0})\right\}$, and $B \equiv \left\{\mathbf{x} \in D^k : f(\mathbf{x}) \neq f(\mathbf{x_0})\right\}$. Clearly, $D^k = A \cup B$, and $A \cap B = \varnothing$. We leave it to the reader to use our special case proved above to show that A and B are each open in $\mathbb{R}^k$. With this, since D^k is connected, we must have that either $A = \varnothing$ or $B = \varnothing$. But $\mathbf{x_0} \in A$, and so $A \neq \varnothing$. Therefore $B = \varnothing$, and $D^k = A$, i.e., $f(\mathbf{x}) = f(\mathbf{x_0})$ for all $\mathbf{x} \in D^k$. ♦

► **6.37** *Answer the (Why?) question in the above proof, and then show that the sets A and B defined in the above proof are each open in $\mathbb{R}^k$.*

► **6.38** *Consider $f : \mathbb{R}^k \to \mathbb{R}$ given by $f(\mathbf{x}) = c$ for all $\mathbf{x} \in \mathbb{R}^k$ where $c \in \mathbb{R}$ is a constant. Prove that $\mathbf{f}'(\mathbf{x}) = \mathbf{0}$ for all $\mathbf{x} \in \mathbb{R}^k$.*

► **6.39** *Can you see why the condition that D^k be connected is necessary in Theorem 2.6?*

2.3 The Gradient and Directional Derivatives

The Gradient

There is a common and convenient notation for collecting all the first-order partial derivatives of a function $f : D^k \to \mathbb{R}$ into one mathematical object. Students of calculus know this to be *the gradient of f*.

Definition 2.7 (Gradient of $\boldsymbol{f} : \boldsymbol{D^k} \to \mathbb{R}$ at the Point a)
Consider a function $f : D^k \to \mathbb{R}$ and an interior point $\mathrm{a} \in D^k$ such that all partial derivatives of f exist at a. The **gradient** of f at a is the $1 \times k$ matrix denoted by $\boldsymbol{\nabla f}(\mathrm{a})$ and defined as

$$\boldsymbol{\nabla f}(\mathrm{a}) = \left[f_{x_1}(\mathrm{a}) \;\; f_{x_2}(\mathrm{a}) \;\; \cdots \;\; f_{x_k}(\mathrm{a}) \right].$$

Note that although we have defined the gradient of $f : D^k \to \mathbb{R}$ as a $1 \times k$ matrix, it can be thought of equivalently as a k-dimensional row vector. The existence of the gradient of $f : D^k \to \mathbb{R}$ at an interior point $\mathrm{a} \in D^k$ depends solely on the existence of the first-order partial derivatives of f at a. That is, if the individual first-order partial derivatives $f_{x_i}(\mathrm{a})$ exist according to Definition 2.5, we could easily compute them and thereby construct the gradient matrix for f at a according to Definition 2.7. Based on our discussion of the derivative $\boldsymbol{f'}(\mathrm{a})$ and its relationship to the first-order partials as described by Definition 2.5 on page 261, a natural question now arises. Does the existence of $\boldsymbol{f'}(\mathrm{a})$ also depend solely on the existence of the first-order partial derivatives? Phrased slightly differently, does the existence of $\boldsymbol{\nabla f}(\mathrm{a})$ imply the existence of $\boldsymbol{f'}(\mathrm{a})$? Although one might be tempted to think so, the answer is *no*. As we have already seen, *if* the derivative $\boldsymbol{f'}(\mathrm{a})$ exists, it is related to the partial derivatives of f according to (6.20). From this and Definition 2.7 we see that *if* the derivative $\boldsymbol{f'}(\mathrm{a})$ exists, it can be represented by the gradient $\boldsymbol{\nabla f}(\mathrm{a})$, since both the derivative and the gradient are the $1 \times k$ matrix whose entries are the first-order partial derivatives of f at a. However, there are situations where the gradient exists while the derivative does not. Rather than determining conditions under which $\boldsymbol{\nabla f}$ *is* in fact the derivative, we postpone this topic until we reach the higher-dimensional situations, $\boldsymbol{f} : D^k \to \mathbb{R}^p$. The result stated and proved there will also apply for the case where $k > 1$ and $p = 1$. As stated above, we will find that when $\boldsymbol{f'}$ exists it *is* equal to the gradient $\boldsymbol{\nabla f}$, but there are situations where the gradient $\boldsymbol{\nabla f}$ exists while the derivative $\boldsymbol{f'}$ does not. This possibility points to a very important fact. *Differentiability for functions of more than one variable entails more than just the existence of the first-order partial derivatives.*

Example 2.8 *We will find the partial derivatives and the gradient of the function given by*

$$f(x,y) = \begin{cases} \frac{xy}{x^2+y^2} & (x,y) \neq (0,0) \\ 0 & (x,y) = (0,0) \end{cases}.$$

We will also show that despite the existence of $\boldsymbol{\nabla f}$ at the origin, the function f is not differentiable there. If $(x,y) \neq (0,0)$, then at (x,y),

$$\frac{\partial f}{\partial x} = \frac{y^3 - x^2 y}{(x^2+y^2)^2}, \quad \frac{\partial f}{\partial y} = \frac{x^3 - y^2 x}{(x^2+y^2)^2},$$

and

$$\nabla f(x,y) = \left[\frac{y^3 - x^2 y}{(x^2+y^2)^2} \;\; \frac{x^3 - y^2 x}{(x^2+y^2)^2}\right].$$

To find the partial derivatives at $(0,0)$*, we use the difference quotient version to obtain*

$$\frac{\partial f}{\partial x}(0,0) = \lim_{x\to 0} \frac{f(x,0) - f(0,0)}{x - 0} = 0, \quad \frac{\partial f}{\partial y}(0,0) = \lim_{y\to 0} \frac{f(0,y) - f(0,0)}{y - 0} = 0,$$

and therefore, $\nabla f(0,0) = [0\;\;0]$*. It follows from this that if* $f'(0,0)$ *were to exist, then it must satisfy* $f'(0,0) = \nabla f(0,0) = \mathbf{0}$*. This in turn implies that for any given* $\epsilon > 0$*, we can find* $\delta > 0$ *such that*

$$\sqrt{x^2+y^2} < \delta \;\Rightarrow\; \left|f(x,y) - f(0,0) - f'(0,0)\,(x,y)\right| \le \epsilon\,\sqrt{x^2+y^2}.$$

In particular, this obtains

$$0 < \sqrt{x^2+y^2} < \delta \;\Rightarrow\; \left|\frac{xy}{x^2+y^2}\right| \le \epsilon\,\sqrt{x^2+y^2},$$

or

$$0 < \sqrt{x^2+y^2} < \delta \;\Rightarrow\; \frac{|xy|}{\left(x^2+y^2\right)^{3/2}} \le \epsilon.$$

But letting $y = 5\,x$ *in the last expression, i.e., having* (x,y) *approach* $(0,0)$ *along the line given by* $y = 5\,x$*, yields*

$$0 < \sqrt{26}\,|x| < \delta \;\Rightarrow\; \frac{5}{(26)^{3/2}|x|} \le \epsilon,$$

a contradiction. (Why?) ◀

▶ **6.40** *In fact, the function in the previous example is not even continuous at the origin. Can you show this?*

Directional Derivatives

When f' exists, what exactly is the relationship between f' and the individual partial derivatives that make up its components? Is there another way to look at each f_{x_i} other than as the components of f'? As we have already mentioned, we can think of the $1 \times k$ matrix f' as the *total* derivative of f. If one wants the rate of change of f in the $\mathbf{e_i}$ direction, i.e., $\frac{\partial f}{\partial x_i}$, one can simply consider the ith entry of the matrix f'. But one can also think of the ith entry of f' as the projection of the row vector f' in the $\mathbf{e_i}$ direction, i.e., $f_{x_i} = f'\,\mathbf{e_i}$. That is, the rate of change of f in the $\mathbf{e_i}$ direction is found by the inner product of f' and the unit vector pointing in the direction of interest. This is true for directions other than the coordinate directions as well.

Definition 2.9 (Directional Derivative of $f : D^k \to \mathbb{R}$ at the Point a) Consider a function $f : D^k \to \mathbb{R}$ having derivative $f'(\mathbf{a})$ at an interior point $\mathbf{a} \in D^k$. Then the derivative of f in the direction $\mathbf{u}$ (where $\mathbf{u}$ is a unit column vector) at $\mathbf{a} \in D^k$ is denoted by $f'_{\mathbf{u}}(\mathbf{a})$, is given by

$$f'_{\mathbf{u}}(\mathbf{a}) = f'(\mathbf{a})\,\mathbf{u},$$

and is referred to as the **directional derivative** of f at $\mathbf{a}$ in the direction $\mathbf{u}$.

Notice that the directional derivative of such an f in the $\mathbf{u}$ direction at a point $\mathbf{a}$ is a *scalar* quantity. It is the rate of change in f (a scalar quantity) when moving from point $\mathbf{a}$ in the $\mathbf{u}$ direction. In fact, for $\mathbf{u} = [u_i]$, we have

$$f'_{\mathbf{u}}(\mathbf{a}) = \boldsymbol{f}'(\mathbf{a})\,\mathbf{u} = \sum_{i=1}^{k} u_i\, f_{x_i}(\mathbf{a}).$$

Example 2.10 *Consider $f(x,y) = \frac{1}{x-y}$ on $D^2 = \left\{(x,y) \in \mathbb{R}^2 : x \neq y\right\}$. We will*

a) Find $\nabla f(2,1)$.

b) Show that $\boldsymbol{f}'(2,1) = \nabla f(2,1)$.

c) Find $f'_{\mathbf{u}}(2,1)$ where $\mathbf{u} = \left(\frac{1}{\sqrt{2}}, -\frac{1}{\sqrt{2}}\right)$.

For the solution to part a), note that for all $(x,y) \in D^2$, we have $\frac{\partial f}{\partial x} = -\frac{1}{(x-y)^2}$, $\frac{\partial f}{\partial y} = \frac{1}{(x-y)^2}$, and so the gradient at $(2,1)$ is given by $\nabla f(2,1) = [-1 \;\; 1]$. To prove b) we need to show that for any $\epsilon > 0$ there exists $\delta > 0$ such that $\sqrt{(x-2)^2 + (y-1)^2} < \delta \;\;\Rightarrow$

$$|f(x,y) - f(2,1) - [-1 \;\; 1]\,(x-2, y-1)| \le \epsilon\sqrt{(x-2)^2 + (y-1)^2}.$$

To see this, note that

$$|f(x,y) - f(2,1) - [-1 \;\; 1]\,(x-2, y-1)| = \left|\frac{(x-2)^2 - 2\,(x-2)\,(y-1) + (y-1)^2}{x-y}\right|.$$

Note that here we have written the numerator of the right-hand side in terms of $(x-2)$ and $(y-1)$. If we now assume that $\sqrt{(x-2)^2 + (y-1)^2} < \frac{1}{10}$, then it follows that $|x-2| < \frac{1}{10}$ and $|y-1| < \frac{1}{10}$, which in turn implies that $x-y > \frac{8}{10}$. (Why?) Finally,

$$\begin{aligned}
&|f(x,y) - f(2,1) - [-1 \;\; 1]\,(x-2, y-1)| \\
&\le \left(\tfrac{10}{8}\right)\left[|x-2|^2 + 2\,|x-2|\,|y-1| + |y-1|^2\right] \\
&\le \frac{5}{4}\left[4\left(\sqrt{(x-2)^2 + (y-1)^2}\right)^2\right] \quad \text{since } |a| \le \sqrt{a^2+b^2}, \\
&\le \epsilon\sqrt{(x-2)^2 + (y-1)^2} \quad \text{if } \sqrt{(x-2)^2 + (y-1)^2} < \tfrac{\epsilon}{5}.
\end{aligned}$$

Letting $\delta = \min\left(\frac{1}{10}, \frac{\epsilon}{5}\right)$ gives the result. To establish c) simply compute the value of $f'_{\mathbf{u}}(2,1) = \boldsymbol{f}'(2,1)\,\mathbf{u} = [-1 \;\; 1]\left(\frac{1}{\sqrt{2}}, -\frac{1}{\sqrt{2}}\right) = -\sqrt{2}$. ◀

▶ **6.41** *As you might expect, there is a difference quotient definition for the directional derivative that is very similar to Definition 2.9. Write this limit expression and verify that it is consistent with Definition 2.9.*

► **6.42** *For a function $f : D^k \to \mathbb{R}$ differentiable at interior point $\mathbf{a} \in D^k$, show that the partial derivatives $f_{x_i}(\mathbf{a})$ are just the directional derivatives $f'_{\mathbf{e}_i}(\mathbf{a})$.*

► **6.43** *Suppose $f : D^2 \to \mathbb{R}$ has directional derivatives in every direction at $\mathbf{x} = \mathbf{0}$. Must f be differentiable at $\mathbf{x} = \mathbf{0}$?*

2.4 Higher-Order Partial Derivatives

Of course, any partial derivative f_{x_i} may exist for many values of $\mathbf{x}$ in the domain of $f : D^k \to \mathbb{R}$. We may refer to the set of such values of $\mathbf{x}$ as the domain of f_{x_i} and denote it by $D^k_{f_{x_i}}$. That is, f_{x_i} may itself be considered a function of $\mathbf{x} = (x_1, x_2, \ldots, x_k)$, and so might be differentiated again with respect to one of the components $x_1, x_2, \ldots, x_k$ at an interior point of its domain. When the differentiation is possible, say with respect to the jth variable, the result is denoted by $f_{x_i x_j}$, or $\frac{\partial^2 f}{\partial x_j \partial x_i}$. In the case where $i \neq j$, this is called the *second-order partial derivative of f with respect to x_i and x_j*. Note that what is implied by the two different notations, and the verbal description, is that the x_i derivative is taken *first*. Under appropriate conditions (which we will see later), we will have the convenience of being able to assume that $f_{x_i x_j} = f_{x_j x_i}$. That is, under the right conditions it does not matter in what order you take the partial derivatives of f. In general, however, $f_{x_i x_j} \neq f_{x_j x_i}$. In the case where $i = j$, the result is $f_{x_i x_i}$ or $\frac{\partial^2 f}{\partial x_i^2}$, the *second-order partial derivative of f with respect to x_i*. Third, and other higher-order derivatives are defined similarly. Of course, each of these higher-order partial derivatives may be considered as a function of $\mathbf{x} = (x_1, x_2, \ldots, x_k)$. The domain of each higher-order partial derivative is the set of $\mathbf{x} \in D^k$ at which the higher-order partial derivative is well defined.

Example 2.11 *Let $f : \mathbb{R}^2 \to \mathbb{R}$ be given by $f(x, y) = \frac{3}{5}x^{5/3}(y + 1)$. It is easy to determine that $\frac{\partial f}{\partial x} = x^{2/3}(y + 1)$, and $\frac{\partial f}{\partial y} = \frac{3}{5}x^{5/3}$, and that these partial derivatives exist for every $(x, y) \in \mathbb{R}^2$. It is also easy to see that $\frac{\partial^2 f}{\partial x^2} = \frac{2\,(y+1)}{3\,x^{1/3}}$ for all (x, y) such that $x \neq 0$. Does the second-order partial derivative $\frac{\partial^2 f}{\partial x^2}$ exist at the origin? To determine this, we need to evaluate the following limit,*

$$\lim_{x \to 0} \frac{\frac{\partial f}{\partial x}(x, 0) - \frac{\partial f}{\partial x}(0, 0)}{x} = \lim_{x \to 0} \frac{x^{2/3} - 0}{x} = \lim_{x \to 0} \frac{1}{x^{1/3}},$$

and we see that the limit does not exist. Therefore, $\frac{\partial^2 f}{\partial x^2}(0, 0)$ does not exist. Note, however, that $\frac{\partial^2 f}{\partial y^2} = 0$ for every $(x, y) \in \mathbb{R}^2$. Also note that for every $(x, y) \in \mathbb{R}^2$,

$$\frac{\partial^2 f}{\partial x \partial y} = \frac{\partial^2 f}{\partial y \partial x} = x^{2/3}.$$

◄

Example 2.12 *Let $f : \mathbb{R}^2 \to \mathbb{R}$ be given by*

$$f(x,y) = \begin{cases} \frac{x^3y - y^3x}{x^2+y^2} & \text{for } (x,y) \neq (0,0) \\ 0 & \text{for } (x,y) = 0 \end{cases}.$$

We will show that, for this function, the second-order mixed partial derivatives are not equal at the origin. Note that

$$f_x(0,0) = \lim_{x\to 0} \frac{f(x,0) - f(0,0)}{x} = \lim_{x\to 0} \frac{0}{x} = 0,$$

while $f_x(0,y)$ is given by

$$f_x(0,y) = \lim_{x\to 0} \frac{f(x,y) - f(0,y)}{x} = \lim_{x\to 0} \frac{x^3y - y^3x}{x\,(x^2+y^2)} = \lim_{x\to 0} \frac{x^2y - y^3}{x^2+y^2} = -y.$$

The value of $f_{xy}(0,0)$ is then found to be

$$f_{xy}(0,0) = \lim_{y\to 0} \frac{f_x(0,y) - f_x(0,0)}{y} = \lim_{y\to 0} \frac{-y - 0}{y} = -1.$$

A similar determination of $f_{yx}(0,0)$ via $f_y(0,0)$ and $f_y(x,0)$, or by recognizing the antisymmetric nature of f relative to switching the roles of x and y, yields the result $f_{yx}(0,0) = 1$. ◄

The following theorem gives sufficient conditions under which $f_{xy} = f_{yx}$ for functions $f : D^2 \to \mathbb{R}$. This allows for algebraic simplifications when dealing with mixed higher-order derivatives of functions of more than one variable. The more general case of $f : D^k \to \mathbb{R}$ is left to the reader as an exercise.

Theorem 2.13 (The Mixed Derivative Theorem for $f : D^2 \to \mathbb{R}$)
Suppose $f : D^2 \to \mathbb{R}$ is such that f, f_x, f_y, f_{xy}, and f_{yx} exist and are continuous on a neighborhood of the interior point $(a,b) \in D^2$. Then $f_{yx}(a,b) = f_{xy}(a,b)$.

PROOF [6] Choose $\epsilon > 0$ small enough so that the open rectangle $(a-\epsilon, a+\epsilon) \times (b-\epsilon, b+\epsilon)$ is contained in the neighborhood where f, f_x, f_y, f_{xy}, and f_{yx} are presumed continuous. Then let $h, k \in \mathbb{R}$ be such that $|h| < \epsilon$ and $|k| < \epsilon$. Define $g : (a-\epsilon, a+\epsilon) \to \mathbb{R}$ to be the function given by

$$g(x) \equiv f(x, b+k) - f(x,b).$$

Then by two applications of the mean value theorem, there exists ξ between

[6] We borrow the idea from [Ful78].

a and $a+h$ and η between b and $b+k$ such that

$$\begin{aligned} g(a+h)-g(a) &= g'(\xi)h \\ &= \left[\frac{\partial f}{\partial x}(\xi, b+k) - \frac{\partial f}{\partial x}(\xi, b)\right] h \\ &= \left(\frac{\partial^2 f}{\partial y \partial x}(\xi, \eta)\right) hk. \end{aligned}$$

Therefore,

$$\begin{aligned} \left(\frac{\partial^2 f}{\partial y \partial x}(\xi, \eta)\right) hk &= g(a+h)-g(a) \\ &= f(a+h, b+k) - f(a+h, b) - f(a, b+k) + f(a, b). \end{aligned} \tag{6.21}$$

Finally, dividing the above by hk and taking the limit as (h,k) approaches $(0,0)$ obtains

$$\frac{\partial^2 f}{\partial y \partial x}(a,b) = \lim_{(h,k)\to(0,0)} \frac{f(a+h, b+k) - f(a+h, b) - f(a, b+k) + f(a, b)}{hk}. \tag{6.22}$$

In taking the limit on the left-hand side of (6.21) above we have exploited the continuity of f_{xy} at (a,b). We leave it to the reader to show that $\frac{\partial^2 f}{\partial x \partial y}(a,b)$ equals the same limit expression on the right-hand side of (6.22). ♦

► **6.44** *Complete the above proof. To do so, let $q : (b-\epsilon, b+\epsilon) \to \mathbb{R}$ be given by $q(y) \equiv f(a+h, y) - f(a, y)$. Then apply the mean value theorem to q.*

► **6.45** *How would you generalize the statement of the above theorem to handle the more general case $f : D^k \to \mathbb{R}$? How would you prove it? What about higher-order mixed partial derivatives? For example, under the right circumstances can we conclude that $f_{yxx} = f_{xyx} = f_{xxy}$?*

2.5 Geometric Interpretation of Partial Derivatives

It is worth a bit of effort to understand partial derivatives geometrically. For convenience, we consider the case $f : D^2 \to \mathbb{R}$, leaving the general case of $f : D^k \to \mathbb{R}$ as an exercise for the reader. Let $f : D^2 \to \mathbb{R}$ be differentiable at the interior point $\mathbf{a} = (a,b) \in D^2$, with graph $S = \{(x, y, f(x,y)) : (x,y) \in D^2\}$ as shown in the figure below. If we "slice" S with the plane given by $H_1 = \{(x,y,z) \in \mathbb{R}^3 : y = b\}$, more compactly denoted by $H_1 = \{(x, b, z)\}$, we obtain the curve

$$C_1 = S \cap H_1 = \left\{(x, b, f(x,b)) : (x,b) \in D^2\right\}.$$

This curve can be represented by the function

$$z = f_1(x) = f(x, b),$$

which describes how the value of f varies when the first component of $\mathbf{a}$ is allowed to change while the other component of $\mathbf{a}$ is held fixed at b. (See

Figure 6.1.) Differentiating this function at $x = a$ obtains

$$f_1'(a) = \frac{\partial f}{\partial x}(a, b),$$

a value representing the slope of the tangent line to the curve represented by $f_1(x)$ at $x = a$. Similarly, we may slice the graph of f with the plane $H_2 = \{(a, y, z)\}$, to obtain the curve

$$C_2 = S \cap H_2 = \left\{(a, y, f(a, y)) : (a, y) \in D^2\right\}.$$

This curve can be represented by the function $f_2(y) = f(a, y)$, a function that describes how the value of f varies when the second component of a is allowed to change while the first component of a is held fixed at a. Differentiating this function at $y = b$ obtains $f_2'(b) = \frac{\partial f}{\partial y}(a, b)$, which represents the slope of the tangent line to $f_2(y)$ at $y = b$.

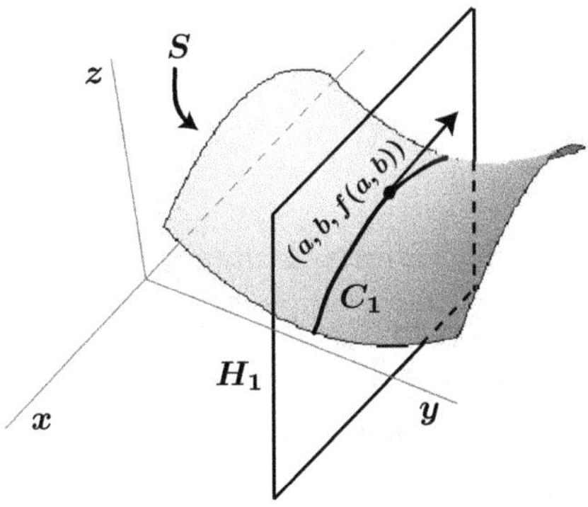

Figure 6.1 *The geometric interpretation of* $\frac{\partial f}{\partial x}(a, b)$.

► **6.46** *Consider a function* $f : D^k \to \mathbb{R}$ *that is differentiable at the interior point* $\mathbf{a} = (a_1, a_2, \ldots, a_k) \in D^k$. *Discuss what happens if you slice the graph of* f *with the planes* $H_i = \{(a_1, a_2, \ldots, a_{i-1}, x_i, a_{i+1}, \ldots, a_k, x_{k+1})\}$. *Relate the partial derivatives of* f *to slopes of tangents to the curves* $C_i = S \cap H_i$.

2.6 Some Useful Results

As in the previous section, we now develop some practical results involving the derivative for functions $f : D^k \to \mathbb{R}$.

Algebraic Results

The following result, analogous to Theorem 1.9 on page 244, provides some convenient rules for manipulating derivatives of real-valued functions of a vector. The proof is left to the exercises.

Proposition 2.14 (Algebraic Properties of the Derivative)
Suppose the functions $f : D^k \to \mathbb{R}$ and $g : D^k \to \mathbb{R}$ are differentiable at the interior point $\mathbf{a} \in D^k$. Then

a) $h \equiv f \pm g$ *is differentiable at* $\mathbf{a}$*, and* $\boldsymbol{h}'(\mathbf{a}) = \boldsymbol{f}'(\mathbf{a}) \pm \boldsymbol{g}'(\mathbf{a})$.

b) $h \equiv fg$ *is differentiable at* $\mathbf{a}$*, and* $\boldsymbol{h}'(\mathbf{a}) = g(\mathbf{a})\boldsymbol{f}'(\mathbf{a}) + f(\mathbf{a})\boldsymbol{g}'(\mathbf{a})$.

c) $h \equiv cf$ *is differentiable at* $\mathbf{a}$*, and* $\boldsymbol{h}'(\mathbf{a}) = c\boldsymbol{f}'(\mathbf{a})$ *for* $c \in \mathbb{R}$.

d) $h \equiv (f/g)$ *is differentiable at* $\mathbf{a}$ *provided* $g(\mathbf{a}) \neq 0$. *In this case,*

$$\boldsymbol{h}'(\mathbf{a}) = \frac{g(\mathbf{a})\boldsymbol{f}'(\mathbf{a}) - f(\mathbf{a})\boldsymbol{g}'(\mathbf{a})}{(g(\mathbf{a}))^2}.$$

► **6.47** *Prove the above result. Also determine the domains of the functions defined in parts a) through d).*

► **6.48** *Prove that the following functions $f : \mathbb{R}^2 \to \mathbb{R}$ are differentiable on $\mathbb{R}^2$.*
a) $f(x, y) = x$ b) $f(x, y) = y$ c) $f(x, y) = x^n$ for $n \in \mathbb{Z}^+$
d) $f(x, y) = y^n$ for $n \in \mathbb{Z}^+$ e) $f(x, y) = c_{ij}\, x^i\, y^j$ for $i, j \in \mathbb{Z}^+$ and $c_{ij} \in \mathbb{R}$
f) $f(x, y) = \sum_{1 \leq i,j \leq m} c_{ij}\, x^i\, y^j$

► **6.49** *Generalize the previous exercise to the case of $f : \mathbb{R}^k \to \mathbb{R}$.*

► **6.50** *Let $f : D^2 \to \mathbb{R}$ be given by $f(x, y) = \frac{p(x,y)}{q(x,y)}$ where $p : \mathbb{R}^2 \to \mathbb{R}$ and $q : \mathbb{R}^2 \to \mathbb{R}$ are of the form described in part f) of the previous exercise, and $D^2 = \{(x, y) \in \mathbb{R}^2 : q(x, y) \neq 0\}$. Show that f is differentiable on D^2.*

► **6.51** *Generalize the previous exercise to the case of the rational function $f : D^k \to \mathbb{R}$ given by $f(x_1, \ldots, x_k) = \frac{p(x_1,\ldots,x_k)}{q(x_1,\ldots,x_k)}$ defined on D^k given by $D^k = \{(x_1, \ldots, x_k) \in \mathbb{R}^k : q(x_1, \ldots, x_k) \neq 0\}$.*

Rather than prove the chain rule for functions $f : D^k \to \mathbb{R}$, we simply point out that the higher-dimensional cases that we will prove in the next section for functions $\boldsymbol{f} : D^k \to \mathbb{R}^p$ will also apply in the case $p = 1$.

Local Extrema Results

The following result is the higher-dimensional version of Theorem 1.13 on page 247.

Proposition 2.15 *Suppose $f : D^k \to \mathbb{R}$ has a local maximum or minimum at the interior point $\mathbf{a} = (a_1, \ldots, a_k)$. If f is differentiable at $\mathbf{a} \in D^k$, then $\boldsymbol{f}'(\mathbf{a}) = \mathbf{0}$. That is, $f_{x_i}(\mathbf{a}) = 0$ for each $i = 1, \ldots, k$.*

PROOF We prove the result for the case $k = 2$. Suppose $f : D^2 \to \mathbb{R}$ has a local maximum or minimum at interior point $\mathbf{a} = (a_1, a_2) \in D^2$, and that f is differentiable at $\mathbf{a}$. Note that

$$\boldsymbol{f}'(\mathbf{a}) = \begin{bmatrix} f_x(\mathbf{a}) & f_y(\mathbf{a}) \end{bmatrix},$$

so establishing that the derivative is zero at the extremum means establishing that each entry of the derivative *matrix* is zero at the extremum. Define $g(x) \equiv f(x, a_2)$ on an open interval containing $x = a_1$. Then according to Theorem 1.13 on page 247, g must have a local maximum (minimum) at $x = a_1$, and so $0 = g'(a_1) = f_x(\mathbf{a})$. To see that $f_y(\mathbf{a}) = 0$ we apply the same argument to the function $h(y) \equiv f(a_1, y)$. ♦

▶ **6.52** *Can you generalize the above proof to the cases $k > 2$?*

Example 2.16 *We will find the local extrema of the function $f : \mathbb{R}^2 \to \mathbb{R}$ given by*

$$f(x, y) = 4x^2 + xy + y^2 + 12$$

and determine their nature. We leave it to the reader to show that this function is, in fact, differentiable on all of $\mathbb{R}^2$. Assuming the function is differentiable, the only points at which extrema can occur are those for which both

$$\frac{\partial f}{\partial x} = 8x + y = 0, \quad \textit{and} \quad \frac{\partial f}{\partial y} = x + 2y = 0,$$

i.e., $(x, y) = (0, 0)$. To determine whether $f(0, 0) = 12$ is a local maximum or minimum, suppose $\epsilon_1, \epsilon_2 \in \mathbb{R}$ are not both zero, and note that

$$\begin{aligned} f(\epsilon_1, \epsilon_2) &= 4\epsilon_1^2 + \epsilon_1\epsilon_2 + \epsilon_2^2 + 12 \\ &= \tfrac{7}{2}\epsilon_1^2 + \left(\tfrac{1}{2}\epsilon_1^2 + \epsilon_1\epsilon_2 + \tfrac{1}{2}\epsilon_2^2\right) + \tfrac{1}{2}\epsilon_2^2 + 12 \\ &= \tfrac{7}{2}\epsilon_1^2 + \left(\tfrac{1}{\sqrt{2}}\epsilon_1 + \tfrac{1}{\sqrt{2}}\epsilon_2\right)^2 + \tfrac{1}{2}\epsilon_2^2 + 12 \\ &> 12. \end{aligned}$$

Therefore, $f(0, 0) = 12$ is the only extremum, and is a minimum. But note that since the above argument actually holds for all *ϵ_1 and ϵ_2 (not just* small *ϵ_1 and ϵ_2), the minimum is in fact a* global *minimum.* ◀

Finally, rounding out our local extrema results for functions $f : D^k \to \mathbb{R}$, we consider the higher-dimensional analog to Corollary 1.14 on page 248.

Proposition 2.17 *Let D^k be a compact subset of $\mathbb{R}^k$, and suppose that $f : D^k \to \mathbb{R}$ is differentiable on* Int(D^k) *and continuous on D^k. Then f achieves a maximum value and a minimum value somewhere in D^k. If either extreme value is attained at an interior point, $\boldsymbol{\xi}$, then it must be true that $\boldsymbol{f}'(\boldsymbol{\xi}) = \mathbf{0}$.*

► **6.53** *Prove the above result.*

Rolle's Theorem

We would like to consider a higher-dimensional version of Rolle's theorem next. One might wonder if it is even possible to generalize Rolle's theorem to higher dimensions. In fact, if one generalizes in the wrong way, the result will not hold. Rolle's theorem for a real-valued function of one real variable says that if the function vanishes at two different points, then under appropriate conditions there is a point between them where the derivative of the function vanishes. Suppose we consider the real valued function of *two* variables, $f(x, y) = \frac{x^3}{3} + x + y^2$. It is easy to verify that this f vanishes at the two points $(0, 0)$ and $(-1, \frac{2}{\sqrt{3}})$. But clearly $\boldsymbol{f}'(x, y) = [(x^2 + 1) \;\; 2y] \neq [0 \;\; 0]$ for *any* $(x, y) \in \mathbb{R}^2$. We must therefore be careful in generalizing such results to higher dimensions, as we now do.

Proposition 2.18 (Rolle's Theorem in $\mathbb{R}^k$)
Let D^k be an open and bounded subset of $\mathbb{R}^k$, and suppose that $f : \overline{D^k} \to \mathbb{R}$ is differentiable on D^k and continuous on $\overline{D^k}$. If $f \equiv 0$ on the boundary of D^k, then there exists an interior point $\mathbf{q} \in D^k$ such that $\boldsymbol{f}'(\mathbf{q}) = \mathbf{0}$.

PROOF First, notice that the correct generalization here involves requiring the function to be zero on the entire boundary of D^k (convince yourself that this idea is, in fact, consistent with our one-dimensional case). The proof of this higher-dimensional version of Rolle's theorem goes just as in the one-dimensional case, and the reader is encouraged to write out the details as an exercise. ♦

► **6.54** *Write out the proof of the above theorem.*

Example 2.19 *What kind of function would satisfy the conditions of the above version of Rolle's theorem? Consider the top half of the unit sphere centered at the origin of $\mathbb{R}^3$. For any point in $D^2 = \{(x, y) \in \mathbb{R}^2 : x^2 + y^2 < 1\}$, the z-coordinate of a point on this half-sphere surface, i.e., its height above the xy-plane, is given by the formula $z = \sqrt{1 - x^2 - y^2}$. Let $f : D^2 \to \mathbb{R}$ be the square of this height, so that $f(x, y) = 1 - x^2 - y^2$. Clearly f is defined on $\overline{D^2}$, differentiable on D^2, continuous on $\overline{D^2}$, and $f \equiv 0$ on $\partial D^2 = \{(x, y) \in \mathbb{R}^2 : x^2 + y^2 = 1\}$. It is also clear that $\boldsymbol{f}'(0, 0) = \mathbf{0}$.* ◄

▶ **6.55** *Verify the claims made in the above example.*

▶ **6.56** *Why not state the conditions of Proposition 2.18 in terms of $f: D^k \to \mathbb{R}$ with D^k closed, f differentiable on the interior of D^k, and f continuous on D^k?*

There is a mean value theorem and a version of Taylor's theorem with remainder for functions $f: D^k \to \mathbb{R}$, but we postpone their statement and proof until the next section. Our proofs of these results will make use of the chain rule associated with the more general class of real functions $\boldsymbol{f}: D^k \to \mathbb{R}^p$, and so we must establish this result first. We begin our discussion of the $\boldsymbol{f}: D^k \to \mathbb{R}^p$ case next.

3 THE DERIVATIVE FOR $\boldsymbol{f}: D^k \to \mathbb{R}^p$

The functions we now consider are those that take k-dimensional vectors to p-dimensional vectors. In this case we are dealing with vector-valued functions of several real variables, or vector-valued functions of a vector. We remind the reader that such functions will be denoted as

$$\boldsymbol{f}(\mathbf{x}), \quad \boldsymbol{f}(x_1, \dots, x_k), \quad \text{or} \quad \big(f_1(\mathbf{x}), f_2(\mathbf{x}), \dots, f_p(\mathbf{x})\big).$$

As described previously, a vector $\mathbf{x}$ having components x_i for $i = 1, \dots, k$ will also be simply denoted by $[x_i]$ where the context is clear regarding its dimension, as well as its being a column or row vector. Similarly, a $p \times k$ matrix with components $A_{i\,j}$ will sometimes be succinctly denoted by $[A_{i\,j}]$.

3.1 Definition

We begin with our formal definition, once again restricting ourselves to interior points of D^k.

Definition 3.1 (The Derivative of $\boldsymbol{f}: D^k \to \mathbb{R}^p$ at the Point a)
Consider $\boldsymbol{f}: D^k \to \mathbb{R}^p$ and an interior point $\mathbf{a} \in D^k$. We say that $\boldsymbol{f}$ is **differentiable** at a if there exists a $p \times k$ matrix $\boldsymbol{A}$ of real numbers, A_{ij}, such that the mapping $\mathbf{p}: D^k \to \mathbb{R}^p$ given by

$$\mathbf{p}(\mathbf{x}) = \boldsymbol{f}(\mathbf{a}) + \boldsymbol{A}\,(\mathbf{x} - \mathbf{a}) \quad \text{satisfies} \quad \lim_{\mathbf{x} \to \mathbf{a}} \frac{\boldsymbol{f}(\mathbf{x}) - \mathbf{p}(\mathbf{x})}{|\mathbf{x} - \mathbf{a}|} = \mathbf{0}.$$

In this case we refer to the matrix $\boldsymbol{A}$ as the **derivative** of $\boldsymbol{f}$ at a, and denote it thereafter by $\boldsymbol{f}'(\mathbf{a})$.

It might seem strange that the derivative of a function $\boldsymbol{f}: D^k \to \mathbb{R}^p$ is a $p \times k$ matrix, but students of linear algebra will recall that this choice of mathematical object, when multiplying the k-dimensional vector $(\mathbf{x} - \mathbf{a})$, will return a p-dimensional vector. (Note that the vector $(\mathbf{x} - \mathbf{a})$ must be thought of here as a *column* vector in order for the product to be well defined.) The reader can verify that this last, most general derivative definition is consistent with the two previous cases we have considered (namely, $k = p = 1$, and $k > 1$, $p = 1$). Just as in the cases we have already considered, there is an equivalent ϵ, δ definition for differentiability in the case of $\boldsymbol{f} : D^k \to \mathbb{R}^p$:

The function $\boldsymbol{f} : D^k \to \mathbb{R}^p$ is differentiable at interior point $\mathbf{a} \in D^k$ and has the derivative $\boldsymbol{f}'(\mathbf{a})$ if and only if for each $\epsilon > 0$ there exists $\delta > 0$ such that

$$|\mathbf{x} - \mathbf{a}| < \delta \quad \Rightarrow \quad \big|\boldsymbol{f}(\mathbf{x}) - \boldsymbol{f}(\mathbf{a}) - \boldsymbol{f}'(\mathbf{a})(\mathbf{x} - \mathbf{a})\big| \le \epsilon\,|\mathbf{x} - \mathbf{a}|. \tag{6.23}$$

► **6.57** *Prove the equivalence of Definition 3.1 and its corresponding ϵ, δ version.*

► **6.58** *For $\boldsymbol{f} : D^k \to \mathbb{R}^p$, consider a difference quotient version for our derivative definition. What difficulties do you encounter?*

► **6.59** *We will have need to define the derivative of a function $\boldsymbol{f} : \mathbb{R} \to \mathbb{R}^p$ in later chapters. Show that Definition 3.1 extends naturally to this case. That is, show that if $k = 1$, Definition 3.1 applies. Show too that if $\boldsymbol{f}$ is given by $\boldsymbol{f}(x) = \big(f_1(x), \ldots, f_p(x)\big)$, then $\boldsymbol{f}'(a)$ is given by $\big(f_1'(a), \ldots, f_p'(a)\big)$.*

Of course, as in the previous cases we have considered, when the derivative exists at $\mathbf{a} \in D^k$ for $\boldsymbol{f} : D^k \to \mathbb{R}^p$, it is unique.

Proposition 3.2 *Consider $\boldsymbol{f} : D^k \to \mathbb{R}^p$ and an interior point $\mathbf{a} \in D^k$. If $\boldsymbol{f}'(\mathbf{a})$ exists, then it is unique.*

PROOF Suppose $\boldsymbol{A}$ and $\boldsymbol{B}$ are matrices such that $\mathbf{p_A}(\mathbf{x}) = \boldsymbol{f}(\mathbf{a}) + \boldsymbol{A}\,(\mathbf{x} - \mathbf{a})$ and $\mathbf{p_B}(\mathbf{x}) = \boldsymbol{f}(\mathbf{a}) + \boldsymbol{B}\,(\mathbf{x} - \mathbf{a})$ satisfy

$$\lim_{\mathbf{x}\to\mathbf{a}} \frac{\boldsymbol{f}(\mathbf{x}) - \mathbf{p_A}(\mathbf{x})}{|\mathbf{x} - \mathbf{a}|} = \mathbf{0} \text{ and } \lim_{\mathbf{x}\to\mathbf{a}} \frac{\boldsymbol{f}(\mathbf{x}) - \mathbf{p_B}(\mathbf{x})}{|\mathbf{x} - \mathbf{a}|} = \mathbf{0},$$

respectively. Subtracting these limit expressions from each other obtains

$$\lim_{\mathbf{x}\to\mathbf{a}} \frac{(\boldsymbol{B} - \boldsymbol{A})\,(\mathbf{x} - \mathbf{a})}{|\mathbf{x} - \mathbf{a}|} = \mathbf{0}.$$

This in turn implies that for any $\epsilon > 0$, there exists $\delta > 0$ such that

$$0 < |\mathbf{x} - \mathbf{a}| < \delta \quad \Rightarrow \quad \frac{|(\boldsymbol{B} - \boldsymbol{A})\,(\mathbf{x} - \mathbf{a})|}{|\mathbf{x} - \mathbf{a}|} < \epsilon. \tag{6.24}$$

Letting $\mathbf{x} = \mathbf{a} + h\,\mathbf{e_i}$ for $i = 1, \ldots, k$ in the inequality (6.24) above obtains

$$0 < |h| < \delta \quad \Rightarrow \quad \frac{|(\boldsymbol{B} - \boldsymbol{A})\,h\,\mathbf{e_i}|}{|h|} < \epsilon.$$

But in fact, $\frac{|(\boldsymbol{B}-\boldsymbol{A})\,h\,\mathbf{e_i}|}{|h|} = |(\boldsymbol{B} - \boldsymbol{A})\,\mathbf{e_i}|$, and so $|(\boldsymbol{B} - \boldsymbol{A})\,\mathbf{e_i}| < \epsilon$ must be true for all $\epsilon > 0$ and for all i, *independent of* h. Therefore, $(\boldsymbol{B} - \boldsymbol{A})\,\mathbf{e_i} = \mathbf{0}$ for all i, and this implies $\boldsymbol{B} - \boldsymbol{A} = \mathbf{0}$ (Why?). That is, $\boldsymbol{A} = \boldsymbol{B}$. ♦

► **6.60** *Answer the (Why?) question in the above proof.*

As one might expect, differentiability implies continuity for functions $\boldsymbol{f} : D^k \to \mathbb{R}^p$ just as it does for functions $f : D \to \mathbb{R}$. We would like to establish this fact next, and to more easily do so, we introduce the notion of a *norm* for matrices. [7]

Definition 3.3 Let $\boldsymbol{A}$ be a $p \times k$ matrix of real numbers A_{ij}. The real number denoted by $|\boldsymbol{A}|$ is called the **norm** of $\boldsymbol{A}$, and is defined by

$$|\boldsymbol{A}| = \max_{i,j} |A_{ij}|.$$

Example 3.4 *For* $\boldsymbol{A} = \begin{bmatrix} 2 & 4 & -12 \\ 0 & 1 & \pi \end{bmatrix}$, *we have* $|\boldsymbol{A}| = 12$. ◄

Our matrix norm also has the following nice property, which we state as a proposition.

Proposition 3.5 *Let* $\boldsymbol{A}$ *be a* $p \times k$ *matrix of real numbers* A_{ij}. *Then there exists a real number* $C > 0$, *such that*

$$|\boldsymbol{A}\mathbf{x}| \le C\,|\boldsymbol{A}|\,|\mathbf{x}| \text{ for every } \mathbf{x} \in \mathbb{R}^k.$$

PROOF Suppose the matrix $\boldsymbol{A}$ has entries A_{ij} for $1 \le i \le p$, and $1 \le j \le k$. Then $(\boldsymbol{A}\mathbf{x})_i = \sum_{j=1}^k A_{ij}x_j$, and so

$$|(\boldsymbol{A}\mathbf{x})_i| \le \sum_{j=1}^k |A_{ij}||x_j| \le |\boldsymbol{A}| \sum_{j=1}^k |x_j| \le |\boldsymbol{A}| \sum_{j=1}^k |\mathbf{x}| = k\,|\boldsymbol{A}|\,|\mathbf{x}|.$$

From this,

$$(\boldsymbol{A}\mathbf{x})_i^2 = |(\boldsymbol{A}\mathbf{x})_i|^2 \le (k\,|\boldsymbol{A}|\,|\mathbf{x}|)^2.$$

Now consider that

$$|\boldsymbol{A}\mathbf{x}| = \sqrt{\left(\sum_{i=1}^p (\boldsymbol{A}\mathbf{x})_i^2\right)} \le \sqrt{\sum_{i=1}^p (k\,|\boldsymbol{A}|\,|\mathbf{x}|)^2} = \sqrt{(k\,|\boldsymbol{A}|\,|\mathbf{x}|)^2\,p} = k\sqrt{p}\,|\boldsymbol{A}|\,|\mathbf{x}|.$$

Letting $C \equiv k\sqrt{p}$ gives the result. ♦

Note in the above proof that C does not depend on $\mathbf{x}$, but it *does* depend on the size of the matrix $\boldsymbol{A}$.

► **6.61** *Prove the following.*

a) $|\boldsymbol{A}| \ge 0$ *with equality if and only if* $\boldsymbol{A} = \boldsymbol{O}$, *the zero matrix.*

b) $|-\boldsymbol{A}| = |\boldsymbol{A}|$

c) $|c\,\boldsymbol{A}| = |c|\,|\boldsymbol{A}|$, *for* $c \in \mathbb{R}$

d) $|\boldsymbol{A} + \boldsymbol{B}| \le |\boldsymbol{A}| + |\boldsymbol{B}|$, *for* $\boldsymbol{A}$ *and* $\boldsymbol{B}$ *both* $p \times k$

e) $|\boldsymbol{A}\,\boldsymbol{B}| \le c\,|\boldsymbol{A}|\,|\boldsymbol{B}|$, *for some constant* c *where* A *is* $p \times k$ *and* B *is* $k \times r$

[7] There are actually many norms one could define on the set of $p \times k$ matrices. We choose one here that is simple and convenient for our needs.

Note that a), c), d), and e) together imply that $|\cdot|$ is a norm as defined in Chapter 1 on the space of all $p \times k$ matrices.

We are now ready to prove the following theorem.

Theorem 3.6 *Suppose $\boldsymbol{f} : D^k \to \mathbb{R}^p$ is differentiable at interior point $\mathbf{a} \in D^k$. Then $\boldsymbol{f}$ is continuous at an $\mathbf{a}$.*

PROOF Remember in what follows that $\boldsymbol{f}'(\mathrm{a})$ is a matrix. Since $\boldsymbol{f}$ is differentiable at a, for $\epsilon = 1$ there exists $\delta_1 > 0$ such that

$$|\mathrm{x} - \mathrm{a}| < \delta_1 \quad \Rightarrow \quad |\boldsymbol{f}(\mathrm{x}) - \boldsymbol{f}(\mathrm{a}) - \boldsymbol{f}'(\mathrm{a})\,(\mathrm{x} - \mathrm{a})| \le |\mathrm{x} - \mathrm{a}|.$$

Therefore,

$$\begin{aligned}
|\boldsymbol{f}(\mathrm{x}) - \boldsymbol{f}(\mathrm{a})| &= |\boldsymbol{f}(\mathrm{x}) - \boldsymbol{f}(\mathrm{a}) - \boldsymbol{f}'(\mathrm{a})\,(\mathrm{x} - \mathrm{a}) + \boldsymbol{f}'(\mathrm{a})\,(\mathrm{x} - \mathrm{a})| \\
&\le |\boldsymbol{f}(\mathrm{x}) - \boldsymbol{f}(\mathrm{a}) - \boldsymbol{f}'(\mathrm{a})\,(\mathrm{x} - \mathrm{a})| + |\boldsymbol{f}'(\mathrm{a})\,(\mathrm{x} - \mathrm{a})| \\
&\le |\mathrm{x} - \mathrm{a}| + C\,|\boldsymbol{f}'(\mathrm{a})|\,|\mathrm{x} - \mathrm{a}| \\
&= (1 + C\,|\boldsymbol{f}'(\mathrm{a})|)\,|\mathrm{x} - \mathrm{a}| \\
&< \epsilon \quad \text{if } |\mathrm{x} - \mathrm{a}| < \delta \equiv \min\left(\delta_1, \frac{\epsilon}{(1 + C\,|\boldsymbol{f}'(\mathrm{a})|)}\right),
\end{aligned}$$

and the theorem is proved. ♦

When $\boldsymbol{f} : D^k \to \mathbb{R}^p$ is differentiable at $\mathrm{a} \in D^k$, what do the entries of the derivative matrix $\boldsymbol{f}'(\mathrm{a})$ look like? How do they depend on $\boldsymbol{f}$ and on a? To answer these questions, recall that for a function $\boldsymbol{f} : D^k \to \mathbb{R}^p$ we may consider each of the p components of the vector $\boldsymbol{f}$ as a function $f_i : D^k \to \mathbb{R}$ for $i = 1, \dots, p$. That is,

$$\boldsymbol{f}(\mathrm{x}) = \big(f_1(\mathrm{x}), f_2(\mathrm{x}), \dots, f_p(\mathrm{x})\big),$$

where each $f_i(\mathrm{x}) = f_i(x_1, \dots, x_k)$ for $i = 1, \dots, p$. Now, suppose that $\boldsymbol{f}$ is known to be differentiable at a and has the derivative given by the matrix $\boldsymbol{A}$ there. Then according to our equivalent ϵ, δ version of Definition 3.1 (as given by expression (6.23) on page 274), for any $\epsilon > 0$ there exists $\delta > 0$ such that

$$|\mathrm{x} - \mathrm{a}| < \delta \quad \Rightarrow \quad |\boldsymbol{f}(\mathrm{x}) - \boldsymbol{f}(\mathrm{a}) - \boldsymbol{A}\,(\mathrm{x} - \mathrm{a})| \le \epsilon\,|\mathrm{x} - \mathrm{a}|.$$

The ith component of the vector $\boldsymbol{f}(\mathrm{x}) - \boldsymbol{f}(\mathrm{a}) - \boldsymbol{A}\,(\mathrm{x} - \mathrm{a})$ is given by

$$f_i(\mathrm{x}) - f_i(\mathrm{a}) - [A_{i1} \;\; A_{i2} \;\; \cdots \;\; A_{ik}]\,(\mathrm{x} - \mathrm{a}),$$

and so, for $\mathrm{x} \in D^k$ and $|\mathrm{x} - \mathrm{a}| < \delta$,

$$\begin{aligned}
|f_i(\mathrm{x}) - f_i(\mathrm{a}) - [A_{i1} \;\; A_{i2} \;\; \cdots \;\; A_{ik}]\,(\mathrm{x} - \mathrm{a})| &\le |\boldsymbol{f}(\mathrm{x}) - \boldsymbol{f}(\mathrm{a}) - \boldsymbol{A}\,(\mathrm{x} - \mathrm{a})| \\
&\le \epsilon\,|\mathrm{x} - \mathrm{a}|.
\end{aligned}$$

This implies that $[A_{i1} \;\; \cdots \;\; A_{ik}] = \boldsymbol{f}_i'(\mathrm{a})$ according to the ϵ, δ version of our derivative definition given by expression (6.17) on page 258. But we have already learned that if $f_i : D^k \to \mathbb{R}$ is differentiable at $\mathrm{a} \in D^k$, then $\boldsymbol{f}_i'(\mathrm{a}) = \boldsymbol{\nabla} \boldsymbol{f}_i(\mathrm{a})$, and so $\boldsymbol{f}'(\mathrm{a}) = \boldsymbol{A}$ is given by

$$\boldsymbol{A} = \begin{bmatrix} \leftarrow & \boldsymbol{f}_1'(\mathrm{a}) & \rightarrow \\ \leftarrow & \boldsymbol{f}_2'(\mathrm{a}) & \rightarrow \\ & \vdots & \\ \leftarrow & \boldsymbol{f}_p'(\mathrm{a}) & \rightarrow \end{bmatrix} = \begin{bmatrix} \leftarrow & \boldsymbol{\nabla} \boldsymbol{f}_1(\mathrm{a}) & \rightarrow \\ \leftarrow & \boldsymbol{\nabla} \boldsymbol{f}_2(\mathrm{a}) & \rightarrow \\ & \vdots & \\ \leftarrow & \boldsymbol{\nabla} \boldsymbol{f}_p(\mathrm{a}) & \rightarrow \end{bmatrix}, \tag{6.25}$$

where the horizontal arrows serve to remind the reader that the entries $\boldsymbol{f}_i'(\mathrm{a})$ and $\boldsymbol{\nabla} \boldsymbol{f}_i(\mathrm{a})$ are row vectors for $1 \le i \le p$.

Example 3.7 *Let $\boldsymbol{f} : \mathbb{R}^2 \to \mathbb{R}^3$ be given by $\boldsymbol{f}(x, y) = (x + y, xy, e^{xy})$, and suppose it is known that $\boldsymbol{f}$ is differentiable on $\mathbb{R}^2$. We will calculate $\boldsymbol{f}'(x, y)$. Denoting each component function of $\boldsymbol{f}$ by f_1, f_2, and f_3, we have for all $(x, y) \in \mathbb{R}^2$,*

$$\frac{\partial f_1}{\partial x} = \frac{\partial f_1}{\partial y} = 1, \quad \frac{\partial f_2}{\partial x} = y, \quad \frac{\partial f_2}{\partial y} = x, \quad \frac{\partial f_3}{\partial x} = y\, e^{x\,y}, \quad \frac{\partial f_3}{\partial y} = x\, e^{x\,y}.$$

From this we obtain for each $(x, y) \in \mathbb{R}^2$ that $\boldsymbol{f}'(x, y) = \begin{bmatrix} 1 & 1 \\ y & x \\ y\, e^{x\,y} & x\, e^{x\,y} \end{bmatrix}$. ◄

If $\boldsymbol{f}'(\mathrm{a})$ is given by the $p \times k$ matrix $\boldsymbol{A}$, what is the rate of change of $\boldsymbol{f}$ at $\mathbf{x} = \mathrm{a}$ with respect to a change in $\mathbf{x}$ *in the x_i th coordinate only?* Recall that the matrix $\boldsymbol{A}$ represents the change in $\boldsymbol{f}$ at $\mathbf{x} = \mathrm{a}$ associated with a *general* change in $\mathbf{x}$ (along *any* direction). In particular, in terms of the ϵ, δ version of Definition 3.1 we have that for any $\epsilon > 0$ there exists $\delta > 0$ such that

$$|\mathbf{x} - \mathrm{a}| < \delta \Rightarrow |\boldsymbol{f}(\mathbf{x}) - \boldsymbol{f}(\mathrm{a}) - \boldsymbol{A}\,(\mathbf{x} - \mathrm{a})| \le \epsilon\, |\mathbf{x} - \mathrm{a}|. \tag{6.26}$$

If we let $\mathbf{x}$ vary from $\mathbf{x} = \mathrm{a}$ along the x_i direction only, that is, let $\mathbf{x} = \mathrm{a} + h\,\mathbf{e_i}$, then (6.26) becomes

$$|h| < \delta \Rightarrow |\boldsymbol{f}(\mathrm{a} + h\,\mathbf{e_i}) - \boldsymbol{f}(\mathrm{a}) - \boldsymbol{A}\,(h\,\mathbf{e_i})| \le \epsilon\, |h|,$$

or

$$|h| < \delta \Rightarrow |\boldsymbol{f}(\mathrm{a} + h\,\mathbf{e_i}) - \boldsymbol{f}(\mathrm{a}) - h\,\boldsymbol{A_i}| \le \epsilon\, |h|, \tag{6.27}$$

where $\boldsymbol{A_i} = \boldsymbol{A}\,\mathbf{e_i}$ is the ith column of the matrix $\boldsymbol{A}$. That is, according to (6.25),

$$\boldsymbol{A_i} = \begin{bmatrix} A_{1\,i} \\ A_{2\,i} \\ \vdots \\ A_{p\,i} \end{bmatrix} = \begin{bmatrix} \frac{\partial f_1}{\partial x_i}(\mathrm{a}) \\ \frac{\partial f_2}{\partial x_i}(\mathrm{a}) \\ \vdots \\ \frac{\partial f_p}{\partial x_i}(\mathrm{a}) \end{bmatrix}.$$

But (6.27) is equivalent to

$$\lim_{h \to 0} \frac{\boldsymbol{f}(\mathrm{a} + h\,\mathbf{e_i}) - \boldsymbol{f}(\mathrm{a})}{h} = \boldsymbol{A_i},$$

and so the column vector $\boldsymbol{A_i}$ is called *the partial derivative of* $\boldsymbol{f}$ *with respect to* x_i *at* a, and is denoted by

$$\boldsymbol{A_i} = \boldsymbol{f}_{x_i}(\mathrm{a}) = \frac{\partial \boldsymbol{f}}{\partial x_i}(\mathrm{a}).$$

That is,

$$\boldsymbol{A} = \begin{bmatrix} \uparrow & \uparrow & & \uparrow \\ \boldsymbol{A_1} & \boldsymbol{A_2} & \cdots & \boldsymbol{A_k} \\ \downarrow & \downarrow & & \downarrow \end{bmatrix} = \begin{bmatrix} \uparrow & \uparrow & & \uparrow \\ \frac{\partial \boldsymbol{f}}{\partial x_1}(\mathrm{a}) & \frac{\partial \boldsymbol{f}}{\partial x_2}(\mathrm{a}) & \cdots & \frac{\partial \boldsymbol{f}}{\partial x_k}(\mathrm{a}) \\ \downarrow & \downarrow & & \downarrow \end{bmatrix},$$

where each column is given more explicitly by

$$\frac{\partial \boldsymbol{f}}{\partial x_i}(\mathrm{a}) = \begin{bmatrix} \frac{\partial f_1}{\partial x_i}(\mathrm{a}) \\ \frac{\partial f_2}{\partial x_i}(\mathrm{a}) \\ \vdots \\ \frac{\partial f_p}{\partial x_i}(\mathrm{a}) \end{bmatrix}.$$

Example 3.8 *Let* $\boldsymbol{f} : \mathbb{R}^3 \to \mathbb{R}^2$ *be given by* $\boldsymbol{f}(x_1, x_2, x_3) = \big(\sin(x_1 x_2),\, x_1 + e^{x_2 x_3}\big)$, *and suppose it is known that* $\boldsymbol{f}$ *is differentiable on* $\mathbb{R}^3$. *The reader can easily confirm that for each* $(x_1, x_2, x_3) \in \mathbb{R}^3$,

$$\frac{\partial \boldsymbol{f}}{\partial x_1} = \begin{bmatrix} x_2 \cos(x_1 x_2) \\ 1 \end{bmatrix}, \quad \frac{\partial \boldsymbol{f}}{\partial x_2} = \begin{bmatrix} x_1 \cos(x_1 x_2) \\ x_3\, e^{x_2 x_3} \end{bmatrix}, \text{ and } \frac{\partial \boldsymbol{f}}{\partial x_3} = \begin{bmatrix} 0 \\ x_2\, e^{x_2 x_3} \end{bmatrix}.$$

◀

Now that we have characterized the matrix $\boldsymbol{A}$ representing the derivative of a function $\boldsymbol{f}$ at a point by both its rows and its columns, we see that when $\boldsymbol{f}'(\mathrm{a})$ exists its matrix representation $\boldsymbol{A}$ has the form given by

$$\boldsymbol{A} = \begin{bmatrix} \frac{\partial f_1}{\partial x_1}(\mathrm{a}) & \frac{\partial f_1}{\partial x_2}(\mathrm{a}) & \cdots & \frac{\partial f_1}{\partial x_k}(\mathrm{a}) \\ \frac{\partial f_2}{\partial x_1}(\mathrm{a}) & \frac{\partial f_2}{\partial x_2}(\mathrm{a}) & \cdots & \frac{\partial f_2}{\partial x_k}(\mathrm{a}) \\ \vdots & \vdots & \ddots & \vdots \\ \frac{\partial f_p}{\partial x_1}(\mathrm{a}) & \frac{\partial f_p}{\partial x_2}(\mathrm{a}) & \cdots & \frac{\partial f_p}{\partial x_k}(\mathrm{a}) \end{bmatrix}.$$

This matrix of partial derivatives is commonly known as *the Jacobian matrix of* $\boldsymbol{f}$ *evaluated at* a.

Definition 3.9 (The Jacobian Matrix of $\boldsymbol{f} : D^k \to \mathbb{R}^p$ at the Point a) Consider a function $\boldsymbol{f} : D^k \to \mathbb{R}^p$, and an interior point a in the domain of $\boldsymbol{f}$ such that all partial derivatives of the component functions f_i of $\boldsymbol{f}$ exist at a.

The **Jacobian matrix** of f at a is the $p \times k$ matrix denoted by $J(\mathbf{a})$ and defined as

$$J(\mathbf{a}) = \begin{bmatrix} \frac{\partial f_1}{\partial x_1}(\mathbf{a}) & \frac{\partial f_1}{\partial x_2}(\mathbf{a}) & \cdots & \frac{\partial f_1}{\partial x_k}(\mathbf{a}) \\ \frac{\partial f_2}{\partial x_1}(\mathbf{a}) & \frac{\partial f_2}{\partial x_2}(\mathbf{a}) & \cdots & \frac{\partial f_2}{\partial x_k}(\mathbf{a}) \\ \vdots & \vdots & \ddots & \vdots \\ \frac{\partial f_p}{\partial x_1}(\mathbf{a}) & \frac{\partial f_p}{\partial x_2}(\mathbf{a}) & \cdots & \frac{\partial f_p}{\partial x_k}(\mathbf{a}) \end{bmatrix}.$$

Notice that the above definition does *not* require the differentiability of f at a, but only that the individual partial derivatives exist at a. Just as for the gradient in the case $f : D^k \to \mathbb{R}$, the Jacobian matrix for any $f : D^k \to \mathbb{R}^p$ can be constructed as long as all k first-order partial derivatives *exist* for each component function f_i for $i = 1, \dots, p$. Just as with the gradient, the existence of the Jacobian matrix for a mapping f at a point a is *not* enough to ensure the existence of the derivative f' there. What exactly is the relationship between the Jacobian matrix and the derivative of a function f at a point a? What we have seen so far is that *if* the function is already known to be differentiable at a, then the derivative at a is in fact the Jacobian matrix evaluated at a. From this we can conclude that *if* the Jacobian matrix J fails to exist at a point a, the derivative of f will not exist there either. In other words, the existence of $J(\mathbf{a})$ is a *necessary* condition for the existence of $f'(\mathbf{a})$. In general, this necessary condition is not a sufficient condition, however. That is, there exist functions $f : D^k \to \mathbb{R}^p$ for which the Jacobian $J(\mathbf{a})$ exists at a, but the derivative $f'(\mathbf{a})$ fails to exist. Fortunately, there is an extra condition which, when combined with the existence of the Jacobian matrix at a, guarantees the existence of f' there. When these two conditions are satisfied, we then will have $f'(\mathbf{a}) = J(\mathbf{a})$. The moral of the story once again is: *The differentiability of a function of more than one variable requires more than just the existence of the first-order partial derivatives.*

Theorem 3.10 (Sufficient Conditions for Existence of f' at the Point a)
Let $f : D^k \to \mathbb{R}^p$ have component functions $f_i : D^k \to \mathbb{R}$ for $i = 1, \dots, p$, and let a be an interior point of D^k. If the partial derivatives $\frac{\partial f_i}{\partial x_j}$ for $i = 1, \dots, p$ and $j = 1, \dots, k$ exist in a neighborhood of a and are continuous at a, then f is differentiable at a and $f'(\mathbf{a}) = J(\mathbf{a})$.

PROOF Consider the simpler case of $f : D^2 \to \mathbb{R}$ first, and assume that f_x and f_y satisfy the hypotheses of the theorem at and near the interior point $\mathbf{a} = (a, b) \in D^2$. We will show that $f'(\mathbf{a}) = \nabla f(\mathbf{a})$. To do this, we must establish that for any given $\epsilon > 0$ we can find $\delta > 0$ such that $|\mathbf{x} - \mathbf{a}| < \delta$ implies

$$|\boldsymbol{f}(\mathbf{x}) - f(\mathbf{a}) - \nabla \boldsymbol{f}(\mathbf{a})\,(\mathbf{x}-\mathbf{a})| \le \epsilon\,|\mathbf{x}-\mathbf{a}|\,.$$

This vector version of the ϵ, δ definition of differentiability is not quite as useful in this particular instance, and so we translate it into its corresponding componentwise form as follows. We will show that, for $\mathbf{x} = (x, y) \in D^2$, whenever $\sqrt{(x-a)^2 + (y-b)^2} < \delta$, we have

$$\big|f(x,y) - f(a,b) - (x-a)f_x(a,b) - (y-b)f_y(a,b)\big| \le \epsilon\sqrt{(x-a)^2 + (y-b)^2}.$$

To this end, we consider

$$\begin{aligned} & f(x,y) - f(a,b) - (x-a)\,f_x(a,b) - (y-b)\,f_y(a,b) \\ &= f(x,y) - f(a,y) - f_x(a,y)\,(x-a) \qquad (6.28) \\ &\quad + f(a,y) - f(a,b) - f_y(a,b)\,(y-b) + [f_x(a,y) - f_x(a,b)]\,(x-a). \end{aligned}$$

Since $f_x(a,y)$ exists for y near b, there exists $\delta_1 > 0$ such that

$$|x-a| < \delta_1 \;\Rightarrow\; |f(x,y) - f(a,y) - f_x(a,y)\,(x-a)| \le \tfrac{\epsilon}{3}\,|x-a|.$$

Similarly, since $f_y(a,b)$ exists, there exists $\delta_2 > 0$ such that

$$|y-b| < \delta_2 \;\Rightarrow\; |f(a,y) - f(a,b) - f_y(a,b)\,(y-b)| \le \tfrac{\epsilon}{3}\,|y-b|.$$

Finally, since f_x is continuous at (a,b), there exists $\delta_3 > 0$ such that

$$|y-b| < \delta_3 \;\Rightarrow\; |f_x(a,y) - f_x(a,b)| < \tfrac{\epsilon}{3}.$$

Overall, letting $\delta = \min(\delta_1, \delta_2, \delta_3)$, we have that $\sqrt{(x-a)^2 + (y-b)^2} < \delta$ implies both $|x-a| < \delta$ and $|y-b| < \delta$, which in turn imply that

$$\begin{aligned} & |f(x,y) - f(a,b) - f_x(a,b)\,(x-a) - f_y(a,b)\,(y-b)| \\ &\qquad \le \tfrac{\epsilon}{3}|x-a| + \tfrac{\epsilon}{3}|y-b| + \tfrac{\epsilon}{3}|x-a| \qquad (6.29) \\ &\qquad \le \epsilon\sqrt{(x-a)^2 + (y-b)^2}. \end{aligned}$$

The first inequality in expression (6.29) is obtained by substituting (6.28) into the absolute value bars on the left-hand side, followed by two applications of the triangle inequality. With this, the special case of the theorem is proved. The case for $f : D^k \to \mathbb{R}$ with $k > 2$ is left as an exercise.

We consider now another case, namely, $\boldsymbol{f} : D^2 \to \mathbb{R}^2$. Suppose that for $\mathbf{x} = (x, y) \in D^2$, and $\boldsymbol{f} : D^2 \to \mathbb{R}^2$, the function $\boldsymbol{f}(\mathbf{x})$ is given by

$$\boldsymbol{f}(\mathbf{x}) = (g(x,y),\, h(x,y))\,,$$

where g_x, g_y, h_x, and h_y satisfy the hypotheses of the theorem at and near the point $\mathrm{a} = (a, b) \in D^2$. We will show that $\boldsymbol{f}$ is differentiable at a with

$$\boldsymbol{f}'(\mathrm{a}) = \boldsymbol{J}(\mathrm{a}) = \begin{bmatrix} g_x(a,b) & g_y(a,b) \\ h_x(a,b) & h_y(a,b) \end{bmatrix}.$$

Since g_x, g_y, h_x, and h_y exist in a neighborhood of the point $\mathrm{a} = (a,b)$ and are continuous there, it follows from the case already proved that each of the component functions g and h are differentiable at the point $\mathbf{a} = (a, b)$. Therefore, for $\epsilon > 0$ there exists $\delta > 0$ such that whenever $|\mathbf{x} - \mathbf{a}| < \delta$ we have

$$\left|g(\mathbf{x}) - g(\mathbf{a}) - \boldsymbol{g}'(\mathbf{a})\,(\mathbf{x} - \mathbf{a})\right| \le \tfrac{\epsilon}{2}\,|\mathbf{x} - \mathbf{a}|,$$

and

$$\left|h(\mathbf{x}) - h(\mathbf{a}) - \boldsymbol{h}'(\mathbf{a})\,(\mathbf{x} - \mathbf{a})\right| \le \tfrac{\epsilon}{2}\,|\mathbf{x} - \mathbf{a}|.$$

From this we obtain that whenever $|\mathbf{x} - \mathbf{a}| < \delta$, we have

$$\begin{aligned}
&|\boldsymbol{f}(\mathbf{x}) - \boldsymbol{f}(\mathbf{a}) - \boldsymbol{J}(\mathbf{a})\,(\mathbf{x} - \mathbf{a})| \\
&\quad= \left|\left(g(\mathbf{x}) - g(\mathbf{a}) - \boldsymbol{g}'(\mathbf{a})\,(\mathbf{x} - \mathbf{a}),\ h(\mathbf{x}) - h(\mathbf{a}) - \boldsymbol{h}'(\mathbf{a})\,(\mathbf{x} - \mathbf{a})\right)\right| \\
&\quad\le \left|g(\mathbf{x}) - g(\mathbf{a}) - \boldsymbol{g}'(\mathbf{a})\,(\mathbf{x} - \mathbf{a})\right| + \left|h(\mathbf{x}) - h(\mathbf{a}) - \boldsymbol{h}'(\mathbf{a})\,(\mathbf{x} - \mathbf{a})\right| \qquad \text{(Why?)} \\
&\quad\le \tfrac{\epsilon}{2}\,|\mathbf{x} - \mathbf{a}| + \tfrac{\epsilon}{2}\,|\mathbf{x} - \mathbf{a}| \\
&\quad= \epsilon\,|\mathbf{x} - \mathbf{a}|.
\end{aligned}$$

This completes the proof for $\boldsymbol{f} : D^2 \to \mathbb{R}^2$. The more general case is handled similarly and is left to the exercises. ♦

► **6.62** *Prove the above theorem for the case $f : D^3 \to \mathbb{R}$. Can you prove it for the more general $f : D^k \to \mathbb{R}$?*

► **6.63** *Prove the above theorem for the case $\boldsymbol{f} : D^3 \to \mathbb{R}^2$. Can you prove it for the more general $\boldsymbol{f} : D^k \to \mathbb{R}^p$?*

Theorem 3.10 is a very practical theorem. It allows one to avoid direct application of the derivative definition when trying to determine $\boldsymbol{f}'$ for a function $\boldsymbol{f} : D^k \to \mathbb{R}^p$. More specifically, it says that if one constructs the Jacobian matrix of first-order partial derivatives, and the partial derivatives all exist in a neighborhood of $\mathbf{a}$ and are all continuous at $\mathbf{a}$, then $\boldsymbol{f}'$ exists at $\mathbf{a}$ and is given by $\boldsymbol{J}(\mathbf{a})$. That is, in many cases we can simply construct the Jacobian matrix and then confirm that it is the derivative!

Example 3.11 *Consider the function $\boldsymbol{g} : \mathbb{R} \to \mathbb{R}^2$ given by*

$$\boldsymbol{g}(t) = \left(g_1(t),\, g_2(t)\right) = \left(t^2,\, e^t\right).$$

It is easy to see that $\boldsymbol{g}'(t)$ is given by

$$\boldsymbol{g}'(t) = \begin{bmatrix} g_1'(t) \\ g_2'(t) \end{bmatrix} = \begin{bmatrix} 2t \\ e^t \end{bmatrix},$$

for all t. (Why?) ◄

Example 3.12 *Consider the function $\boldsymbol{f} : D^2 \to \mathbb{R}^2$ given by*

$$\boldsymbol{f}(x, y) = \left(\frac{e^{xy}}{x},\, \sqrt{\sin y}\right)$$

defined on the strip $D^2 = (-\infty, 0) \times (0, \pi)$. If we denote the two component functions of $\boldsymbol{f}$ by f_1 and f_2 we have for all $(x, y) \in D^2$,

$$\frac{\partial f_1}{\partial x} = \frac{(xy-1)e^{xy}}{x^2}, \qquad \frac{\partial f_1}{\partial y} = e^{xy},$$

and

$$\frac{\partial f_2}{\partial x} = 0, \qquad \frac{\partial f_2}{\partial y} = \frac{\cos y}{2\sqrt{\sin y}},$$

all of which are continuous on D^2. Therefore, $\boldsymbol{f}$ is differentiable on D^2 and the Jacobian matrix gives us the derivative as

$$\boldsymbol{f}'(x, y) = \begin{bmatrix} \frac{(xy-1)e^{xy}}{x^2} & e^{xy} \\ 0 & \frac{\cos y}{2\sqrt{\sin y}} \end{bmatrix}.$$

◀

Example 3.13 Consider the function given by

$$f(x, y) = \begin{cases} \frac{xy}{x^2+y^2} & (x, y) \neq (0, 0) \\ 0 & (x, y) = (0, 0) \end{cases}.$$

Clearly the first-order partial derivatives of f with respect to x and y exist at the origin, since

$$\frac{\partial f}{\partial x}(0, 0) = \lim_{x \to 0} \frac{f(x, 0) - f(0, 0)}{x - 0} = 0,$$

and

$$\frac{\partial f}{\partial y}(0, 0) = \lim_{y \to 0} \frac{f(0, y) - f(0, 0)}{y - 0} = 0.$$

But this function can't be differentiable at the origin, since

$$f(x, mx) = \frac{mx^2}{x^2 + m^2x^2} = \frac{m}{1 + m^2}$$

is enough to show that f isn't even continuous there. It is left as an exercise to show that $\partial f/\partial x$ and $\partial f/\partial y$ are not continuous at the origin. ◀

▶ **6.64** In the previous example, $\partial f/\partial x$ and $\partial f/\partial y$ can't both be continuous at the origin. Why? Show that both $\partial f/\partial x$ and $\partial f/\partial y$ are not continuous at the origin.

The following result is a generalization of Theorem 2.6 on page 262.

Proposition 3.14 *Let D^k be a connected open subset of $\mathbb{R}^k$, and suppose that $\boldsymbol{f} : D^k \to \mathbb{R}^p$ is differentiable on D^k with $\boldsymbol{f}'(\mathbf{x}) = \mathbf{0}$ for all $\mathbf{x} \in D^k$. Then there exists a constant vector $\mathbf{c} \in \mathbb{R}^p$ such that $\boldsymbol{f}(\mathbf{x}) = \mathbf{c}$ for all $\mathbf{x} \in D^k$.*

▶ **6.65** Prove the above theorem.

▶ **6.66** Consider $\boldsymbol{f} : D^k \to \mathbb{R}^p$. If there exists a constant vector $\mathbf{c} \in \mathbb{R}^p$ such that $\boldsymbol{f}(\mathbf{x}) = \mathbf{c}$ for all $\mathbf{x} \in D^k$, show that $\boldsymbol{f}'(\mathbf{x}) = \mathbf{0}$ for all interior points $\mathbf{x} \in D^k$.

3.2 Some Useful Results

In this subsection we develop the higher-dimensional analogues of *some* of our earlier "useful results." Note that functions $f : D^k \to \mathbb{R}^p$ have no "local extrema results," nor is there a version of Rolle's theorem that applies to them. This is because higher-dimensional Euclidean spaces lack the special order properties possessed by $\mathbb{R}$.

Algebraic Results

The following result is analogous to Theorem 1.9 on page 244 and Proposition 2.14 on page 270.

Proposition 3.15 (Algebraic Properties of the Derivative)
Suppose the functions $f : D^k \to \mathbb{R}^p$ and $g : D^k \to \mathbb{R}^p$ are differentiable at the interior point $\mathbf{a} \in D^k$. Then

a) $f \pm g$ *is differentiable at* $\mathbf{a}$, *and* $(f \pm g)'(\mathbf{a}) = f'(\mathbf{a}) \pm g'(\mathbf{a})$.

b) cf *is differentiable at* $\mathbf{a}$, *and* $(cf)'(\mathbf{a}) = cf'(\mathbf{a})$ *for* $c \in \mathbb{R}$.

c) $f \cdot g$ *is differentiable at* $\mathbf{a}$, *and* $(f \cdot g)'(\mathbf{a}) = f(\mathbf{a})g'(\mathbf{a}) + g(\mathbf{a})f'(\mathbf{a})$.

► **6.67** *Prove the above theorem. Note that in part c) the terms $f(\mathbf{a})g'(\mathbf{a})$ and $g(\mathbf{a})f'(\mathbf{a})$ are products of a p-dimensional row vector with a $p \times k$ matrix. That is, $(f \cdot g)'(\mathbf{a})$ is a $1 \times k$ matrix, as it should be. Why is the corresponding part d) of Theorem 1.9 and Proposition 2.14 not represented here?*

The higher-dimensional version of the chain rule for functions $f : D^k \to \mathbb{R}^p$ is a generalization of the more familiar one corresponding to one dimensional Euclidean space, as it must be. We prove this important theorem next.

Theorem 3.16 (The Chain Rule for $f : D^k \to \mathbb{R}^p$)
Suppose $f : D^k \to D^p$ is differentiable at the interior point $\mathbf{a} \in D^k$, and $g : D^p \to \mathbb{R}^m$ is differentiable at the interior point $\mathbf{b} = f(\mathbf{a}) \in D^p$. Then the composite function $h \equiv g \circ f : D^k \to \mathbb{R}^m$ is differentiable at $\mathbf{a} \in D^k$ and $h'(\mathbf{a}) = g'(\mathbf{b})\, f'(\mathbf{a})$.

Note that $h'(\mathbf{a}) = g'(\mathbf{b})\, f'(\mathbf{a})$ in the above theorem is a product of *matrices*. In particular, $g'(\mathbf{b})$ is an $m \times p$ matrix and $f'(\mathbf{a})$ is a $p \times k$ matrix.

PROOF [8] We begin the proof with a rather convenient choice of function. We define the function $\mathbf{G} : D^p \to \mathbb{R}^m$ as follows:

$$\mathbf{G}(\mathbf{y}) \equiv \begin{cases} \frac{g(\mathbf{y})-g(\mathbf{b})-g'(\mathbf{b})(\mathbf{y}-\mathbf{b})}{|\mathbf{y}-\mathbf{b}|} & \text{for } \mathbf{y} \neq \mathbf{b} \text{ and } \mathbf{y} \in D^p \\ \mathbf{0} & \text{for } \mathbf{y} = \mathbf{b} \end{cases}.$$

[8]We follow [Fit95] for the proof.

This function $\mathbf{G}$ is continuous at $\mathbf{y} = \mathbf{b}$ (Why?). Rearranging the expression for $\mathbf{G}(\mathbf{y})$ gives

$$\boldsymbol{g}(\mathbf{y}) - \boldsymbol{g}(\mathbf{b}) = \boldsymbol{g}'(\mathbf{b})(\mathbf{y} - \mathbf{b}) + \mathbf{G}(\mathbf{y})|\mathbf{y} - \mathbf{b}| \quad \text{for all } \mathbf{y} \in D^p.$$

Letting $\mathbf{y} = \boldsymbol{f}(\mathbf{x})$ and substituting $\boldsymbol{f}(\mathbf{a})$ for $\mathbf{b}$ yields

$$\boldsymbol{h}(\mathbf{x}) - \boldsymbol{h}(\mathbf{a}) = \boldsymbol{g}'(\mathbf{b})(\boldsymbol{f}(\mathbf{x}) - \boldsymbol{f}(\mathbf{a})) + \mathbf{G}\left(\boldsymbol{f}(\mathbf{x})\right)|\boldsymbol{f}(\mathbf{x}) - \boldsymbol{f}(\mathbf{a})|. \tag{6.30}$$

Subtracting $\boldsymbol{g}'(\mathbf{b})\boldsymbol{f}'(\mathbf{a})(\mathbf{x} - \mathbf{a})$ from both sides of (6.30), and dividing both sides by $|\mathbf{x} - \mathbf{a}|$ under the assumption $\mathbf{x} \neq \mathbf{a}$ obtains the lengthy expression

$$\frac{\boldsymbol{h}(\mathbf{x})-\boldsymbol{h}(\mathbf{a})-\boldsymbol{g}'(\mathbf{b})\boldsymbol{f}'(\mathbf{a})(\mathbf{x}-\mathbf{a})}{|\mathbf{x}-\mathbf{a}|} = \boldsymbol{g}'(\mathbf{b})\left[\frac{\boldsymbol{f}(\mathbf{x})-\boldsymbol{f}(\mathbf{a})-\boldsymbol{f}'(\mathbf{a})(\mathbf{x}-\mathbf{a})}{|\mathbf{x}-\mathbf{a}|}\right] + \mathbf{G}\left(\boldsymbol{f}(\mathbf{x})\right)\frac{|\boldsymbol{f}(\mathbf{x})-\boldsymbol{f}(\mathbf{a})|}{|\mathbf{x}-\mathbf{a}|}. \tag{6.31}$$

Now, since $\boldsymbol{f}$ is differentiable at $\mathbf{a}$, the following is true:

1. $\lim_{\mathbf{x}\to\mathbf{a}}\left[\frac{\boldsymbol{f}(\mathbf{x})-\boldsymbol{f}(\mathbf{a})-\boldsymbol{f}'(\mathbf{a})(\mathbf{x}-\mathbf{a})}{|\mathbf{x}-\mathbf{a}|}\right] = \mathbf{0}$.
2. There exists $\delta_1 > 0$ and a constant $C > 0$ such that $0 < |\mathbf{x} - \mathbf{a}| < \delta_1 \Rightarrow$

$$\frac{|\boldsymbol{f}(\mathbf{x})-\boldsymbol{f}(\mathbf{a})|}{|\mathbf{x}-\mathbf{a}|} \leq C\left(1 + |\boldsymbol{f}'(\mathbf{a})|\right). \qquad \text{(Why?)}$$

Also, by continuity of $\boldsymbol{f}$ and $\mathbf{G}$, we have that

3. $\lim_{\mathbf{x}\to\mathbf{a}}\mathbf{G}\left(\boldsymbol{f}(\mathbf{x})\right) = \mathbf{G}\left(\boldsymbol{f}(\mathbf{a})\right) = \mathbf{G}(\mathbf{b}) = \mathbf{0}$.

Considering the lengthy expression (6.31), along with these last three facts, yields

$$\lim_{\mathbf{x}\to\mathbf{a}} \frac{\boldsymbol{h}(\mathbf{x}) - \boldsymbol{h}(\mathbf{a}) - \boldsymbol{g}'(\mathbf{b})\,\boldsymbol{f}'(\mathbf{a})(\mathbf{x} - \mathbf{a})}{|\mathbf{x} - \mathbf{a}|} = \mathbf{0}.$$

That is,

$$\boldsymbol{h}'(\mathbf{a}) = \boldsymbol{g}'(\mathbf{b})\,\boldsymbol{f}'(\mathbf{a}).$$

◆

► **6.68** *Show that the function* $\mathbf{G} : D^p \to \mathbb{R}^m$ *in the above proof is continuous at* $\mathbf{y} = \mathbf{b}$. *Also, answer the (Why?) question following remark number 2.*

Note again that the derivative $\boldsymbol{h}'(\mathbf{a})$ is the product of matrices $\boldsymbol{g}'(\mathbf{b})$ and $\boldsymbol{f}'(\mathbf{a})$. This is found to be either strange or convenient, depending on one's mathematical experience.

Mean Value Theorems and Taylor's Theorems with Remainder

Recall that we postponed stating and proving the mean value theorem and Taylor's theorem with remainder for functions $f : D^k \to \mathbb{R}$ in the previous section. This was because we needed the chain rule in the $\boldsymbol{f} : D^k \to \mathbb{R}^p$ case. Now that this chain rule is established, we can tie up these loose ends. Before stating and proving these results, we take the time to clarify what is meant by the phrase "between two points" in higher-dimensional spaces. One natural interpretation of this phrase is "on the line segment joining" the two points. In the discussion that follows, we denote the line segment joining two points $\mathbf{p}_1$ and $\mathbf{p}_2$ in $\mathbb{R}^k$ by $[\mathbf{p}_1, \mathbf{p}_2]$ or $(\mathbf{p}_1, \mathbf{p}_2)$. Square brackets will be used to indicate that the segment includes the endpoints $\mathbf{p_1}$ and $\mathbf{p_2}$, while parentheses will be used to indicate that the segment does not include the endpoints $\mathbf{p_1}$ and $\mathbf{p_2}$. For example, the line segment joining $\mathbf{p_1}$ and $\mathbf{p_2}$ but not containing either is the set $(\mathbf{p_1}, \mathbf{p_2}) \equiv \big\{\mathbf{p} \in \mathbb{R}^k : \mathbf{p} = \mathbf{p_1} + t\,(\mathbf{p_2} - \mathbf{p_1}),\text{ for } 0 < t < 1\big\}$.

Theorem 3.17 (The Mean Value Theorem for $f : D^k \to \mathbb{R}$)
Let D^k be an open subset of $\mathbb{R}^k$ and suppose that $f : D^k \to \mathbb{R}$ is differentiable on D^k. Suppose $\mathbf{p_1}$ and $\mathbf{p_2}$ are two distinct interior points in D^k such that the segment $(\mathbf{p_1}, \mathbf{p_2})$ is entirely contained within D^k. Then, there exists a point $\mathbf{q}$ on the segment $(\mathbf{p_1}, \mathbf{p_2})$ such that

$$f(\mathbf{p_2}) - f(\mathbf{p_1}) = \boldsymbol{f}'(\mathbf{q})\,(\mathbf{p_2} - \mathbf{p_1}).$$

PROOF [9] To prove the result, we define a real-valued function of a single real variable, $g(t) \equiv f\left(\mathbf{p_1} + t\,(\mathbf{p_2} - \mathbf{p_1})\right)$. It is easy to see that g is continuous on $[0, 1]$ and differentiable on $(0, 1)$, with derivative

$$g'(t) = \boldsymbol{f}'\left(\mathbf{p_1} + t\,(\mathbf{p_2} - \mathbf{p_1})\right)(\mathbf{p_2} - \mathbf{p_1}).$$

Then there exists $\tau \in (0, 1)$ such that $g(1) - g(0) = g'(\tau)$. (Why?) Defining $\mathbf{q}$ by $\mathbf{q} \equiv \mathbf{p_1} + \tau\,(\mathbf{p_2} - \mathbf{p_1})$, we obtain

$$f(\mathbf{p_2}) - f(\mathbf{p_1}) = \boldsymbol{f}'(\mathbf{q})\,(\mathbf{p_2} - \mathbf{p_1}),$$

and the theorem is proved. ♦

► **6.69** *Can you use this version of the mean value theorem to prove the version of Rolle's theorem given by Proposition 2.18 on page 272?*

To round out our list of "useful results" for functions $f : D^k \to \mathbb{R}$, we now develop a version of Taylor's theorem with remainder for real-valued functions of several variables. For convenience, we will handle the $f : D^2 \to \mathbb{R}$ case and leave the more general $f : D^k \to \mathbb{R}$ case to the reader. For convenience, assume $f : D^2 \to \mathbb{R}$ has as many mixed partial derivatives as needed. Fix interior point $\mathbf{a} \in D^2$ and $\mathbf{x} \in D^2$ near $\mathbf{a}$. As in the proof of the mean value theorem, we define

$$g(t) \equiv f\left(\mathbf{a} + t\,(\mathbf{x} - \mathbf{a})\right), \tag{6.32}$$

where $\mathbf{a} = (a, b)$, and $\mathbf{x} = (x, y)$. Applying the one-dimensional version of Taylor's theorem with remainder to g, centered at $t = 0$, gives

$$g(t) = \sum_{j=0}^{n} \frac{g^{(j)}(0)}{j!}\,t^j + R_n(t).$$

Setting $t = 1$ obtains

[9] We follow [Ful78].

$$g(1) = \sum_{j=0}^{n} \frac{g^{(j)}(0)}{j!} + R_n(1), \tag{6.33}$$

where, for some $\xi \in (0, 1)$, $R_n(1)$ is given by

$$R_n(1) = \frac{g^{(n+1)}(\xi)}{(n+1)!}. \tag{6.34}$$

But (6.32) yields that $g(1) = f(\mathbf{x})$, and so (6.33) becomes

$$f(\mathbf{x}) = \sum_{j=0}^{n} \frac{g^{(j)}(0)}{j!} + \frac{g^{(n+1)}(\xi)}{(n+1)!},$$

where we have used (6.34) to replace $R_n(1)$. We denote the finite sum on the right-hand side of the above formula for $f(\mathbf{x})$ by $P_n(\mathbf{x})$, that is, we define $P_n(\mathbf{x}) \equiv \sum_{j=0}^{n} \frac{g^{(j)}(0)}{j!}$. Though a bit difficult to see, $P_n(\mathbf{x})$ is actually the nth-degree Taylor polynomial for $f(\mathbf{x})$ centered at $\mathbf{a}$. To see this more clearly, we use (6.32) to compute the first few terms in the sum:

$$\begin{aligned} g(0) &= f(\mathbf{a}) \\ g'(0) &= (x-a) f_x(\mathbf{a}) + (y-b) f_y(\mathbf{a}) \\ g''(0) &= (x-a)^2 f_{xx}(\mathbf{a}) + (x-a)(y-b) f_{xy}(\mathbf{a}) \\ &\quad + (x-a)(y-b) f_{yx}(\mathbf{a}) + (y-b)^2 f_{yy}(\mathbf{a}). \end{aligned}$$

Notice that $g'(0)$ is a first-degree polynomial in x and y, and $g''(0)$ is a second-degree polynomial in x and y, where x and y are the components of the vector $\mathbf{x}$. Of course, just as in the one-dimensional case, we can approximate $f(\mathbf{x})$ by its Taylor polynomial $P_n(\mathbf{x})$.[10]

► **6.70** *Compute $g'''(0)$. Can you find a pattern for $g^{(j)}(0)$?*

► **6.71** *Find $P_2(\mathbf{x})$ for the function $f(\mathbf{x}) = \sin x \sin y$.*

In order to state Taylor's theorem with remainder in a more compact form, we take the time now to develop a bit of informal operator notation. It will greatly simplify the form of the derivative terms $g^{(j)}(0)$, which can otherwise seem rather complicated. We do this in the $f : D^2 \to \mathbb{R}$ case for convenience, although it is readily generalizable to the $f : D^k \to \mathbb{R}$ case. Define the differential operator L by[11]

[10]Just as in the one-dimensional case, if $P_n(\mathbf{x})$ exists for all n within a neighborhood of $\mathbf{a}$, and if $R_n(\mathbf{x}) \to 0$ as $n \to \infty$ within that neighborhood, we refer to the resulting associated "infinite degree Taylor polynomial" as *the Taylor series of f centered at* $\mathbf{a}$. Again, Taylor series will be discussed more fully in Chapter 9.

[11]Note that the differential operator could have been more explicitly denoted by $L_{\mathbf{a}}$, in that the point $\mathbf{a} = (a, b) \in \mathbb{R}^2$ is specific to L. However, to retain a less cluttered notation we will dispense with the subscript and work under the assumption that the context will make it clear about which point we are expanding our Taylor polynomial.

$$L \equiv (x-a)\frac{\partial}{\partial x} + (y-b)\frac{\partial}{\partial y}.$$

Applying this operator to the function f, we have

$$Lf = (x-a)\frac{\partial f}{\partial x} + (y-b)\frac{\partial f}{\partial y}.$$

We also define "powers" of L in the following formal, nonrigorous manner,[12] namely,

$$\begin{aligned} L_2 &\equiv \left((x-a)\tfrac{\partial}{\partial x} + (y-b)\tfrac{\partial}{\partial y}\right)^2 \\ &\equiv \left((x-a)\tfrac{\partial}{\partial x} + (y-b)\tfrac{\partial}{\partial y}\right)\left((x-a)\tfrac{\partial}{\partial x} + (y-b)\tfrac{\partial}{\partial y}\right) \\ &= (x-a)^2\tfrac{\partial^2}{\partial x^2} + (x-a)(y-b)\tfrac{\partial^2}{\partial x\partial y} + (x-a)(y-b)\tfrac{\partial^2}{\partial y\partial x} + (y-b)^2\tfrac{\partial^2}{\partial y^2}. \end{aligned}$$

Likewise, we define L_j for any $j \in \mathbb{N}$ according to

$$L_j \equiv \left((x-a)\tfrac{\partial}{\partial x} + (y-b)\tfrac{\partial}{\partial y}\right)^j.$$

It is not hard to show that

$$g^{(j)}(0) = \left(L_j f\right)(\mathbf{a}), \tag{6.35}$$

and so

$$f(\mathbf{x}) = \sum_{j=0}^{n} \frac{\left(L_j f\right)(\mathbf{a})}{j!} + R_n(\mathbf{x}), \tag{6.36}$$

where $R_n(\mathbf{x})$ is given by

$$R_n(\mathbf{x}) = \frac{\left(L_{(n+1)} f\right)(\boldsymbol{\xi})}{(n+1)!} \quad \text{for some} \quad \boldsymbol{\xi} \in (\mathbf{a}, \mathbf{x}). \tag{6.37}$$

► **6.72** *Verify (6.35) and therefore that equations (6.36) and (6.37) can be obtained from the one-dimensional Taylor's theorem with remainder, Theorem 1.22 on page 252, applied to the function g.*

We now formally state the theorem implied by our special-case derivation.

Theorem 3.18 (Taylor's Theorem with Remainder for $f : D^k \to \mathbb{R}$)
Suppose all partial derivatives of order n of the function $f : D^k \to \mathbb{R}$ are continuous on D^k, and suppose all partial derivatives of order $n+1$ exist on D^k. Fix interior point $\mathbf{a} \in D^k$ and fix $r > 0$ such that $N_r(\mathbf{a}) \subset D^k$. Define the differential operator L by

$$L \equiv \sum_{i=1}^{k} (x_i - a_i)\frac{\partial}{\partial x_i}.$$

Then for $\mathbf{x} \in N_r(\mathbf{a})$,

$$f(\mathbf{x}) = P_n(\mathbf{x}) + R_n(\mathbf{x}),$$

[12]We use a subscript rather than an exponent to remind the reader that the resulting operator is not actually the associated operator power of L. To see this, simply consider the case of L applied to Lf, i.e., $L(Lf)$. The result is *not* given by what we refer to as $L_2 f$.

where

$$P_n(\mathbf{x}) = \sum_{j=0}^{n} \frac{(L_j f)(\mathbf{a})}{j!} \quad \text{and} \quad R_n(\mathbf{x}) = \frac{\left(L_{(n+1)} f\right)(\boldsymbol{\xi})}{(n+1)!}$$

for some $\boldsymbol{\xi} \in (\mathbf{a}, \mathbf{x})$.

► **6.73** *How does the expression for* $L_2 f$ *simplify for a function* $f : D^2 \to \mathbb{R}$ *that satisfies* $f_{xy} = f_{yx}$?

► **6.74** *Write the derivation of Taylor's theorem with remainder for* $f : D^k \to \mathbb{R}$.

► **6.75** *Can you derive the associated mean value theorem from Taylor's theorem with remainder for* $f : D^2 \to \mathbb{R}$? *How about for* $f : D^k \to \mathbb{R}$?

Now that we have established the mean value theorem and Taylor's theorem for functions $f : D^k \to \mathbb{R}$, results that were left unfinished in the previous section, we can refocus our attention on the functions specific to this section, namely, $\boldsymbol{f} : D^k \to \mathbb{R}^p$. There are versions of the mean value theorem and Taylor's theorem with remainder in this more general case as well, although they are a bit different from the previous cases we have seen.

Theorem 3.19 (The Mean Value Theorem for $\boldsymbol{f} : D^k \to \mathbb{R}^p$)
Let D^k *be an open subset of* $\mathbb{R}^k$, *and suppose that* $\boldsymbol{f} : D^k \to \mathbb{R}^p$ *is differentiable on* D^k. *Suppose* $\mathbf{p}_1$ *and* $\mathbf{p}_2$ *are two distinct interior points in* D^k *such that the segment* $(\mathbf{p}_1, \mathbf{p}_2)$ *is entirely contained within* D^k. *Then, there exist points* $\mathbf{q}_1, \mathbf{q}_2, \ldots, \mathbf{q}_p$ *on the segment* $(\mathbf{p}_1, \mathbf{p}_2)$ *such that*

$$\boldsymbol{f}(\mathbf{p}_2) - \boldsymbol{f}(\mathbf{p}_1) = \begin{bmatrix} \leftarrow & f_1'(\mathbf{q}_1) & \rightarrow \\ \leftarrow & f_2'(\mathbf{q}_2) & \rightarrow \\ & \vdots & \\ \leftarrow & f_p'(\mathbf{q}_p) & \rightarrow \end{bmatrix} (\mathbf{p}_2 - \mathbf{p}_1).$$

► **6.76** *Prove the above theorem along the lines of Theorem 3.17 on page 285. In what significant way does it differ from previous versions of the mean value theorem?*

► **6.77** *Let* $\boldsymbol{f} : \mathbb{R}^2 \to \mathbb{R}^2$ *be given by* $\boldsymbol{f}(x, y) = \left(x^2 + 4y, x^3 - y^2\right)$. *If* $\mathbf{a} = (0, 0)$, *and* $\mathbf{b} = (1, 0)$, *show that there does not exist a point* $\boldsymbol{\xi} \in \mathbb{R}^2$ *such that* $\boldsymbol{f}(\mathbf{b}) - \boldsymbol{f}(\mathbf{a}) = \boldsymbol{f}'(\boldsymbol{\xi})(\mathbf{b} - \mathbf{a})$.

Corresponding to the above mean value theorem for functions $\boldsymbol{f} : D^k \to \mathbb{R}^p$ is the following Taylor's theorem with remainder.

Theorem 3.20 (Taylor's Theorem with Remainder for $\boldsymbol{f} : \boldsymbol{D^k} \to \mathbb{R}^{\boldsymbol{p}}$) *Consider the function $\boldsymbol{f} : D^k \to \mathbb{R}^p$ given by $\boldsymbol{f}(\mathbf{x}) = \big(f_1(\mathbf{x}), f_2(\mathbf{x}), \ldots, f_p(\mathbf{x})\big)$ for all $\mathbf{x} \in D^k$, where each f_i is a real-valued function on D^k. Suppose all partial derivatives of order n of the functions f_i are continuous on D^k for $i = 1, 2, \ldots p$, and suppose all partial derivatives of order $n+1$ of the functions f_i exist on D^k for $i = 1, 2, \ldots p$. Fix $\mathbf{a} \in D^k$ and $r > 0$ such that $N_r(\mathbf{a}) \subset D^k$. Then, for $\mathbf{x} \in N_r(\mathbf{a})$,*

$$\boldsymbol{f}(\mathbf{x}) = \boldsymbol{P_n}(\mathbf{x}) + \boldsymbol{R_n}(\mathbf{x})$$

where

$$\boldsymbol{P_n}(\mathbf{x}) = \left(\sum_{j=0}^{n} \frac{\big(L_j f_1\big)(\mathbf{a})}{j!}, \ldots, \sum_{j=0}^{n} \frac{\big(L_j f_p\big)(\mathbf{a})}{j!} \right),$$

and

$$\boldsymbol{R_n}(\mathbf{x}) = \left(\frac{\big(L_{(n+1)} f_1\big)(\boldsymbol{\xi_1})}{(n+1)!}, \ldots, \frac{\big(L_{(n+1)} f_p\big)(\boldsymbol{\xi_p})}{(n+1)!} \right)$$

for some $\boldsymbol{\xi_1}, \boldsymbol{\xi_2}, \ldots, \boldsymbol{\xi_p} \in (\mathbf{a}, \mathbf{x})$.

► **6.78** *Prove the above theorem.*

► **6.79** *Consider the function $\boldsymbol{f} : \mathbb{R}^2 \to \mathbb{R}^2$ given by $\boldsymbol{f}(x, y) = \big(e^{xy},\ x \sin y\big)$. Find $\boldsymbol{P_3}(\mathbf{x})$ and $\boldsymbol{R_3}(\mathbf{x})$ centered at $\mathbf{0}$.*

3.3 Differentiability Classes

Functions can be conveniently categorized according to how many *continuous* derivatives they possess, i.e., how many derivatives they possess that are themselves continuous functions. For example, one such class of functions of interest in analysis is the class of functions having a continuous first derivative. Such functions are often referred to as *continuously differentiable* functions. The following definition generalizes this classification idea.

Definition 3.21 A function $f : D^1 \to \mathbb{R}$ is called a $\boldsymbol{C^n}$ **function** on D^1, or is considered to be in the class $C^n(D^1)$, if f has n derivatives that are themselves continuous functions on D^1.

If f has continuous derivatives of all orders on D^1 we say that f is a $\boldsymbol{C^\infty}$ **function** on D^1 and consider it to be in the class $C^\infty(D^1)$.

Example 3.22 *Consider $f : D^1 \to \mathbb{R}$ given by $f(x) = x^{7/2}$, where $D^1 = (-1, 1)$. Then $f \in C^3(D^1)$. (Why?)* ◄

► **6.80** *Verify the claim made in the above example.*

Note that a function $f : D^1 \to \mathbb{R}$ that is $C^n(D^1)$ is also $C^m(D^1)$ for $m < n$. In fact, the differentiability classes have a nested relationship. Often, the domain of interest in using the above terminology is clear from the context of the problem, in which case D^1 won't be specified explicitly and the function will simply be referred to as a C^n function. For example, if f is $C^3\,((0,1))$, and the interval $(0,1)$ is unambiguously understood to be the domain of interest, we will simply say that f is C^3. It is worth pointing out that identifying f as C^3 does not preclude the possibility that f might in fact be C^4 or even C^n for some other $n > 3$. While one usually categorizes a function into its appropriate C^n class according to the largest n for which the function qualifies, it is not necessary, and in certain instances not very convenient to do so. Also, if f is C^n, but is not C^{n+1}, this does not imply that f does not *have* a derivative of order $(n+1)$, but only that such a derivative (*if* it exists) is not continuous. It is worth mentioning that, in general, the higher the differentiability class to which it belongs, the "better behaved" a function can be expected to be. A clue as to why this is so is the fact, proved in Chapter 5, that if a function is continuous at a point then the function remains bounded in some neighborhood of that point. Therefore, if a function f is C^1, for example, its derivative f' remains bounded in some neighborhood of each point of its domain. If the function is differentiable but not C^1, then its derivative may be unbounded at any point in its domain where f' is not continuous. This possibility can lead to the function f having unusual behavior at or near such a point.

Just as we categorized a function $f : D^1 \to \mathbb{R}$ having n continuous derivatives on D^1 as being $C^n(D^1)$, we can likewise categorize functions $f : D^k \to \mathbb{R}$ as $C^n(D^k)$ according to how many continuous *partial* derivatives they have.

Definition 3.23 A function $f : D^k \to \mathbb{R}$ is called a $\boldsymbol{C^n}$ **function** on D^k, or is considered to be in the class $C^n(D^k)$, if all partial derivatives (including mixed) of order n exist and are continuous on D^k. We say that f is a $\boldsymbol{C^\infty}$ **function** on D^k if it has continuous partial derivatives of *all orders* on D^k. In this latter case, we consider f to be in the class $C^\infty(D^k)$.

Example 3.24 *Let $f(x,y) = x(x^2+y^2)^{3/2}$ on all of $\mathbb{R}^2$. Then it is not hard to verify that $f_x(x,y) = 3x^2(x^2+y^2)^{1/2} + (x^2+y^2)^{3/2}$, and $f_y(x,y) = 3xy(x^2+y^2)^{1/2}$. From this we determine that f is at least $C^1(\mathbb{R}^2)$. In determining the second-order partial derivatives, one can show that f_{xy} does not exist at $(0,0)$. Therefore, f cannot be in $C^2(\mathbb{R}^2)$.* ◀

▶ **6.81** *In the above example, verify that f_{xy} does not exist at $(0,0)$. Also, find the other second-order partial derivatives of f, and verify that the origin is the only discontinuity among them.*

▶ **6.82** *For the same function as in the last example, suppose we restrict the domain to $D^2 \equiv \{(x,y) : xy \neq 0\}$. The result is a new function (since the domain is dif-*

ferent from the function in the example); call it g. What is the largest n for which $g \in C^n(D^2)$?

► **6.83** *Can you extend Definition 3.23 in a natural way to be able to similarly categorize functions $\boldsymbol{f} : D^k \to \mathbb{R}^p$?*

The concepts of derivative and differentiability play an even more significant role for functions of a complex variable than for functions of real variables. In fact, we will ultimately discover that if a complex function $f : D \to \mathbb{C}$ is differentiable on $D \subset \mathbb{C}$, then f *must* be C^∞ on D. This is a significant result. It means that differentiable complex functions can't exhibit the unusual behaviors that some differentiable real functions might exhibit. We begin discussing this very important case next.

4 THE DERIVATIVE FOR $f : D \to \mathbb{C}$

In this section we investigate the concept of differentiation for complex valued functions of a complex variable, commonly referred to as complex functions. There are in fact two ways to view such functions. From a geometric point of view, since the complex variable z can be thought of as a vector having components x and y, a function $f : D \to \mathbb{C}$ can be thought of as a special case of $\boldsymbol{f} : D^2 \to \mathbb{R}^2$. That is, a complex function is a function from the (complex) plane to itself. From an algebraic point of view, since $\mathbb{C}$ is a field, a complex function can be thought of as a function that takes an element of the field $\mathbb{C}$ to another element of the field $\mathbb{C}$. That is, just like real-valued functions of a single real variable, a complex function is a function from a field to itself. The notation we adopt for complex functions follows from this algebraic perspective. Since we are discussing complex-valued functions of a single complex variable, it is common practice to use the same style font as for a real-valued function of a single real variable, namely $f(z)$ (rather than $\boldsymbol{f}(\mathbf{z})$). Although the algebraic view "wins out" notationally, we will see that complex functions can be more rewardingly viewed from the geometric perspective. (See, for example, the discussion of complex functions as mappings in Chapter 10.) In the end, we hope to reconcile both points of view as they pertain to the world of complex functions. As already mentioned in Chapter 4, we note that in the context of complex functions $f : D \to \mathbb{C}$, the domain will be denoted by D rather than D^2 to distinguish this rather special class of functions from the real case(s). Finally, before we begin our discussion of derivatives of complex functions, we must clarify our position on the terminology common to the theory of functions of a complex variable. In most texts on complex function theory, a complex function that is differentiable on an open set $D \subset \mathbb{C}$ is referred to as *analytic* (or, somewhat less commonly nowadays, *holomorphic*) on D. In some such texts, a complex function is even defined to be analytic on such a set D exactly when it is differentiable there. It is in fact true that such differentiable complex functions are analytic, but in our development of the subject we will not define analyticity until later in the text.[13] Once we have defined the term *analytic*, we will show that a complex function that is differentiable on an open set D must be analytic there. The reader may then return to this section and replace each instance of the phrase "complex function differentiable on the open set $D \subset \mathbb{C}$" with "analytic complex function on the open set $D \subset \mathbb{C}$," thereby gaining consistency between the statements of theorems in this work and the corresponding statements in other, more traditional treatments. It should be noted, however, that *real* functions that are differentiable on an open set are *not* necessarily analytic there, no matter what text you are reading. This is one of the significant ways in which complex functions and real functions differ from each other.

4.1 Three Derivative Definitions Again

Let us look more carefully now at the algebraic representation of the derivative. Just as in the real case, we will present an "official" derivative definition, as well as two equivalent, alternative versions that will prove to be quite practical in our work. As in the real case, our "official" version is the linear approximation version.

The Linear Approximation Version

Definition 4.1 (Derivative of $f : D \to \mathbb{C}$ at the point z_0) For a complex function $f : D \to \mathbb{C}$ and $z_0 \in \text{Int}(D)$, we say that f is **differentiable** at z_0 if there exists a complex number A such that the linear function $p : D \to \mathbb{C}$ given by

$$p(z) = f(z_0) + A\,(z - z_0) \text{ satisfies } \lim_{z \to z_0} \frac{f(z) - p(z)}{|z - z_0|} = 0.$$

In this case, we refer to the complex number A as the **derivative** of f at z_0, and denote it thereafter by $f'(z_0)$ or $\frac{df}{dz}(z_0)$.

Note that the absolute value bars in the denominator of the above limit expression are not really necessary, just as in the case of a real-valued function $f : D^1 \to \mathbb{R}$. This is because the numerator $f(z) - p(z)$ and the denominator $z - z_0$ are members of the same field, and so their ratio (without absolute value bars) is well defined as long as the denominator is nonzero. We have retained the absolute value bars only to be consistent with the corresponding real function versions of the definition.

The Difference Quotient Version

Omitting the absolute value bars mentioned above, and a bit of rearrangement, leads us naturally to the convenient, practical, and equivalent difference quotient version of the derivative of a complex function.

[13] Our definition of *analyticity*, in fact, will rely on the existence of a convergent Taylor series for the function in question. This will provide us with a definition of analyticity that extends consistently to real functions.

A complex function $f : D \to \mathbb{C}$ is differentiable at $z_0 \in \text{Int}(D)$ if and only if

$$\lim_{z \to z_0} \frac{f(z) - f(z_0)}{z - z_0}$$

exists. In this case, we denote the limit by $f'(z_0)$ or $\frac{df}{dz}(z_0)$.

We emphasize again that this version of the derivative definition is only available in this case and that of a real-valued function of a single real variable. In each of these cases, the function is a mapping from a field to itself, and so the difference quotient (without absolute value bars) is well defined.

The ϵ, δ Version

Likewise, the ϵ, δ version of the definition of the derivative is often convenient to apply.

The function $f : D \to \mathbb{C}$ is differentiable at $z_0 \in \text{Int}(D)$ and has derivative $f'(z_0)$ there if and only if for each $\epsilon > 0$ there exists $\delta > 0$ such that

$$|z - z_0| < \delta \quad \Rightarrow \quad |f(z) - f(z_0) - f'(z_0)(z - z_0)| \le \epsilon\, |z - z_0|.$$

The equivalence of these three definitions is stated in the following result, an analog to Theorem 1.2 on page 238.

Proposition 4.2 *Consider $f : D \to \mathbb{C}$ and $z_0 \in \text{Int}(D)$. The following three statements are mathematically equivalent.*

a) $\lim_{z \to z_0} \frac{f(z) - f(z_0)}{z - z_0} = A.$

b) There exists $A \in \mathbb{C}$ such that for any $\epsilon > 0$ there exists $\delta > 0$ such that

$$|z - z_0| < \delta \quad \Rightarrow \quad |f(z) - f(z_0) - A\,(z - z_0)| \le \epsilon\, |z - z_0|.$$

c) There exists $A \in \mathbb{C}$ such that the function $p : D \to \mathbb{C}$ given by

$$p(z) = f(z_0) + A\,(z - z_0) \quad \textit{satisfies} \quad \lim_{z \to z_0} \frac{f(z) - p(z)}{|z - z_0|} = 0.$$

When any one (and therefore all three) of these statements is true, the function f is differentiable at $z_0 \in \text{Int}(D)$ with derivative $f'(z_0) = A$.

▶ **6.84** *Prove the above result.*

▶ **6.85** *For each $f(z)$ below, use the difference quotient formulation to verify that the given $f'(z_0)$ is as claimed. Then use the ϵ, δ version to prove that the derivative is the* *given* $f'(z_0)$.

a) $f(z) = c$ for $D = \mathbb{C}$, constant $c \in \mathbb{C}$, $f'(z_0) = 0$ for all $z_0 \in D$

b) $f(z) = z^3$ for $D = \mathbb{C}$, $f'(z_0) = 3z_0^2$ for all $z_0 \in D$

c) $f(z) = z^{-1}$ for $D = \mathbb{C} \setminus \{0\}$, $f'(z_0) = -z_0^{-2}$ for all $z_0 \in D$

d) $f(z) = z^n$ for $D = \mathbb{C}$, $n \in \mathbb{Z}^+$, $f'(z_0) = nz_0^{n-1}$ for all $z_0 \in D$

e) $f(z) = z^n$ for $D = \mathbb{C} \setminus \{0\}$, $n \in \mathbb{Z}^-$, $f'(z_0) = nz_0^{n-1}$ for all $z_0 \in D$

f) $f(z) = \sin z$ for $D = \mathbb{C}$, $f'(z_0) = \cos z_0$, for all $z_0 \in D$

g) $f(z) = \cos z$ for $D = \mathbb{C}$, $f'(z_0) = -\sin z_0$, for all $z_0 \in D$

A hint for part f) is to consider the trigonometric identity

$$\sin\theta - \sin\phi = 2\cos\left(\tfrac{\theta+\phi}{2}\right)\sin\left(\tfrac{\theta-\phi}{2}\right).$$

Just as for real functions, when the derivative of a complex function f at a point z_0 exists, it is unique. This fact is given in the following proposition, the proof of which is left to the reader.

Proposition 4.3 *Consider $f : D \to \mathbb{C}$ and $z_0 \in \text{Int}(D)$. If $f'(z_0)$ exists, then it is unique.*

► **6.86** *Prove the above result.*

It is also true for complex functions that differentiability implies continuity. We state this too as a separate result, mirroring Proposition 1.7 on page 242 in the real case almost exactly. Once again, the proof is left to the reader.

Proposition 4.4 *Let the function $f : D \to \mathbb{C}$ be differentiable at $z_0 \in \text{Int}(D)$. Then f is continuous at z_0 as well.*

► **6.87** *Prove the above result.*

Higher-Order Derivatives

Of course, just as for real functions, a function $f : D \to \mathbb{C}$ might be differentiable at more than just a single point of its domain. In fact, we may consider all the points $D_{f'} \subset D$ at which f is differentiable as the domain of the function $f' : D_{f'} \to \mathbb{C}$. That is, the derivative of f is itself a complex function, and we say that *f is differentiable on $D_{f'}$*. If z_0 is an interior point of $D_{f'}$, we may consider whether f' has a derivative at z_0. We can determine this by checking whether the limit

$$\lim_{z \to z_0} \frac{f'(z) - f'(z_0)}{z - z_0} = B$$

for some complex number B. If so, then $B = f''(z_0)$ is called *the second derivative of f at z_0*. In similar fashion, one may consider whether the third derivative $f'''(z_0)$ exists at z_0, and more generally, whether the nth derivative $f^{(n)}(z_0)$ exists at z_0. Just as we found for real-valued functions of a single real variable, this derivative-generating scheme shows that it is necessary for the $(n-1)^{st}$ derivative to exist before one can consider the existence of the nth derivative. It is also necessary for the point z_0 to be an interior point of the domain of $f^{(n-1)}$ to consider the existence of $f^{(n)}(z_0)$. Once again, this last fact implies that the domain of the nth derivative will be contained in the domain of the $(n-1)^{st}$.

4.2 Some Useful Results

Just as in the real function case, we list a few of the more convenient results that allow for easier manipulation of derivatives for functions $f : D \to \mathbb{C}$. We begin with the algebraic properties.

Algebraic Results

Proposition 4.5 (Algebraic Properties of the Derivative)
Suppose the functions $f : D \to \mathbb{C}$ and $g : D \to \mathbb{C}$ are differentiable at $z_0 \in \text{Int}(D)$. Then

a) $f \pm g$ is differentiable at z_0, and $(f \pm g)'(z_0) = f'(z_0) \pm g'(z_0)$.

b) fg is differentiable at z_0, and $(fg)'(z_0) = g(z_0)f'(z_0) + f(z_0)g'(z_0)$.

c) cf is differentiable at z_0, and $(cf)'(z_0) = cf'(z_0)$ for $c \in \mathbb{C}$.

d) (f/g) is differentiable at z_0 provided $g(z_0) \neq 0$. In this case,

$$(f/g)'(z_0) = \frac{g(z_0)f'(z_0) - f(z_0)g'(z_0)}{(g(z_0))^2}.$$

► **6.88** *Prove the above result.*

► **6.89** *Consider $f : D \to \mathbb{C}$. If $f' \equiv 0$ on D, what can you conclude about f? Does it matter what kind of set D is?*

► **6.90** *Prove that the following functions $f : \mathbb{C} \to \mathbb{C}$ are differentiable on $\mathbb{C}$.*
a) $f(z) = z$ *b)* $f(z) = z^n$ for $n \in \mathbb{Z}^+$
c) $f(z) = c_0 + c_1 z + c_2 z^2 + \cdots + c_n z^n$ for $c_j \in \mathbb{C}$ and $n \in \mathbb{N}$ with $c_n \neq 0$.

► **6.91** *Suppose $p : \mathbb{C} \to \mathbb{C}$ is given by $p(z) = a_0 + a_1 z + a_2 z^2 + \cdots + a_n z^n$ for $n \in \mathbb{N}$, $a_j \in \mathbb{C}$ for all $0 \leq j \leq n$, and $a_n \neq 0$, and $q : \mathbb{C} \to \mathbb{C}$ is given by $q(z) = b_0 + b_1 z + b_2 z^2 + \cdots + b_m z^m$ for $m \in \mathbb{N}$, $b_j \in \mathbb{C}$ for all $0 \leq j \leq m$, and $b_m \neq 0$. Then, for $f : D \to \mathbb{C}$ given by $f(z) = \frac{p(z)}{q(z)}$, where $D = \{z \in \mathbb{C} : q(z) \neq 0\}$, show that f is differentiable on D.*

Proposition 4.6 (The Chain Rule for $f : D \to \mathbb{C}$)
Suppose $f : D_f \to \mathbb{C}$ is differentiable at $z_0 \in \text{Int}(D_f)$, and let $\mathcal{R}_f \equiv f(D_f)$. Suppose $\mathcal{R}_f \subset D_g$. If $g : D_g \to \mathbb{C}$ is differentiable at $w_0 = f(z_0) \in \text{Int}(D_g)$, then $h \equiv g \circ f : D_f \to \mathbb{C}$ is differentiable at $z_0 \in \text{Int}(D_f)$, and

$$h'(z_0) = g'(w_0)f'(z_0) = g'\left(f(z_0)\right)f'(z_0).$$

▶ **6.92** *Prove the chain rule for complex functions.*

Lack of Local Extrema Results

Just as in the real case $\boldsymbol{f} : D^k \to \mathbb{R}^p$, there are no local extrema results for functions $f : D \to \mathbb{C}$, nor is there a comparable version of Rolle's theorem. After all, $\mathbb{C}$ is geometrically similar to $\mathbb{R}^2$, and both lack the special order properties possessed by $\mathbb{R}$ that are required. Also lacking in the complex function case is a version of the mean value theorem that would in some way depend on Rolle's theorem (recall that in the case $f : D^1 \to \mathbb{R}$ one could show that Rolle's theorem implied the mean value theorem and vice versa). In fact, the reader is encouraged to work the following exercise, which provides a counterexample to this type of mean value theorem for complex functions. Later we will see that a different kind of mean value theorem holds in this case.

▶ **6.93** *Let $f : \mathbb{C} \to \mathbb{C}$ be given by $f(z) = z^3$, and consider the points $z_1 = 1$ and $z_2 = i$ as $\mathbf{p}_1$ and $\mathbf{p}_2$ in the statement of Theorem 3.17 on page 285. Show that there does not exist a complex number playing the role of $\mathbf{q}$ as described in that theorem.*

As a consequence of the above discussion, there is no version of Taylor's theorem with remainder that would have followed directly from the mean value theorem as was possible for $f : D^1 \to \mathbb{R}$ and for $f : D^k \to \mathbb{R}$. This is not to say that there is *no* Taylor's theorem with remainder for complex functions—there is! However, we will see in a later chapter that it is in a form that differs from the versions we have seen thus far. In particular, its remainder is expressible in a way that does not involve a derivative of the function in question.

There *is* one very significant derivative result that is special to the class of complex functions $f : D \to \mathbb{C}$. In fact, this result is named after not one, but *two* mathematicians worthy of note. Just as the Jacobian provided a convenient way to determine whether a real function $\boldsymbol{f} : D^k \to \mathbb{R}^p$ was differentiable at a point, the *Cauchy-Riemann equations* provide a means for doing the same thing in the complex function case. We investigate this important set of equations next.

4.3 The Cauchy-Riemann Equations

The Cauchy-Riemann equations of complex function theory are arguably among the most significant results in analysis, providing a convenient characterization of differentiability for complex functions. We begin with an informal motivation, followed by a more careful, mathematically rigorous development.

An Informal Motivation for the Cauchy-Riemann Equations

For the moment, consider $f(z) = u(x,y) + i\,v(x,y)$ from the geometric point of view. At any point $z = x + i\,y$ in the domain of f, we have, with a slight abuse of notation,

$$f(z) = u(x,y) + i\,v(x,y) = \big(u(x,y),\, v(x,y)\big) = \boldsymbol{f}(x,y).$$

Here, u and v can be thought of as the real-valued component functions of a vector-valued function $\boldsymbol{f} : D^2 \to \mathbb{R}^2$. The function $\boldsymbol{f}$ maps elements of D^2 to $\mathbb{R}^2$, and so it clearly lends itself to the derivative definition from the previous subsection. That is, if the derivative $f'(z_0)$ exists, it should take the form of a 2×2 matrix $\boldsymbol{A} = \boldsymbol{f}'(x_0, y_0)$, which must be the Jacobian matrix,[14]

$$\boldsymbol{J}(x_0, y_0) = \begin{bmatrix} u_x(x_0, y_0) & u_y(x_0, y_0) \\ & \\ v_x(x_0, y_0) & v_y(x_0, y_0) \end{bmatrix}.$$

There is more here, however, than meets the eye. We will see that for a complex function $f : D \to \mathbb{C}$ to be differentiable implies much more than for a vector valued function $\boldsymbol{f} : D^2 \to \mathbb{R}^2$. Since a significant difference between the two situations is that $\mathbb{C}$ is a *field*, whereas $\mathbb{R}^2$ is not, the field property must be what gives complex functions their additional structure.

While the above 2×2 matrix description of the derivative of $f(z)$ at z_0 seems reasonable enough, now consider the function $f(z)$ the way one considers a real-valued function of a single real variable, i.e., algebraically. The reader can readily verify that, consistent with (6.1) on page 235, the derivative of f at $z_0 \in \mathbb{C}$ might also be viewed as the limit of the appropriate difference quotient,

$$\lim_{z \to z_0} \frac{f(z) - f(z_0)}{z - z_0}.$$

This limit, when it exists, must be a member of the field of which the difference quotient is a member, i.e., the limit, when it exists, is a complex number. But just as with the geometric point of view, there is something of substance yet hidden from us. In the comparison of complex functions $f : D \to \mathbb{C}$ to

[14]Note that $\boldsymbol{J}(x, y)$ is a function of x and y, and not z, since each of its entries is a function of x and y, and not z. When the context is clear, we will sometimes suppress the explicit x and y dependence in the entries of the Jacobian matrix to avoid a cluttered appearance in our equations.

real-valued functions $f : D^1 \to \mathbb{R}$, the only difference is that of *dimension.* Since $\mathbb{C}$ is two-dimensional, the theory of complex functions is a richer theory than that of real-valued functions, as we will see.

How do we reconcile these seemingly different representations for the derivative $f'(z_0)$ when it exists? We will "follow our noses" and argue rather informally first, then we will more rigorously justify what we have found in our intuitive investigations. In the geometric point of view, the derivative is a 2×2 matrix. In the algebraic point of view, the derivative is a complex number. As we've seen in Chapter 2, there is a subset of 2×2 matrices that is naturally isomorphic to the complex numbers, indicated by

$$a + ib \quad \leftrightarrow \quad \begin{bmatrix} a & -b \\ b & a \end{bmatrix}.$$

Comparing this to our 2×2 Jacobian matrix $\boldsymbol{J}(x_0, y_0)$, and viewing $\boldsymbol{J}(x_0, y_0)$ as a complex number via the above-mentioned isomorphism, yields

$$\boldsymbol{J}(x_0, y_0) = \begin{bmatrix} u_x(x_0, y_0) & u_y(x_0, y_0) \\ & \\ v_x(x_0, y_0) & v_y(x_0, y_0) \end{bmatrix} = \begin{bmatrix} a & -b \\ b & a \end{bmatrix},$$

giving us a pair of equations that seemingly must hold true in order to make sense of our notion of derivative for complex functions. That is,

$$u_x(x_0, y_0) = v_y(x_0, y_0), \quad \text{and} \quad u_y(x_0, y_0) = -v_x(x_0, y_0).$$

In fact, these equalities relating the first-order partial derivatives of u and v at (x_0, y_0) are the *Cauchy-Riemann equations* that lie at the heart of complex function theory. Applying them to our Jacobian representation for $f'(z_0)$, and recognizing the complex number represented by the resulting 2×2 matrix yields the correspondence

$$\boldsymbol{f}'(x_0, y_0) = \begin{bmatrix} u_x(x_0, y_0) & u_y(x_0, y_0) \\ & \\ v_x(x_0, y_0) & v_y(x_0, y_0) \end{bmatrix} \longleftrightarrow f'(z_0) = u_x(x_0, y_0) + i\, v_x(x_0, y_0).$$

That is, the matrix representation of the derivative at $z_0 = x_0 + i\, y_0$ is equivalent to the complex number having real part $u_x(x_0, y_0)$, and imaginary part $v_x(x_0, y_0)$. So, just as for a differentiable real function $f : D^1 \to \mathbb{R}$, a differentiable complex function can be seen to have as its derivative a member of the field it maps from and into, even when starting from the geometric point of view. This is not true for a general real function $\boldsymbol{f} : D^2 \to \mathbb{R}^2$ whose derivative, if it exists, is *not* a member of $\mathbb{R}^2$, which itself is not a field. Also, while we can represent the derivative of a differentiable complex function in matrix form, the set of possible candidates is restricted to a very special class of 2×2

matrices, namely, a subset of matrices isomorphic to the complex numbers.[15] This last remark deserves further comment, since it is directly related to one of the key ways in which complex functions differ from their real counterparts. Recall that Theorem 3.10 from the last section gave us the following result. For a *real* function $\boldsymbol{f} : D^2 \to \mathbb{R}^2$ given by $\boldsymbol{f}(x, y) = (f_1(x, y), f_2(x, y))$, if each partial derivative, $\frac{\partial f_1}{\partial x}, \frac{\partial f_1}{\partial y}, \frac{\partial f_2}{\partial x}$, and $\frac{\partial f_2}{\partial y}$ exists in a neighborhood of the point $(x_0, y_0) \in D^2$ and is continuous at (x_0, y_0), then the function $\boldsymbol{f}$ is differentiable at (x_0, y_0) and $\boldsymbol{f}'(x_0, y_0)$ is given by $\boldsymbol{J}(x_0, y_0)$. Although a complex function $f(z) = u(x, y) + i\, v(x, y)$ can also be viewed in certain ways as a mapping from a subset of the plane to itself, the similarity runs only so deep. In fact, for such a complex function Theorem 3.10 is *not enough* to guarantee differentiability at a point z_0. As already described, the Cauchy-Riemann equations impose an extra set of conditions that must be satisfied by a complex function in order for the derivative $f'(z_0)$ to exist, an extra set of conditions that effectively restrict the set of 2×2 Jacobian matrices available for a complex function to qualify as differentiable at a point z_0.

Example 4.7 *Suppose $f : \mathbb{C} \to \mathbb{C}$ is given by $f(z) = z + \overline{z} = 2x$. Viewing f as $\boldsymbol{f} : D^2 \to \mathbb{R}^2$, we have $\boldsymbol{f}(x, y) = (2x, 0)$, and $\boldsymbol{J}(x, y) = \begin{bmatrix} 2 & 0 \\ 0 & 0 \end{bmatrix}$. From this we see that the Jacobian matrix exists, and each entry is in fact a continuous function on all of $\mathbb{R}^2$. Therefore $\boldsymbol{f}$ is differentiable as a function from $\mathbb{R}^2$ to $\mathbb{R}^2$. Furthermore, in this case the Jacobian matrix is the derivative. However, the Jacobian matrix is not of the form $\begin{bmatrix} a & -b \\ b & a \end{bmatrix}$, and so we infer that f is not differentiable as a function from $\mathbb{C}$ to $\mathbb{C}$.* ◀

Example 4.8 *Suppose $f : \mathbb{C} \to \mathbb{C}$ is given by*

$$f(z) = z^2 = \left(x^2 - y^2\right) + i\, 2xy.$$

Then $u(x, y) = x^2 - y^2$ and $v(x, y) = 2xy$. If we view f as $\boldsymbol{f} : \mathbb{R}^2 \to \mathbb{R}^2$, then $\boldsymbol{f}(x, y) = \left(x^2 - y^2, 2xy\right)$, and the Jacobian is given by

$$\boldsymbol{J}(x, y) = \begin{bmatrix} 2x & -2y \\ 2y & 2x \end{bmatrix}.$$

It is clear that all of the entries of $\boldsymbol{J}(x, y)$ are continuous for all values of x and y, and so the derivative $\boldsymbol{f}'$ exists and is $\boldsymbol{J}(x, y)$ at all points of $\mathbb{R}^2$. Since the Jacobian matrix has the form $\boldsymbol{J}(x, y) = \begin{bmatrix} a & -b \\ b & a \end{bmatrix}$, we can identify $\boldsymbol{f}'$ with

[15]The practical significance of the matrix representation for $f'(z_0)$ lies in its interpretation as a linear transformation. When applied to an element of the complex plane, the matrix has the effect of scaling and rotating as discussed in subsection 3.6 of Chapter 1. The derivative of a complex function at a point z_0, when it exists, must be a scaling and a rotation at the point z_0 in this sense. We will consider this idea more carefully when we discuss conformal mappings in Chapter 10.

the complex number $a + i\,b$, or $2x + i\,2y = 2z$ in this case. That is, according to the matrix $\leftrightarrow$ complex number isomorphism,

$$\boldsymbol{f}'(x,y) = \boldsymbol{J}(x,y) = \begin{bmatrix} 2x & -2y \\ 2y & 2x \end{bmatrix} \quad \longleftrightarrow \quad f'(z) = 2x + i\,2y = 2\,z,$$

and so we can infer that when considered as a function from $\mathbb{C}$ to $\mathbb{C}$, f' exists and will be given by $f'(z) = 2\,z$. Note that this yields a derivative formula for the complex function $f(z) = z^2$ that is analogous to that of the real function $f(x) = x^2$. ◄

► **6.94** *Using the same line of reasoning as in the last example, and recalling the fact that $e^z = e^x \cos y + i\,e^x \sin y$, show that $\frac{d}{dz}e^z$ can be inferred to be e^z.*

► **6.95** *Assuming that $\frac{d}{dz}e^z = e^z$, show that $\frac{d}{dz}\sin z$ can be inferred to be $\cos z$, and that $\frac{d}{dz}\cos z$ can be inferred to be $-\sin z$. (Hint: Use the exponential forms of $\sin z$ and $\cos z$.)*

► **6.96** *Consider $f : \mathbb{C} \to \mathbb{C}$ given by $f(z) = u(x,y) + i\,v(x,y)$ where $u(x,y) = x^2 + y$ and $v(x,y) = x^2 + y^2$. Infer that the only $z \in \mathbb{C}$ at which $f'(z)$ might exist is $z = -\frac{1}{2} - i\,\frac{1}{2}$.*

► **6.97** *Consider $f : \mathbb{C} \to \mathbb{C}$ given by*

$$f(z) = \left(5x + \tfrac{1}{2}x^2 + xy - y^2\right) + i\left(5y + 2xy + \tfrac{1}{2}y^2 - \tfrac{1}{2}x^2\right).$$

Use the method of this section to infer where f might be differentiable.

Rigorous Derivation of the Cauchy-Riemann Equations

Suppose $f : D \to \mathbb{C}$ is differentiable at $z_0 = x_0 + i\,y_0 \in \text{Int}(D)$ with derivative $f'(z_0) = A_1 + iA_2$, where $A_1, A_2 \in \mathbb{R}$. What are the values of A_1 and A_2? To determine this we will force z to approach z_0 in the complex plane first horizontally and then vertically. To this end, substitute $z = x + i\,y_0$ into Definition 4.1 given on page 292, and note that y_0 is fixed. The definition then yields that for any $\epsilon > 0$ there exists $\delta > 0$ such that
$|x - x_0| < \delta \ \Rightarrow$

$$\left|\left(u(x,y_0) - u(x_0,y_0) - A_1(x - x_0)\right) + i\left(v(x,y_0) - v(x_0,y_0) - A_2(x - x_0)\right)\right| \le \epsilon\,|x - x_0|.$$

From this, we find that $|x - x_0| < \delta \ \Rightarrow$

$$\begin{aligned} |u(x,y_0) - u(x_0,y_0) - A_1(x - x_0)| &\le \epsilon\,|x - x_0|, \quad \text{and} \\ |v(x,y_0) - v(x_0,y_0) - A_2(x - x_0)| &\le \epsilon\,|x - x_0|, \quad \text{(Why?)} \end{aligned}$$

which in turn implies that

$$\frac{\partial u}{\partial x}(x_0,y_0) = A_1, \qquad \frac{\partial v}{\partial x}(x_0,y_0) = A_2. \tag{6.38}$$

A similar calculation associated with the substitution of $z = x_0 + i\,y$ into Definition 4.1 (and hence, forcing z to approach z_0 vertically in the complex plane) yields

$$\frac{\partial v}{\partial y}(x_0,y_0) = A_1, \qquad \frac{\partial u}{\partial y}(x_0,y_0) = -A_2. \tag{6.39}$$

Together, (6.38) and (6.39) give us the Cauchy-Riemann equations. We summarize these results formally in the following theorem.

Theorem 4.9 (The Cauchy-Riemann Equations (Necessity))
Consider $f : D \to \mathbb{C}$ where $f(z) = u(x, y) + i\,v(x, y)$ is differentiable at $z_0 = x_0 + i\,y_0 \in \text{Int}(D)$. Then the Cauchy-Riemann equations given by

$$u_x = v_y, \quad u_y = -v_x,$$

must hold at $z_0 = (x_0, y_0)$. In this case, the derivative at z_0 is given by

$$f'(z_0) = u_x(x_0, y_0) + i\,v_x(x_0, y_0) \quad \textit{or} \quad f'(z_0) = v_y(x_0, y_0) - i\,u_y(x_0, y_0).$$

This theorem tells us that for a complex function $f : D \to \mathbb{C}$ given by $f(z) = u(x, y) + i\,v(x, y)$ to be differentiable at a point z_0, the Cauchy-Riemann equations must hold there. They are *necessary* conditions for the derivative $f'(z_0)$ to exist. This means that in order for f to be a differentiable complex function, its imaginary part v must be related to its real part u in a very special way. Not just any old $u(x, y)$ and $v(x, y)$ will do, even if they are (individually) differentiable real-valued functions of x and y. It is worth emphasizing that the two key ways in which the world of complex functions differs from its closest real function counterparts are what give rise to the Cauchy-Riemann equations in the first place. If one starts with the geometric point of view, as in our informal discussion earlier, the field property possessed by $\mathbb{C}$ imposes conditions on the entries of the 2×2 matrix representation of the derivative, giving rise to the Cauchy-Riemann equations. If one starts with the algebraic point of view as represented by Definition 4.1, the two-dimensional geometry of $\mathbb{C}$ imposes conditions on the resulting limit. In particular, the same limiting value must be obtained regardless of the direction of approach, again giving rise to the Cauchy-Riemann equations. These very properties that give rise to the Cauchy-Riemann equations also act as a severe constraint on the function f. Differentiable complex functions comprise a very special class of functions.

► **6.98** *Consider Example 4.8 and the complex functions from the two exercises immediately following it. Verify that for each of these (complex) differentiable functions, the Cauchy-Riemann equations hold.*

Example 4.10 *Recall the function* $\boldsymbol{f}$ *from Example 3.12 on page 281 in the previous section, given by*

$$\boldsymbol{f}(x, y) = \left(\frac{e^{xy}}{x}, \sqrt{\sin y}\right)$$

defined on the strip $D^2 = (-\infty, 0) \times (0, \pi)$. This function was found to be *differentiable on D^2, and the Jacobian matrix gave the derivative as*

$$\boldsymbol{f}'(x,y) = \begin{bmatrix} \frac{(xy-1)e^{xy}}{x^2} & e^{xy} \\ 0 & \frac{\cos y}{2\sqrt{\sin y}} \end{bmatrix}.$$

Denoting the complex plane counterpart of $D^2 \subset \mathbb{R}^2$ by D, if we consider the corresponding complex function given by $f(z) = u(x,y) + i\,v(x,y)$ with

$$u(x,y) = \frac{e^{xy}}{x} \quad \text{and} \quad v(x,y) = \sqrt{\sin y},$$

we might wonder whether $f : D \to \mathbb{C}$ is differentiable as a complex function. A simple comparison of the partial derivatives u_x, u_y, v_x, and v_y clearly shows that the Cauchy-Riemann equations are not satisfied at any point in the plane, and so, although $\boldsymbol{f}$ is differentiable as a function from $D^2 \subset \mathbb{R}^2$ to $\mathbb{R}^2$, f is not differentiable as a function from $D \subset \mathbb{C}$ to $\mathbb{C}$. ◀

Example 4.11 *Suppose $f(z)$ is given by*

$$f(z) = \frac{1}{z} = \frac{x}{x^2+y^2} - i\,\frac{y}{x^2+y^2}.$$

Then f can be seen to be differentiable at $z \neq 0$ (Why?), i.e., x and y not both 0. Therefore, the functions

$$u(x,y) = \frac{x}{x^2+y^2} \quad \text{and} \quad v(x,y) = \frac{-y}{x^2+y^2}$$

must satisfy the Cauchy-Riemann equations. To check this, we compute

$$u_x = \frac{y^2-x^2}{\left(x^2+y^2\right)^2}, \qquad u_y = \frac{-2xy}{\left(x^2+y^2\right)^2}$$

and

$$v_x = \frac{2xy}{\left(x^2+y^2\right)^2}, \qquad v_y = \frac{y^2-x^2}{\left(x^2+y^2\right)^2}$$

to see that the Cauchy-Riemann equations are indeed satisfied for $z \neq 0$. ◀

► **6.99** *Show that $f : \mathbb{C} \setminus \{0\} \to \mathbb{C}$ given by $f(z) = \frac{1}{z}$ is differentiable as suggested in the previous example.*

The Cauchy-Riemann equations by themselves are not enough to ensure differentiability of a function at a point.[16] However, the Cauchy-Riemann equations along with a continuity condition are enough to do the trick. We now establish a set of *sufficient* conditions for a complex function $f(z) = u(x,y) + i\,v(x,y)$ to be differentiable at $z_0 \in \text{Int}(D)$. Note in the statement

[16] Our results focus on the question of differentiability at a point. If instead we were to consider differentiability on an open set, it can be shown that the Cauchy-Riemann equations are sufficient to ensure differentiability for all points of the set.

of this result that the sufficient conditions that the function $f : D \to \mathbb{C}$ must satisfy in order to be differentiable at a point $z_0 \in \text{Int}(D)$ are exactly those of Theorem 3.10 on page 279 supplemented by the Cauchy-Riemann equations.

Theorem 4.12 (The Cauchy-Riemann Equations (Sufficiency))
Suppose $f : D \to \mathbb{C}$ is given by $f(z) = u(x,y) + i\,v(x,y)$, and consider $z_0 = x_0 + i\,y_0 \in \text{Int}(D)$. If the first-order partial derivatives u_x, u_y, v_x, and v_y exist in a neighborhood of (x_0, y_0) and are continuous at (x_0, y_0), and if the Cauchy-Riemann equations given by

$$u_x = v_y, \quad u_y = -v_x,$$

hold at (x_0, y_0), then the derivative of f exists at z_0 and is given by

$$f'(z_0) = u_x(x_0, y_0) + i\,v_x(x_0, y_0).$$

PROOF Suppose the hypotheses of the theorem hold. It follows from Theorem 3.10 that $u(x,y)$ and $v(x,y)$ are differentiable at (x_0, y_0). Therefore,

$$\begin{aligned}\sqrt{(x-x_0)^2 + (y-y_0)^2} < \delta, \;\Rightarrow \\ \big|u(x,y) - u(x_0,y_0) - u_x(x_0,y_0)(x-x_0) - u_y(x_0,y_0)(y-y_0)\big| \\ \le \tfrac{\epsilon}{2}\sqrt{(x-x_0)^2 + (y-y_0)^2},\end{aligned} \tag{6.40}$$

and

$$\begin{aligned}\big|v(x,y) - v(x_0,y_0) - v_x(x_0,y_0)(x-x_0) - v_y(x_0,y_0)(y-y_0)\big| \\ \le \tfrac{\epsilon}{2}\sqrt{(x-x_0)^2 + (y-y_0)^2}.\end{aligned} \tag{6.41}$$

Consider now

$$\begin{aligned}&\Big|f(z) - f(z_0) - [u_x(x_0,y_0) + i\,v_x(x_0,y_0)]\,(z-z_0)\Big| \\ &= \Big|u(x,y) + i\,v(x,y) - [u(x_0,y_0) + i\,v(x_0,y_0)] \\ &\qquad - [u_x(x_0,y_0) + i\,v_x(x_0,y_0)]\,[(x-x_0) + i\,(y-y_0)]\Big| \\ &= \Big|u(x,y) - u(x_0,y_0) - u_x(x_0,y_0)(x-x_0) + v_x(x_0,y_0)(y-y_0) \\ &\qquad + i\,[v(x,y) - v(x_0,y_0) - v_x(x_0,y_0)(x-x_0) - u_x(x_0,y_0)(y-y_0)]\Big| \\ &\le \Big|u(x,y) - u(x_0,y_0) - u_x(x_0,y_0)(x-x_0) + v_x(x_0,y_0)(y-y_0)\Big| \\ &\qquad + \Big|v(x,y) - v(x_0,y_0) - v_x(x_0,y_0)(x-x_0) - u_x(x_0,y_0)(y-y_0)\Big|.\end{aligned}$$

To complete the proof, we use the Cauchy-Riemann equations at (x_0, y_0) and inequalities (6.40) and (6.41) to obtain for $|z - z_0| < \delta$,

$$\begin{aligned}&\Big|f(z) - f(z_0) - [u_x(x_0,y_0) + i\,v_x(x_0,y_0)]\,(z-z_0)\Big| \\ &\qquad \le \tfrac{\epsilon}{2}\sqrt{(x-x_0)^2 + (y-y_0)^2} + \tfrac{\epsilon}{2}\sqrt{(x-x_0)^2 + (y-y_0)^2} \\ &\qquad = \epsilon\,|z - z_0|.\end{aligned}$$

♦

It is worth reiterating that for complex functions $f : D \to \mathbb{C}$ given by $f(z) = u(x, y) + i\,v(x, y)$, in order for the derivative f' to exist at a point $z_0 = x_0 + i\,y_0$, it is *not enough* that u_x, u_y, v_x, and v_y all exist in a neighborhood of the point and be continuous at the point. Although this was a sufficient set of conditions for the existence of the derivative of a *real*-valued function $f(x, y) = (u(x, y), v(x, y))$ at the point $(x_0, y_0) \in \mathbb{R}^2$, it is *not enough* in the otherwise analogous complex function case. A complex function must satisfy a bit more in order for the derivative to exist. As indicated by the previous two theorems, *satisfaction of the Cauchy-Riemann equations is a definitive property of any differentiable complex function.*

Example 4.13 *Suppose $f(z) = \sinh z$. Recall from Chapter 4 that*

$$u(x, y) = \tfrac{1}{2}\left(e^x - e^{-x}\right)\cos y \;\; \textit{and} \;\; v(x, y) = \tfrac{1}{2}\left(e^x + e^{-x}\right)\sin y.$$

From this we easily obtain

$$u_x = \tfrac{1}{2}\left(e^x + e^{-x}\right)\cos y, \qquad u_y = -\tfrac{1}{2}\left(e^x - e^{-x}\right)\sin y,$$

and

$$v_x = \tfrac{1}{2}\left(e^x - e^{-x}\right)\sin y, \qquad v_y = \tfrac{1}{2}\left(e^x + e^{-x}\right)\cos y,$$

and so it follows that the Cauchy-Riemann equations hold at every point of the complex plane. Moreover, the partial derivatives are clearly continuous for all x and y, and hence are continuous in the entire complex plane as well. Hence, $f(z) = \sinh z$ is differentiable everywhere, and

$$f'(z) = \tfrac{1}{2}\left(e^x + e^{-x}\right)\cos y + i\,\tfrac{1}{2}\left(e^x - e^{-x}\right)\sin y = \cosh z.$$

Notice that the derivative of $\sinh z$ is $\cosh z$. The reader will recall that the derivative of $\sinh x$ is $\cosh x$ in the real function case, and so our complex function theory has retained another well-known derivative identity. ◄

Example 4.14 *Let $f : D \to \mathbb{C}$ be differentiable on the open and connected set $D \subset \mathbb{C}$, such that $|f(z)| = c$ on D for some $c \in \mathbb{R}$. Then f must be constant on D. To see this, first note that if $|f(z)| = 0$ on D, then $f(z) = 0$ and we are done; so assume*

$$|f(z)| = \sqrt{u^2 + v^2} = c \neq 0 \;\text{ on } D.$$

Then, by differentiation, we have

$$2uu_x + 2vv_x = 0, \tag{6.42}$$

$$2uu_y + 2vv_y = 0. \tag{6.43}$$

Applying the Cauchy-Riemann equations to (6.42) and (6.43) above, we obtain

$$(a) \quad uv_y + vv_x = 0,$$
$$(b) \quad -uv_x + vv_y = 0.$$

Since $u^2 + v^2 \neq 0$, it follows that either $u \neq 0$ or $v \neq 0$. If $u \neq 0$ then (a) yields $v_y = -\frac{v}{u}v_x$, which can be subbed into (b) to obtain $v_x = 0$. This in turn yields $v_y = u_x = u_y = 0$. If $v \neq 0$ then (a) yields $v_x = -\frac{u}{v}v_y$, which can be subbed into

(b) to obtain $v_y = 0$. This in turn yields $v_x = u_x = u_y = 0$. From this we may conclude that each of u and v are constant, and so therefore is $f = u + iv$. ◀

Before leaving this section, we give an example of a function for which the Cauchy-Riemann equations apply at a point, and yet that lacks a derivative at that point. In fact, the function is not differentiable anywhere in its domain.

Example 4.15 *Let $f : \mathbb{C} \to \mathbb{C}$ be given by $f(z) = u(x, y) + i\, v(x, y)$, where*

$$u(x,y) = \begin{cases} \frac{x^3 - 3xy^2}{x^2+y^2} & \text{for } (x,y) \neq (0,0) \\ 0 & \text{for } (x,y) = (0,0) \end{cases},$$

and

$$v(x,y) = \begin{cases} \frac{y^3 - 3x^2 y}{x^2+y^2} & \text{for } (x,y) \neq (0,0) \\ 0 & \text{for } (x,y) = (0,0) \end{cases}.$$

We leave it to the reader to show that the Cauchy-Riemann equations only hold at $(0, 0)$, yet despite this, the function f is still not differentiable there. ◀

► **6.100** *As claimed in the previous example, show that the Cauchy-Riemann equations only hold at $(0, 0)$ for the function f described there. Also show why the function is not differentiable anywhere in its domain, even at $(0, 0)$, despite the fact that the Cauchy-Riemann equations hold there.*

4.4 The z and $\overline{z}$ Derivatives

In this subsection we consider an interesting and convenient way of characterizing differentiable complex functions. In this section we consider functions $f : D \to \mathbb{C}$ with D open where $f(z) = u(x, y) + iv(x, y)$ is such that u and v have continuous x and y partial derivatives. Then it is not unreasonable to define $\frac{\partial f}{\partial x}$ and $\frac{\partial f}{\partial y}$ according to

$$\frac{\partial f}{\partial x} \equiv \frac{\partial u}{\partial x} + i \frac{\partial v}{\partial x} \quad \text{and} \quad \frac{\partial f}{\partial y} \equiv \frac{\partial u}{\partial y} + i \frac{\partial v}{\partial y}.$$

We now define new operators $\frac{\partial}{\partial z}$ and $\frac{\partial}{\partial \overline{z}}$ in terms of these operators $\frac{\partial}{\partial x}$ and $\frac{\partial}{\partial y}$ on f.

Definition 4.16 (The z and $\overline{z}$ Derivatives)
Suppose $f : D \to \mathbb{C}$ is given by $f(z) = u(x, y) + iv(x, y)$. We define $\frac{\partial f}{\partial z}$ and $\frac{\partial f}{\partial \overline{z}}$ according to

$$\frac{\partial f}{\partial z} \equiv \frac{1}{2}\left(\frac{\partial f}{\partial x} - i \frac{\partial f}{\partial y}\right) \quad \text{and} \quad \frac{\partial f}{\partial \overline{z}} \equiv \frac{1}{2}\left(\frac{\partial f}{\partial x} + i \frac{\partial f}{\partial y}\right).$$

Initially, the definition of these operators might seem arbitrary. We now elaborate on two motivations behind it, one of them rigorous and the other less so. We'll begin with the nonrigorous, symbolic motivation. Note that

$$x = \tfrac{1}{2}(z + \overline{z}) \quad \text{and} \quad y = \tfrac{1}{2i}(z - \overline{z}). \tag{6.44}$$

This implies that any $f(z) = u(x, y) + i\,v(x, y)$ given in terms of the independent variables x and y can be written in terms of z and $\overline{z}$ by simply substituting via the formulas in (6.44). However, it must be remembered that z and $\overline{z}$ are *not* actually independent, since knowing the value of z uniquely determines that of $\overline{z}$. Yet, by a symbolic but nonrigorous application of the chain rule to the complex function $f(x, y)$, we obtain

$$\begin{aligned} \frac{\partial f}{\partial z} &= \frac{\partial f}{\partial x}\frac{\partial x}{\partial z} + \frac{\partial f}{\partial y}\frac{\partial y}{\partial z} \\ &= \frac{\partial f}{\partial x}\frac{1}{2} + \frac{\partial f}{\partial y}\frac{1}{2i} \qquad \text{by (6.44)} \\ &= \frac{1}{2}\left(\frac{\partial f}{\partial x} - i\,\frac{\partial f}{\partial y}\right). \end{aligned}$$

A motivation for $\frac{\partial f}{\partial \overline{z}}$ can be obtained similarly. Yet, it is important to emphasize that Definition 4.16 is valid even without the above justification. Definitions, of course, do not need to be proved. Applying Definition 4.16 to a complex function $f(z)$, then, is a perfectly valid mathematical calculation. In fact, there is no need to interpret the operators $\frac{\partial}{\partial z}$ and $\frac{\partial}{\partial \overline{z}}$ as partial derivatives (which they are not), and there is no need to interpret the variables z and $\overline{z}$ as independent (which they are not). The great convenience of the method described in this section lies in a very happy coincidence associated with the function $f(z)$ and the operators $\frac{\partial}{\partial z}$ and $\frac{\partial}{\partial \overline{z}}$. That coincidence is at the heart of the more rigorous motivation for using these operators, which we present shortly. But if these operators are going to be intuitively useful, they should at least provide what one would expect when applied to certain functions. In particular, the values of $\frac{\partial f}{\partial z}$ and $\frac{\partial f}{\partial \overline{z}}$ for certain simple functions should be what one expects. The reader is urged to verify, in fact, that

$$\frac{\partial}{\partial z} z = 1, \quad \frac{\partial}{\partial \overline{z}} z = 0, \quad \frac{\partial}{\partial z}\overline{z} = 0, \text{ and } \frac{\partial}{\partial \overline{z}}\overline{z} = 1.$$

► **6.101** *Verify the above.*

► **6.102** *Show that* $\frac{\partial}{\partial z}(z^2\overline{z}) = 2z\overline{z}$ *and that* $\frac{\partial}{\partial \overline{z}}(z^2\overline{z}) = z^2$.

For the real motivation behind Definition 4.16, and its true significance, consider $f(x, y) = u(x, y) + i\,v(x, y)$ on $D \subset \mathbb{C}$. Then

$$\begin{aligned} \frac{\partial}{\partial \overline{z}} f &= \frac{1}{2}\left(\frac{\partial f}{\partial x} + i\,\frac{\partial f}{\partial y}\right) \\ &= \tfrac{1}{2}\left(u_x + i\,v_x\right) + i\,\tfrac{1}{2}\left(u_y + i\,v_y\right) \\ &= \tfrac{1}{2}\left(u_x - v_y\right) + i\,\tfrac{1}{2}\left(v_x + u_y\right). \end{aligned}$$

From this we see that $\frac{\partial}{\partial \overline{z}} f = 0$ on $D \subset \mathbb{C}$ *if and only if* the Cauchy-Riemann equations hold on $D \subset \mathbb{C}$. By Theorem 4.12 on page 303, it follows that f is differentiable on D. We have just proved the following result.

Theorem 4.17 *Let $f : D \to \mathbb{C}$ be given by $f(z) = u(x, y) + i\, v(x, y)$. Then f is differentiable on D if and only if $\frac{\partial}{\partial \overline{z}} f = 0$ on D.*

This is a very useful theorem. It provides an extremely simple way to determine, for certain functions $f : D \to \mathbb{C}$, whether or not f is differentiable without necessarily computing the derivative itself or applying its definition.

Example 4.18 *Consider the complex function given by $f(z) = \overline{z}$. We have already seen that the $\overline{z}$-derivative of this function is 1, and so is never zero. By the above theorem, this function cannot be differentiable anywhere in the complex plane.* ◀

The above theorem is certainly a good justification for our new operator $\frac{\partial}{\partial \overline{z}}$. We now establish an equally valid justification for $\frac{\partial}{\partial z}$. We will show that for a differentiable complex function f on $D \subset \mathbb{C}$, the z derivative and f' are one and the same, i.e., $\frac{\partial f}{\partial z} = f'$. To show this, consider that for such a function we have

$$\begin{aligned}
\frac{\partial}{\partial z} f = \frac{1}{2}\left(\frac{\partial f}{\partial x} - i\,\frac{\partial f}{\partial y}\right) &= \tfrac{1}{2}(u_x + i\, v_x) - i\,\tfrac{1}{2}(u_y + i\, v_y) \\
&= \tfrac{1}{2}(u_x + v_y) + i\,\tfrac{1}{2}(v_x - u_y) \\
&= \tfrac{1}{2}\left(2u_x + i\, 2v_x\right) \text{ by the Cauchy-Riemann equations} \\
&= u_x + i\, v_x = f'.
\end{aligned}$$

We have just established the following theorem.

Theorem 4.19 *If $f : D \to \mathbb{C}$ is a differentiable function on D, then $f' = \frac{\partial}{\partial z} f$ on D.*

Note that for an arbitrary function $f : D \to \mathbb{C}$, one may calculate $\frac{\partial f}{\partial z}$ even if f' fails to exist, that is, even if f is not differentiable. The function $f(z) = \overline{z}$ is such a function.

Of course, to apply our new operators efficiently we need to develop some simple rules for their manipulation. In this way, a calculus for manipulating $\frac{\partial}{\partial z}$ and $\frac{\partial}{\partial \overline{z}}$ is born. While we will not establish all the details of this development here, it can be shown that our new operators satisfy many of the convenient properties of other differential operators.[17] We leave the details to the reader.

► **6.103** *Show that the operators $\frac{\partial}{\partial x}$, $\frac{\partial}{\partial y}$, $\frac{\partial}{\partial z}$, and $\frac{\partial}{\partial \overline{z}}$ are linear operators. That is, for constant $c \in \mathbb{C}$ and for $f, g : D \to \mathbb{C}$ complex functions, show that $\frac{\partial}{\partial x}(c\,f) = c\,\frac{\partial}{\partial x} f$ and $\frac{\partial}{\partial x}(f \pm g) = \frac{\partial}{\partial x} f \pm \frac{\partial}{\partial x} g$. Likewise for $\frac{\partial}{\partial y}$, $\frac{\partial}{\partial z}$, and $\frac{\partial}{\partial \overline{z}}$.*

► **6.104** *The product rule. For $f, g : D \to \mathbb{C}$, show that $\frac{\partial}{\partial x}(f\,g) = f\,\frac{\partial}{\partial x} g + g\,\frac{\partial}{\partial x} f$. Do likewise for $\frac{\partial}{\partial y}$, $\frac{\partial}{\partial z}$, and $\frac{\partial}{\partial \overline{z}}$*

► **6.105** *Is there a quotient rule for $\frac{\partial}{\partial x}$, $\frac{\partial}{\partial y}$, $\frac{\partial}{\partial z}$, and $\frac{\partial}{\partial \overline{z}}$? (Hint: Consider $\frac{1}{f}$ instead of f.)*

► **6.106** *Use induction to show that $\frac{\partial}{\partial z}(\overline{z}^n) = 0$ and $\frac{\partial}{\partial \overline{z}}(z^n) = 0$, and that $\frac{\partial}{\partial z}(z^n) = nz^{n-1}$ and $\frac{\partial}{\partial \overline{z}}(\overline{z}^n) = n\overline{z}^{n-1}$.*

► **6.107** *Show that $\frac{\partial}{\partial z}(z^2\overline{z}) = 2z\overline{z}$ and that $\frac{\partial}{\partial \overline{z}}(z^2\overline{z}) = z^2$ more efficiently than by use of the definition for $\frac{\partial}{\partial z}$ and $\frac{\partial}{\partial \overline{z}}$.*

► **6.108** *Show that $\frac{\partial f}{\partial \overline{z}} = \overline{\left(\frac{\partial \overline{f}}{\partial z}\right)}$, and that $\frac{\partial \overline{f}}{\partial \overline{z}} = \overline{\left(\frac{\partial f}{\partial z}\right)}$.*

► **6.109** *The chain rule. Let w and h be two complex functions as described in this section. Show that $\frac{\partial}{\partial z}(h \circ w) = \frac{\partial h}{\partial w}\frac{\partial w}{\partial z} + \frac{\partial h}{\partial \overline{w}}\frac{\partial \overline{w}}{\partial z}$, and $\frac{\partial}{\partial \overline{z}}(h \circ w) = \frac{\partial h}{\partial w}\frac{\partial w}{\partial \overline{z}} + \frac{\partial h}{\partial \overline{w}}\frac{\partial \overline{w}}{\partial \overline{z}}$.*

► **6.110** *Show that any polynomial in z is differentiable everywhere in $\mathbb{C}$.*

► **6.111** *Let f and g be two complex functions as described in this section. Then $\frac{\partial}{\partial z} g\big(f(z)\big) = g'\big(f(z)\big)\frac{\partial f}{\partial z}$ and $\frac{\partial}{\partial \overline{z}} g\big(f(z)\big) = g'\big(f(z)\big)\frac{\partial f}{\partial \overline{z}}$.*

► **6.112** *Find the z and $\overline{z}$ derivatives of the following functions.*

a) $k \in \mathbb{C}$ *b)* $z\overline{z}$ *c)* $\frac{1}{2}(z + \overline{z})$ *d)* $|z|$ *e)* $e^{|z|}$

► **6.113** *Show that e^z is differentiable on all of $\mathbb{C}$ by using the definition of $\frac{\partial}{\partial \overline{z}}$ to compute $\frac{\partial}{\partial \overline{z}} e^z = 0$. Now do it via the chain rule. Similarly, show that $e^{\overline{z}}$ is not differentiable at any point in $\mathbb{C}$.*

► **6.114** *Show that $\frac{\partial}{\partial \overline{z}} \sin z = 0$ and that $\frac{\partial}{\partial z} \sin z = \cos z$.*

► **6.115** *Show that $\frac{\partial}{\partial \overline{z}} \sin|z| = \big(\cos|z|\big)\frac{z}{2|z|}$.*

► **6.116** *Show that a first-degree Taylor polynomial with remainder can be written for f as $f(z) = f(z_0) + \frac{\partial f}{\partial z}(z_0)\,(z - z_0) + \frac{\partial f}{\partial \overline{z}}(z_0)\,\overline{(z - z_0)} + R_1(|z - z_0|^2)$, where $R_1(|z - z_0|^2) \to 0$ as $z \to z_0$ in such a way that $\frac{R_1}{|z - z_0|^2} \to C$ as $z \to z_0$ for some constant $C \in \mathbb{C}$.*

[17]It is worth emphasizing again that the operators $\frac{\partial}{\partial z}$ and $\frac{\partial}{\partial \overline{z}}$ are *not* partial derivative operators. To be so, the function f would have to justifiably be interpretable as depending on z and $\overline{z}$ independently. Clearly this is not the case, since specifying z completely specifies $\overline{z}$. One cannot, then, think of $\frac{\partial f}{\partial \overline{z}}$ as holding z fixed and varying $\overline{z}$ to see how f varies. These operators are, in fact, called *derivations*, a sort of generalized derivative operator that only needs to satisfy the product rule, namely, that $\frac{\partial}{\partial \overline{z}}(fg) = f\frac{\partial g}{\partial \overline{z}} + g\frac{\partial f}{\partial \overline{z}}$. This alone, it turns out, provides much of what a true derivative operator provides, but not all.

► **6.117** *Consider the function given by $f(z) = (x^2 + y^2) + i\,(x - y)$. Find the z and $\overline{z}$ derivatives both by Definition 4.16 and formally through the associated $F(z, \overline{z})$. Does $f'(z) = \frac{\partial f}{\partial z}$ in this case?*

► **6.118** *For each function in the exercise just prior to Definition 4.16, find (formally) $\partial F/\partial z$ and $\partial F/\partial \overline{z}$ from the F found in each case by treating z and $\overline{z}$ as independent variables. Then find the derivatives by writing each function as a function of x and y and using Definition 4.16.*

► **6.119** *Show that $\frac{\partial f}{\partial x}$ and $\frac{\partial f}{\partial y}$ can be written in terms of $\frac{\partial f}{\partial z}$ and $\frac{\partial f}{\partial \overline{z}}$ as*

$$\frac{\partial f}{\partial x} = \left(\frac{\partial f}{\partial z} + \frac{\partial f}{\partial \overline{z}}\right) \quad \text{and} \quad \frac{\partial f}{\partial y} = i\left(\frac{\partial f}{\partial z} - \frac{\partial f}{\partial \overline{z}}\right).$$

► **6.120** *For each function $f = u + i\,v$ given below, verify that u and v are $C^1(D)$, and that the partial derivative of f with respect to $\overline{z}$ is zero. Conclude from this that each f is a complex-differentiable function on the specified domain. Also, find the derivative f' in each case.*

a) $f(z) = z^3$, *for all* z *b)* $f(z) = e^{-z}$, *for all* z *c)* $f(z) = 1/z$, *for* $z \neq 0$

► **6.121** *Determine whether the following functions are differentiable on all or part of the complex plane.*

a) $f(z) = 4xy + i\,2(y^2 - x^2)$ *b)* $f(z) = e^x + i\,e^y$ *c)* $f(z) = z^2 + |z|^2$

5 THE INVERSE AND IMPLICIT FUNCTION THEOREMS

For a given function, we would like to know under what circumstances the function is guaranteed to have an inverse. Clearly, in the simple case where $f : D^1 \to \mathbb{R}$ is linear with nonzero slope, say, $f(x) = mx + b$ for $m, b \in \mathbb{R}$, the function will be invertible on all of $\mathbb{R}$, since $y = mx + b$ and simple algebra yields

$$x = \frac{1}{m}(y - b) \equiv f^{-1}(y).$$

Indeed, the inverse is itself a linear function, and $(f^{-1})'(y) = \frac{1}{f'(x)}$. But the simplicity reflected in this special case is not to be expected in general. What can we expect for a nonlinear function f? And what of functions on higher-dimensional spaces? The reader might recall from a first-year calculus course that any function $f : D^1 \to \mathbb{R}$ that is continuously differentiable on D^1 and whose derivative does not vanish there is one-to-one on D^1, and therefore has a continuously differentiable inverse there as well. Also, if $f(x) = y$, the derivative of the inverse at y is given by $(f^{-1})'(y) = \frac{1}{f'(x)}$. The key here is that such a function is adequately approximated by its *linear* Taylor polynomial near x, and so the nice result we obtained for linear functions will apply for nonlinear functions too, at least *locally*, that is, in some neighborhood of x. In this section we prove a generalization of this result to higher-dimensional spaces. In fact, a vector-valued function $\boldsymbol{f} : D^k \to \mathbb{R}^p$ with a derivative at $\mathbf{x}_0 \in D^k$ can be similarly approximated locally near $\mathbf{x}_0$ by a matrix (a linear transformation). If this matrix is invertible, i.e., if the determinant of the derivative at $\mathbf{x}_0$ is nonzero, the function will have a local inverse near $\mathbf{x}_0$. We begin by developing some necessary machinery.

5.1 Some Technical Necessities

Matrix Norms Revisited

While the matrix norm defined earlier in this chapter was convenient for its initial purpose, we shall use a different norm here, again for convenience.

Definition 5.1 If A is a $p \times k$ matrix, we define the **sup norm** of A by

$$\|A\| \equiv \sup_{\mathbf{x}\neq\mathbf{0}} \frac{|A\mathbf{x}|}{|\mathbf{x}|}.$$

We leave it to the reader to prove the following.

Proposition 5.2 *Let A and B be $k \times k$ matrices, with I the $k \times k$ identity matrix and O the $k \times k$ zero matrix. Then*

a) $\|A\| \geq 0$ *with equality if and only if* $A = O$,

b) $\|cA\| = |c|\,\|A\|$ *for* $c \in \mathbb{R}$,

c) $\|A + B\| \leq \|A\| + \|B\|$,

d) $\|AB\| \leq \|A\|\,\|B\|$,

e) $\|I\| = 1$.

▶ **6.122** *Prove the above.*

▶ **6.123** *Show that the defining condition given above for $\|A\|$, namely, that $\|A\| = \sup_{\mathbf{x}\neq\mathbf{0}} \frac{|A\mathbf{x}|}{|\mathbf{x}|}$, is equivalent to $\max_{|\mathbf{x}|=1} |A\mathbf{x}|$. (Hint: Use the fact that $|A\mathbf{x}|$ is a continuous function on the compact set $\{\mathbf{x} \in \mathbb{R}^k : |\mathbf{x}| = 1\}$, and therefore a maximum exists.)*

Example 5.3 *In this example we will find $\|A\|$ for $A = \begin{bmatrix} 1 & 0 \\ 0 & 2 \end{bmatrix}$. Note that if $|\mathbf{x}| = 1$, then*

$$|A\mathbf{x}| = \left| \begin{bmatrix} 1 & 0 \\ 0 & 2 \end{bmatrix} \begin{bmatrix} x_1 \\ x_2 \end{bmatrix} \right| = \left| \begin{bmatrix} x_1 \\ 2x_2 \end{bmatrix} \right| = \sqrt{x_1^2 + 4x_2^2} = \sqrt{1 + 3x_2^2},$$

and since $|x_2| \leq 1$, we see that $\max_{|\mathbf{x}|=1} |A\mathbf{x}| = 2$. Using the result of the previous exercise, we have that $\|A\| = 2$. ◀

▶ **6.124** *Show that $\|A\| = \sqrt{2}$ for $A = \begin{bmatrix} 1 & -1 \\ 1 & 1 \end{bmatrix}$.*

▶ **6.125** *Show that for $a, b \in \mathbb{R}$ if $\boldsymbol{A} = \begin{bmatrix} a & 0 \\ 0 & b \end{bmatrix}$ then $\|\boldsymbol{A}\| = \max\big(|a|, |b|\big)$.*

We now establish that the two matrix norms we have introduced (recall the matrix norm used in Chapter 6) are in fact equivalent. That is, convergence in one implies convergence in the other.

Proposition 5.4 *Let $\boldsymbol{A} = [a_{ij}]$ be a $p \times k$ matrix and recall that $|\boldsymbol{A}| = \max_{i,j} |a_{ij}|$ where the maximum is taken over all $1 \le i \le p$ and $1 \le j \le k$. Then*

$$|\boldsymbol{A}| \le \|\boldsymbol{A}\| \le k\sqrt{p}\,|\boldsymbol{A}|.$$

PROOF Note that for fixed j such that $1 \le j \le k$, we have

$$\|\boldsymbol{A}\| \ge \big|\boldsymbol{A}\mathbf{e_j}\big| = \left|\begin{bmatrix} a_{ij} \\ \vdots \\ a_{pj} \end{bmatrix}\right| \ge \big|a_{ij}\big| \quad \text{for } 1 \le i \le p.$$

Therefore $\|\boldsymbol{A}\| \ge |\boldsymbol{A}|$, and the left side of the double inequality is established. We will now establish the right side. To do so, note that

$$\begin{aligned} |\boldsymbol{A}\mathbf{x}|^2 &= \left|\begin{bmatrix} a_{11}x_1 + \cdots + a_{1k}x_k \\ \vdots \\ a_{p1}x_1 + \cdots + a_{pk}x_k \end{bmatrix}\right|^2 \\ &= (a_{11}x_1 + \cdots + a_{1k}x_k)^2 + \cdots + (a_{p1}x_1 + \cdots + a_{pk}x_k)^2 \\ &\le \big(|a_{11}||x_1| + \cdots + |a_{1k}||x_k|\big)^2 + \cdots + \big(|a_{p1}||x_1| + \cdots + |a_{pk}||x_k|\big)^2 \\ &\le (k|\boldsymbol{A}|\,|\mathbf{x}|)^2 + \cdots + (k|\boldsymbol{A}|\,|\mathbf{x}|)^2 \\ &= pk^2|\boldsymbol{A}|^2\,|\mathbf{x}|^2, \end{aligned}$$

which implies the right side of the result. ♦

Once we have the notion of a norm on the set of all $p \times k$ matrices we can define topological properties as well as convergence properties. For example, if $\{\boldsymbol{A_n}\}$ is a sequence of $p \times k$ matrices then we say $\lim \boldsymbol{A_n} = \boldsymbol{A}$ exists if

(1) $\boldsymbol{A}$ is a $p \times k$ matrix,

(2) For all $\epsilon > 0$ there exists $N \in \mathbb{N}$ such that $n > N \;\Rightarrow\; |\boldsymbol{A_n} - \boldsymbol{A}| < \epsilon$.

Proposition 5.4 implies that the convergence of the sequence $\{\boldsymbol{A_n}\}$ is independent of the two norms $|\cdot|$ and $\|\cdot\|$.

▶ **6.126** *Show that the convergence referred to above is independent of the two norms $|\cdot|$ and $\|\cdot\|$.*

▶ **6.127** *Can you define what it means for $\sum_{j=1}^{\infty} \boldsymbol{A_j}$ to converge where the $\boldsymbol{A_j}$ are $p \times k$ matrices?*

▶ **6.128** *Suppose the norm of A is less than 1. Guess at $\sum_{j=1}^{\infty} A^j$. Can you verify whether your guess is correct?*

The following result shows that the set of invertible matrices forms an open set, in that if A is invertible and B is "close enough" to A then B is invertible too.

Proposition 5.5 *Suppose A^{-1} exists and that $\|B - A\| < \frac{1}{\|A^{-1}\|}$. Then B^{-1} exists.*

PROOF Suppose B^{-1} does not exist. We will establish a contradiction. Since B is singular, there exists a vector $\mathbf{v} \neq \mathbf{0}$ such that $B\mathbf{v} = \mathbf{0}$. Then

$$-\mathbf{v} = \mathbf{0} - \mathbf{v} = A^{-1}B\mathbf{v} - \mathbf{v} = A^{-1}(B - A)\mathbf{v}.$$

From this we obtain

$$|\mathbf{v}| = |-\mathbf{v}| = \left|A^{-1}(B - A)\mathbf{v}\right| \leq \|A^{-1}\|\,\|B - A\|\,|\mathbf{v}| < |\mathbf{v}|.$$

Since $\mathbf{v} \neq \mathbf{0}$, this is a contradiction, and therefore B^{-1} exists. ♦

A Fixed Point Theorem

The following result is known as a fixed point theorem for reasons that will be readily apparent. It will be useful in what follows.

Theorem 5.6 *Let $f : V^k \to V^k$ be continuous on the closed set V^k, and suppose there exists $c \in \mathbb{R}$ such that $0 < c < 1$ and*

$$|f(\mathbf{x}) - f(\mathbf{y})| \leq c\,|\mathbf{x} - \mathbf{y}| \quad \textit{for all} \quad \mathbf{x}, \mathbf{y} \in V^k.$$

Then there exists a unique point $\boldsymbol{\xi} \in V^k$ such that $f(\boldsymbol{\xi}) = \boldsymbol{\xi}$.

The point $\boldsymbol{\xi}$ in the above theorem is called a *fixed point* of the mapping f on V^k since f maps $\boldsymbol{\xi}$ to itself.

PROOF Fix $\boldsymbol{\xi}_0 \in V^k$ and define $\boldsymbol{\xi}_{n+1} \equiv f(\boldsymbol{\xi}_n)$ for $n \geq 0$. Since

$$|\boldsymbol{\xi}_{n+1} - \boldsymbol{\xi}_n| = |f(\boldsymbol{\xi}_n) - f(\boldsymbol{\xi}_{n-1})| \leq c\,|\boldsymbol{\xi}_n - \boldsymbol{\xi}_{n-1}| \quad \text{for } n \geq 1,$$

it follows from Exercise 3.35 on page 106 in Chapter 3 that $\{\boldsymbol{\xi}_n\}$ is a Cauchy sequence and therefore there exists $\boldsymbol{\xi}$ such that $\lim \boldsymbol{\xi}_n = \boldsymbol{\xi}$. Since $\{\boldsymbol{\xi}_n\} \subset V^k$ and V^k is closed, it follows that $\boldsymbol{\xi} \in V^k$. Since f is continuous on V^k we have

$$f(\boldsymbol{\xi}) = f(\lim \boldsymbol{\xi}_n) = \lim f(\boldsymbol{\xi}_n) = \lim \boldsymbol{\xi}_{n+1} = \boldsymbol{\xi}.$$

To establish uniqueness, suppose there are two fixed points, $\boldsymbol{\xi}, \boldsymbol{\eta} \in V^k$. Then

$$|\boldsymbol{\xi} - \boldsymbol{\eta}| = |f(\boldsymbol{\xi}) - f(\boldsymbol{\eta})| \leq c\,|\boldsymbol{\xi} - \boldsymbol{\eta}|.$$

This implies $|\boldsymbol{\xi} - \boldsymbol{\eta}| = 0$, i.e., $\boldsymbol{\xi} = \boldsymbol{\eta}$. ♦

5.2 The Inverse Function Theorem

With all of the necessary technical results now at our disposal, we may prove, in stages, the inverse function theorem. We begin with the statement of the theorem.

Theorem 5.7 (The Inverse Function Theorem)
Let $\boldsymbol{f} : D^k \to \mathbb{R}^k$ be C^1 and suppose $\boldsymbol{f}'(\mathbf{a})$ is invertible for some $\mathbf{a} \in D^k$. Then there exist open sets $U \subset D^k$ and $V \subset \mathbb{R}^k$ with $\mathbf{a} \in U$ such that $\boldsymbol{f}$ maps U one-to-one and onto V. Moreover, the function $\boldsymbol{g} : V \to U$ given by $\boldsymbol{g} = \boldsymbol{f}^{-1}$ is differentiable at each $\mathbf{y}_0 \in V$ with $\boldsymbol{g}'(\mathbf{y}_0) = \left[\boldsymbol{f}'(\mathbf{x}_0)\right]^{-1}$ where $\mathbf{x}_0 = \boldsymbol{g}(\mathbf{y}_0)$.

The theorem is surprisingly difficult to prove. To do so we will need a series of five lemmas, which we now establish. The first one gives a convenient means, in a different but equivalent manner to that given in Chapter 6, for characterizing when a function is continuously differentiable at a point.

Lemma 5.8 *Consider $\boldsymbol{f} : D^k \to \mathbb{R}^k$ where $\boldsymbol{f}(\mathbf{x}) \equiv \big(f_1(\mathbf{x}), f_2(\mathbf{x}), \ldots, f_k(\mathbf{x})\big)$ for $f_j : D^k \to \mathbb{R}$. Then $\boldsymbol{f}'$ is continuous at $\mathbf{a} \in D^k$ if and only if for any $\epsilon > 0$ there exists a $\delta > 0$ such that*

$$|\mathbf{x} - \mathbf{a}| < \delta \;\Rightarrow\; \|\boldsymbol{f}'(\mathbf{x}) - \boldsymbol{f}'(\mathbf{a})\| < \epsilon.$$

PROOF Given any $\epsilon > 0$, suppose $\|\boldsymbol{f}'(\mathbf{x}) - \boldsymbol{f}'(\mathbf{a})\| < \epsilon$ whenever $|\mathbf{x} - \mathbf{a}| < \delta$. Then we have for any $1 \le i, j \le k$ that

$$|\mathbf{x} - \mathbf{a}| < \delta \;\Rightarrow\; \left|\frac{\partial f_i}{\partial x_j}(\mathbf{x}) - \frac{\partial f_i}{\partial x_j}(\mathbf{a})\right| \le \|\boldsymbol{f}'(\mathbf{x}) - \boldsymbol{f}'(\mathbf{a})\| < \epsilon,$$

which implies that all partial derivatives $\frac{\partial f_i}{\partial x_j}$ are continuous at $\mathbf{a}$, and hence that $\boldsymbol{f}'$ is continuous at $\mathbf{a}$. Now suppose $\boldsymbol{f}$ is C^1 at $\mathbf{a} \in D^k$. Then all partial derivatives $\frac{\partial f_i}{\partial x_j}$ are continuous at $\mathbf{a}$. That is, for any $\epsilon > 0$ there exists a single $\delta > 0$ (Why?) such that for any $1 \le i, j \le k$ we have

$$|\mathbf{x} - \mathbf{a}| < \delta \;\Rightarrow\; \left|\frac{\partial f_i}{\partial x_j}(\mathbf{x}) - \frac{\partial f_i}{\partial x_j}(\mathbf{a})\right| < \frac{\epsilon}{k^{3/2}}.$$

By Proposition 5.4 on page 311 we have

$$|\mathbf{x} - \mathbf{a}| < \delta \;\Rightarrow\; \|\boldsymbol{f}'(\mathbf{x}) - \boldsymbol{f}'(\mathbf{a})\| \le k^{3/2} \max_{i,j}\left[\left|\frac{\partial f_i}{\partial x_j}(\mathbf{x}) - \frac{\partial f_i}{\partial x_j}(\mathbf{a})\right|\right] < \epsilon,$$

and the result is proved. ♦

The second lemma establishes a certain topological property about the image $\mathbf{b} = \boldsymbol{f}(\mathbf{a})$ when $\boldsymbol{f}'(\mathbf{a})$ is an invertible matrix.

Lemma 5.9 *Let $\boldsymbol{f} : D^k \to \mathbb{R}^k$ be C^1 and suppose $\mathbf{a} \in D^k$ with $\boldsymbol{f}(\mathbf{a}) = \mathbf{b}$ and $\boldsymbol{f}'(\mathbf{a})$ invertible. Then $\mathbf{b}$ is an interior point to $\boldsymbol{f}\big(N_r(\mathbf{a})\big)$ for every neighborhood $N_r(\mathbf{a}) \subset D^k$.*

PROOF Let $\boldsymbol{f}'(\mathbf{a}) = \boldsymbol{A}$ and let $\lambda = \frac{1}{4}\|\boldsymbol{A}^{-1}\|^{-1}$. Then there exists $\delta > 0$ such that

$$\mathbf{x} \in N_\delta(\mathbf{a}) \ \Rightarrow\ |\boldsymbol{f}(\mathbf{x}) - \boldsymbol{f}(\mathbf{a}) - \boldsymbol{A}(\mathbf{x} - \mathbf{a})| \le \lambda|\mathbf{x} - \mathbf{a}| \quad \text{and} \quad \|\boldsymbol{f}'(\mathbf{x}) - \boldsymbol{A}\| < \lambda.$$

The first part of above conclusion follows from the differentiability of $\boldsymbol{f}$ at a. The second part of the conclusion follows from Lemma 5.8. Note too that by Proposition 5.5 on page 312 the function $\boldsymbol{f}'$ is invertible on all of $N_\delta(\mathbf{a})$. Given the above, we will show that if $r \le \delta$ then $N_{r\lambda}(\mathbf{b}) \subset \boldsymbol{f}\big(N_r(\mathbf{a})\big)$. To see this, let $\mathbf{y_0} \in N_{r\lambda}(\mathbf{b})$, i.e., $|\mathbf{y_0} - \mathbf{b}| < r\lambda$, and let $\boldsymbol{G} : D^k \to \mathbb{R}^k$ be given by

$$\begin{aligned} \boldsymbol{G}(\mathbf{x}) &= \boldsymbol{A}^{-1}(\mathbf{y_0} - \mathbf{b}) - \boldsymbol{A}^{-1}\big(\boldsymbol{f}(\mathbf{x}) - \boldsymbol{f}(\mathbf{a}) - \boldsymbol{A}(\mathbf{x} - \mathbf{a})\big) + \mathbf{a} \\ &= \boldsymbol{A}^{-1}\big(\mathbf{y_0} - \boldsymbol{f}(\mathbf{x})\big) + \mathbf{x}. \end{aligned} \tag{6.45}$$

It follows that for $\mathbf{x} \in \overline{N_{r/2}(\mathbf{a})}$ we have

$$\begin{aligned} |\boldsymbol{G}(\mathbf{x}) - \mathbf{a}| &\le \|\boldsymbol{A}^{-1}\|\,|\mathbf{y_0} - \mathbf{b}| + \|\boldsymbol{A}^{-1}\|\,|\boldsymbol{f}(\mathbf{x}) - \boldsymbol{f}(\mathbf{a}) - \boldsymbol{A}(\mathbf{x} - \mathbf{a})| \\ &< \|\boldsymbol{A}^{-1}\|\,r\lambda + \|\boldsymbol{A}^{-1}\|\,\lambda\,|\mathbf{x} - \mathbf{a}| \\ &< \tfrac{1}{4}r + \tfrac{1}{4}\tfrac{1}{2}r < \tfrac{1}{2}r, \end{aligned}$$

that is, $\boldsymbol{G}\Big(\overline{N_{r/2}(\mathbf{a})}\Big) \subset \overline{N_{r/2}(\mathbf{a})}$. Now, from (6.45) we have $\boldsymbol{G}'(\mathbf{x}) = \boldsymbol{I} - \boldsymbol{A}^{-1}\boldsymbol{f}'(\mathbf{x})$, and so

$$\|\boldsymbol{G}'(\mathbf{x})\| = \|\boldsymbol{A}^{-1}(\boldsymbol{A} - \boldsymbol{f}'(\mathbf{x}))\| \le \|\boldsymbol{A}^{-1}\|\,\lambda = \tfrac{1}{4}.$$

By the mean value theorem for functions of several variables, this implies $|\boldsymbol{G}(\mathbf{x}) - \boldsymbol{G}(\mathbf{a})| \le \frac{1}{4}|\mathbf{x} - \mathbf{a}|$, and so $\boldsymbol{G}$ has a fixed point $\mathbf{x_0} \in \overline{N_{r/2}(\mathbf{a})}$. From (6.45), this fixed point has the property that

$$\mathbf{y_0} = \boldsymbol{f}(\mathbf{x_0}) \in \boldsymbol{f}\big(\overline{N_{r/2}(\mathbf{a})}\big) \subset \boldsymbol{f}\big(N_r(\mathbf{a})\big),$$

and the lemma is established if $r \le \delta$. Finally, if $r > \delta$ we have $N_{\delta\lambda}(\mathbf{b}) \subset \boldsymbol{f}\big(N_\delta(\mathbf{a})\big)$ by the argument just established, and clearly $\boldsymbol{f}\big(N_\delta(\mathbf{a})\big) \subset \boldsymbol{f}\big(N_r(\mathbf{a})\big)$. Overall then we have shown that $N_{\delta\lambda}(\mathbf{b}) \subset \boldsymbol{f}\big(N_r(\mathbf{a})\big)$ and the lemma is proved. ♦

The third lemma gives sufficient conditions under which $\boldsymbol{f} : D^k \to \mathbb{R}^k$ is one-to-one at and near $\mathbf{a} \in D^k$.

Lemma 5.10 *Let $\boldsymbol{f} : D^k \to \mathbb{R}^k$ be C^1 and suppose $\boldsymbol{f}'(\mathbf{a})$ is invertible for some $\mathbf{a} \in D^k$. Then there exists $\delta > 0$ such that $\boldsymbol{f}$ is one-to-one on $N_\delta(\mathbf{a})$.*

PROOF Choose $\delta > 0$ as in the proof of Lemma 5.9, and suppose $\boldsymbol{f}(\mathbf{x}_1) = \boldsymbol{f}(\mathbf{x}_2) = \mathbf{y}_0$ for $\mathbf{x}_1, \mathbf{x}_2 \in N_\delta(\mathbf{a})$. Once again let $\boldsymbol{G} : D^k \to \mathbb{R}^k$ be given by

$$\boldsymbol{G}(\mathbf{x}) = \boldsymbol{A}^{-1}(\mathbf{y}_0 - \mathbf{b}) - \boldsymbol{A}^{-1}(\boldsymbol{f}(\mathbf{x}) - \boldsymbol{f}(\mathbf{a}) - \boldsymbol{A}(\mathbf{x} - \mathbf{a})) + \mathbf{a}.$$

Then for $\mathbf{x} \in N_\delta(\mathbf{a})$ we have that $\|\boldsymbol{G}'(\mathbf{x})\| \le \frac{1}{4}$, which implies

$$|\boldsymbol{G}(\mathbf{x}_2) - \boldsymbol{G}(\mathbf{x}_1)| \le \tfrac{1}{4} |\mathbf{x}_2 - \mathbf{x}_1| . \tag{6.46}$$

But we also have that $\boldsymbol{G}(\mathbf{x_j}) = \boldsymbol{A}^{-1}\big(\mathbf{y}_0 - \boldsymbol{f}(\mathbf{x_j})\big) + \mathbf{x_j}$ for $j = 1, 2$. Subbing this into (6.46) obtains

$$|\boldsymbol{G}(\mathbf{x}_2) - \boldsymbol{G}(\mathbf{x}_1)| = |\mathbf{x}_2 - \mathbf{x}_1| \le \tfrac{1}{4} |\mathbf{x}_2 - \mathbf{x}_1|.$$

This implies $\mathbf{x}_1 = \mathbf{x}_2$, and therefore $\boldsymbol{f}$ is one-to-one on $N_\delta(\mathbf{a})$. ♦

For a C^1 function $\boldsymbol{f} : D^k \to \mathbb{R}^k$ with invertible $\boldsymbol{f}'$ at $\mathbf{a} \in D^k$, the fourth lemma establishes that $\boldsymbol{f}$ is an *open mapping* at and near $\mathbf{a}$; that is, it takes an open neighborhood of $\mathbf{a}$ to an open set in $\mathbb{R}^k$.

Lemma 5.11 *Let $\boldsymbol{f} : D^k \to \mathbb{R}^k$ be C^1 and suppose $\boldsymbol{f}'(\mathbf{a})$ is invertible for some $\mathbf{a} \in D^k$. Then there exists $\delta > 0$ such that $\boldsymbol{f}\big(N_\delta(\mathbf{a})\big)$ is open in $\mathbb{R}^k$.*

PROOF Again, choose $\delta > 0$ as in the proof of Lemma 5.9 above, and suppose $\mathbf{y}_0 \in \boldsymbol{f}\big(N_\delta(\mathbf{a})\big)$. Then $\mathbf{y}_0 = \boldsymbol{f}(\mathbf{x}_0)$ for some $\mathbf{x}_0 \in N_\delta(\mathbf{a})$. Now let $N_\rho(\mathbf{x}_0) \subset N_\delta(\mathbf{a})$. By Lemma 5.9 the point $\mathbf{y}_0$ is an interior point to $\boldsymbol{f}\big(N_\rho(\mathbf{x}_0)\big)$, so there exists $N_s(\mathbf{y}_0) \subset \boldsymbol{f}\big(N_\rho(\mathbf{x}_0)\big) \subset \boldsymbol{f}\big(N_\delta(\mathbf{a})\big)$, which shows that $\boldsymbol{f}\big(N_\delta(\mathbf{a})\big)$ is open. ♦

For a C^1 function $\boldsymbol{f} : D^k \to \mathbb{R}^p$ with invertible $\boldsymbol{f}'$ at $\mathbf{a} \in D^k$, our fifth and last lemma establishes the existence of a continuous inverse function for $\boldsymbol{f}$ on a neighborhood of $\boldsymbol{f}(\mathbf{a})$.

Lemma 5.12 *Let $\boldsymbol{f} : D^k \to \mathbb{R}^k$ be C^1 and suppose $\boldsymbol{f}'(\mathbf{a}) = \boldsymbol{A}$ is invertible for some $\mathbf{a} \in D^k$. Then there exists $\delta > 0$ such that for $U = N_\delta(\mathbf{a})$ and $V = \boldsymbol{f}\big(N_\delta(\mathbf{a})\big)$, the function $\boldsymbol{g} : V \to U$ given by $\boldsymbol{g} = \boldsymbol{f}^{-1}$ is continuous on V.*

PROOF Yet again choose $\delta > 0$ as in the proof of Lemma 5.9 above. Fix $\mathbf{y}_0 \in V$ and let $\mathbf{y}_0 = \boldsymbol{f}(\mathbf{x}_0)$ for $\mathbf{x}_0 \in U$. Define $\boldsymbol{h} : U \to \mathbb{R}^k$ by $\boldsymbol{h}(\mathbf{x}) = \mathbf{x} - \boldsymbol{A}^{-1}\boldsymbol{f}(\mathbf{x})$. Then $\boldsymbol{h}'(\mathbf{x}) = \boldsymbol{I} - \boldsymbol{A}^{-1}\boldsymbol{f}'(\mathbf{x}) = \boldsymbol{A}^{-1}\big(\boldsymbol{A} - \boldsymbol{f}'(\mathbf{x})\big)$, and so

$$\|\boldsymbol{h}'(\mathbf{x})\| \le \left\|\boldsymbol{A}^{-1}\right\| \, \|\boldsymbol{A} - \boldsymbol{f}'(\mathbf{x})\| \le \tfrac{1}{4},$$

as argued in the proof of Lemma 5.9. As was argued in the proof of Lemma 5.10 with the function $\boldsymbol{G}$ there, this is enough to imply

$$|\boldsymbol{h}(\mathbf{x}) - \boldsymbol{h}(\mathbf{x}_0)| \le \tfrac{1}{4}|\mathbf{x} - \mathbf{x}_0|.$$

But we also have

$$|h(\mathbf{x}) - h(\mathbf{x}_0)| = \left|(\mathbf{x} - \mathbf{x}_0) - A^{-1}(f(\mathbf{x}) - f(\mathbf{x}_0))\right|,$$

from which the reverse triangle inequality obtains

$$|\mathbf{x} - \mathbf{x}_0| \le \tfrac{4}{3}\,\|A^{-1}\|\,|f(\mathbf{x}) - f(\mathbf{x}_0)|,$$

and if we let $\mathbf{x} = g(\mathbf{y})$, we get

$$|g(\mathbf{y}) - g(\mathbf{y}_0)| \le \tfrac{4}{3}\,\|A^{-1}\|\,|\mathbf{y} - \mathbf{y}_0|. \tag{6.47}$$

This last expression implies that g is (Lipschitz) continuous. ♦

We are now ready to prove the inverse function theorem.

PROOF (of the inverse function theorem)
We assume that $f : D^k \to \mathbb{R}^k$ is C^1 and that $f'(\mathbf{a})$ is invertible for some $\mathbf{a} \in D^k$. Let $U = N_\delta(\mathbf{a})$ with $\delta > 0$ as chosen in the proofs of the previous lemmas, and let $V = f\big(N_\delta(\mathbf{a})\big)$. For any $\mathbf{y}_0 = f(\mathbf{x}_0) \in V$ we will show that $g : V \to U$ given by $g = f^{-1}$ is differentiable at $\mathbf{y}_0$ and that $g'(\mathbf{y}_0) = \big[f'(\mathbf{x}_0)\big]^{-1}$. To this end, let $\epsilon > 0$ and fix $\mathbf{y}_0 \in V$. Let $f'(\mathbf{x}_0) = A$. By Lemma 5.12 we have that $g : V \to U$ given by $g = f^{-1}$ is continuous on V. For $\mathbf{y} \in V$ with $g(\mathbf{y}) = \mathbf{x}$, let $\epsilon > 0$ be given. Then there exists $\delta_1 > 0$ such that if $|\mathbf{x} - \mathbf{x}_0| < \delta_1$ we have

$$\begin{aligned}
\left|g(\mathbf{y}) - g(\mathbf{y}_0) - A^{-1}(\mathbf{y} - \mathbf{y}_0)\right| &= \left|\mathbf{x} - \mathbf{x}_0 - A^{-1}(f(\mathbf{x}) - f(\mathbf{x}_0))\right| \\
&= \left|A^{-1}(f(\mathbf{x}) - f(\mathbf{x}_0) - A(\mathbf{x} - \mathbf{x}_0))\right| \\
&\le \|A^{-1}\|\,\frac{\frac{3}{4}\epsilon}{\|A^{-1}\|^2}\,|\mathbf{x} - \mathbf{x}_0| \\
&= \tfrac{3}{4}\epsilon\,\frac{1}{\|A^{-1}\|}\,|g(\mathbf{y}) - g(\mathbf{y}_0)| \\
&\le \epsilon\,|\mathbf{y} - \mathbf{y}_0|.
\end{aligned} \tag{6.48}$$

The first inequality in the above follows from the differentiability of f at $\mathbf{x}_0$, while the second follows from (6.47) in the proof of Lemma 5.12. Note that $|\mathbf{x} - \mathbf{x}_0| < \delta_1$ if and only if $|g(\mathbf{y}) - g(\mathbf{y}_0)| < \delta_1$, which will be true if

$$\tfrac{4}{3}\,\|A\|\,|\mathbf{y} - \mathbf{y}_0| < \delta_1,$$

i.e., if $|\mathbf{y} - \mathbf{y}_0| < \delta \equiv \frac{3}{4}\,\frac{1}{\|A\|}\,\delta_1$. We have shown that if $\mathbf{y}$ is within δ of $\mathbf{y}_0$, then (6.48) holds, and therefore g is differentiable at $\mathbf{y}_0 \in V$ with $g'(\mathbf{y}_0) = A^{-1} = \big[f'(\mathbf{x}_0)\big]^{-1}$. ♦

It will be convenient to have the differentiability of the inverse function of a complex differentiable function. Fortunately, this is a simple corollary of the result already proved.

Theorem 5.13 *Suppose $f : D \to \mathbb{C}$ is differentiable with $f'(a) \neq 0$. Then there exist open sets $U \subset D$, and $V \subset \mathbb{C}$ such that $a \in U$, and such that f maps U one-to-one and onto V. Moreover, the function $g : V \to U$ given by $g = f^{-1}$ is differentiable at each $w_0 \in V$ with $g'(w_0) = (f'(z_0))^{-1}$ where $z_0 = g(w_0)$.*

PROOF Writing $z_0 = x_0 + iy_0$ and $f = u + iv$, and noting that $f'(z_0) = u_x(x_0, y_0) + iv_x(x_0, y_0) \neq 0$, we have by the Cauchy-Riemann equations

$$u_x(x_0, y_0) = v_y(x_0, y_0) \neq 0 \quad \text{and} \quad u_y(x_0, y_0) = -v_x(x_0, y_0) \neq 0.$$

Now consider f as a function $\boldsymbol{f} : \mathbb{R}^2 \to \mathbb{R}^2$. Then the Jacobian matrix of $\boldsymbol{f}$ at (x_0, y_0) is

$$\boldsymbol{J}(x_0, y_0) = \begin{bmatrix} u_x(x_0, y_0) & u_y(x_0, y_0) \\ v_x(x_0, y_0) & v_y(x_0, y_0) \end{bmatrix},$$

whose determinant is $u_x^2(x_0, y_0) + u_y^2(x_0, y_0) > 0$, and so $\boldsymbol{J}(x_0, y_0)$ is invertible. We thus know, by the inverse function theorem, that there are open sets $U, V \subset \mathbb{R}^2$ such that $\boldsymbol{f}$ maps U onto V in a one-to-one correspondence and such that the inverse $\boldsymbol{g}$ is continuously differentiable *as a real function*. We need to show that it is also differentiable as a complex function, i.e., that it satisfies the Cauchy-Riemann equations. We know that $f'(z_0)$ is of the form

$$\begin{bmatrix} A & B \\ -B & A \end{bmatrix} \quad \text{with } A^2 + B^2 > 0,$$

and we must show that its inverse matrix, which is the derivative of f^{-1} at $f(z_0)$, is of that same form. This will show that the inverse function satisfies the Cauchy-Riemann equations. But this can be seen directly; the inverse is given by

$$\frac{1}{A^2 + B^2} \begin{bmatrix} A & -B \\ B & A \end{bmatrix},$$

and so clearly the inverse of $\boldsymbol{f}$ satisfies the Cauchy-Riemann equations. By the inverse function theorem the inverse is differentiable, and therefore continuous, and so A and B in the above matrix expression are continuous partial derivatives. Therefore, going back to the complex plane, the inverse g of f is complex differentiable with the appropriate derivative, and the proof is complete. ♦

Corollary 5.14 *Suppose $f : D \to \mathbb{C}$ is differentiable on the open set $D \subset \mathbb{C}$. Suppose that $f(D) = V$ is open and f is one-to-one, and suppose f' does not vanish. Then the inverse map $g = f^{-1} : V \to D$ is also differentiable.*

PROOF By the previous theorem, g is differentiable in a neighborhood of each point, and since g may be defined everywhere since f is a one-to-one correspondence, g is also differentiable everywhere. ♦

Some examples will illustrate the inverse function theorem's implications and limitations.

Example 5.15 *Let $f : \mathbb{R} \to \mathbb{R}$ be given by $f(x) = x^2$. Then if $a \neq 0$ we have $f'(a) \neq 0$ and therefore there exists a neighborhood $N_r(a)$ such that $f : N_r(a) \to V$ is one-to-one and onto, and $f^{-1} : V \to N_r(a)$ is differentiable on V. Moreover, $\left(f^{-1}\right)'(b) = \frac{1}{f'(a)}$ where $f(a) = b$. In fact, if $a > 0$, the neighborhood $N_r(a)$ can be replaced by the interval $(0, \infty)$. That is, an inverse function for f with the stated properties exists on all of $(0, \infty)$. Likewise, one can find an inverse for f with the stated properties on all of $(-\infty, 0)$. But one cannot find a single inverse for f with all the stated properties on all of $\mathbb{R}$.* ◀

► **6.129** *What are the inverse functions defined on $(0, \infty)$ and $(-\infty, 0)$ referred to in the above example?*

Example 5.16 *Maps that are "dimension reducing" are not invertible anywhere. To see an example of this, let $\boldsymbol{f} : \mathbb{R}^2 \to \mathbb{R}^2$ be given by $\boldsymbol{f}(x, y) = (\sec xy, \tan xy)$. We will show that $\boldsymbol{f}$ is not invertible in a neighborhood of any point. To see this, note that*

$$\boldsymbol{f}'(x, y) = \begin{bmatrix} y \sec xy \tan xy & x \sec xy \tan xy \\ y \sec^2 xy & x \sec^2 xy \end{bmatrix},$$

and $\det\left(\boldsymbol{f}'\right)(x, y) = xy \sec^3 xy \tan xy - xy \sec^3 xy \tan xy \equiv 0$ for every $(x, y) \in \mathbb{R}^2$. In this case, where $\boldsymbol{f}'$ is singular, the function $\boldsymbol{f}$ effectively maps the set $\mathbb{R}^2$ to a curve, a set of lower dimension. To see this, let $u = \sec xy$ and $v = \tan xy$. Then for all $(x, y) \in \mathbb{R}^2$ we have $\boldsymbol{f}(x, y) = (u, v)$ where $u^2 - v^2 = 1$. Therefore $\boldsymbol{f}$ takes all points in the plane to the points on a hyperbola in the plane. ◀

► **6.130** *Let $\boldsymbol{f} : \mathbb{R}^2 \to \mathbb{R}^2$ be given by $\boldsymbol{f}(x, y) = (x^2 + xy, x^2 + y^2) = (u, v)$.*

a) At which points in $\mathbb{R}^2$ can you find an inverse for this mapping?

b) Describe the set of points where no inverse to this mapping exists.

Example 5.17 *Let $f : \mathbb{C} \to \mathbb{C}$ given by $f(z) = e^z$. Then for every $a \in \mathbb{C}$, we have $f'(a) = e^a \neq 0$. Therefore there exists a neighborhood $N_r(a)$ such that $f : N_r(a) \to V$ is one-to-one and onto, and the inverse function $f^{-1} : V \to N_r(a)$ is differentiable on V. Note that f^{-1} is a logarithm of z defined on V. Moreover, $\left(f^{-1}\right)'(w) = \frac{1}{f'(a)} = \frac{1}{e^a} = \frac{1}{w}$ where $e^a = w$.* ◀

► **6.131** *For a given nonzero $a \in \mathbb{C}$, what is the inverse of f referred to in the previous example? What is its domain? What if a is replaced by $-a$? What if $a = 0$?*

5.3 The Implicit Function Theorem

Suppose we have an equation in two variables $F(x, y) = 0$, and that the pair $(x, y) = (a, b)$ is a solution. It is convenient to be able to solve the equation

$F(x,y)=0$ for y as a function of x, say $y=h(x)$, near (a,b) such that $F(x,y)=F(x,h(x))=0$. The implicit function theorem tells us that under the right conditions we can do this *locally*, even in higher dimensions. Before stating and proving the theorem, however, we introduce some helpful notation and make some preliminary comments.

Notation

Suppose $\mathbf{x}=(x_1,x_2,\ldots,x_k)\in\mathbb{R}^k$ and $\mathbf{y}=(y_1,y_2,\ldots,y_l)\in\mathbb{R}^l$. Then we write $(\mathbf{x},\mathbf{y})\equiv(x_1,x_2,\ldots,x_k,y_1,y_2,\ldots,y_l)\in\mathbb{R}^{k+l}$, and thus identify $\mathbb{R}^{k+l}$ with the *Cartesian product* $\mathbb{R}^k\times\mathbb{R}^l$. If $(\mathbf{x},\mathbf{y})\in\mathbb{R}^{k+l}$, we will assume that $\mathbf{x}\in\mathbb{R}^k$ and that $\mathbf{y}\in\mathbb{R}^l$ unless otherwise noted. For two sets $A\subset\mathbb{R}^k$ and $B\subset\mathbb{R}^l$, we write $A\times B\equiv\{(\mathbf{x},\mathbf{y}):\mathbf{x}\in A,\ \ \mathbf{y}\in B\}$.

▶ **6.132** *Show that* $(\mathbf{x}_1+\mathbf{x}_2,\mathbf{y})=(\mathbf{x}_1,\mathbf{y})+(\mathbf{x}_2,\mathbf{y})$. *Also, show that a similar identity holds for scalar products. If $A\times B=\varnothing$, then what are A and B? What if $A\times B=\mathbb{R}^{k+l}$?*

Finally, suppose we are given a differentiable $\boldsymbol{F}:D^{k+l}\to\mathbb{R}^l$, with $\boldsymbol{F}=(F_1,\ldots,F_l)$ and $F_i:D^{k+l}\to\mathbb{R}$ for $1\le i\le l$. The derivative $\boldsymbol{F}'(\mathbf{x},\mathbf{y})$ is given by the usual matrix:

$$\boldsymbol{F}'(\mathbf{x},\mathbf{y})=\begin{bmatrix}\frac{\partial F_1}{\partial x_1} & \cdots & \frac{\partial F_1}{\partial x_k} & \frac{\partial F_1}{\partial y_1} & \cdots & \frac{\partial F_1}{\partial y_l}\\ \frac{\partial F_2}{\partial x_1} & \cdots & \frac{\partial F_2}{\partial x_k} & \frac{\partial F_2}{\partial y_1} & \cdots & \frac{\partial F_2}{\partial y_l}\\ \vdots & \ddots & \vdots & \vdots & \ddots & \vdots\\ \frac{\partial F_l}{\partial x_1} & \cdots & \frac{\partial F_l}{\partial x_k} & \frac{\partial F_l}{\partial y_1} & \cdots & \frac{\partial F_l}{\partial y_l}\end{bmatrix}$$

(The only difference is that there are entries for both the x and y coordinates.) Define the matrix $\boldsymbol{F}_{\mathbf{x}}$ by

$$\boldsymbol{F}_{\mathbf{x}}\equiv\begin{bmatrix}\frac{\partial F_1}{\partial x_1} & \frac{\partial F_1}{\partial x_2} & \cdots & \frac{\partial F_1}{\partial x_k}\\ \frac{\partial F_2}{\partial x_1} & \frac{\partial F_2}{\partial x_2} & \cdots & \frac{\partial F_2}{\partial x_k}\\ \vdots & \vdots & \ddots & \vdots\\ \frac{\partial F_l}{\partial x_1} & \frac{\partial F_l}{\partial x_2} & \cdots & \frac{\partial F_l}{\partial x_k}\end{bmatrix},$$

and similarly for $\boldsymbol{F}_{\mathbf{y}}$. This notational convention allows us to write

$$\boldsymbol{F}'(\mathbf{x},\mathbf{y})=\begin{bmatrix}\boldsymbol{F}_{\mathbf{x}} & \boldsymbol{F}_{\mathbf{y}}\end{bmatrix}.$$

▶ **6.133** *Show that all the standard linearity properties hold for $\boldsymbol{F}_{\mathbf{x}}$ and $\boldsymbol{F}_{\mathbf{y}}$. What about a function $\boldsymbol{F}:D^{k+l+m}\to\mathbb{R}^l$ given by $\boldsymbol{F}(\mathbf{x},\mathbf{y},\mathbf{z})$ in three variables? Define what $(\mathbf{x},\mathbf{y},\mathbf{z})$ means and define the expressions $\boldsymbol{F}_{\mathbf{x}}$, $\boldsymbol{F}_{\mathbf{y}}$, and $\boldsymbol{F}_{\mathbf{z}}$.*

The Implicit Function Theorem

We now use the above-developed notation to state and prove the implicit function theorem, which itself is an application of the inverse function theorem. Our proof of the result is from [Rud76].

Theorem 5.18 (The Implicit Function Theorem)
Suppose $\boldsymbol{F} : D^{k+l} \to \mathbb{R}^l$ *is continuously differentiable on the open set* $D^{k+l} \subset \mathbb{R}^{k+l}$. *Suppose* $(\mathbf{a}, \mathbf{b}) \in D^{k+l}$ *satisfies* $\boldsymbol{F}(\mathbf{a}, \mathbf{b}) = \mathbf{0}$. *Assume, moreover, that the square matrix* $\boldsymbol{F}_{\mathbf{y}}(\mathbf{a}, \mathbf{b})$ *is invertible. Then there exists a neighborhood* $N_r(\mathbf{a})$ *of* $\mathbf{a}$ *and a continuously differentiable function* $\boldsymbol{h} : N_r(\mathbf{a}) \to \mathbb{R}^l$ *satisfying* $\boldsymbol{h}(\mathbf{a}) = \mathbf{b}$ *such that if* $\mathbf{x} \in N_r(\mathbf{a})$, *then* $\boldsymbol{F}(\mathbf{x}, \boldsymbol{h}(\mathbf{x})) = \mathbf{0}$. *Moreover, on* $N_r(\mathbf{a})$ *we have*

$$\boldsymbol{h}'(\mathbf{x}) = -\left[\boldsymbol{F}_{\mathbf{y}}(\mathbf{x}, \boldsymbol{h}(\mathbf{x}))\right]^{-1} \boldsymbol{F}_{\mathbf{x}}(\mathbf{x}, \boldsymbol{h}(\mathbf{x})).$$

PROOF Consider the function $\boldsymbol{G} : D^{k+l} \to \mathbb{R}^{k+l}$ given by $\boldsymbol{G}(\mathbf{x}, \mathbf{y}) = (\mathbf{x}, \boldsymbol{F}(\mathbf{x}, \mathbf{y}))$. Then we have $\boldsymbol{G}(\mathbf{a}, \mathbf{b}) = (\mathbf{a}, \boldsymbol{F}(\mathbf{a}, \mathbf{b})) = (\mathbf{a}, \mathbf{0})$. The definition of $\boldsymbol{G}$ makes it clear that if the output of $\boldsymbol{G}$ has the second coordinate zero, then the input represents a solution of $\boldsymbol{F}(\mathbf{x}, \mathbf{y}) = \mathbf{0}$. Hence it is for such points that we search. To do this, we will apply the inverse function theorem to $\boldsymbol{G}$. Note that

$$\boldsymbol{G}'(\mathbf{a}, \mathbf{b}) = \begin{bmatrix} \boldsymbol{I} & \boldsymbol{O} \\ \boldsymbol{F}_{\mathbf{x}}(\mathbf{a}, \mathbf{b}) & \boldsymbol{F}_{\mathbf{y}}(\mathbf{a}, \mathbf{b}) \end{bmatrix}, \quad \text{(Why?)}$$

with all partial derivatives evaluated at $(\mathbf{a}, \mathbf{b})$. On the right, we have written the matrix in what is often referred to as "block form," with $\boldsymbol{I}$ representing the $k \times k$ identity matrix and $\boldsymbol{O}$ representing the $l \times l$ zero matrix. We assumed that $\boldsymbol{F}_{\mathbf{y}}(\mathbf{a}, \mathbf{b})$ was invertible, and obviously $\boldsymbol{I}$ is invertible. We now use a fact from linear algebra, namely, if a matrix is given in block form

$$\boldsymbol{M} = \begin{bmatrix} \boldsymbol{A} & \boldsymbol{O} \\ \boldsymbol{B} & \boldsymbol{C} \end{bmatrix},$$

where $\boldsymbol{A}$ and $\boldsymbol{C}$ are square matrices, then $\det \boldsymbol{M} = \det \boldsymbol{A} \det \boldsymbol{C}$. In particular, $\boldsymbol{M}$ is invertible if and only if $\boldsymbol{A}$ and $\boldsymbol{C}$ are. This obtains that $\boldsymbol{G}'(\mathbf{a}, \mathbf{b})$ is invertible. Since $\boldsymbol{F}$ is continuously differentiable it follows that $\boldsymbol{G}$ is continuously differentiable, and therefore the inverse function theorem applies to $\boldsymbol{G}$. Let $U \subset D^{k+l}$ and $V \subset \mathbb{R}^{k+l}$ be open sets such that $(\mathbf{a}, \mathbf{b}) \in U$ and $(\mathbf{a}, \mathbf{0}) = \boldsymbol{F}(\mathbf{a}, \mathbf{b}) \in V$. Then $\boldsymbol{G} : U \to V$ has inverse function $\boldsymbol{G}^{-1} : V \to U$, which is differentiable on V. Since V is open, there exists $N_\rho(\mathbf{a}, \mathbf{0}) \subset V$. We leave it to the reader to show that there exists $N_{r_1}(\mathbf{a}) \subset \mathbb{R}^k$ and $N_{r_2}(\mathbf{0}) \subset \mathbb{R}^l$ such that $(\mathbf{a}, \mathbf{0}) \in N_{r_1}(\mathbf{a}) \times N_{r_2}(\mathbf{0}) \subset N_\rho(\mathbf{a}, \mathbf{0}) \subset V$. It follows that there exists a function $\boldsymbol{H} : V \to U$ differentiable on V such that $\boldsymbol{G}^{-1}(\mathbf{x}, \mathbf{y}) = (\mathbf{x}, \boldsymbol{H}(\mathbf{x}, \mathbf{y}))$. If we let $\boldsymbol{h} : N_{r_1}(\mathbf{a}) \to \mathbb{R}^l$ be given by the differentiable function $\boldsymbol{h}(\mathbf{x}) = \boldsymbol{H}(\mathbf{x}, \mathbf{0})$. Then the function $\boldsymbol{h}$ has the desired property that we seek, namely, for $\mathbf{x} \in N_{r_1}(\mathbf{a})$

we have

$$(\mathbf{x}, \mathbf{0}) = \boldsymbol{G}(\boldsymbol{G}^{-1}(\mathbf{x}, \mathbf{0})) = \boldsymbol{G}(\mathbf{x}, \boldsymbol{H}(\mathbf{x}, \mathbf{0})) = \boldsymbol{G}(\mathbf{x}, \boldsymbol{h}(\mathbf{x})) = \big(\mathbf{x}, \boldsymbol{F}(\mathbf{x}, \boldsymbol{h}(\mathbf{x}))\big).$$

This means that $\boldsymbol{F}(\mathbf{x}, \boldsymbol{h}(\mathbf{x})) = \mathbf{0}$ on $N_{r_1}(\mathbf{a})$. Finally, we leave it to the reader to verify that

$$\begin{aligned} \mathbf{0} &= \boldsymbol{F}'(\mathbf{x}, \boldsymbol{h}(\mathbf{x})) \begin{bmatrix} \mathbf{I} \\ \boldsymbol{h}'(\mathbf{x}) \end{bmatrix} \\ &= \begin{bmatrix} \boldsymbol{F}_{\mathbf{x}}(\mathbf{x}, \boldsymbol{h}(\mathbf{x})) & \boldsymbol{F}_{\mathbf{y}}(\mathbf{x}, \boldsymbol{h}(\mathbf{x})) \end{bmatrix} \begin{bmatrix} \mathbf{I} \\ \boldsymbol{h}'(\mathbf{x}) \end{bmatrix} \\ &= \boldsymbol{F}_{\mathbf{x}}(\mathbf{x}, \boldsymbol{h}(\mathbf{x})) + \boldsymbol{F}_{\mathbf{y}}(\mathbf{x}, \boldsymbol{h}(\mathbf{x}))\boldsymbol{h}'(\mathbf{x}). \end{aligned}$$

This implies that

$$\boldsymbol{h}'(\mathbf{x}) = -\big[\boldsymbol{F}_{\mathbf{y}}(\mathbf{x}, \boldsymbol{h}(\mathbf{x}))\big]^{-1}\boldsymbol{F}_{\mathbf{x}}(\mathbf{x}, \boldsymbol{h}(\mathbf{x})),$$

and the theorem is proved. ♦

► **6.134** *Answer the (Why?) question in the above proof. Also, given a neighborhood $N_\rho(\mathbf{x}_0, \mathbf{y}_0) \subset \mathbb{R}^{k+l}$, show that it contains a product of two neighborhoods $N_{r_1}(\mathbf{x}_0) \times N_{r_2}(\mathbf{y}_0)$.*

Example 5.19 *Consider the system of equations given by*

$$\begin{aligned} 3x + 6y + z^3 &= 0 \\ x + 2y + 4z &= 0. \end{aligned}$$

We will show that it is possible to solve this system of equations for x and z in terms of y. To do so, let $\boldsymbol{f} : \mathbb{R}^3 \to \mathbb{R}^2$ be given by $\boldsymbol{f}(x, y, z) = (3x + 6y + z^3, x + 2y + 4z)$. Then $\boldsymbol{f}'(x, y, z) = \begin{bmatrix} 3 & 3z^2 & 6 \\ 1 & 4 & 2 \end{bmatrix}$. Since $\boldsymbol{f}(0, 0, 0) = (0, 0)$, we examine $\boldsymbol{f}'(0, 0, 0) = \begin{bmatrix} 3 & 0 & 6 \\ 1 & 4 & 2 \end{bmatrix}$. Since $\begin{bmatrix} 3 & 0 \\ 1 & 4 \end{bmatrix}$ is invertible, we may solve for y in terms of x and z. ◄

► **6.135** *In the above example, can we solve for z in terms of x and y?*

6 Supplementary Exercises

1. *Consider the function $f : \mathbb{R} \to \mathbb{R}$ given by $f(x) = 2x^2 - 3$. Use the ϵ, δ version of derivative to show that $f'(1) \neq 3$. Hint: Use the method of proof by contradiction.*

2. *Let $f : \mathbb{R} \to \mathbb{R}$ be given by $f(x) = \begin{cases} x^2 & \text{for } x \text{ irrational} \\ 0 & \text{for } x \text{ rational} \end{cases}$. Find the points $x \in \mathbb{R}$ where f is differentiable.*

3. *Can you find a function $f : \mathbb{R} \to \mathbb{R}$ that is differentiable at exactly two points?*

4. *The function $f : \mathbb{R} \to \mathbb{R}$ given by $f(x) = |x|$ is differentiable everywhere in $\mathbb{R}$ except at $x = 0$. Can you find a function $f : \mathbb{R} \to \mathbb{R}$ that is differentiable everywhere except at exactly two points?*

5. *For a fixed positive integer N, show that the function $f : \mathbb{R} \to \mathbb{R}$ given by $f(x) = \sum_{j=1}^{N} |x - j|$ is differentiable everywhere in $\mathbb{R}$ except at $x = j$ for $j = 1, 2, \ldots, N$. What if each of the points $x = j$ is replaced by a more generic $x_j \in \mathbb{R}$?*

6. *Consider the function $f : \mathbb{R} \to \mathbb{R}$ given by $f(x) = \sum_{j=0}^{\infty} \frac{|x-j|}{j!}$. Show that f is differentiable everywhere in $\mathbb{R}$ except at $j = 0, 1, 2, \ldots$.*

7. *Recall from a previous exercise that $f : \mathbb{R} \to \mathbb{R}$ is called* Lipschitz continous *if there exists a nonnegative real number M independent of x and y where*

$$|f(x) - f(y)| \leq M\,|x - y| \text{ for all } x, y \in \mathbb{R}.$$

Is a function that is Lipschitz continuous on $\mathbb{R}$ necessarily differentiable on $\mathbb{R}$? Is a function that is differentiable on $\mathbb{R}$ necessarily Lipschitz continuous on $\mathbb{R}$?

8. *Suppose $f : \mathbb{R} \to \mathbb{R}$ is such that there exists a nonnegative real number M and a positive real number ϵ each independent of x and y where*

$$|f(x) - f(y)| \leq M\,|x - y|^{1+\epsilon} \text{ for all } x, y \in \mathbb{R}.$$

Show that f is differentiable on $\mathbb{R}$. Can you say anything else about f?

9. *Consider $f : D^1 \to \mathbb{R}$ and interior point $x_0 \in D^1$. Suppose that*

$$\lim_{\substack{x \to x_0 \\ x \in \mathbb{Q}}} \frac{f(x) - f(x_0)}{x - x_0}$$

exists. That is, for any sequence of rational x that converges to x_0, the limit of the difference quotient always gives the same finite value. Show by counterexample that despite this, $f'(x_0)$ does not necessarily exist.

10. *Suppose $f : D^1 \to \mathbb{R}$ is given by $f(x) = \tan x$ where $D^1 = \left\{ x \in \mathbb{R} : x \neq \frac{2n+1}{2}\,\pi \text{ for } n \in \mathbb{Z} \right\}$. Show that $f'(a) = \sec^2 a$ for all $a \in D^1$. Use the quotient rule to find the derivative for each of the functions $\sec x, \csc x$, and $\cot x$, remembering to specify each on its appropriate domain.*

11. *Consider the function $f(x) : (-1, 1) \to \mathbb{R}$ given by $f(x) = x^{2/3}$. What is the domain of $f'(x)$?*

12. *For the following functions, find the local extrema and determine their nature as in Example 1.15 on page 248.*

a) $f(x) = \frac{1}{4}x^4 - x^3 + \frac{3}{2}x^2 - x - 5, \quad x \in \mathbb{R}$ *b)* $f(x) = \frac{x}{x+1}, \quad x \in [1, 4]$

c) $f(x) = \frac{x^2-1}{x^2+1}, \quad x \in \mathbb{R}$ *d)* $f(x) = \frac{x}{\sqrt{2}} - \sin x, \quad x \in [0, \pi)$

13. *Consider the function $f : [-1, 1] \to \mathbb{R}$ given by $f(x) = |x|$. Find and characterize any extrema. Does this example contradict Corollary 1.14?*

14. *We used Rolle's theorem to prove the mean value theorem. See if you can use the mean value theorem to prove Rolle's theorem instead.*

15. *Why do you suppose Theorem 1.17 is called the mean value theorem?*

16. *Suppose $f : (a,b) \cup (c,d) \to \mathbb{R}$ is differentiable on the disjoint union $(a,b) \cup (c,d)$, and $f'(x) = 0$ for all $x \in (a,b) \cup (c,d)$. Is it still true that there exists a constant c such that $f(x) \equiv c$ for all $x \in (a,b) \cup (c,d)$? What if $c = b$?*

17. *As students of calculus are well aware, the second derivative test can be of great practical value in determining the nature of a critical point. If the critical point corresponds to an extrema of the function f, the second derivative test described below can determine whether it is a local minimum or a local maximum. Of course, not all critical points correspond to extrema, as the example of $x = 0$ for the function $f(x) = x^3$ illustrates. In this case, $x = 0$ corresponds to an inflection point, a point where the concavity of the curve associated with the graph of $f(x)$ in the xy-plane changes sign. At such points, the second derivative test is inconclusive. Use the mean value theorem to prove the following:*

Suppose $f : N_r(a) \to \mathbb{R}$ for some $r > 0$ is such that f'' exists and is continuous on $N_r(a)$, and $f'(a) = 0$.

a) If $f''(a) > 0$, then $f(a)$ is a local minimum.

b) If $f''(a) < 0$, then $f(a)$ is a local maximum.

c) If $f''(a) = 0$, the test is inconclusive.

18. *Prove the following: If $f : (a,b) \to \mathbb{R}$ is differentiable, then at each $x \in (a,b)$, the derivative f' is either continuous or has an essential discontinuity, i.e., f' has no removable or jump discontinuities.*

19. *Consider the function $f : \mathbb{R} \to \mathbb{R}$ given by $f(x) = \begin{cases} \frac{1}{24}\,x^4 & \text{for } x \geq 0 \\ -\frac{1}{24}\,x^4 & \text{for } x < 0 \end{cases}$.*

a) Find the highest possible degree Taylor polynomial at $x = 0$ having a remainder according to Theorem 1.22.

b) What is the expression for its remainder term?

c) Suppose you are interested in evaluating $f(x)$ at $x = -\frac{1}{2}$ via your Taylor approximation with remainder as described by your results from parts a) and b). How small might the remainder be? How large? Can you make it arbitrarily small?

d) Suppose you are interested in evaluating $f(x)$ at $x = \frac{1}{2}$ via your Taylor approximation with remainder as described by your results from parts a) and b). How small might the remainder be? How large? Can you make it arbitrarily small?

20. *Suppose f and g are differentiable on $N_r'(a) \subset \mathbb{R}$ for some $r > 0$ such that the following all hold:*

(i) $\lim_{x \to a} f(x) = \lim_{x \to a} g(x) = \pm\infty$
(ii) $g'(x) \neq 0$ for all $x \in N_r'(a)$
(iii) $\lim_{x \to a} \frac{f'(x)}{g'(x)} = L$ exists

Prove that $\lim_{x \to a} \frac{f(x)}{g(x)} = L$.

21. *For the following functions, find the largest* n *for which* $f \in C^n(D^1)$.

a) $f(x) = x^{-1}, \ D^1 = (0,2)$ b) $f(x) = |x|, \ D^1 = (-1,0)$
c) $f(x) = |x|, \ D^1 = (-1,1)$ d) $f(x) = \frac{x}{x^2+1}, \ D^1 = \mathbb{R}$
e) $f(x) = \sin\left(x^2\right), \ D^1 = \mathbb{R}$ f) $f(x) = \begin{cases} x^2/2 & \text{if } x \geq 0 \\ -x^2/2 & \text{if } x \leq 0 \end{cases}, \ D^1 = \mathbb{R}$

22. *Let* $f_n : \mathbb{R} \to \mathbb{R}$ *be given by* $f_n(x) = \begin{cases} x^n \sin\left(\frac{1}{x}\right) & x \neq 0 \\ 0 & x = 0 \end{cases}$.

a) *Show that* f_1 *is continuous, but not differentiable at* $x = 0$.
b) *What can you say about* f_2?
c) *What can you say about* f_n?

23. *Let* $f : [0, \infty) \to \mathbb{R}$ *be given by* $f(x) = \sqrt{x}$. *Find* $D_{f'}$.

24. *Let* $f : \mathbb{R} \to \mathbb{R}$ *be given by* $f(x) = |x|$. *Find* $D_{f'}$.

25. *Let* $f : \mathbb{R} \to \mathbb{R}$ *be given by* $f(x) = \begin{cases} x^2 & \text{if } x \leq 0 \\ x & \text{if } x > 0 \end{cases}$. *Find* $D_{f'}$ *and* $D_{f''}$.

26. *Discuss the critical point of the function* $f : [-1,1] \to \mathbb{R}$ *given by* $f(x) = x^{2/3}$. *Does the (interior) minimum happen at a point where* $f' = 0$? *Does this example contradict Corollary 1.14 on page 248?*

27. *Consider* $f : \mathbb{R} \to \mathbb{R}$ *given by* $f(x) = \begin{cases} x^2 \sin\left(\frac{1}{x}\right) & \text{if } x \neq 0 \\ 0 & \text{if } x = 0 \end{cases}$.

a) *Show that* f *is differentiable on* $\mathbb{R}$, *and that* $f'(0) = 0$. *Therefore* $x = 0$ *is a critical point.*

b) *Show that* $f'(x)$ *is discontinuous at* $x = 0$.

c) *Show that* $f'(x)$ *takes values that are greater than 0 and less than 0 in any neighborhood of* $x = 0$, *and hence* $x = 0$ *is not a local extremum.*

d) *Show that* $f''(0)$ *doesn't exist, and so* $x = 0$ *is not an inflection point.*

28. *Confirm that* $f : \mathbb{R} \to \mathbb{R}$ *given by* $f(x) = \begin{cases} x^4 \sin\left(\frac{1}{x}\right) & \text{if } x \neq 0 \\ 0 & \text{if } x = 0 \end{cases}$ *has the property that* $f'(0) = f''(0) = 0$, *but* $x = 0$ *is not an extremum or an inflection point.*

29. *Consider the function* $f : \mathbb{R} \to \mathbb{R}$ *given by* $f(x) = \begin{cases} \frac{x}{2} + x^2 \sin\left(\frac{1}{x}\right) & \text{for } x \neq 0 \\ 0 & \text{for } x = 0. \end{cases}$

a) *Show that* f *is differentiable in a neighborhood of* $x = 0$*, and that* $f'(0)$ *is positive.*

b) *Can you find an interval containing* $x = 0$ *throughout which* f *is increasing?*

c) *Find a* $\delta > 0$ *such that* $f(x) < 0$ *on* $(-\delta, 0)$ *and* $f(x) > 0$ *on* $(0, \delta)$.

30. *Consider* $f : \mathbb{R} \to \mathbb{R}$ *given by* $f(x) = \begin{cases} x^2 & \text{if } x \le 2 \\ a\,x + b & \text{if } x > 2 \end{cases}$ *where* a *and* b *are constants. Can you find values of* a *and* b *that make* f' *continuous on* $\mathbb{R}$?

31. *Suppose* $f : \mathbb{R} \to \mathbb{R}$ *has the following properties.*

a) f *is continuous at* $x = 0$,

b) $f(a + b) = f(a) + f(b)$ *for all* $a, b \in \mathbb{R}$,

c) $f'(0) = C$ *for some constant* C.

Show that $f(x) = Cx$ *for all* $x \in \mathbb{R}$.

32. *For* $f : \mathbb{R} \to \mathbb{R}$ *given by* $f(x) = \begin{cases} e^{-1/x} & \text{if } x > 0 \\ 0 & \text{if } x \le 0 \end{cases}$ *find* $f'(x)$ *and* $f''(x)$ *for all* $x \in \mathbb{R}$.

33. *Suppose* $f : (a, b) \to \mathbb{R}$ *is differentiable on* (a, b). *Find a nonconstant sequence* $\{x_n\} \subset (a, b)$ *satisfying both of the following (Hint: Use the mean value theorem.):*

a) $\lim x_n = c \in (a, b)$,

b) $\lim f'(x_n) = f'(c)$.

34. *Suppose* $f : \mathbb{R} \to \mathbb{R}$ *is differentiable at the point* x_0. *Compute*

$$\lim_{h \to 0} \frac{f(x_0 + \alpha\, h) - f(x_0 + \beta\, h)}{h}$$

where α *and* β *are constants.*

35. *Can you use the Cauchy mean value theorem on page 250 to prove the mean value theorem?*

36. *Consider* $f : \mathbb{R} \to \mathbb{R}$ *and suppose there exists some constant* $K \in \mathbb{R}$ *such that for all* x *and* y *in* $\mathbb{R}$, $|f(x) - f(y)| \le K\,(x - y)^2$. *Prove that there exists some constant* $c \in \mathbb{R}$ *such that* $f \equiv c$ *on* $\mathbb{R}$.

37. *In the previous exercise, what if the exponent 2 is replaced by* α? *For what values of* α *does the conclusion still hold?*

38. *Let* $a_1, \ldots, a_n$ *be a set of nonnegative real numbers. The* arithmetic-geometric mean inequality *states that the arithmetic mean is at least the geometric mean, i.e.,*

$$\frac{1}{n} \sum_{j=1}^{n} a_j \ge (a_1 \ldots a_n)^{1/n}.$$

The inequality is obvious if one of the a_js is zero, so assume all of them are nonzero.

a) *Let $c > 0$. If the inequality is true for $a_1, \ldots, a_n$, then it is true for $ca_1, \ldots, ca_n$. Thus show that it is enough to consider the case when $a_1 \ldots a_n = 1$, so the geometric mean is 1.*

b) *Let $x > 0$. Prove that $x - 1 \geq \log x$. Use the mean value theorem, considering the cases $x > 1$, $x < 1$, and $x = 1$ separately.*

c) *Now suppose $a_1 \ldots a_n = 1$. Use the previous inequality to get*

$$\sum_{j=1}^{n} a_j - n \geq 0, \text{ hence } \frac{1}{n}\sum_{j=1}^{n} a_j \geq 1.$$

This proves the inequality. This proof, due to Schaumberger, is from [Sch85].

39. *Consider $f : \mathbb{R} \to \mathbb{R}$ such that $\lim_{x\to\infty} f(x) = \infty$.*

a) *Can you find a function $h : \mathbb{R} \to \mathbb{R}$ such that $\lim_{x\to\infty} h(x) = \infty$, and for any integer $k \geq 1$ $\lim_{x\to\infty}\left[(f(x))^{1/k}(h(x))^{-1}\right] = \infty$?*

b) *Can you find a function $H : \mathbb{R} \to \mathbb{R}$ such that $\lim_{x\to\infty} H(x) = \infty$, and for any integer $k \geq 1$ $\lim_{x\to\infty}\left[H(x)(f(x))^{-k}\right] = \infty$?*

40. *Suppose $f : \mathbb{R} \to \mathbb{R}$ is C^2 on its whole domain. Show that for any $h > 0$ and any x there exists a point ξ between $x - h$ and $x + h$ such that*

$$\frac{f(x+h) - 2f(x) + f(x-h)}{h^2} = f''(\xi).$$

41. *Find the positive real values x for which $\sin x \geq x - \frac{x^3}{6}$.*

42. *Suppose $f : \mathbb{R} \to \mathbb{R}$ is continuously differentiable on $\mathbb{R}$, and assume $a < b$ such that*

(i) $f(b) \geq f(a) > 0$

(ii) *f and f' never have the same sign at any point in $[a, b]$.*

Show that there exists a constant c such that $f(x) = c$ on $[a, b]$.

43. *Suppose $f : \mathbb{R} \to \mathbb{R}$ is continuously differentiable on $\mathbb{R}$ and that f' is never zero at the roots of f. Show that any bounded interval $[a, b]$ contains no more than a finite number of roots of f.*

44. *Suppose f and $g : \mathbb{R} \to \mathbb{R}$ are differentiable on $\mathbb{R}$ such that $f\,g' - f'\,g \neq 0$ on $\mathbb{R}$. If r_1 and r_2 are consecutive roots of f, show there is a root of g between r_1 and r_2.*

45. *Suppose in the previous exercise that f' and g' are also continuous. Show that there is only one root of g between r_1 and r_2.*

46. *Suppose $f : \mathbb{R} \to \mathbb{R}$ is differentiable on $\mathbb{R}$, and assume $a < b$ such that $f'(a) < 0 < f'(b)$. Show that there exists a number c between a and b such that $f'(c) = 0$. (Note that f' need not be continuous!)*

47. *Determine whether the following limit exists: $\lim_{x\to 0} x\,e^{1/x}$.*

48. *Show that* $\frac{1}{20} \le \frac{x}{16+x^3} \le \frac{1}{12}$ *for* $1 \le x \le 4$.

49. *Suppose* $f : [a,b] \to \mathbb{R}$ *is continuous on* $[a,b]$ *and differentiable on* $(a,c) \cup (c,b)$ *for some* $c \in (a,b)$. *Also, suppose* $\lim_{x\to c} f'(x) = L$. *Use the mean value theorem to show that* $f'(c) = L$.

50. *Establish the following inequalities for nonnegative values of* x.
a) $e^{-x} \le 1$ *b)* $1 - x \le e^{-x}$ *c)* $e^{-x} \le 1 - x + \frac{x^2}{2}$
Determine a general rule consistent with the above, and prove it.

51. *Consider a differentiable function* $f : [0,\infty) \to \mathbb{R}$ *such that* $\lim_{x\to\infty} f(x) = M$ *and* $\lim_{x\to\infty} f'(x) = L$. *Can you use the mean value theorem to determine* L?

52. *Consider the function* $f : \mathbb{R}\setminus\{0\} \to \mathbb{R}$ *given by* $f(x) = \frac{x}{1+e^{1/x}}$. *Can you extend this function's domain to include* $x = 0$? *Is* f *differentiable there? (Hint: Use one-handed limits.)*

53. *Suppose* $f_n : [0,1] \to \mathbb{R}$ *is given by* $f_n(x) = \frac{1}{n} e^{-n^2x^2}$ *for* $n = 1, 2, \ldots$.
a) Show that the sequence of functions $\{f_n\}$ *converges uniformly on* $[0,1]$.
b) Is the same true for the sequence $\{f_n'\}$? *Why or why not?*

54. *Suppose* $f : \mathbb{R} \to \mathbb{R}$ *and* $g : \mathbb{R} \to \mathbb{R}$ *are* $C^1(-\infty,\infty)$, *and that* $g(x) > 0$ *for all* x. *For each real number* ξ *define the function* $W_\xi : \mathbb{R} \to \mathbb{R}$ *by*

$$W_\xi(x) = g(x)f'(x) - \xi\, f(x)g'(x).$$

If a *and* b *are consecutive zeros of the function* f, *show that for each real number* ξ *the function* W_ξ *has a zero between* a *and* b.

55. *Suppose* $f : \mathbb{R} \to \mathbb{R}$ *is differentiable on* $\mathbb{R}$, *and that* $f'(a) < 0 < f'(b)$ *for some* $a < b$. *Show that there exists* $c \in (a,b)$ *such that* $f'(c) = 0$ *even though* f' *is not necessarily continuous.*

56. *In a supplementary exercise from the previous chapter, we considered the function* $f : \mathbb{R} \to \mathbb{R}$ *given by* $f(x) = \sum_{j=0}^{\infty} \frac{|x-j|}{j!}$. *There, you were asked to show that this function is well defined, i.e., converges, for any* $x \in \mathbb{R}$, *and that* f *is continuous on* $\mathbb{R}$. *Show that this function is differentiable at any* $x \ne N \in \mathbb{Z}$. *To get started, what is the derivative of* f *at* $x \in (N, N+1)$ *for* $N \in \mathbb{Z}$? *Compare it to the derivative of* f *at* $x \in (N-1, N)$.

57. *Consider the function* $f : \mathbb{R} \setminus \{0\} \to \mathbb{R}$ *given by* $f(x) = (\cos x)^{1/x}$. *Can* f *be assigned a value at* $x = 0$ *so as to be continuous there?*

58. *A Continuous Nowhere Differentiable Function.*
Such "pathological" functions are actually more typical than originally realized. In fact, if $C[a,b]$ *is the space of continuous functions on the interval* $[a,b]$, *and if* $C_{ND}[a,b]$ *is the subset of continuous nowhere differentiable functions on* $[a,b]$, *one can show that* $C_{ND}[a,b]$ *is dense in* $C[a,b]$ *when* $C[a,b]$ *is viewed as a normed complete vector space with the sup norm. That is, every continuous function in* $C[a,b]$ *has a continuous nowhere differentiable function arbitrarily near to it. Also,* $C_{ND}[a,b]$ *functions*

are what is known as "prevalent" in $C[a,b]$, a measure-theoretic concept meaning roughly that "almost all" functions in $C[a,b]$ are, in fact, members of $C_{ND}[a,b]$. That is, they are to $C[a,b]$ what the irrationals are to $\mathbb{R}$! While there are many examples of functions that are continuous everywhere and differentiable nowhere on their domains, one of the earliest examples known was constructed by Bolzano in the 1830s. A more famous example was constructed by Weierstrass some 40 years later. The relatively recent example presented here is by McCarthy (published in the American Mathematical Monthly, Vol. LX, No.10, December 1953). It has the benefit of simplicity over its historical predecessors. Our presentation is inspired by [Thi03]. Define the function $f : \mathbb{R} \to \mathbb{R}$ according to

$$f(x) = \sum_{k=1}^{\infty} 2^{-k} \phi(2^{2^k} x),$$

where $\phi : \mathbb{R} \to \mathbb{R}$ is the "saw-tooth" function with period 4 given by

$$\phi(x) = \begin{cases} 1 + x & \text{for } x \in [-2,0] \\ 1 - x & \text{for } x \in [0,2] \end{cases},$$

and $\phi(x + 4) = \phi(x)$ for all $x \in \mathbb{R}$.

a) Use the Weierstrass M-test to see that f is continuous on $\mathbb{R}$.

b) To show that f is nowhere differentiable, fix an arbitrary $x \in \mathbb{R}$. We will construct a sequence $\{x_n\}$ of real numbers approaching x such that the limit $\lim\limits_{n\to\infty} \frac{|f(x_n)-f(x)|}{|x_n-x|}$ does not exist. To this end, let $x_n = x \pm 2^{-2^n}$ where the sign is selected so that x_n and x are on the same straight-line segment of the saw-tooth function ϕ. Show the following:

1. For $k \in \mathbb{N}$ such that $k > n$, $\phi(2^{2^k} x_n) - \phi(2^{2^k} x) = 0$.

2. For $k \in \mathbb{N}$ such that $k = n$, $\left|\phi(2^{2^k} x_n) - \phi(2^{2^k} x)\right| = 1$.

3. For $k \in \mathbb{N}$ such that $k < n$, $\sup\limits_{k\in\{1,2,\ldots,n-1\}} \left|\phi(2^{2^k} x_n) - \phi(2^{2^k} x)\right| \le 2^{-2^{n-1}}$.

c) Show that

$$\begin{aligned} \frac{|f(x_n) - f(x)|}{|x_n - x|} &= 2^{2^n} \left| \sum_{k=1}^{\infty} 2^{-k} \left[\phi(2^{2^k} x_n) - \phi(2^{2^k} x)\right]\right| \\ &\ge 2^{2^n} \left(1 - \left| \sum_{k=1}^{n-1} 2^{-k} \left[\phi(2^{2^k} x_n) - \phi(2^{2^k} x)\right]\right|\right) \\ &\ge 2^{2^n} \left(1 - (n-1) 2^{-2^{n-1}}\right) \\ &\ge 2^{2^n} \left(1 - 2^n 2^{-2^{n-1}}\right), \quad \text{which goes to } \infty \text{ as } n \to \infty. \end{aligned}$$

59. *For each of the following functions, use the ϵ, δ definition to verify that the derivative is as claimed.*

a) Consider the function $f : D^2 \to \mathbb{R}$ with $D^2 = \{(x,y) \in \mathbb{R}^2 : xy \ge 0\}$ and f given by $f(x,y) = \sqrt{2xy}$. At the point $(1,1)$, show that f has the derivative $\left[\frac{1}{\sqrt{2}}\ \frac{1}{\sqrt{2}}\right]$.

b) Consider the function $f : D^2 \to \mathbb{R}$ with $D^2 = \{(x,y) \in \mathbb{R}^2 : xy \ne 0\}$ and f given by $f(x,y) = (xy)^{-1}$. At the point $(1,1)$, show that f has the derivative $[-1\ \ -1]$.

c) Consider the function $f : D^3 \to \mathbb{R}$ with $D^3 = \{(x,y,z) \in \mathbb{R}^3 : x + y + z \ne 0\}$

and f given by $f(x,y) = \frac{x}{x+y+z}$. At the point $(0,0,1)$, show that f has derivative given by $[1\ \ 0\ \ 0]$.

60. *Can you construct a function $f : \mathbb{R}^2 \to \mathbb{R}$ that is differentiable at exactly one point? Exactly two points?*

61. *Can you construct a function $f : \mathbb{R}^2 \to \mathbb{R}$ that is differentiable everywhere on its domain except at exactly one point? Exactly two points?*

62. *Show that the function $f : \mathbb{R}^2 \to \mathbb{R}$ given by $f(x,y) = 4x^2 + xy + y^2 + 12$ in Example 2.16 on page 271 is, in fact, differentiable on its whole domain.*

63. *For the following functions, find all extrema and determine their nature.*

a) $f(x,y) = x^3/3 - x^2y + xy^2 + y^2/2, \quad (x,y) \in \mathbb{R}^2$

b) $f(x,y) = \sin(xy), \quad (x,y) \in \mathbb{R}^2$

64. *Just as in the case of $f : D^1 \to \mathbb{R}$, a critical point for a function $f : D^2 \to \mathbb{R}$ might not correspond to a local extremum. Such a point in the higher-dimensional case might be what is called a* saddle point, *a higher-dimensional analog of the one-dimensional inflection point. A* saddle point *corresponding to a function $f : D^2 \to \mathbb{R}$ can be defined as a critical point that is neither a local maximum nor a local minimum. Along at least one direction of approach to the saddle point, the behavior is that of a local minimum, while along at least one other direction the behavior is that of a local maximum. A simple example of such a saddle point is the origin corresponding to the graph of the function $f : \mathbb{R}^2 \to \mathbb{R}$ given by $f(x,y) = y^2 - x^2$. For $f : \mathbb{R}^2 \to \mathbb{R}$ given by $f(x,y) = y^2 - x^2$, find a direction along which f attains what appears to be a local minimum as you approach $(0,0)$. Similarly, find a direction along which this function attains what appears to be a local maximum as you approach $(0,0)$.*

65. *As we have already seen in the case of $f : D^1 \to \mathbb{R}$, determining the type of extremum associated with a given critical point is often made easier by applying a second derivative test. We now describe such a test for the special case of functions $f : D^2 \to \mathbb{R}$. Implementation of the test in this case involves a function $\mathfrak{D} : D^2 \to \mathbb{R}$ called the* discriminant of f, *a function not referred to in the case of $f : D^1 \to \mathbb{R}$. In particular, we define $\mathfrak{D}(\mathbf{x})$ to be the determinant of the two-by-two array of second-order partial derivatives of f evaluated at $\mathbf{x}$, i.e.,*

$$\mathfrak{D}(\mathbf{x}) \equiv \begin{vmatrix} f_{xx}(\mathbf{x}) & f_{xy}(\mathbf{x}) \\ f_{yx}(\mathbf{x}) & f_{yy}(\mathbf{x}) \end{vmatrix} = f_{xx}(\mathbf{x})f_{yy}(\mathbf{x}) - f_{xy}(\mathbf{x})f_{yx}(\mathbf{x}).$$

Our test is then stated as follows:

Consider $f : D^2 \to \mathbb{R}$ and suppose $\mathbf{a}$ is an interior point of D^2 such that $\boldsymbol{f}'(\mathbf{a}) = \mathbf{0}$. Suppose also that f_{xx}, f_{xy}, f_{yx}, and f_{yy} exist and are continuous on $N_r(\mathbf{a})$ for some $r > 0$.

a) If $\mathfrak{D}(\mathbf{a}) > 0$ and $f_{xx}(\mathbf{a}) > 0$, then $f(\mathbf{a})$ is a local minimum.

If $\mathfrak{D}(\mathbf{a}) > 0$ and $f_{xx}(\mathbf{a}) < 0$, then $f(\mathbf{a})$ is a local maximum.

b) If $\mathfrak{D}(\mathbf{a}) < 0$, then $\mathbf{a}$ corresponds to a saddle point of the function f.

c) If $\mathfrak{D}(\mathbf{a}) = 0$, the test is inconclusive.

Prove the above. To establish part a), consider the second degree Taylor polynomial of the function $G(t) \equiv f(\mathbf{a}+t\mathbf{u})$ for all possible unit vectors $\mathbf{u} = (h, k)$. Make clever use of the resulting quadratic expression involving the second-order partial derivatives that results. To establish part b), analyze how the saddle point case relates to the quadratic expression alluded to in the hint for part a). A similar second derivative test can be derived for functions $f : D^k \to \mathbb{R}$, where the function $\mathfrak{D} : D^k \to \mathbb{R}$ is the determinant of the matrix of second-order partial derivatives of f at $\mathbf{x}$, called the Hessian matrix.

66. *Use the above second derivative test to confirm your findings in Supplementary Exercise 64 above.*

67. *Let $\boldsymbol{f} : \mathbb{R}^2 \to \mathbb{R}^2$ be given by $\boldsymbol{f}(x, y) = \left(x^2 - y^2, 2xy\right)$. Use the ϵ, δ definition of differentiability to confirm directly that $\boldsymbol{f}'(1, 2) = \begin{bmatrix} 2 & -4 \\ 4 & 2 \end{bmatrix}$.*

68. *Let $\boldsymbol{f} : D^2 \to \mathbb{R}^3$ be given by $\boldsymbol{f}(x, y) = \left(xy, x + y, x^{-1}\right)$ for $D^2 = \left\{(x, y) \in \mathbb{R}^2 : x \neq 0\right\}$. Use the ϵ, δ definition of differentiability to confirm directly that $\boldsymbol{f}'(1, 0) = \begin{bmatrix} 0 & 1 \\ 1 & 1 \\ -1 & 0 \end{bmatrix}$.*

69. *Find $\partial \boldsymbol{f}/\partial x_i$ for every i for the following functions, assuming that $\boldsymbol{f} : D^k \to \mathbb{R}^p$ is differentiable on its domain in each case.*

a) $k = 2, p = 4$, $\quad \boldsymbol{f}(x, y) = (x - y,\ x + y,\ xy,\ x/y)$

b) $k = 3, p = 2$, $\quad \boldsymbol{f}(x, y, z) = \left(xe^y \sin z,\ (x - y - z)^3\right)$

c) $k = p$, $\quad \boldsymbol{f}(x_1, x_2, \ldots, x_k) = (x_1,\ x_1 + x_2,\ \ldots, x_1 + x_2 + \cdots + x_k)$

70. *Consider $f : \mathbb{R}^2 \to \mathbb{R}$ given by* $f(x, y) = \begin{cases} (x^2 + y^2) \sin \frac{1}{\sqrt{x^2+y^2}} & \text{if } (x, y) \neq (0, 0) \\ 0 & \text{if } (x, y) = (0, 0) \end{cases}$.

a) Show that f is differentiable on $\mathbb{R}^2$.
b) Show that $\frac{\partial f}{\partial x}$ and $\frac{\partial f}{\partial y}$ are discontinuous at $(0, 0)$.
Does this contradict Theorem 3.10 on page 279?

71. *Suppose $\boldsymbol{f} : [a, b] \to \mathbb{R}^p$ is continuous on $[a, b]$ and differentiable on (a, b). Then there exists a point $c \in (a, b)$ such that $|\boldsymbol{f}(b) - \boldsymbol{f}(a)| \leq |\boldsymbol{f}'(c)|\,(b - a)$. (Hint: Let $h(x) = \left(\boldsymbol{f}(b) - \boldsymbol{f}(a)\right) \cdot \boldsymbol{f}(x)$.)*

72. *Let D^k be an open, convex subset of $\mathbb{R}^k$, and $\boldsymbol{f} : D^k \to \mathbb{R}^p$ be differentiable on D^k. Suppose there exists a constant M such that $|\boldsymbol{f}'(\mathbf{x})| \leq M$ for all $\mathbf{x} \in D^k$. Prove that $|\boldsymbol{f}(\mathbf{b}) - \boldsymbol{f}(\mathbf{a})| \leq M\,|\mathbf{b} - \mathbf{a}|$ for all $\mathbf{a}, \mathbf{b} \in D^k$. (Hint: Let $\boldsymbol{g}(x) = \boldsymbol{f}\left((1 - x)\,\mathbf{a} + x\,\mathbf{b}\right)$.)*

73. *Suppose $\boldsymbol{f} : \mathbb{R}^k \to \mathbb{R}^p$ is differentiable on $\mathbb{R}^k$ and that there exists a nonnegative constant $c < 1$ such that $|\boldsymbol{f}'(\mathbf{x})| \leq c$ for all $\mathbf{x} \in \mathbb{R}^k$. Prove that there exists a unique point $\mathbf{p} \in \mathbb{R}^k$ such that $\boldsymbol{f}(\mathbf{p}) = \mathbf{p}$. Such a point $\mathbf{p}$ is called a* fixed point *of the function*

f. *Hints:*

1. *Show that* $|f(\mathbf{x}) - f(\mathbf{y})| \le c\,|\mathbf{x} - \mathbf{y}|$ *for all* $\mathbf{x}, \mathbf{y} \in \mathbb{R}^k$.
2. *Let* $\mathbf{x}_0$ *be a point in* $\mathbb{R}^k$ *and define* $\mathbf{x}_{n+1} = f(\mathbf{x}_n)$ *for all* $n \ge 1$. *Show that* $\lim \mathbf{x}_n = \mathbf{p}$ *exists.*
3. *Show that* $f(\mathbf{p}) = \mathbf{p}$.

74. *Let* D^2 *be an open disk in* $\mathbb{R}^2$, *and* $f : D^2 \to \mathbb{R}$ *differentiable on* D^2 *such that* $\frac{\partial f}{\partial x}(x, y) = 0$ *for all* $(x, y) \in D^2$. *Show that* $f(x, y)$ *depends only on* y.

75. *Let* $D^2 = R_1 \cup R_2 \cup R_3$, *where* $R_1 = (0, 1) \times (0, 6)$, $R_2 = [1, 2] \times (5, 6)$, *and* $R_3 = (2, 3) \times (0, 6)$. *If* $f : D^2 \to \mathbb{R}$ *is differentiable on the interior of* D^2 *such that* $\frac{\partial f}{\partial x}(x, y) = 0$ *for all interior points* $(x, y) \in D^2$, *does it follow that* $f(x, y)$ *depends only on* y?

76. *Suppose* $f : \mathbb{R}^2 \to \mathbb{R}$ *has the property that there exist two constants* M_1 *and* M_2 *such that for all* $(x, y) \in \mathbb{R}^2$, $\left|\frac{\partial f}{\partial x}(x, y)\right| \le M_1$ *and* $\left|\frac{\partial f}{\partial y}(x, y)\right| \le M_2$. *Show that* f *is continuous on* $\mathbb{R}^2$.

77. *Suppose* D^2 *is open in* $\mathbb{R}^2$ *and that* $f : D^2 \to \mathbb{R}$ *is such that* f *and all mixed partial derivatives of order* ≤ 3 *are continuous on* D^2. *Show that on* D^2, $f_{xxy} = f_{xyx} = f_{yxx}$.

78. *Suppose* $f : \mathbb{R}^2 \to \mathbb{R}$ *has continuous first- and second-order partial derivatives. If* $g_{ab}(t) \equiv f(at, bt)$ *has a local minimum at* $t = 0$ *for every choice of real numbers* a *and* b, *does* f *have a local minimum at* $(0, 0)$?

79. *Consider the function* $f : \mathbb{R}^2 \to \mathbb{R}$ *given by* $f(x, y) = \begin{cases} \frac{x^2 y}{x^4 + y^2} & \text{for } (x, y) \ne (0, 0) \\ 0 & \text{for } (x, y) = (0, 0) \end{cases}$.

a) *Is* f *continuous at* $(0, 0)$?

b) *Do the partial derivatives* f_x *and* f_y *exist at* $(0, 0)$?

c) *Is* f *differentiable at* $(0, 0)$?

80. *If* $f : \mathbb{R}^2 \to \mathbb{R}$ *is given by* $f(x, y) = \begin{cases} (x + y)^2 \sin(1/y) & \text{for } (x, y) \ne (0, 0) \\ 0 & \text{for } (x, y) = (0, 0) \end{cases}$, *is* f_y *continuous at* $(0, 0)$?

81. *Suppose* $f : \mathbb{R}^2 \to \mathbb{R}$ *has continuous partial derivatives up to order* $k \ge 2$, *and for all* $a, b \in \mathbb{R}$ *the function* $g_{a,b} : \mathbb{R} \to \mathbb{R}$ *given by* $g_{a,b}(t) \equiv f(at, bt)$ *has a local minimum at* $t = 0$. *Show that* f *has a local minimum at* $(0, 0)$. *(The order* k *may be chosen to be any value greater than or equal to 2 for convenience in justifying your answer.)*

82. *Consider the function* $f : \mathbb{C} \to \mathbb{C}$ *given by* $f(z) = 2z^2 - 3$. *Use the* ϵ, δ *version of the derivative to show that* $f'(1) \ne 3$. *Hint: Use the method of proof by contradiction.*

83. *In each of the following, consider the function $f : D \to \mathbb{C}$ that is given, and find as large a domain D as is possible for the function f. Also find the corresponding domain of f'.* a) $f(z) = z^5$ b) $f(z) = |z|$ c) $f(z) = z^{2/3}$

84. *Determine the values of z for which the following functions are differentiable (presume the natural domain for f in each case). First, use the Jacobian to infer what the derivative should be, and then use the Cauchy-Riemann equations to find the derivative in each case.*

a) $f(z) = (z-1)^{-1}$ b) $f(z) = z^3$ c) $f(z) = \sinh z$

d) $f(z) = z^2 + (\overline{z})^2$ e) $f(z) = -2xy + i\left(x^2 - y^2\right)$ f) $f(z) = 2xy + i\,x^2$

g) $f(z) = e^x + i\,e^y$ h) $f(z) = x^3y - y^3x + i\left(\frac{3}{2}x^2y^2 - \frac{1}{4}y^4 - \frac{1}{4}x^4\right)$

85. *Suppose $u : \mathbb{R}^2 \to \mathbb{R}$ is given by $u(x,y) = xy$. Can you find a (complex) differentiable function $f : \mathbb{C} \to \mathbb{C}$ such that* Re $(f) = u$?

86. *Suppose $v : \mathbb{R}^2 \to \mathbb{R}$ is given by $v(x,y) = x^2y$. Can you find a (complex) differentiable function $f : D \to \mathbb{C}$ such that $Im(f) = v$?*

87. *Suppose $u : D^2 \to \mathbb{R}$ is such that its second-order partial derivatives exist and are continuous on D^2, and $\Delta u \equiv u_{xx} + u_{yy} = 0$ on D^2. Such a function u is called* harmonic *on D^2. What would you need to assume in order to find a function $f : D \to \mathbb{C}$ such that* Re$(f) = u$? *What if you require f to be differentiable?*

88. *Suppose $f : D \to \mathbb{C}$ given by $f(z) = u(x,y) + i\,v(x,y)$ is differentiable on $D \subset \mathbb{C}$, and that the functions u and v satisfy the conditions of the mixed derivative theorem on page 267. Compute Δu and Δv. What can you conclude? We will see in a later chapter that differentiablity of the complex function f implies these other conditions on its real and imaginary parts u and v, and so the conclusion you come to will apply to the real and imaginary parts of any differentiable complex function.*

89. *Compare your results from the last exercise to what you find for the real function $\mathbf{g} : D^2 \to \mathbb{R}^2$ given by $\mathbf{g}(x,y) = \left(x^2 + y,\, x^2 + y^2\right)$.*

90. *Suppose the function $f : D \to \mathbb{C}$ is given in polar coordinates as $f(z) = u(r,\theta) + i\,v(r,\theta)$, and that f is defined in a neighborhood of the point $z_0 = r_0 e^{i\,\theta_0} \neq 0$. Under the right conditions (which?), show that the Cauchy-Riemann equations in polar coordinates are given by*

$$u_r = \frac{1}{r}\,v_\theta \quad \text{and} \quad \frac{1}{r}\,u_\theta = -v_r.$$

91. *Show that the Laplacian operator $\Delta \equiv \frac{\partial^2}{\partial x^2} + \frac{\partial^2}{\partial y^2}$ can be expressed in terms of $\frac{\partial}{\partial z}$ and $\frac{\partial}{\partial \overline{z}}$ as $\Delta = 4\,\frac{\partial}{\partial \overline{z}}\,\frac{\partial}{\partial z}$.*

92. *Suppose we have a function of two complex variables, $f(z_1, z_2)$. Show that such a function can be written in the form $f(z_1, z_2) = F(z_1, \overline{z_1}, z_2, \overline{z_2})$, where $z_1 = x_1 + i\,y_1$, $\overline{z_1} = x_1 - i\,y_1$, $z_2 = x_2 + i\,y_2$, and $\overline{z_2} = x_2 - i\,y_2$.*

93. *Find F for the following functions.*
a) $f(z_1, z_2) = z_1^2 + z_1 z_2$ *b)* $f(z_1, z_2) = |z_1|e^{z_2}$ *c)* $f(z_1, z_2) = Im(z_1\overline{z_2})$

94. *For the functions you found in the previous exercise, formally determine the following:* $\partial F/\partial z_1$, $\partial F/\partial z_2$, $\partial F/\partial \overline{z_1}$, *and* $\partial F/\partial \overline{z_2}$.

95. *One could define $f(z_1, z_2)$ to be differentiable if after finding its associated function $F(z_1, \overline{z_1}, z_2, \overline{z_2})$ it was determined that $\partial F/\partial \overline{z_1} = \partial F/\partial \overline{z_2} = 0$. Which functions in the previous exercise would be differentiable according to this definition?*

96. *Given the above definition of differentiability for functions of more than one complex variable, is there a corresponding set of "Cauchy-Riemann equations¿' Explore this possibility.*

97. *Let $f : \mathbb{R}^2 \to \mathbb{R}$ be given by $f(x, y) = x^2 + y^2 - 1$. Use the implicit function theorem to solve for $y = g(x)$ in a neighborhood of $(\frac{1}{\sqrt{2}}, \frac{1}{\sqrt{2}})$. Can this be done in a neighborhood of $(1, 0)$?*

98. *Consider the equation $x^3 + y \cos xy = 0$. Can you solve for $y = f(x)$ in a neighborhood of $(0, 0)$? Can you solve for $x = g(y)$ in a neighborhood of $(0, 0)$?*

99. *Consider the equation $x^2 y - z \ln x + e^z = 1$. Can you solve for $z = f(x, y)$ in a neighborhood of $(1, 0, 0)$? Can you solve for $x = g(y, z)$ in a neighborhood of $(1, 0, 0)$? Can you solve for $y = h(x, z)$ in a neighborhood of $(1, 0, 0)$?*

100. *Consider the system of equations given below and determine whether you can solve for z in terms of $x, y,$ and w:*

$$\begin{aligned} x - y + z + w^8 &= 0 \\ 2x + y - z + w &= 0 \\ y + z - 3w &= 0. \end{aligned}$$

7

REAL INTEGRATION

God does not care about our mathematical difficulties. He integrates empirically.

Albert Einstein

We begin by studying what will be defined as the Riemann integral of a function $f : [a, b] \to \mathbb{R}$ over a closed interval. Since the Riemann integral is the only type of integral we will discuss in detail,[1] any reference to the integral of a function should be understood to be the Riemann integral of that function. One motivation for the development of the Riemann integral corresponds to the special case where such a function f is continuous and nonnegative on its domain. We may then determine the area under the curve described by the graph of f in the xy-plane lying above the x-axis. The key idea, so familiar to all students of calculus, is to approximate the area by a finite sum of rectangular subareas, a so-called *Riemann sum*, and then take the appropriate limit to obtain the exact area under the curve, the value of which corresponds to the value of the integral in this case. Even this special case prompts some natural questions. What other kinds of functions will obtain a limit in this process? Must the associated Riemann sum be set up in a particular way? Of course, we will be able to define such a Riemann integral even when the function f is negative over part of, or even all of, its domain. However, in such cases we usually forgo the interpretation of the integral as an area, instead relying on the more abstract notion of the integral as the limit of a Riemann sum.

1 THE INTEGRAL OF $f : [a, b] \to \mathbb{R}$

1.1 Definition of the Riemann Integral

We begin with the following definition.

[1]Other types include the Riemann-Stieltjes integral, the Lebesgue integral, the Lebesgue-Stieltjes integral, and the Daniell integral. Each was developed, in some measure, to amend perceived shortcomings of the integral type that preceded it historically.

Definition 1.1 A **partition** of $[a, b] \subset \mathbb{R}$ is an ordered collection of $n+1$ points

$$\mathcal{P} = \{x_0, x_1, x_2, \ldots, x_n\} \subset [a, b],$$

such that $a = x_0 < x_1 < x_2 < \cdots < x_n = b$. Each partition $\mathcal{P}$ determines a collection of n **subintervals** $\{I_j\}$ of $[a, b]$, each given by

$$I_j = [x_{j-1}, x_j] \text{ for } 1 \le j \le n.$$

The length of each subinterval I_j is denoted by

$$\Delta x_j \equiv x_j - x_{j-1}.$$

Finally, the **norm of the partition** $\mathcal{P}$ is the length of the largest subinterval determined by $\mathcal{P}$, and is given by

$$\|\mathcal{P}\| \equiv \max_{1 \le j \le n} \Delta x_j.$$

We now introduce the useful concept of a *refinement* of a partition $\mathcal{P}$.

Definition 1.2 A **refinement** of a partition $\mathcal{P}$ on $[a, b]$ is any partition $\mathcal{P}'$ on $[a, b]$ satisfying

$$\mathcal{P} \subset \mathcal{P}'.$$

A simple example that illustrates the above definition is to consider the interval $[0, 1]$ and the partition $\mathcal{P} = \{0, \frac{1}{2}, 1\}$. One refinement of $\mathcal{P}$ is $\mathcal{P}' = \{0, \frac{1}{4}, \frac{1}{2}, \frac{3}{4}, 1\}$. The partition given by $\mathcal{P}'' = \{0, \frac{1}{4}, \frac{3}{4}, 1\}$ is not a refinement of $\mathcal{P}$. (Why?) Finally, we will sometimes have need to form the union of two partitions, and we do this in a very natural way. For example, $\mathcal{P}_1 = \{0, \frac{1}{2}, 1\}$ and $\mathcal{P}_2 = \{0, \frac{1}{4}, \frac{1}{3}, 1\}$ are partitions of $[0, 1]$, and $\mathcal{P}_3 = \{1, \frac{3}{2}, 2\}$ is a partition of $[1, 2]$. The partition $\mathcal{P} = \mathcal{P}_1 \cup \mathcal{P}_2 \equiv \{0, \frac{1}{4}, \frac{1}{3}, \frac{1}{2}, 1\}$ is also a partition of $[0, 1]$, and $\widetilde{\mathcal{P}} = \mathcal{P}_1 \cup \mathcal{P}_3 \equiv \{0, \frac{1}{2}, 1, \frac{3}{2}, 2\}$ is a partition of $[0, 2]$. With the concept of a partition of an interval established, we now define a *Riemann sum* associated with a function $f : [a, b] \to \mathbb{R}$.

Definition 1.3 Consider $f : [a, b] \to \mathbb{R}$ and suppose $\mathcal{P} = \{x_0, x_1, x_2, \ldots, x_n\}$ is a partition of $[a, b]$. Let $\mathcal{C} = \{c_1, c_2, \ldots, c_n\}$ be a collection of points such that $c_j \in I_j$ for $1 \le j \le n$. Then the **Riemann sum** of f associated with the partition $\mathcal{P}$ and the collection $\mathcal{C}$ is given by

$$\mathcal{S}_{\mathcal{P}}(f, \mathcal{C}) \equiv \sum_{j=1}^{n} f(c_j) \Delta x_j.$$

It is important to note that the value of the Riemann sum $\mathcal{S}_{\mathcal{P}}(f, \mathcal{C})$ described in the above definition will, in general, depend on the partition $\mathcal{P}$ and the choice of points $\mathcal{C}$, as well as on the function f.

Example 1.4 *Let $f : [0, 1] \to \mathbb{R}$ be given by $f(x) = x^2+1$, and let $\mathcal{P} = \{0, \frac{1}{3}, 1\}$ be a partition of $[0, 1]$. The subintervals associated with $\mathcal{P}$ are $I_1 = [0, \frac{1}{3}]$ and $I_2 = [\frac{1}{3}, 1]$, and the norm of $\mathcal{P}$ is $\|\mathcal{P}\| = \frac{2}{3}$. Computing $\mathcal{S}_\mathcal{P}(f, \mathcal{C})$ for $\mathcal{C} = \{0, \frac{1}{2}\}$, we obtain*

$$\mathcal{S}_\mathcal{P}(f, \mathcal{C}) = f(0)\,\tfrac{1}{3} + f\left(\tfrac{1}{2}\right)\,\tfrac{2}{3} = \tfrac{7}{6}.$$

◀

► **7.1** *Consider the function from the previous example.*

a) Using the same partition as given in the example, compute the Riemann sum $\mathcal{S}_\mathcal{P}(f, \mathcal{C})$ associated with $\mathcal{C} = \{\frac{1}{4}, \frac{3}{4}\}$.

b) Using the same $\mathcal{C}$ as in the example, compute the Riemann sum $\mathcal{S}_\mathcal{P}(f, \mathcal{C})$ associated with the partition $\mathcal{P} = \{0, \frac{1}{8}, 1\}$.

c) Compute the Riemann sum $\mathcal{S}_{\mathcal{P}'}(f, \mathcal{C}')$ associated with the refinement $\mathcal{P}' = \{0, \frac{1}{8}, \frac{1}{3}, 1\}$ of the partition from the example, and $\mathcal{C}' = \{0, \frac{1}{8}, \frac{1}{3}\}$.

d) Compute the Riemann sums $\mathcal{S}_{\mathcal{P}_n}(f, \mathcal{C}_n)$ associated with the partitions $\mathcal{P}_n = \{0, \frac{1}{n}, \frac{2}{n}, \dots, 1\}$ and associated $\mathcal{C}_n = \{\frac{1}{n}, \frac{2}{n}, \dots, 1\}$ for $n = 3, 5$, 8, and 10. What seems to be happening here?

If $f(x) \geq 0$ is continuous on $[a, b]$, it is easy to see that $\mathcal{S}_\mathcal{P}(f, \mathcal{C})$ approximates the area under the curve above the x-axis between a and b. For a fixed partition $\mathcal{P}$ of $[a, b]$ we may adjust $\mathcal{C} = \{c_1, \dots, c_n\}$ to obtain a better approximation to the area, but in general it is easier to improve the approximation by reducing $\|\mathcal{P}\|$. This can be accomplished through the use of a "finer" partition, achieved by increasing n. Ultimately, and under suitable conditions, letting $n \to \infty$ forces $\|\mathcal{P}\| \to 0$, and the associated Riemann sum approaches the exact area. This idea is summarized in the following definition, which also specifies what it means for a function $f : [a, b] \to \mathbb{R}$ to be *integrable* on $[a, b]$.

Definition 1.5 We say that

$$\lim_{\|\mathcal{P}\| \to 0} \mathcal{S}_\mathcal{P}(f, \mathcal{C}) = L$$

exists if and only if for each $\epsilon > 0$ there exists $\delta > 0$ such that for any partition $\mathcal{P}$ of $[a, b]$ satisfying $\|\mathcal{P}\| < \delta$, and for any corresponding choice of $\mathcal{C}$, we obtain

$$|\mathcal{S}_\mathcal{P}(f, \mathcal{C}) - L| < \epsilon.$$

We say that $f : [a, b] \to \mathbb{R}$ is **integrable** on $[a, b]$ if and only if

$$\lim_{\|\mathcal{P}\| \to 0} \mathcal{S}_\mathcal{P}(f, \mathcal{C}) = L$$

exists. In this case we refer to L as **the integral of** f **from** a **to** b, and we write

$$\int_a^b f(x)\,dx = L.$$

Note that the above definition implies that when $f : [a, b] \to \mathbb{R}$ is integrable on $[a, b]$, then $\int_a^b f(x)\,dx \equiv \lim_{\|\mathcal{P}\|\to 0} \mathcal{S}_{\mathcal{P}}(f, \mathcal{C})$. Also note that when we refer to the integral of a function "on $[a, b]$," we are referring to the integral from a to b. Finally, it is worth mentioning that what we call an integrable function as described by Definition 1.5 is more specifically called a *Riemann integrable function* in texts that treat other types of integrals. Since we will only develop the Riemann integral in what follows, we simply refer to such functions as integrable.

To maintain consistency in our development, we define the following quantity for an integrable function $f : [a, b] \to \mathbb{R}$, namely,

$$\int_b^a f(x)\,dx \equiv -\int_a^b f(x)\,dx. \tag{7.1}$$

This effectively defines the integral of f from b to a in terms of the integral of f from a to b. Also note that for any integrable $f : [a, b] \to \mathbb{R}$, and any $c \in [a, b]$, we define

$$\int_c^c f(x)\,dx \equiv 0. \tag{7.2}$$

The somewhat arbitrary nature of the identities (7.1) and (7.2) can be explained through a more detailed development of the Riemann sum concept, which the reader may explore in the supplementary exercises section of this chapter.

Note that according to Definition 1.5, if $f : [a, b] \to \mathbb{R}$ is integrable on $[a, b]$, then for each $\epsilon > 0$ there exists $\delta > 0$ such that for *any* partition $\mathcal{P}$ of $[a, b]$ satisfying $\|\mathcal{P}\| < \delta$, and any corresponding choice of $\mathcal{C}$, we obtain

$$\left| \mathcal{S}_{\mathcal{P}}(f, \mathcal{C}) - \int_a^b f(x)\,dx \right| < \epsilon.$$

Therefore, for such an integrable f, we may *choose* a convenient sequence of partitions, say, $\mathcal{P}_n = \{a = x_0, x_1, \ldots, x_n = b\}$ where $x_j = a + \frac{j\,(b-a)}{n}$ for $0 \le j \le n$, and associated points $\mathcal{C}_n = \{c_1, \ldots, c_n\}$ with $c_j = x_j \in I_j$ for $1 \le j \le n$, for which it follows that $\|\mathcal{P}_n\| = \frac{b-a}{n}$ and $\mathcal{S}_{\mathcal{P}_n}(f, \mathcal{C}_n) = \sum_{j=1}^n f\big(a + \frac{j\,(b-a)}{n}\big)\left(\frac{b-a}{n}\right)$. Note that with such a choice, we have that $\|\mathcal{P}_n\| = \frac{b-a}{n} < \delta$ as long as $n > \left(\frac{b-a}{\delta}\right)$, and so choosing $N \in \mathbb{N}$ such that $N > \left(\frac{b-a}{\delta}\right)$ obtains the result

$$n > N \;\Rightarrow\; \left| \mathcal{S}_{\mathcal{P}_n}(f, \mathcal{C}_n) - \int_a^b f(x)\,dx \right| < \epsilon,$$

or more succinctly,

$$\lim_{n\to\infty} \left(\tfrac{b-a}{n}\right) \sum_{j=1}^n f\left(a + \tfrac{j\,(b-a)}{n}\right) = \int_a^b f(x)\,dx. \tag{7.3}$$

Such a convenient choice of partitions is common in practice when a function is known to be integrable on the interval in question, allowing the integral to be expressed as the limit of a Riemann sum with the limit being taken as $n \to \infty$. We now give our first, and simplest, example of a subclass of integrable functions on a closed and bounded interval.

Example 1.6 *Let $f : [a,b] \to \mathbb{R}$ be given by $f(x) = k$, where $k \in \mathbb{R}$ is a constant. If $\mathcal{P}$ is any partition of $[a,b]$, then*

$$\mathcal{S}_{\mathcal{P}}(f, C) = \sum_j k\,\Delta x_j = k\,(b-a),$$

for all choices of C. Hence,

$$\int_a^b k\,dx = \lim_{\|\mathcal{P}\|\to 0} \mathcal{S}_{\mathcal{P}}(f, C) = k\,(b-a),$$

and so any function $f : [a,b] \to \mathbb{R}$ that is constant on $[a,b]$ is integrable on $[a,b]$. ◀

In the remainder of this section we will determine other subclasses of integrable functions that are especially convenient and common and that are easier to identify than the more general class of integrable functions given by Definition 1.5. Also, we will derive the important properties of integrals that allow for easier manipulation of them in various settings, and we will see some examples of functions that are not integrable. While the definition of integrability in terms of Riemann sums given by Definition 1.5 is relatively straightforward and conveys the relevant ideas in an uncluttered way, it is not always easily applied. To remedy this, we will now recast integrability in terms of what are commonly referred to as *upper and lower Riemann sums*. This will require us to restrict our focus to bounded functions $f : [a,b] \to \mathbb{R}$.

1.2 Upper and Lower Sums and Integrals

Upper and Lower Sums and Their Properties

For all that follows in this subsection, we will presume that $f : [a,b] \to \mathbb{R}$ is a bounded function. We begin with a definition.

Definition 1.7 Consider the bounded function $f : [a,b] \to \mathbb{R}$, and let $\mathcal{P}$ be a partition of $[a,b]$ having associated subintervals $\{I_j\}$. Finally, let $M_j \equiv \sup_{I_j} f$ and let $m_j \equiv \inf_{I_j} f$. Then the **upper sum** associated with f on $[a,b]$ is given by

$$\overline{\mathcal{S}}_{\mathcal{P}}(f) \equiv \sum_{j=1}^{n} M_j \Delta x_j,$$

and the **lower sum** associated with f on $[a, b]$ is given by

$$\underline{S}_{\mathcal{P}}(f) \equiv \sum_{j=1}^{n} m_j \Delta x_j.$$

Note in the above that $\sup_{I_j} f \equiv \sup\{f(x) : x \in I_j\}$, and likewise $\inf_{I_j} f \equiv \inf\{f(x) : x \in I_j\}$. We now give an example.

Example 1.8 *Consider $f : [0,1] \to \mathbb{R}$ given by $f(x) = x^2 + 1$, and let $\mathcal{P} = \{0, \frac{1}{3}, 1\}$ be a partition of $[0,1]$. Clearly f is bounded on $[a, b]$ since f is continuous there. We will find the associated upper and lower sums. Note that $I_1 = \left[0, \frac{1}{3}\right]$, and so*

$$M_1 = \sup_{I_1} f = \left(\tfrac{1}{3}\right)^2 + 1 = \tfrac{10}{9},$$

while $I_2 = \left[\frac{1}{3}, 1\right]$, and so

$$M_2 = \sup_{I_2} f = 1^2 + 1 = 2.$$

Therefore, we have that

$$\overline{S}_{\mathcal{P}}(f) = \tfrac{10}{9}\left(\tfrac{1}{3}\right) + 2\left(\tfrac{2}{3}\right) = \tfrac{46}{27}.$$

In a similar fashion, we find that

$$m_1 = \inf_{I_1} f = 0^2 + 1 = 1, \quad \text{and} \quad m_2 = \inf_{I_2} f = \left(\tfrac{1}{3}\right)^2 + 1 = \tfrac{10}{9}.$$

Therefore, we have that

$$\underline{S}_{\mathcal{P}}(f) = 1\left(\tfrac{1}{3}\right) + \tfrac{10}{9}\left(\tfrac{2}{3}\right) = \tfrac{29}{27}.$$ ◀

► **7.2** *In what follows, let $f : [a, b] \to \mathbb{R}$ be bounded.*

a) *Suppose $f : [a, b] \to \mathbb{R}$ is bounded, and that $\mathcal{P} = \{a = x_0, x_1, \ldots, x_n = b\}$ is a partition of $[a, b]$. Fix any $0 < j < n$, and let $\mathcal{P}_1 = \{a = x_0, \ldots, x_j\}$ and $\mathcal{P}_2 = \{x_j, \ldots, x_n = b\}$, so that $\mathcal{P} = \mathcal{P}_1 \cup \mathcal{P}_2$. Show that $\overline{S}_{\mathcal{P}_1 \cup \mathcal{P}_2}(f) = \overline{S}_{\mathcal{P}_1}(f) + \overline{S}_{\mathcal{P}_2}(f)$, and that $\underline{S}_{\mathcal{P}_1 \cup \mathcal{P}_2}(f) = \underline{S}_{\mathcal{P}_1}(f) + \underline{S}_{\mathcal{P}_2}(f)$.*

b) *Consider any bounded function $f : [a, b] \to \mathbb{R}$ and suppose $c \in (a, b)$. If $\mathcal{P}_1 = \{a = x_0, x_1, \ldots, x_m = c\}$ partitions $[a, c]$ and $\mathcal{P}_2 = \{c = x_m, \ldots, x_{m+n} = b\}$ partitions $[c, b]$, then $\mathcal{P} = \mathcal{P}_1 \cup \mathcal{P}_2$ is a partition of $[a, b]$. Show that $\overline{S}_{\mathcal{P}_1 \cup \mathcal{P}_2}(f) = \overline{S}_{\mathcal{P}_1}(f) + \overline{S}_{\mathcal{P}_2}(f)$, and that $\underline{S}_{\mathcal{P}_1 \cup \mathcal{P}_2}(f) = \underline{S}_{\mathcal{P}_1}(f) + \underline{S}_{\mathcal{P}_2}(f)$.*

Recall that in Example 1.4 on page 337 the value obtained for the Riemann sum associated with the same function and partition as in the above example was $\frac{7}{6}$, a value lying *between* the lower and upper sum values of $\frac{29}{27}$ and $\frac{46}{27}$, respectively. This illustrates a general fact, and we state the result in the following lemma.

Lemma 1.9 *For any bounded function $f : [a,b] \to \mathbb{R}$, any partition $\mathcal{P}$ of $[a,b]$, and any associated choice of points $\mathcal{C} = \{c_1,\dots,c_n\}$,*

$$\underline{\mathcal{S}}_{\mathcal{P}}(f) \le \mathcal{S}_{\mathcal{P}}(f,\mathcal{C}) \le \overline{\mathcal{S}}_{\mathcal{P}}(f).$$

► **7.3** *Prove the above lemma.*

Note that the above lemma also establishes that for any partition $\mathcal{P}$, the lower sum associated with f is less than or equal to the upper sum associated with f, i.e.,

$$\underline{\mathcal{S}}_{\mathcal{P}}(f) \le \overline{\mathcal{S}}_{\mathcal{P}}(f).$$

The above lemma will also enable us to study the integrability of bounded functions in terms of upper and lower sums, which are often easier to work with than Riemann sums. In particular, we will ultimately show that the integral of a bounded function f exists as described in Definition 1.5 on page 337 if and only if the supremum of the associated lower sums equals the infimum of the associated upper sums. Before formally stating and proving this result, however, we will prove a series of lemmas establishing various properties of upper and lower sums.

Lemma 1.10 *Suppose $f : [a,b] \to \mathbb{R}$ is a bounded function and $\mathcal{P}$ is a partition of $[a,b]$. If $\mathcal{P}'$ is a refinement of $\mathcal{P}$, then*

$$\underline{\mathcal{S}}_{\mathcal{P}}(f) \le \underline{\mathcal{S}}_{\mathcal{P}'}(f) \quad \textit{and} \quad \overline{\mathcal{S}}_{\mathcal{P}'}(f) \le \overline{\mathcal{S}}_{\mathcal{P}}(f).$$

The lemma says that under refinements of a given partition, lower sums never decrease, and upper sums never increase.

PROOF We prove that $\overline{\mathcal{S}}_{\mathcal{P}'}(f) \le \overline{\mathcal{S}}_{\mathcal{P}}(f)$ and leave the proof that $\underline{\mathcal{S}}_{\mathcal{P}}(f) \le \underline{\mathcal{S}}_{\mathcal{P}'}(f)$ to the reader. Note that it is only necessary to consider a "one point" refinement, since the general case can then be done by induction. To this end, suppose $\mathcal{P} = \{x_0, x_1, \dots, x_n\}$ and $\mathcal{P}' = \mathcal{P} \cup \{\xi\}$, where $x_{k-1} < \xi < x_k$ for some $1 \le k \le n$. Then,

$$\overline{\mathcal{S}}_{\mathcal{P}}(f) = \sum_{j=1}^{n} M_j \Delta x_j = M_1\Delta x_1 + M_2\Delta x_2 + \cdots + M_n\Delta x_n,$$

and

$$\overline{\mathcal{S}}_{\mathcal{P}'}(f) = M_1\Delta x_1 + \cdots + M_{k-1}\Delta x_{k-1} + M_k^{\text{left}}(\xi - x_{k-1}) + M_k^{\text{right}}(x_k - \xi) \\ + M_{k+1}\Delta x_{k+1} + \cdots + M_n\Delta x_n,$$

where

$$M_k^{\text{left}} \equiv \sup_{[x_{k-1},\xi]} f \quad \text{and} \quad M_k^{\text{right}} \equiv \sup_{[\xi, x_k]} f.$$

Therefore,

$$\overline{\mathcal{S}}_{\mathcal{P}}(f) - \overline{\mathcal{S}}_{\mathcal{P}'}(f) = M_k\Delta x_k - \left(M_k^{\text{left}}(\xi - x_{k-1}) + M_k^{\text{right}}(x_k - \xi)\right). \tag{7.4}$$

To complete the proof, we note that

$$M_k^{\text{left}} = \sup_{[x_{k-1},\xi]} f \le \sup_{I_k} f = M_k, \quad \text{and similarly,} \quad M_k^{\text{right}} \le M_k.$$

This combined with (7.4) above yields

$$\begin{aligned}\overline{S}_{\mathcal{P}}(f) - \overline{S}_{\mathcal{P}'}(f) &\ge M_k \Delta x_k - \Big(M_k(\xi - x_{k-1}) + M_k(x_k - \xi)\Big) \\ &= M_k\Delta x_k - M_k \Delta x_k \\ &= 0.\end{aligned}$$

That is,

$$\overline{S}_{\mathcal{P}'}(f) \le \overline{S}_{\mathcal{P}}(f). \qquad \blacklozenge$$

► **7.4** *Complete the proof of the above lemma by showing that* $\underline{S}_{\mathcal{P}}(f) \le \underline{S}_{\mathcal{P}'}(f)$.

The following lemma differs from the previous one in that it considers two possibly different partitions of $[a, b]$. Basically, it says that lower sums are never greater than upper sums.

Lemma 1.11 *Suppose* $f : [a, b] \to \mathbb{R}$ *is a bounded function, and* $\mathcal{P}_1$ *and* $\mathcal{P}_2$ *are two partitions of* $[a, b]$. *Then,*

$$\underline{S}_{\mathcal{P}_1}(f) \le \overline{S}_{\mathcal{P}_2}(f).$$

PROOF Let $\mathcal{P} = \mathcal{P}_1 \cup \mathcal{P}_2$. Then $\mathcal{P}$ is a refinement of $\mathcal{P}_1$ and of $\mathcal{P}_2$. By Lemmas 1.9 and 1.10 on page 341, we have that

$$\underline{S}_{\mathcal{P}_1}(f) \le \underline{S}_{\mathcal{P}_1\cup\mathcal{P}_2}(f) \le \overline{S}_{\mathcal{P}_1\cup\mathcal{P}_2}(f) \le \overline{S}_{\mathcal{P}_2}(f). \qquad \blacklozenge$$

Upper and Lower Integrals and Their Properties

For a given bounded function $f : [a, b] \to \mathbb{R}$, since every lower sum is less than or equal to every upper sum, and Riemann sums are caught between them, it is reasonable to consider the supremum of the lower sums and the infimum of the upper sums. If these values are the same, the common value will be shown to equal the limit of the Riemann sums, and hence the integral of f over $[a, b]$. In fact, for $m \le f \le M$ on $[a, b]$, it is easy to see that lower sums are bounded above, since

$$\underline{S}_{\mathcal{P}}(f) \le \overline{S}_{\mathcal{P}}(f) = \sum_{j=1}^{n} M_j \Delta x_j \le \sum_{j=1}^{n} M\, \Delta x_j = M\,(b - a).$$

Therefore, the supremum of the lower sums exists. Similarly, one can show that upper sums are bounded below, and hence, that the infimum of the upper sums exists.

► **7.5** *Show that, for a bounded function* $f : [a, b] \to \mathbb{R}$, *the upper sums are bounded below.*

With the above facts in mind, we make the following definition.

Definition 1.12 For any bounded function $f : [a, b] \to \mathbb{R}$, we define the **lower integral** of f over $[a, b]$ by

$$\underline{\int_a^b} f(x)\, dx \equiv \sup_{\mathcal{P}} \underline{S}_{\mathcal{P}}(f).$$

Similarly, we define the **upper integral** of f over $[a, b]$ by

$$\overline{\int_a^b} f(x)\, dx \equiv \inf_{\mathcal{P}} \overline{S}_{\mathcal{P}}(f).$$

Note in the above that

$$\sup_{\mathcal{P}} \underline{S}_{\mathcal{P}}(f) \equiv \sup \left\{ \underline{S}_{\mathcal{P}}(f) : \mathcal{P} \text{ is a partition of } [a, b] \right\}.$$

Similarly,

$$\inf_{\mathcal{P}} \overline{S}_{\mathcal{P}}(f) \equiv \inf \left\{ \overline{S}_{\mathcal{P}}(f) : \mathcal{P} \text{ is a partition of } [a, b] \right\}.$$

As expected, the upper integral of a bounded function is never less than the lower integral of the function.

Theorem 1.13 *Suppose $f : [a, b] \to \mathbb{R}$ is a bounded function. Then,*

$$\underline{\int_a^b} f(x)\, dx \le \overline{\int_a^b} f(x)\, dx.$$

PROOF Fix a partition $\mathcal{P}_1$ of $[a, b]$, and note that for any other partition $\mathcal{P}$ of $[a, b]$ we have

$$\underline{S}_{\mathcal{P}_1}(f) \le \overline{S}_{\mathcal{P}}(f).$$

Therefore,

$$\underline{S}_{\mathcal{P}_1}(f) \le \overline{\int_a^b} f(x)\, dx. \quad \text{(Why?)}$$

Since $\mathcal{P}_1$ was arbitrary, it follows that $\underline{S}_{\mathcal{P}}(f) \le \overline{\int_a^b} f(x)\, dx$ for all partitions $\mathcal{P}$ of $[a, b]$. This yields

$$\underline{\int_a^b} f(x)\, dx \le \overline{\int_a^b} f(x)\, dx,$$

and the theorem is proved. ♦

► **7.6** *Suppose $f, g : [a, b] \to \mathbb{R}$ are both bounded functions such that $f(x) \le g(x)$ on $[a, b]$. Prove that $\underline{\int_a^b} f(x)\, dx \le \underline{\int_a^b} g(x)\, dx$, and $\overline{\int_a^b} f(x)\, dx \le \overline{\int_a^b} g(x)\, dx$.*

The following lemmas will be useful to us later.

Lemma 1.14 *Suppose $f : [a, b] \to \mathbb{R}$ is a bounded function. Then, for any $\epsilon > 0$ there exists $\delta > 0$ such that if $\mathcal{P}$ is a partition of $[a, b]$,*

$$\|\mathcal{P}\| < \delta \;\Rightarrow\; \overline{S}_{\mathcal{P}}(f) < \overline{\int_a^b} f(x)\,dx + \epsilon \text{ and } \underline{S}_{\mathcal{P}}(f) > \underline{\int_a^b} f(x)\,dx - \epsilon.$$

PROOF We establish the inequality $\overline{S}_{\mathcal{P}}(f) < \overline{\int_a^b} f(x)\,dx + \epsilon$ and leave the inequality involving the lower sum to the reader. For $\epsilon > 0$ we may choose a partition $\mathcal{P}_1$ of $[a, b]$ such that

$$\overline{S}_{\mathcal{P}_1}(f) < \overline{\int_a^b} f(x)\,dx + \tfrac{\epsilon}{2}.$$

Now suppose $\mathcal{P}$ is a partition of $[a, b]$ such that $\|\mathcal{P}\| < \delta$, where $\delta > 0$ is to be determined so that $\overline{S}_{\mathcal{P}}(f) < \overline{\int_a^b} f(x)\,dx + \epsilon$. If $\mathcal{P}_2 = \mathcal{P} \cup \mathcal{P}_1$, then $\overline{S}_{\mathcal{P}_2}(f) \le \overline{S}_{\mathcal{P}_1}(f)$ so that

$$\begin{aligned}
\overline{S}_{\mathcal{P}}(f) &= \left(\overline{S}_{\mathcal{P}}(f) - \overline{S}_{\mathcal{P}_2}(f)\right) + \overline{S}_{\mathcal{P}_2}(f) \\
&\le \left(\overline{S}_{\mathcal{P}}(f) - \overline{S}_{\mathcal{P}_2}(f)\right) + \overline{S}_{\mathcal{P}_1}(f) \\
&< \left(\overline{S}_{\mathcal{P}}(f) - \overline{S}_{\mathcal{P}_2}(f)\right) + \overline{\int_a^b} f(x)\,dx + \tfrac{\epsilon}{2}.
\end{aligned} \tag{7.5}$$

If we can show that $\left(\overline{S}_{\mathcal{P}}(f) - \overline{S}_{\mathcal{P}_2}(f)\right) < \frac{\epsilon}{2}$, we are done. To this end, we describe the partitions $\mathcal{P}$, $\mathcal{P}_1$, and $\mathcal{P}_2$ more explicitly by

$$\begin{aligned}
\mathcal{P}_1 &= \{x_0', x_1', \ldots, x_n'\}, \\
\mathcal{P} &= \{x_0, x_1, \ldots, x_m\}, \\
\mathcal{P}_2 &= \{x_0'', x_1'', \ldots, x_r''\},
\end{aligned}$$

and illustrate these partitions in Figure 7.1. Let $\delta_1 \equiv \min\limits_{\mathcal{P}_1}(\Delta x_j')$, and assume that $\|\mathcal{P}\| < \delta < \delta_1$ so that every subinterval associated with $\mathcal{P}_2 = \mathcal{P} \cup \mathcal{P}_1$ is smaller than the smallest subinterval associated with $\mathcal{P}_1$, i.e., between any two points in $\mathcal{P}$ there is at most one point of $\mathcal{P}_1$. Then, denoting

$$M_j \equiv \sup_{[x_{j-1}, x_j]} f, \text{ and } M_j'' \equiv \sup_{\left[x_{j-1}'', x_j''\right]} f,$$

we have

$$\overline{S}_{\mathcal{P}}(f) - \overline{S}_{\mathcal{P}_2}(f) = \sum M_j \Delta x_j - \sum M_j'' \Delta x_j''. \tag{7.6}$$

Note that for any subinterval that is common to $\mathcal{P}$ and $\mathcal{P}_2$ there is no contribution to $\overline{S}_{\mathcal{P}}(f) - \overline{S}_{\mathcal{P}_2}(f)$. The only contributions come from subintervals $I_j = [x_{j-1}, x_j]$ of $\mathcal{P}$ that have been subdivided by a point x_k' from $\mathcal{P}_1$, i.e., $I_j = [x_{j-1}, x_j] = [x_{j-1}, x_k'] \cup [x_k', x_j]$, where $\Delta x_j^{\text{left}} \equiv x_k' - x_{j-1}$, and $\Delta x_j^{\text{right}} \equiv x_j - x_k'$. Recall that $\Delta x_j = x_j - x_{j-1}$. There are no more than n such

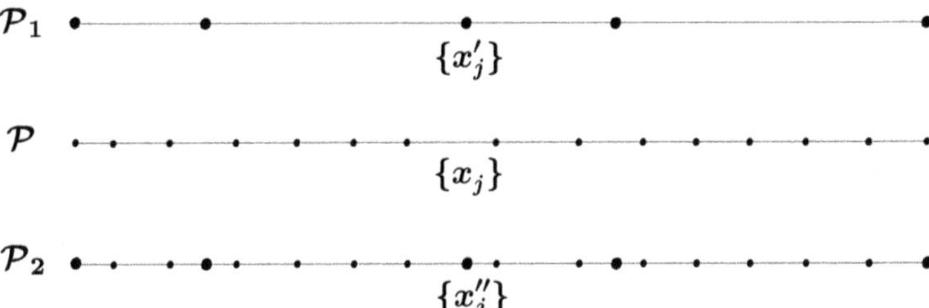

Figure 7.1 *The partitions $\mathcal{P}_1, \mathcal{P}$, and $\mathcal{P}_2$.*

subintervals from $\mathcal{P}$, and each such subdivided subinterval I_j gives rise to a contribution to the right-hand side of (7.6) of the form

$$E_j = \Big(\sup_{[x_{j-1},x_j]} f \Big) \Delta x_j - \left[\Big(\sup_{[x_{j-1},x'_k]} f \Big) \Delta x_j^{\text{left}} + \Big(\sup_{[x'_k,x_j]} f \Big) \Delta x_j^{\text{right}} \right].$$

If we let $\Lambda \equiv \sup_{[a,b]} |f|$, then we have

$$|E_j| \le 3\,\Lambda\,\|\mathcal{P}\| < 3\,\Lambda\,\delta.$$

All of this yields

$$\begin{aligned} \overline{S}_{\mathcal{P}}(f) - \overline{S}_{\mathcal{P}_2}(f) &= \left|\overline{S}_{\mathcal{P}}(f) - \overline{S}_{\mathcal{P}_2}(f)\right| \\ &\le 3\,\Lambda\,\delta\,n \\ &< \tfrac{\epsilon}{2}, \end{aligned}$$

provided that $\delta < \frac{\epsilon}{6\,\Lambda\,n}$. Combining this with (7.5) obtains

$$\overline{S}_{\mathcal{P}}(f) < \overline{\int_a^b} f(x)\,dx + \epsilon,$$

provided that $\|\mathcal{P}\| < \delta \equiv \min\left(\delta_1, \frac{\epsilon}{6\,\Lambda\,n}\right)$. ♦

► **7.7** *Complete the above proof by writing up the similar argument for the lower integral. What δ must ultimately be used to establish the overall result?*

Lemma 1.15 *Suppose $f : [a,b] \to \mathbb{R}$ is bounded on $[a,b]$, and $a < c < b$. Then*

$$\underline{\int_a^b} f(x)\,dx = \underline{\int_a^c} f(x)\,dx + \underline{\int_c^b} f(x)\,dx,$$

and

$$\overline{\int_a^b} f(x)\,dx = \overline{\int_a^c} f(x)\,dx + \overline{\int_c^b} f(x)\,dx.$$

PROOF We will establish the equality for upper integrals. To this end, let $\mathcal{P}$ be a partition of $[a,b]$. Then $\mathcal{P}\cup\{c\}=\mathcal{P}_1\cup\mathcal{P}_2$ where $\mathcal{P}_1$ is a partition of $[a,c]$, and $\mathcal{P}_2$ is a partition of $[c,b]$. Now, since

$$\begin{aligned}\overline{S}_{\mathcal{P}}(f) &\geq \overline{S}_{\mathcal{P}\cup\{c\}}(f)\\ &= \overline{S}_{\mathcal{P}_1}(f)+\overline{S}_{\mathcal{P}_2}(f) && \text{by Exercise 7.2 on page 340,}\\ &\geq \overline{\int_a^c} f(x)\,dx+\overline{\int_c^b} f(x)\,dx,\end{aligned}$$

it follows that

$$\overline{\int_a^b} f(x)\,dx \geq \overline{\int_a^c} f(x)\,dx+\overline{\int_c^b} f(x)\,dx. \tag{7.7}$$

We now establish the reverse inequality. To this end, for any $\epsilon>0$ there exists a partition $\mathcal{P}_1$ of $[a,c]$, and a partition $\mathcal{P}_2$ of $[c,b]$ such that

$$\overline{S}_{\mathcal{P}_1}(f)<\overline{\int_a^c} f(x)\,dx+\tfrac{\epsilon}{2},\quad\text{and}\quad \overline{S}_{\mathcal{P}_2}(f)<\overline{\int_c^b} f(x)\,dx+\tfrac{\epsilon}{2}.\quad\text{(Why?)}$$

Now, since $\mathcal{P}_1\cup\mathcal{P}_2$ is a partition of $[a,b]$, we have

$$\begin{aligned}\overline{\int_a^b} f(x)\,dx &\leq \overline{S}_{\mathcal{P}_1\cup\mathcal{P}_2}(f)\\ &= \overline{S}_{\mathcal{P}_1}(f)+\overline{S}_{\mathcal{P}_2}(f) && \text{by Exercise 7.2 on page 340,}\\ &< \overline{\int_a^c} f(x)\,dx+\overline{\int_c^b} f(x)\,dx+\epsilon.\end{aligned}$$

This is true for every $\epsilon>0$, and therefore

$$\overline{\int_a^b} f(x)\,dx \leq \overline{\int_a^c} f(x)\,dx+\overline{\int_c^b} f(x)\,dx. \tag{7.8}$$

Inequalities (7.7) and (7.8) together obtain

$$\overline{\int_a^b} f(x)\,dx = \overline{\int_a^c} f(x)\,dx+\overline{\int_c^b} f(x)\,dx.$$

We leave the proof of the lower integral equality to the reader. ♦

► **7.8** *Answer the (Why?) question in the above proof, and prove the lower integral result.*

1.3 Relating Upper and Lower Integrals to Integrals

The following theorem is the key result of this section. It provides a means for determining the integrability of a function $f:[a,b]\to\mathbb{R}$, and its value when it is integrable, via upper and lower integrals.

Theorem 1.16 *Suppose $f : [a, b] \to \mathbb{R}$ is a bounded function. Then,*

$$f : [a, b] \to \mathbb{R} \text{ is integrable if and only if } \underline{\int_a^b} f(x)\, dx = \overline{\int_a^b} f(x)\, dx,$$

and in either case,

$$\int_a^b f(x)\, dx = \underline{\int_a^b} f(x)\, dx = \overline{\int_a^b} f(x)\, dx.$$

PROOF Suppose $\underline{\int_a^b} f(x)\, dx = \overline{\int_a^b} f(x)\, dx$. We will show that this implies that f is integrable. To this end, consider $\epsilon > 0$. By Lemma 1.14, there exists $\delta > 0$ such that if $\mathcal{P}$ is a partition of $[a, b]$,

$$\|\mathcal{P}\| < \delta \quad \Rightarrow \quad \overline{S}_{\mathcal{P}}(f) < \overline{\int_a^b} f(x)\, dx + \epsilon \ \text{ and } \ \underline{S}_{\mathcal{P}}(f) > \underline{\int_a^b} f(x)\, dx - \epsilon.$$

From this we obtain, for $\|\mathcal{P}\| < \delta$ and any $\mathcal{C}$ associated with $\mathcal{P}$,

$$\underline{\int_a^b} f(x)\, dx - \epsilon < \underline{S}_{\mathcal{P}}(f) \le S_{\mathcal{P}}(f, \mathcal{C}) \le \overline{S}_{\mathcal{P}}(f) < \overline{\int_a^b} f(x)\, dx + \epsilon. \qquad (7.9)$$

Denoting the common value of $\underline{\int_a^b} f(x)\, dx$ and $\overline{\int_a^b} f(x)\, dx$ by L, we obtain from expression (7.9) that

$$\|\mathcal{P}\| < \delta \quad \Rightarrow \quad |S_{\mathcal{P}}(f, \mathcal{C}) - L| < \epsilon,$$

i.e.,

$$\lim_{\|\mathcal{P}\| \to 0} S_{\mathcal{P}}(f, \mathcal{C}) = L, \qquad (7.10)$$

and hence f is integrable on $[a, b]$. Note also that equality (7.10) establishes that $\int_a^b f(x)\, dx = \underline{\int_a^b} f(x)\, dx = \overline{\int_a^b} f(x)\, dx$. Now assume that f is integrable on $[a, b]$, i.e., $\lim_{\|\mathcal{P}\| \to 0} S_{\mathcal{P}}(f, \mathcal{C}) = \int_a^b f(x)\, dx$ exists. Then, for $\epsilon > 0$ there exists $\delta > 0$ such that if $\mathcal{P}$ is a partition of $[a, b]$,

$$\|\mathcal{P}\| < \delta \quad \Rightarrow \quad \left| S_{\mathcal{P}}(f, \mathcal{C}) - \int_a^b f(x)\, dx \right| < \frac{\epsilon}{2}.$$

Choose a partition $\mathcal{P}$ of $[a, b]$ such that $\|\mathcal{P}\| < \delta$, and select $c_j \in I_j$ such that for $M_j \equiv \sup_{I_j} f$,

$$f(c_j) > M_j - \frac{\epsilon}{2\,(b - a)} \quad \text{for } j = 1, \ldots, n.$$

Then for $\mathcal{C} = \{c_1, \ldots, c_n\}$,

$$S_{\mathcal{P}}(f, \mathcal{C}) > \overline{S}_{\mathcal{P}}(f) - \tfrac{\epsilon}{2} \ge \overline{\int_a^b} f(x)\, dx - \tfrac{\epsilon}{2}.$$

Since $\mathcal{S}_{\mathcal{P}}(f,\mathcal{C}) < \int_a^b f(x)\,dx + \frac{\epsilon}{2}$, it follows that $\overline{\int_a^b} f(x)\,dx < \int_a^b f(x)\,dx + \epsilon$ for all $\epsilon > 0$. Hence,

$$\overline{\int_a^b} f(x)\,dx \le \int_a^b f(x)\,dx. \tag{7.11}$$

If we choose $c_j \in I_j$ such that, for $m_j \equiv \inf_{I_j} f$,

$$f(c_j) < m_j + \frac{\epsilon}{2\,(b-a)},$$

we may similarly conclude that

$$\int_a^b f(x)\,dx \le \underline{\int_a^b} f(x)\,dx. \tag{7.12}$$

Combining inequalities (7.11) and (7.12), we have

$$\overline{\int_a^b} f(x)\,dx \le \int_a^b f(x)\,dx \le \underline{\int_a^b} f(x)\,dx,$$

which implies

$$\underline{\int_a^b} f(x)\,dx = \int_a^b f(x)\,dx = \overline{\int_a^b} f(x)\,dx.$$

♦

Example 1.17 *Consider the bounded function $f : [0,2] \to \mathbb{R}$ given by*

$$f(x) = \begin{cases} 0 & \text{if } 0 \le x \le 1 \\ 1 & \text{if } 1 < x \le 2 \end{cases}.$$

We will show that f is integrable on $[0,2]$, and determine the value of the integral. To this end, let $\mathcal{P} = \{x_0, x_1, \dots, x_n\}$ be a partition of $[0,2]$. We consider two cases: Case 1 with $x_{m-1} < 1 < x_m$, and case 2 with $x_m = 1$. Note that in case 1,

$$\overline{\mathcal{S}}_{\mathcal{P}}(f) = \sum_{j=m}^{n} \Delta x_j = x_n - x_{m-1} = 2 - x_{m-1},$$

and

$$\underline{\mathcal{S}}_{\mathcal{P}}(f) = \sum_{j=m+1}^{n} \Delta x_j = x_n - x_m = 2 - x_m.$$

In case 2 we similarly have

$$\overline{\mathcal{S}}_{\mathcal{P}}(f) = 1, \quad \text{and} \quad \underline{\mathcal{S}}_{\mathcal{P}}(f) = 2 - x_{m+1}.$$

We leave it to the reader to show that

$$\inf_{\mathcal{P}} \overline{\mathcal{S}}_{\mathcal{P}}(f) = \overline{\int_0^2} f(x)\,dx = 1, \quad \text{and} \quad \sup_{\mathcal{P}} \underline{\mathcal{S}}_{\mathcal{P}}(f) = \underline{\int_0^2} f(x)\,dx = 1,$$

and therefore that f is integrable on $[0,2]$ and $\int_0^2 f(x)\,dx = 1$. ◄

▶ **7.9** *Complete the above example.*

Example 1.18 *Consider the bounded function $f : [0, 1] \to \mathbb{R}$ given by*

$$f(x) = \begin{cases} 0 & \text{if } x \in \mathbb{Q} \cap [0, 1] \\ 1 & \text{if } x \in \mathbb{I} \cap [0, 1] \end{cases}.$$

We will show that f is not integrable on $[0, 1]$ by showing that $\underline{\int_0^1} f(x)\,dx \neq \overline{\int_0^1} f(x)\,dx$. To this end, consider the upper integral and let $\mathcal{P}$ be any partition of $[0, 1]$. Then $M_j \equiv \sup_{I_j} f = 1$ for each $1 \leq j \leq n$, so

$$\overline{S}_{\mathcal{P}}(f) = \sum_{j=1}^{n} \Delta x_j = 1 \text{ for all partitions } \mathcal{P} \text{ of } [0, 1].$$

This implies that

$$\overline{\int_0^1} f(x)\,dx = 1.$$

We leave it to the reader to show that $\underline{\int_0^1} f(x)\,dx = 0$, and therefore f is not integrable on $[0, 1]$. ◀

▶ **7.10** *For the function of the previous example, show that $\underline{\int_0^1} f(x)\,dx = 0$, and hence, that f is not integrable on $[0, 1]$.*

2 Properties of the Riemann Integral

2.1 Classes of Bounded Integrable Functions

Identifying classes or families of integrable functions is of great use in many applications. It also allows us to avoid checking each function we encounter individually for integrability. In order to prove our first such result, we will need the following lemma.

Lemma 2.1 *Suppose $f : [a, b] \to \mathbb{R}$ is a bounded function. Then, f is integrable on $[a, b]$ if and only if for any $\epsilon > 0$, there exists a partition $\mathcal{P}$ of $[a, b]$ such that*

$$\overline{S}_{\mathcal{P}}(f) - \underline{S}_{\mathcal{P}}(f) < \epsilon.$$

PROOF Suppose f is integrable on $[a, b]$, and consider any $\epsilon > 0$. There exist partitions $\mathcal{P}_1$ and $\mathcal{P}_2$ of $[a, b]$ such that

$$\overline{S}_{\mathcal{P}_1}(f) < \overline{\int_a^b} f(x)\,dx + \tfrac{\epsilon}{2} \quad \text{(Why?)},$$

and

$$\underline{\int_a^b} f(x)\,dx - \tfrac{\epsilon}{2} < \underline{S}_{\mathcal{P}_2}(f) \quad \text{(Why?)}.$$

Now let $\mathcal{P} = \mathcal{P}_1 \cup \mathcal{P}_2$. Then,

$$\overline{S}_{\mathcal{P}}(f) \le \overline{S}_{\mathcal{P}_1}(f) < \overline{\int_a^b} f(x)\,dx + \tfrac{\epsilon}{2} = \int_a^b f(x)\,dx + \tfrac{\epsilon}{2},$$

and

$$\underline{S}_{\mathcal{P}}(f) \ge \underline{S}_{\mathcal{P}_2}(f) > \underline{\int_a^b} f(x)\,dx - \tfrac{\epsilon}{2} = \int_a^b f(x)\,dx - \tfrac{\epsilon}{2},$$

which together imply

$$\overline{S}_{\mathcal{P}}(f) - \underline{S}_{\mathcal{P}}(f) < \epsilon.$$

Now suppose that for any $\epsilon > 0$ there exists a partition $\mathcal{P}$ of $[a, b]$ such that $\overline{S}_{\mathcal{P}}(f) - \underline{S}_{\mathcal{P}}(f) < \epsilon$. To prove that f is integrable on $[a, b]$, note that

$$0 \le \overline{\int_a^b} f(x)\,dx - \underline{\int_a^b} f(x)\,dx \le \overline{S}_{\mathcal{P}}(f) - \underline{S}_{\mathcal{P}}(f) < \epsilon.$$

Since this is true for every $\epsilon > 0$, it follows that $\overline{\int_a^b} f(x)\,dx = \underline{\int_a^b} f(x)\,dx$, and the lemma is proved. ♦

▶ **7.11** *Answer the two (Why?) questions in the above proof.*

The following important example makes critical use of the above lemma.

Example 2.2 *Suppose the bounded function $f : [a, b] \to \mathbb{R}$ is nonnegative and integrable on $[a, b]$. The reader can easily verify that $\sqrt{f}$ is then bounded on $[a, b]$. We will show that $\sqrt{f}$ is also integrable on $[a, b]$. To this end, consider $\delta > 0$ and let $\mathcal{P}$ be a partition of $[a, b]$ such that*

$$\overline{S}_{\mathcal{P}}(f) - \underline{S}_{\mathcal{P}}(f) < \delta^2. \tag{7.13}$$

Note that for $M_j \equiv \sup_{I_j} f$ and $m_j \equiv \inf_{I_j} f$,

$$\overline{S}_{\mathcal{P}}(\sqrt{f}) - \underline{S}_{\mathcal{P}}(\sqrt{f}) = \sum_{j=1}^{n} \left(\sqrt{M_j} - \sqrt{m_j}\right) \Delta x_j. \tag{7.14}$$

We now split up the sum in (7.14) according to

$$A \equiv \{j : M_j - m_j < \delta\} \quad \text{and} \quad B \equiv \{j : M_j - m_j \ge \delta\}.$$

That is,

$$\overline{S}_{\mathcal{P}}(\sqrt{f}) - \underline{S}_{\mathcal{P}}(\sqrt{f}) = \sum_{j\in A} \left(\sqrt{M_j} - \sqrt{m_j}\right) \Delta x_j + \sum_{j\in B} \left(\sqrt{M_j} - \sqrt{m_j}\right) \Delta x_j. \tag{7.15}$$

Note that

$$j \in A \;\Rightarrow\; \sqrt{M_j} - \sqrt{m_j} \le \sqrt{M_j - m_j} < \sqrt{\delta}, \quad \textit{(Why?)}$$

and therefore

$$\sum_{j\in A} \left(\sqrt{M_j} - \sqrt{m_j}\right) \Delta x_j < \sqrt{\delta}\,(b - a). \tag{7.16}$$

Also note that inequality (7.13) *yields*

$$\begin{aligned} \delta^2 > \overline{S}_{\mathcal{P}}(f) - \underline{S}_{\mathcal{P}}(f) &= \sum_{j=1}^{n} (M_j - m_j)\, \Delta x_j \\ &\geq \sum_{j \in B} (M_j - m_j)\, \Delta x_j \\ &\geq \delta \sum_{j \in B} \Delta x_j, \end{aligned}$$

and so

$$\sum_{j \in B} \Delta x_j \leq \delta. \tag{7.17}$$

Inequality (7.17) *now yields that*

$$\sum_{j \in B} \left(\sqrt{M_j} - \sqrt{m_j} \right) \Delta x_j \leq \sqrt{M} \sum_{j \in B} \Delta x_j \leq \sqrt{M}\, \delta, \tag{7.18}$$

where $M \equiv \sup_{[a,b]} f$. *Finally, we have that*

$$\begin{aligned} \overline{S}_{\mathcal{P}}(\sqrt{f}\,) - \underline{S}_{\mathcal{P}}(\sqrt{f}\,) &= \sum_{j \in A} \left(\sqrt{M_j} - \sqrt{m_j} \right) \Delta x_j + \sum_{j \in B} \left(\sqrt{M_j} - \sqrt{m_j} \right) \Delta x_j \\ &< \sqrt{\delta}\,(b-a) + \sqrt{M}\,\delta \qquad \text{by (7.16) and (7.18),} \\ &< \epsilon \quad \text{if } \delta < \min\left\{ \frac{\epsilon}{2\left(\sqrt{M}+1\right)}, \left[\frac{\epsilon}{2\,(b-a)}\right]^2 \right\}. \end{aligned}$$

Therefore, $\sqrt{f}$ *is integrable on* $[a, b]$. ◀

► **7.12** *Verify that* $\sqrt{f}$ *is bounded on* $[a, b]$, *and answer the (Why?) question in the above example.*

Using Lemma 2.1, we can now prove that continuous functions on $[a, b]$ are integrable on $[a, b]$.

Theorem 2.3 *If* $f : [a, b] \to \mathbb{R}$ *is continuous on* $[a, b]$, *then* f *is integrable on* $[a, b]$.

PROOF Since f is continuous on the closed interval $[a, b]$, it is also bounded there. Consider any $\epsilon > 0$. We will find a partition $\mathcal{P}$ of $[a, b]$ such that $\overline{S}_{\mathcal{P}}(f) - \underline{S}_{\mathcal{P}}(f) < \epsilon$. To this end, note that since f is continuous on the compact set $[a, b]$, it is uniformly continuous there. Therefore, there exists $\delta > 0$ such that for $\xi, \eta \in [a, b]$,

$$|\xi - \eta| < \delta \quad \Rightarrow \quad |f(\xi) - f(\eta)| < \frac{\epsilon}{(b-a)}.$$

Now choose partition $\mathcal{P}$ of $[a, b]$ such that $\|\mathcal{P}\| < \delta$, and consider

$$\overline{S}_{\mathcal{P}}(f) - \underline{S}_{\mathcal{P}}(f) = \sum_{j=1}^{n} \left(M_j - m_j\right) \Delta x_j. \tag{7.19}$$

By continuity of f on compact I_j, we have for each $1 \le j \le n$,

$$M_j - m_j = f(a_j) - f(b_j) \text{ where } a_j, b_j \in I_j. \text{ (Why?)}$$

Therefore,

$$M_j - m_j = |M_j - m_j| = |f(a_j) - f(b_j)| < \frac{\epsilon}{(b-a)},$$

which obtains from (7.19) that

$$\overline{S}_{\mathcal{P}}(f) - \underline{S}_{\mathcal{P}}(f) < \frac{\epsilon}{(b-a)} \sum_{j=1}^{n} \Delta x_j = \epsilon,$$

and the theorem is proved. ♦

▶ **7.13** *Answer the (Why?) question in the above proof.*

Example 2.4 *Consider $f : [0,1] \to \mathbb{R}$ given by $f(x) = x^2$. Since f is continuous on $[0,1]$, we know that it is integrable there. We will now determine the value of $\int_0^1 f(x)\,dx$. Note that the method we use establishes the integrability of f while it determines the value of the integral. Note also that since $f \ge 0$ on $[a,b]$, this value represents the area under the graph of f between $x = 0$ and $x = 1$ lying above the x-axis. To determine the value of this integral, consider the sequence of partitions of $[0,1]$ given by $\{\mathcal{P}_n\}$, where*

$$\mathcal{P}_n = \left\{0, \tfrac{1}{n}, \tfrac{2}{n}, \ldots, \tfrac{n-1}{n}, 1\right\}.$$

Then[2]

$$\begin{aligned} \overline{S}_{\mathcal{P}_n}(f) = \sum_{j=1}^{n} M_j \Delta x_j &= \sum_{j=1}^{n} \left(\tfrac{j}{n}\right)^2 \tfrac{1}{n} \\ &= \frac{1}{n^3} \sum_{j=1}^{n} j^2 \\ &= \frac{1}{n^3} \left[\frac{n\,(n+1)\,(2\,n+1)}{6}\right] \\ &= \frac{(n+1)\,(2\,n+1)}{6\,n^2}, \end{aligned} \qquad (7.20)$$

and, similarly,

$$\underline{S}_{\mathcal{P}_n}(f) = \frac{(n-1)\,(2\,n-1)}{6\,n^2}. \qquad (7.21)$$

Since $\underline{S}_{\mathcal{P}_n}(f) \le \underline{\int_0^1} f(x)\,dx \le \overline{\int_0^1} f(x)\,dx \le \overline{S}_{\mathcal{P}_n}(f)$, identities (7.20) and (7.21) give us

$$\frac{(n-1)\,(2\,n-1)}{6\,n^2} \le \underline{\int_0^1} f(x)\,dx \le \overline{\int_0^1} f(x)\,dx \le \frac{(n+1)\,(2\,n+1)}{6\,n^2}. \qquad (7.22)$$

[2]The reader should note the formula $\sum_{j=1}^{n} j^2 = \frac{n(n+1)(2n+1)}{6}$, which can be established by a simple inductive argument.

Letting $n \to \infty$ in (7.22) then yields

$$\frac{1}{3} \le \underline{\int_0^1} f(x)\, dx \le \overline{\int_0^1} f(x)\, dx \le \frac{1}{3},$$

and therefore

$$\int_0^1 f(x)\, dx = \underline{\int_0^1} f(x)\, dx = \overline{\int_0^1} f(x)\, dx = \frac{1}{3}.$$

◀

► **7.14** *Consider $f : [0,1] \to \mathbb{R}$ given by $f(x) = x$. Is f integrable on $[0,1]$? Find $\underline{\int_0^1} f(x)\, dx$ and $\overline{\int_0^1} f(x)\, dx$, and show they are equal. From this, determine $\int_0^1 f(x)\, dx$. Do the same for $g : [0,1] \to \mathbb{R}$ given by $g(x) = x^3$. You will need the formula $\sum_{j=1}^n j^3 = \frac{n^2(n+1)^2}{4}$.*

The following theorem establishes another important class of integrable bounded functions.

Theorem 2.5 *If $f : [a,b] \to \mathbb{R}$ is bounded and monotone on $[a,b]$, then f is integrable on $[a,b]$.*

PROOF We consider the case where f is nondecreasing, and leave the nonincreasing case to the reader. If $f(a) = f(b)$ we are done (Why?), so suppose $f(a) < f(b)$. Consider any $\epsilon > 0$, and choose a partition $\mathcal{P}$ of $[a,b]$ such that $\|\mathcal{P}\| < \frac{\epsilon}{f(b)-f(a)}$. Then

$$\begin{aligned}
\overline{S}_{\mathcal{P}}(f) - \underline{S}_{\mathcal{P}}(f) &= \sum_{j=1}^n (M_j - m_j)\, \Delta x_j \\
&= \sum_{j=1}^n \big(f(x_j) - f(x_{j-1})\big)\, \Delta x_j \\
&< \frac{\epsilon}{f(b) - f(a)} \sum_{j=1}^n \big(f(x_j) - f(x_{j-1})\big) \\
&= \frac{\epsilon}{f(b) - f(a)} \big(f(b) - f(a)\big) \\
&= \epsilon,
\end{aligned}$$

and the theorem is proved in the nondecreasing case. ♦

► **7.15** *Answer the (Why?) in the above proof, and finish the proof by handling the case where f is nonincreasing.*

The following theorem establishes yet another important class of integrable bounded functions.

Theorem 2.6 *If $f : [a,b] \to \mathbb{R}$ is bounded and has finitely many discontinuities on $[a,b]$, then f is integrable on $[a,b]$.*

PROOF We consider the case where f has exactly one discontinuity on $[a,b]$, and we assume it is also in (a,b). In particular, we leave the case of this discontinuity being at one of the endpoints, and the more general case of f having more than one discontinuity, to the reader. To this end, let $x_0 \in (a,b)$ be the only discontinuity of f on $[a,b]$, and consider any $\epsilon > 0$. We will apply Lemma 2.1 on page 349. For $\Lambda \equiv \sup_{[a,b]} |f(x)|$, choose $\delta > 0$ such that $(x_0 - \delta, x_0 + \delta) \subset (a,b)$. Then f is integrable on each of $[a, x_0 - \delta]$ and $[x_0 + \delta, b]$, since f is continuous on each of these intervals. Therefore, there exists a partition $\mathcal{P}_1$ of $[a, x_0 - \delta]$ such that

$$\overline{S}_{\mathcal{P}_1}(f) - \underline{S}_{\mathcal{P}_1}(f) < \tfrac{\epsilon}{3},$$

and there exists a partition $\mathcal{P}_2$ of $[x_0 + \delta, b]$ such that

$$\overline{S}_{\mathcal{P}_2}(f) - \underline{S}_{\mathcal{P}_2}(f) < \tfrac{\epsilon}{3}.$$

Now, $\mathcal{P} = \mathcal{P}_1 \cup \mathcal{P}_2$ is a partition of all of $[a,b]$, and

$$\begin{aligned} \overline{S}_{\mathcal{P}}(f) - \underline{S}_{\mathcal{P}}(f) &= \overline{S}_{\mathcal{P}_1}(f) - \underline{S}_{\mathcal{P}_1}(f) + (2\,\delta)\Big(\sup_{[x_0-\delta,\, x_0+\delta]} f\Big) \\ &\quad - (2\,\delta)\Big(\inf_{[x_0-\delta,\, x_0+\delta]} f\Big) + \overline{S}_{\mathcal{P}_2}(f) - \underline{S}_{\mathcal{P}_2}(f) \\ &< \tfrac{\epsilon}{3} + 4\,\delta\,\Lambda + \tfrac{\epsilon}{3}. \end{aligned}$$

Choose $\delta < \frac{\epsilon}{12\,\Lambda}$ to obtain the result. ♦

► **7.16** *Prove the case with $x_0 = a$ or $x_0 = b$, as well as the case of more than one discontinuity in the above theorem. What can happen if there are infinitely many discontinuities in $[a,b]$?*

2.2 Elementary Properties of Integrals

We now prove several convenient and familiar properties of integrals of bounded functions.

Algebraic Properties

The following theorem establishes two properties that together are often referred to as *the linearity property* of the integral.

Theorem 2.7 *Suppose $f : [a,b] \to \mathbb{R}$ and $g : [a,b] \to \mathbb{R}$ are both bounded and integrable on $[a,b]$. Then* *a)* $kf : [a,b] \to \mathbb{R}$ *is bounded and integrable on* $[a,b]$ *for all* $k \in \mathbb{R}$*, and*

$$\int_a^b k\, f(x)\, dx = k \int_a^b f(x)\, dx.$$

b) $f \pm g : [a,b] \to \mathbb{R}$ *is bounded and integrable on* $[a,b]$*, and*

$$\int_a^b (f \pm g)(x)\, dx = \int_a^b f(x)\, dx \pm \int_a^b g(x)\, dx.$$

PROOF We leave it to the reader to verify that kf and $f \pm g$ are bounded on $[a,b]$. To prove integrability in part a), let $\mathcal{P}$ be any partition of $[a,b]$ with associated $\mathcal{C}$. Then,

$$\mathcal{S}_\mathcal{P}(k\, f, \mathcal{C}) = \sum_{j=1}^n k\, f(c_j)\, \Delta x_j = k\, \mathcal{S}_\mathcal{P}(f, \mathcal{C}).$$

Since f is integrable on $[a,b]$, for any $\epsilon > 0$ there exists $\delta > 0$ such that

$$\|\mathcal{P}\| < \delta \;\Rightarrow\; \left| \mathcal{S}_\mathcal{P}(f, \mathcal{C}) - \int_a^b f(x)\, dx \right| < \frac{\epsilon}{|k|+1}.$$

Therefore, $\|\mathcal{P}\| < \delta$ implies

$$\begin{aligned} \left| \mathcal{S}_\mathcal{P}(kf, \mathcal{C}) - k\int_a^b f(x)\, dx \right| &= \left| k\, \mathcal{S}_\mathcal{P}(f, \mathcal{C}) - k\int_a^b f(x)\, dx \right| \\ &= |k| \left| \mathcal{S}_\mathcal{P}(f, \mathcal{C}) - \int_a^b f(x)\, dx \right| \\ &< |k| \left(\frac{\epsilon}{|k|+1} \right) < \epsilon. \end{aligned}$$

That is,

$$\int_a^b k\, f(x)\, dx = \lim_{\|\mathcal{P}\| \to 0} \mathcal{S}_\mathcal{P}(k\, f, \mathcal{C}) = k \int_a^b f(x)\, dx.$$

To prove integrability in part b), let $\mathcal{P}$ be any partition of $[a,b]$. Then,

$$\mathcal{S}_\mathcal{P}(f \pm g, \mathcal{C}) = \sum_{j=1}^n \big(f(c_j) \pm g(c_j) \big)\, \Delta x_j = \mathcal{S}_\mathcal{P}(f, \mathcal{C}) \pm \mathcal{S}_\mathcal{P}(g, \mathcal{C}).$$

It is left to the reader to show that

$$\lim_{\|\mathcal{P}\| \to 0} \mathcal{S}_\mathcal{P}(f \pm g, \mathcal{C}) = \lim_{\|\mathcal{P}\| \to 0} \mathcal{S}_\mathcal{P}(f, \mathcal{C}) \pm \lim_{\|\mathcal{P}\| \to 0} \mathcal{S}_\mathcal{P}(g, \mathcal{C}),$$

and hence,

$$\int_a^b (f \pm g)(x)\, dx = \int_a^b f(x)\, dx \pm \int_a^b f(x)\, dx.$$

♦

► **7.17** *Complete the proof of part a) by showing that* kf *is bounded on* $[a,b]$ *and that* $\lim_{\|\mathcal{P}\| \to 0} \mathcal{S}_\mathcal{P}(k\, f, \mathcal{C}) = k \lim_{\|\mathcal{P}\| \to 0} \mathcal{S}_\mathcal{P}(f, \mathcal{C})$.

► **7.18** *Complete the proof of part b) by showing that $f \pm g$ is bounded on $[a,b]$ and that* $\lim_{\|\mathcal{P}\|\to 0} \mathcal{S}_{\mathcal{P}}(f \pm g, \mathcal{C}) = \lim_{\|\mathcal{P}\|\to 0} \mathcal{S}_{\mathcal{P}}(f, \mathcal{C}) \pm \lim_{\|\mathcal{P}\|\to 0} \mathcal{S}_{\mathcal{P}}(g, \mathcal{C})$.

We would next like to prove that if two functions are bounded and integrable over a closed interval, their product is also bounded and integrable over that interval. To do so we will need the following two technical lemmas.

Lemma 2.8 *Suppose $f : [a,b] \to \mathbb{R}$ is bounded and integrable on $[a,b]$. Then $f^+ : [a,b] \to \mathbb{R}$ given by*

$$f^+(x) = \begin{cases} 0 & \text{when } f(x) < 0 \\ f(x) & \text{when } f(x) \geq 0 \end{cases},$$

and $f^- : [a,b] \to \mathbb{R}$ given by

$$f^-(x) = \begin{cases} -f(x) & \text{when } f(x) < 0 \\ 0 & \text{when } f(x) \geq 0 \end{cases},$$

are also bounded and integrable on $[a,b]$.

PROOF We consider the case of f^+, and leave the f^- case to the reader. Clearly f^+ is bounded on $[a,b]$. Let $\mathcal{P}$ be a partition of $[a,b]$ and define $M_j^+ \equiv \sup_{I_j} f^+$ and $m_j^+ \equiv \inf_{I_j} f^+$. Then $M_j^+ - m_j^+ \leq M_j - m_j$ (Why?). Therefore, for $\epsilon > 0$ there exists a partition $\mathcal{P}$ of $[a,b]$ such that

$$\overline{\mathcal{S}}_{\mathcal{P}}(f^+) - \underline{\mathcal{S}}_{\mathcal{P}}(f^+) \leq \overline{\mathcal{S}}_{\mathcal{P}}(f) - \underline{\mathcal{S}}_{\mathcal{P}}(f) < \epsilon.$$

This establishes the integrability of f^+ on $[a,b]$. ♦

► **7.19** *Answer the (Why?) question in the above proof, and prove the case for f^-.*

Lemma 2.9 *Suppose $f : [a,b] \to \mathbb{R}$ and $g : [a,b] \to \mathbb{R}$ are both bounded, nonnegative and integrable on $[a,b]$. Then $fg : [a,b] \to \mathbb{R}$ is also bounded and integrable on $[a,b]$.*

PROOF We leave it to the reader to verify that fg is bounded on $[a,b]$. Let $\mathcal{P}$ be a partition of $[a,b]$ and define $M_j^{fg} \equiv \sup_{I_j}(fg)$ and $m_j^{fg} \equiv \inf_{I_j}(fg)$. Then, since

$$M_j^{fg} \leq M_j^f M_j^g \quad \text{and} \quad m_j^{fg} \geq m_j^f m_j^g, \qquad \text{(Why?)}$$

we have that

$$\begin{aligned}\overline{S}_{\mathcal{P}}(fg) - \underline{S}_{\mathcal{P}}(fg) &= \sum_j \left(M_j^{fg} - m_j^{fg}\right) \Delta x_j \\ &\le \sum_j \left(M_j^f\, M_j^g - m_j^f\, m_j^g\right) \Delta x_j \\ &= \sum_j \left[M_j^f \left(M_j^g - m_j^g\right) + m_j^g \left(M_j^f - m_j^f\right)\right] \Delta x_j \\ &\le M^f \left(\overline{S}_{\mathcal{P}}(g) - \underline{S}_{\mathcal{P}}(g)\right) + M^g \left(\overline{S}_{\mathcal{P}}(f) - \underline{S}_{\mathcal{P}}(f)\right), \end{aligned} \tag{7.23}$$

where $M^f = \sup_{[a,b]} f$ and $M^g = \sup_{[a,b]} g$. Now, since f and g are integrable on $[a,b]$, there exists a partition $\mathcal{P}$ of $[a,b]$ such that

$$\overline{S}_{\mathcal{P}}(g) - \underline{S}_{\mathcal{P}}(g) < \frac{\epsilon}{2\,(M^f+1)}, \quad \text{and} \quad \overline{S}_{\mathcal{P}}(f) - \underline{S}_{\mathcal{P}}(f) < \frac{\epsilon}{2\,(M^g+1)},$$

which together with (7.23) imply

$$\overline{S}_{\mathcal{P}}(fg) - \underline{S}_{\mathcal{P}}(fg) < \epsilon,$$

and the result is proved. ♦

► **7.20** *Verify that fg is bounded on $[a,b]$, and answer the (Why?) question in the above proof. In particular, show that $M_j^{fg} \le M_j^f\, M_j^g$, and that $m_j^{fg} \ge m_j^f\, m_j^g$.*

We can now state the following more general result, the proof of which is left to the reader.

Theorem 2.10 *Suppose $f : [a,b] \to \mathbb{R}$ and $g : [a,b] \to \mathbb{R}$ are both bounded and integrable on $[a,b]$. Then $fg : [a,b] \to \mathbb{R}$ is also bounded and integrable on $[a,b]$.*

► **7.21** *Prove the above theorem, noting that $f(x) = f^+(x) - f^-(x)$, and $g(x) = g^+(x) - g^-(x)$.*

Theorem 2.11 *Suppose $f : [a,b] \to \mathbb{R}$ and $g : [a,b] \to \mathbb{R}$ are both bounded and integrable on $[a,b]$.*

a) If $f(x) \le g(x)$ on $[a,b]$, then $\int_a^b f(x)\,dx \le \int_a^b g(x)\,dx$.

b) $|f(x)|$ is integrable on $[a,b]$, and $\left|\int_a^b f(x)\,dx\right| \le \int_a^b |f(x)|\,dx$.

PROOF To prove *a*), define $h : [a,b] \to \mathbb{R}$ by $h(x) = g(x) - f(x)$, and note that $h(x) \ge 0$ on $[a,b]$ and that h is integrable according to part *b*) of Theorem 2.7 on page 354. Also, for any partition $\mathcal{P}$ of $[a,b]$ we have

$$S_{\mathcal{P}}(h, \mathcal{C}) = \sum_{j=1}^{n} h(c_j)\, \Delta x_j \ge 0.$$

It is left to the reader to show that $\lim_{\|\mathcal{P}\|\to 0} S_{\mathcal{P}}(h, C) \geq 0$, and therefore,

$$\int_a^b h(x)\,dx \geq 0.$$

Again by part *b*) of Theorem 2.7, we then have

$$\int_a^b f(x)\,dx \leq \int_a^b g(x)\,dx.$$

To establish part *b*) of the present theorem, we note that if f is integrable then f^+ and f^- are integrable. Since $|f| = f^+ + f^-$, it follows that $|f|$ is integrable. The details of the rest of the proof of part *b*) are left to the reader. ♦

► **7.22** *Complete the proof of part a) by considering the bounded function $h : [a,b] \to \mathbb{R}$ satisfying $h(x) \geq 0$ on $[a,b]$. Show that $\lim_{\|\mathcal{P}\|\to 0} S_{\mathcal{P}}(h, C) \geq 0$. Complete the proof of part b).*

The following corollary to the above theorem is a useful result on its own.

Corollary 2.12 *If $f : [a,b] \to \mathbb{R}$ is bounded and integrable on $[a,b]$ with $\Lambda \equiv \sup_{[a,b]} |f(x)|$, then*

$$\left|\int_a^b f(x)\,dx\right| \leq \Lambda\,(b-a).$$

► **7.23** *Prove the above corollary.*

► **7.24** *In this exercise we will establish a result known as* ***Jordan's Inequality****. It will be useful to us in Chapter 9. In particular, we will establish that $\int_0^\pi e^{-R\sin\theta}\,d\theta \leq \frac{\pi}{R}$. From this, we will be able to conclude that $\lim_{R\to\infty} \int_0^\pi e^{-R\sin\theta}\,d\theta = 0$. To begin, note that $\int_0^\pi e^{-R\sin\theta}\,d\theta = 2\int_0^{\pi/2} e^{-R\sin\theta}\,d\theta$. Draw a graph of the sine function on $[0, \pi/2]$ to see that $\sin\theta \geq \frac{2\theta}{\pi}$ on that interval. Verify this inequality analytically by considering the function $f(\theta) = \sin\theta - \frac{2}{\pi}$ on the interval $[0, \pi/2]$. Where does it achieve its maximum and minimum values? With this inequality established, note that*

$$\int_0^\pi e^{-R\sin\theta}\,d\theta = 2\int_0^{\pi/2} e^{-R\sin\theta}\,d\theta \leq 2\int_0^{\pi/2} e^{-2R\theta/\pi}\,d\theta = \frac{\pi}{R}(1-e^{-R}) \leq \frac{\pi}{R}.$$

Integration over Subintervals

One more result in this subsection rounds out our list of convenient integral properties for bounded functions. To prove it, we will need the following lemma.

Lemma 2.13 *Suppose $f : [a,b] \to \mathbb{R}$ is bounded and integrable on $[a,b]$. If $[c,d] \subset [a,b]$, then f is also integrable on $[c,d]$.*

PROOF Note that after two applications of Lemma 1.15 on page 345, we have

$$\overline{\int_a^b} f(x)\,dx = \overline{\int_a^c} f(x)\,dx + \overline{\int_c^d} f(x)\,dx + \overline{\int_d^b} f(x)\,dx, \tag{7.24}$$

and similarly,

$$\underline{\int_a^b} f(x)\,dx = \underline{\int_a^c} f(x)\,dx + \underline{\int_c^d} f(x)\,dx + \underline{\int_d^b} f(x)\,dx. \tag{7.25}$$

Subtracting (7.25) from (7.24), we obtain

$$\begin{aligned} 0 &= \overline{\int_a^b} f(x)\,dx - \underline{\int_a^b} f(x)\,dx \\ &= \left(\overline{\int_a^c} f(x)\,dx - \underline{\int_a^c} f(x)\,dx\right) + \left(\overline{\int_c^d} f(x)\,dx - \underline{\int_c^d} f(x)\,dx\right) + \\ &\qquad \left(\overline{\int_d^b} f(x)\,dx - \underline{\int_d^b} f(x)\,dx\right) \\ &\geq \left(\overline{\int_c^d} f(x)\,dx - \underline{\int_c^d} f(x)\,dx\right) \\ &\geq 0, \end{aligned}$$

which implies

$$\overline{\int_c^d} f(x)\,dx = \underline{\int_c^d} f(x)\,dx.$$

Therefore, f is integrable on $[c, d]$. ♦

► **7.25** *For an alternative proof of the above theorem, consider $\epsilon > 0$ and let $\mathcal{P}$ be a partition of $[a, b]$ such that $\overline{S}_{\mathcal{P}}(f) - \underline{S}_{\mathcal{P}}(f) < \epsilon$. Let $\mathcal{P}' \equiv \mathcal{P} \cup \{c, d\}$, and verify that $\overline{S}_{\mathcal{P}'}(f) - \underline{S}_{\mathcal{P}'}(f) < \epsilon$ still holds. Finally, for $\mathcal{P}'' \equiv \mathcal{P}' \cap [c, d]$, show that $\overline{S}_{\mathcal{P}''}(f) - \underline{S}_{\mathcal{P}''}(f) \leq \overline{S}_{\mathcal{P}'}(f) - \underline{S}_{\mathcal{P}'}(f) < \epsilon$, and so f is integrable on $[c, d]$.*

The following important property can now be proved, a task that is left to the reader.

Theorem 2.14 *Suppose $f : [a, b] \to \mathbb{R}$ is bounded and integrable on $[a, b]$, and $c \in [a, b]$. Then*

$$\int_a^b f(x)\,dx = \int_a^c f(x)\,dx + \int_c^b f(x)\,dx.$$

► **7.26** *Prove the above theorem. Show also that if $c_1 \leq c_2 \leq \cdots \leq c_m$ are points each lying within $[a, b]$, then $\int_a^b f(x)\,dx = \int_a^{c_1} f(x)\,dx + \int_{c_1}^{c_2} f(x)\,dx + \cdots + \int_{c_m}^b f(x)\,dx$.*

► **7.27** *Suppose $f : [a, b] \to \mathbb{R}$ is bounded and integrable on $[a, b]$, and let c_1, c_2, and c_3 be any three points in $[a, b]$. Show that $\int_{c_1}^{c_2} f(x)\,dx + \int_{c_2}^{c_3} f(x)\,dx = \int_{c_1}^{c_3} f(x)\,dx$. Note that you need not presume $c_1 \leq c_2 \leq c_3$.*

2.3 The Fundamental Theorem of Calculus

We now will establish one of the most important results in all of mathematics, the fundamental theorem of calculus. To do so, we note that if $f : [a, b] \to \mathbb{R}$ is bounded and integrable on $[a, b]$, then it is integrable on every subinterval $[a, u] \subset [a, b]$, and therefore the integral $\int_a^u f(x)\,dx$ determines a function $F(u)$ defined on $[a, b]$. This function has nice properties, as the following theorem establishes.

Theorem 2.15 *If $f : [a, b] \to \mathbb{R}$ is bounded and integrable on $[a, b]$, let $F : [a, b] \to \mathbb{R}$ be given by*

$$F(u) = \int_a^u f(x)\,dx.$$

The function F has the following properties:

a) F is continuous on $[a, b]$.

b) F is differentiable at points $u_0 \in [a, b]$ where f is continuous, and in this case $F'(u_0) = f(u_0)$.

Note that the conclusion to part *b)* of the theorem can be understood as

$$\left(\frac{d}{du}\int_a^u f(x)\,dx\right)\bigg|_{u=u_0} = f(u_0),$$

a result with great significance.

PROOF Let $\Lambda \equiv \sup_{[a,b]} |f|$. If $\Lambda = 0$, then $F \equiv 0$ on $[a, b]$, and so the theorem holds. Assume $\Lambda \neq 0$. Consider any $\epsilon > 0$ and fix $u_0 \in [a, b]$. For any $u \in [a, b]$, we have

$$\begin{aligned}
|F(u) - F(u_0)| &= \left|\int_a^u f(x)\,dx - \int_a^{u_0} f(x)\,dx\right| \\
&= \left|\int_{u_0}^u f(x)\,dx\right| \\
&\leq \Lambda\,|u - u_0| \\
&< \epsilon \quad \text{if} \quad |u - u_0| < \tfrac{\epsilon}{\Lambda},
\end{aligned}$$

establishing continuity of F at u_0. To prove part *b)* of the theorem, consider $u_0 \in [a, b]$ such that f is continuous at u_0. Then for $\epsilon > 0$ we must show there exists $\delta > 0$ such that for $u \in (a, b)$,

$$|u - u_0| < \delta \;\Rightarrow\; |F(u) - F(u_0) - f(u_0)\,(u - u_0)| \leq \epsilon\,|u - u_0|.$$

We have

$$\begin{aligned}
|F(u) - F(u_0) - f(u_0)\,(u - u_0)| &= \left|\int_a^u f(x)\,dx - \int_a^{u_0} f(x)\,dx - f(u_0)\,(u - u_0)\right| \\
&= \left|\int_{u_0}^u \big(f(x) - f(u_0)\big)\,dx\right|. \qquad (7.26)
\end{aligned}$$

Since f is continuous at u_0, there exists $\delta > 0$ such that for $x \in [a, b]$,

$$|x - u_0| < \delta \quad \Rightarrow \quad |f(x) - f(u_0)| < \epsilon.$$

If we also impose the condition that $|u - u_0| < \delta$, then (7.26) becomes

$$|F(u) - F(u_0) - f(u_0)(u - u_0)| \leq \epsilon\, |u - u_0|,$$

proving that F is differentiable at u_0 and that

$$F'(u_0) = f(u_0).$$

♦

Up to this point we have studied integrals and their properties without providing a practical method for calculating them. For complicated functions, calculating Riemann sums or even upper and lower sums can be difficult. Our next theorem, besides being of deep theoretical significance, is therefore an extremely practical result. Before stating and proving it, we give a definition.

Definition 2.16 For any function $f : [a, b] \to \mathbb{R}$, we define an **antiderivative** of f on $[a, b]$ to be any differentiable function $F : [a, b] \to \mathbb{R}$ satisfying

$$F'(x) = f(x) \quad \text{on} \quad [a, b].$$

► **7.28** *If F is an antiderivative for $f : [a, b] \to \mathbb{R}$ on $[a, b]$, is it unique? If F and G are two antiderivatives for f on $[a, b]$, how might they differ? Show this by proper use of Definition 2.16.*

Theorem 2.17 (The Fundamental Theorem of Calculus)
Suppose $f : [a, b] \to \mathbb{R}$ is bounded and integrable on $[a, b]$ and has antiderivative $F : [a, b] \to \mathbb{R}$ on $[a, b]$. Then

$$\int_a^b f(x)\, dx = F(b) - F(a).$$

PROOF Let $\mathcal{P} = \{a = x_0, x_1, \ldots, x_n = b\}$ be any partition of $[a, b]$. Then,

$$\begin{aligned} F(b) - F(a) &= \sum_{j=1}^{n} \big(F(x_j) - F(x_{j-1})\big) \\ &= \sum_{j=1}^{n} F'(c_j)\, \Delta x_j \quad \text{for points} \quad c_j \in I_j \quad \text{(Why?)} \\ &= \sum_{j=1}^{n} f(c_j)\, \Delta x_j \\ &= \mathcal{S}_{\mathcal{P}}(f, \mathcal{C}) \quad \text{where} \quad \mathcal{C} = \{c_j\}. \end{aligned}$$

From this we obtain,

$$\underline{S}_{\mathcal{P}}(f) \leq F(b) - F(a) \leq \overline{S}_{\mathcal{P}}(f) \quad \text{for all partitions } \mathcal{P} \text{ of } [a,b].$$

Therefore,

$$\underline{\int_a^b} f(x)\,dx \leq F(b) - F(a) \leq \overline{\int_a^b} f(x)\,dx.$$

Since $\int_a^b f(x)\,dx = \underline{\int_a^b} f(x)\,dx = \overline{\int_a^b} f(x)\,dx$, it follows that

$$\int_a^b f(x)\,dx = F(b) - F(a).$$

♦

► **7.29** *Answer the (Why?) question in the above proof. If f has more than one antiderivative on $[a,b]$, does it matter which one you use in evaluating the integral of f over $[a,b]$?*

► **7.30** *Show that, for $R > 0$, $\int_0^\pi e^{-R\sin\theta}\,d\theta < \frac{\pi}{R}$. (Hint: To do so, first prove Jordan's inequality, namely, $\frac{2\theta}{\pi} \leq \sin\theta$ for $\theta \in [0, \frac{\pi}{2}]$. This in turn yields $\int_0^{\frac{\pi}{2}} e^{-R\sin\theta}\,d\theta \leq \int_0^{\frac{\pi}{2}} e^{\frac{-2R\theta}{\pi}}\,d\theta$.)*

In practice, the expression of the conclusion of the fundamental theorem of calculus is often given as

$$\int_a^b f(x)\,dx = F(x)\Big|_a^b \equiv F(b) - F(a),$$

a notation with which students of calculus should be familiar.

Average Values

For a function $f : [a,b] \to \mathbb{R}$, the *average value* of f on the interval $[a,b]$, denoted by $\langle f \rangle$, might be of interest. If $\mathcal{P}$ is a partition of $[a,b]$, then the average value of f over $[a,b]$ might be estimated by

$$\frac{1}{n}\sum_{j=1}^n f(c_j) \quad \text{where} \quad c_j \in I_j. \tag{7.27}$$

If the partition $\mathcal{P}$ is chosen so that $\Delta x_j = \frac{b-a}{n}$ for $1 \leq j \leq n$, then $\frac{1}{n} = \frac{\Delta x_j}{b-a}$ for every $1 \leq j \leq n$, and the estimate (7.27) becomes

$$\frac{1}{b-a}\sum_{j=1}^n f(c_j)\,\Delta x_j. \tag{7.28}$$

If the function f is integrable on $[a,b]$, letting $\|\mathcal{P}\| \to 0$ in (7.28) obtains the value

$$\lim_{\|\mathcal{P}\|\to 0}\left(\frac{1}{b-a}\sum_{j=1}^n f(c_j)\,\Delta x_j\right) = \frac{1}{b-a}\int_a^b f(x)\,dx.$$

This discussion motivates the following definition.

Definition 2.18 Suppose $f : [a, b] \to \mathbb{R}$ is integrable on $[a, b]$. Then the **average value** of f over $[a, b]$ is given by

$$\langle f \rangle \equiv \frac{1}{b-a} \int_a^b f(x)\, dx.$$

It is an interesting fact that if f is continuous on $[a, b]$, then f assumes its average value over the interval $[a, b]$ at some point $\xi \in (a, b)$. We prove this result next.

Theorem 2.19 *Suppose $f : [a, b] \to \mathbb{R}$ is continuous on $[a, b]$. Then there exists $\xi \in (a, b)$ such that*

$$f(\xi) = \frac{1}{b-a} \int_a^b f(x)\, dx.$$

PROOF Let $F : [a, b] \to \mathbb{R}$ be given by

$$F(u) = \int_a^u f(x)\, dx.$$

Then, according to Theorem 2.15 on page 360, we know that F is continuous and differentiable on $[a, b]$ such that $F'(x) = f(x)$ on $[a, b]$. Applying the mean value theorem to F, we obtain

$$F(b) - F(a) = F'(\xi)\,(b-a) \quad \text{for some} \quad \xi \in (a, b),$$

i.e.,

$$\int_a^b f(x)\, dx = f(\xi)\,(b-a).$$

◆

► **7.31** *For an alternative proof of the above theorem, show that the quantity given by $\frac{1}{b-a} \int_a^b f(x)\, dx$ is between the maximum and the minimum of f on $[a, b]$, and apply the intermediate value theorem.*

3 Further Development of Integration Theory

3.1 Improper Integrals of Bounded Functions

We sometimes wish to investigate the integrability of bounded functions over unbounded intervals. To begin, we define what it means for a function to be integrable on an interval of the form $[a, \infty)$, or $(-\infty, b]$.

Definition 3.1 Suppose $f : [a, \infty) \to \mathbb{R}$ is bounded and integrable on $[a, b]$ for every $b > a$. If the limit $\lim\limits_{b\to\infty} \int_a^b f(x)\, dx$ exists, then we say that f is **integrable on** $[a, \infty)$ and we denote the integral by

$$\int_a^\infty f(x)\, dx \equiv \lim_{b\to\infty} \int_a^b f(x)\, dx.$$

Similarly, suppose $f : (-\infty, b] \to \mathbb{R}$ is bounded and integrable on $[a, b]$ for every $a < b$. If the limit $\lim_{a\to-\infty} \int_a^b f(x)\,dx$ exists, then we say that f is **integrable on** $(-\infty, b]$ and we denote the integral by

$$\int_{-\infty}^{b} f(x)\,dx \equiv \lim_{a\to-\infty} \int_a^b f(x)\,dx.$$

Example 3.2 *Consider $f : [1, \infty) \to \mathbb{R}$ given by $f(x) = \frac{1}{x^2}$. Then,*

$$\int_1^b \frac{1}{x^2}\,dx = -\frac{1}{x}\bigg|_1^b = -\tfrac{1}{b} + 1,$$

and so

$$\int_1^\infty \frac{1}{x^2}\,dx = \lim_{b\to\infty} \left(-\tfrac{1}{b} + 1\right) = 1.$$

◀

▶ **7.32** *Evaluate the integral $\int_{-\infty}^{-1} \frac{1}{x^2}\,dx$.*

If each of the integrals described in Definition 3.1 exists, then we consider the function to be integrable on the whole real line. We define this formally below.

Definition 3.3 Suppose $f : \mathbb{R} \to \mathbb{R}$ is bounded and integrable on $[a, b]$ for every choice of $a < b \in \mathbb{R}$, and for every $c \in \mathbb{R}$ the integrals $\int_c^\infty f(x)\,dx$ and $\int_{-\infty}^c f(x)\,dx$ exist. Then we say that f is **integrable on** $\mathbb{R}$, and we denote the integral by

$$\int_{-\infty}^{\infty} f(x)\,dx \equiv \int_{-\infty}^{c} f(x)\,dx + \int_c^\infty f(x)\,dx.$$

▶ **7.33** *For a function $f : \mathbb{R} \to \mathbb{R}$ satisfying the conditions of the above definition, show that the sum $\int_{-\infty}^c f(x)\,dx + \int_c^\infty f(x)\,dx$ is independent of the choice of c.*

▶ **7.34** *Suppose $f : [a, b] \to \mathbb{R}$ is integrable on $[a, b]$ for every choice of $a, b \in \mathbb{R}$, and for a particular choice of $c \in \mathbb{R}$ the integrals $\int_c^\infty f(x)\,dx$ and $\int_{-\infty}^c f(x)\,dx$ exist. Show that this is enough to conclude that Definition 3.3 holds.*

Example 3.4 *Consider $f : \mathbb{R} \to \mathbb{R}$ given by $f(x) = x\,e^{-x^2}$. Then since f is continuous on all of $\mathbb{R}$, it follows that f is integrable on $[a, b]$ for all $a, b \in \mathbb{R}$. Note that*

$$\begin{aligned}
\int_{-\infty}^{c} x\,e^{-x^2}\,dx &= \lim_{a\to-\infty} \int_a^c x\,e^{-x^2}\,dx \\
&= \lim_{a\to-\infty} \left(-\tfrac{1}{2}\,e^{-x^2}\Big|_a^c\right) \\
&= \lim_{a\to-\infty} \left(-\tfrac{1}{2}\,e^{-c^2} + \tfrac{1}{2}\,e^{-a^2}\right) \\
&= -\tfrac{1}{2}\,e^{-c^2}.
\end{aligned}$$

Similarly,

$$\int_c^\infty x\, e^{-x^2}\, dx = \tfrac{1}{2}\, e^{-c^2}.$$

Therefore, we have that

$$\int_{-\infty}^\infty x\, e^{-x^2}\, dx = \int_{-\infty}^c x\, e^{-x^2}\, dx + \int_c^\infty x\, e^{-x^2}\, dx = -\tfrac{1}{2}\, e^{-c^2} + \tfrac{1}{2}\, e^{-c^2} = 0.$$

◀

Example 3.5 *Consider $f : \mathbb{R} \to \mathbb{R}$ given by $f(x) = x$. Then since f is continuous on all of $\mathbb{R}$, it follows that f is integrable on $[a, b]$ for all $a, b \in \mathbb{R}$. In this case, note that*

$$\begin{aligned}\int_{-\infty}^c x\, dx &= \lim_{a\to-\infty} \int_a^c x\, dx \\ &= \lim_{a\to-\infty} \left(\tfrac{1}{2}\, x^2\Big|_a^c\right) \\ &= \lim_{a\to-\infty} \left(\tfrac{1}{2}\, c^2 - \tfrac{1}{2}\, a^2\right),\end{aligned}$$

which does not exist. Therefore, $\int_{-\infty}^\infty x\, dx$ does not exist. ◀

► **7.35** *Verify that the integral $\int_c^\infty x\, dx$ does not exist either, although we do not need this additional information in order to come to the conclusion of the previous example.*

Note that if we had computed the improper integral $\int_{-\infty}^\infty x\, dx$ from the previous example somewhat differently, we would have arrived at a different conclusion. Namely, consider the limit given by

$$\lim_{R\to\infty} \int_{-R}^R x\, dx = \lim_{R\to\infty} \tfrac{1}{2}\, x^2\Big|_{-R}^R = \lim_{R\to\infty} \left(\tfrac{1}{2}\, R^2 - \tfrac{1}{2}\, (-R)^2\right) = 0.$$

In computing the limit in this way, we are mistakenly led to believe that the improper integral $\int_{-\infty}^\infty x\, dx$ exists and equals 0. Note that in carrying out this "different" limit process, we have not used our definition for evaluating such "doubly improper" integrals as $\int_{-\infty}^\infty x\, dx$, namely, Definition 3.3. According to Definition 3.3, this integral does not exist. And yet this "different" limit procedure is sometimes useful. We define it more formally below.

Definition 3.6 Suppose $f : \mathbb{R} \to \mathbb{R}$ is bounded and integrable on $[a, b]$ for every choice of $a < b \in \mathbb{R}$. Then, if it exists, the **Cauchy principal value** of the integral $\int_{-\infty}^\infty f(x)\, dx$ is given by the following limit,

$$\text{CPV} \int_{-\infty}^\infty f(x)\, dx \equiv \lim_{R\to\infty} \int_{-R}^R f(x)\, dx.$$

The reader is cautioned to note, as illustrated by Example 3.5, that the value CPV $\int_{-\infty}^{\infty} f(x)\,dx$ may exist for a given function f even if $\int_{-\infty}^{\infty} f(x)\,dx$ does not.

► **7.36** *Calculate CPV $\int_{-\infty}^{\infty} e^{-|x|}\,dx$. Does $\int_{-\infty}^{\infty} e^{-|x|}\,dx$ exist? What is its relationship to CPV $\int_{-\infty}^{\infty} e^{-|x|}\,dx$?*

► **7.37** *Show that if $f : \mathbb{R} \to \mathbb{R}$ is an even function then $\int_{-\infty}^{\infty} f(x)\,dx$ exists if and only if CPV $\int_{-\infty}^{\infty} f(x)\,dx$ gives the same value.*

► **7.38** *Show that if $\int_{-\infty}^{\infty} f(x)\,dx$ converges, then CPV $\int_{-\infty}^{\infty} f(x)\,dx$ must converge to the same value.*

3.2 Recognizing a Sequence as a Riemann Sum

We sometimes can use our acquired knowledge of integrals, paticularly integrable functions and their associated Riemann sums, to determine the limiting value of an otherwise difficult sequence. To this end, we will make use of equality (7.3) on page 338. For example, consider the expression given by

$$\lim_{n\to\infty} \sum_{j=1}^{n} \frac{\sqrt{j}}{n^{3/2}} = \lim_{n\to\infty} \left(\frac{1}{n^{3/2}} + \frac{\sqrt{2}}{n^{3/2}} + \frac{\sqrt{3}}{n^{3/2}} + \cdots + \frac{\sqrt{n}}{n^{3/2}} \right).$$

If we can recognize the finite sum as being a Riemann sum associated with some integrable function $f : [a,b] \to \mathbb{R}$, we can evaluate the limit as the integral of f over $[a,b]$, a much simpler task in most cases than determining the sum directly. In our example, note that the finite sum can be rewritten as

$$\sum_{j=1}^{n} \frac{\sqrt{j}}{n^{3/2}} = \sum_{j=1}^{n} \sqrt{\frac{j}{n}}\,\frac{1}{n}.$$

This corresponds to a Riemann sum of the function $f : [0,1] \to \mathbb{R}$ given by $f(x) = \sqrt{x}$, with $a = 0$, $b = 1$, and $\Delta x_j = \frac{1}{n}$ for each $1 \le j \le n$. Since we know this f is integrable over $[0,1]$, the limit of the sum is equal to the integral $\int_0^1 \sqrt{x}\,dx$, that is,

$$\lim_{n\to\infty} \sum_{j=1}^{n} \frac{\sqrt{j}}{n^{3/2}} = \lim_{n\to\infty} \sum_{j=1}^{n} \sqrt{\frac{j}{n}}\,\frac{1}{n} = \int_0^1 \sqrt{x}\,dx = \tfrac{2}{3}\,x^{3/2}\Big|_0^1 = \tfrac{2}{3}.$$

► **7.39** *What partition $\mathcal{P}$ of $[0,1]$ is implicitly used in the above Riemann sum for $f(x) = \sqrt{x}$? Also, what set of points C is implicitly used there?*

► **7.40** *Find $\lim\limits_{n\to\infty} \sum_{j=1}^{n} \frac{1}{n} e^{j/n}$.*

► **7.41** *Show that $\ln u = \lim\limits_{n\to\infty} \sum_{j=1}^{n} \frac{u-1}{n+j\,(u-1)}$.*

3.3 Change of Variables Theorem

The following theorem is especially useful, as every student of calculus knows.

Theorem 3.7 *Suppose $f : [a,b] \to \mathbb{R}$ is continuous on $[a,b]$. Suppose $g : [c,d] \to [a,b]$ is a function such that*

$$(i)\ \ g(c) = a \ \text{ and } \ g(d) = b, \tag{7.29}$$

$$(ii)\ \ g' \text{ exists on } [c,d] \text{ and is continuous on } [c,d]. \tag{7.30}$$

Then,

$$\int_a^b f(x)\,dx = \int_c^d f(g(y))\,g'(y)\,dy.$$

PROOF The proof makes explicit use of the chain rule. Let $F(u) \equiv \int_a^u f(x)\,dx$. Then $F'(u) = f(u)$ for all $u \in [a,b]$. Note that

$$\frac{d}{dy}F(g(y)) = F'(g(y))\,g'(y) = f(g(y))\,g'(y),$$

i.e., $F(g(y))$ is an antiderivative of $f(g(y))\,g'(y)$. Therefore,

$$\begin{aligned}\int_c^d f(g(y))\,g'(y)\,dy &= F(g(d)) - F(g(c)) \\ &= F(b) - F(a) \\ &= \int_a^b f(x)\,dx.\end{aligned}$$

◆

In applying the above theorem, we implicitly make use of the relationship between x and y given by $x = g(y)$. The following example illustrates the technique, so familiar to calculus students.

Example 3.8 *Consider $\int_1^4 x\,e^{x^2}\,dx$. If we let $g : [1,2] \to [1,4]$ be given by $g(y) = \sqrt{y}$, then*

$$\int_1^4 x\,e^{x^2}\,dx = \int_1^2 \sqrt{y}\,e^y\,\frac{1}{2\sqrt{y}}\,dy = \tfrac{1}{2}\int_1^2 e^y\,dy = \tfrac{1}{2}\,(e^2 - e).$$

◄

3.4 Uniform Convergence and Integration

Theorem 3.9 *Suppose $f_n : [a,b] \to \mathbb{R}$ is a sequence of bounded integrable functions on $[a,b]$ that converges uniformly to $f : [a,b] \to \mathbb{R}$. Then f is bounded and integrable on $[a,b]$.*

PROOF We leave it to the reader to prove that f is bounded on $[a,b]$. We will establish that f is integrable. By the uniform convergence of $\{f_n\}$ to f on $[a,b]$, for any $\epsilon > 0$ there exists $N \in \mathbb{N}$ such that for all $x \in [a,b]$,

$$n > N \ \Rightarrow\ |f_n(x) - f(x)| < \frac{\epsilon}{2\,(b-a)} \equiv \tilde{\epsilon}.$$

This implies that for all $x \in [a, b]$ and for all $n > N$,

$$f_n(x) - \widetilde{\epsilon} < f(x) < f_n(x) + \widetilde{\epsilon},$$

and therefore

$$\int_a^b \left(f_n(x) - \widetilde{\epsilon}\right) dx = \underline{\int_a^b} \left(f_n(x) - \widetilde{\epsilon}\right) dx \le \underline{\int_a^b} f(x)\, dx$$
$$\le \overline{\int_a^b} f(x)\, dx \le \overline{\int_a^b} \left(f_n(x) + \widetilde{\epsilon}\right) dx = \int_a^b \left(f_n(x) + \widetilde{\epsilon}\right) dx. \tag{7.31}$$

A simple rearrangement of (7.31) yields

$$0 \le \overline{\int_a^b} f(x)\, dx - \underline{\int_a^b} f(x)\, dx \le \int_a^b \left(f_n(x) + \widetilde{\epsilon}\right) dx - \int_a^b \left(f_n(x) - \widetilde{\epsilon}\right) dx = \epsilon.$$

Since this holds for all $\epsilon > 0$, it follows that $\underline{\int_a^b} f(x)\, dx = \overline{\int_a^b} f(x)\, dx$, and so f is integrable on $[a, b]$. ♦

▶ **7.42** *Complete the above proof by establishing that f is bounded on $[a, b]$.*

Theorem 3.10 *Suppose $f_n : [a, b] \to \mathbb{R}$ is a sequence of bounded integrable functions on $[a, b]$ that converges uniformly to $f : [a, b] \to \mathbb{R}$. Then,*

$$\lim_{n\to\infty} \int_a^b f_n(x)\, dx = \int_a^b \left(\lim_{n\to\infty} f_n(x)\right) dx = \int_a^b f(x)\, dx.$$

PROOF By Theorem 3.9 we know that f is bounded and integrable on $[a, b]$. By the uniform convergence of $\{f_n\}$ to f on $[a, b]$, for any $\epsilon > 0$ there exists $N \in \mathbb{N}$ such that for all $x \in [a, b]$,

$$n > N \ \Rightarrow\ |f_n(x) - f(x)| < \frac{\epsilon}{2\,(b-a)}.$$

Therefore,

$$n > N \ \Rightarrow\ \left|\int_a^b \left(f_n(x) - f(x)\right) dx\right| \le \frac{\epsilon}{2\,(b-a)}\,(b-a) < \epsilon. \tag{7.32}$$

Expression (7.32) is equivalent to

$$n > N \ \Rightarrow\ \left|\int_a^b f_n(x)\, dx - \int_a^b f(x)\, dx\right| < \epsilon,$$

which implies that

$$\lim_{n\to\infty} \int_a^b f_n(x)\, dx = \int_a^b f(x)\, dx.$$

♦

Example 3.11 *Consider the sequence of functions $f_n : [1, 2] \to \mathbb{R}$ given by*

$f_n(x) = \frac{n^2 x}{1+n^3 x^2}$. We know from a previous exercise that $f_n(x) \to 0$ uniformly on $[1, 2]$. Therefore, it must be true that

$$\lim_{n\to\infty} \int_1^2 \frac{n^2 x}{1+n^3 x^2}\, dx = \int_1^2 \left(\lim_{n\to\infty} \frac{n^2 x}{1+n^3 x^2} \right) dx = \int_1^2 0\, dx = 0.$$

◀

► **7.43** *Change the domain in the previous example to $[0, 1]$. What happens now?*

► **7.44** *Let $f_n : [-1, 1] \to \mathbb{R}$ be a sequence of functions given by $f_n(x) = \frac{n x^2}{n+1+x}$.*

a) Find $\lim_{n\to\infty} f_n(x)$.
b) Is the convergence uniform?
c) Compute $\lim_{n\to\infty} \int_{-1}^1 f_n(x)\, dx$ and $\int_{-1}^1 \left(\lim_{n\to\infty} f_n(x)\right) dx$. Are they equal? Why?

Corollary 3.12 *Suppose $f_j : [a, b] \to \mathbb{R}$ is a sequence of bounded integrable functions on $[a, b]$ such that $f(x) = \sum_{j=1}^{\infty} f_j(x)$ converges uniformly on $[a, b]$. Then $f : [a, b] \to \mathbb{R}$ is bounded and integrable on $[a, b]$, and,*

$$\int_a^b f(x)\, dx = \int_a^b \Big(\sum_{j=1}^{\infty} f_j(x) \Big) dx = \sum_{j=1}^{\infty} \left(\int_a^b f_j(x)\, dx \right).$$

► **7.45** *Prove the above corollary.*

Example 3.13 *Consider the series of functions $\sum_{j=1}^{\infty} \frac{\sin(jx)}{j^5}$. We know from an exercise in Chapter 3 that this series converges uniformly on $\mathbb{R}$; therefore it must converge uniformly on $[-A, A]$ for every choice of real number $A > 0$. From this, we conclude that*

$$\int_{-A}^{A} \left(\sum_{j=1}^{\infty} \frac{\sin(jx)}{j^5} \right) dx = \sum_{j=1}^{\infty} \left(\int_{-A}^{A} \frac{\sin(jx)}{j^5}\, dx \right) = \sum_{j=1}^{\infty} \frac{1}{j^5} \left(\int_{-A}^{A} \sin(jx)\, dx \right) = 0.$$

◀

4 Vector-Valued and Line Integrals

4.1 The Integral of $f : [a, b] \to \mathbb{R}^p$

We now consider the integral of a function $\boldsymbol{f} : [a, b] \to \mathbb{R}^p$, easily extending many of the results from the real-valued case. We begin with the following definitions.

Definition 4.1 Consider $\boldsymbol{f} : [a, b] \to \mathbb{R}^p$, given by $\boldsymbol{f}(x) = \big(f_1(x), \ldots, f_p(x)\big)$ where $f_j : [a, b] \to \mathbb{R}$ for $1 \le j \le p$. Then, if f_j is integrable on $[a, b]$ for each $1 \le j \le p$, we say that $\boldsymbol{f}$ is **integrable** on $[a, b]$, and

$$\int_a^b \boldsymbol{f}(x)\, dx = \left(\int_a^b f_1(x)\, dx, \ldots, \int_a^b f_p(x)\, dx \right).$$

Definition 4.2 Suppose $\boldsymbol{f} : [a, b] \to \mathbb{R}^p$ is given by $\boldsymbol{f}(x) = \big(f_1(x), \ldots, f_p(x)\big)$ where $f_j : [a, b] \to \mathbb{R}$ for $1 \le j \le p$. Then we define **an antiderivative** of $\boldsymbol{f}$ on $[a, b]$ to be any differentiable function $\boldsymbol{F} : [a, b] \to \mathbb{R}^p$ satisfying

$$\boldsymbol{F}'(x) = \boldsymbol{f}(x) \quad \text{on } [a, b].$$

► **7.46** *If the function* $\boldsymbol{F} : [a, b] \to \mathbb{R}^p$ *in the above definition is given by* $\boldsymbol{F}(x) = \big(F_1(x), \ldots, F_p(x)\big)$ *where* $F_j : [a, b] \to \mathbb{R}$ *for* $1 \le j \le p$, *show that each* F_j *satisfies* $F_j' = f_j$ *on* $[a, b]$, *and is therefore an antiderivative of* f_j *on* $[a, b]$ *for* $1 \le j \le p$.

With these definitions in hand, we state the following theorems, which are analogous to Theorems 2.15 and 2.17 on pages 360 and 361.

Theorem 4.3 *If* $\boldsymbol{f} : [a, b] \to \mathbb{R}^p$ *is bounded and integrable on* $[a, b]$, *let* $\boldsymbol{F} : [a, b] \to \mathbb{R}^p$ *be the function given by*

$$\boldsymbol{F}(u) = \int_a^u \boldsymbol{f}(x)\, dx.$$

Then $\boldsymbol{F}$ *has the following properties:*

a) $\boldsymbol{F}$ *is continuous on* $[a, b]$,

b) $\boldsymbol{F}$ *is differentiable at points* $u_0 \in [a, b]$ *where* $\boldsymbol{f}$ *is continuous, and in this case* $\boldsymbol{F}'(u_0) = \boldsymbol{f}(u_0)$.

Note that the conclusion to part *b)* of the theorem can be written

$$\frac{d}{du}\left(\int_a^u \boldsymbol{f}(x)\, dx \right)\bigg|_{u_0} = \boldsymbol{f}(u_0).$$

► **7.47** *Prove the above theorem.*

Theorem 4.4 *Suppose* $\boldsymbol{f} : [a,b] \to \mathbb{R}^p$ *is bounded and integrable on* $[a,b]$, *and has antiderivative* $\boldsymbol{F} : [a,b] \to \mathbb{R}^p$. *Then*

$$\int_a^b \boldsymbol{f}(x)\,dx = \boldsymbol{F}(b) - \boldsymbol{F}(a).$$

► **7.48** *Prove the above theorem.*

Theorem 4.5 *If* $\boldsymbol{f} : [a,b] \to \mathbb{R}^p$ *is bounded and integrable on* $[a,b]$, *then* $|\boldsymbol{f}| : [a,b] \to \mathbb{R}$ *is integrable on* $[a,b]$, *and*

$$\left|\int_a^b \boldsymbol{f}(x)\,dx\right| \le \int_a^b |\boldsymbol{f}(x)|\,dx.$$

PROOF Let $\boldsymbol{f}(x) = \big(f_1(x), \ldots, f_p(x)\big)$, where $f_j(x)$ is integrable on $[a,b]$ for $1 \le j \le p$. Then, by Theorems 2.7 and 2.10 on pages 354 and 357 and Example 2.2 on page 350, $|\boldsymbol{f}|(x) = \sqrt{f_1^2(x) + \cdots + f_p^2(x)}$ is integrable on $[a,b]$. Define $I_j \equiv \int_a^b f_j(x)\,dx$ for $1 \le j \le p$, $I \equiv \int_a^b \boldsymbol{f}(x)dx = (I_1, I_2, \ldots, I_p)$, and note that

$$\begin{aligned}
\left|\int_a^b \boldsymbol{f}(x)\,dx\right|^2 &= \sum_{j=1}^p \left(\int_a^b f_j(x)\,dx\right)^2 \\
&= \sum_{j=1}^p I_j^2 \\
&= \sum_{j=1}^p \int_a^b I_j\, f_j(x)\,dx \\
&= \int_a^b \sum_{j=1}^p I_j\, f_j(x)\,dx \\
&= \int_a^b I \cdot \boldsymbol{f}(x)\,dx \\
&\le \int_a^b |I|\, |\boldsymbol{f}(x)|\,dx \\
&= \left|\int_a^b \boldsymbol{f}(x)\,dx\right| \int_a^b |\boldsymbol{f}(x)|\,dx.
\end{aligned}$$

From this we have that

$$\left|\int_a^b \boldsymbol{f}(x)\,dx\right| \le \int_a^b |\boldsymbol{f}(x)|\,dx.$$

♦

We would now like to generalize our notion of integral to include paths of integration other than segments of the real axis. In general, we would like to be able to integrate a real-valued function of more than one variable $f : D^k \to \mathbb{R}$ over some curve $C \subset D^k$. There are many types of functions f to integrate, and there are many kinds of curves C to integrate over. The properties possessed by f and C will play a significant role in determining the value of such an integral.

4.2 Curves and Contours

We begin with some definitions.

Definition 4.6 Let C be a subset of D^k. If there exists a continuous function $\mathbf{x} : [a, b] \to D^k$ such that $C = \mathbf{x}([a, b])$, then C is called a **curve** in D^k. In this case, the function $\mathbf{x} : [a, b] \to D^k$ is called a **parametrization** of C.

Example 4.7 *Let* $C = \{(x, y) \in \mathbb{R}^2 : \frac{x^2}{4} + \frac{y^2}{9} = 1\}$. *To see that* C *is a curve in* $\mathbb{R}^2$, *let* $\mathbf{x} : [0, 2\pi] \to \mathbb{R}^2$ *be given by* $\mathbf{x}(t) = (2\cos t, 3\sin t)$ *and note that* $\mathbf{x}$ *is continuous on* $[0, 2\pi]$. *(Why?) We will now show that* $\mathbf{x}([0, 2\pi]) = C$. *To see that* $\mathbf{x}([0, 2\pi]) \subset C$, *note that* $\frac{(2\cos t)^2}{4} + \frac{(3\sin t)^2}{9} = 1$. *To see that* $C \subset \mathbf{x}([0, 2\pi])$, *fix* $(x_0, y_0) \in \mathbb{R}^2$ *such that* $\frac{x_0^2}{4} + \frac{y_0^2}{9} = 1$. *We will find at least one value of* t *such that* $\mathbf{x}(t) = (x_0, y_0)$. *Since* $\frac{x_0^2}{4} = 1 - \frac{y_0^2}{9} \le 1$, *it follows that* $-1 \le \frac{x_0}{2} \le 1$ *and there exists* t_0 *such that* $\frac{x_0}{2} = \cos t_0$. *From this we have that*

$$\frac{y_0^2}{9} = 1 - \frac{x_0^2}{4} = 1 - \cos^2 t_0 = \sin^2 t_0,$$

so that $\frac{y_0}{3} = \pm \sin t_0$. *This then yields*

$$x_0 = 2\cos t_0 \text{ and } y_0 = \pm 3\sin t_0.$$

If $y_0 = 3\sin t_0$ *we are done, but if* $y_0 = -3\sin t_0$ *we have*

$$x_0 = 2\cos t_0 = 2\cos(-t_0) \text{ and } y_0 = -3\sin t_0 = 3\sin(-t_0),$$

and again we are done. ◀

It is important to note that while a given parametrization specifies a curve $C \subset D^k$, a given curve $C \subset D^k$ has many possible parametrizations associated with it. When we refer to a *parametrized curve* $C \subset D^k$, we are referring to the curve along with a particular choice of parametrization $\mathbf{x} : [a, b] \to D^k$. In such a situation, the point $\mathbf{x}(a)$ is called the *initial point* of the curve, while the point $\mathbf{x}(b)$ is called the *terminal point* of the curve. If $\mathbf{x}(a) = \mathbf{x}(b)$, the curve is called *closed*. If the curve does not cross itself, i.e., $\mathbf{x}(t_1) \neq \mathbf{x}(t_2)$ for all $a < t_1 < t_2 < b$, the curve is called *simple*. A *simple closed curve* is a closed curve whose only point of intersection is its common initial/terminal point. As an aid to visualizing a parametrized curve, each component function $x_j : [a,b] \to \mathbb{R}$ of the parametrization $\mathbf{x}$ can be thought of as describing the j-th coordinate of some particle traversing the curve C, starting from the initial point $\mathbf{x}(a)$ and ending at the terminal point $\mathbf{x}(b)$. While certain parametrizations could allow for the particle to "backtrack" as it moves along the curve, we will typically not use such parametrizations. For any parametrized curve, the parametrization will specify an initial and a terminal point and will simultaneously specify a direction in which the curve is "traversed." Changing from one parametrization to another for a given curve is referred to as *reparametrizing* the curve.

Example 4.8 *Consider the circle of radius R centered in the plane $\mathbb{R}^2$. One parametrization of this simple closed curve is given by $\mathbf{x} : [0, 2\pi] \to \mathbb{R}^2$ where $\mathbf{x}(t) = (R\cos t,\, R\sin t)$. The initial point is $\mathbf{x}(0) = (R, 0)$, the terminal point is $\mathbf{x}(2\pi) = (R, 0)$, and the circle is traversed in the counterclockwise direction.* ◄

► **7.49** *Parametrize the circle described in the previous example so that it is traversed in the clockwise direction.*

► **7.50** *Suppose C in $\mathbb{R}^k$ is the straight line segment that connects two points $\mathbf{p} \neq \mathbf{q}$ in $\mathbb{R}^k$. Find a parametrization of this "curve."*

► **7.51** *Let C be a curve in $\mathbb{R}^k$ parametrized by $\mathbf{x} : [a,b] \to \mathbb{R}^k$. Let $\widetilde{\mathbf{x}} : [c,d] \to \mathbb{R}^k$ be given by $\widetilde{\mathbf{x}}(s) = \mathbf{x}\Big(a + \frac{b-a}{d-c}(s-c)\Big)$. Show that $\widetilde{\mathbf{x}}$ is a parametrization of C on the parameter interval $[c,d]$.*

It is often convenient to catenate curves to form new curves.

Definition 4.9 Let $\{C_1, C_2, \dots, C_n\}$ be a finite collection of curves in $\mathbb{R}^k$ with C_j parametrized by $\mathbf{x_j} : [j-1, j] \to \mathbb{R}^k$ for $1 \le j \le n$, and $\mathbf{x_{j-1}}(j-1) = \mathbf{x_j}(j-1)$ for $2 \le j \le n$. Let C be the curve in $\mathbb{R}^k$ given by

$$C \equiv (C_1, C_2, \dots, C_n) \equiv \bigcup_{j=1}^{n} C_j.$$

When referred to as the **parametrized curve** $C = (C_1, C_2, \dots, C_n)$, unless specified otherwise, we will assume C to be parametrized by $\mathbf{x} : [0,n] \to \mathbb{R}^k$ where

$$\mathbf{x}(t) = \begin{cases} \mathbf{x_1}(t) & \text{for } 0 \le t \le 1, \\ \mathbf{x_2}(t) & \text{for } 1 \le t \le 2, \\ \vdots & \vdots \\ \mathbf{x_n}(t) & \text{for } n-1 \le t \le n. \end{cases}$$

► **7.52** *Verify that the parametrized $C = (C_1, C_2, \dots, C_n)$ is a curve.*

It is also often convenient to subdivide a given curve into smaller curves.

Definition 4.10 Let C be a curve in $\mathbb{R}^k$ with parametrization $\mathbf{x} : [a, b] \to \mathbb{R}^k$. Let $S = \{\boldsymbol{\xi}_1, \boldsymbol{\xi}_2, \dots, \boldsymbol{\xi}_{n-1}\}$ be any finite collection of ordered (via the parametrization) points of C, i.e., there exists a partition $\mathcal{P} = \{a = t_0, t_1, \dots, t_{n-1}, t_n = b\}$, of $[a, b]$ such that $\boldsymbol{\xi}_j = \mathbf{x}(t_j)$ for $1 \leq j \leq n - 1$. Define the **subcurve** $C_j \subset C$ for each $1 \leq j \leq n$ to be the curve parametrized by $\mathbf{x_j} : [t_{j-1}, t_j] \to \mathbb{C}$ where $\mathbf{x_j}(t) \equiv \mathbf{x}(t)$. We say that S **subdivides** C into the n subcurves $C_1, C_2, \dots, C_n$, which form a **subdivision** of C.

Note from the above definition that C_1 is that part of C from $\mathbf{x}(a)$ to $\boldsymbol{\xi}_1$, C_2 is that part of C from $\boldsymbol{\xi}_1$ to $\boldsymbol{\xi}_2$, and so on. Note too, that any partition specified for a given parametrized curve implicitly provides a subdivision of that curve. Also, by selecting a set of (ordered) points $S = \{\boldsymbol{\xi}_1, \dots, \boldsymbol{\xi}_{n-1}\}$ from a given parametrized curve C, one has also selected a partition $\mathcal{P}$ of $[a, b]$.

Both S and the associated $\mathcal{P}$ subdivide C into $C_1, C_2, \dots, C_n$, and clearly we may write $C = (C_1, C_2, \dots, C_n)$.

Of particular importance in our work will be the kinds of curves known as *smooth* curves.

Definition 4.11 Let C be a curve in D^k. If there exists a parametrization $\mathbf{x} : [a, b] \to D^k$ such that

(i) $\mathbf{x}$ is continuously differentiable on $[a, b]$,

(ii) $\mathbf{x}'(t) \neq \mathbf{0}$ for all $t \in [a, b]$,

then C is called a **smooth** curve in D^k.

Condition 2, that $\mathbf{x}'(t)$ not vanish on $[a, b]$, distinguishes a smooth curve from one that is merely continuously differentiable. Recall the definition of the unit tangent vector to a curve from calculus, and the imposition of this condition becomes a bit clearer. Consider a curve in $\mathbb{R}^2$ as an example, with parametrization $\mathbf{x}(t) = (x(t), y(t))$ for $a \leq t \leq b$. If we think of $\mathbf{x}(t)$ as indicating a particle's position in the plane at any time $t \in [a, b]$, the unit tangent vector to the path of this particle is given by

$$\widehat{\boldsymbol{T}} = \frac{\mathbf{x}'(t)}{|\mathbf{x}'(t)|},$$

which would clearly be undefined if $\mathbf{x}'(t)$ were zero. A smooth curve is one whose unit tangent vector is well defined along the entire curve and that changes its direction, that is, its angle of inclination with the positive real axis, in a continuous manner.

We now define what we will call a *contour*.

Definition 4.12 Let C be a curve in $\mathcal{D}^k$. If there exists a parametrization $\mathbf{x} : [a, b] \to \mathcal{D}^k$ and a partition $\mathcal{P} = \{a = t_0, t_1, \ldots, t_n = b\}$ of $[a, b]$ such that $C_j \equiv \mathbf{x}\big([t_{j-1}, t_j]\big)$ is a smooth curve for $1 \le j \le n$, then C is called a **contour**.

A contour C in $\mathcal{D}^k$ is a curve that is the union of a finite number of smooth curves $C_j \subset \mathcal{D}^k$. If $\mathbf{x} : [a, b] \to \mathcal{D}^k$ is a parametrization of the contour C consistent with Definition 4.12, then for each $1 \le j \le n$ the function $\mathbf{x_j} : [t_{j-1}, t_j] \to \mathcal{D}^k$ given by $\mathbf{x_j}(t) \equiv \mathbf{x}(t)$ is the *induced parametrization* of the smooth curve C_j. Note that the initial point of C_j coincides with the terminal point of C_{j-1} for every $1 \le j \le n$. Clearly, the initial point of C_1 and the terminal point of C_n are the initial and terminal points of the contour C, and so, $C = (C_1, C_2, \ldots, C_n)$. Also, there are such things as *closed contours* (those for which the initial point of C_1 coincides with the terminal point of C_n), *simple contours* (those that have no points of self-intersection other than possibly the initial and terminal points coinciding), and *smooth contours* (those that are smooth curves). It is worth noting that with respect to this latter category, "patching" smooth curves together to form contours will not necessarily result in a smooth contour. The "patch points" will not in general be points with well-defined derivatives. The following example illustrates how two smooth curves can be patched together to form a contour.

Example 4.13 *Consider the contour C in $\mathbb{R}^2$ shown in Figure 7.2 below. Let C_1 be the smooth curve parametrized by $\mathbf{x_1} : [0, \pi] \to \mathbb{R}^2$ where $\mathbf{x_1}(t) = (R\cos t,\, R\sin t)$, and let C_2 be the smooth curve parametrized by $\mathbf{x_2} : [\pi, \pi + 2R] \to \mathbb{R}^2$ where $\mathbf{x_2}(t) = (t - \pi - R, 0)$. Then the contour $C = (C_1, C_2)$ is parametrized by $\mathbf{x} : [0, \pi + 2R] \to \mathbb{R}^2$ given by*

$$\mathbf{x}(t) = \begin{cases} \mathbf{x_1}(t) & \text{for } 0 \le t \le \pi, \\ \mathbf{x_2}(t) & \text{for } \pi \le t \le \pi + 2R. \end{cases}$$

Note that C is not smooth.

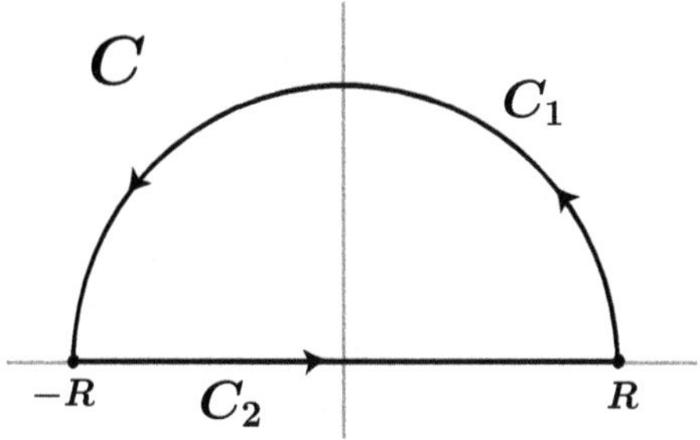

◀

Figure 7.2 *The semicircular contour C.*

▶ **7.53** *Find another parametrization for the contour C described above such that the initial and terminal points are both $(-R, 0)$.*

For a given contour C in $\mathcal{D}^k$ it will be useful to define when two parametrizations of C are *equivalent* in a certain sense. We begin by defining when two parametrizations of a smooth curve are equivalent.

Definition 4.14 Let C be a smooth curve in $\mathcal{D}^k$ with parametrization $\mathbf{x} : [a, b] \to \mathcal{D}^k$. Another parametrization of C given by $\widetilde{\mathbf{x}} : [c, d] \to \mathcal{D}^k$ is called **equivalent** to $\mathbf{x}$ if there exists a mapping $g : [c, d] \to [a, b]$ such that

$$(i)\ \widetilde{\mathbf{x}} = \mathbf{x} \circ g,$$

$$(ii)\ g \text{ is } C^1 \text{ on } [c, d],$$

$$(iii)\ g' > 0 \text{ on } [c, d],$$

$$(iv)\ g(c) = a \text{ and } g(d) = b.$$

Loosely speaking, two parametrizations of a contour C are equivalent if they associate the same initial and terminal points, and if they preserve the direction in which the contour is traversed as the parameter increases. In terms of an imagined particle traversing the contour, the difference between two different but equivalent parametrizations can be likened to the particle traversing the contour at different speeds. It is useful to know how to *reparametrize* a given contour, that is, to change from a given parametrization to a different one.

▶ **7.54** *Let C be a smooth curve in $\mathcal{D}^k$ with parametrization $\mathbf{x} : [a, b] \to \mathcal{D}^k$. Let $g : [c, d] \to [a, b]$ be such that properties 2–4 of Definition 4.14 hold. Show that $\widetilde{\mathbf{x}} : [c, d] \to \mathcal{D}^k$ given by $\widetilde{\mathbf{x}} = \mathbf{x} \circ g$ is a parametrization of C and is therefore equivalent to $\mathbf{x}$.*

We can easily define two *contour* parametrizations to be *equivalent* in the same way that two curve parametrizations are defined to be equivalent, namely, $\mathbf{x}$ and $\widetilde{\mathbf{x}}$ are equivalent if $\widetilde{\mathbf{x}} = \mathbf{x} \circ g$ for a function g as above.

Example 4.15 *Consider again the circle of radius R centered in the plane $\mathbb{R}^2$ and parametrized by $\mathbf{x} : [0, 2\pi] \to \mathbb{R}^2$ where $\mathbf{x}(t) = (R\cos t,\ R\sin t)$. Let $g : [0, \pi] \to [0, 2\pi]$ be given by $g(s) = 2\,s$. Then $\widetilde{\mathbf{x}} : [0, \pi] \to \mathbb{R}^2$ given by $\widetilde{\mathbf{x}}(s) = \mathbf{x} \circ g(s) = \big(R\cos(2s), R\sin(2s)\big)$ is a reparametrization of C. Recall from vector calculus that the instantaneous velocity of an imagined particle as it traverses the curve C according to the original parametrization is given by $\mathbf{x}'(t)$. Its instantaneous speed is the magnitude of this velocity, which in this case is just R. We leave it to the reader to verify that a particle traversing this curve according to the reparametrization does so twice as fast as with the original parametrization, but traverses it in the same direction, and with the same initial and terminal points. Therefore, this new parametrization is equivalent to the original one.* ◀

► **7.55** *Confirm the claim in the above example. Then reparametrize the curve so that an imagined particle traverses it three times more slowly than the original parametrization.*

► **7.56** *Consider the contour C described in Example 4.13 above. Find an equivalent reparametrization of C.*

It can also be useful to traverse a given parametrized contour C "in reverse." To this end, let C be a given parametrized contour having initial and terminal points $\mathbf{x}(a)$ and $\mathbf{x}(b)$, respectively. We define the parametrized contour $-C$ to be the same set of points as C, but with initial point $\mathbf{x}(b)$, terminal point $\mathbf{x}(a)$, and parametrized so as to be traversed in the "opposite" direction from the parametrized contour C. The following result establishes a convenient way to obtain a parametrization for $-C$ with these properties.

Theorem 4.16 *Suppose C is a contour in D^k with parametrization $\mathbf{x}_C : [a,b] \to D^k$. Then $\mathbf{x}_{-C} : [-b,-a] \to D^k$ given by*

$$\mathbf{x}_{-C}(t) = \mathbf{x}_C(-t)$$

parametrizes the contour $-C$.

► **7.57** *Prove the above theorem. Begin by considering the case where C is a smooth curve.*

Example 4.17 *Let $C \subset \mathbb{R}^2$ be the circle parametrized by $\mathbf{x}_C : [0,2\pi] \to \mathbb{R}^2$ where $\mathbf{x}_C(t) = (r \cos t, r \sin t)$. Then $-C$ is the circle parametrized by $\mathbf{x}_{-C} : [-2\pi, 0] \to \mathbb{R}^2$ where $\mathbf{x}_{-C}(t) = \big(r \cos(-t), r \sin(-t)\big) = (r \cos t, -r \sin t)$. The reader should verify that $\mathbf{x}_{-C}$ parametrizes the circle in the clockwise direction, whereas $\mathbf{x}_C$ parametrized the circle in the counterclockwise direction.*◄

► **7.58** *Verify the above.*

4.3 Line Integrals

We begin our discussion of line integrals with the following definition.

Definition 4.18 Let $f : D^k \to \mathbb{R}$ be continuous on the open set $D^k \subset \mathbb{R}^k$ and let C be a smooth curve in D^k with parametrization $\mathbf{x} : [a,b] \to D^k$ given by $\mathbf{x}(t) = \big(x_1(t), \ldots, x_k(t)\big)$. For each $1 \leq j \leq k$, the **line integral of f with respect to x_j** is given by

$$\int_C f(x_1, \ldots, x_k)\, dx_j \equiv \int_a^b f\big(x_1(t), \ldots, x_k(t)\big)\, x_j'(t)\, dt.$$

If $C = (C_1, \ldots, C_n)$ is a contour in D^k, then for each $1 \le j \le k$,

$$\int_C f(x_1, \ldots, x_k)\, dx_j \equiv \sum_{\ell=1}^{n} \int_{C_\ell} f(x_1, \ldots, x_k)\, dx_j.$$

► **7.59** *Let $f, g : D^k \to \mathbb{R}$ be continuous on the open set D^k, and suppose C is a contour in D^k. Show the following:*

a) $\int_C (f \pm g)(\mathbf{x})\, dx_j = \int_C f(\mathbf{x})\, dx_j \pm \int_C g(\mathbf{x})\, dx_j$ *for each* $1 \le j \le k$.

b) For any $\alpha \in \mathbb{R}$, $\int_C \alpha\, f(\mathbf{x})\, dx_j = \alpha \int_C f(\mathbf{x})\, dx_j$ *for each* $1 \le j \le k$.

Example 4.19 *Let $C = (C_1, C_2)$ be the piecewise-smooth contour described in Example 4.13, and let $f : \mathbb{R}^2 \to \mathbb{R}$ be given by $f(x, y) = x^2 + y^2$. Then,*

$$\int_{C_1} f(x, y)\, dx = \int_0^{\pi} R^2\,(-R\,\sin t)dt = -2\,R^3,$$

and

$$\int_{C_2} f(x, y)\, dx = \int_{\pi}^{\pi+2\,R} (t - \pi - R)^2\, dt = \tfrac{2}{3}\,R^3.$$

Therefore,

$$\int_C f(x, y)\, dx = \int_{C_1} f(x, y)\, dx + \int_{C_2} f(x, y)\, dx = -2\,R^3 + \tfrac{2}{3}\,R^3 = -\tfrac{4}{3}\,R^3.$$ ◄

► **7.60** *Find $\int_C f(x, y)\, dy$ for the previous example.*

The following theorem is an especially useful result. We leave its proof to the reader.

Theorem 4.20 *Suppose $f : D^k \to \mathbb{R}$ is continuous on the open set $D^k \subset \mathbb{R}^k$ and C is a contour in D^k with parametrization $\mathbf{x} : [a, b] \to D^k$ given by $\mathbf{x}(t) = \big(x_1(t), \ldots, x_k(t)\big)$. Suppose $\widetilde{\mathbf{x}} : [c, d] \to D^k$ given by $\widetilde{\mathbf{x}}(t) = \big(\widetilde{x}_1(t), \ldots, \widetilde{x}_k(t)\big)$ is a parametrization of C equivalent to $\mathbf{x}$. Then for each $1 \le j \le k$,*

$$\int_C f(x_1, \ldots, x_k)\, dx_j = \int_C f(\widetilde{x}_1, \ldots, \widetilde{x}_k)\, d\widetilde{x}_j.$$

► **7.61** *Prove the above theorem, considering first the case where C is a smooth curve.*

Example 4.21 *Let $f : \mathbb{R}^2 \to \mathbb{R}$ be given by $f(x, y) = x^2 + xy + 1$, and let $C \subset \mathbb{R}^2$ be the contour parametrized by $\mathbf{x} : [0, 1] \to \mathbb{R}^2$ where $\mathbf{x}(t) = (t, t^2)$. Then,*

$$\int_C f(x,y)\,dx = \int_0^1 (t^2 + t^3 + 1)\,dt = \tfrac{19}{12}.$$

If we let $g : [-1,0] \to [0,1]$ *be given by* $g(s) = s^3 + 1$*, then by Definition 4.14 on page 376 it follows that* $\widetilde{\mathbf{x}} : [-1,0] \to \mathbb{R}^2$ *given by*

$$\widetilde{\mathbf{x}}(s) \equiv \mathbf{x}\big(g(s)\big) = \big(s^3+1, (s^3+1)^2\big)$$

is an equivalent parametrization of C*. Recalculating* $\int_C f(x,y)\,dx$ *with the equivalent parametrization* $\widetilde{\mathbf{x}}$ *obtains*

$$\int_C f(\widetilde{x},\widetilde{y})\,d\widetilde{x} = \int_{-1}^0 \big[(s^3+1)^2 + (s^3+1)^3 + 1\big]\,3s^2\,ds = \tfrac{19}{12},$$

the same value as with the original parametrization. ◀

▶ **7.62** *Show that* $\int_C f(x,y)\,dy = \int_C f(\widetilde{x},\widetilde{y})\,d\widetilde{y}$ *in the above example.*

For the most general line integral, consider the following definition.

Definition 4.22 Let $\boldsymbol{f} : D^k \to \mathbb{R}^k$ be continuous on the open set $D^k \subset \mathbb{R}^k$ with $\boldsymbol{f} = (f_1, \dots, f_k)$, and let C be a smooth curve in D^k with parametrization $\mathbf{x} : [a,b] \to D^k$ given by $\mathbf{x}(t) = \big(x_1(t), \dots, x_k(t)\big)$. Then the **line integral of $\boldsymbol{f}$ with respect to x along C** is given by

$$\int_C \boldsymbol{f}(\mathbf{x}) \cdot d\mathbf{x} \equiv \int_C f_1(\mathbf{x})\,dx_1 + \cdots + f_k(\mathbf{x})\,dx_k \equiv \int_C f_1(\mathbf{x})\,dx_1 + \cdots + \int_C f_k(\mathbf{x})\,dx_k.$$

Note that the above, in combination with the chain rule, yields

$$\begin{aligned}
\int_C \boldsymbol{f}(\mathbf{x}) \cdot d\mathbf{x} &\equiv \int_C f_1(\mathbf{x})\,dx_1 + \cdots + f_k(\mathbf{x})\,dx_k \\
&= \int_C f_1(\mathbf{x})\,dx_1 + \cdots + \int_C f_k(\mathbf{x})\,dx_k \\
&= \int_a^b f_1(\mathbf{x}(t))\,\frac{dx_1}{dt}\,dt + \cdots + \int_a^b f_k(\mathbf{x}(t))\,\frac{dx_k}{dt}\,dt \\
&= \int_a^b \left(f_1(\mathbf{x}(t))\,\frac{dx_1}{dt} + \cdots + f_k(\mathbf{x}(t))\,\frac{dx_k}{dt} \right) dt \\
&= \int_a^b \boldsymbol{f}(\mathbf{x}) \cdot \frac{d\mathbf{x}}{dt}\,dt. \qquad (7.33)
\end{aligned}$$

The formula for the line integral indicated by (7.33) is often useful in applications. It also leads to the following nice result.

Theorem 4.23 (The Fundamental Theorem of Calculus for Line Integrals) *Suppose* $\boldsymbol{f} : D^k \to \mathbb{R}^k$ *is continuous on the open set* $D^k \subset \mathbb{R}^k$ *with* $\boldsymbol{f} = (f_1, \dots, f_k)$, *and let* C *be a contour in* D^k *with parametrization* $\mathbf{x} : [a, b] \to D^k$ *given by* $\mathbf{x}(t) = \big(x_1(t), \dots, x_k(t)\big)$. *Suppose also that there exists* $F : D^k \to \mathbb{R}$ *such that* $\boldsymbol{f} = \nabla F$ *on* D^k. *Then,*

$$\int_C \boldsymbol{f}(\mathbf{x}) \cdot d\mathbf{x} = F\big(\mathbf{x}(b)\big) - F\big(\mathbf{x}(a)\big).$$

► **7.63** *Prove the above theorem.*

The above theorem is a generalization of the fundamental theorem of calculus, and it is a great convenience. It implies two very interesting facts. First, under the right circumstances, the value of the line integral of a vector-valued function $\boldsymbol{f}$ along a contour C can be obtained simply by evaluating the function's antiderivative *at the endpoints of the contour.* Second, if the contour C is *closed*, then the value of the line integral is *zero*.

► **7.64** *Is it necessary that the function* $F : D^k \to \mathbb{R}$ *referred to in Theorem 4.23 be differentiable, i.e., is it necessary that* $\boldsymbol{f}$ *be the derivative of* F?

Example 4.24 Let $\boldsymbol{f} : \mathbb{R}^3 \to \mathbb{R}^3$ be given by $\boldsymbol{f}(x, y, z) = (yz, xz, xy)$, and let $C \subset \mathbb{R}^3$ be the contour parametrized by $\mathbf{x} : [0, 1] \to \mathbb{R}^3$ where $\mathbf{x}(t) = (t, t^2, t^3)$. Then,

$$\begin{aligned}\int_C \boldsymbol{f}(\mathbf{x}) \cdot d\mathbf{x} &= \int_C yz\, dx + xz\, dy + xy\, dz \\ &= \int_0^1 t^5\, dt + \int_0^1 t^4\,(2t)\, dt + \int_0^1 t^3\,(3t^2)\, dt \\ &= \int_0^1 6t^5\, dt = 1.\end{aligned}$$

Note, however, that since $\boldsymbol{f}$ has an antiderivative $F : \mathbb{R}^3 \to \mathbb{R}$ given by $F(x, y, z) = xyz$, we have

$$\int_C \boldsymbol{f}(\mathbf{x}) \cdot d\mathbf{x} = F\big(\mathbf{x}(1)\big) - F\big(\mathbf{x}(0)\big) = F(1, 1, 1) - F(0, 0, 0) = 1,$$

which is much easier. ◄

► **7.65** *Verify that the function* F *in the above example is an antiderivative for the function* $\boldsymbol{f}$ *given there.*

The following theorem tells us how a line integral changes when we traverse a contour C in the opposite direction.

Theorem 4.25 *Suppose $f : D^k \to \mathbb{R}$ is continuous on D^k and that C is a contour in D^k. Then,*

$$\int_{-C} f(x_1,\ldots,x_k)\,dx_j = -\int_C f(x_1,\ldots,x_k)\,dx_j.$$

PROOF We consider the case where C is a smooth curve and leave the more general case to the reader. If $\mathbf{x}_C : [a,b] \to D^k$ given by $\mathbf{x}_C(t) = \big(x_1(t),\ldots,x_k(t)\big)$ is a parametrization of C, then $\mathbf{x}_{-C} : [-b,-a] \to D^k$ given by $\mathbf{x}_{-C}(t) = \big(x_1(-t),\ldots,x_k(-t)\big)$ is a parametrization of $-C$. From this we have that

$$\int_{-C} f(x_1,\ldots,x_k)\,dx_j = \int_{-b}^{-a} f\big(x_1(-t),\ldots,x_k(-t)\big)\,\big(-x_j'(-t)\big)\,dt.$$

Changing variables by letting $s = -t$, we obtain

$$\begin{aligned}\int_{-C} f(x_1,\ldots,x_k)\,dx_j &= \int_b^a f\big(x_1(s),\ldots,x_k(s)\big)\,\big(-x_j'(s)\big)\,(-ds)\\ &= -\int_a^b f\big(x_1(s),\ldots,x_k(s)\big)\,x_j'(s)\,ds\\ &= -\int_C f(x_1,\ldots,x_k)\,dx_j.\end{aligned}$$

♦

► **7.66** *Prove the above theorem for contours.*

Example 4.26 *Let $f : \mathbb{R}^2 \to \mathbb{R}$ be given by $f(x,y) = x^2 + y^2$, and C be the curve in $\mathbb{R}^2$ parametrized by $\mathbf{x}_C(t) = (R\cos t, R\sin t)$ for $t \in [0,\pi]$. Then $-C$ is parametrized by $\mathbf{x}_{-C}(t) = (R\cos t, -R\sin t)$ for $t \in [-\pi, 0]$, and*

$$\int_C (x^2+y^2)\,dx = \int_0^\pi R^2\,(-R\sin t)\,dt = -2\,R^3,$$

while

$$\int_{-C} (x^2+y^2)\,dx = \int_{-\pi}^0 R^2\,(-R\sin t)\,dt = 2\,R^3.$$

◄

5 SUPPLEMENTARY EXERCISES

1. *Recall that we appended our definition of the integrability of $f : [a,b] \to \mathbb{R}$ given in Definition 1.5 on page 337 with the seemingly arbitrary definition (7.1), repeated here for convenience:*

$$\int_b^a f(x)\,dx \equiv -\int_a^b f(x)\,dx. \qquad (7.1)$$

This exercise will help the reader to justify this definition. In particular, note that in defining the Riemann sum associated with the integral of f from a to b, one can think of the Riemann sum, and hence the associated integral, as accumulating values as you traverse the interval $[a,b]$ from a to b, left-to-right along the x-axis. We had defined the length of each subinterval I_j as $\Delta x_j = x_j - x_{j-1}$, a natural definition that assigns a positive length to each subinterval I_j since $x_{j-1} < x_j$. Using the same points $\mathcal{C}$ and subintervals $\{I_j\}$ defined by $\mathcal{P} = \{a = x_0, x_1, \ldots, x_n = b\}$ from before, one could also determine the integral obtained by traversing the interval $[a,b]$ in the opposite direction, from b to a, or right-to-left along the x-axis. Since a partition is an ordered set of points from the interval $[a,b]$, we are now effectively making use of the "reverse" partition $\mathcal{P}^r = \{x_0^r, x_1^r, \ldots, x_n^r\} = \{b = x_n, x_{n-1} \ldots, x_0 = a\}$. From this, we see that $\Delta x_j^r = x_j^r - x_{j-1}^r = x_{n-(j-1)} - x_{n-j} = -(x_{n-j} - x_{n-(j-1)}) = -\Delta x_{n-j}$. The reader should verify that the resulting Riemann sum associated with $\mathcal{P}^r$ is the same in magnitude as that associated with $\mathcal{P}$, but opposite in sign. This results in the formula given by (7.1).

2. *Use the result of the previous exercise to justify the definition given by (7.2) on page 338, i.e., $\int_c^c f(x)\,dx \equiv 0$, for any $c \in [a,b]$.*

3. *Suppose $f : [a,b] \to \mathbb{R}$ is nonnegative and continuous on $[a,b]$ such that $\int_a^b f(x)\,dx = 0$. Show that $f \equiv 0$ on $[a,b]$.*

4. *Suppose $f \in C^2([a,b])$, and that $f(x) \geq 0$ and $f''(x) \leq 0$ on $[a,b]$. Show that*

$$\tfrac{1}{2}(b-a)\left(f(a)+f(b)\right) \leq \int_a^b f(x)\,dx \leq (b-a)\, f\left(\tfrac{a+b}{2}\right).$$

5. Integration by parts. *Suppose $u, v : [a,b] \to \mathbb{R}$ are differentiable on $[a,b]$, and u' and v' are integrable on $[a,b]$. Show that*

$$\int_a^b u(x)\, v'(x)\,dx = u(x)\, v(x)\big|_a^b - \int_a^b v(x)\, u'(x)\,dx.$$

6. *Evaluate $\int_1^2 \sqrt[3]{x}\,dx$ using Definition 1.5 on page 337.*

7. *Evaluate $\int_0^2 (x + |x-1|)\,dx$.*

8. *Let $f : [0,3] \to \mathbb{R}$ be given by $f(x) = \begin{cases} x & \text{on } [0,1] \\ x-1 & \text{on } (1,2] \\ 0 & \text{on } (2,3] \end{cases}$. Define $F : [0,3] \to \mathbb{R}$ by $F(u) = \int_0^u f(x)\,dx$. Is F continuous? Does $F' = f$ on all of $[0,3]$?*

9. *Let $f : [a,b] \to \mathbb{R}$ be bounded and integrable, and suppose $g : [a,b] \to \mathbb{R}$ is such that $g(x) = f(x)$ on $[a,b]$ except at a finite number of points of $[a,b]$. Show that g is also integrable on $[a,b]$, and find $\int_a^b g(x)\,dx$.*

10. *The function g to which we referred in the previous exercise is not necessarily integrable if $g(x) \neq f(x)$ on $\mathbb{Q} \cap [a,b]$. To see this, construct such a $g : [a,b] \to \mathbb{R}$ and* *show that it is not integrable on* $[a, b]$*. Suppose* $g(x) \neq f(x)$ *on some infinite collection of points within* $[a, b]$*. Is it necessarily true that* $\int_a^b f(x)\,dx \neq \int_a^b g(x)\,dx$*?*

11. *Suppose* $f : [0, 1] \to \mathbb{R}$ *is continuous on* $[0, 1]$ *and* $\int_0^u f(x)\,dx = \int_u^1 f(x)\,dx$ *for every* $u \in [0, 1]$*. Find* $f(x)$*.*

12. *Suppose* $f : [a, b] \to \mathbb{R}$ *is continuous on* $[a, b]$ *and* $\int_a^b f(x)\,dx = 0$*. Show that there exists* $c \in [a, b]$ *such that* $f(c) = 0$*.*

13. *Find an example of a function* $f : [a, b] \to \mathbb{R}$ *such that* f *is not integrable on* $[a, b]$*, but* $|f|$ *and* f^{2j} *are for any* $j \in \mathbb{N}$*.*

14. *Suppose* $f : [a, b] \to \mathbb{R}$ *is continuous on* $[a, b]$ *such that* $|f(x)| \leq 1$ *on* $[a, b]$*, and* $\int_a^b f(x)\,dx = b - a$*. Show that* $f(x) \equiv 1$ *on* $[a, b]$*.*

15. *Find an integrable function* $f : [0, 1] \to \mathbb{R}$ *such that* $F : [a, b] \to \mathbb{R}$ *given by* $F(u) = \int_0^u f(x)\,dx$ *does not have a derivative at some point of* $[a, b]$*. Can you find an* f *such that* F *is not continuous at some point of* $[a, b]$*?*

16. *Let* $f : [0, 1] \to \mathbb{R}$ *be given by* $f(x) = \begin{cases} x^2 \sin\left(\frac{1}{x^2}\right) & \text{if } x \neq 0 \text{ and } x \in [0, 1] \\ 0 & \text{if } x = 0 \end{cases}$*. Show that* f *is differentiable on* $[0, 1]$*, but* f' *is not integrable on* $[0, 1]$*. Clearly* f' *can't be continuous on* $[0, 1]$*, right?*

17. *Let* C *be the Cantor set, and define* $\chi_C : [0, 1] \to \mathbb{R}$*, the* **characteristic function** *of* C*, by* $\chi_C(x) = \begin{cases} 1 & \text{for } x \in C \cap [0, 1] \\ 0 & \text{for } x \notin C \cap [0, 1] \end{cases}$*. Show that* χ_C *is integrable and that* $\int_0^1 \chi_C(x)\,dx = 0$*. Contrast this with Example 1.18 on page 349. How does* χ_C *differ from the function described there? How are they similar?*

18. *Suppose* $f : [a, b] \to \mathbb{R}$ *is continuous on* $[a, b]$*, and that* $\int_a^b f(x)\,h(x)\,dx = 0$ *for every continuous* $h : [a, b] \to \mathbb{R}$*. Show that* $f \equiv 0$ *on* $[a, b]$*.*

19. *Suppose* $f : [a, b] \to \mathbb{R}$ *is integrable on* $[a, b]$*, and that* $\int_a^b f(x)\,h(x)\,dx = 0$ *for every continuous* $h : [a, b] \to \mathbb{R}$*. Show that* $f(x) = 0$ *at all points* $x \in [a, b]$ *where* f *is continuous.*

20. *Suppose* $f : [a, b] \to \mathbb{R}$ *is integrable on* $[a, b]$*, and* $g : \mathbb{R} \to \mathbb{R}$ *is continuous on* $\mathbb{R}$*. Is* $g \circ f : [a, b] \to \mathbb{R}$ *necessarily integrable?*

21. *Show that* $\lim_{n\to\infty} \int_0^b e^{-nx}\,dx = 0$ *for any choice of* $b > 0$*. Now answer the following with a valid explanation: The sequence of functions* $f_n(x) = e^{-nx}$ *converges uniformly on* $[0, b]$*, true or false?*

22. *Suppose* $0 < a < 2$*. Show that* $\lim_{n\to\infty} \int_a^2 e^{-nx}\,dx = 0$*. Now consider the case* $a = 0$*.*

23. *Leibniz's rule*
Let $f : [a, b] \to \mathbb{R}$ be a bounded and integrable function on $[a, b]$, and let $g, h : D^1 \to [a, b]$ be differentiable on D^1. Show the following.

a) *The function $F : D^1 \to \mathbb{R}$ given by $F(u) = \int_a^{g(u)} f(x)\,dx$ is differentiable at points $u_0 \in D^1$ where f is continuous, and $F'(u_0) = f\big(g(u_0)\big)\, g'(u_0)$.*

b) *The function $G : D^1 \to \mathbb{R}$ given by $G(u) = \int_{h(u)}^{g(u)} f(x)\,dx$ is differentiable at points $u_0 \in D^1$ where f and g are continuous. Find $G'(u)$.*

c) *Generalize the above results to the case of $\boldsymbol{f} : [a, b] \to \mathbb{R}^p$, where $\boldsymbol{f}(x) = \big(f_1(x), \ldots, f_p(x)\big)$ and $f_j : [a, b] \to \mathbb{R}$ for $1 \le j \le p$.*

24. *Let $\sum_{n=1}^{\infty} x_n$ be a series of positive nonincreasing real numbers, and suppose $f : [N, \infty) \to \mathbb{R}$ is a function such that $f(n) = x_n$ for $n \ge N$. Show that $\sum_{n=1}^{\infty} x_n$ and $\int_N^{\infty} f(x)\,dx$ both converge or both diverge.*

25. *Improper integrals of unbounded functions*
Let $f : (a, b] \to \mathbb{R}$ be such that f is integrable on $[c, b]$ for every $a < c < b$, and $\lim_{x \to a^+} f(x) = \pm\infty$. We define the improper integral $\int_a^b f(x)\,dx = \lim_{c \to a^+} \int_c^b f(x)\,dx$ when this limit exists. Similarly, suppose $g : [a, b) \to \mathbb{R}$ is integrable on $[a, c]$ for every $a < c < b$, and $\lim_{x \to b^-} g(x) = \pm\infty$. Define the improper integral $\int_a^b g(x)\,dx = \lim_{c \to b^-} \int_a^c g(x)\,dx$ when this limit exists. Show that $\int_0^1 \frac{1}{\sqrt{x}}\,dx$ exists, but $\int_0^1 \frac{1}{x}\,dx$ does not.

26. *Integrals depending on a parameter*
Let $D^2 = [a, b] \times [c, d]$ and suppose $f(x, t)$ is continuous on D^2. Define $F : [c, d] \to \mathbb{R}$ by $F(t) \equiv \int_a^b f(x, t)\,dx$.

a) *Show that F is continuous on $[c, d]$.*

b) *Suppose $\frac{\partial f}{\partial t}$ and $\frac{\partial f}{\partial x}$ are continuous on D^2. Show that F is differentiable on $[c, d]$, and $F'(t_0) = \int_a^b \frac{\partial f}{\partial t}(x, t_0)\,dx$.*

27. *The Cauchy-Schwarz inequality for integrals*

a) *If $a, b \in \mathbb{R}$, show that $ab \le \frac{a^2+b^2}{2}$.*

b) *Given Riemann-integrable functions $f, g : [a, b] \to \mathbb{R}$, show that*

$$\int_a^b |f(x)g(x)|\,dx \le \frac{1}{2}\left(\int_a^b f(x)^2 dx + \int_a^b g(x)^2 dx\right).$$

c) *Show that if $\int_a^b f(x)^2 dx = \int_a^b g(x)^2 = 1$, then $\int_a^b |f(x)g(x)|\,dx \le 1$.*

d) *Show that for any f, g,*

$$\int_a^b |f(x)g(x)|\,dx \le \left(\int_a^b f(x)^2 dx\right)^{1/2} \left(\int_a^b g(x)^2 dx\right)^{1/2}.$$

To do this, note that it is true when both factors on the right are 1 (Why?). Observe that the inequality is homogeneous in f and g, i.e., you can divide f or g by a constant factor and it will still hold. And when one of those is zero, this is trivial (Why?).

e) *Look back at Chapter 1 at the section on inner products. Prove that*

$$\langle f, g\rangle = \int_a^b f(x)g(x)dx$$

is an inner product on the vector space of continuous functions on $[a, b]$. *Does the proof of the Cauchy-Schwarz inequality there still work? What does the triangle inequality state here, if it works?*

28. *A function* $\varphi : \mathbb{R} \to \mathbb{R}$ *is called* convex *when* $\varphi(w_1x_1 + w_2x_2 + \cdots + w_nx_n) \leq w_1\varphi(x_1) + w_2\varphi(x_2) + \cdots + w_n\varphi(x_n)$. *Suppose* $f : [0, 1] \to \mathbb{R}$ *is continuous. Show that*

$$\varphi\left(\int_0^1 f(x)dx\right) \leq \int_0^1 \varphi(f(x))dx,$$

a statement known as Jensen's inequality *by considering Riemann sums whose norms are very small, and applying the definition of a convex function.*

29. *This exercise and the next illustrate useful techniques in analysis, cf. [Zyg77]. Let* $f : [a, b] \to \mathbb{R}$ *be continuous. Define the sequence* $\{f_n\}$ *as*

$$f_n = \int_a^b f(x) \sin(nx)dz. \tag{7.34}$$

In this exercise you will prove a version of the Riemann-Lebesgue lemma, *i.e., that* $\lim f_n = 0$.

a) *Suppose* g *is the* characteristic function *of a subinterval* I' *of* $[a, b]$, *i.e.,* $g(x) = 1$ *if* $x \in I'$, *and* $g(x) = 0$ *otherwise. Define the sequence* g_n *correspondingly, as in* (7.34). *Prove by direct calculation that* $\lim g_n = 0$.

b) *Prove the same when* g *is a* step function, *or a linear combination of characteristic functions of intervals.*

c) *Given* $\epsilon > 0$ *and* f *continuous, prove that there exists a step function* g *such that* $|f(x) - g(x)| < \epsilon$ *for all* $x \in [a, b]$. *Deduce that* $|f_n| \leq |g_n| + \epsilon(b - a)$.

d) *Show that* $\limsup |f_n| \leq \epsilon(b - a)$ *and finish the proof.*

30. *Let* $f : [0, 1] \to \mathbb{R}$ *be nonnegative and continuous. Write* $M = \sup_{[0,1]} f$. *Consider the expressions*

$$\mathfrak{M}_n = \left(\int_a^b f^n dx\right)^{1/n}$$

for each $n \in \mathbb{N}$.

a) *Prove that* $\mathfrak{M}_n \leq M$ *for all* n.

b) *Fix* $\epsilon > 0$. *Prove that there exists an interval* $J = [c - \eta, c + \eta] \subset [0, 1]$ *such that* $f(x) > M - \epsilon$ *for* $x \in J$.

c) *Prove that (with the same notation)* $\mathfrak{M}_n \geq (2\eta)^{1/n}(M - \epsilon)$. *Deduce that* $\liminf \mathfrak{M}_n \geq M - \epsilon$.

d) *Prove that* $\lim \mathfrak{M}_n = M$.

8

COMPLEX INTEGRATION

The shortest path between two truths in the real domain passes through the complex domain.

Jacques Hadamard

We have already seen how complex functions can have significantly different properties than real-valued or vector-valued functions. Not the least of these are the very different and far-reaching implications associated with complex differentiability. The theory of complex integration is no less rich. Some of the results we ultimately develop will rely on subtle topological properties of certain subsets of the plane, as well as the curves along which we will integrate. In fact a proper development of the main theorem of this chapter, Cauchy's integral theorem, will require quite a bit of careful effort due to such subtleties. Despite this, we will find that integration is as useful an analytical tool as differentiation for investigating complex functions and their key properties. In fact it is in the theory of complex integration where we will discover some of the most interesting and useful results in all of analysis, results that will apply even to real functions of a real variable.

1 INTRODUCTION TO COMPLEX INTEGRALS

1.1 Integration over an Interval

We begin by considering the integral of a continuous complex-valued function of a single real variable, $w : [a, b] \to \mathbb{C}$. We assume w has the form $w(t) = u(t) + iv(t)$, and so w being continuous implies that each of u and v is continuous. We define the integral of $w(t)$ on $[a, b]$ according to

$$\int_a^b w(t)\,dt \equiv \int_a^b u(t)\,dt + i\int_a^b v(t)\,dt. \tag{8.1}$$

This definition is consistent with the linearity property possessed by the integral of a real-valued function of a single real variable, extended in the natural way to include pure imaginary constants. We will build on this basic definition, developing for our complex integral many of the properties possessed by the more familiar integral of real-valued functions. Consistent with this, we define the complex integral from b to a where $a < b$ according to

$$\int_b^a w(t)\,dt \equiv -\int_a^b w(t)\,dt,$$

and the complex integral from c to c for any $c \in [a,b]$ according to

$$\int_c^c w(t)\,dt \equiv 0.$$

The following result should also be familiar from the real case.

Proposition 1.1 *Suppose $w, w_1, w_2 : [a,b] \to \mathbb{C}$ are continuous on $[a,b]$.*

a) If $z_1, z_2 \in \mathbb{C}$, then $\int_a^b \big(z_1\, w_1(t) + z_2\, w_2(t)\big)dt = z_1 \int_a^b w_1(t)\,dt + z_2 \int_a^b w_2(t)\,dt$.

b) $\mathrm{Re}\left(\int_a^b w(t)\,dt\right) = \int_a^b \mathrm{Re}\big(w(t)\big)\,dt$, and $\mathrm{Im}\left(\int_a^b w(t)\,dt\right) = \int_a^b \mathrm{Im}\big(w(t)\big)\,dt$.

c) If c_1, c_2, and c_3 are any three points from $[a,b]$, then,

$$\int_{c_1}^{c_2} w(t)\,dt = \int_{c_1}^{c_3} w(t)\,dt + \int_{c_3}^{c_2} w(t)\,dt.$$

► **8.1** *Prove the above proposition.*

Some of the most useful results in analysis aren't identities or equalities, they are inequalities. Such results are particularly useful in proving theorems, and one such inequality, a sort of "triangle inequality for line integrals" follows.

Theorem 1.2 *Suppose $w : [a,b] \to \mathbb{C}$ is continuous on $[a,b]$, with $\Lambda \equiv \max_{[a,b]} |w(t)|$. Then,*

$$\left|\int_a^b w(t)\,dt\right| \le \int_a^b |w(t)|\,dt \le \Lambda\,(b-a).$$

PROOF If $\int_a^b w(t)\,dt = 0$, we are done. Therefore assume that $\int_a^b w(t)\,dt = r_0\, e^{i\theta} \neq 0$. Then we have

$$\begin{aligned}\left|\int_a^b w(t)\,dt\right| = r_0 = \mathrm{Re}(r_0) &= \mathrm{Re}\left(e^{-i\theta}\, r_0\, e^{i\theta}\right)\\ &= \mathrm{Re}\left(e^{-i\theta}\int_a^b w(t)\,dt\right)\\ &= \mathrm{Re}\left(\int_a^b e^{-i\theta}\, w(t)\,dt\right)\\ &= \int_a^b \mathrm{Re}\big(e^{-i\theta}\, w(t)\big)\,dt\end{aligned}$$

$$\leq \int_a^b \left|e^{-i\theta}\, w(t)\right| dt$$
$$= \int_a^b \left|w(t)\right| dt.$$

Clearly, $\int_a^b \left|w(t)\right| dt \leq \Lambda\,(b-a)$, and the theorem is proved. ♦

In the above theorem it is often more convenient to replace Λ with any $M \geq \Lambda$. This is particularly true when any upper bound is all that is needed, or when the maximum of $|w(t)|$ over $[a,b]$ is inconveniently obtained.

The following result is the familiar change-of-variables formula from calculus in terms of our new complex integral.

Proposition 1.3 *Suppose $w : [a,b] \to \mathbb{C}$ is continuous on $[a,b]$, and let $t = g(s)$ where $g : [c,d] \to [a,b]$ has the following properties:*

(i) $g(c) = a$ and $g(d) = b$,

(ii) g is continuously differentiable on $[c,d]$.

Then,

$$\int_a^b w(t)\, dt = \int_c^d w(g(s))\, g'(s)\, ds.$$

► **8.2** *Prove the above proposition.*

What would our complex integral be worth if the fundamental theorem of calculus didn't apply? It does, as described by the following results.

Proposition 1.4 *Suppose $w : [a,b] \to \mathbb{C}$ is continuous on $[a,b]$ and $W : [a,b] \to \mathbb{C}$ is given by*

$$W(s) = \int_a^s w(t)\, dt.$$

Then W is differentiable on $[a,b]$, and $W'(s) = w(s)$ for all $s \in [a,b]$.

The careful reader will notice that we never formally defined what it means for a function $W : [a,b] \to \mathbb{C}$ to be differentiable. But the definition is transparent. If we write $W = U + iV$ for real-valued functions U, V, then $W'(t) = w(t)$ just means $U'(t) = \mathrm{Re}(w(t))$ and $V'(t) = \mathrm{Im}(w(t))$.

► **8.3** *Prove the above proposition.*

Theorem 1.5 (The Fundamental Theorem of Calculus)
Suppose $w : [a,b] \to \mathbb{C}$ is continuous on $[a,b]$, and there exists a differentiable complex-valued function $W : [a,b] \to \mathbb{C}$ such that $W'(t) = w(t)$ on $[a,b]$. Then,

$$\int_a^b w(t)\,dt = W(t)\Big|_a^b = W(b) - W(a).$$

PROOF Let $w(t) = u(t) + iv(t)$ and $W(t) = U(t) + iV(t)$, where $U'(t) = u(t)$ and $V'(t) = v(t)$ for all $t \in [a,b]$. Then by definition of our complex integral and the fundamental theorem of calculus for real functions, we have

$$\begin{aligned}\int_a^b w(t)\,dt &= \int_a^b u(t)\,dt + i\int_a^b v(t)\,dt \\ &= \big(U(b) - U(a)\big) + i\,\big(V(b) - V(a)\big) \\ &= W(b) - W(a).\end{aligned}$$

♦

As a simple illustration of the above, consider $w : [0,1] \to \mathbb{R}$ given by $w(t) = t^2 + i\,t^3$. Then,

$$\int_0^1 w(t)\,dt = \int_0^1 (t^2 + i\,t^3)\,dt = \left(\tfrac{1}{3}\,t^3 + i\,\tfrac{1}{4}\,t^4\right)\Big|_0^1 = \tfrac{1}{3} + i\,\tfrac{1}{4}.$$

1.2 Curves and Contours

In general, we would like to be able to integrate a complex-valued function of a complex variable over a curve C in the complex plane. To begin, we extend our notion of a parametrized curve from $\mathbb{R}^2$ to $\mathbb{C}$ in the natural way. Recall that a curve C in $\mathbb{R}^2$ can be parametrized by a continuous map $\mathbf{x} : [a,b] \to \mathbb{R}^2$. Similarly, we will say that a curve C in $\mathbb{C}$ can be parametrized by a continuous map $z : [a,b] \to \mathbb{C}$. In this case, $z(a)$ is referred to as the *initial point* of the curve C and $z(b)$ is referred to as the *terminal point* of the curve C. The curve $C = (C_1, \ldots, C_n) = \bigcup_{j=1}^n C_j$ is to be interpreted in $\mathbb{C}$ as in $\mathbb{R}^2$, as a union of curves C_j for $1 \le j \le n$, or, when parametrized, as a catenation of curves C_j for $1 \le j \le n$. The notion of *equivalent parametrizations* for a given curve C in $\mathbb{C}$ is also defined as in the real case, as are the definitions for *closed* curve, *simple* curve, and *smooth* curve. In particular, the curve C in $\mathbb{C}$ is smooth if there exists a parametrization $z : [a,b] \to \mathbb{C}$ such that z is continuously differentiable on $[a,b]$ and $z'(t) \neq 0$ for all $t \in [a,b]$. Finally, *contours* are also defined as in the real case. A contour $C' \subset \mathbb{C}$ will be called a *subcontour* of the contour $C \subset \mathbb{C}$ if every point of C' is also a point of C, i.e., $C' \subset C$.

Circles

The following definitions and conventions will also be of use to us in what follows.

Definition 1.6 For any $r > 0$ and $z_0 \in \mathbb{C}$, let $C_r(z_0)$ be the circle in $\mathbb{C}$ having radius r and center z_0, i.e.,

$$C_r(z_0) \equiv \{z \in \mathbb{C} : |z - z_0| = r\}.$$

When referred to as **the parametrized circle** $C_r(z_0)$, unless specified otherwise, we will assume $C_r(z_0)$ to be parametrized by $z : [0, 2\pi] \to \mathbb{C}$ where

$$z(t) = z_0 + r\, e^{i\,t}.$$

▶ **8.4** *Verify that the parametrized circle $C_r(z_0)$ is a smooth curve, and that the default parametrization described in Definition 1.6 associates a counterclockwise direction to the traversal of $C_r(z_0)$.*

▶ **8.5** *Fix $\theta_0 \in \mathbb{R}$. Let $\tilde{z} : [\theta_0, \theta_0 + 2\pi] \to \mathbb{C}$ be given by $\tilde{z}(t) = z_0 + r\, e^{i\,t}$. Show that $\tilde{z}$ is also a parametrization of $C_r(z_0)$. How does $C_r(z_0)$ parametrized by $\tilde{z}$ differ from $C_r(z_0)$ parametrized by z?*

Line Segments and Polygonal Contours

Definition 1.7 For any $z_1, z_2 \in \mathbb{C}$ with $z_1 \neq z_2$, let $[z_1, z_2] \subset \mathbb{C}$ be the closed line segment in $\mathbb{C}$ determined by z_1 and z_2. When referred to as **the parametrized segment** $[z_1, z_2]$, unless specified otherwise, we will assume $[z_1, z_2]$ to be parametrized by $z : [0, 1] \to \mathbb{C}$ where

$$z(t) = z_1 + t\,(z_2 - z_1).$$

▶ **8.6** *Verify that the parametrized segment $[z_1, z_2]$ is a smooth curve with initial point z_1 and terminal point z_2.*

▶ **8.7** *Let $\tilde{z} : [a, b] \to \mathbb{C}$ be given by $\tilde{z}(t) = z_1 + \frac{t-a}{b-a}\,(z_2 - z_1)$. Show that $\tilde{z}$ is a parametrization of $[z_1, z_2]$ that is equivalent, according to Definition 4.14 in Chapter 7, to the parametrization z given in Definition 1.7.*

Definition 1.8 The collection of points $\{z_0, z_1, z_2, \ldots, z_N\} \subset \mathbb{C}$ are called **vertices** if they have the following properties:

1. $z_{j-1} \neq z_j$ for $1 \leq j \leq N$,
2. z_{j-1}, z_j, z_{j+1} for $1 \leq j \leq N - 1$ are not collinear.

Let P be the polygonal contour in $\mathbb{C}$ given by

$$P \equiv [z_0, z_1, \ldots, z_N] \equiv \bigcup_{j=1}^{N} [z_{j-1}, z_j].$$

When referred to as **the parametrized polygonal contour** P, unless specified otherwise, we will assume P to be parametrized by $z : [0, N] \to \mathbb{C}$ where

$$z(t) = \begin{cases} z_0 + t\,(z_1 - z_0) & \text{for } 0 \le t \le 1, \\ z_1 + (t-1)\,(z_2 - z_1) & \text{for } 1 \le t \le 2, \\ \vdots & \vdots \\ z_{N-1} + \big(t - (N-1)\big)\,(z_N - z_{N-1}) & \text{for } N-1 \le t \le N. \end{cases}$$

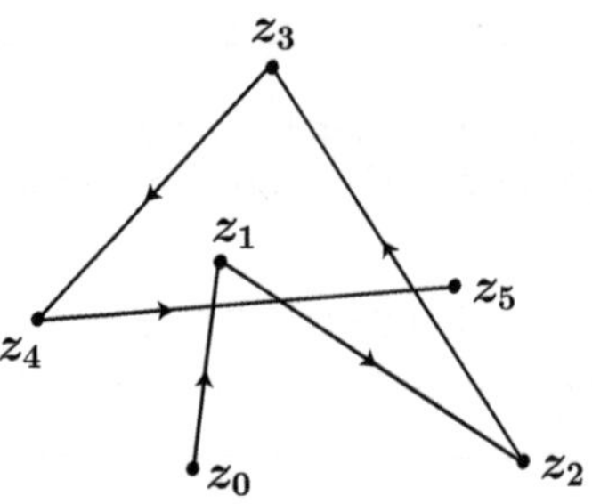

Figure 8.1 *A polygonal contour.*

The condition that $z_{j-1} \neq z_j$ for $1 \le j \le N$ in the above definition simply ensures that no point is repeated consecutively in the collection $\{z_0, z_1, z_2, \ldots, z_N\}$. Also, note that when $N = 1$ the polygonal contour P described above reduces to a single segment. Finally, if $z_0 = z_N$, then P is called *closed,* otherwise P is called *open.* The following special case will also be of importance to us.

Definition 1.9 A **rectangular contour** is a polygonal contour whose segments are horizontal or vertical.

▶ **8.8** *Verify that the parametrized polygonal contour* $P = [z_0, z_1, \ldots, z_N]$ *is a contour.*

The notion of a *connected set* was defined in Chapter 2. A particular type of connected set is a *path-connected set,* one such that any pair of points from it can be connected to each other by a curve lying completely in the set. Path-connected sets are nice for many reasons, but we will have need of a bit more. We will usually require that the curve connecting any two such points be a contour. If, in fact, any two points within the set can always be connected by a contour, we will sometimes refer to the set as *contour connected,* indicating a special type of path-connected set. As it turns out, within all of the Euclidean spaces of interest to us, an open connected set must be contour connected. The proof of this fact, and some other related results, is left to the reader.

► **8.9** *Let $E \subset \mathbb{X}$ be open and connected. Show that E is contour connected. To do this, fix any point $w_0 \in E$, and consider the set A defined as follows:*

$$A = \{w \in E : \text{There exists a contour } x : [a,b] \to E \text{ with } x(a) = w_0 \text{ and } x(b) = w\}.$$

(In fact, one can choose a rectangular contour.) Show that A is open. If A is all of E, there is nothing else to prove, so assume there exists $w_1 \in B \equiv E \setminus A$. Show that B is also open, that $\overline{A} \cap B = A \cap \overline{B} = \varnothing$, and hence that $E = A \cup B$ is therefore disconnected, a contradiction. Hence $E = A$ is contour connected.

► **8.10** *In this exercise, you will show that if a set $E \subset \mathbb{X}$ is contour connected, then it is connected. To do this, suppose $E \subset \mathbb{X}$ has the property that for every pair of points $w_1, w_2 \in E$ there exists a contour $x : [a,b] \to E$ such that $x(a) = w_1$ and $x(b) = w_2$. Assume $E = A \cup B$ where A and B are nonempty and $\overline{A} \cap B = A \cap \overline{B} = \varnothing$. Choose $w_1 \in A$ and $w_2 \in B$, and a contour $x : [a,b] \to E$ such that $x(a) = w_1$ and $x(b) = w_2$. Argue that $x([a,b])$ is a connected subset of E, and that $x([a,b])$ is completely contained in either A or B, a contradiction.*

► **8.11** *Suppose $D \subset \mathbb{C}$ is open and connected. Define the open set $D_0 \equiv D \setminus \{z_0\}$ where $z_0 \in D$. Show that for any pair of points $w_1, w_2 \in D_0$, there exists a contour $z : [a,b] \to D_0$ such that $z(a) = w_1$ and $z(b) = w_2$. Therefore D_0 is connected. Convince yourself that this is also true for $\mathbb{R}^k$ for $k \geq 2$, but not for $\mathbb{R}$.*

► **8.12** *Suppose $D \subset \mathbb{C}$ is open and connected. Define the open set $D_n \equiv D \setminus \{z_1, z_2, \ldots, z_n\}$ where $z_j \in D$ for $1 \leq j \leq n$. Show that for any pair of points $w_1, w_2 \in D_n$, there exists a contour $z : [a,b] \to D_n$ such that $z(a) = w_1$ and $z(b) = w_2$. Therefore D_n is connected. Convince yourself that this is also true for $\mathbb{R}^k$ for $k \geq 2$, but not for $\mathbb{R}$.*

Finally, we end this section with the statement of a significant result from topology that will be of use to us. For a proof, we refer the reader to the book [Ful97] by Fulton.

Theorem 1.10 (The Jordan Curve Theorem)
If C is a simple closed curve in the plane, then the complement of C consists of two components, the interior of the curve C, *and* the exterior of the curve C, *denoted by* $\text{Int}(C)$ *and* $Ext(C)$, *respectively.* $\text{Int}(C)$ *and* $Ext(C)$ *have the following properties:*

a) $\text{Int}(C)$ *and* $Ext(C)$ *are open, connected, and disjoint.*

b) $\text{Int}(C)$ *is bounded and* $Ext(C)$ *is unbounded.*

c) The curve C is the boundary of $\text{Int}(C)$ *and* $Ext(C)$.

1.3 Complex Line Integrals

We would ultimately like to consider integrals such as $\int_C f(z)\,dz$ for complex functions f and parametrized contours C. We begin by parallelling the work done in the real case, and so it will be convenient to assume that f is continuous on the contour C. We will take this to mean that, for C with parametrization $z : [a, b] \to \mathbb{C}$, the function $f\big(z(t)\big)$ is continuous for all $t \in [a, b]$. With this, the following definition is analogous to that of the previous chapter for real-valued functions. Recall that if $C = (C_1, \ldots, C_n)$ is a given contour in $\mathbb{C}$, then each C_j for $1 \leq j \leq n$ is a smooth curve in $\mathbb{C}$.

Definitions and Examples

We begin with a definition.

Definition 1.11 Consider $f : D \to \mathbb{C}$ on the open set $D \subset \mathbb{C}$. Suppose C is a smooth curve in D with parametrization $z : [a, b] \to D$, and suppose f is continuous on C. Then,

$$\int_C f(z)\, dz \equiv \int_a^b f(z(t))\, z'(t)\, dt.$$

If $C = (C_1, \ldots, C_n)$ is a parametrized contour in D, then

$$\int_C f(z)\, dz \equiv \sum_{j=1}^{n} \int_{C_j} f(z)\, dz.$$

Closed contours are especially significant in the context of complex integration, and in cases where we integrate a complex function f around a parametrized contour C that is known to be closed we often use a special notation to emphasize this fact:

$$\oint_C f(z)\, dz.$$

Example 1.12 *Consider the two parametrized contours P and C illustrated in Figure 8.2, each having initial point 0 and terminal point $1 + i$ in $\mathbb{C}$, and given by the parametrizations $z_P : [0, 1] \to \mathbb{C}$ and $z_C : [0, 1] \to \mathbb{C}$ where $z_P(t) = t + i\,t$ and $z_C(t) = t + i\,t^2$, respectively. Then,*

$$\int_P z\, dz = \int_0^1 (t + i\,t)\,(1 + i)\, dt = (1 + i)^2\, \tfrac{1}{2}\, t^2 \Big|_0^1 = \tfrac{1}{2}\,(1 + i)^2 = i,$$

while

$$\int_C z\, dz = \int_0^1 (t + i\,t^2)\,(1 + i\,2t)\, dt = \int_C \left((t - 2t^3) + i\,3t^2\right) dt$$

$$= \tfrac{1}{2}\,(t^2 - t^4) + i\,t^3 \Big|_0^1 = i. \qquad \blacktriangleleft$$

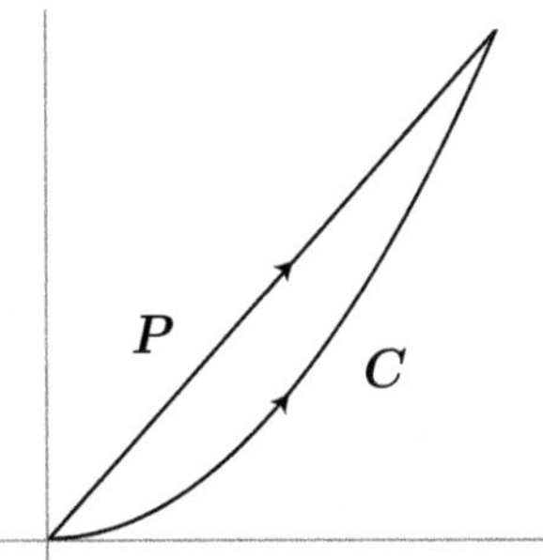

Figure 8.2 *The smooth parametrized contours of Example 1.12.*

► **8.13** *Using the same contours P and C as parametrized in the previous example, compute $\int_P \overline{z}\,dz$ and $\int_C \overline{z}\,dz$. What is different here from what we found in the example?*

► **8.14** *Let $P = [1, i, -1, -i, 1]$ be the parametrized square whose interior contains the origin, and let C_1 be the parametrized unit circle centered at the origin. Compute $\oint_P z\,dz$ and $\oint_{C_1} z\,dz$, and compare the results. Then compute $\oint_P \frac{1}{z}\,dz$ and $\oint_{C_1} \frac{1}{z}\,dz$ and do the same. By the way, what relationship exists between $\frac{1}{z}$ and $\overline{z}$ on C_1?*

The following example will be of great use to us later.

Example 1.13 *We will evaluate the integral $\oint_C \frac{dz}{z-z_0}$ where $C = C_r(z_0)$ is the parametrized circle of radius r centered at z_0. To do so, let $z : [0, 2\pi] \to \mathbb{C}$ be given by $z(t) = z_0 + r\,e^{it}$. Then,*

$$\oint_C \frac{dz}{z - z_0} = \int_0^{2\pi} \frac{i\,r\,e^{it}}{r\,e^{it}}\,dt = 2\pi\,i.$$

◄

► **8.15** *Let C be the circle described in the previous example. Show that for any integer $n \neq 1$, $\oint_C \frac{dz}{(z-z_0)^n} = 0$.*

► **8.16** *Let $\mathrm{Log} : \mathbb{C} \setminus \{0\} \to \mathbb{C}$ be the principal branch of the logarithm, i.e., $\mathrm{Log}(z) = \ln|z| + i\,\mathrm{Arg}(z)$, where $-\pi < \mathrm{Arg}(z) \leq \pi$. Let C_1 be the parametrized unit circle centered at the origin, and let Γ_ϵ be the contour parametrized by $z_\epsilon : [-\pi+\epsilon, \pi-\epsilon] \to \mathbb{C}$ where $z_\epsilon(t) = e^{i\,t}$ as shown in Figure 8.3. Note that since $\Gamma_\epsilon \subset \mathbb{C}\setminus(-\infty, 0]$ for every $\epsilon > 0$, the function $\mathrm{Log}(z)$ is continuous on Γ_ϵ. Compute $\oint_{C_1} \mathrm{Log}(z)\,dz \equiv \lim_{\epsilon\to 0} \int_{\Gamma_\epsilon} \mathrm{Log}(z)\,dz$. What happens to the integral defined above if a different branch of the logarithm is used, say, $\log : \mathbb{C}\setminus\{0\} \to \mathbb{C}$ given by $\log(z) = \ln|z| + i\,\widetilde{\mathrm{Arg}}(z)$, where $0 \leq \widetilde{\mathrm{Arg}}(z) < 2\pi$?*

Elementary Properties

The following proposition establishes some useful properties of our complex integral.

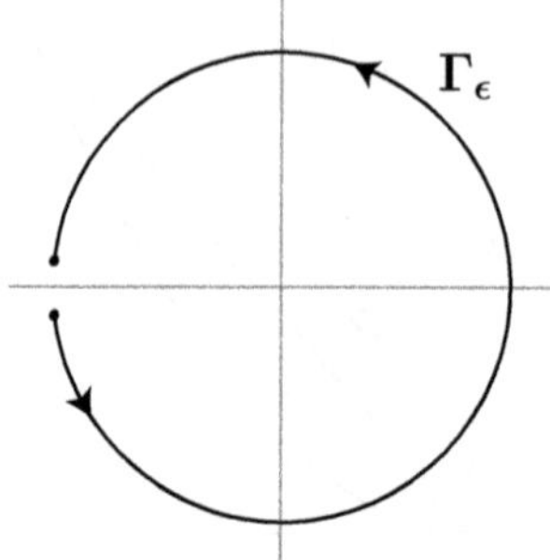

Figure 8.3 *The contour Γ_ϵ in Exercise 8.16, which is "almost a circle."*

Proposition 1.14 *Let C be a parametrized contour in the open set $D \subset \mathbb{C}$, and suppose $f, g : D \to \mathbb{C}$ are continuous on D.*

a) For constants $w_1, w_2 \in \mathbb{C}$, then $\int\limits_C \big(w_1 f(z) \pm w_2\, g(z)\big)\, dz = w_1 \int\limits_C f(z)\, dz \pm w_2 \int\limits_C g(z)\, dz$.

b) If $S = \{z_1, \ldots, z_{N-1}\} \subset C$ subdivides C into the N subcontours $C_1, C_2, \ldots, C_N$, then

$$\int\limits_C f(z)\, dz = \sum_{j=1}^{N} \Bigg(\int\limits_{C_j} f(z)\, dz \Bigg).$$

PROOF We prove the special case of *b)* where C is a smooth curve and S consists of a single point in C. We leave the more general case, and the proof of *a)* to the reader. Suppose $S = \{z^*\} \subset C$, and let $z : [a, b] \to D$ be a parametrization of the smooth curve C. Then $z^* = z(t^*)$ for some $t^* \in [a, b]$. In this case $C = (C_1, C_2)$ where C_1 is parametrized by $z_1 : [a, t^*] \to D$ and $z_1(t) \equiv z(t)$, and C_2 is parametrized by $z_2 : [t^*, b] \to D$ where $z_2(t) \equiv z(t)$. Finally,

$$\begin{aligned}
\int\limits_C f(z)\, dz &= \int_a^b f\big(z(t)\big)\, z'(t)\, dt \\
&= \int_a^{t^*} f\big(z_1(t)\big)\, z_1'(t)\, dt + \int_{t^*}^b f\big(z_2(t)\big)\, z_2'(t)\, dt \\
&= \int\limits_{C_1} f(z)\, dz + \int\limits_{C_2} f(z)\, dz.
\end{aligned}$$

♦

▶ **8.17** *Complete the proof of the above proposition by proving a), and proving b) for contours and more general subdivisions of C.*

Example 1.15 *Let C be the parametrized contour given by $z : [0, 2\pi] \to \mathbb{C}$ where*

$$z(t) = \begin{cases} e^{it} & \text{for } 0 \le t \le \pi \\ -1 + \frac{2}{\pi}(t - \pi) & \text{for } \pi \le t \le 2\pi \end{cases}.$$

Let $S = \{-1, 1\}$ subdivide C into two subcontours C_1, C_2 as shown in Figure 8.4 where $C = (C_1, C_2)$. It is easy to show that $\int_C z^2\, dz = \sum_{j=1}^{2} \left(\int_{C_j} z^2\, dz \right)$. To see

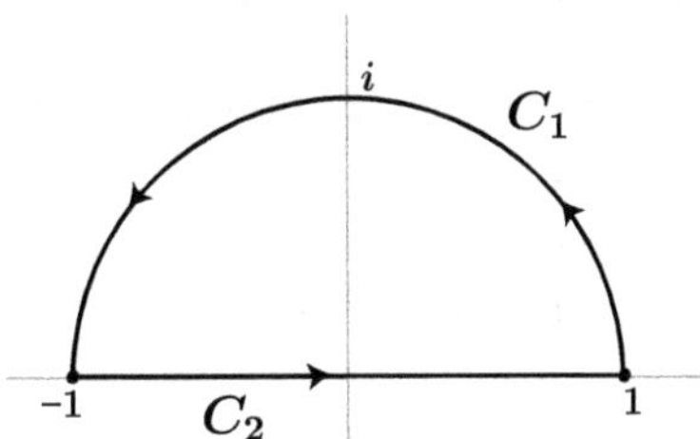

Figure 8.4 *The semicircular contour C.*

this, note that consistent with the given parametrization of C, the subcontour C_1 is parametrized by $z_1 : [0, \pi] \to \mathbb{C}$ where $z_1(t) = e^{it}$. Likewise, the subcontour C_2 is parametrized by $z_2 : [\pi, 2\pi] \to \mathbb{C}$ where $z_2(t) = -1 + \frac{2}{\pi}(t - \pi)$. We leave it to the reader to show that $\int_C z^2\, dz = 0$, $\int_{C_1} z^2\, dz = -\frac{2}{3}$, $\int_{C_2} z^2\, dz = \frac{2}{3}$, and that therefore $\int_C z^2\, dz = \sum_{j=1}^{3} \left(\int_{C_j} z^2\, dz \right)$. ◀

▶ **8.18** *Verify the claim made in the previous example.*

Lengths of Contours

It is often useful to determine the length of a smooth curve C in the complex plane. Suppose C is a smooth curve parametrized by the map $z : [a, b] \to \mathbb{C}$ for $a \le t \le b$. Then the length of C will be denoted by L_C, and is given by

$$L_C = \int_a^b \left| z'(t) \right| dt. \tag{8.2}$$

To see this, recall from multivariable calculus that for a smooth curve in the plane parametrized by $\mathbf{r}(t) = x(t)\hat{\imath} + y(t)\hat{\jmath}$,

$$L_C = \int_a^b \sqrt{\left(\frac{dx}{dt}\right)^2 + \left(\frac{dy}{dt}\right)^2}\, dt,$$

which is equivalent to (8.2) when the plane is $\mathbb{C}$ and the curve is given by $z : [a, b] \to \mathbb{C}$ where $z(t) = x(t) + iy(t)$. If $\mathcal{P} = \{a = t_0, t_1, \ldots, t_N = b\}$ is a partition of $[a, b]$ corresponding to the parametrized contour $C \subset \mathbb{C}$, then $\mathcal{P}$

subdivides C into subcontours $C_1, C_2, \ldots, C_N$, and

$$L_C = \sum_{j=1}^{N} L_{C_j}.$$

► **8.19** *Prove the above claim.*

► **8.20** *Consider C parametrized by $z : (0,1] \to \mathbb{C}$ where $z(t) = t + i \sin\left(\frac{1}{t}\right)$. Find L_C. Is C a contour according to Definition 4.12 in Chapter 7? Why or why not?*

The following result is the complex analog to Corollary 2.12 in Chapter 7.

Proposition 1.16 *Suppose C is a parametrized contour contained in the open set $D \subset \mathbb{C}$, and that $f : D \to \mathbb{C}$ is continuous on D. If $\Lambda \equiv \max_{z\in C} |f(z)|$, then*

$$\left| \int_C f(z)\, dz \right| \leq \Lambda\, L_C.$$

► **8.21** *Prove the above proposition. Clearly it also holds if Λ is replaced by any $M > \Lambda$.*

Just as in the real line integral case, it is an extremely useful fact that if two equivalent parametrizations of $C \subset \mathbb{C}$ are given, the value of the integral of a complex function f along C will be the same with respect to either of the equivalent parametrizations used to describe C. We state this result more formally in the following theorem, the proof of which is left to the reader.

Proposition 1.17 *Suppose $f : D \to \mathbb{C}$ is continuous on the open set $D \subset \mathbb{C}$, and C is a contour in D with parametrization $z : [a,b] \to D$. Suppose $\widetilde{z} : [c,d] \to D$ is a parametrization of C equivalent to z. Then,*

$$\int_C f(z)\, dz \equiv \int_a^b f(z(t))\, z'(t)\, dt = \int_c^d f(\widetilde{z}(s))\, \widetilde{z}'(s)\, ds.$$

► **8.22** *Prove the above proposition. Refer to Proposition 1.3 on page 389.*

Example 1.18 Consider the segment $[1,i]$ and the integral $\int_{[1,i]} \overline{z}\, dz$. With the standard parametrization $z_1 : [0,1] \to \mathbb{C}$ given by $z_1(t) = 1 + t\,(i-1)$, we have

$$\int_{[1,i]} \overline{z}\, dz = \int_0^1 (1 + t\,(-i-1))\,(i-1)\, dt = 2 + i.$$

With the equivalent parametrization $z_2 : [-1,1] \to \mathbb{C}$ given by $z_2(t) = 1 +$

$\frac{1}{2}(t+1)(i-1)$, *we have*

$$\int_{[1,i]} \overline{z}\, dz = \int_{-1}^{1} \left(1 + \tfrac{1}{2}(t+1)(-i-1)\right) \tfrac{1}{2}(i-1)\, dt = 2+i,$$

and the integral has the same value for both parametrizations. ◀

Based on our results for contours in $\mathbb{R}^2$, for any contour $C \subset \mathbb{C}$ with parametrization $z_C : [a,b] \to \mathbb{C}$, the associated parametrized curve or contour $-C$ consisting of the same set of points in $\mathbb{C}$ but traversed in the opposite way can be parametrized by $z_{-C} : [-b,-a] \to \mathbb{C}$ where $z_{-C}(t) = z_C(-t)$. We leave it to the reader to show that if $f : D \to \mathbb{C}$ is continuous in the open set D, and if C is a parametrized contour in D, then

$$\int_{-C} f(z)\, dz = -\int_{C} f(z)\, dz.$$

▶ **8.23** *Prove the above claim.*

Example 1.19 *Suppose C is a simple closed contour contained in an open set $D \subset \mathbb{C}$. Connect two points of C by a contour $S \subset D$ that passes only through the interior of C. More specifically, consider Figure 8.5 with C_L the contour parametrized so as to be traversed from z_1 to z_3 via z_2, and with C_R the contour parametrized so as to be traversed from z_3 to z_1 via z_4. Take the*

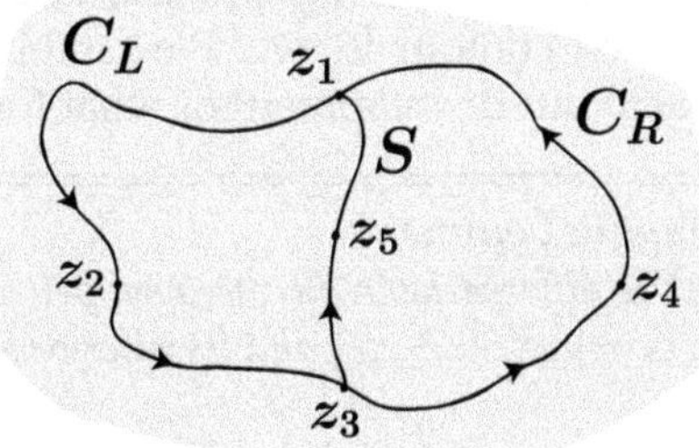

Figure 8.5 *The situation in Example 1.19.*

closed contour C to be (C_L, C_R), and let S be the contour parametrized from z_3 to z_1 via z_5. Then note that for any function $f : D \to \mathbb{C}$ that is continuous on $(C \cup S) \subset D$, we have

$$\begin{aligned}\oint_C f(z)\, dz &= \oint_{C_L} f(z)\, dz + \oint_{C_R} f(z)\, dz \\ &= \oint_{C_L} f(z)\, dz + \int_S f(z)\, dz - \int_S f(z)\, dz + \oint_{C_R} f(z)\, dz\end{aligned}$$

$$= \oint_{(C_L,S)} f(z)\,dz + \oint_{(C_R,-S)} f(z)\,dz.$$

That is, the integral of f around the the closed contour C is equal to the sum of the integrals of f around the two closed contours (C_L, S) and $(C_R, -S)$, and so we may interpret the integral of f along S and $-S$ as "canceling each other out." ◀

▶ **8.24** *In the above example, suppose S connects two points of C by going through the exterior of C. Does the conclusion of the example still hold? That is, does the integral of f along S still cancel out the integral of f along $-S$?*

▶ **8.25** *In the above example, suppose the contour C is not simple and S is a contour that connects two points of C. Does the conclusion of the example still hold now?*

▶ **8.26** *Suppose $C \subset \mathbb{C}$ is a closed contour parametrized by $z : [a, b] \to \mathbb{C}$. For any $t_0 \in \mathbb{R}$ show that if C is reparametrized by $\tilde{z} : [a+t_0, b+t_0] \to \mathbb{C}$ where $\tilde{z}(t) \equiv z(t-t_0)$, then the value of $\oint_C f(z)\,dz$ is unchanged.*

2 Further Development of Complex Line Integrals

2.1 The Triangle Lemma

One of the most important concepts in the theory of complex integration is that, under the right conditions, the integral of a differentiable complex function around a closed contour is *zero*. We establish our first such result next. It will be critical to our development of what follows.

Lemma 2.1 (The Triangle Lemma)
Suppose $f : D \to \mathbb{C}$ is differentiable on the open set $D \subset \mathbb{C}$, and suppose the parametrized triangle $\triangle = [z_0, z_1, z_2, z_0]$ and its interior are contained in D. Then

$$\oint_{\triangle} f(z)\,dz = 0.$$

Proof Let P_0 be the perimeter of $\triangle$, and let L_0 be the length of the longest side of $\triangle$. Let M_1 be the midpoint of $[z_0, z_1]$, let M_2 be the midpoint of $[z_1, z_2]$, and let M_3 be the midpoint of $[z_2, z_0]$. Connect these midpoints to form four parametrized triangles $T_1 = [z_0, M_1, M_3, z_0]$, $T_2 = [M_1, z_1, M_2, M_1]$, $T_3 = [M_2, z_2, M_3, M_2]$, and $T_4 = [M_3, M_1, M_2, M_3]$, as shown below. As in Example 1.19 on page 399, it is easy to see that by cancelation of contour contributions,

$$\oint_{\triangle} f(z)\,dz = \oint_{T_1} f(z)\,dz + \oint_{T_2} f(z)\,dz + \oint_{T_3} f(z)\,dz + \oint_{T_4} f(z)\,dz.$$

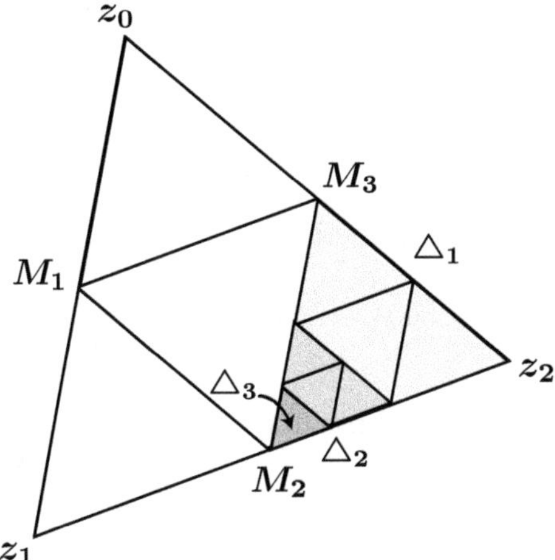

Figure 8.6 *Demonstration of the proof of the triangle lemma.*

One of the smaller triangles T_1, T_2, T_3, or T_4, hereafter denoted by $\triangle_1$, has the property that

$$\left|\oint_{\triangle} f(z)\,dz\right| \leq 4\left|\oint_{\triangle_1} f(z)\,dz\right|. \tag{8.3}$$

(Why?) Consider the same procedure with $\triangle_1$ as the initial triangle (shaded in Figure 8.6), and of the four smaller triangles coming from $\triangle_1$, let the triangle having property (8.3) be denoted by $\triangle_2$. Then we have

$$\left|\oint_{\triangle} f(z)\,dz\right| \leq 4\left|\oint_{\triangle_1} f(z)\,dz\right| \leq 4^2\left|\oint_{\triangle_2} f(z)\,dz\right|.$$

Continuing this procedure indefinitely, we generate a sequence of triangles $\triangle, \triangle_1, \triangle_2, \ldots$ such that

1. $\left|\oint_{\triangle} f(z)\,dz\right| \leq 4^n\left|\oint_{\triangle_n} f(z)\,dz\right|$,
2. P_n = (perimeter of $\triangle_n$) = $\frac{1}{2^n}\,P_0$,
3. L_n = (length of the longest side of $\triangle_n$) = $\frac{1}{2^n}\,L_0$.

Now denote the closure of the interior of each triangle in this sequence by $\blacktriangle$ and by $\blacktriangle_j$ for $j = 1, 2, \ldots$. Then, $\blacktriangle \supset \blacktriangle_1 \supset \blacktriangle_2 \supset \cdots$, forms a nested sequence of closed sets, and therefore by Theorem 4.9 in Chapter 2 there exists a unique $z_0 \in \blacktriangle \cap \left(\bigcap_{j=1}^{\infty} \blacktriangle_j\right)$. Since f is differentiable at z_0, for any $\epsilon > 0$ there exists $\delta > 0$ such that

$$0 < |z - z_0| < \delta \;\Rightarrow\; \left|\frac{f(z) - f(z_0)}{z - z_0} - f'(z_0)\right| < \frac{\epsilon}{2\,L_0\,P_0}.$$

Now choose N large enough that $\blacktriangle_N \subset N_\delta(z_0)$. Then it follows that

$$\begin{aligned}\big|f(z) - f(z_0) - f'(z_0)\,(z-z_0)\big| &\le \frac{\epsilon}{2\,L_0\,P_0}\,|z-z_0| \\ &\le \frac{\epsilon}{2\,L_0\,P_0}\,L_N \qquad \text{for all } z \in \blacktriangle_N.\end{aligned}$$

From this last inequality we obtain

$$\left|\oint\limits_{\triangle_N} \big(f(z) - f(z_0) - f'(z_0)\,(z-z_0)\big)\,dz\right| \le \frac{\epsilon}{2\,L_0\,P_0}\,L_N\,P_N = \frac{\epsilon}{2\,(4^N)}. \qquad (8.4)$$

We leave it to the reader to show that $\oint\limits_{\triangle_N} f(z_0)\,dz = \oint\limits_{\triangle_N} f'(z_0)\,(z-z_0)\,dz = 0$, and therefore that $\oint\limits_{\triangle_N} f(z)\,dz = \oint\limits_{\triangle_N} \big(f(z) - f(z_0) - f'(z_0)\,(z-z_0)\big)\,dz$. This in turn, along with (8.4), leads to

$$\begin{aligned}\left|\oint\limits_{\triangle} f(z)\,dz\right| &\le 4^N \left|\oint\limits_{\triangle_N} f(z)\,dz\right| \\ &= 4^N \left|\oint\limits_{\triangle_N} \big(f(z) - f(z_0) - f'(z_0)\,(z-z_0)\big)\,dz\right| \\ &\le \tfrac{\epsilon}{2} < \epsilon,\end{aligned}$$

and so, $\oint\limits_{\triangle} f(z)\,dz = 0$. ♦

► **8.27** *Show that* $\oint\limits_{\triangle_N} f(z_0)\,dz = \oint\limits_{\triangle_N} f'(z_0)\,(z-z_0)\,dz = 0$ *by parametrizing* $\triangle_N$ *and computing.*

The above result can clearly be applied to integrals along nontriangular contours, as the following example illustrates.

Example 2.2 *Suppose* $f : D \to \mathbb{C}$ *is differentiable on* D *and that the parametrized rectangle* $R = [z_0, z_1, z_2, z_3, z_0]$ *and its interior is contained in* D. *Clearly* R *can be carved up into parametrized triangular pieces* $\triangle_1$ *and* $\triangle_2$ *such that* $\oint\limits_R f(z)\,dz = \oint\limits_{\triangle_1} f(z)\,dz + \oint\limits_{\triangle_2} f(z)\,dz$. *Since the integral over each triangle is zero, it follows that* $\oint\limits_R f(z)\,dz = 0$. ◄

► **8.28** *Fill in the details of the above example.*

An important refinement of the triangle lemma follows next. It says that the differentiability of f is allowed to be relaxed at a single point $p \in D$, and yet the result still holds.

Corollary 2.3 *Suppose $f : D \to \mathbb{C}$ is continuous on D, and suppose the parametrized triangle $\triangle = [z_0, z_1, z_2, z_0]$ and its interior are contained in D. Assume that f is differentiable on $D \setminus \{p\}$ for some $p \in D$. Then*

$$\oint_{\triangle} f(z)\, dz = 0.$$

PROOF [1] Note that if p is outside the triangle, the proof is the same as for the previous theorem. There are three other cases to consider (see Figure 8.7).

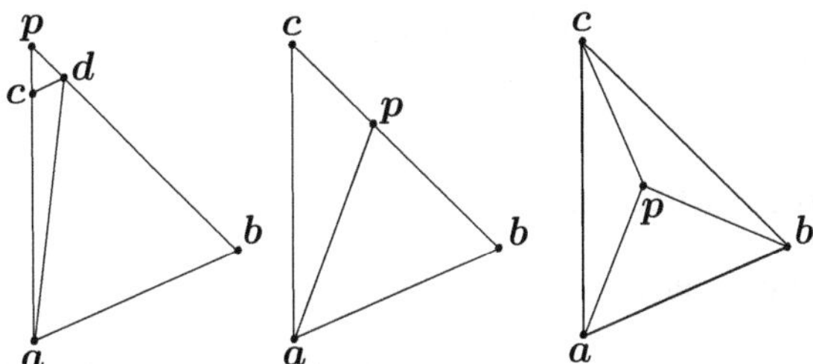

Figure 8.7 *Demonstration of the proof of Corollary 2.3 in Cases 1–3, from left to right.*

Case 1: p is a vertex of $\triangle$.

Denote the other two vertices of $\triangle$ by a and b, and let $\epsilon > 0$ be given. Let $\Lambda \equiv \max |f(z)|$, the maximum value of f on (and within) $\triangle$. Consider the figure above, where we choose points c and d on $\triangle$ so that the perimeter of the triangle formed by c, d, and p, denoted by $P_{\triangle_{cdp}}$, is less than $\frac{\epsilon}{\Lambda}$. Then we have

$$\begin{aligned}\oint_{\triangle} f(z)\, dz &= \oint_{[a,d,c,a]} f(z)\, dz + \oint_{[a,b,d,a]} f(z)\, dz + \oint_{[c,d,p,c]} f(z)\, dz \\ &= 0 + 0 + \oint_{[c,d,p,c]} f(z)\, dz,\end{aligned}$$

and so,

$$\left|\oint_{\triangle} f(z)\, dz\right| = \left|\oint_{[c,d,p,c]} f(z)\, dz\right| \le \Lambda\, P_{\triangle_{cdp}} < \epsilon,$$

which yields the result.

Case 2: p lies on a side of $\triangle$, but not on a vertex.

[1] We follow [Rud87] in the proof.

Denote the vertices of $\triangle$ by a, b, and c. Then,

$$\oint_{\triangle} f(z)\,dz = \oint_{[a,b,p,a]} f(z)\,dz + \oint_{[a,p,c,a]} f(z)\,dz = 0 + 0 = 0.$$

Case 3: p lies in the interior of $\triangle$.

Again, denote the vertices of $\triangle$ by a, b, and c. Then,

$$\oint_{\triangle} f(z)\,dz = \oint_{[a,b,p,a]} f(z)\,dz + \oint_{[b,c,p,b]} f(z)\,dz + \oint_{[a,p,c,a]} f(z)\,dz = 0 + 0 + 0 = 0.$$

♦

► **8.29** *In the above proof, where in cases 2 and 3 is the result of case 1 implicitly used? Extend the corollary to the case where $f : D \to \mathbb{C}$ is differentiable on $D \setminus \{p_1, p_2, \ldots, p_n\}$ for $p_1, p_2, \ldots, p_n \in D$.*

2.2 Winding Numbers

To motivate what follows, consider the integral $\frac{1}{2\pi i}\oint_C \frac{dz}{z}$ for some parametrized closed curve C in $\mathbb{C}$ where $0 \notin C$. Since $0 \notin C$, we may parametrize C according to $z(t) = r(t)e^{i\theta(t)}$ where $r(t)$ and $\theta(t)$ are continuously differentiable functions on $[a, b]$. Following our noses in a nonrigorous way, we obtain

$$\begin{aligned}
\frac{1}{2\pi i}\oint_C \frac{dz}{z} &= \frac{1}{2\pi i}\int_a^b \frac{z'(t)}{z(t)}\,dt \\
&= \frac{1}{2\pi i}\int_a^b \frac{\big(r'(t) + i\,r(t)\theta'(t)\big)e^{i\theta(t)}}{r(t)e^{i\theta(t)}}\,dt \\
&= \frac{1}{2\pi i}\int_a^b \left(\frac{r'(t)}{r(t)} + i\,\theta'(t)\right)dt \\
&= \frac{1}{2\pi i}\left(\ln|r(t)|\Big|_a^b + i\,\theta(t)\Big|_a^b\right) \\
&= \frac{\theta(b) - \theta(a)}{2\pi} \\
&= \frac{\Delta\theta}{2\pi}.
\end{aligned}$$

Here, $\Delta\theta$ is the net change in the argument of $z(t)$ as C is traversed once from $t = a$ to $t = b$. As we rigorously show in Appendix A, the value of $\Delta\theta$ is an integer (not necessarily zero!) multiple of 2π. Hence, $\frac{1}{2\pi i}\oint_C \frac{dz}{z}$ can be thought of as the number of times the contour C "winds around" the origin. The definition below exploits this conclusion and characterizes a topological concept that will be of surprisingly great use to us. In fact, the statements and proofs of many of our most potent results will rely on it.

Definition 2.4 Let C be a closed parametrized contour in $\mathbb{C}$, and consider any $z_0 \notin C$. The **winding number of C about** z_0, denoted by $n_C(z_0)$, is given by

$$n_C(z_0) \equiv \frac{1}{2\pi i} \oint_C \frac{d\zeta}{\zeta - z_0}.$$

As described above, the number $n_C(z_0)$ can be seen to be the number of times the contour C "winds around" the point z_0. A rigorous justification for this geometric interpretation can be found in Appendix A. To illustrate this geometric connection to the formula in the above definition, we consider the simple case where $C = C_r(z_0)$ is a circle of radius r centered at z_0.

Example 2.5 *Consider the circle $C = C_r(z_0)$ centered at z_0 in the complex plane and parametrized by $\zeta : [0, 2\pi] \to \mathbb{C}$ where $\zeta(t) = z_0 + r\,e^{i\,t}$. Then,*

$$n_C(z_0) = \frac{1}{2\pi i} \oint_C \frac{d\zeta}{\zeta - z_0} = \frac{1}{2\pi i} \int_0^{2\pi} \frac{i\,r\,e^{i\,t}\,dt}{r\,e^{i\,t}} = 1.$$

◀

▶ **8.30** *Show that if you traverse the circle in the above example in the same counterclockwise direction, but n times rather than once, then $n_C(z_0) = n$. (Hint: What is the parametrization for this new curve?)*

▶ **8.31** *Parametrize the circle from the above example in the opposite direction and show that $n_C(z_0) = -1$.*

Proposition 2.6 *$n_C(z_0)$ is an integer.*

PROOF We prove the case where C is a smooth closed curve, and leave the more general case where C is a contour to the exercises. For C parametrized by $z : [a,b] \to \mathbb{C}$, let $F : [a,b] \to \mathbb{C}$ be defined by

$$F(s) \equiv \int_a^s \frac{z'(t)}{z(t) - z_0}\,dt \quad \text{for } s \in [a,b].$$

Note that $F(b) = 2\pi i\, n_C(z_0)$. Also, by Proposition 1.4 on page 389 we have $F'(s) = \frac{z'(s)}{z(s)-z_0}$. Now consider the following derivative,

$$\frac{d}{ds}\left[\frac{e^{F(s)}}{z(s) - z_0}\right] = \frac{\big(z(s) - z_0\big)\, e^{F(s)}\, F'(s) - e^{F(s)}\, z'(s)}{\big(z(s) - z_0\big)^2} = 0.$$

This yields that $\frac{e^{F(s)}}{z(s)-z_0}$ is a constant for all $s \in [a,b]$, and therefore in particular,

$$\frac{e^{F(a)}}{z(a) - z_0} = \frac{e^{F(b)}}{z(b) - z_0}.$$

Since C is closed we have that $z(a) = z(b)$, and so $e^{F(a)} = e^{F(b)}$. This last equality implies that $F(b) = F(a) + 2\pi i\, m$ for some integer m. But since $F(a) = 0$, we obtain

$$n_C(z_0) = \frac{F(b)}{2\pi i} = \frac{2\pi i\, m}{2\pi i} = m,$$

proving the result. ♦

▶ **8.32** *Prove the above proposition for a general closed contour C.*

For a fixed closed parametrized contour $C \subset \mathbb{C}$, we can allow the point $z_0 \notin C$ to vary continuously, thereby denoting it by z, and investigate how the value of $n_C(z)$ varies as a consequence. In light of this, for a fixed closed parametrized contour C in $\mathbb{C}$, we may view $n_C : \mathbb{C} \setminus C \to \mathbb{Z}$ as a function, given by

$$n_C(z) \equiv \frac{1}{2\pi i} \oint_C \frac{d\zeta}{\zeta - z}.$$

We will now try to characterize what integer values $n_C(z)$ may take, and how $n_C(z)$ may vary as z is varied for a given closed parametrized contour C.

Proposition 2.7 *Let C be a closed parametrized contour in $\mathbb{C}$. Then $n_C(z)$ is constant (and therefore a single integer value) on any connected subset of $\mathbb{C} \setminus C$.*

▶ **8.33** *Prove the above theorem by establishing the following:*
a) Show that n_C is continuous on $\mathbb{C} \setminus C$.
b) Show that if $A \subset \mathbb{C} \setminus C$ is connected, then $n_C(A)$ must also be connected.

Proposition 2.8 *Let C be a closed parametrized contour in $\mathbb{C}$. Then $n_C(z) = 0$ on any unbounded connected set in $\mathbb{C} \setminus C$.*

PROOF We handle the case where C is a smooth curve and leave the proof of the more general case where C is a contour to the reader. To this end, let A be an unbounded connected set in $\mathbb{C} \setminus C$. Let $z_0 \in A$ be far enough from C such that $|z_0 - \zeta(t)| \geq L_C$ for all $t \in [a, b]$. Then

$$|n_C(z_0)| = \frac{1}{2\pi} \left| \oint_C \frac{d\zeta}{\zeta - z_0} \right|.$$

Since $|\zeta(t) - z_0| \geq L_C$, it follows that

$$|n_C(z_0)| \leq \frac{1}{2\pi\, L_C} L_C < 1,$$

which implies that $n_C(z_0) = 0$. (Why?) Finally, since A is connected, it follows from the previous proposition that $n_C(z) = 0$ for all $z \in A$. ♦

▶ **8.34** *Answer the (Why?) question in the above proof, and complete the proof by establishing the result for the case where C is a closed parametrized contour.*

▶ **8.35** *In Example 2.5 on page 405, we found that the winding number $n_C(z_0)$ for a counterclockwise circle C around its center z_0 was 1. Consider this same contour again. Let a be any point from* Int(C). *Show that $n_C(a) = 1$. What if $a \in Ext(C)$?*

Example 2.9 *Let R be a rectangle that is parametrized counterclockwise, and let $z_0 \in$* Int(R). *We will show that $n_R(z_0) = 1$ using a technique sometimes referred to as "the horseshoe method." Without any loss of generality, let z_0 be the origin of the complex plane, and choose $C_r \subset$* Int(R) *where $C_r \equiv C_r(0)$ is the parametrized circle of radius r centered at 0. Consider the simple closed contours C_1 and C_2 as shown in Figure 8.8. (C_1 has initial point z_1, and passes through z_2, z_3, z_4, z_5, z_6, and z_7 as t increases, terminating at z_1. Likewise, C_2 starts at z_1 and passes through $z_8, z_6, z_5, z_9, z_{10}$, and z_2 before terminating at z_1.) Clearly, $0 \in Ext(C_1)$ and $0 \in Ext(C_2)$, and so*

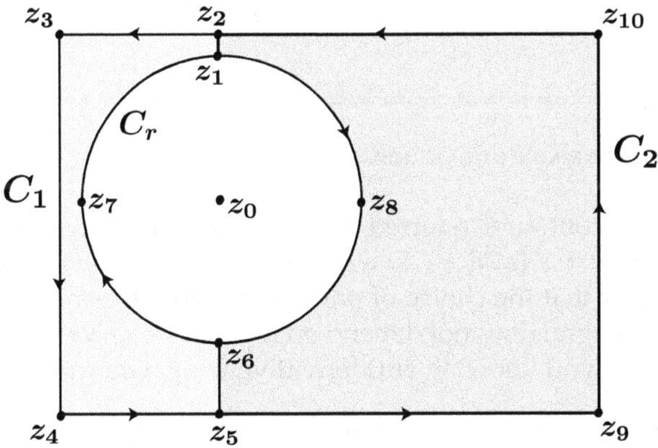

Figure 8.8 *The horseshoe method of computing the winding number of a rectangular contour. Note the shading of the interiors of C_1 and C_2.*

$n_{C_1}(0) = n_{C_2}(0) = 0$. But since

$$\begin{aligned} 0 &= n_{C_1}(0) + n_{C_2}(0) \\ &= \frac{1}{2\pi i}\oint_{C_1} \frac{dz}{z} + \frac{1}{2\pi i}\oint_{C_2} \frac{dz}{z} \\ &= \frac{1}{2\pi i}\oint_{R} \frac{dz}{z} - \frac{1}{2\pi i}\oint_{C_r} \frac{dz}{z} \qquad \textit{(Why?)} \\ &= n_R(0) - n_{C_r}(0), \end{aligned}$$

it follows that $n_R(0) = n_{C_r}(0)$. Since $n_{C_r}(0) = 1$, we have that $n_R(0) = 1$. ◀

▶ **8.36** *Answer the (Why?) question in the above example. See Example 1.19 on page 399 for a clue.*

2.3 Antiderivatives and Path-Independence

An *antiderivative* associated with a complex function f is defined as in the real function case. That is, for open $D \subset \mathbb{C}$, consider $f : D \to \mathbb{C}$ and suppose there exists a differentiable function $F : D \to \mathbb{C}$ such that $F'(z) = f(z)$ for all $z \in D$. Then F is called an *antiderivative* of f on D. As you might expect, antiderivatives play a useful role in evaluating line integrals of complex functions.

Proposition 2.10 (The Fundamental Theorem of Calculus)
Suppose $f : D \to \mathbb{C}$ is continuous on the open set $D \subset \mathbb{C}$ and let C be a parametrized contour in D with initial and terminal points z_I and z_T, respectively. If f has an antiderivative F on D, then

$$\int_C f(z)\, dz = F(z_T) - F(z_I).$$

► **8.37** *Prove the above proposition.*

In practice, the contour C referred to in Proposition 2.10 will have a particular parametrization $z : [a, b] \to \mathbb{C}$ with $z_I = z(a)$ and $z_T = z(b)$. However, it should be clear that the choice of parametrization does not matter, since the value of the integral does not depend on it. This is a very practical theorem. It says that when you know an antiderivative for f, you can evaluate $\int_C f(z)\, dz$ without applying Definition 1.11, and the result depends only on the initial and terminal points of C. The following corollary, partly illustrated in Figure 8.9, is a useful consequence of Proposition 2.10.

Corollary 2.11 *Suppose $f : D \to \mathbb{C}$ is continuous on the open set $D \subset \mathbb{C}$ and has antiderivative F there.*

a) If C is a closed parametrized contour in D, then $\oint_C f(z)\, dz = 0$.

b) Let z_1 and z_2 be any two points in D. If C_1 and C_2 are any two parametrized contours in D with initial point z_1 and terminal point z_2, then

$$\int_{C_1} f(z)\, dz = \int_{C_2} f(z)\, dz.$$

► **8.38** *Prove the above corollary.*

► **8.39** *If $\tilde{z}$ is another parametrization for C, is the result of part a) of the above corollary still true? Does it matter if $\tilde{z}$ is equivalent to z or not? Does the result of part b) depend on the parametrizations used for C_1 and C_2?*

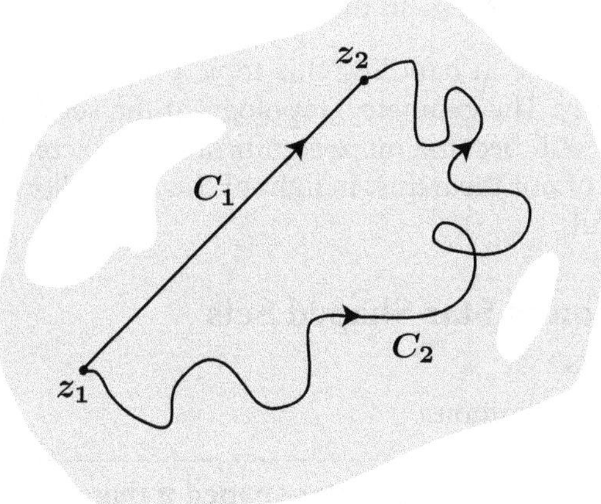

Figure 8.9 *The situation in Corollary 2.11, part b): path independence.*

Corollary 2.11 tells us that the class of complex functions that are continuous and have an antiderivative on D is an especially nice class. If $f : D \to \mathbb{C}$ is such a function, then not only can we conclude that $\oint\limits_C f(z)\,dz = 0$ for any closed contour $C \subset D$, we can also conclude that the integral of f from one point to another in D is *path independent*. That is, the value of the integral depends only on the initial point and the terminal point, not on the path (within D) taken between them. This is significant. After all, in general for $f : D \to \mathbb{C}$ there are infinitely many paths within D that connect two points in D. The answer to the question *When will the integral of a complex function f from one point to another not depend on the choice of contour between them?* is important for both practical and theoretical reasons. In particular, the property of path independence is convenient, since it allows one to choose a different contour than the one specified (if specified) connecting the initial point to the terminal point.

Example 2.12 *We will evaluate the integral $\int\limits_C e^z\,dz$ where C is the curve parametrized by $z : [0, 2\pi] \to \mathbb{C}$ where $z(t) = t + i \sin t$. Note that since $f(z) = e^z$ has antiderivative $F(z) = e^z$ on $\mathbb{C}$, we may change the path C to $\widetilde{C}$ parametrized by $\widetilde{z} : [0, 2\pi] \to \mathbb{C}$ where $\widetilde{z}(t) = t$. Then we have*

$$\int\limits_C e^z\,dz = \int\limits_{\widetilde{C}} e^{\widetilde{z}}\,d\widetilde{z} = \int_0^{2\pi} e^t\,dt = e^{2\pi} - 1.$$

◀

The results indicated in Corollary 2.11 need deeper exploration. In particular, we would like to identify exactly those properties a function f, a set D, and a closed contour C need to possess so that f has an antiderivative in D. More generally, we would like to identify the properties f, D, and the closed contour C need to possess so that $\oint_C f(z)\,dz = 0$. There are many technical hurdles to overcome in handling this topic properly, and so we will move forward carefully. The geometry/topology of the set D in which our function is defined will become more significant as we continue to "fine-tune" the statements of our theorems. In light of this, the following concept will be especially useful.

2.4 Integration in Star-Shaped Sets

Star-Shaped Sets

We begin with a definition.

Definition 2.13 A set $D \subset \mathbb{C}$ is **star-shaped** if there exists $z_0 \in D$ such that $[z_0, z] \subset D$ for every $z \in D$. Such a point $z_0 \in D$ is sometimes called a **star center**.

Figure 8.10 provides examples of star-shaped sets and a set that is not star-shaped.

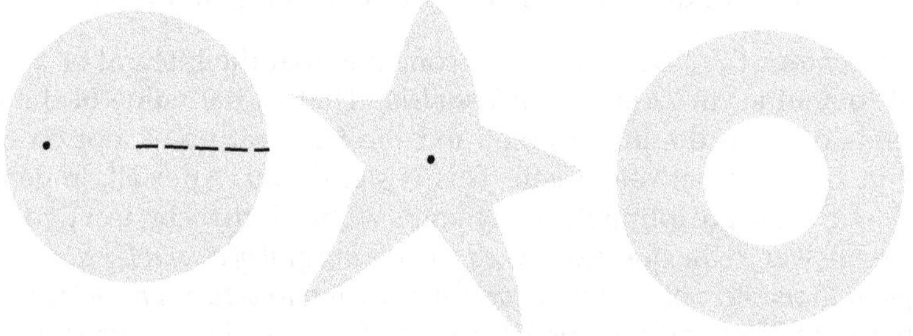

Figure 8.10 *Two star-shaped sets (to the left) with star centers shown, and one set that is not star-shaped (to the right).*

A special subclass of star-shaped sets, and one that is probably more familiar, is the class of *convex* sets. A set is convex if the line segment connecting any pair of points of the set is completely contained in the set. A star-shaped set is one that contains at least one star center. A convex set is one for which every point is a star center. Simple examples of convex sets, and therefore star-shaped sets, are disks, rectangles, and ellipses. The following, nonconvex set is star-shaped. Consider a point $z^* \in \mathbb{C}$, and let R be any ray beginning at z^*. Then $D \equiv \mathbb{C} \setminus R$ is star-shaped, but not convex.

► **8.40** *Let $D \subset \mathbb{C}$ be star-shaped and suppose $f : D \to \mathbb{C}$ is such that $f' \equiv 0$ on D. Show that $f \equiv c$ on D for some $c \in \mathbb{C}$.*

The star-shaped condition on D allows us to rephrase one of our earlier results. Recall that Corollary 2.11 on page 408 stated that if the continuous function $f : D \to \mathbb{C}$ had an antiderivative on the open set $D \subset \mathbb{C}$, then the integral of f around any closed contour contained in D is zero and the integral of f between any two points $z_1, z_2 \in D$ is path independent. For the function f, the key to the path-independence property holding is the existence of an antiderivative for f on D. The following result establishes convenient sufficient conditions for the existence of such an antiderivative in the case where the open set D is star-shaped.

Theorem 2.14 *Suppose $f : D \to \mathbb{C}$ is differentiable on the open star-shaped set $D \subset \mathbb{C}$. Then f has an antiderivative on D.*

Of course, the significant consequences of having an antiderivative on D are precisely those listed in the conclusion of Corollary 2.11, namely, if z_1 and z_2 are any two points in D, then the integral of f within D along contours starting from z_1 and ending at z_2 are path independent. Also, integrals of f along closed contours within D are zero.

PROOF Let $z^* \in D$ be such that $[z^*, z] \subset D$ for all $z \in D$, i.e., z^* is a star center for D. We define $F : D \to \mathbb{C}$ for the parametrized $[z^*, z]$ by

$$F(z) \equiv \int_{[z^*,z]} f(\zeta)\, d\zeta.$$

We will show that for each $z_0 \in D$,

$$F'(z_0) = \lim_{z \to z_0} \frac{F(z) - F(z_0)}{z - z_0} = f(z_0).$$

To establish this, we note that since f is continuous at z_0, given $\epsilon > 0$ there exists $\delta > 0$ such that

$$|\zeta - z_0| < \delta \quad \Rightarrow \quad |f(\zeta) - f(z_0)| < \tfrac{\epsilon}{2}.$$

Since we wish to consider points near z_0 that are also elements of D, we take δ small enough that $N_\delta(z_0)$ is entirely contained in D. (See Figure 8.11.) For

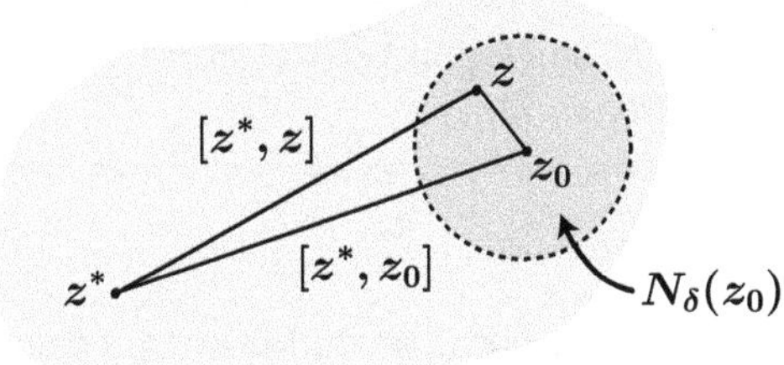

Figure 8.11 *The situation in the proof of Theorem 2.14.*

such a δ, it follows that for z satisfying $0 < |z - z_0| < \delta$ we have

$$\begin{aligned}\left|\frac{F(z)-F(z_0)}{z-z_0} - f(z_0)\right| &= \left|\frac{1}{z-z_0}\int_{[z_0,z]} f(\zeta)\,d\zeta - f(z_0)\right| \\ &= \left|\frac{1}{z-z_0}\int_{[z_0,z]} f(\zeta)\,d\zeta - \frac{1}{z-z_0}\int_{[z_0,z]} f(z_0)\,d\zeta\right| \\ &= \frac{1}{z-z_0}\left|\int_{[z_0,z]} \big(f(\zeta)-f(z_0)\big)\,d\zeta\right| \\ &\le \frac{1}{|z-z_0|}\,\tfrac{\epsilon}{2}\,|z-z_0| \\ &= \tfrac{\epsilon}{2} \\ &< \epsilon, \end{aligned} \tag{8.5}$$

and therefore F is an antiderivative of f on D. By Corollary 2.11 on page 408 integrals of f on contours $C \subset D$ are path independent. ♦

Note that in the above proof, the first equality in expression (8.5) implicitly makes use of the identity

$$\frac{1}{z-z_0}\left(\int_{[z^*,z]} f(\zeta)\,d\zeta - \int_{[z^*,z_0]} f(\zeta)\,d\zeta\right) = \frac{1}{z-z_0}\left(\int_{[z_0,z]} f(\zeta)\,d\zeta\right). \tag{8.6}$$

Here, the existence of the difference on the left-hand side requires that the integrals along the parametrized segments $[z^*, z]$ and $[z^*, z_0]$ be well defined, i.e., that these segments are contained in D. That D is star-shaped with star center z^* ensures this. Equality (8.6) then follows from the triangle lemma.

The following corollary to the above theorem weakens the conditions on f just a bit, and yet the conclusion still holds. This corollary will be needed when we prove the Cauchy integral theorem in the next section.

Corollary 2.15 *Suppose $f : D \to \mathbb{C}$ is differentiable on the open star-shaped set $D \subset \mathbb{C}$, except possibly at a single point $p \in D$ where f is merely continuous. Then f still has an antiderivative on D.*

▶ **8.41** *Prove the above corollary.*

The conclusion to the above corollary holds even if the number of excluded points $p_1, \ldots, p_n$, is greater than one, but finite. We leave the proof of this to the reader.

▶ **8.42** *Prove that the conclusion to Corollary 2.15 still holds even for a finite collection $p_1, \ldots, p_n$ of excluded points.*

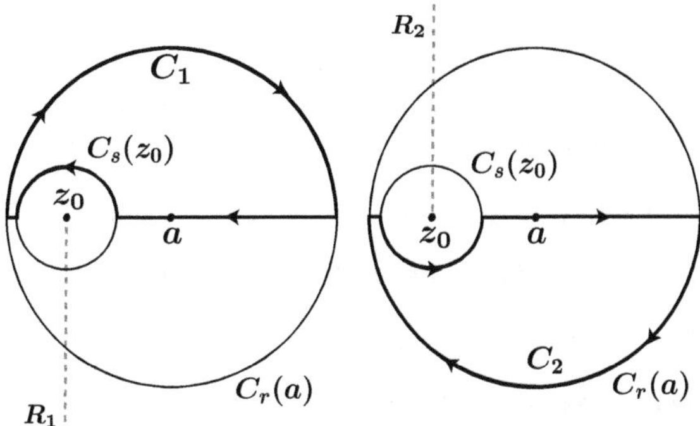

Figure 8.12 *The contours in the horseshoe construction of Example 2.16, with rays shown.*

Example 2.16 *Let $C_r \equiv C_r(a)$ and suppose $z_0 \in \mathbb{C}$ is such that $|z_0 - a| < r$. We will show that $\oint_{C_r} \frac{dz}{z-z_0} = 2\pi\, i$. To this end, choose $s > 0$ so that $C_s \equiv C_s(z_0) \subset$ Int(C_r). Define the parametrized contours C_1 and C_2 as shown in Figure 8.12, and such that*

$$\oint_{C_1} \frac{dz}{z-z_0} + \oint_{C_2} \frac{dz}{z-z_0} = \oint_{C_r} \frac{dz}{z-z_0} - \oint_{C_s} \frac{dz}{z-z_0}.$$

In fact, $\oint_{C_j} \frac{dz}{z-z_0} = 0$ for $j = 1, 2$. To see this, embed C_1 in a star-shaped region $D_1 \equiv \mathbb{C} \setminus R_1$, formed by omitting the ray R_1 emanating from z_0 from $\mathbb{C}$. Now note that $\frac{1}{z-z_0}$ is differentiable on D_1. Similarly, embed C_2 in a star-shaped region $D_2 \equiv \mathbb{C} \setminus R_2$ on which $\frac{1}{z-z_0}$ is differentiable. Since $\oint_{C_s} \frac{dz}{z-z_0} = 2\pi\, i$, it must also be true that $\oint_{C_r} \frac{dz}{z-z_0} = 2\pi\, i$. ◀

Example 2.17 *Let $D \equiv \mathbb{C} \setminus R$ where $R = (-\infty, 0]$ is the ray from the origin along the negative real axis. Note that $z^* = 1$ is a star center for D and so D is star-shaped. Since $\frac{1}{z}$ is differentiable on D, Theorem 2.14 implies that there exists a function $F : D \to \mathbb{C}$ such that F is differentiable on D and $F'(z) = \frac{1}{z}$. We will show that the function F is related to a branch of the logarithm function. To see this, consider $F : D \to \mathbb{C}$ given by $F(z) \equiv \int_{[1,z]} \frac{d\zeta}{\zeta}$. Since F is differentiable on D, it follows that $F = U + i\, V$ where $\frac{\partial U}{\partial x} = \frac{\partial V}{\partial y}$, and $\frac{\partial U}{\partial y} = -\frac{\partial V}{\partial x}$. We also have*

$$F'(z) = \frac{\partial U}{\partial x} + i\,\frac{\partial V}{\partial x} = \frac{1}{z} = \frac{x}{x^2+y^2} - i\,\frac{y}{x^2+y^2},$$

and therefore,

$$\frac{\partial U}{\partial x} = \frac{x}{x^2+y^2} \quad \text{and} \quad \frac{\partial V}{\partial x} = -\frac{y}{x^2+y^2}. \tag{8.7}$$

Integrating each equation in (8.7) with respect to x yields

$$U(x,y) = \ln\sqrt{x^2+y^2} + c_1, \quad \text{and} \quad V(x,y) = \tan^{-1}\left(\tfrac{y}{x}\right) + c_2,$$

where c_1 and c_2 are constants. From this, we have

$$F(z) = U + i\,V = \ln\sqrt{x^2+y^2} + i\,\tan^{-1}\left(\tfrac{y}{x}\right) + (c_1 + i\,c_2),$$

and since $F(1) = 0$, we see that $c_1 + i\,c_2 \equiv 0$. Therefore,

$$F(z) = \ln|z| + i\,\widetilde{\text{Arg}}(z) \equiv \log(z),$$

for $\widetilde{\text{Arg}}(z) \in (-\pi,\pi)$. Here, $\widetilde{\text{Arg}} : D \to [-\pi,\pi)$ is a (nonprincipal) branch of the argument of z, and $\log(z)$ is its associated (nonprincipal) branch of the logarithm function as described in Chapter 4. The function F is the restriction of $\log(z)$ to D. ◀

▶ **8.43** *Consider the situation described in the previous example, but with $D \equiv \mathbb{C}\setminus R$ where R is some other ray emanating from $z_0 \in \mathbb{C}$. How does the result differ in this case?*

Example 2.18 *Suppose $D \subset \mathbb{C}$ is open and star-shaped, and $f : D \to \mathbb{C}\setminus\{0\}$ is twice differentiable on D. We will show that f has a differentiable logarithm on D, i.e., that there exists $g : D \to \mathbb{C}$ such that g is differentiable on D and $e^{g(z)} = f(z)$ on D. By Theorem 2.14, since $\frac{f'(z)}{f(z)}$ is differentiable on D it must have an antiderivative $F(z)$ on D, i.e., $F'(z) = \frac{f'(z)}{f(z)}$. Now, since $\left(f(z)\,e^{-F(z)}\right)' = f'(z)\,e^{-F(z)} - f(z)\,e^{-F(z)}\,F'(z) = 0$, it follows that there exists some $z_0 \in D$ such that $f(\dot{z})\,e^{-F(z)} = c = e^{z_0}$ on D, and $f(z) = e^{F(z)+z_0}$. The function $F(z) + z_0$ is sometimes called a differentiable logarithm of f.* ◀

Example 2.19 *Suppose $D \subset \mathbb{C}$ is open and star-shaped and $f : D \to \mathbb{C}\setminus\{0\}$ is twice differentiable on D. Suppose also that g is a differentiable logarithm of f on D. If $C \subset D$ is any contour parametrized by $z : [a,b] \to \mathbb{C}$, we will show that $\int\limits_C \frac{f'(z)}{f(z)}\,dz = g\big(z(b)\big) - g\big(z(b)\big)$. To see this, note that $f(z) = e^{g(z)}$, and so it follows that $f'(z) = e^{g(z)}\,g'(z)$, or $g'(z) = \frac{f'(z)}{e^{g(z)}} = \frac{f'(z)}{f(z)}$, i.e., $g(z)$ is an antiderivative of $\frac{f'(z)}{f(z)}$. This implies that $\int\limits_C \frac{f'(z)}{f(z)}\,dz = g\big(z(b)\big) - g\big(z(a)\big)$.* ◀

The following theorem establishes a useful equivalence that summarizes many of the significant results of this section.

Theorem 2.20 *Suppose $f : D \to \mathbb{C}$ is continuous on the open connected set D. Then the following are equivalent:*

(i) Integrals of f in D are path independent.

(ii) Integrals of f around closed contours in D are 0.

(iii) There is an antiderivative $F : D \to \mathbb{C}$ for f on D.

PROOF By Corollary 2.11 we see that *(iii)* $\Rightarrow$ *(i)*, and that *(iii)* $\Rightarrow$ *(ii)*. We leave it to the reader to show that *(i)* $\Leftrightarrow$ *(ii)*. We now prove that *(i)* $\Rightarrow$ *(iii)*. Fix $a \in D$. Then, for any contour $C \subset D$ that has initial point $a \in D$ and terminal point $z \in D$, define $F : D \to \mathbb{C}$ by

$$F(z) \equiv \int_C f(\zeta)\, d\zeta.$$

We now show that $F'(z_0) = f(z_0)$ for every point $z_0 \in D$. Note that for any contour C_0 in D with initial point a and terminal point z_0 there exists $N_r(z_0) \subset D$, and for $z \in N_r(z_0)$ we have that $[z_0, z] \subset N_r(z_0) \subset D$. (Why?) Therefore,

$$F(z) - F(z_0) = \int_C f(\zeta)\, d\zeta - \int_{C_0} f(\zeta)\, d\zeta = \int_{[z_0,z]} f(\zeta)\, d\zeta.$$

This yields

$$F(z) - F(z_0) - f(z_0)\,(z - z_0) = \int_{[z_0,z]} \big(f(\zeta) - f(z_0)\big)\, d\zeta.$$

Since f is continuous, for each $\epsilon > 0$ there exists $\delta > 0$ such that

$$|\zeta - z_0| < \delta \;\Rightarrow\; |f(\zeta) - f(z_0)| < \epsilon.$$

Choosing $\delta < r$, we obtain

$$|z - z_0| < \delta \;\Rightarrow\; |F(z) - F(z_0) - f(z_0)\,(z - z_0)| \le \epsilon\, |z - z_0|,$$

which implies $F'(z_0) = f(z_0)$. ◆

► **8.44** *Prove (i) $\Leftrightarrow$ (ii) in the above theorem.*

► **8.45** *Show that connectedness is not necessary in the above theorem.*

3 CAUCHY'S INTEGRAL THEOREM AND ITS CONSEQUENCES

We will now generalize the triangle lemma to more general simple closed contours in D. The resulting Cauchy's integral theorem is as fundamental to the theory of complex functions as the Cauchy-Riemann equations. Many

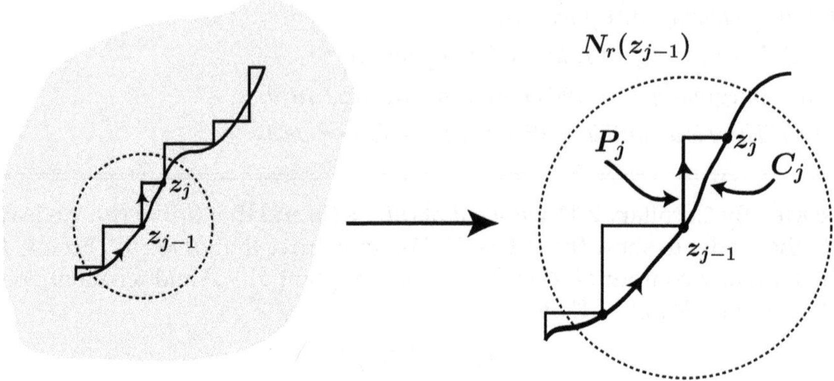

Figure 8.13 *The proof of Lemma 3.1.*

far-reaching results are a consequence of it, which we will see in a later section of this chapter. For now, it can be seen as a generalization of the circumstances under which differentiable functions $f : D \to \mathbb{C}$ that don't necessarily have an antiderivative on D still satisfy $\oint_C f(z)\,dz = 0$ for closed contours $C \subset D$.

3.1 Auxiliary Results

In order to prove Cauchy's integral theorem we will need several technical results. In the first of these we establish that for any function $f : D \to \mathbb{C}$ that is differentiable on the open set D, and any parametrized contour $C \subset D$, the integral of f along C is equal to the integral of f along a rectangular parametrized contour $P \subset D$ consisting of only vertical and horizontal segments.

Lemma 3.1 *Let C be a contour in the open set $D \subset \mathbb{C}$, and let C be parametrized by $z_C : [a, b] \to \mathbb{C}$. Then there exists a rectangular contour P in D with parametrization $z_P : [a, b] \to \mathbb{C}$ satisfying $z_P(a) = z_C(a)$ and $z_P(b) = z_C(b)$ such that if $f : D \to \mathbb{C}$ is any differentiable function on D,*

$$\int_C f(z)\,dz = \int_P f(z)\,dz.$$

PROOF Choose $r > 0$ small enough so that $|z_C(t) - w| \geq r$ for all $w \in D^C$ and for all $t \in [a, b]$. Since $z_C(t)$ is continuous on the closed interval $[a, b]$, it is uniformly continuous there, and so there exists $\delta > 0$ such that for $t, s \in [a, b]$,

$$|t - s| < \delta \Rightarrow |z_C(t) - z_C(s)| < r,$$

and this δ is independent of t and s. Now, choose a partition $\mathcal{P} = \{a = t_0, t_1, \ldots, t_N = b\}$ of $[a, b]$ with $\|\mathcal{P}\| < \delta$ and such that $z_C(t_{j-1}) \neq z_C(t_j)$ for $1 \leq j \leq N$. Also, for notational convenience, let $z_j \equiv z_C(t_j)$ for $0 \leq j \leq N$. If C_j is the smooth piece of C parametrized by $z_{C_j} : [t_{j-1}, t_j] \to \mathbb{C}$ where $z_{C_j}(t) \equiv z_C(t)$ for $t \in [t_{j-1}, t_j]$, then it follows that $|z_{C_j}(t) - z_{j-1}| < r$ for $t \in [t_{j-1}, t_j]$, i.e., $C_j \subset N_r(z_{j-1})$ for $j = 1, 2, \ldots, N$. By our choice of r we have that $N_r(z_{j-1}) \subset D$ for $j = 1, 2, \ldots, N$. Now, since each $N_r(z_{j-1})$ is star-shaped, we can replace C_j by a rectangular contour P_j having parametrization $z_{P_j} : [t_{j-1}, t_j] \to \mathbb{C}$ that lies in $N_r(z_{j-1})$ and that connects z_{j-1} to z_j (see Figure 8.13). By Corollary 2.11 on page 408 and Theorem 2.14 on page 411, we know that $\int\limits_{C_j} f(z)\,dz = \int\limits_{P_j} f(z)\,dz$. Let $P \equiv (P_1, P_2, \ldots, P_N)$. Then P is rectangular, and part *b)* of Proposition 1.14 on page 396 yields

$$\int\limits_C f(z)\,dz = \int\limits_P f(z)\,dz. \qquad \blacklozenge$$

Clearly the above lemma holds for both open and closed contours C.

Now that we have established that integrals along contours can be adequately replaced by integrals along rectangular contours, we state and prove a key lemma relating to integrals along rectangular contours.

Lemma 3.2 *Suppose P is a closed rectangular parametrized contour in $\mathbb{C}$. Then there exist counterclockwise-parametrized rectangles $R_1, \ldots, R_N$ in $\mathbb{C}$ such that for arbitrary points $w_j \in \text{Int}(R_j)$ for $1 \leq j \leq N$, and for $g : \bigcup_{j=1}^N R_j \to \mathbb{C}$ continuous on $\bigcup_{j=1}^N R_j$,*

$$\oint\limits_P g(z)\,dz = \sum_{j=1}^N \left(n_P(w_j) \oint\limits_{R_j} g(z)\,dz \right).$$

PROOF [2] Let $P = [z_0, z_1, \ldots z_{M-1}, z_M, z_0]$ for some $M \in \mathbb{N}$. We begin by choosing a rectangle $R \subset \mathbb{C}$ that contains all of P in its interior. Form a grid of rectangles R_j for $1 \leq j \leq N$ by extending each segment of P in both directions until it intersects R. Note that $P \subset \bigcup_{j=1}^N R_j$, where each R_j is parametrized counterclockwise. For each R_j let w_j be an arbitrary point from $\text{Int}(R_j)$. Note that $\bigcup_{j=1}^N R_j$ consists of parametrized segments of two types, each of which is a parametrized side of some rectangle R_j. A segment of the

[2] The proof of this technical lemma, and this approach to Cauchy's integral theorem, are taken from [AN07].

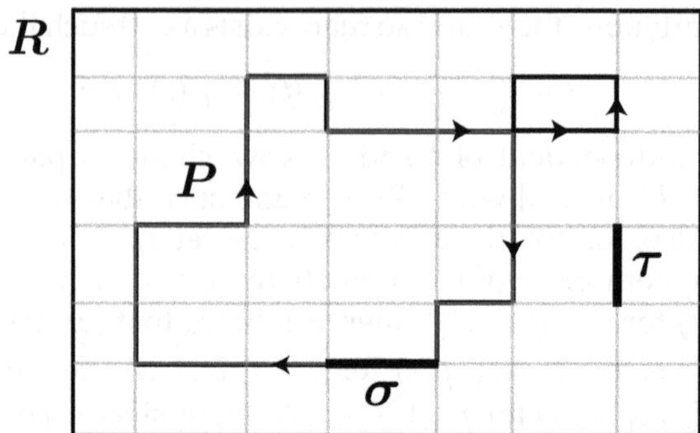

Figure 8.14 *The proof of Lemma 3.2, with a sigma (σ) and a tau (τ) segment shown.*

first type, hereafter referred to as a *tau segment*, is *not* a segment of P. A segment of the second type, hereafter referred to as a *sigma segment*, *is* a segment of P. We will show that

$$\oint_P g(z)\,dz = \sum_{j=1}^{N}\left(n_P(w_j)\oint_{R_j} g(z)\,dz\right) \tag{8.8}$$

for any function $g : \bigcup_{j=1}^{N} R_j \to \mathbb{C}$ that is continuous on $\bigcup_{j=1}^{N} R_j$. To do this, we will prove that the parametrized tau segments make no net contribution to the right-hand side of equality (8.8), and that each parametrized sigma segment, $\sigma = (P \cap R_j)$ for some j, contributes an equal amount to each side of equality (8.8). The situation is illustrated in Figure 8.14.

Let τ be a parametrized tau segment of rectangle R_I where τ is not contained in P. We will show that τ makes no net contribution to the right-hand side of (8.8). There are two cases to consider, each illustrated in Figure 8.15:

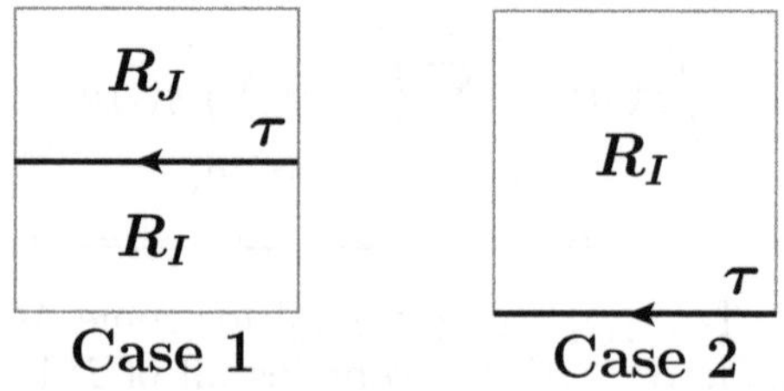

Figure 8.15 *The two cases in the proof for τ segments.*

Case 1: τ is a common side for adjacent rectangles R_I and R_J, i.e., τ is not part of R.

Since τ is not contained in P, it follows that the segment $[w_I, w_J]$ does not intersect P, and therefore

$$n_P(w_I) = n_P(w_J). \quad \text{(Why?)}$$

The only contributions from τ to the right-hand side of (8.8) are from the terms

$$n_P(w_I)\oint_{R_I} g(z)\,dz + n_P(w_J)\oint_{R_J} g(z)\,dz,$$

whose τ-related contributions are, more explicitly,

$$n_P(w_I)\int_{\tau} g(z)\,dz + n_P(w_J)\int_{-\tau} g(z)\,dz = \big(n_P(w_I) - n_P(w_J)\big)\int_{\tau} g(z)\,dz = 0.$$

Case 2: τ is a side of exactly one rectangle R_I, i.e., τ is part of R.

In this case, since τ is not contained in P it follows that $n_P(w_I) = 0$. To see this, note that τ is part of R but not a part of P. Therefore, there exists a ray from w_I that crosses τ and therefore exits $\overline{\text{Int}(R)}$ but never intersects P. Proposition 2.8 on page 406 then yields that $n_P(w_I) = 0$, and hence the only contribution from τ to the right-hand side of equation (8.8) is $n_P(w_I)\int_{\tau} g(z)\,dz = 0$.

Now suppose $\big(P \cap R_I\big) \neq \varnothing$ for some I. Then $\big(P \cap R_I\big)$ consists of one or more segments. Fix one. It must be a common side of two adjacent rectangles, R_I and R_J. Parametrize the segment from one endpoint to the other (it will not matter which endpoint is considered the initial point and which the terminal point) and denote the parametrized segment by σ. We will show that the contribution of σ to each side of equality (8.8) is the same. Begin by considering the left-hand side of (8.8), and let $\alpha \geq 0$ be the number of times the parametrized segment σ is traversed in a single traversal of P. Similarly, let $\beta \geq 0$ be the number of times the parametrized segment $-\sigma$ is traversed in a single traversal of P (see Figure 8.16). Then the net contribution of σ to

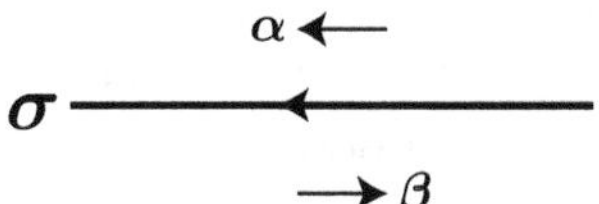

Figure 8.16 *The meaning of the quantities α and β.*

$\oint_P g(z)\,dz$ is $(\alpha - \beta)\int_{\sigma} g(z)\,dz$. We will now show that the parametrized segment σ contributes the same amount, $(\alpha - \beta)\int_{\sigma} g(z)\,dz$, to the right side of (8.8). To see this, we note that the only place the term $\int_{\sigma} g(z)\,dz$ can appear in the right side of (8.8) is in

$$n_P(w_I)\oint_{R_I} g(z)\,dz + n_P(w_J)\oint_{R_J} g(z)\,dz. \tag{8.9}$$

The σ contributions to (8.9) are given by

$$n_P(w_I)\int_\sigma g(z)\,dz + n_P(w_J)\int_{-\sigma} g(z)\,dz = \big(n_P(w_I) - n_P(w_J)\big)\int_\sigma g(z)\,dz.$$

We will show that $\big(n_P(w_I) - n_P(w_J)\big) = (\alpha - \beta)$, which will complete the proof. To show this, consider

$$\oint_P g(z)\,dz - (\alpha-\beta)\oint_{R_I} g(z)\,dz = \oint_{\widetilde{P}} g(z)\,dz. \tag{8.10}$$

Here, $\widetilde{P}$ is the closed rectangular contour obtained by traversing P but detouring around the σ segment of R_I along the other three sides of R_I. Obtained in this way, $\widetilde{P}$ does not contain σ, and the points w_I and w_J end up in the same unbounded connected region of the plane determined by the closed contour $\widetilde{P}$ (see Figure 8.17). Since σ is not contained in $\widetilde{P}$, it fol-

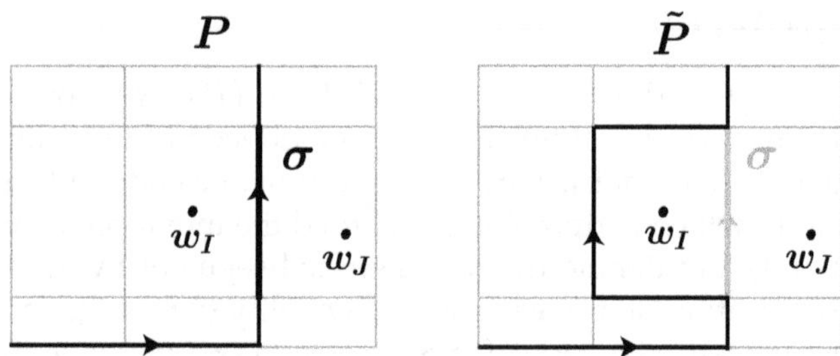

Figure 8.17 *An example of the replacement of P by $\tilde{P}$, where the whole grid of rectangles is not shown in either case.*

lows that $n_{\widetilde{P}}(w_I) = n_{\widetilde{P}}(w_J)$. Since equation (8.10) holds for any function $g : \bigcup_{j=1}^N R_j \to \mathbb{C}$ that is continuous on $\bigcup_{j=1}^N R_j$, we may choose g to be given by $g(z) = \frac{1}{z-w_I}$. We may also choose g to be given by $g(z) = \frac{1}{z-w_J}$. Making these substitutions into 8.10 and multiplying each integral in that equality by $\frac{1}{2\pi i}$ yields that

$$n_P(w_I) - (\alpha-\beta) = n_{\widetilde{P}}(w_I) = n_{\widetilde{P}}(w_J) = n_P(w_J),$$

i.e.,

$$n_P(w_I) - n_P(w_J) = \alpha - \beta.$$

This completes the proof. ♦

3.2 Cauchy's Integral Theorem

The following result is the main result of this section and one of the key results of complex function theory. Its proof relies on Lemma 3.2.

Theorem 3.3 (Cauchy's Integral Theorem)
Let C be a closed parametrized contour in the open set $D \subset \mathbb{C}$, and suppose $n_C(w) = 0$ for all $w \in D^C$. If $f : D \to \mathbb{C}$ is differentiable on D, then

$$\oint_C f(z)\,dz = 0.$$

PROOF[3] By Lemma 3.1, there exists a closed rectangular parametrized contour P in D (see Figure 8.18) such that for any $g : D \to \mathbb{C}$ differentiable on D,

$$\oint_C g(z)\,dz = \oint_P g(z)\,dz.$$

If we choose $g = f$ we obtain $\oint_C f(z)\,dz = \oint_P f(z)\,dz$, and if we choose $g(z) = \frac{1}{z-w}$ for $w \in D^C$ it follows that

$$n_P(w) = n_C(w) = 0 \quad \text{for all } w \in D^C.$$

Therefore, we need only establish the theorem for P. By Lemma 3.2, we can find rectangles $R_1, \ldots, R_N$ in $\mathbb{C}$ and points $w_j \in \text{Int}(R_j)$ for $1 \le j \le N$ such that

$$\oint_P g(z)\,dz = \sum_{j=1}^{N} \left(n_P(w_j) \oint_{R_j} g(z)\,dz \right)$$

for all functions g continuous on $\bigcup_{j=1}^N R_j$. Let $S_j \equiv \overline{\text{Int}(R_j)}$. Since f is differ-

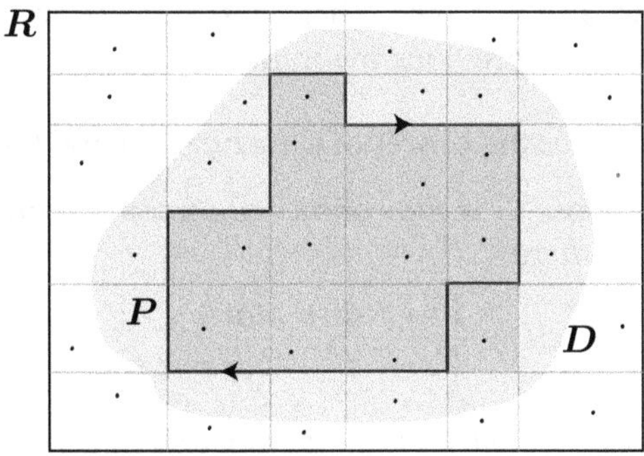

Figure 8.18 *The proof of Cauchy's integral theorem, with the points z_j shown and the rectangles contained in D shaded darker.*

entiable on D, it must be continuous on $P \cup \bigcup_{S_j \subset D} S_j$. By the Tietze extension

[3]We follow [AN07] in the proof.

theorem, we may extend f to a continuous function $\tilde{f}$ on all of $\mathbb{C}$ such that $\tilde{f} = f$ on $P \cup \bigcup_{S_j \subset D} S_j$. Therefore,

$$\oint_P \tilde{f}(z)\,dz = \sum_{j=1}^{N} \left(n_P(w_j) \oint_{R_j} \tilde{f}(z)\,dz \right). \tag{8.11}$$

There are two cases to consider, $S_j \subset D$ and $S_j \not\subset D$. If $S_j \subset D$, then by the triangle lemma $\oint_{R_j} \tilde{f}(z)\,dz = \oint_{R_j} f(z)\,dz = 0$. If $S_j \not\subset D$, there exists $w \in S_j \cap D^C$ such that $[w_j, w]$ does not intersect P, and therefore

$$n_P(w_j) = n_P(w) = 0.$$

In either case we have $n_P(w_j) \oint_{R_j} \tilde{f}(z)\,dz = 0$, and therefore,

$$\oint_P f(z)\,dz = \oint_P \tilde{f}(z)\,dz = \sum_{j=1}^{N} \left(n_P(w_j) \oint_{R_j} \tilde{f}(z)\,dz \right) = 0.$$ ♦

Corollary 3.4 *Let C be a simple closed parametrized contour in the open set $D \subset \mathbb{C}$ such that* Int$(C) \subset D$. *If $f : D \to \mathbb{C}$ is differentiable on D, then*

$$\oint_C f(z)\,dz = 0.$$

PROOF Consider any $w \in D^C$. Then $w \in Ext(C)$, and so $n_C(w) = 0$. The result follows from the previous theorem. ♦

An example illustrates the power of Cauchy's integral theorem.

Example 3.5 *Suppose C is the contour shown in Figure 8.19 parametrized by $z : [0, 2\pi] \to \mathbb{C}$ where*

$$z(t) = \begin{cases} t + i \sin t & \text{for } 0 \le t \le \pi \\ 2\pi - t & \text{for } \pi < t \le 2\pi \end{cases}.$$

It then follows from Cauchy's integral theorem that $\oint_C e^{\cos z}\,dz = 0$. ◄

The conclusion of the corollary to Cauchy's integral theorem, namely, that $\oint_C f(z)\,dz = 0$ for a simple closed contour $C \subset D$ such that Int$(C) \subset D$, clearly holds if f has an antiderivative F on D. From this we see that Cauchy's integral theorem gives more general conditions under which the integral of a function f around a closed contour equals zero.

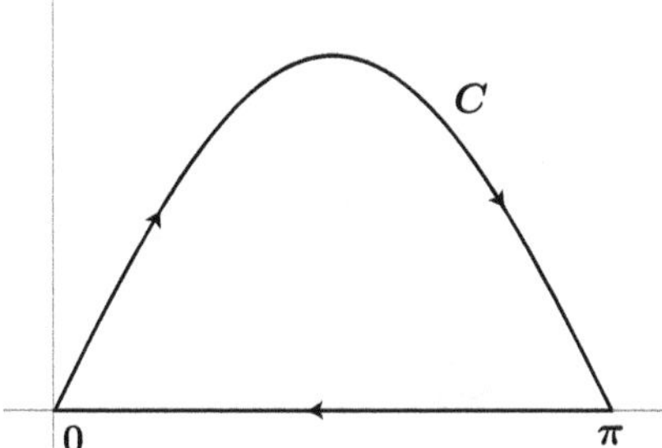

Figure 8.19 *The contour of Example 3.5.*

3.3 Deformation of Contours

The following results provide a convenient means for determining the value of the integral of $f : D \to \mathbb{C}$ along a closed contour $C_1 \subset D$ by giving conditions under which C_1 can be replaced by a simpler contour C_2. The effect is that, under the right circumstances, one can think of "deforming" the contour C_1 into the simpler contour C_2, and thereby determine $\oint\limits_{C_1} f(z)\,dz$ by computing $\oint\limits_{C_2} f(z)\,dz$ instead. With experience, these results provide a useful visual intuition for manipulating such integrals in $\mathbb{C}$.

Theorem 3.6 *Suppose $D \subset \mathbb{C}$ is open and connected, and that $f : D \to \mathbb{C}$ is differentiable on D.*

a) Let C_1 and C_2 be closed parametrized contours in D such that for all $w \in D^C$ $n_{C_1}(w) = n_{C_2}(w)$. Then $\oint\limits_{C_1} f(z)\,dz = \oint\limits_{C_2} f(z)\,dz$.

b) Let $\{C, C_1, C_2, \ldots, C_N\}$ be a collection of closed parametrized contours in D, and suppose $n_C(w) = \sum_{j=1}^N n_{C_j}(w)$ for all $w \in D^C$. Then $\oint\limits_C f(z)\,dz = \sum_{j=1}^N \oint\limits_{C_j} f(z)\,dz$.

PROOF We prove part *a)* and leave the proof of part *b)* to the reader. Let C_1 and C_2 be parametrized by $z_1 : [a,b] \to D$ and $z_2 : [c,d] \to D$, respectively. Figure 8.20 illustrates one possible situation. Since D is connected, we may find a contour C_3 in D that is parametrized by $z_3 : [\alpha, \beta] \to D$ such that $z_3(\alpha) = z_1(a)$ and $z_3(\beta) = z_2(c)$. Now let $C = (C_1, C_3, -C_2, -C_3)$. Note that C is closed and contained in D.

Let $w \in D^C$. Then $w \notin C \subset D$, and since

$$\oint\limits_C \frac{dz}{z-w} = \oint\limits_{C_1} \frac{dz}{z-w} + \int\limits_{C_3} \frac{dz}{z-w} + \oint\limits_{-C_2} \frac{dz}{z-w} + \int\limits_{-C_3} \frac{dz}{z-w},$$

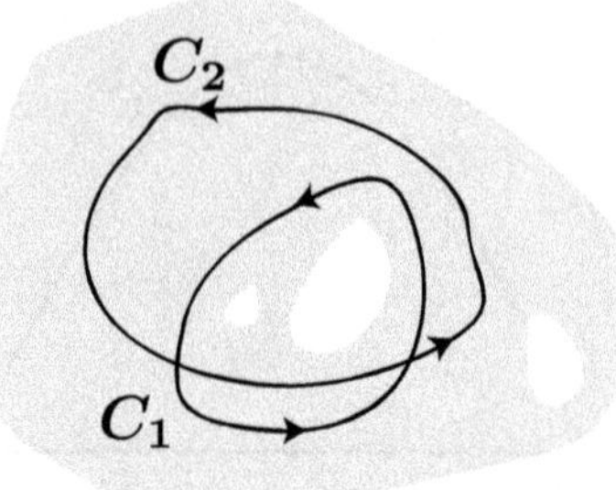

Figure 8.20 *The situation in case a) of Theorem 3.6.*

we obtain after multiplying by $\frac{1}{2\pi i}$,

$$n_C(w) = n_{C_1}(w) + \frac{1}{2\pi i}\int_{C_3} \frac{dz}{z-w} - n_{C_2}(w) - \frac{1}{2\pi i}\int_{C_3} \frac{dz}{z-w} = 0.$$

Therefore by Cauchy's integral theorem, for any $f : D \to \mathbb{C}$ where f is differentiable on D, we have that $\oint_C f(z)\,dz = 0$. Since for any such f we also have

$$\begin{aligned}\oint_C f(z)\,dz &= \oint_{C_1} f(z)\,dz + \int_{C_3} f(z)\,dz + \oint_{-C_2} f(z)\,dz + \int_{-C_3} f(z)\,dz \\ &= \oint_{C_1} f(z)\,dz - \oint_{C_2} f(z)\,dz,\end{aligned}$$

the result is proved. ♦

▶ **8.46** *Prove part b) of the above theorem.*

Example 3.7 *Let C_1 be the parametrized nonsimple closed contour indicated in Figure 8.21. By the geometric interpretation suggested in Appendix A, this contour has winding number +1 around the origin. Let C_2 be the parametrized circle given by $z_2 : [0, 2\pi] \to \mathbb{C}$ where $z_2(t) = e^{it}$. Then, according to Theorem 3.6, we have $\oint_{C_1} \frac{\cos z}{z}\,dz = \oint_{C_2} \frac{\cos z}{z}\,dz$.* ◀

Example 3.8 *Let $z_1 = \frac{1}{2}(\pi + i)$, $z_2 = \frac{1}{2}(3\pi - i)$, and consider the function $f : D \to \mathbb{C}$ given by $f(z) = \frac{1}{(z-z_1)(z-z_2)}$ where $D \equiv \mathbb{C} \setminus \{z_1, z_2\}$. Let $C = (C_\alpha, C_\beta)$ where C_α is the parametrized cycle of the sine curve given by $z_\alpha : [0, 2\pi] \to \mathbb{C}$ where $z_\alpha(t) = (t, \sin t)$, and C_β is the parametrized segment of the real axis from 2π to the origin given by $z_\beta : [2\pi, 4\pi] \to \mathbb{C}$ where $z_\beta(t) = (4\pi - t, 0)$. By the geometric interpretation suggested in Appendix A, the contour C has winding number -1 around z_1, and winding number +1 around z_2. Let C_1*

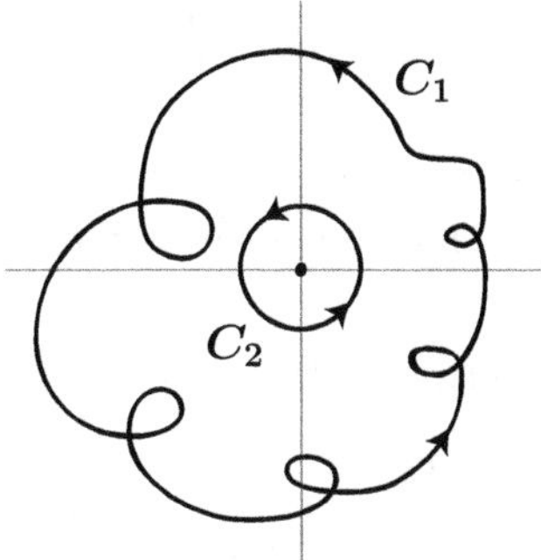

Figure 8.21 *The nonsimple closed contour winding once around the origin in Example 3.7.*

be a circle of small enough radius $\epsilon_1 > 0$ centered at z_1 and parametrized clockwise, and let C_2 be a circle of small enough radius $\epsilon_2 > 0$ centered at z_2 and parametrized counterclockwise. (See Figure 8.22.) By Theorem 3.6, we

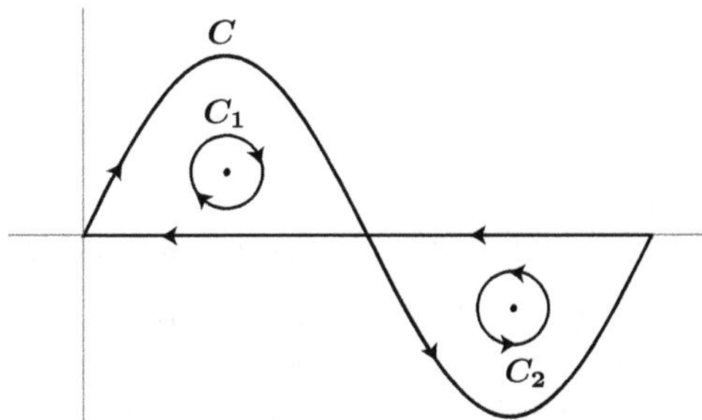

Figure 8.22 *The contours of Example 3.8.*

see that

$$\oint_C \frac{dz}{(z-z_1)(z-z_2)} = \oint_{C_1} \frac{dz}{(z-z_1)(z-z_2)} + \oint_{C_2} \frac{dz}{(z-z_1)(z-z_2)}.$$

◀

Theorem 3.6 is a very useful result. The following corollaries make this plainer. The first is illustrated in Figure 8.23.

Corollary 3.9 *Let $f : D \to \mathbb{C}$ be differentiable on the open connected set D, and suppose C_1 and C_2 are simple closed contours in D such that*

(i) $C_2 \subset \text{Int}(C_1)$,

(ii) $Ext(C_2) \cap \text{Int}(C_1) \subset D$,

(iii) $n_{C_2}(z) = n_{C_1}(z)$ *for all* $z \in \text{Int}(C_2)$.

Then,

$$\oint_{C_1} f(z)\, dz = \oint_{C_2} f(z)\, dz.$$

PROOF By Theorem 3.6 it suffices to show that $n_{C_2}(w) = n_{C_1}(w)$ for all $w \in D^C$. From *(ii)*,

$$D^C \subset \left(Ext(C_2)\right)^C \cup \left(\text{Int}(C_1)\right)^C = \left(C_2 \cup \text{Int}(C_2)\right) \cup \left(C_1 \cup Ext(C_1)\right),$$

and since $C_1, C_2 \subset D$, we have

$$D^C \subset \text{Int}(C_2) \cup Ext(C_1),$$

where $\text{Int}(C_2) \cap Ext(C_1) = \varnothing$. Now suppose $w_0 \in D^C$. Then there are two

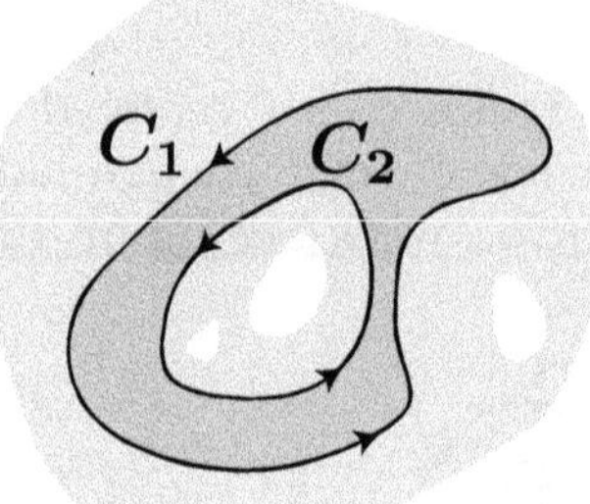

Figure 8.23 *The situation in Corollary 3.9.*

cases, $w_0 \in \text{Int}(C_2)$ and $w_0 \in Ext(C_1)$. If $w_0 \in \text{Int}(C_2)$, then by condition *(iii)* $n_{C_2}(w_0) = n_{C_1}(w_0)$. For the case $w_0 \in Ext(C_1)$, we leave it to the reader to show that $Ext(C_1) \subset Ext(C_2)$ and therefore $w_0 \in Ext(C_2)$, which implies

$$n_{C_2}(w_0) = n_{C_1}(w_0) = 0.$$ ♦

▶ **8.47** *Complete the proof of the above corollary by showing that if $w_0 \in Ext(C_1)$ then $w_0 \in Ext(C_2)$, implying $n_{C_2}(w_0) = n_{C_1}(w_0) = 0$.*

Example 3.10 *Consider the parametrized contour C shown in Figure 8.24, and consider $f : D \to \mathbb{C}$ given by $f(z) = \frac{\text{Log}(z)}{z-i}$ where $D \equiv \mathbb{C} \setminus \left((-\infty, 0] \cup \{i\}\right)$. The reader should verify, via the argument given in Appendix A, that the winding number of C around $z = i$ is $+1$, and that f is differentiable on the open connected set D. We would like to evaluate the integral $\oint_C f(z)\, dz$. To do so, imagine "deforming" the contour C into the circle $C_{\frac{1}{3}} \equiv C_{\frac{1}{3}}(i)$. This can be done in such a way that as C deforms into $C_{\frac{1}{3}}$, it does so within D where f is differentiable. Note that the contour C and the contour $C_{\frac{1}{3}}$ into which we have chosen to have C deform satisfy the conditions of Corollary 3.9. Therefore, $\oint_C f(z)\, dz = \oint_{C_{\frac{1}{3}}} f(z)\, dz$, and so*

$$\oint_C \frac{\text{Log}(z)}{z-i}\,dz = \oint_{C_{\frac{1}{3}}} \frac{\text{Log}(z)}{z-i}\,dz.$$

◀

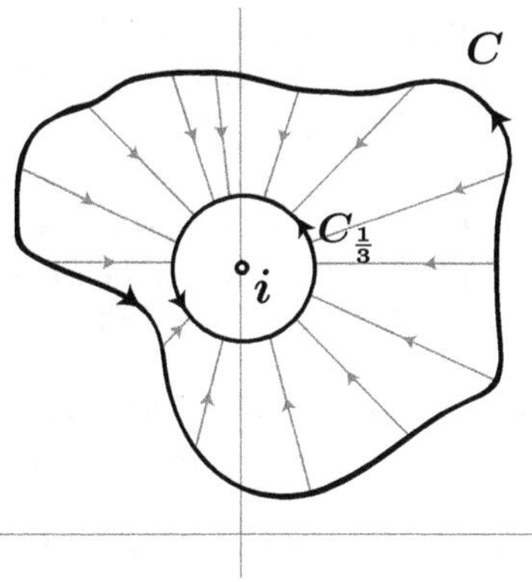

Figure 8.24 *The contour for Example 3.10.*

The following corollary extends the result of the previous one to a finite collection of closed contours.

Corollary 3.11 *Let $f : D \to \mathbb{C}$ be differentiable on the open connected set D, and suppose $C, C_1, C_2, \ldots, C_n$ are simple closed contours in D such that*

(i) $C_j \subset \text{Int}(C)$ *for* $j = 1, 2 \ldots, n$,

(ii) $\text{Int}(C) \cap \left(\bigcap_{j=1}^n Ext(C_j) \right) \subset D$,

(iii) $\text{Int}(C_j) \cap \text{Int}(C_k) = \varnothing$ *for* $j \neq k$,

(iv) $n_C(z) = n_{C_j}(z)$ *for all* $z \in \text{Int}(C_j)$ *for each of* $j = 1, 2, \ldots, n$.

Then,

$$\oint_C f(z)\,dz = \sum_{j=1}^n \left(\oint_{C_j} f(z)\,dz \right).$$

PROOF By Theorem 3.6 it suffices to prove that $n_C(w) = \sum_{j=1}^n n_{C_j}(w)$ on D^C. Since

$$D^C \subset \left(\text{Int}(C)\right)^C \cup \left(\bigcup_{j=1}^n \left(Ext(C_j)\right)^C \right) = (C \cup Ext(C)) \cup \bigcup_{j=1}^n (C_j \cup \text{Int}(C_j)),$$

and since $C \subset D$, and $C_j \subset D$ for $j = 1, \ldots, n$, we have

$$D^C \subset Ext(C) \cup \left(\bigcup_{j=1}^n \text{Int}(C_j) \right).$$

As in the proof of the previous corollary, if $w_0 \in D^C$ there are two cases to consider. If $w_0 \in Ext(C)$, we leave it to the reader to show that $Ext(C) \subset Ext(C_j)$ for $j = 1, \ldots, n$, and hence that $n_C(w_0) = n_{C_j}(w_0) = 0$ so that $n_C(w_0) = \sum_{j=1}^n n_{C_j}(w_0)$. If instead $w_0 \in \left(\bigcup_{j=1}^n \text{Int}(C_j)\right)$, then $w_0 \in \text{Int}(C_j)$ for some $1 \le j \le n$, and for this j we have $n_C(w_0) = n_{C_j}(w_0)$. Since $w_0 \notin \text{Int}(C_k)$ for $k \neq j$ (Why?), we have $n_{C_k}(w_0) = 0$ for $k \neq j$. Overall, this obtains in this second case that $n_C(w_0) = \sum_{j=1}^n n_{C_j}(w_0)$. ♦

► **8.48** *Complete the proof of the above corollary by showing that when $w_0 \in Ext(C)$ we have $Ext(C) \subset Ext(C_j)$ for $j = 1, \ldots, n$, and hence that $n_C(w_0) = n_{C_j}(w_0) = 0$ so that $n_C(w_0) = \sum_{j=1}^n n_{C_j}(w_0)$. Also, answer the (Why?) question in the above proof.*

Example 3.12 *For the contour C shown in Figure 8.25, we will show that $\oint\limits_C \frac{dz}{z^2-z} = \oint\limits_{C_1} \frac{dz}{z^2-z} + \oint\limits_{C_2} \frac{dz}{z^2-z}$ where C_1 is the parametrized circle of radius ϵ_1 centered at the origin, and C_2 is the parametrized circle of radius ϵ_2 centered at $z = 1$. Here, ϵ_1 and ϵ_2 are to be taken small enough to ensure that C_1 and C_2*

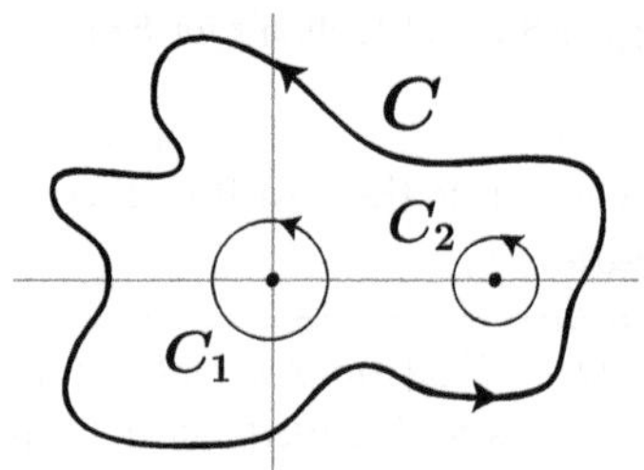

Figure 8.25 *The situation in Example 3.12.*

are contained in Int(C). *Verify via the geometric intuition established in Appendix A that $n_C(0) = n_{C_1}(0) = 1$ and that $n_C(1) = n_{C_2}(1) = 1$. By confirming too that the conditions of Corollary 3.11 hold, we have the result.* ◄

4 CAUCHY'S INTEGRAL FORMULA

It is an interesting and useful property of differentiable complex functions that they can be appropriately represented by integrals around closed contours. Cauchy's integral formula and the variations of it that follow are not only of great practical use in computations, they are also fundamental in establishing other results of deep significance in complex function theory.

4.1 The Various Forms of Cauchy's Integral Formula

We begin with a version of Cauchy's integral formula on star-shaped sets.

Proposition 4.1 (Cauchy's Integral Formula on Star-Shaped Sets)
Let $f : D \to \mathbb{C}$ be differentiable on the open star-shaped set D, and suppose C is a closed parametrized contour in D. Then for $z_0 \in D \setminus C$,

$$\frac{1}{2\pi i} \oint_C \frac{f(\zeta)}{\zeta - z_0}\, d\zeta = n_C(z_0)\, f(z_0).$$

PROOF Let $g : D \to \mathbb{C}$ be given by $g(\zeta) = \begin{cases} \frac{f(\zeta)-f(z_0)}{\zeta - z_0} & \text{for } \zeta \in D \setminus \{z_0\} \\ f'(z_0) & \text{for } \zeta = z_0 \end{cases}$. Note that g is continuous on D and differentiable on $D \setminus \{z_0\}$, and therefore by Corollary 2.11 on page 408 and Theorem 2.14 on page 411 we have that

$$0 = \oint_C g(\zeta)\, d\zeta = \oint_C \frac{f(\zeta)}{\zeta - z_0}\, d\zeta - 2\pi i\, n_C(z_0)\, f(z_0).$$

A simple rearrangement gives the result. ♦

► **8.49** *Show that the function g described in the proof of the above result is continuous on D and differentiable on $D \setminus \{z_0\}$.*

The following corollary is a useful consequence of the above result.

Corollary 4.2 (Cauchy's Integral Formula for Circles)
Suppose $f : D \to \mathbb{C}$ is differentiable on the open set D, and let C_r be a parametrized circle of radius r such that it and its interior are contained in D. Then for $z_0 \in D \setminus C_r$,

$$\frac{1}{2\pi i} \oint_{C_r} \frac{f(\zeta)}{\zeta - z_0}\, d\zeta = \begin{cases} 0 & \text{if } z_0 \in Ext(C_r) \\ f(z_0) & \text{if } z_0 \in \mathrm{Int}(C_r) \end{cases}.$$

PROOF We first consider the case where z_0 is outside C_r. Let D' be an open disk in D that contains C_r but excludes z_0, i.e., $C_r \subset D' \subset D$ and $z_0 \notin D'$. The function $F : D' \to \mathbb{C}$ given by $F(\zeta) \equiv \frac{f(\zeta)}{\zeta - z_0}$ is differentiable on D', and so

$$\oint_{C_r} F(\zeta)\, d\zeta = \oint_{C_r} \frac{f(\zeta)}{\zeta - z_0}\, d\zeta = 0. \quad \text{(Why?)}$$

Now consider the case where z_0 is inside C_r. Again, let D' be a disk such that $C_r \subset D' \subset D$. Apply the previous proposition, and the result is proved. ♦

► **8.50** *How do we know a disk such as D' as described in each of the cases of the above proof exists? What type of region is D' in each case? To answer these questions, let C_r be a circle such that it and its interior are contained in the open set $D \subset \mathbb{C}$. If r and a are the circle's radius and center, respectively, show that there* *exists a neighborhood $N_\rho(a)$ such that $C_r \subset N_\rho(a) \subset D$. (Hint: Assume otherwise, that is, that for all $\rho > r$ we have $N_\rho(a) \cap D^C \neq \varnothing$. In particular, show that there exists $w_n \in N_{r+\frac{1}{n}}(a) \cap D^C$.)*

► **8.51** *Suppose the circle in the statement of the above corollary is parametrized so that it is traversed $n \in \mathbb{N}$ times in the counterclockwise direction, rather than just once. Show that this yields the result $\frac{1}{2\pi i} \oint\limits_{C_r} \frac{f(\zeta)}{\zeta - z_0}\, d\zeta = \begin{cases} 0 & \text{if } z_0 \in Ext(C_r) \\ n\, f(z_0) & \text{if } z_0 \in \text{Int}(C_r) \end{cases}$. What if the circle is traversed in the clockwise direction?*

► **8.52** *Use Corollary 4.2 to prove Gauss's mean value theorem. If $f : D \to \mathbb{C}$ is differentiable on the open set $D \subset \mathbb{C}$, and $C_r(z_0)$ and its interior are contained within D, then $f(z_0) = \frac{1}{2\pi} \int_0^{2\pi} f(z_0 + r\, e^{i\,t})\, dt$.*

Let $D_r \equiv \text{Int}(C_r)$. Then Corollary 4.2 (and also Exercise 8.52) tells us that if $\overline{D_r} \subset D$ and if f is differentiable on D, the values of f inside the disk D_r are completely determined by the values of f *on the boundary of the disk*, an important insight about differentiable complex functions. That is, for $z_0 \in D_r$,

$$f(z_0) = \frac{1}{2\pi i} \oint\limits_{C_r} \frac{f(\zeta)}{\zeta - z_0}\, d\zeta. \tag{8.12}$$

Of course, this expression can also be rearranged as

$$\oint\limits_{C_r} \frac{f(\zeta)}{\zeta - z_0}\, d\zeta = 2\pi i\, f(z_0), \tag{8.13}$$

highlighting its computational utility in evaluating integrals of this type.

Example 4.3 Note that, for $C_2 \equiv C_2(i)$, the parametrized circle of radius 2 centered at i, the integrals $\oint\limits_{C_2} \frac{\sin z}{z}\, dz$ and $\oint\limits_{C_2} \frac{\sin z}{z - 5i}\, dz$ can both readily be seen to equal 0 by virtue of formula (8.13) and Corollary 4.2. ◄

► **8.53** *Use (8.13) to calculate the results of the integrals obtained in Examples 3.7 and 3.8 on page 424, and Example 3.10 on page 426.*

Next, a generalization of Cauchy's integral formula to open connected sets.

Theorem 4.4 (Cauchy's Integral Formula on Open Connected Sets)
Let $f : D \to \mathbb{C}$ be differentiable on the open connected set $D \subset \mathbb{C}$. Suppose C is a closed parametrized contour in D and suppose $n_C(w) = 0$ for all $w \in D^C$. Then for $z_0 \in D \setminus C$,

$$\frac{1}{2\pi i} \oint\limits_{C} \frac{f(\zeta)}{\zeta - z_0}\, d\zeta = n_C(z_0)\, f(z_0).$$

PROOF Choose $r > 0$ small enough so that $N_r(z_0) \subset D \setminus C$ and let $\widetilde{C}_{\frac{r}{2}}$ be the contour parametrized by $\widetilde{z} : [0, 2\pi] \to \mathbb{C}$, where $\widetilde{z}(t) = z_0 + \frac{r}{2} \exp\left(i\, n_C(z_0)\, t\right)$ (see Figure 8.26). Then as shown in Exercise 8.30 on page 405, $n_{\widetilde{C}_{\frac{r}{2}}}(z_0) = n_C(z_0)$. Now, if $D_0 \equiv D \setminus \{z_0\}$ it follows that $\frac{f(\zeta)}{\zeta - z_0}$ is differentiable on the open connected set D_0, and C and $\widetilde{C}_{\frac{r}{2}}$ are closed contours in D_0. If we can

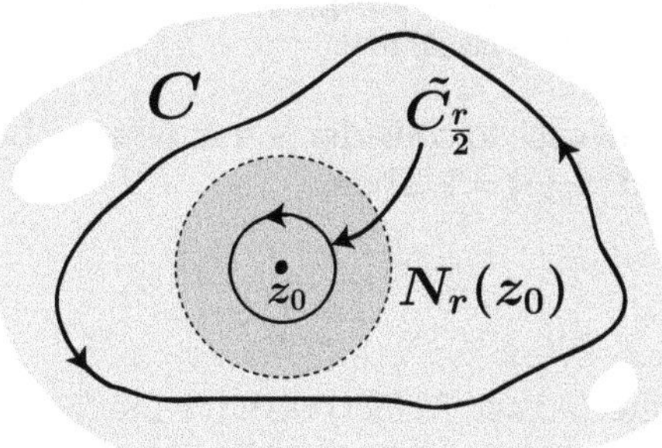

Figure 8.26 *The situation described in the proof of Cauchy's integral formula on open connected sets.*

show that $n_{\widetilde{C}_{\frac{r}{2}}}(w) = n_C(w)$ for all $w \in D_0^C$, then by Theorem 3.6 on page 423, we have

$$\oint_C \frac{f(\zeta)}{\zeta - z_0}\, d\zeta = \oint_{\widetilde{C}_{\frac{r}{2}}} \frac{f(\zeta)}{\zeta - z_0}\, d\zeta.$$

Since f is differentiable on $N_r(z_0)$, Cauchy's integral formula on star-shaped sets yields that

$$\oint_{\widetilde{C}_{\frac{r}{2}}} \frac{f(\zeta)}{\zeta - z_0}\, d\zeta = 2\pi i\, n_{\widetilde{C}_{\frac{r}{2}}}(z_0)\, f(z_0) = 2\pi i\, n_C(z_0)\, f(z_0).$$

It remains only to show that $n_{\widetilde{C}_{\frac{r}{2}}}(w) = n_C(w)$ for all $w \in D_0^C$. To see this, consider $w \in D_0^C$. Then $w \in D^C \cup \{z_0\}$. If $w \in D^C$, then $n_{\widetilde{C}_{\frac{r}{2}}}(w) = 0$ and $n_C(w) = 0$, and hence, $n_{\widetilde{C}_{\frac{r}{2}}}(w) = n_C(w)$. If $w = z_0$, then $n_{\widetilde{C}_{\frac{r}{2}}}(w) = n_{\widetilde{C}_{\frac{r}{2}}}(z_0) = n_C(z_0) = n_C(w)$, and the theorem is proved. ♦

Example 4.5 *Recall that in Example 3.7 on page 424 we established that* $\oint_{C_1} \frac{\cos z}{z}\, dz = \oint_{C_2} \frac{\cos z}{z}\, dz$ *for the contours* C_1 *and* C_2 *depicted there. Applying Corollary 4.2 yields the result that* $\oint_{C_1} \frac{\cos z}{z}\, dz = \oint_{C_2} \frac{\cos z}{z}\, dz = \frac{1}{2\pi i}$. *Likewise, in Example 3.8 we established that* $\oint_C \frac{dz}{(z-z_1)(z-z_2)} = \oint_{C_1} \frac{dz}{(z-z_1)(z-z_2)} + \oint_{C_2} \frac{dz}{(z-z_1)(z-z_2)}$.

Once again, by applying Corollary 4.2 we obtain

$$\oint_C \frac{dz}{(z-z_1)(z-z_2)} = \oint_{C_1} \frac{dz}{(z-z_1)(z-z_2)} + \oint_{C_2} \frac{dz}{(z-z_1)(z-z_2)}$$

$$= (-1)\,\frac{1}{2\pi i}\,\frac{1}{(z_1-z_2)} + (-1)\,\frac{1}{2\pi i}\,\frac{1}{(z_2-z_1)}. \quad \blacktriangleleft$$

► **8.54** *Reconsider Example 3.12 on page 428 , and show via Cauchy's integral formula that the integral there is equal to 0.*

In practice, a frequent case to consider is that where z belongs to a connected subset of $D \setminus C$, and $n_C(z) = 1$. Then,

$$f(z) = \frac{1}{2\pi i} \oint_C \frac{f(\zeta)}{\zeta - z}\, d\zeta$$

for all z in this connected set. In fact, we have the following result.

Corollary 4.6 (Cauchy's Integral Formula for Simple Closed Contours)
Let $f: D \to \mathbb{C}$ be differentiable on the open set $D \subset \mathbb{C}$, and suppose C is a simple closed parametrized contour such that C and $\mathrm{Int}(C)$ *are contained in D. If $n_C(z) = 1$ for all $z \in \mathrm{Int}(C)$, then for any $z_0 \in \mathrm{Int}(C)$,*

$$\frac{1}{2\pi i} \oint_C \frac{f(\zeta)}{\zeta - z_0}\, d\zeta = f(z_0).$$

► **8.55** *Prove the above corollary.*

Note that determining that $n_C(z) = 1$ for a particular simple closed contour C and a point $z \in \mathrm{Int}(C)$ is usually accomplished through the geometric intuition developed in the Appendices section of this chapter (see the section preceding the Supplementary Exercises). An actual computation of $n_C(z)$ for such a simple closed contour is not generally necessary. For a single traversal of a specific contour such as that described here, one can trace the contour starting from the initial/terminal point and note that the continuously changing argument has a net increase of 2π when one arrives back at the initial/terminal point. This is enough to imply that the associated winding number for the contour is 1. The reader is encouraged to read the Appendices section now.

Example 4.7 *We will use Corollary 4.6 to evaluate $\oint_C \frac{e^z}{z}\, dz$ where C is the parametrized square $[2+2i, -2+2i, -2-2i, 2-2i, 2+2i]$. Simply note that e^z is differentiable on $\mathbb{C}$, and therefore by the above corollary the integral is equal to $2\pi i$.* ◄

Example 4.8 *We will evaluate* $\oint\limits_{C_{\frac{1}{2}}} \frac{dz}{z^3-z}$ *where* $C_{\frac{1}{2}} \equiv C_{\frac{1}{2}}(1)$ *is the parametrized circle centered at* $z = 1$ *of radius* $\frac{1}{2}$*. If we consider* $f : D \to \mathbb{C}$ *where* $f(z) = \frac{1}{z\,(z+1)}$ *and* $D = \{z \in \mathbb{C} : |z-1| < \frac{3}{4}\}$*, then* f *is differentiable on* D*, and* $C_{\frac{1}{2}}$ *and its interior are contained in* D*. By Cauchy's integral formula for circles, we have*

$$\oint\limits_{C_{\frac{1}{2}}} \frac{dz}{z^3 - z} = 2\pi\, i\, f(1) = 2\pi\, i\, \left(\tfrac{1}{2}\right) = \pi\, i.$$

◀

4.2 The Maximum Modulus Theorem

We present one more important property of differentiable complex functions. Our proof of it relies on the application of Cauchy's integral formula. Called the maximum modulus theorem, it establishes that if a differentiable complex function on $D \subset \mathbb{C}$ is such that $|f|$ attains a local maximum at $a \in D$, then $|f|$ must equal that maximum value *throughout some neighborhood* $N_r(a) \subset D$. Even more than this is true, namely, there exists some constant $c \in \mathbb{C}$ such that $|c| = |f(a)|$ *and for which* $f(z) = c$ *throughout* $N_r(a)$. This result is *not* true generally for differentiable functions in any of the *real* cases we have considered.[4]

Theorem 4.9 (Maximum Modulus Theorem—First Version)
Suppose $f : D \to \mathbb{C}$ *is differentiable on the open connected set* $D \subset \mathbb{C}$ *and that for some* $a \in D$ *there exists* $N_R(a) \subset D$ *such that* $|f(z)| \leq |f(a)|$ *for all* $z \in N_R(a)$*. Then* $f(z) = f(a)$ *for all* $z \in N_R(a)$*.*

PROOF Choose $0 < r < R$, and define $C_r \equiv C_r(a)$ to be the parametrized circle of radius r centered at a. Then

$$\begin{aligned} f(a) &= \frac{1}{2\pi\, i} \oint\limits_{C_r} \frac{f(z)}{z-a}\, dz \\ &= \frac{1}{2\pi\, i} \int_0^{2\pi} \frac{f(a + r\, e^{it})}{r\, e^{it}}\, i\, r\, e^{it}\, dt \\ &= \frac{1}{2\pi} \int_0^{2\pi} f(a + r\, e^{it})\, dt. \end{aligned}$$

Therefore,

$$|f(a)| \leq \frac{1}{2\pi} \int_0^{2\pi} |f(a + r\, e^{it})|\, dt \leq \frac{1}{2\pi} \int_0^{2\pi} |f(a)|\, dt = |f(a)|,$$

[4]In fact, it is true for the special case of *harmonic* functions $u : D^2 \to \mathbb{R}$, but not more generally among real differentiable functions. Of course, a differentiable complex function $f = u + iv$ has harmonic functions u and v as its real and imaginary parts!

which implies that

$$\frac{1}{2\pi}\int_0^{2\pi} |f(a+r\,e^{it})|\,dt = |f(a)| = \frac{1}{2\pi}\int_0^{2\pi} |f(a)|\,dt.$$

From this we have,

$$\int_0^{2\pi}\Big(|f(a+r\,e^{it})| - |f(a)|\Big)\,dt = 0.$$

But since $|f(a+r\,e^{it})| \le |f(a)|$, it follows that $|f(a+r\,e^{it})| - |f(a)| \le 0$, and hence that

$$|f(a+r\,e^{it})| - |f(a)| = 0. \quad \text{(Why?)}$$

Overall then, we have that

$$|f(a+r\,e^{it})| = |f(a)| \quad \text{for all } 0 < r < R, \quad \text{and for all } t \in [0, 2\pi],$$

which means that

$$|f(z)| = |f(a)| \quad \text{for all } z \in N_R(a).$$

Finally, we see that f maps a small neighborhood of z into a circle. Taking derivatives in different directions, it follows from the Cauchy-Riemann equations that $f(z) = c$ on $N_R(a)$ for some constant $c \in \mathbb{C}$. The value of c must clearly be $f(a)$. ♦

► **8.56** *Justify the conclusion of the proof of the above theorem.*

Actually, it can be shown that if f satisfies the conditions of the above theorem, it must equal c *on all of D*. This is even more surprising, but requires a bit more work to establish.[5] An extension of the above theorem that we *will* provide is the following.

Theorem 4.10 (Maximum Modulus Theorem—Second Version)
Let $D \subset \mathbb{C}$ be bounded, open, and connected. Suppose $f : \overline{D} \to \mathbb{C}$ is differentiable on D and continuous on $\overline{D}$. Then the maximum value of $|f|$ on $\overline{D}$ is attained on the boundary ∂D. If the maximum of $|f|$ is also attained on the interior of D, then f must be constant on $\overline{D}$.

PROOF We follow a proof idea as described in [MH98]. Let $M = \max_{\overline{D}} |f(z)| = |f(a)|$ for some $a \in \overline{D}$. If $a \in \partial D$ we are done. So assume $a \in D$. Define

$$D_1 \equiv \{z \in D : f(z) = f(a)\} \quad \text{and} \quad D_2 \equiv D \setminus D_1.$$

We will establish the following:

1. $D \subset D_1 \cup D_2$.
2. $D_1 \cap D_2 = \varnothing$.
3. D_1 is open.

[5] The proof involves the idea of *analytic continuation* of a differentiable complex function. See [MH98].

4. D_2 is open.

To establish 1, let z_0 be an arbitrary point in D. If $z_0 \in D_2$, we are done. If $z_0 \notin D_2$, then $z_0 \in \overline{D_1}$. (Why?) Choosing a sequence $\{z_n\} \subset D_1$ such that $\lim z_n = z_0$, continuity of f then implies that $\lim f(z_n) = f(z_0)$. But $f(z_n) = f(a)$ for all $n \in \mathbb{N}$, and so it follows that $f(z_0) = f(a)$ and therefore $z_0 \in D_1$. To prove 2, suppose there exists $z_0 \in D_1 \cap D_2$. Then $z_0 \in \overline{D_1}$ (Why?) and $z_0 \in D_2 = D \setminus \overline{D_1}$, a contradiction. Therefore, $D_1 \cap D_2 = \varnothing$. To establish 3, let z_0 be an element of D_1. Then $|f(z_0)| = |f(a)| = M$. Since D is open there exists a neighborhood $N_R(z_0) \subset D$, and

$$|f(z)| \leq M = |f(a)| = |f(z_0)| \quad \text{on } N_R(z_0).$$

From the first version of the maximum modulus theorem it follows that $f(z) = f(z_0) = f(a)$ on $N_R(z_0)$, that is, $N_R(z_0) \subset D_1$. Therefore D_1 is open. Finally to prove 4, note that $D_2 = D \setminus \overline{D_1} = D \cap (\overline{D_1})^C$, the intersection of open sets. Therefore D_2 is open. With 1–4 established, since D is connected, it must follow that either $D_1 = \varnothing$ or $D_2 = \varnothing$. But $a \in D_1$, so it must be the case that $D_2 = \varnothing$. This and 1 then implies that $D \subset D_1$, and therefore $f(z) = f(a)$ for all $z \in D$. By continuity, this implies $f(z) = f(a)$ for all $z \in \overline{D}$. This says that $|f(z)| = M$ on D as well as on $\overline{D}$, and the proof is complete. ♦

► **8.57** *Let $D \subset \mathbb{C}$ be bounded, open and connected. Suppose the complex function $f : \overline{D} \to \mathbb{C}$ is nonconstant and differentiable on D and continuous on $\overline{D}$. What does the above theorem allow you to conclude about the location of the maximum value of f on $\overline{D}$? How does this differ from the case where f is a real-valued function of two variables?*

4.3 Cauchy's Integral Formula for Higher-Order Derivatives

The following proposition is a valuable tool.

Proposition 4.11 *Let $f : D \to \mathbb{C}$ be continuous on the open set $D \subset \mathbb{C}$, and let C be a parametrized contour in D. For each $n \in \mathbb{N}$ define $F_n : \mathbb{C} \setminus C \to \mathbb{C}$ according to*

$$F_n(z) \equiv \int_C \frac{f(\zeta)}{(\zeta - z)^n}\, d\zeta.$$

Then for each $n \in \mathbb{N}$ the function F_n is differentiable on $\mathbb{C} \setminus C$, and

$$F_n'(z) = n \int_C \frac{f(\zeta)}{(\zeta - z)^{n+1}}\, d\zeta = n\, F_{n+1}(z) \quad \text{for all } z \in \mathbb{C} \setminus C.$$

Note that this result says that under the conditions of the proposition, we may differentiate under the integral sign, i.e.,

$$F_n'(z) = n \int_C \frac{f(\zeta)}{(\zeta - z)^{n+1}}\, d\zeta = \int_C \frac{d}{dz}\left(\frac{f(\zeta)}{(\zeta - z)^n}\right) d\zeta.$$

PROOF We will establish the result for $n = 1$ and leave the general case to the reader. Let $z_0 \in D \setminus C$ and choose $\rho > 0$ such that $N_\rho(z_0) \subset D \setminus C$. For any $\epsilon > 0$, we will find $\delta > 0$ such that

$$z \in D \cap N_\delta(z_0) \Rightarrow \left| F_1(z) - F_1(z_0) - \left(\int_C \frac{f(\zeta)}{(\zeta - z_0)^2}\, d\zeta \right) (z - z_0) \right| \le \epsilon\, |z - z_0|.$$

To see this, note that for $z \in N_{\frac{\rho}{2}}(z_0)$ we have

$$\begin{aligned}
&\left| F_1(z) - F_1(z_0) - \left(\int_C \frac{f(\zeta)}{(\zeta - z_0)^2}\, d\zeta \right) (z - z_0) \right| \\
&= \left| \int_C \frac{f(\zeta)}{\zeta - z}\, d\zeta - \int_C \frac{f(\zeta)}{\zeta - z_0}\, d\zeta - \left(\int_C \frac{f(\zeta)}{(\zeta - z_0)^2}\, d\zeta \right) (z - z_0) \right| \\
&= \left| \left(\int_C \frac{f(\zeta)}{(\zeta - z)(\zeta - z_0)}\, d\zeta - \int_C \frac{f(\zeta)}{(\zeta - z_0)^2}\, d\zeta \right) (z - z_0) \right| \\
&= \left| (z - z_0) \int_C \frac{(\zeta - z_0)(z - z_0) f(\zeta)}{(\zeta - z_0)^3 (\zeta - z)}\, d\zeta \right| \\
&= (z - z_0)^2 \left| \int_C \frac{f(\zeta)}{(\zeta - z_0)^2 (\zeta - z)}\, d\zeta \right|.
\end{aligned}$$

Now for $\zeta \in C$ and $z \in N_{\frac{\rho}{2}}(z_0)$, it follows that

1. $|f(\zeta)| \le \Lambda \equiv \max_{\zeta \in C} |f(\zeta)|$,
2. $|\zeta - z_0| \ge \rho$,
3. $|\zeta - z| \ge \frac{\rho}{2}$,

and therefore for $|z - z_0| < \frac{\epsilon\, \rho^3}{2 \Lambda L_C} \equiv \delta$,

$$\begin{aligned}
\left| F_1(z) - F_1(z_0) - \left(\int_C \frac{f(\zeta)}{(\zeta - z_0)^2}\, d\zeta \right) (z - z_0) \right| &\le |z - z_0|^2 \frac{2 \Lambda L_C}{\rho^3} \\
&\le \epsilon\, |z - z_0|.
\end{aligned}$$

Therefore, $F_1'(z) = \int_C \frac{f(\zeta)}{(\zeta - z_0)^2}\, d\zeta$ and the proof of the $n = 1$ case is complete. ♦

► **8.58** *Prove the general case.*

Using the above proposition, one can now prove the following result, a kind of converse to Corollary 4.6.

Corollary 4.12 *Let $f : D \to \mathbb{C}$ be continuous on the open connected set $D \subset \mathbb{C}$. Suppose that for every $z_0 \in D$ and every simple closed parametrized contour C*

satisfying the following conditions,

(i) $z_0 \in \text{Int}(C)$,

(ii) C *and* $\text{Int}(C)$ *are contained in* D,

we have

$$f(z_0) = \frac{1}{2\pi i} \oint\limits_C \frac{f(\zeta)}{\zeta - z_0}\, d\zeta.$$

Then f *is differentiable on* D.

► **8.59** *Prove the above corollary to Proposition 4.11*

The following corollary is a surprising and potent extension of the idea represented by equation (8.12) on page 430.

Corollary 4.13 *Suppose* $f : D \to \mathbb{C}$ *is differentiable on the open connected set* D, *and let* C *be a closed parametrized contour contained in* D. *Then for any* $z_0 \in D \setminus C$,

$$n_C(z_0)\, f^{(n)}(z_0) = \frac{n!}{2\pi i} \oint\limits_C \frac{f(\zeta)}{(\zeta - z_0)^{n+1}}\, d\zeta \quad \textit{for } n = 0, 1, 2, \ldots. \tag{8.14}$$

Also, for Λ *and* r *defined by* $\Lambda \equiv \max\limits_{\zeta \in C} |f(\zeta)|$ *and* $r \equiv \min\limits_{\zeta \in C} |\zeta - z_0|$,

$$\left| n_C(z_0)\, f^{(n)}(z_0) \right| \le \frac{n!\, \Lambda\, L_C}{2\pi\, r^{n+1}}. \tag{8.15}$$

Expressions (8.14) and (8.15) in Corollary 4.13 are known as *Cauchy's integral formulas for the derivatives* and *Cauchy's inequalities*, respectively.

► **8.60** *Prove the above corollary using Proposition 4.11.*

► **8.61** *In the special case where* $C = C_r(z_0)$, *a circle of radius* r *centered at* z_0, *show that* (8.15) *in the above corollary reduces to* $\left|f^{(n)}(z_0)\right| \le \frac{n!\,\Lambda}{r^n}$.

The above corollary tells us some very significant things about a complex function $f : D \to \mathbb{C}$ that is differentiable on the open connected set D.

1. It tells us that such a function *has* derivatives of all orders inside D, and provides a formula for computing any derivative of f inside D.
2. It tells us that for any $z_0 \in \text{Int}(C) \subset D$, the values of f and its derivatives of all orders at z_0 are determined by the values of f *on the boundary of the disk* $D_r \subset D$ *centered at* z_0.

The first fact is significant enough to state as a separate corollary.

Corollary 4.14 *If $f : D \to \mathbb{C}$ is differentiable at z_0 in the open set D, then f is infinitely differentiable at $z_0 \in D$.*

It is important to note that while the above corollary is true for differentiable complex functions $f : D \to \mathbb{C}$, it does not hold for any of the real function classes we have studied thus far. For example, it is not true that all differentiable functions $f : \mathbb{R} \to \mathbb{R}$ are necessarily infinitely differentiable, nor is it true that all differentiable functions $\boldsymbol{f} : \mathbb{R}^k \to \mathbb{R}^m$ are necessarily infinitely differentiable, and so on. This is one of the significant ways in which complex functions distinguish themselves from real functions.

5 Further Properties of Complex Differentiable Functions

5.1 Harmonic Functions

There *is* a special subclass of real functions $u : D^2 \to \mathbb{R}$ called *harmonic functions* to which Corollary 4.14 does apply. And it is no coincidence that this subclass of real-valued functions is directly related to the class of differentiable complex functions, as the next corollary shows.

Corollary 5.1 *If $f : D \to \mathbb{C}$ given by $f = u + i\,v$ is differentiable on the open set $D \subset \mathbb{C}$, then $u, v : D^2 \to \mathbb{R}$, where D^2 is the set D as represented in the xy-plane, are each twice differentiable and satisfy*

$$\Delta u \equiv \frac{\partial^2 u}{\partial x^2} + \frac{\partial^2 u}{\partial y^2} = 0 \quad \text{and} \quad \Delta v \equiv \frac{\partial^2 v}{\partial x^2} + \frac{\partial^2 v}{\partial y^2} = 0$$

on D^2. That is, u and v are **harmonic functions**.

PROOF Since $f = u + i\,v$ is differentiable on D, the Cauchy-Riemann equations yield $\frac{\partial u}{\partial x} = \frac{\partial v}{\partial y}$ and $\frac{\partial u}{\partial y} = -\frac{\partial v}{\partial x}$. Since $f' = \frac{\partial u}{\partial x} + i\,\frac{\partial v}{\partial x} = \frac{\partial v}{\partial y} - i\,\frac{\partial u}{\partial y}$ is also differentiable on D, the Cauchy-Riemann equations applied once again yield

$$\frac{\partial^2 u}{\partial x^2} = \frac{\partial^2 v}{\partial y \partial x}, \quad \frac{\partial^2 u}{\partial y^2} = -\frac{\partial^2 v}{\partial x \partial y},$$

and

$$\frac{\partial^2 v}{\partial x^2} = -\frac{\partial^2 u}{\partial y \partial x}, \quad \frac{\partial^2 v}{\partial y^2} = \frac{\partial^2 u}{\partial x \partial y}.$$

These equations together, along with the mixed derivative theorem, imply $\Delta u = \Delta v = 0$. ♦

A special relationship exists between the functions u and v in the case where $f : D \to \mathbb{C}$ given by $f = u + iv$ is a differentiable complex function. From

Corollary 5.1 we know that u and v must both be harmonic on D^2, the version of D in $\mathbb{R}^2$. But this is not all. The functions u and v must also satisfy the Cauchy-Riemann equations, a condition that forces them to be what are called *harmonic conjugates* of each other. This is illustrated in the following example.

Example 5.2 *Suppose $u : \mathbb{R}^2 \to \mathbb{R}$ is given by $u(x, y) = x^2 - y^2$. Can we find a function $v : \mathbb{R}^2 \to \mathbb{R}$ such that the complex function $f : \mathbb{C} \to \mathbb{C}$ where $f = u + iv$ is differentiable? If so, is the associated v unique? Before looking for such a v, note that for the given u we have that $\Delta u \equiv 0$ on $\mathbb{R}^2$. Therefore according to Corollary 5.1, the function u might be the associated real part of a differentiable complex function $f = u + iv$. Of course, any v we seek as the imaginary part of f must also satisfy $\Delta v \equiv 0$ on $\mathbb{R}^2$, but if f is to be differentiable u and v must also satisfy the Cauchy-Riemann equations, namely, $u_x = v_y$ and $u_y = -v_x$ on $\mathbb{R}^2$. Therefore, since $u_x = 2\,x$, we must have $v_y = 2\,x$. Integrating this last equation with respect to y obtains $v = 2\,xy + k(x)$. Note that the constant of integration associated with the y partial integration might actually be a function of x. From this candidate for v we see that $v_x = 2\,y + k'(x)$. And since $-u_y = 2\,y$ the Cauchy-Riemann equations imply that $2\,y + k'(x) = 2\,y$, or $k'(x) = 0$. That is, $k(x) \equiv c \in \mathbb{R}$, a constant. Overall then, any function $v : \mathbb{R}^2 \to \mathbb{R}$ of the form $v = 2\,xy + c$ will work. It is easy to see that our discovered form for v satisfies the condition $\Delta v \equiv 0$ as well. Any such v is called a* ***harmonic conjugate*** *of $u = x^2 - y^2$ on $\mathbb{R}^2$. The reader should check that u is a harmonic conjugate of v on $\mathbb{R}^2$.* ◀

► **8.62** *Given $v : \mathbb{R}^2 \to \mathbb{R}$ where $v(x, y) = 2xy$, show that there exists a harmonic function $u : \mathbb{R}^2 \to \mathbb{R}$ such that $f : \mathbb{C} \to \mathbb{C}$ given by $f = u + iv$ is a differentiable complex function on $\mathbb{C}$, and that u has the form $u(x, y) = x^2 - y^2 + c$ for arbitrary constant $c \in \mathbb{R}$.*

► **8.63** *If $u : D^2 \to \mathbb{R}$ is harmonic on D^2, then u is $C^\infty(D^2)$. In fact, $u = \mathrm{Re}(f)$ for some differentiable complex function $f : D \to \mathbb{C}$, where D is D^2 as represented in $\mathbb{C}$. Note that this result only holds locally, even if D is connected. To see this, consider $u : D^2 \to \mathbb{R}$ given by $u(x, y) = \ln(x^2 + y^2)$ where $D^2 = \mathbb{R}^2 \setminus \{0\}$. Determine the harmonic conjugates of u. How do they differ from each other? What is the domain of the resulting differentiable complex function f?*

5.2 A Limit Result

The following result will be of use to us in the next chapter. Its proof, perhaps surprisingly, requires the use of Cauchy's integral formula.

Theorem 5.3 *Let $f_n : D \to \mathbb{C}$ be differentiable on D for $n \geq 1$ and suppose there exists a function $f : D \to \mathbb{C}$ such that $\lim\limits_{n\to\infty} f_n(z) = f(z)$ uniformly on D. Then f is differentiable on D and $f'(z) = \lim\limits_{n\to\infty} f_n'(z)$ on D.*

PROOF Fix $a \in D$, and choose the parametrized circle $C_r \equiv C_r(a)$ such that C_r and its interior belong to D. Then Cauchy's integral formula obtains

$$f_n(z) = \frac{1}{2\pi i} \oint_{C_r} \frac{f_n(\zeta)}{\zeta - z}\, d\zeta \quad \text{for } |z-a| < r.$$

We ask the reader to show that for fixed z,

$$\lim_{n\to\infty} \oint_{C_r} \frac{f_n(\zeta)}{\zeta - z}\, d\zeta = \oint_{C_r} \frac{f(\zeta)}{\zeta - z}\, d\zeta.$$

This implies

$$f(z) = \frac{1}{2\pi i} \oint_{C_r} \frac{f(\zeta)}{\zeta - z}\, d\zeta \quad \text{for } |z-a| < r,$$

and so by Proposition 4.11 we have

$$f'(z) = \frac{1}{2\pi i} \oint_{C_r} \frac{f(\zeta)}{(\zeta - z)^2}\, d\zeta \quad \text{for } |z-a| < r.$$

Finally, we know that

$$f_n'(z) = \frac{1}{2\pi i} \oint_{C_r} \frac{f_n(\zeta)}{(\zeta - z)^2}\, d\zeta \quad \text{for } |z-a| < r,$$

which yields

$$\lim_{n\to\infty} f_n'(z) = \frac{1}{2\pi i} \oint_{C_r} \frac{f(\zeta)}{(\zeta - z)^2}\, d\zeta = f'(z) \quad \text{for } |z-a| < r.$$

◆

▶ **8.64** *For fixed z, show that $\lim_{n\to\infty} \oint_{C_r} \frac{f_n(\zeta)}{\zeta - z}\, d\zeta = \oint_{C_r} \frac{f(\zeta)}{\zeta - z}\, d\zeta$, and thereby complete the proof of the above theorem.*

5.3 Morera's Theorem

The following result is a kind of converse to Cauchy's integral theorem. Its proof requires Cauchy's integral formula, however, and so we present it here rather than in the previous section.

Theorem 5.4 (Morera's Theorem)
Let $f : D \to \mathbb{C}$ be continuous on the open set $D \subset \mathbb{C}$ such that $\oint_C f(z)\, dz = 0$ for every closed contour $C \subset D$. Then f is differentiable on D.

PROOF Fix $z_0 \in D$. We will show that f is differentiable at z_0. To this end, consider $N_r(z_0) \subset D$, and define $F : N_r(z_0) \to \mathbb{C}$ by

$$F(z) \equiv \int\limits_{[z_0,z]} f(\zeta)\, d\zeta.$$

By the path-independence of integrals of continuous functions in star-shaped regions, for any triangle $\triangle \subset N_r(z_0)$ we have that $\oint\limits_{\triangle} f(\zeta)\, d\zeta = 0$. It then follows by an argument similar to that in the proof of Theorem 2.14 on page 411 that F is differentiable on $N_r(z_0)$, and $F'(z) = f(z)$. By Corollary 4.14 on page 438, F is infinitely differentiable on $N_r(z_0)$, which implies that f is also infinitely differentiable on $N_r(z_0)$, and hence at z_0. ♦

► **8.65** *Let $f : D \to \mathbb{C}$ be continuous on the open set $D \subset \mathbb{C}$, and suppose f is differentiable on $D \setminus \{p\}$ for some $p \in D$. Show that f is differentiable on all of D. What if f is presumed initially to only be differentiable on $D \setminus \{z_1, z_2, \ldots, z_n\}$ for $z_1, z_2, \ldots, z_n \in D$?*

5.4 Liouville's Theorem

The next result pertains to complex functions that are differentiable in the entire complex plane. Such functions are called, appropriately enough, *entire* functions. The result follows the formal definition.

Definition 5.5 A function $f : \mathbb{C} \to \mathbb{C}$ that is differentiable on all of $\mathbb{C}$ is called an **entire** function.

Constant functions and polynomials are entire. The following theorem may seem surprising at first.

Theorem 5.6 (Liouville's Theorem)
If $f : \mathbb{C} \to \mathbb{C}$ is entire and bounded, then $f(z) \equiv c$ for some constant $c \in \mathbb{C}$.

PROOF The proof of this result is almost as surprising as the result itself. In fact, it is surprisingly simple. If f is to be shown to be constant, then we should be able to show that its derivative is identically zero. Suppose $|f(z)| \leq M$ for all $z \in \mathbb{C}$. Then fix an arbitrary $z_0 \in \mathbb{C}$ and let C be a circle of radius R centered at z_0. By Corollary 4.13 on page 437, we have

$$f'(z_0) = \frac{1}{2\,\pi\, i} \oint\limits_C \frac{f(z)}{(z - z_0)^2}\, dz,$$

and

$$|f'(z_0)| \leq \frac{1}{2\,\pi} \frac{M}{R^2} (2\,\pi\, R) = \frac{M}{R}. \tag{8.16}$$

Note that (8.16) is true for any $R > 0$. If $\epsilon > 0$ is given, choose R large enough such that

$$|f'(z_0)| \le \frac{M}{R} < \epsilon,$$

which implies that $f'(z_0) = 0$. Since $z_0 \in \mathbb{C}$ was arbitrary, $f'(z) = 0$ for all $z \in \mathbb{C}$, and so f must be constant on $\mathbb{C}$. ♦

5.5 The Fundamental Theorem of Algebra

Liouville's theorem is useful in proving the following famous result.

Theorem 5.7 (The Fundamental Theorem of Algebra)
If $p : \mathbb{C} \to \mathbb{C}$ is a nonconstant polynomial given by

$$p(z) = a_0 + a_1\, z + \cdots + a_N\, z^N, \text{ where } a_N \ne 0 \text{ and } N \ge 1,$$

then there exists $z_0 \in \mathbb{C}$ such that $p(z_0) = 0$, i.e., the polynomial $p(z)$ has at least one root.

PROOF Assume $p(z)$ has no roots. Then the function $f : \mathbb{C} \to \mathbb{C}$ given by $f(z) = \frac{1}{p(z)}$ is entire. We will show that f is bounded, and hence, by Liouville's theorem, constant (a contradiction). To see that f must be bounded, note

$$|f(z)| = \frac{1}{|p(z)|} \le \frac{1}{\left| |a_N\, z^N| - |a_0 + a_1\, z + \cdots + a_{N-1}\, z^{N-1}| \right|}. \tag{8.17}$$

But,

$$\begin{aligned} |a_0 + a_1\, z + \cdots + a_{N-1}\, z^{N-1}| & \\ &\le \left(\max_k |a_k|\right)\left(1 + |z| + \cdots + |z|^{N-1}\right) \\ &= \left(\max_k |a_k|\right)\frac{1 - |z|^N}{1 - |z|} \\ &\le |a_N|\left(|z| - 1\right)\frac{|z|^N - 1}{|z| - 1} \quad \text{for } |z| > R \text{ sufficiently large,} \\ &= |a_N|\,|z|^N - |a_N| \\ &= |a_N\, z^N| - |a_N|. \end{aligned} \tag{8.18}$$

The "sufficiently large" R referred to above can be any value greater than $\frac{\max_k |a_k|}{|a_N|} + 1$. Subbing (8.18) into (8.17) obtains

$$|f(z)| = \frac{1}{|p(z)|} \le \frac{1}{|a_N|} \equiv c \quad \text{for } |z| > R.$$

From this, we have that f is bounded by c outside the disk of radius R. But since the closed disk of radius R is a compact set, there exists M such that $|f| < M$ on the closed disk. Overall then, we have $|f| \le \max(M, c)$, and hence, f is bounded. By Liouville's theorem f must be constant, a contradiction. ♦

The following corollary to the fundamental theorem of algebra is a frequently exploited fact, namely, that any polynomial of degree $N \in \mathbb{N}$ with real or complex coefficients has exactly N roots, including multiplicities. We preface its more formal statement with some terminology. Note that if z_1 is a root of the polynomial $p(z)$, we may write $p(z)$ as follows,[6]

$$p(z) = (z - z_1)\, q_1(z),$$

where $q_1(z)$ is a polynomial of degree one less than that of $p(z)$. If z_1 is a root of $q_1(z)$, we may repeat the procedure and obtain

$$p(z) = (z - z_1)^2\, q_2(z),$$

where $q_2(z)$ is a polynomial of degree one less than that of $q_1(z)$. We may continue in this way until z_1 is no longer a root of q_{m_1} for some $m_1 \in \mathbb{N}$, yielding

$$p(z) = (z - z_1)^{m_1}\, q_{m_1}(z).$$

We then say that $p(z)$ has root z_1 with *multiplicity* m_1. Doing this for each root of $p(z)$, and accounting for each root's multiplicity, we obtain the following result.

Corollary 5.8 *Suppose $p : \mathbb{C} \to \mathbb{C}$ is a nonconstant polynomial given by*

$$p(z) = a_0 + a_1 z + \cdots + a_N z^N, \quad \textit{where } a_N \neq 0 \textit{ and } N \in \mathbb{N}.$$

If $z_1, z_2, \ldots, z_r$ are the distinct roots of p, with associated multiplicities $m_1, m_2, \ldots, m_r$, then $\sum_{j=1}^r m_j = N$, and $p(z)$ can be written as

$$p(z) = c\,(z - z_1)^{m_1}\,(z - z_2)^{m_2} \cdots (z - z_r)^{m_r}$$

for some constant $c \in \mathbb{C}$.

► **8.66** *Prove the above corollary.*

6 Appendices: Winding Numbers Revisited

6.1 A Geometric Interpretation

In this appendix, we will establish a useful geometric interpretation of the winding number $n_C(z_0)$ of a curve $C \subset \mathbb{C}$ about a fixed point $z_0 \in \mathbb{C}$. In particular, $n_C(z_0)$ can be interpreted as the number of times the curve C "winds around" the point z_0. To establish this, we will need to prove several technical results.

Continuous Logarithms and Arguments

For the following four lemmas, we follow [AN07]. The results are actually special cases of more general theorems about "covering spaces" in topology, for which we refer the reader to [Hat02].

[6]See any introductory text in abstract algebra.

Lemma 6.1 *Let $z : [a, b] \to \mathbb{C}$ be continuous on $[a, b]$ such that $z(t) \neq 0$ for all $t \in [a, b]$. There exists a continuous function $g : [a, b] \to \mathbb{C}$ such that $z(t) = e^{g(t)}$ for all $t \in [a, b]$.*

The function g is sometimes called a *continuous logarithm* of z. This result would probably seem obvious if the complex z were replaced by real $x > 0$.

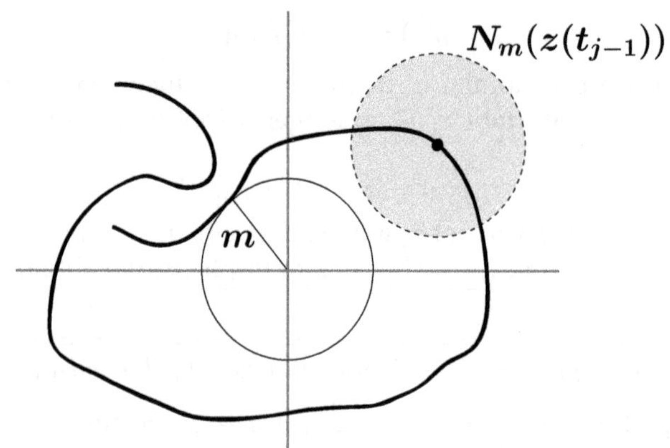

Figure 8.27 *The proof of Lemma 6.1.*

PROOF As depicted in Figure 8.27, let $m \equiv \min_{t\in[a,b]} |z(t)| > 0$. Since z is uniformly continuous on $[a, b]$, there exists $\delta > 0$ such that

$$|t - t'| < \delta \Rightarrow |z(t) - z(t')| < m.$$

Let $\mathcal{P} = \{a = t_0, t_1, \ldots, t_n = b\}$ be a partition of $[a, b]$ with $\|\mathcal{P}\| < \delta$. It follows that for each $1 \leq j \leq n$,

$$z(t) \in N_m\big(z(t_{j-1})\big) \quad \text{for} \quad t \in [t_{j-1}, t_j]. \quad \text{(Why?)}$$

Since $0 \notin N_m\big(z(t_{j-1})\big)$ for $1 \leq j \leq n$, there exists a continuous function $h_j : N_m\big(z(t_{j-1})\big) \to \mathbb{C}$ such that

$$\zeta = e^{h_j(\zeta)} \quad \text{for all} \quad \zeta \in N_m\big(z(t_{j-1})\big).$$

Note here that h_j is a branch of the complex logarithm (and is therefore differentiable on $N_m\big(z(t_{j-1})\big)$), and that it takes complex arguments; the g we seek will take real arguments. We'll now restrict this h_j to points on the contour within $N_m\big(z(t_{j-1})\big)$. Since $z(t) \in N_m\big(z(t_{j-1})\big)$ for all $t \in [t_{j-1}, t_j]$, we have

$$z(t) = e^{h_j(z(t))} \quad \text{for} \quad t \in [t_{j-1}, t_j].$$

Now define $g_j : [t_{j-1}, t_j] \to \mathbb{C}$ for $1 \leq j \leq n$ according to $g_j(t) \equiv h_j\big(z(t)\big)$ for $t \in [t_{j-1}, t_j]$. Then,

$$z(t) = e^{g_1(t)} \quad \text{for} \quad t \in [a, t_1],$$
$$z(t) = e^{g_2(t)} \quad \text{for} \quad t \in [t_1, t_2],$$
$$\vdots \qquad\qquad \vdots$$
$$z(t) = e^{g_n(t)} \quad \text{for} \quad t \in [t_{n-1}, b].$$

Note that $g_2(t_1) = g_1(t_1) + 2\pi\, i\, N$ for some $N \in \mathbb{Z}$ (Why?), so that we may replace g_2 by $g_2 - 2\pi\, i\, N$ and define $g : [a, t_2] \to \mathbb{C}$ by

$$g(t) = \begin{cases} g_1(t) & \text{for } t \in [a, t_1] \\ g_2(t) - 2\pi\, i\, N & \text{for } t \in [t_1, t_2] \end{cases}.$$

Then g is continuous on $[a, t_2]$ (Why?), and we have that $z(t) = e^{g(t)}$ for $t \in [a, t_2]$. Continuing in this way, we can define g on $[a, t_3]$, and eventually on all of $[a, b]$ such that g is continuous on $[a, b]$, and $z(t) = e^{g(t)}$ on $[a, b]$. ♦

If g is a continuous logarithm of z, it is reasonable to consider $g(z)$ as $\log z = \ln|z| + i \arg z$, and to therefore expect that $\arg z$ can be chosen to be continuous. We prove this next.

Lemma 6.2 *Let $z : [a, b] \to \mathbb{C}$ be a continuous function on $[a, b]$ such that $z(t) \neq 0$ for all $t \in [a, b]$. Then there exists $\theta : [a, b] \to \mathbb{R}$ continuous on $[a, b]$ such that $z(t) = |z(t)|\, e^{i\theta(t)}$ for all $t \in [a, b]$.*

PROOF By the previous lemma there exists $g : [a, b] \to \mathbb{C}$ continuous on $[a, b]$ such that $z(t) = e^{g(t)}$ for all $t \in [a, b]$. From this we have

$$z(t) = e^{g(t)} = e^{\text{Re}(g(t)) + i\,\text{Im}(g(t))} = e^{\text{Re}(g(t))}\, e^{i\,\text{Im}(g(t))},$$

and therefore, $|z(t)| = e^{\text{Re}(g(t))}$. Defining $\theta : [a, b] \to \mathbb{R}$ by $\theta(t) \equiv \text{Im}(g(t))$, we have

$$z(t) = |z(t)|\, e^{i\,\theta(t)}.$$ ♦

Two more technical results are required. The first one says that if you have two continuous logarithms of z, then they can differ by only a multiple of $2\pi\, i$. Likewise, their arguments can differ only by a multiple of 2π.

Lemma 6.3 *Let $z : [a, b] \to \mathbb{C}$ be a continuous function on $[a, b]$ such that $z(t) \neq 0$ for all $t \in [a, b]$. Suppose $g_1, g_2 : [a, b] \to \mathbb{C}$ and $\theta_1, \theta_2 : [a, b] \to \mathbb{R}$ are continuous on $[a, b]$ such that*

$$z(t) = e^{g_1(t)} = e^{g_2(t)} = |z(t)|\, e^{i\,\theta_1(t)} = |z(t)|\, e^{i\,\theta_2(t)} \quad \text{for all} \quad t \in [a, b].$$

Then, for some $m, n \in \mathbb{Z}$,

$$g_2(t) = g_1(t) + 2\pi\, i\, n, \quad \text{and} \quad \theta_2(t) = \theta_1(t) + 2\pi\, m.$$

PROOF Note that since $e^{g_1(t)} = e^{g_2(t)}$ for all $t \in [a,b]$, we have $e^{g_1(t)-g_2(t)} = 1 = e^{2\pi i n}$ for some $n \in \mathbb{Z}$, and so $g_1(t) - g_2(t) = 2\pi i n$ or $\frac{g_1(t)-g_2(t)}{2\pi i} = n$. Therefore the function $h : [a,b] \to \mathbb{C}$ given by $h(t) \equiv \frac{g_2(t)-g_1(t)}{2\pi i}$ is integer-valued, i.e., $h(t) = n(t)$. The function h is also continuous on $[a,b]$. Since $[a,b]$ is connected, $\frac{g_2(t)-g_1(t)}{2\pi i} \equiv n$ on $[a,b]$ for some fixed $n \in \mathbb{Z}$. A similar argument obtains the conclusion that $\frac{\theta_2(t)-\theta_1(t)}{2\pi} \equiv m$ on $[a,b]$ for some fixed $m \in \mathbb{Z}$, and the lemma is proved. ♦

Lemma 6.4 *Let $z : [a,b] \to \mathbb{C}$ be a continuous function on $[a,b]$ such that $z(t) \neq 0$ for all $t \in [a,b]$. Let $g : [a,b] \to \mathbb{C}$ and $\theta : [a,b] \to \mathbb{R}$ be continuous on $[a,b]$ such that $z(t) = e^{g(t)} = |z(t)|\, e^{i\,\theta(t)}$. Then, for $t_1, t_2 \in [a,b]$,*

$$g(t_2) - g(t_1) = \ln|z(t_2)| - \ln|z(t_1)| + i\left(\theta(t_2) - \theta(t_1)\right).$$

PROOF Let $h : [a,b] \to \mathbb{C}$ be given by $h(t) \equiv \ln|z(t)| + i\,\theta(t)$. Then,

$$z(t) = e^{h(t)} = |z(t)|\, e^{i\,\theta(t)},$$

and by Lemma 6.3 there exists $n \in \mathbb{N}$ such that

$$g(t) = h(t) + 2\pi\, i\, n = \ln|z(t)| + i\,\theta(t) + 2\pi\, i\, n.$$

The result follows. ♦

Winding Numbers and the Logarithm

And now the key result. We continue to follow [AN07].

Theorem 6.5 *Let C be a closed contour in $\mathbb{C}$ parametrized by $z : [a,b] \to \mathbb{C}$, and suppose $z_0 \notin C$. Choose $\theta : [a,b] \to \mathbb{R}$ to be a continuous argument of $z(t) - z_0$. Then,*

$$n_C(z_0) = \frac{1}{2\pi}\left(\theta(b) - \theta(a)\right).$$

PROOF Let $r \equiv \text{dist}(z_0, C) = \min_{t\in[a,b]} |z(t) - z_0| > 0$. Since $z(t)$ is uniformly continuous on $[a,b]$, there exists $\delta > 0$ such that

$$|t - t'| < \delta \;\Rightarrow\; |z(t) - z(t')| < r.$$

Choose a partition $\mathcal{P} = \{a = t_0, t_1, \ldots, t_n = b\}$ of $[a,b]$ such that $\|\mathcal{P}\| < \delta$. Then, for $t \in [t_{j-1}, t_j]$, we have that $z(t) \in N_r\left(z(t_{j-1})\right)$ while $z_0 \notin N_r\left(z(t_{j-1})\right)$. For all $\zeta \in N_r\left(z(t_{j-1})\right)$ we have $\zeta - z_0 \neq 0$. Therefore, there exists $h_j : N_r\left(z(t_{j-1})\right) \to \mathbb{C}$ differentiable on $N_r\left(z(t_{j-1})\right)$, where h_j is the restriction of a branch of the logarithm function onto $N_r\left(z(t_{j-1})\right)$. Then

$$\zeta - z_0 = e^{h_j(\zeta)} \quad \text{for all} \quad \zeta \in N_r\left(z(t_{j-1})\right).$$

It then follows that

$$z(t) - z_0 = e^{h_j(z(t))} \quad \text{for} \quad t \in [t_{j-1}, t_j]. \tag{8.19}$$

If we define $g_j : [t_{j-1}, t_j] \to \mathbb{C}$ for $1 \le j \le n$ according to $g_j(t) \equiv h_j(z(t))$ for $t \in [t_{j-1}, t_j]$, equality (8.19) becomes

$$z(t) - z_0 = e^{g_j(t)} \quad \text{for} \quad t \in [t_{j-1}, t_j].$$

Therefore, $g_j'(t) = \frac{z'(t)}{z(t)-z_0}$. From this and Lemma 6.4, we have

$$\begin{aligned}
\int_{C_j} \frac{dz}{z - z_0} &= \int_{t_{j-1}}^{t_j} \frac{z'(t)}{z(t) - z_0}\, dt \\
&= g_j(t_j) - g_j(t_{j-1}) \\
&= \ln|z(t_j) - z_0| - \ln|z(t_{j-1}) - z_0| + i\left(\theta(t_j) - \theta(t_{j-1})\right),
\end{aligned}$$

and

$$\oint_C \frac{dz}{z - z_0} = \sum_{j=1}^{n} \left(\int_{C_j} \frac{dz}{z - z_0} \right) = i\left(\theta(b) - \theta(a)\right).$$

The result follows. ♦

6.2 Winding Numbers of Simple Closed Contours

In this subsection, we establish that when traversing a simple closed contour in the plane, the winding number around any interior point is either +1 or −1. We begin with a lemma corresponding to a special case. In its statement, we denote the upper half-plane of $\mathbb{C}$ by $H_U \equiv \{z \in \mathbb{C} : \text{Im}(z) > 0\}$, and the lower half-plane by $H_L \equiv \{z \in \mathbb{C} : \text{Im}(z) < 0\}$. We follow [Ful97] in the proofs.

Lemma 6.6 *Suppose $C \subset \mathbb{C}$ is a parametrized closed contour that does not contain the point P. Suppose also that there exist points $z_U \in C \cap H_U$ and $z_L \in C \cap H_L$. Let A be the subcontour of C that begins at z_L and terminates at z_U, and let B be the subcontour of C that begins at z_U and terminates at z_L. Suppose A and B have the following properties:*

(i) $A \cap (-\infty, P) = \varnothing$,

(ii) $B \cap (P, \infty) = \varnothing$.

Then $n_C(P) = +1$.

PROOF Without loss in generality, assume $P \in \mathbb{R}$. Parametrize a circle C' around P such that C does not intersect C' or its interior. Let z_1 and z_2 be points on C' as shown in Figure 8.28, with $\widetilde{A}$ the clockwise arc from z_1 to z_2,

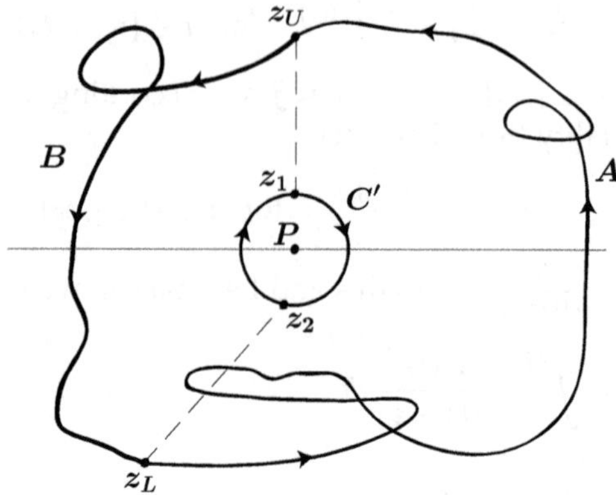

Figure 8.28 *The situation in Lemma 6.6.*

and $\widetilde{B}$ the clockwise arc from z_2 to z_1. Define the contours γ_A and γ_B by

$$\gamma_A \equiv \big(A, [z_U, z_1], \widetilde{A}, [z_2, z_L]\big),$$
$$\gamma_B \equiv \big(B, [z_L, z_2], \widetilde{B}, [z_1, z_U]\big).$$

It follows that

$$n_{\gamma_A}(P) + n_{\gamma_B}(P) = n_C(P) + n_{C'}(P).$$

Since $\gamma_A \cap (-\infty, P) = \varnothing$, we conclude that $n_{\gamma_A}(P) = 0$. Similarly, since $\gamma_B \cap (P, \infty) = \varnothing$, it follows that $n_{\gamma_B}(P) = 0$. Therefore,

$$n_C(P) = -n_{C'}(P) = 1.$$

◆

And now we present the key result of this subsection.

Theorem 6.7 *Let C be a parametrized simple closed contour. Then $n_C(P) = -1$ for every $P \in \operatorname{Int}(C)$, or $n_C(P) = +1$ for every $P \in \operatorname{Int}(C)$.*

PROOF Without loss of generality, we may assume that C is positioned so that the following all hold for some parametrization $z : [a, b] \to \mathbb{C}$ (see Figure 8.29).

1. C lies in the right half-plane, i.e., $\operatorname{Re}\big(z(t)\big) > 0$ for all $t \in [a, b]$,
2. $S \equiv \min C \cap (0, \infty)$, and $T \equiv \max C \cap (0, \infty)$,
3. $M = z(a)$ is in the lower half-plane, and $N = z(u)$ is in the upper half-plane for some u with $a < u < b$,
4. $[0, M] \cap C = \{M\}$ and $[0, N] \cap C = \{N\}$,
5. $A \equiv z\big([a, u]\big)$, and $B \equiv z\big([u, b]\big)$,
6. $\gamma_A \equiv \big([0, M], A, [N, 0]\big)$ and $\gamma_B \equiv \big([0, M], -B, [N, 0]\big)$.

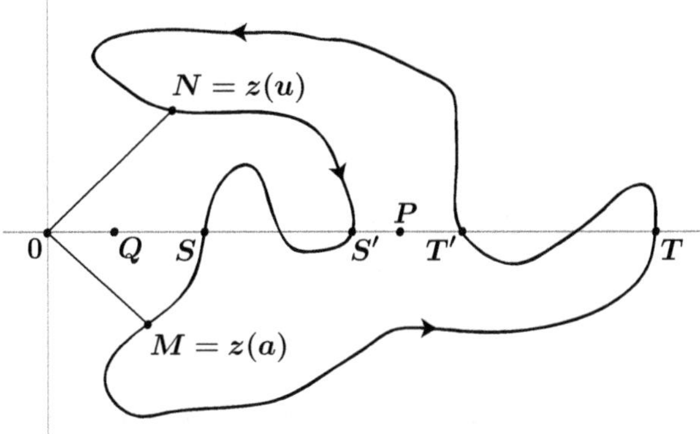

Figure 8.29 *The conditions imposed on the contour C in the proof of Theorem 6.7.*

Since $C = (A, B)$, and $A \cap B = \{M, N\}$, it follows that $T \in A$ or $T \in B$. We will consider the case where $T \in A$ and leave the other case, which is handled similarly, to the reader. Assuming $T \in A$, we note that $[T, \infty)$ does not intersect γ_B, and therefore,

$$n_{\gamma_B}(T) = 0.$$

If $0 < Q < S$, then by Lemma 6.6 we have

$$n_{\gamma_A}(Q) = n_{\gamma_B}(Q) = 1.$$

We now show that $S \in B$. To do so, suppose instead that $S \in A$. Note that the subcontour of A that connects S to T never intersects γ_B, and therefore

$$n_{\gamma_B}(S) = n_{\gamma_B}(T) = 0.$$

Since $[Q, S]$ also never intersects γ_B, we have

$$n_{\gamma_B}(S) = n_{\gamma_B}(Q) = 1,$$

a contradiction. Therefore, $S \in B$. We now define

$$T' \equiv \min A \cap (0, \infty) \quad \text{and} \quad S' \equiv \max B \cap (0, T').$$

Note that the interval (S', T') contains no points of $A \cup B$. If we take any point $P \in (S', T')$, then

$$n_C(P) = n_{\gamma_A}(P) - n_{\gamma_B}(P). \tag{8.20}$$

But note also that

$$n_{\gamma_A}(P) = n_{\gamma_A}(S') = n_{\gamma_A}(S) = n_{\gamma_A}(Q) = 1, \quad \text{(Why?)}$$

and

$$n_{\gamma_B}(P) = n_{\gamma_B}(T') = n_{\gamma_B}(T) = 0. \quad \text{(Why?)}$$

From this, and (8.20), we have

$$n_C(P) = n_{\gamma_A}(P) - n_{\gamma_B}(P) = 1.$$

Note that P must be in the interior of C, since $n_C(W) = 0$ for all $W \in Ext(C)$. ♦

▶ **8.67** *Complete the above proof by considering the case where $T \in B$. In this case, show that $n_C(P) = -1$ for $P \in \text{Int}(C)$.*

7 SUPPLEMENTARY EXERCISES

1. Evaluate $\int_1^2 \left(\frac{1}{t} - i\,t\right)^2 dt$.

2. Suppose $w : [-a, a] \to \mathbb{C}$ is an even function, i.e., $w(-t) = w(t)$ for all $t \in [-a, a]$. Show that $\int_{-a}^{a} w(t)\,dt = 2\int_0^a w(t)\,dt$. What if $w(t)$ is odd, i.e., $w(-t) = -w(t)$ for all $t \in [-a, a]$?

3. Evaluate $\int_0^1 e^{z_0 t}\,dt$, where $z_0 \neq 0$ is some complex constant.

4. Let $z_0, z_1, z_2 \in \mathbb{C}$ form a triangle, $T = [z_0, z_1, z_2, z_0]$, in $\mathbb{C}$. For the parametrized T, write down the explicit parametrization z as described in Definition 1.8.

5. Let $z_0, z_1, z_2, z_3 \in \mathbb{C}$ form a rectangle, $R = [z_0, z_1, z_2, z_3, z_0]$, in $\mathbb{C}$. For the parametrized R, write down the explicit parametrization z as described in Definition 1.8.

6. Let $D^2 = \{(x, y) \in \mathbb{R}^2 : x > 0, y = \sin\left(\frac{1}{x}\right)\} \cup \{(0, y) \in \mathbb{R}^2 : -1 \leq y \leq 1\}$. Show that D^2 is connected but is not contour connected. In fact, D^2 is not even path connected. Does this contradict Exercise 8.9 on page 393?

7. Evaluate $\oint\limits_C \overline{z}\,dz$ if C is the parametrized circle $C_r(0)$.

8. Let $f : \mathbb{C} \to \mathbb{C}$ be given by $f(z) = \text{Im}(z) + i\,\text{Re}(z)$. Evaluate $\int\limits_{C_1} f(z)\,dz$ and $\int\limits_{C_2} f(z)\,dz$ where C_1 and C_2 are the parametrized polygonal contours given by $C_1 = [0, i, 1 + i]$ and $C_2 = [0, 1 + i]$. Comment.

9. Evaluate $\int\limits_C \sin z\,dz$ for the parametrized polygonal contour $C = [1, i, -1]$.

10. Evaluate $\oint\limits_C e^{\overline{z}}\,dz$ for the parametrized polygonal contour $C = [0, 1, 1 + i, i, 0]$.

11. Evaluate the integral $\oint\limits_P e^z\,dz$ for the parametrized polygonal contour $P = [0, 1, i, 0]$ in $\mathbb{C}$, and compare it to the value of the integral $\oint\limits_C e^z\,dz$ where $C \equiv C_1(-1)$ is the circle parametrized by $z_C : [0, 2\pi] \to \mathbb{C}$ and $z_C(t) = -1 + e^{i\,t}$.

12. Let $f : \mathbb{C} \to \mathbb{C}$ be given by $f(z) = \begin{cases} z & \text{if } \text{Im}(z) \geq 0 \\ \overline{z} & \text{if } \text{Im}(z) < 0 \end{cases}$. Let C be the portion of the parabola $y = x^2$ from the point $(-1, 1)$ to the point $(2, 4)$ in the plane. Find $\int\limits_C f(z)\,dz$.

13. *Let C be the circular arc from the point $(5,0)$ to the point $(0,5)$ in the first quadrant. Show that* $\left|\int\limits_C \frac{dz}{z^3-1}\, dz\right| \le \frac{5\pi}{248}$.

14. *Let C be the parametrized polygonal contour $[-1,i,1]$. Show that* $\left|\int\limits_C \frac{dz}{z}\, dz\right| \le 4$.

15. *Show by direct calculation that* $\oint\limits_{\triangle} z^2\, dz = 0$ *where* $\triangle = [0,1,i,0]$.

16. *Let $C_1 \equiv C_1(1)$ and $C_2 \equiv C_2(2)$ be circles in $\mathbb{C}$, and let $C = (C_1, C_2)$ be the nonsimple closed contour (see Figure 8.30) in $\mathbb{C}$ with C_1 and C_2 as subcontours. Parametrize C so that the origin is the initial and terminal points, and such that C_1 is traversed counterclockwise first, followed by counterclockwise traversal of C_2. Use Proposition 1.14 part b) on page 396 to evaluate* $\oint\limits_C \overline{z}\, dz$.

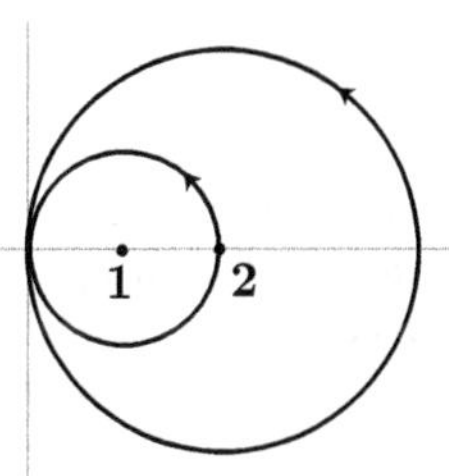

Figure 8.30 *The contour C comprised of two circular contours in Exercise 16.*

17. *Show by direct calculation that* $\int\limits_{-C_1(i)} z^2\, dz = -\int\limits_{C_1(i)} z^2\, dz$.

18. *Let C_1 be the parametrized circular arc from $(R,0)$ to $(-R,0)$ in the upper half-plane, and let C be the parametrized contour given by $C = (C_1, [-R,R])$. Evaluate* $\lim\limits_{R\to\infty}\left|\oint\limits_C \frac{z}{3\,z^8+1}\, dz\right|$.

19. *Use Proposition 2.10 on page 408 to evaluate* $\int\limits_P \cos z\, dz$ *where* $P = [\pi, \frac{3\pi\, i}{2}]$.

20. *Let P_1, P_2, and P_3 be the parametrized segments $[0,1]$, $[1,i]$, and $[0,i]$, respectively. If $P = (P_1, P_2)$, is it true that* $\int\limits_P \overline{z}\, dz = \int\limits_{P_3} \overline{z}\, dz$? *Does this contradict Corollary 2.11?*

21. *Let $C \subset \mathbb{C}$ be the parametrized contour given by $z : [0, 2\pi] \to \mathbb{C}$ where*

$$z(t) = \begin{cases} R\, e^{it} & \text{for } 0 \le t \le \pi \\ -R + \left(\frac{t}{\pi} - 1\right) 2R & \text{for } \pi < t \le 2\pi \end{cases}.$$

Show that $n_C(z_0) = 1$ for all $z_0 \in \mathrm{Int}(C)$.

22. *The conditions of Cauchy's integral theorem can be generalized a bit more. In fact, suppose C is a simple closed parametrized contour in the open set $D \subset \mathbb{C}$ such that*

$\text{Int}(C) \subset D$, just as in Cauchy's integral theorem. But now assume that $f : D \to \mathbb{C}$ is differentiable on $D \setminus \{p\}$ but only continuous at p for some $p \in \mathbb{C}$. Show that it is still true that $\oint\limits_C f(z)\,dz = 0$.

23. *Show that the condition of continuity of f at p in the previous exercise can be weakened even further. One need only presume that $\lim\limits_{z\to p} f(z)$ exists. In fact, if $p \notin C$, one can presume the even weaker condition that $\lim\limits_{z\to p} \big((z-p)\,f(z)\big)$ exists.*

24. *Evaluate $\oint\limits_C \frac{e^z}{z^5}\,dz$ where C is the parametrized square $[2+2\,i, -2+2\,i, -2-2\,i, 2-2\,i, 2+2\,i]$.*

25. *Let C be the ellipse given by $z : [0, 2\pi] \to \mathbb{C}$ where $z(t) = (3\cos t, 2\sin t)$. Let $C_1 \equiv C_{\frac{1}{2}}(1)$, $C_2 \equiv C_{\frac{1}{2}}(-1)$, $C_3 \equiv C_{\frac{1}{2}}(i)$, and $C_4 \equiv C_{\frac{1}{2}}(-i)$. Show that $\oint\limits_C \frac{dz}{z^4-1} = \sum_{j=1}^4 \oint\limits_{C_j} \frac{dz}{z^4-1}$.*

26. *This and the following exercise relate to uniformly convergent sequences of complex functions and the interchanging of limit operations. Let $f_n : D \to \mathbb{C}$ for $n = 1, 2, \ldots$, be a sequence of functions continuous on the open set $D \subset \mathbb{C}$, such that $\{f_n\}$ converges uniformly to $f : D \to \mathbb{C}$ on D. Show that for any parametrized contour $C \subset D$,*

$$\lim_{n\to\infty}\left(\int\limits_C f_n(z)\,dz\right) = \int\limits_C f(z)\,dz.$$

To do so, let C be a parametrized contour in D, and consider $\int\limits_C \big(f_n(z) - f(z)\big)\,dz$ along with Proposition 1.16 on page 398.

27. *Let $f_n : D \to \mathbb{C}$ for $n = 1, 2, \ldots$, be a sequence of functions differentiable on the open set $D \subset \mathbb{C}$, and suppose $\{f_n\}$ converges uniformly to $f : D \to \mathbb{C}$ on D. Show that f is differentiable on D, and that*

$$\lim_{n\to\infty} f_n'(z) = f'(z) \quad \text{on } D.$$

To do so, establish the following:

a) Show that f is continuous on D.

b) Fix $z_0 \in D$, and let $C \subset D$ be a simple closed parametrized contour such that $\text{Int}(C) \subset D$. Let $r \equiv \text{dist}(z_0, C)$, and show that for $\epsilon > 0$ there exists $N \in \mathbb{N}$ such that $n > N \;\Rightarrow\; |f_n(z) - f(z)| < \frac{\epsilon\, r}{2\,L_C}$ for all $z \in D$.

c) Note that $\oint\limits_C \frac{f_n(\zeta)}{\zeta - z_0}\,d\zeta = 2\pi\,i\,f_n(z_0)$ for each $n = 1, 2, \ldots$, and show from this

$$n > N \;\Rightarrow\; \left|\oint\limits_C \left(\frac{f_n(\zeta)}{\zeta-z_0} - \frac{f(\zeta)}{\zeta-z_0}\right)d\zeta\right| < \epsilon, \text{ and so } \lim\left(\oint\limits_C \frac{f_n(\zeta)}{\zeta-z_0}\,d\zeta\right) = \oint\limits_C \frac{f(\zeta)}{\zeta-z_0}\,d\zeta.$$

d) Apply the assumption of part c) to the conclusion of part c) to obtain the result that $f(z_0) = \frac{1}{2\pi\,i}\oint\limits_C \frac{f(\zeta)}{\zeta-z_0}\,d\zeta$. By Corollary 4.12 on page 436, it follows that f is differentiable on D.

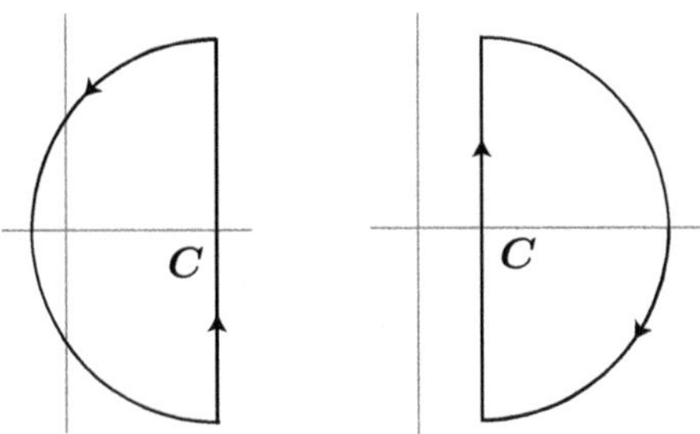

Figure 8.31 *Two semicircular contours for Exercise 28.*

e) *Show that* $\lim f_n'(z_0) = f'(z_0)$ *by noting that* $f_n'(z_0) = \frac{1}{2\pi i}\oint\limits_C \frac{f_n(\zeta)}{(\zeta-z_0)^2}\,d\zeta$, *and* $f'(z_0) = \frac{1}{2\pi i}\oint\limits_C \frac{f(\zeta)}{(\zeta-z_0)^2}\,d\zeta$.

28. *Compute*

$$\lim_{t\to\infty}\int_{[c-it,c+it]} \frac{e^{\pm z}}{z^{n+1}}dz \quad \text{where } c \neq 0.$$

Hint: Consider the semicircular contours shown in Figure 8.31.

29. *Evaluate* $\oint_C \frac{dz}{z^2-1}$ *where* C *is the contour shown in Figure 8.32.*

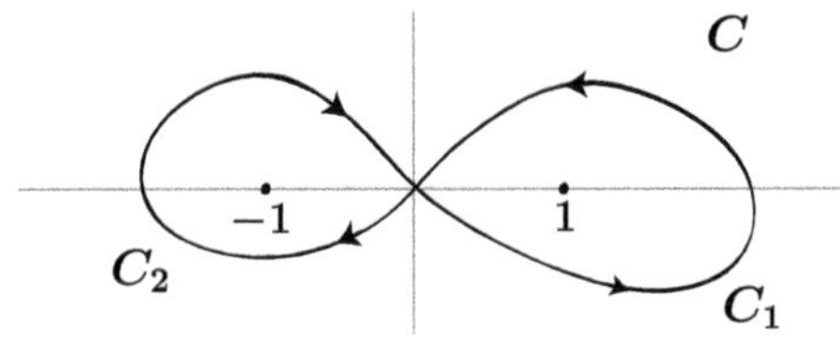

Figure 8.32 *The contour for Exercise 29.*

30. *In this exercise, you will prove the fundamental theorem of algebra using only techniques of winding numbers. Start with a monic*[7] *nonconstant polynomial* $F(z) = z^n + a_1 z^{n-1} + \cdots + a_n$. *Assume it vanishes nowhere. We will define the closed curve* Γ_r *parametrized by* $\gamma_r : [0, 2\pi] \to \mathbb{C}$ *given by* $\gamma_r(t) = \frac{F(re^{it})}{|F(re^{it})|}$.

a) *Prove that* $n(\Gamma_0, 0) = 0$.

b) *Fix* $r > 0$. *Show that if* r' *is sufficiently close to* r, *then* $n(\Gamma_{r'}, 0) = n(\Gamma_r, 0)$. *(The two curves are very close, and their winding numbers are thus very close integers.)*

c) *Deduce that*

$$n(\Gamma_r, 0) = \int_0^{2\pi} \frac{\gamma_r'(t)}{\gamma_r(t)}dt$$

is continuous. Show that it is constant.

d) *Take* r *very large. Show that* $\gamma_r(t)$ *is close to* e^{int}, *where* n *is the degree of* $F(z)$. *If* r *is large enough, then* $n(\Gamma_r, 0) = n$, *right? (Note that the dominant term in* $F(z)$ *is* z^n.*)*

[7]This means that the leading coefficient is 1. Why can we assume this?

e) Deduce that $n(\Gamma_r, 0) = n$ *for all* n*, and derive a contradiction.*

The above exercise is from r [Hat02], where it is a theorem (though proved using the fundamental group, not winding numbers).

31. *Suppose the open connected set* D *is simply connected,*[8] *i.e., for any contour* $C \subset D$ *and* $w \notin D$*, we have* $n_C(w) = 0$.

a) Prove that any differentiable $f : D \to \mathbb{C}$ *has an antiderivative on* D.

b) Let $u : D^2 \to \mathbb{R}$ *be harmonic on* D^2*, the corresponding region to* D *in* $\mathbb{R}^2$*. Prove that* $f(z) = \frac{\partial u}{\partial x} - i\frac{\partial u}{\partial y}$ *is differentiable on* D *(as a complex function).*

c) Prove that on a simply connected open set, any harmonic function has a harmonic conjugate.

d) Prove that $D \subset \mathbb{C}$ *is simply connected if and only if every differentiable* $g : D \to \mathbb{C} - \{0\}$ *has a differentiable logarithm. (Cf. the Appendix for a hint.)*

e) Show $D \subset \mathbb{C}$ *is simply connected if and only if every harmonic function on* D^2 *has a harmonic conjugate. (Think of* $\log|g(z)|$ *for* g *differentiable.)*

[8]The standard definition of a simply connected set, that (intuitively) a closed curve can be deformed into a point within that set, can be proved to be equivalent with this one.

9

Taylor Series, Laurent Series, and the Residue Calculus

A good calculator does not need artificial aids.

Lao Tze

In the first section of this chapter we will develop the theory of real and complex power series. While such series can be defined for some of the other spaces we have previously discussed, we will restrict our attention to the most important cases, namely $\mathbb{R}$ and $\mathbb{C}$. For results that apply to both $\mathbb{R}$ and $\mathbb{C}$, we will denote these fields more generally by $\mathbb{F}$, and an arbitrary element of $\mathbb{F}$ by ξ. Results that are specific to either $\mathbb{R}$ or $\mathbb{C}$ will be stated explicitly in terms of either, and we will denote an element of $\mathbb{R}$ and an element of $\mathbb{C}$ by the usual x and z, respectively. In section 2 we define the important type of power series known as Taylor series and develop its key properties. In section 3 we use Taylor series to characterize what it means for a function to be *analytic*. In the case of a complex function defined on an open connected set $D \subset \mathbb{C}$, we will see that f is analytic on D if and only if f is differentiable on D, a significant identification that does not hold in the case of real functions. We then define a kind of series representation for a function near a point at which the function is not differentiable, also called a *singularity* of the function. The resulting *Laurent series* is extremely useful in characterizing such singularities. Finally, in the last section of the chapter, we show that a particular coefficient of the Laurent series called the *residue* can be exploited in evaluating complex contour integrals of the function. This so-called *residue calculus* is an extremely useful technique, even for evaluating certain real integrals.

1 POWER SERIES

1.1 Definition, Properties, and Examples

We begin with a definition.

Definition 1.1 Consider a fixed $\xi_0 \in \mathbb{F}$ and constants $a_j \in \mathbb{F}$ for $j \geq 0$. A series of the form

$$\sum_{j=0}^{\infty} a_j \, (\xi - \xi_0)^j,$$

for $\xi \in \mathbb{F}$ is called a **power series centered at** ξ_0. The **radius of convergence** of the power series is defined as

$$R \equiv \sup \left\{ |\xi - \xi_0| : \sum_{j=0}^{\infty} a_j (\xi - \xi_0)^j \text{ converges} \right\},$$

when the supremum exists. When R exists and is nonzero, the neighborhood $N_R(\xi_0)$ is called the **neighborhood of convergence**.

When $R = 0$, the series converges only at the point $\xi = \xi_0$, to the value a_0. If the supremum fails to exist the radius of convergence is considered infinite, and we will often write $R = \infty$ and refer to $N_R(\xi_0)$ even in this case. When $R = \infty$ it is not hard to show that the series converges for all $\xi \in \mathbb{F}$. In the case where $\mathbb{F} = \mathbb{R}$, we sometimes refer to $N_R(x_0) = (x_0 - R, x_0 + R)$ as *the interval of convergence* rather than as the neighborhood of convergence. In the case where $\mathbb{F} = \mathbb{C}$ we may refer to the circle $C_R(z_0) \equiv \partial N_R(z_0)$ as *the circle of convergence*, whereas in the case $\mathbb{F} = \mathbb{R}$ the set $C_R(x_0)$ consists of the endpoints of the interval of convergence. While any power series will converge for at least $\xi = \xi_0$, it is not clear from the definition alone for what other points $\xi \in \mathbb{F}$ the series might converge. Even the definition of the radius of convergence R only tells us (roughly) the furthest distance from ξ_0 at which convergence occurs. Fortunately, the situation is not as complicated as it might otherwise be, as the following theorem establishes.

Theorem 1.2 *For any power series $\sum_{j=0}^{\infty} a_j \, (\xi - \xi_0)^j$ centered at ξ_0 with radius of convergence $R > 0$, the following all hold:*

a) The series converges absolutely on $N_R(\xi_0)$.

b) The series converges uniformly on $\overline{N_r(\xi_0)}$ for every $r < R$.

c) The series diverges for $\xi \notin \overline{N_R(\xi_0)}$.

Convergence or divergence of the power series at points on $C_R(\xi_0)$ must be studied individually.

PROOF Part *c*) is immediate from the definition of R. To prove *a*) and *b*), consider $r < R$. There exists $\xi_1 \in \mathbb{F}$ such that $r < |\xi_1 - \xi_0| \leq R$ and

$\sum_{j=0}^{\infty} a_j(\xi_1 - \xi_0)^j$ converges. Since $\lim_{j\to\infty} a_j(\xi_1 - \xi_0)^j = 0$, there exists $M \geq 0$ such that

$$|a_j(\xi_1 - \xi_0)^j| \leq M \quad \text{for all } j \geq 0.$$

Now suppose $\xi \in \overline{N_r(\xi_0)}$. Then,

$$|a_j(\xi - \xi_0)^j| = |a_j(\xi_1 - \xi_0)^j| \left|\frac{\xi - \xi_0}{\xi_1 - \xi_0}\right|^j \leq M \left(\frac{r}{|\xi_1 - \xi_0|}\right)^j \quad \text{for all } j \geq 0,$$

and since $\frac{r}{|\xi_1 - \xi_0|} < 1$, it follows from the Weierstrass M-test that $\sum_{j=0}^{\infty} a_j(\xi - \xi_0)^j$ converges absolutely and uniformly on $\overline{N_r(\xi_0)}$. ♦

▶ **9.1** *Show that when $R = 0$ the associated power series centered at ξ_0 converges absolutely at ξ_0, but diverges for $\xi \neq \xi_0$.*

▶ **9.2** *Suppose $\sum_{j=0}^{\infty} a_j (\xi - \xi_0)^j$ is a power series centered at ξ_0 that converges for $\xi_1 \neq \xi_0$. Show that the series converges absolutely for any ξ such that $|\xi - \xi_0| < |\xi_1 - \xi_0|$, that is, for any $\xi \in N_r(\xi_0)$ where $r \equiv |\xi_1 - \xi_0|$.*

▶ **9.3** *Show that a power series converges uniformly on any compact subset of its neighborhood of convergence.*

To see that "anything can happen" at points lying on the circle of convergence of a power series, consider the particular complex example of $\sum_{j=1}^{\infty} \frac{1}{j} z^j$. Note that the series converges at $z = -1$, but it diverges at $z = 1$. The radius of convergence must be $R = 1$ (Why?), and so the circle of convergence in this case contains at least one point of convergence and one point of divergence for the series.

▶ **9.4** *Consider the complex power series given by $\sum_{j=0}^{\infty} \frac{1}{j^2} z^j$. Use the ratio test to show that $R = 1$, and that the series converges absolutely at every point on its circle of convergence $C_1(0)$.*

▶ **9.5** *Can you find an example of a power series that does not converge at any point on its circle of convergence?*

The following corollary to Theorem 1.2 establishes a convenient fact about power series, namely, they are continuous functions on their neighborhoods of convergence.

Corollary 1.3 *A power series $\sum_{j=0}^{\infty} a_j (\xi - \xi_0)^j$ represents a continuous function on its neighborhood of convergence.*

PROOF Fix $\xi_1 \in N_R(\xi_0)$, where R is the radius of convergence for the series. Now choose $r < R$ so that $\xi_1 \in \overline{N_r(\xi_0)} \subset N_R(\xi_0)$. Let $f_n : N_R(\xi_0) \to \mathbb{F}$ be the continuous function defined by $f_n(\xi) = \sum_{j=0}^{n} a_j (\xi - \xi_0)^j$ for $n \geq 1$. By the previous theorem we know that $\sum_{j=0}^{\infty} a_j (\xi - \xi_0)^j$ converges uniformly on

$\overline{N_r(\xi_0)}$. This, along with Theorem 3.8 from Chapter 5, implies that

$$\lim_{n\to\infty} f_n(\xi) = \lim_{n\to\infty} \sum_{j=0}^{n} a_j\,(\xi-\xi_0)^j = \sum_{j=0}^{\infty} a_j\,(\xi-\xi_0)^j$$

is continuous at ξ_1. Since $\xi_1 \in N_R(\xi_0)$ was arbitrary, we have that $f : N_R(\xi_0) \to \mathbb{F}$ given by $f(\xi) \equiv \sum_{j=0}^{\infty} a_j\,(\xi-\xi_0)^j$ is continuous on $N_R(\xi_0)$. ◆

Therefore, any power series centered at ξ_0 is a continuous function of ξ on its neighborhood of convergence. This neighborhood of convergence is the function's domain and can be as small as ξ_0 itself, or as large as all of $\mathbb{F}$.

Examples of Power Series

The simplest type of power series is the *geometric series* introduced in Chapter 3, namely, $\sum_{j=0}^{\infty} \xi^j$. Recall that this series diverges for $|\xi| \geq 1$ and converges to $\frac{1}{1-\xi}$ for $|\xi| < 1$. This fact will be extremely useful to us in what follows, since, as indicated in the proof of Theorem 1.2, determining those values of ξ for which a given power series converges hinges largely on determining those values of ξ for which the tail of the series behaves like that of a convergent geometric series.

Example 1.4 *Since $\sum_{j=0}^{\infty} z^j$ converges for $|z| < 1$, we may define the function $f : N_1(0) \to \mathbb{C}$ according to*

$$f(z) = \sum_{j=0}^{\infty} z^j.$$

Given the discussion above, we see that the function $g : \mathbb{C} \setminus \{1\} \to \mathbb{C}$ given by $g(z) = \frac{1}{1-z}$ is equal to f on $N_1(0)$. That is,

$$f(z) = \sum_{j=0}^{\infty} z^j = 1 + z + z^2 + \cdots = \frac{1}{1-z} \quad \text{on } N_1(0).$$

Note that the function g is actually defined outside of $N_1(0)$ as well; we have simply established that $f(z) = g(z)$ on $N_1(0)$. The radius of convergence of the series is $R = 1$, and the circle of convergence is $C_1(0)$. (See Figure 9.1.) ◀

As the above example illustrates, a power series can sometimes be written as a "closed formula" function (rather than as an infinite series). In the particular example above, we say that the function $g(z) = \frac{1}{1-z}$ *has power series representation* $\sum_{j=0}^{\infty} z^j$ *on* $N_1(0)$. For the same function $g(z) = \frac{1}{1-z}$, we can also find a power series representation centered at a different point.

Example 1.5 *For $g : \mathbb{C} \setminus \{1\} \to \mathbb{C}$ given by $g(z) = \frac{1}{1-z}$, we will find a power series representation for g centered at $z = i$. Note that*

$$\frac{1}{1-z} = \frac{1}{1-i-(z-i)} = \frac{1}{1-i}\,\frac{1}{1-\frac{z-i}{1-i}} = \frac{1}{1-i}\,\frac{1}{1-w},$$

for $w = \frac{z-i}{1-i}$. From this we have

$$\frac{1}{1-z} = \frac{1}{1-i}\sum_{j=0}^{\infty} w^j = \frac{1}{1-i}\sum_{j=0}^{\infty}\left(\frac{z-i}{1-i}\right)^j = \sum_{j=0}^{\infty}\left(\frac{1}{1-i}\right)^{j+1}(z-i)^j,$$

which converges for $|w| = \left|\frac{z-i}{1-i}\right| < 1$, i.e., for $|z-i| < |1-i| = \sqrt{2}$, or on $N_{\sqrt{2}}(i)$. Note that $z = 1$ lies on the circle of convergence $C_{\sqrt{2}}(i)$. (See Figure 9.1 again.) ◀

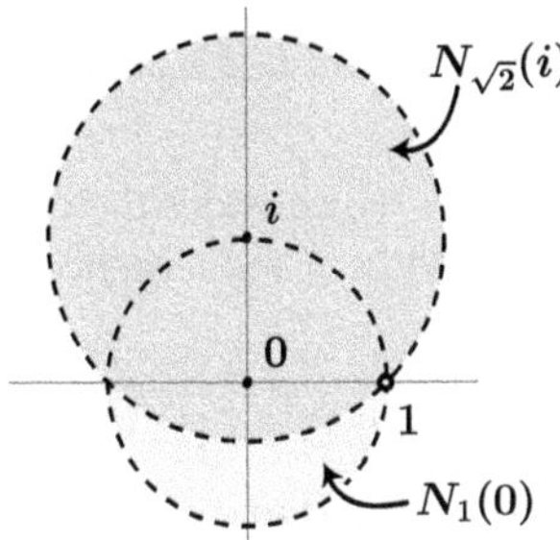

Figure 9.1 *The regions of convergence for the power series in Examples 1.4 and 1.5*

From Examples 1.4 and 1.5 we see that the same function $g(z) = \frac{1}{1-z}$ has two different power series representations,

$$\frac{1}{1-z} = \sum_{j=0}^{\infty} z^j = 1 + z + z^2 + \cdots \quad \text{on } N_1(0),$$

and

$$\frac{1}{1-z} = \frac{1}{1-i} + \left(\frac{1}{1-i}\right)^2(z-i) + \left(\frac{1}{1-i}\right)^3(z-i)^2 + \cdots \quad \text{on } N_{\sqrt{2}}(i).$$

In fact, $g(z) = \frac{1}{1-z}$ has infinitely many different power series representations corresponding to different values of $z_0 \neq 1$.

▶ **9.6** *Suppose $f : \mathbb{C} \setminus \{1\} \to \mathbb{C}$ is given by $f(z) = \frac{1}{z-1}$. Find a power series representation for f centered at $z = 2$, and determine its radius and neighborhood of convergence.*

Other examples can be analyzed in light of the special geometric series case. In each, a function is given by a formula, and an associated power series is sought. The strategy is to express the given function as a multiple of the quantity $\frac{1}{1-w}$ for some function w of ξ, and then recognize that $\frac{1}{1-w} = \sum_{j=0}^{\infty} w^j$ where the series converges for $|w| < 1$. This will yield the radius and neighborhood of convergence of the derived power series representation for the original function.

Example 1.6 *Consider the function $f : \mathbb{C} \setminus \{\frac{1}{3}\} \to \mathbb{C}$ given by $f(z) = \frac{1}{1-3z}$.*

Note that the function is not defined at $z = \frac{1}{3}$*. Letting* $w = 3z$*, we have*

$$\frac{1}{1-3z} = \frac{1}{1-w} = \sum_{j=0}^{\infty} w^j = \sum_{j=0}^{\infty} (3z)^j = \sum_{j=0}^{\infty} 3^j z^j,$$

which converges for $|w| = |3z| < 1$*, or for* $|z| < \frac{1}{3}$*. Therefore the radius of convergence is* $\frac{1}{3}$*, and the neighborhood of convergence is* $N_{\frac{1}{3}}(0)$*. Note that the point* $z = \frac{1}{3}$ *lies on the circle of convergence* $C_{\frac{1}{3}}(0)$*.* ◀

▶ **9.7** *Choose any real* $\alpha > 0$*. Can you find a power series having radius of convergence equal to* α*?*

▶ **9.8** *For fixed* $a, b \in \mathbb{F}$ *and* $m \in \mathbb{N}$*, consider the function given by the rule* $f(\xi) = \frac{1}{a - b\xi^m}$*. Determine the neighborhood of convergence for this function's power series representation centered at* $\xi_0 = 0$*.*

Example 1.7 *Consider the function* $f : \mathbb{C} \setminus \{\pm 2i\} \to \mathbb{C}$ *given by* $f(z) = \frac{1}{z^2+4}$*. Note that the function is not defined at* $z = \pm 2i$*. Factoring out a* $\frac{1}{4}$*, and letting* $w = -\frac{z^2}{4}$*, we obtain*

$$\frac{1}{z^2+4} = \frac{1}{4}\frac{1}{1-\left(-\frac{z^2}{4}\right)} = \frac{1}{4}\frac{1}{1-w} = \frac{1}{4}\sum_{j=0}^{\infty} w^j = \frac{1}{4}\sum_{j=0}^{\infty} \left(\frac{i}{2}\right)^{2j} z^{2j},$$

which converges for $|w| = \left|-\frac{z^2}{4}\right| < 1$*, or* $|z| < 2$*. Hence, the radius of convergence is* $R = 2$*, and the neighborhood of convergence is* $N_2(0)$*. Note that the points* $\pm 2i$ *lie on the circle of convergence* $C_2(0)$*.* ◀

The previous examples illustrate a useful method of exploiting the behavior of geometric series to find power series representations for certain functions. The following two results are also useful in that they provide convenient means for computing the radius of convergence R for a given power series. The first makes use of the ratio test, and the second makes use of the root test. Recall that both tests are based on comparison of the given series to geometric series.

Theorem 1.8 *Let* $\sum_{j=0}^{\infty} a_j (\xi - \xi_0)^j$ *be a power series centered at* ξ_0 *with* $a_j \neq 0$ *for all* $j \geq 0$*, and let*

$$\rho \equiv \lim \left|\frac{a_{j+1}}{a_j}\right|,$$

when the limit exists or is ∞*. Then there are three cases:*

a) If $\rho \neq 0$*, then* $R = \frac{1}{\rho}$*.*

b) If $\rho = 0$*, then* $R = \infty$*.*

c) If $\rho = \infty$*, then* $R = 0$*.*

PROOF We first consider *a)*. Suppose $\rho = \lim \left|\frac{a_{j+1}}{a_j}\right| \neq 0$. Then if $\xi \neq \xi_0$ we may apply the ratio test to obtain

$$\lim_{j\to\infty} \left|\frac{a_{j+1}(\xi - \xi_0)^{j+1}}{a_j(\xi - \xi_0)^j}\right| = \rho\,|\xi - \xi_0|. \tag{9.1}$$

From this we see that the series converges absolutely if $\rho\,|\xi - \xi_0| < 1$ and diverges if $\rho\,|\xi - \xi_0| > 1$. This shows that $R = \frac{1}{\rho}$. Now consider *b)*. In this case, the limit in (9.1) equals zero for all $\xi \in \mathbb{F}$. The ratio test then yields that the series converges absolutely for all $\xi \in \mathbb{F}$, i.e., $R = \infty$. Finally, consider *c)* and suppose $\xi \neq \xi_0$. Then,

$$\lim_{j\to\infty} \left|\frac{a_{j+1}(\xi - \xi_0)^{j+1}}{a_j(\xi - \xi_0)^j}\right| = \lim_{j\to\infty} \left(\left|\frac{a_{j+1}}{a_j}\right| |\xi - \xi_0|\right) = \infty.$$

Therefore, the only case in which convergence holds is $\xi = \xi_0$, i.e., $R = 0$. ♦

Clearly this theorem will be well suited to power series whose a_j involve factorials. On the other hand, the following result is typically more useful for those power series whose a_j involve j-dependent powers. Note too that it is a bit more general than Theorem 1.8 in that, like the root test of Chapter 3, it defines ρ in terms of a lim sup rather than a limit. Of course, in those cases where the limit exists, it will equal the lim sup.

Theorem 1.9 *Let $\sum_{j=0}^{\infty} a_j\,(\xi - \xi_0)^j$ be a power series centered at ξ_0, and let*

$$\rho \equiv \limsup \sqrt[j]{|a_j|},$$

when it exists, otherwise let $\rho = \infty$. Then there are three cases:

a) If $\rho \neq 0$, then $R = \frac{1}{\rho}$.

b) If $\rho = 0$, then $R = \infty$.

c) If $\rho = \infty$, then $R = 0$.

PROOF In cases *a)* and *b)*, consider that

$$\limsup \sqrt[j]{|a_j\,(\xi - \xi_0)^j|} = \left(\limsup \sqrt[j]{|a_j|}\right) |\xi - \xi_0| = \rho\,|\xi - \xi_0|.$$

If $\rho \neq 0$ then $\limsup \sqrt[j]{|a_j\,(\xi - \xi_0)^j|} = \rho\,|\xi - \xi_0| < 1$ if and only if $|\xi - \xi_0| < \frac{1}{\rho}$, and therefore $R = \frac{1}{\rho}$. If $\rho = 0$ then $\limsup \sqrt[j]{|a_j\,(\xi - \xi_0)^j|} = 0 < 1$, and the series converges for all $\xi \in \mathbb{F}$, i.e., $R = \infty$. If $\rho = \infty$ then $\limsup \sqrt[j]{|a_j\,(\xi - \xi_0)^j|} < 1$ only if $\xi = \xi_0$, i.e., $R = 0$. ♦

In applying either Theorem 1.8 or Theorem 1.9, it is convenient to think of R as the reciprocal of ρ, even when $\rho = 0$ or $\rho = \infty$.

Example 1.10 *We look at several examples.*

a) Consider the complex power series $\sum_{j=0}^{\infty} j!z^j$. Applying Theorem 1.8 we compute

$$\rho = \lim \left| \frac{(j+1)!}{j!} \right| = \lim(j+1) = \infty,$$

and so $R = 0$. Therefore the series converges only for $z = 0$.

b) Consider the power series $\sum_{j=0}^{\infty} \frac{1}{j} z^j$ discussed previously. Applying Theorem 1.8 we compute

$$\rho = \lim \left| \frac{j}{j+1} \right| = 1,$$

and so we confirm that $R = 1$ as we had deduced earlier. Recall that $z = 1$ and $z = -1$ each lie on the circle of convergence $C_1(0)$ in this case, and that the series converges at $z = -1$ while it diverges at $z = 1$.

c) Consider the real power series $\sum_{j=0}^{\infty} (-j)^j (x-1)^j$. Applying Theorem 1.9 we see that

$$\rho = \limsup \sqrt[j]{|a_j|} = \limsup \sqrt[j]{j^j} = \limsup j,$$

which does not exist. Therefore, $\rho = \infty$ and so $R = 0$. The series converges only for $x = 1$.

d) Consider the power series $\sum_{j=0}^{\infty} a_j z^j$, with

$$a_j = \begin{cases} \left(\frac{1}{2}\right)^j & \text{for odd } j \geq 1 \\ \left(\frac{1}{3}\right)^j & \text{for even } j \geq 2 \end{cases}.$$

Then it is easy to see that $\lim_{j\to\infty} \sqrt[j]{|a_j|}$ does not exist. But $\limsup \sqrt[j]{|a_j|} = \frac{1}{2}$, and therefore the radius of convergence is $R = 2$.

e) Recall from Chapter 3 that our definition of e^x was

$$e^x \equiv \sum_{j=0}^{\infty} \frac{x^j}{j!}.$$

We saw then, via the ratio test, that the above series converges for every $x \in \mathbb{R}$. In Chapter 4, we naturally defined the exponential function $\exp : \mathbb{R} \to \mathbb{R}$ by

$$\exp(x) = e^x \equiv \sum_{j=0}^{\infty} \frac{x^j}{j!}.$$

From this, we might naturally define a complex function given by

$$f(z) = \sum_{j=0}^{\infty} \frac{z^j}{j!} \tag{9.2}$$

and expect it to be equal to the complex exponential function on its neighborhood of convergence. To determine the radius of convergence for this *complex power series, apply Theorem 1.8 to see that $R = \infty$. We might be tempted at this point to conclude that the complex exponential function has this power series representation on all of $\mathbb{C}$. Recall, though, that in Chapter 4 we defined the complex exponential function as $\exp(z) \equiv e^z \equiv e^x\, e^{iy}$. Later we will show that the power series given in (9.2) is equivalent to our definition from Chapter 4.* ◀

We prove one more result in this subsection, which will be of use in what follows. Referred to as Abel's theorem, it establishes that if a real power series converges at the right endpoint of its interval of convergence, then the power series is left continuous there.

Theorem 1.11 (Abel's Theorem)
Suppose $f(x) = \sum_{j=0}^{\infty} a_j (x - x_0)^j$ is a real power series that converges for $|x - x_0| < R$. If $\sum_{j=0}^{\infty} a_j R^j$ converges then $\lim\limits_{x \to (x_0+R)^-} f(x) = \sum_{j=0}^{\infty} a_j R^j$.

PROOF We consider the case $x_0 = 0$ and $R = 1$, and leave the general case to the reader. Suppose $\sum_{j=0}^{\infty} a_j = s$, let $S_n(x) = \sum_{j=0}^{n} a_j x^j$, and let $S_j \equiv S_j(1) = \sum_{k=0}^{j} a_k$. Then for $|x| < R = 1$,

$$\frac{f(x)}{1-x} = \Big(\sum_{j=0}^{\infty} a_j x^j\Big)\Big(\sum_{j=0}^{\infty} x^j\Big) = \sum_{j=0}^{\infty} S_j x^j,$$

where the second equality above is due to Theorem 5.6 in Chapter 3. From this we have, for $|x| < 1$,

$$\begin{aligned} f(x) &= (1-x)\sum_{j=0}^{\infty} S_j x^j \\ &= (1-x)\sum_{j=0}^{\infty} (S_j - s)x^j + (1-x)\sum_{j=0}^{\infty} s\, x^j \\ &= (1-x)\sum_{j=0}^{\infty} (S_j - s)x^j + s. \end{aligned} \tag{9.3}$$

Now note that since $\lim\limits_{j\to\infty} S_j = s$, for any $\epsilon > 0$ there exists $N \in \mathbb{N}$ such that

$$j > N \quad \Rightarrow \quad |S_j - s| < \tfrac{\epsilon}{2}.$$

Therefore, for $0 \le x < 1$, equality (9.3) obtains

$$\begin{aligned} f(x) - s &= (1-x)\sum_{j=0}^{\infty}(S_j - s)x^j \\ &= (1-x)\sum_{j=0}^{N} (S_j - s)x^j + (1-x)\sum_{j=N+1}^{\infty} (S_j - s)x^j, \end{aligned}$$

which implies that

$$\begin{aligned}|f(x)-s| &\le |1-x|\left|\sum_{j=0}^{N}(S_j-s)x^j\right| + |1-x|\sum_{j=N+1}^{\infty}\frac{\epsilon}{2}\,|x|^j \\ &\le |1-x|\sum_{j=0}^{N}|S_j-s| + \frac{\epsilon}{2}\,|1-x|\,\frac{1}{1-|x|} \\ &= |1-x|\sum_{j=0}^{N}|S_j-s| + \tfrac{\epsilon}{2}. \qquad \text{(Why?)} \end{aligned} \tag{9.4}$$

If we choose $\delta \equiv \epsilon\left(2\sum_{j=0}^{N}|S_j-s|\right)^{-1}$, we obtain from inequality (9.4) that

$$1-\delta < x < 1 \;\Rightarrow\; |f(x)-s| < \epsilon,$$

and the special case of the result is proved. ♦

► **9.9** *Answer the (Why?) question in the above proof. Then prove the general case; to do so, let $y = x - x_0$ so that $f(x) = \sum_{j=0}^{\infty} a_j(x-x_0)^j$ becomes $f(x_0+y) = \sum_{j=0}^{\infty} a_j y^j$. Then rescale by letting $y = Rt$. This gives $f(x_0+Rt) = \sum_{j=0}^{\infty} a_j R^j t^j$.*

► **9.10** *What conditions need be presumed in the above theorem in order to conclude that $\lim_{x\to(x_0-R)^+} f(x) = \sum_{j=0}^{\infty} a_j(-R)^j$?*

1.2 Manipulations of Power Series

Algebraic Manipulations of Power Series

The following results are a great convenience in manipulating power series.

Theorem 1.12 *Let $f(\xi) = \sum_{j=0}^{\infty} a_j\,(\xi-\xi_0)^j$ and $g(\xi) = \sum_{j=0}^{\infty} b_j\,(\xi-\xi_0)^j$ be power series having radii of convergence R_f and R_g, respectively. Then,*

a) $(f\pm g)(\xi) = \sum_{j=0}^{\infty} c_j\,(\xi-\xi_0)^j$ where $c_j = a_j \pm b_j$, and

b) $(fg)(\xi) = \sum_{j=0}^{\infty} c_j\,(\xi-\xi_0)^j$ where $c_j = \sum_{k=0}^{j} a_k\, b_{j-k}$.

In both cases, the resulting series has radius of convergence $R \ge \min(R_f, R_g)$.

PROOF We prove $b)$ and leave the proof of $a)$ to the reader. Since $\sum_{j=0}^{\infty} a_j\,(\xi-\xi_0)^j$ and $\sum_{j=0}^{\infty} b_j\,(\xi-\xi_0)^j$ converge absolutely on $N_r(\xi_0)$ for $r = \min(R_f, R_g)$, it follows from Theorem 5.6 in Chapter 3 (which allows for multiplication of infinite series) that

$$
\begin{aligned}
f(\xi)g(\xi) &= \left(\sum_{j=0}^{\infty} a_j\,(\xi-\xi_0)^j\right)\left(\sum_{j=0}^{\infty} b_j\,(\xi-\xi_0)^j\right) \\
&= \sum_{j=0}^{\infty}\left(\sum_{k=0}^{j} a_k\,(\xi-\xi_0)^k\, b_{j-k}\,(\xi-\xi_0)^{j-k}\right) \\
&= \sum_{j=0}^{\infty}\left(\sum_{k=0}^{j} a_k\, b_{j-k}\right)(\xi-\xi_0)^j \\
&\equiv \sum_{j=0}^{\infty} c_j\,(\xi-\xi_0)^j, \text{ where } c_j \equiv \sum_{k=0}^{j} a_k b_{j-k},
\end{aligned}
$$

and since this series converges on $N_r(\xi_0)$ where $r = \min(R_f, R_g)$, it follows that the radius of convergence R satisfies $R \geq r$. ♦

► **9.11** *Prove part 1 of the above theorem.*

Theorem 1.13 *Let $f(\xi) = \sum_{j=0}^{\infty} a_j\,(\xi-\xi_0)^j$ be a power series having radius of convergence R. If $a_0 \neq 0$, then there exists $r > 0$ such that on $N_r(\xi_0)$ we have*

$$\frac{1}{f(\xi)} = \sum_{j=0}^{\infty} b_j\,(\xi-\xi_0)^j,$$

where

$$b_0 = \frac{1}{a_0}, \quad \textit{and} \quad b_j = -\frac{1}{a_0}\sum_{k=1}^{j} a_k b_{j-k} \;\; \textit{for} \;\; j \geq 1.$$

PROOF [1] If there is such a series $\sum_{j=0}^{\infty} b_j\,(\xi-\xi_0)^j$, then

$$
\begin{aligned}
1 &= \left(\sum_{j=0}^{\infty} a_j\,(\xi-\xi_0)^j\right)\left(\sum_{j=0}^{\infty} b_j\,(\xi-\xi_0)^j\right) \\
&= \sum_{j=0}^{\infty}\left(\sum_{k=0}^{j} a_k\,(\xi-\xi_0)^k\, b_{j-k}\,(\xi-\xi_0)^{j-k}\right) \\
&= \sum_{j=0}^{\infty}\left(\sum_{k=0}^{j} a_k\, b_{j-k}\right)(\xi-\xi_0)^j.
\end{aligned}
$$

This implies that $a_0 b_0 = 1$, and $\sum_{k=0}^{j} a_k\, b_{j-k} = 0$ for $j \geq 1$, i.e., that $b_0 = \frac{1}{a_0}$ and $b_j = -\frac{1}{a_0}\sum_{k=0}^{j} a_k b_{j-k}$ for $j \geq 1$. We must now prove that $\sum_{j=0}^{\infty} b_j\,(\xi - \xi_0)^j$ has a positive radius of convergence. Since $\sum_{j=0}^{\infty} a_j\,(\xi-\xi_0)^j$ converges

[1] We follow [Ful78] in the proof.

for $|\xi - \xi_0| < R$, choose $0 < r < R$ such that $\sum_{j=0}^{\infty} a_j r^j$ converges. This implies there exists $M \geq 1$ such that $|a_j r^j| \leq M |a_0|$ for all $j \geq 0$, i.e.,

$$|a_j| \leq \frac{M|a_0|}{r^j} \quad \text{for all } j \geq 0.$$

Now,

$$\begin{aligned}
|b_0| &= \frac{1}{|a_0|} \leq \frac{M}{|a_0|}, \\
|b_1| &= \frac{1}{|a_0|} |a_1 b_0| \leq \frac{1}{|a_0|} \left(\frac{M|a_0|}{r}\right) \frac{1}{|a_0|} = \frac{1}{|a_0|} \left(\frac{M}{r}\right), \\
|b_2| &= \frac{1}{|a_0|} |a_1 b_1 + a_2 b_0| \leq \frac{1}{|a_0|} \left(\frac{M|a_0|}{r} \frac{M}{|a_0| r} + \frac{M|a_0|}{r^2} \frac{M}{|a_0|}\right) = \frac{2}{|a_0|} \left(\frac{M}{r}\right)^2,
\end{aligned}$$

and by induction it can be shown that

$$|b_j| \leq \frac{2^{j-1}}{|a_0|} \left(\frac{M}{r}\right)^j \quad \text{for } j \geq 1. \tag{9.5}$$

This implies

$$\limsup \sqrt[j]{|b_j|} \leq \limsup \frac{1}{\sqrt[j]{2\,|a_0|}} \left(\frac{2M}{r}\right) = \frac{2M}{r},$$

and therefore,

$$R = \frac{1}{\limsup \sqrt[j]{|b_j|}} \geq \frac{r}{2M} > 0.$$

♦

► **9.12** *Use induction to prove inequality* (9.5) *for* $j \geq 1$.

► **9.13** *Show that the coefficients* b_j *described in Theorem 1.13 may be determined by the "long division" algorithm as illustrated in Figure 9.2. That is, divide 1 by the power series representation for* $f(z)$*, i.e., by* $a_0 + a_1 z + a_2 z^2 + \cdots$.

$$\begin{array}{r|l}
 & a_0^{-1} - a_0^{-2} a_1 z + \ldots \\
\hline
a_0 + a_1 z + a_2 z^2 + \ldots & 1 + a_0^{-1} a_1 z a_0^{-1} a_2 z^2 \\
 & 1 + a_0^{-1} a_1 z + a_0^{-1} a_2 z^2 + \ldots \\
\hline
 & -a_0^{-1} a_1 z - a_0^{-2} a_1^2 z^2 + \ldots \\
 & -a_0^{-1} a_1 z - a_0^{-2} a_1^2 z^2 + \ldots \\
\hline
 & \ddots
\end{array}$$

Figure 9.2 *The "long division" algorithm one can use to find the reciprocal of a power series.*

Example 1.14 *Suppose* $f : \mathbb{C} \to \mathbb{C}$ *is given by* $f(z) = 1 + z^2$*. Then* f *is its own power series representation on all of* $\mathbb{C}$*. Let* $g : \mathbb{C} \setminus \{\pm i\} \to \mathbb{C}$ *be given by* $g(z) = \frac{1}{f(z)} = \frac{1}{1+z^2}$*. Note that* $f(z) \neq 0$ *on* $N_1(0)$*. According to the above theorem,* $g(z)$ *has a power series representation on* $N_r(0)$ *for some* $r > 0$ *given by*

$$g(z) = \frac{1}{f(z)} = \sum_{j=0}^{\infty} b_j\, z^j,$$

where

$$b_0 = 1,$$
$$b_1 = -\frac{1}{a_0}(a_1 b_0) = 0,$$
$$b_2 = -\frac{1}{a_0}(a_1 b_1 + a_2 b_0) = -1,$$
$$b_3 = -\frac{1}{a_0}(a_1 b_2 + a_2 b_1 + a_3 b_0) = 0,$$
$$b_4 = -\frac{1}{a_0}(a_1 b_3 + a_2 b_2 + a_3 b_1 + a_4 b_0) = 1,$$

etc.,

i.e.,

$$g(z) = \frac{1}{f(z)} = \frac{1}{1+z^2} = 1 - z^2 + z^4 - z^6 + \cdots, \quad \text{on } N_r(0).$$

The reader should verify that this power series representation for g has radius of convergence $R = 1$. This is the same conclusion as that based on convergent geometric series, namely,

$$\frac{1}{1+z^2} = \sum_{j=0}^{\infty}(-z^2)^j = \sum_{j=0}^{\infty}(-1)^j\, z^{2j} \quad \text{on } N_1(0).$$

◀

▶ **9.14** *Verify that the derived power series representation for g in the above example has radius of convergence $R = 1$ as claimed.*

▶ **9.15** *Find the first several terms of the power series representation for $f(z) = \sec z$ centered at $z_0 = 0$ using the method described in the previous exercise.*

Corollary 1.15 *Let $f(\xi) = \sum_{j=0}^{\infty} a_j\,(\xi - \xi_0)^j$ and $g(\xi) = \sum_{j=0}^{\infty} c_j\,(\xi - \xi_0)^j$ be power series each having radius of convergence R_f and R_g, respectively. If $a_0 \neq 0$, then for some positive $r < \min(R_f, R_g)$, $\frac{g(\xi)}{f(\xi)}$ has power series representation*

$$\frac{g(\xi)}{f(\xi)} = \sum_{j=0}^{\infty} \gamma_j\,(\xi - \xi_0)^j \quad \text{on } N_r(\xi_0),$$

with

$$\gamma_j = \sum_{k=0}^{j} c_k\, b_{j-k},$$

where

$$b_0 = \frac{1}{a_0}, \quad \text{and} \quad b_j = -\frac{1}{a_0}\sum_{k=1}^{j} a_k b_{j-k} \quad \text{for } j \geq 1$$

as described in Theorem 1.13.

▶ **9.16** *Prove the above corollary and show via "long division" that the coefficients* γ_j *are as claimed in the above theorem.*

▶ **9.17** *Find the first four terms of the Taylor series representation centered at* $z_0 = 0$ *for the complex function* $h(z) = \frac{z+1}{z-1}$ *on the neighborhood* $N_1(0)$.

Term-by-Term Differentiation of Power Series

We have already seen that any power series $\sum_{j=0}^{\infty} a_j\,(\xi - \xi_0)^j$ represents a continuous function on its neighborhood of convergence. We will now show, in fact, that any power series represents a *differentiable* function on its neighborhood of convergence.

Theorem 1.16 *The power series* $\sum_{j=0}^{\infty} a_j\,(\xi - \xi_0)^j$ *with radius of convergence* R *is infinitely differentiable at every* $\xi \in N_R(\xi_0)$, *and its nth derivative for* $n \geq 1$ *is the power series given by*

$$\frac{d^n}{d\xi^n}\sum_{j=0}^{\infty} a_j\,(\xi - \xi_0)^j = \sum_{j=n}^{\infty} j\,(j-1)\cdots(j-(n-1))\,a_j\,(\xi - \xi_0)^{j-n},$$

i.e., the power series can be differentiated term-by-term. The differentiated power series has the same neighborhood of convergence as the original power series.

PROOF Suppose $\sum_{j=0}^{\infty} a_j(z - z_0)^j$ has radius of convergence $R > 0$ and fix $z_1 \in N_R(z_0)$. We will show that

$$\frac{d}{dz}\bigg|_{z_1}\left(\sum_{j=0}^{\infty} a_j(z - z_0)^j\right) = \sum_{j=1}^{\infty} j\,a_j(z_1 - z_0)^{j-1}.$$

We begin by choosing r such that $|z_1 - z_0| < r < R$, and recalling that $\sum_{j=0}^{\infty} a_j(z - z_0)^j$ converges uniformly on $N_r(z_0)$. This means that $S_n(z) \equiv \sum_{j=0}^{n} a_j(z - z_0)^j$ converges uniformly to $S(z) \equiv \sum_{j=0}^{\infty} a_j(z - z_0)^j$ on $N_r(z_0)$. By Theorem 5.3 of section 5 in Chapter 8, we know that $S(z)$ is differentiable, and for all $z \in N_r(z_0)$,

$$\begin{aligned}\frac{d}{dz}\left(\sum_{j=0}^{\infty} a_j(z - z_0)^j\right) &= \frac{d}{dz}\left(\lim_{n\to\infty} S_n(z)\right)\\ &= \lim_{n\to\infty}\left(\frac{d}{dz}S_n(z)\right)\\ &= \lim_{n\to\infty}\sum_{j=1}^{n} j\,a_j(z - z_0)^{j-1}\\ &= \sum_{j=1}^{\infty} j\,a_j(z - z_0)^{j-1}.\end{aligned}$$

Finally note that if we are to consider the real power series $\sum_{j=0}^{\infty} a_j(x-x_0)^j$, we may "complexify" the series and use the result proved above to obtain

$$\frac{d}{dz}\left(\sum_{j=0}^{\infty} a_j(z-x_0)^j\right) = \sum_{j=1}^{\infty} j\, a_j(z-x_0)^{j-1}. \tag{9.6}$$

Now replace z in (9.6) with x to obtain

$$\frac{d}{dx}\left(\sum_{j=0}^{\infty} a_j(x-x_0)^j\right) = \sum_{j=1}^{\infty} j\, a_j(x-x_0)^{j-1}.$$

Extending this result to the case of higher-order derivatives, and the proof that the differentiated series has the same radius of convergence are left to the reader. ♦

► **9.18** *Complete the proof of the above theorem by extending the result to higher-order derivatives. Also, show that the original power series and its claimed derivative have the same radius of convergence by applying Theorem 1.9 on page 461.*

Theorem 1.16 is a very practical result. It says that for a function given as a power series, the function is differentiable inside its associated power series representation's neighborhood of convergence, and the derivative is found by differentiating the power series "term-by-term."

Example 1.17 *We have seen that the power series representation of the function $g(z) = \frac{1}{1-z}$ is given by*

$$\frac{1}{1-z} = \sum_{j=0}^{\infty} z^j \quad \text{on } N_1(0).$$

By Theorem 1.16, the derivative $\frac{dg}{dz} = \frac{1}{(1-z)^2}$ has power series representation given by

$$\frac{1}{(1-z)^2} = \frac{d}{dz}\sum_{j=0}^{\infty} z^j = \sum_{j=0}^{\infty} \frac{d}{dz} z^j = \sum_{j=1}^{\infty} j z^{j-1} \quad \text{on } N_1(0).$$

By differentiating again, we obtain a power series representation for $\frac{d^2g}{dz^2} = \frac{2}{(1-z)^3}$, namely,

$$\frac{2}{(1-z)^3} = \sum_{j=2}^{\infty} j\,(j-1)\, z^{j-2} \quad \text{on } N_1(0).$$

Clearly, there is no limit to the number of times one can differentiate a power series in this way. Note too that the above obtains a power series representation for $\frac{1}{(1-z)^3}$, namely,

$$\frac{1}{(1-z)^3} = \frac{1}{2}\sum_{j=2}^{\infty} j\,(j-1)\, z^{j-2} \quad \text{on } N_1(0).$$

◄

Term-by-Term Integration of Power Series

In our consideration of the integration of power series, we will handle the real and complex cases separately. Focusing on the real case first, we will show that under the right conditions, a power series can be integrated term-by-term.

Recall from Chapter 7 that if $\{f_n\}$ is a sequence of integrable functions $f_n : [a,b] \to \mathbb{R}$ for $n \ge 1$, such that $\lim_{n\to\infty} f_n(x) = f(x)$ uniformly on $[a,b]$ for some $f : [a,b] \to \mathbb{R}$, then f is integrable on $[a,b]$ and

$$\lim_{n\to\infty} \int_a^b f_n(x)\,dx = \int_a^b \lim_{n\to\infty} f_n(x)\,dx = \int_a^b f(x)\,dx. \tag{9.7}$$

Now consider a real power series $\sum_{j=0}^{\infty} a_j\,(x-x_0)^j$ with neighborhood of convergence $N_R(x_0)$. Then by Theorem 1.2 on page 456, we know that this series converges uniformly on every closed interval $[a,b] \subset N_R(x_0)$. Letting $f_n : [a,b] \to \mathbb{R}$ be defined by $f_n(x) \equiv \sum_{j=0}^{n} a_j\,(x-x_0)^j$ for $n \ge 0$, we have a uniformly convergent sequence of functions $\{f_n\}$ on $[a,b]$ to $f : [a,b] \to \mathbb{R}$ where $f(x) \equiv \sum_{j=0}^{\infty} a_j\,(x-x_0)^j$. Subbing into (9.7), we obtain

$$\lim_{n\to\infty} \int_a^b \left(\sum_{j=0}^{n} a_j\,(x-x_0)^j \right) dx = \int_a^b \left(\sum_{j=0}^{\infty} a_j\,(x-x_0)^j \right) dx. \tag{9.8}$$

Since the left-hand side of the above equality involves an integral of a finite sum, we may interchange summation and integration and then take the limit to obtain

$$\sum_{j=0}^{\infty} \left(\int_a^b a_j\,(x-x_0)^j\,dx \right).$$

Overall then, (9.8) and the above yield

$$\int_a^b \left(\sum_{j=0}^{\infty} a_j\,(x-x_0)^j \right) dx = \sum_{j=0}^{\infty} \left(\int_a^b a_j\,(x-x_0)^j\,dx \right). \tag{9.9}$$

Equality (9.9) says that the series can be integrated term-by-term. We have proved the following theorem.

Theorem 1.18 *The real power series $\sum_{j=0}^{\infty} a_j\,(x-x_0)^j$ with neighborhood of convergence $N_R(x_0)$ is integrable over any closed interval $[a,b] \subset N_R(x_0)$, and the integral is given by*

$$\int_a^b \left(\sum_{j=0}^{\infty} a_j\,(x-x_0)^j \right) dx = \sum_{j=0}^{\infty} \left(\int_a^b a_j\,(x-x_0)^j\,dx \right),$$

i.e., the power series can be integrated term-by-term.

Example 1.19 *Consider the function $f : \mathbb{R} \to \mathbb{R}$ given by $f(x) = \frac{1}{1+x^2}$. It has power series representation given by*

$$\frac{1}{1+x^2} = \sum_{j=0}^{\infty}(-x^2)^j = \sum_{j=0}^{\infty}(-1)^j\, x^{2j},$$

with neighborhood of convergence $N_1(0)$. For any $\epsilon > 0$ we have $[0, 1-\epsilon] \subset N_1(0)$, and so

$$\begin{aligned}\int_0^{1-\epsilon} \frac{1}{1+x^2}\, dx &= \int_0^{1-\epsilon}\left(\sum_{j=0}^{\infty}(-1)^j\, x^{2j}\right) dx \\ &= \sum_{j=0}^{\infty}(-1)^j\left(\int_0^{1-\epsilon} x^{2j}\, dx\right) \\ &= \sum_{j=0}^{\infty}(-1)^j\,\frac{(1-\epsilon)^{2j+1}}{2j+1}.\end{aligned}$$

But we also have

$$\int_0^{1-\epsilon} \frac{1}{1+x^2}\, dx = \tan^{-1} x\,\Big|_0^{(1-\epsilon)} = \tan^{-1}(1-\epsilon),$$

and so overall we obtain

$$\tan^{-1}(1-\epsilon) = \sum_{j=0}^{\infty}(-1)^j\,\frac{(1-\epsilon)^{2j+1}}{2j+1}. \tag{9.10}$$

Applying Abel's theorem on page 463, we may take the limit as $\epsilon \to 0^+$ in (9.10). This yields $\tan^{-1}(1) = \frac{\pi}{4}$ on the left-hand side and $\sum_{j=0}^{\infty}\frac{(-1)^j}{2j+1}$ on the right-hand side, obtaining what is known as the Leibniz series,

$$\frac{\pi}{4} = \sum_{j=0}^{\infty}\frac{(-1)^j}{2j+1}.$$

Multiplying the above equality by 4 obtains what is referred to as the Leibniz formula for π,

$$\pi = 4\left(1 - \tfrac{1}{3} + \tfrac{1}{5} - \tfrac{1}{7} + \cdots\right).$$

◄

Example 1.20 *In this example we will find a power series centered at $x_0 = 1$ and convergent on $N_1(1) = (0, 2)$ for the real natural logarithm $\ln x$. To do so, we will exploit term-by-term integration of power series. Note that since*

$$\frac{1}{x} = \frac{1}{1-(1-x)} = \sum_{j=0}^{\infty}(1-x)^j$$

converges on $N_1(1)$, it follows from Theorem 1.18 that for $0 < x < 2$,

$$\begin{aligned} \ln x &= \int_1^x \frac{1}{y}\,dy \\ &= \int_1^x \sum_{j=0}^{\infty}(1-y)^j\,dy \\ &= \sum_{j=0}^{\infty}\left(\int_1^x (1-y)^j\,dy\right) \\ &= \sum_{j=0}^{\infty}\frac{(-1)^j}{j+1}(x-1)^{j+1} \quad \text{for } 0 < x < 2. \end{aligned} \tag{9.11}$$

◀

▶ **9.19** *From the results of the previous example, justify letting $x \to 2^-$ to show that* $\ln 2 = \sum_{j=0}^{\infty}\frac{(-1)^j}{j+1}$.

We now develop a term-by-term integration theorem for complex integrals of complex power series. We will need the following result, analogous to Theorem 3.10 in Chapter 7.

Theorem 1.21 *For each $n \in \mathbb{N}$, let $f_n : D \to \mathbb{C}$ be a continuous function on D and suppose there exists a function $f : D \to \mathbb{C}$ such that $\lim_{n\to\infty} f_n(z) = f(z)$ uniformly on D. Then for any contour $C \subset D$,*

$$\lim_{n\to\infty}\int_C f_n(z)\,dz = \int_C \lim_{n\to\infty} f_n(z)\,dz = \int_C f(z)\,dz.$$

PROOF Since each f_n for $n \geq 1$ is continuous on D, it follows from Theorem 3.8 in Chapter 5 that f is continuous on D. Let L_C be the length of the contour C, and let $\epsilon > 0$. Then there exists $N \in \mathbb{N}$ such that for all $z \in D$,

$$n > N \quad\Rightarrow\quad |f_n(z) - f(z)| < \frac{\epsilon}{2L_C}.$$

Note that

$$n > N \quad\Rightarrow\quad \left|\int_C f_n(z)\,dz - \int_C f(z)\,dz\right| = \left|\int_C \big(f_n(z) - f(z)\big)\,dz\right| \leq \frac{\epsilon}{2L_C}L_C < \epsilon.$$

Therefore, $\lim_{n\to\infty}\int_C f_n(z)\,dz = \int_C f(z)\,dz$. ♦

We now apply this theorem to prove the following term-by-term integration theorem for complex power series.

Theorem 1.22 *Let $\sum_{j=0}^{\infty} a_j\,(z-z_0)^j$ be a complex power series with neighborhood of convergence $N_R(z_0)$. Then for any contour $C \subset N_R(z_0)$,*

$$\int_C \left(\sum_{j=0}^{\infty} a_j\,(z-z_0)^j\right) dz = \sum_{j=0}^{\infty}\left(\int_C a_j\,(z-z_0)^j\,dz\right),$$

i.e., the power series can be integrated term-by-term.

PROOF Let $\sum_{j=0}^{\infty} a_j\,(z-z_0)^j$ be a power series with neighborhood of convergence $N_R(z_0)$. This series is a continuous function on $N_R(z_0)$, and so it is continuous on C. Also, the series converges uniformly on $\overline{N_r(z_0)}$ for all $r < R$, and there exists $r > 0$ such that $r < R$ and $C \subset \overline{N_r(z_0)}$. Since the series $\sum_{j=0}^{n} a_j\,(z-z_0)^j$ converges uniformly to $\sum_{j=0}^{\infty} a_j\,(z-z_0)^j$ on $\overline{N_r(z_0)}$, Theorem 1.21 implies

$$\int_C \left(\sum_{j=0}^{\infty} a_j\,(z-z_0)^j\right) dz = \sum_{j=0}^{\infty}\left(\int_C a_j\,(z-z_0)^j\,dz\right),$$

and the theorem is proved. ♦

Example 1.23 *Let C be any contour in $N_1(0)$ and consider the integral $\int_C \frac{1}{1-z}\,dz$. Recall that $\frac{1}{1-z} = \sum_{j=0}^{\infty} z^j$ on $N_1(0)$, so that Theorem 1.22 obtains*

$$\int_C \frac{dz}{1-z} = \int_C \left(\sum_{j=0}^{\infty} z^j\right) dz = \sum_{j=0}^{\infty}\left(\int_C z^j\,dz\right).$$

If C is parametrized by $z(t)$ for $a \le t \le b$, this yields

$$\int_C \frac{dz}{1-z} = \sum_{j=0}^{\infty}\left(\int_C z^j\,dz\right) = \sum_{j=0}^{\infty} \frac{\big(z(b)\big)^{j+1} - \big(z(a)\big)^{j+1}}{j+1}. \quad \textit{(Why?)}$$

◀

2 Taylor Series

We now develop the theory of Taylor series expansions for real and complex functions. A *Taylor series expansion* (also referred to as a *Taylor series representation*) of a function f centered at a point ξ_0 is a special kind of power series uniquely determined by f and ξ_0. To motivate the idea, suppose a function f is infinitely differentiable at ξ_0 and also suppose that f has a power series representation on $N_R(\xi_0)$ given by

$$f(\xi) = \sum_{j=0}^{\infty} a_j(\xi-\xi_0)^j.$$

Evaluating the series at ξ_0, we see that $a_0 = f(\xi_0)$. If we now differentiate the series we obtain $f'(\xi) = \sum_{j=1}^{\infty} j a_j (\xi - \xi_0)^{j-1}$, and evaluating the differentiated series at ξ_0 then obtains $a_1 = f'(\xi_0)$. Differentiating the series again and evaluating at ξ_0 once more obtains $a_2 = \frac{f''(\xi_0)}{2!}$, and more generally we have

$$a_j = \frac{f^{(j)}(\xi_0)}{j!} \quad \text{for } j \geq 0. \tag{9.12}$$

From this we see that if f is infinitely differentiable at ξ_0 and if f has a power series representation there, then that power series representation is unique since its coefficients must be those given by (9.12). For $\xi \in N_R(\xi_0)$, the neighborhood of convergence of the series, we will investigate the significance of the series and its relation to the defining function f. We now formally define this unique power series associated with f at ξ_0 as *the Taylor series of f centered at* ξ_0.

Definition 2.1 Suppose $f : \mathbb{D} \to \mathbb{F}$ is infinitely differentiable at $\xi_0 \in \mathbb{D} \subset \mathbb{F}$. Then the **Taylor series of f centered at** ξ_0 is defined to be the power series given by

$$\sum_{j=0}^{\infty} \frac{f^{(j)}(\xi_0)}{j!} (\xi - \xi_0)^j. \tag{9.13}$$

The above definition says nothing about where the Taylor series of f centered at ξ_0 converges. Of course, it must converge at $\xi = \xi_0$ (to $f(\xi_0)$ in fact), but where else might it converge? If it converges at some other $\xi \in \mathbb{D}$, is it necessarily true that it converges to the value $f(\xi)$ there? Taylor's theorem answers these questions. Since there are significant differences in the real and complex cases, we will state Taylor's theorem separately for each.

In the real case, we can appeal to Taylor's theorem with remainder from Chapter 6. Recall from this result that if $f : N_r(x_0) \to \mathbb{R}$ has $n+1$ derivatives on $N_r(x_0)$, then

$$f(x) = \sum_{j=0}^{n} \frac{f^{(j)}(x_0)}{j!} (x - x_0)^j + R_n(x) \quad \text{on } N_r(x_0), \tag{9.14}$$

where $R_n(x) \equiv \frac{f^{(n+1)}(c)}{(n+1)!} (x - x_0)^{n+1}$ for some c between x_0 and x. Suppose in addition we know that f is C^∞ on $N_r(x_0)$ (not just at x_0) and that $\lim_{n\to\infty} R_n(x) = 0$ on $N_r(x_0)$. Then, taking the limit as $n \to \infty$ in (9.14) obtains

$$f(x) = \sum_{j=0}^{\infty} \frac{f^{(j)}(x_0)}{j!} (x - x_0)^j \quad \text{on } N_r(x_0),$$

i.e., the Taylor series of f centered at x_0 converges at each $x \in N_r(x_0)$, and it converges to the value $f(x)$. This is the basis for referring to a Taylor series expansion of a function f centered at a point ξ_0 as a Taylor series *representation* of f centered at that point. We have established the following result.

Theorem 2.2 (Taylor's Theorem for Real Functions)
Let $f : \mathcal{D}^1 \to \mathbb{R}$ be C^∞ on some neighborhood $N_r(x_0) \subset \mathcal{D}^1$, and suppose $\lim_{n\to\infty} R_n(x) = 0$ on $N_r(x_0)$. Then f has unique Taylor series representation on $N_r(x_0)$ given by

$$f(x) = \sum_{j=0}^{\infty} \frac{f^{(j)}(x_0)}{j!}\,(x - x_0)^j.$$

▶ **9.20** *Prove the uniqueness in the above theorem.*

While the real case of Taylor's theorem above followed relatively easily from Taylor's theorem with remainder from Chapter 6, the version of Taylor's theorem for complex functions requires a bit more work. In fact, its proof will require key results from our development of complex integration.

Theorem 2.3 (Taylor's Theorem for Complex Functions)
Let $f : D \to \mathbb{C}$ be differentiable on some neighborhood $N_r(z_0) \subset D$. Then f has unique Taylor series representation on $N_r(z_0)$ given by

$$f(z) = \sum_{j=0}^{\infty} \frac{f^{(j)}(z_0)}{j!}\,(z - z_0)^j.$$

Before proving the complex version of Taylor's theorem, we point out how it differs from the real one. Note that in the complex case we require only that f be differentiable on $N_r(z_0)$, since this will imply f is infinitely differentiable on $N_r(z_0)$. Also note in the complex case that we make no reference to a Taylor remainder. It will turn out that in the complex case the remainder will *always* go to zero as the degree of the associated Taylor polynomial goes to ∞, and so we need not include it in the conditions of the theorem. These are two significant ways in which the complex case differs from the real, and they point to just how restrictive it is for a complex function to be differentiable.

PROOF Without loss of generality, we may assume that $z_0 = 0$ for the purposes of the proof. Fix $z \in N_r(0)$. Let $C_s \equiv C_s(0)$ be a circle of radius $s < r$ centered at 0 such that $z \in \text{Int}(C_s)$. Then Cauchy's integral formula yields

$$f(z) = \frac{1}{2\pi i} \oint_{C_s} \frac{f(\zeta)}{\zeta - z}\, d\zeta. \tag{9.15}$$

Now,

$$\begin{aligned}\frac{1}{\zeta - z} &= \frac{1}{\zeta}\left(\frac{1}{1-\frac{z}{\zeta}}\right)\\ &= \frac{1}{\zeta}\left(1+\left(\frac{z}{\zeta}\right)+\left(\frac{z}{\zeta}\right)^2+\dots\right) \quad \text{by geometric series,}\\ &= \frac{1}{\zeta}\left(1+\left(\frac{z}{\zeta}\right)+\left(\frac{z}{\zeta}\right)^2+\dots+\left(\frac{z}{\zeta}\right)^N+\left(\frac{z}{\zeta}\right)^{N+1}\frac{1}{1-\frac{z}{\zeta}}\right). \quad \text{(Why?)}\end{aligned}$$

Multiplying this result by $f(\zeta)$ obtains

$$\frac{f(\zeta)}{\zeta - z} = \sum_{j=0}^{N} z^j \frac{f(\zeta)}{\zeta^{j+1}} + \left(\frac{z}{\zeta}\right)^{N+1}\frac{f(\zeta)}{\zeta - z}.$$

Subbing this back into expression (9.15) yields

$$\begin{aligned}f(z) &= \frac{1}{2\pi i}\oint_{C_s}\sum_{j=0}^{N} z^j \frac{f(\zeta)}{\zeta^{j+1}}d\zeta + \frac{1}{2\pi i}\oint_{C_s}\left(\frac{z}{\zeta}\right)^{N+1}\frac{f(\zeta)}{\zeta - z}\,d\zeta\\ &= \sum_{j=0}^{N} z^j \frac{1}{2\pi i}\oint_{C_s}\frac{f(\zeta)}{\zeta^{j+1}}\,d\zeta + \frac{1}{2\pi i}\oint_{C_s}\left(\frac{z}{\zeta}\right)^{N+1}\frac{f(\zeta)}{\zeta - z}\,d\zeta\\ &= \sum_{j=0}^{N} z^j \frac{f^{(j)}(0)}{j!} + \frac{1}{2\pi i}\oint_{C_s}\left(\frac{z}{\zeta}\right)^{N+1}\frac{f(\zeta)}{\zeta - z}\,d\zeta. \end{aligned} \tag{9.16}$$

Note here that we have a version of Taylor's theorem with remainder for complex functions, in that (9.16) is equivalent to

$$f(z) = \sum_{j=0}^{N} \frac{f^{(j)}(0)}{j!} z^j + \widetilde{R}_N(z), \tag{9.17}$$

where

$$\widetilde{R}_N(z) \equiv \frac{1}{2\pi i}\oint_{C_s}\left(\frac{z}{\zeta}\right)^{N+1}\frac{f(\zeta)}{\zeta - z}\,d\zeta. \tag{9.18}$$

We will now show that it is always the case that $\lim_{N\to\infty} \widetilde{R}_N(z) = 0$, yielding the result we seek. To see this, note that for ζ on C_s there exists $M > 0$ such that

$|f(\zeta)| \le M$ (Why?), and so

$$\left|\widetilde{R}_N(z)\right| = \left|\frac{1}{2\pi i}\oint_{C_s}\left(\frac{z}{\zeta}\right)^{N+1}\frac{f(\zeta)}{\zeta - z}\,d\zeta\right|$$
$$\le \frac{1}{2\pi}\left(\frac{|z|}{s}\right)^{N+1}\frac{M}{s-|z|}\,2\pi s$$
$$= \left(\frac{|z|}{s}\right)^{N+1}\frac{Ms}{s-|z|},$$

which goes to zero as $N \to \infty$ since $|z| < s$. Taking the limit as $N \to \infty$ in (9.17) then yields

$$f(z) = \sum_{j=0}^{\infty}\frac{f^{(j)}(0)}{j!}\,z^j.$$

Since $z \in N_r(0)$ was arbitrarily chosen, the result holds on all of $N_r(0)$. The uniqueness follows from the argument that began this section. ♦

Note that in Theorems 2.2 and 2.3 the neighborhood $N_r(\xi_0)$ on which each series is shown to converge is not necessarily the whole neighborhood of convergence $N_R(\xi_0)$ for the series where R is the series' radius of convergence. In general, $N_r(\xi_0) \subset N_R(\xi_0)$. Of course, the series actually converges on all of $N_R(\xi_0)$, and so it is customary to take r as large as one can so that $N_r(\xi_0) \subset \mathbb{D}$ and the conditions of the theorem are satisfied.

Example 2.4 *Recall in Chapter 6 we had considered the real function $f : \mathbb{R} \to \mathbb{R}$ given by $f(x) = \sin x$ with $x_0 = 0$. We saw there that*

$$\sin x = x - \frac{x^3}{3!} + \frac{x^5}{5!} - \cdots + \frac{(-1)^n x^{2n+1}}{(2n+1)!} + R_{2n+1}(x),$$

where $R_{2n+1}(x) = (\pm \sin c)\,\frac{x^{2n+2}}{(2n+2)!}$ for some real number c between 0 and x. Since $f(x) = \sin x$ is C^∞ on $\mathbb{R}$ and $\lim_{n\to\infty} R_{2n+1}(x) = 0$ for all $x \in \mathbb{R}$, it follows from Taylor's theorem for real functions that

$$\sin x = \sum_{j=0}^{\infty}(-1)^j\,\frac{x^{2j+1}}{(2j+1)!} \quad \text{on } \mathbb{R}.$$

Therefore, $\sin x$ is equal to its Taylor series for every $x \in \mathbb{R}$. ◀

► **9.21** *Show that the complex sine function has a Taylor series representation centered at $z_0 = 0$ of the same form as the real sine function.*

Example 2.5 *Consider the function $f : \mathbb{C} \to \mathbb{C}$ given by $f(z) = \cos z$, differentiable on $\mathbb{C}$. We will find the Taylor series centered at $z_0 = 0$. Note that $f^{(2j+1)}(0) = 0$ for all $j \ge 0$, and so the Taylor series of the even function $\cos z$ will consist of only even powers of z. It is not hard to see that*

$f''(z) = -\cos z$, $f^{(4)}(z) = \cos z$, and in general $f^{(2j)}(z) = (-1)^j \cos z$ for $j \geq 0$. Therefore, $f^{(2j)}(0) = (-1)^j$ for $j \geq 0$, which yields a Taylor series centered at $z_0 = 0$ for $\cos z$ of

$$\cos z = 1 - \frac{z^2}{2!} + \frac{z^4}{4!} - \cdots = \sum_{j=0}^{\infty} \frac{(-1)^j}{(2j)!} z^{2j} \quad \text{on } \mathbb{C}.$$

Note that this Taylor series has the same form as that of the real cosine function $f(x) = \cos x$. ◀

▶ **9.22** *Find the Taylor series for $f(z) = \cos z$ centered at $z_0 = \frac{\pi}{2}$.*

▶ **9.23** *Show that the real cosine function has a Taylor series centered at $x_0 = 0$ of the same form as the complex cosine function.*

Example 2.6 *Consider the complex exponential function $f(z) = e^z$, differentiable on $\mathbb{C}$. Since $f^{(j)}(z) = e^z$ for each $j \geq 0$ (as derived from the definition $e^z \equiv e^x e^{iy}$), we have that $f^{(j)}(0) = 1$ for all $j \geq 0$ and so the Taylor series of $f(z) = e^z$ centered at $z_0 = 0$ is*

$$e^z = \sum_{j=0}^{\infty} \frac{1}{j!} z^j \quad \text{on } \mathbb{C}.$$

Note that this Taylor series has the same form as that of the real exponential function $f(x) = \exp(x)$. Also, compare this result to that of part e) in Example 1.10 on page 462. This result establishes that the complex exponential function has a power series representation on its entire domain of the same form as the real exponential function. ◀

The careful reader might have noted that the previous three examples and their associated exercises hint at an interesting and useful technique. Ordinarily, in investigating convergence of a real Taylor series for a real function $f(x)$, one must consider whether and where the remainder $R_n(x)$ goes to zero; that is, *unless the real function is a restriction of some differentiable complex function to the real line.* In such a case, one may determine a Taylor series representation for the "complexified" version of the function $f(z)$. One can then obtain the corresponding Taylor series representation for the original *real* function by setting $z = x$ in the complex Taylor series representation. This way, no remainders need be considered.

Example 2.7 *Suppose one wishes to find a Taylor series for the real natural logarithm centered at $x_0 = 1$. Instead, consider the principal branch of the complex logarithm (restricted to $\mathbb{C} \setminus (-\infty, 0]$) and find a Taylor series for it centered at $z_0 = 1$. Recall that $\mathrm{Log} z$ is differentiable on $\mathbb{C} \setminus (-\infty, 0]$, and so it is differentiable on a neighborhood $N_r(1)$ of $z_0 = 1$. Also, $\frac{d}{dz} \mathrm{Log} z = \frac{1}{z}$. From this it follows that*

$$\frac{d^j}{dz^j} \mathrm{Log} z = \frac{(-1)^{j-1}(j-1)!}{z^j} \quad \text{for } j \geq 1 \quad \text{on } N_r(1),$$

and so we obtain the Taylor series,

$$\text{Log}\,z = \sum_{j=0}^{\infty} \frac{(-1)^j}{j+1} (z-1)^{j+1} \quad \text{on } N_r(1).$$

We leave it to the reader to show that this series has neighborhood of convergence $N_1(1)$. It must therefore converge for $z = x \in N_1(1)$, that is, we may conclude that the real function $\ln : (0, \infty) \to \mathbb{R}$ *has Taylor series centered at $x_0 = 1$ given by*

$$\ln x = \sum_{j=0}^{\infty} \frac{(-1)^j}{j+1} (x-1)^{j+1},$$

convergent on $N_1(1) = (0, 2)$ in $\mathbb{R}$. Note that this agrees with our result from Example 1.20 on page 471. ◄

► **9.24** *Show that the complex Taylor series centered at $z_0 = 1$ for* Log$z$ *has neighborhood of convergence $N_1(1)$.*

In the following example we see yet another interesting way in which a complex Taylor series can shed some light on the corresponding real version.

Example 2.8 *Consider the real function $f : \mathbb{R} \to \mathbb{R}$ where $f(x) = \frac{1}{4+x^2}$ is defined on the whole real line. We might expect that the Taylor series of this function centered at any x_0 would converge on all of $\mathbb{R}$, but this is not true! In fact, for $x_0 = 0$ we obtain the Taylor series*

$$\frac{1}{4+x^2} = \sum_{j=0}^{\infty} \frac{(-1)^j}{4^{j+1}} x^{2j}$$

with neighborhood of convergence $N_2(0) = (-2, 2)$ in $\mathbb{R}$. (The reader should confirm this.) Why is that? To understand what happens here, consider the corresponding complex function $\frac{1}{4+z^2}$ studied in Example 1.7 on page 460. It's domain consists of $\mathbb{C} \setminus \{\pm 2i\}$, and its Taylor series expansion centered at $z_0 = 0$ is given by $\frac{1}{4+z^2} = \sum_{j=0}^{\infty} \frac{(-1)^j}{4^{j+1}} z^{2j}$ with neighborhood of convergence $N_2(0)$ in $\mathbb{C}$. The bounded neighborhood of convergence suddenly makes sense. The complex version of the function can't include the points $\pm 2i$ in the complex plane, although this limitation was not evident in the real case. The corresponding real Taylor series, a special case of the complex one with $z = x$, must conform to this limitation too! ◄

One can go from the real Taylor series to the corresponding complex one as well. For example, as we saw in Example 2.4 on page 477, the real sine function has a real Taylor series $\sin x = \sum_{j=0}^{\infty} \frac{(-1)^j}{(2j+1)!} x^{2j+1}$ that is convergent on all of $\mathbb{R}$. This allows us to conclude that the complex Taylor series given by $F(z) = \sum_{j=0}^{\infty} \frac{(-1)^j}{(2j+1)!} z^{2j+1}$ converges on all of $\mathbb{C}$ and equals $\sin x$ when $z = x$. In fact, it must equal $\sin z$ for all $z \in \mathbb{C}$. To see this, apply Taylor's theorem for complex functions to $\sin z$ centered at $z = 0$. Since $\sin z$ and all of its derivatives evaluated at the origin must match those of $\sin x$, the Taylor series representation of the complex sine function centered at the origin must have the same form as the Taylor series representation of the real sine function centered at the origin. Hence the complex version must converge for all real z. If it didn't converge at some $z_1 \in \mathbb{C}$, a contradiction would result.

► **9.25** *Derive the contradiction that would result from the above argument. Also, in the above discussion, does it matter whether the Taylor series representations for the real and complex sine functions are centered at the origin? What if they were both centered at some other real value x_0? What if the complex series is centered at some other $z_0 \in \mathbb{C}$?*

We give one more example of interest.

Example 2.9 *For $c \in \mathbb{C}$ fixed, consider the function $f : N_1(0) \to \mathbb{C}$ written as $f(z) = (1+z)^c$ and defined as*

$$f(z) = (1+z)^c \equiv \exp(c\,\mathrm{Log}(1+z)). \tag{9.19}$$

Then f is a differentiable function in the open unit disk. In fact, the chain rule applied to (9.19) obtains $f'(z) = c(1+z)^{c-1}$, and by induction it follows that $f^{(n)}(0) = c(c-1)\dots(c-(n-1))$. If we define the binomial coefficient $\binom{c}{k}$ by

$$\binom{c}{k} \equiv \frac{c(c-1)\dots(c-(k-1))}{k!},$$

then Taylor's theorem for complex functions yields

$$(1+z)^c = 1 + \binom{c}{1} z + \binom{c}{2} z^2 + \cdots = \sum_{j=0}^{\infty} \binom{c}{j} z^j \quad \text{if } |z| < 1. \tag{9.20}$$

◄

► **9.26** *Use induction to show that $f^{(n)}(0) = c(c-1)\dots(c-(n-1))$ for $n \geq 1$.*

► **9.27** *Show directly that the radius of convergence of the series in (9.20) is 1 except in one important case. What is that case?*

► **9.28** *Show that the series in (9.20) gives the ordinary binomial theorem if c is a positive integer.*

It is worth pointing out that we usually don't compute the Taylor series coefficients for a given function $f(\xi)$ by direct application of the formula $a_j = \frac{f^{(j)}(\xi_0)}{j!}$ given in expression (9.13) of Definition 2.1. Instead, we exploit the uniqueness of the Taylor series representation for a given function f centered at a particular point.

Example 2.10 *We have already seen that the function given by $f(z) = \frac{1}{1-z}$ has power series $\sum_{j=0}^{\infty} z^j$ centered at $z_0 = 0$ with neighborhood of convergence $N_1(0)$. Since the Taylor series representation of f is unique, this power series must be the Taylor series of f centered at $z_0 = 0$. Each coefficient $a_j = 1$*

must be $\frac{f^{(j)}(z_0)}{j!} = \frac{f^{(j)}(0)}{j!}$, *that is,* $f^{(j)}(0) = j!$. *We leave it to the reader to verify this.* ◀

▶ **9.29** *Recall that in our discussion of power series in the last section we saw that the function with the formula* $f(z) = \frac{1}{4+z^2}$ *had power series representation given by* $\sum_{j=0}^{\infty} \frac{(-1)^j}{4^{j+1}} z^{2j}$ *centered at* $z_0 = 0$ *and convergent on* $N_2(0)$. *Verify from this that* $f''(0) = -\frac{1}{8}$, *and that* $f^n(0) = 0$ *for all odd* n.

▶ **9.30** *Suppose* $f, g : N_r(\xi_0) \to \mathbb{F}$ *each have Taylor series representations on* $N_r(\xi_0)$. *Suppose also that* $f^{(j)}(\xi_0) = g^{(j)}(\xi_0)$ *for* $j \geq 0$. *Show that* $f(\xi) = g(\xi)$ *on* $N_r(\xi_0)$.

3 Analytic Functions

3.1 Definition and Basic Properties

We now define one of the most important classes of functions in analysis, especially complex analysis.

Definition 3.1 A function $f : \mathbb{D} \to \mathbb{F}$ is called **analytic at** $\xi_0 \in \mathbb{D}$ if and only if f has a Taylor series representation on some neighborhood $N_r(\xi_0) \subset \mathbb{D}$. If f is analytic at every point $\xi \in \mathbb{D}$, we say that f is **analytic on** $\mathbb{D}$.

Note that according to the definition, a function is analytic at a point ξ_0 in its domain only if there exists a *neighborhood* $N_r(\xi_0)$ centered at ξ_0 such that $f(\xi)$ equals its Taylor series expansion *throughout that neighborhood*, i.e.,

$$f(\xi) = \sum_{j=0}^{\infty} \frac{f^{(j)}(\xi_0)}{j!} (\xi - \xi_0)^j \quad \text{on} \quad N_r(\xi_0).$$

If a function equals its Taylor series representation only at the point ξ_0, this is not enough to conclude that f is analytic at ξ_0.

Example 3.2 *Since the complex function* $f : N_1(0) \to \mathbb{C}$ *given by* $f(z) = \frac{1}{1-z}$ *has the Taylor series representation*

$$f(z) = \sum_{j=0}^{\infty} z^j \quad \text{on} \quad N_1(0),$$

it is analytic at $z_0 = 0$. *Similarly, the real function* $f : N_1(0) \to \mathbb{R}$ *given by* $f(x) = \frac{1}{1-x}$ *is analytic at* $x_0 = 0$. ◀

Theorem 3.3 *Suppose* $f : \mathbb{D} \to \mathbb{F}$ *has a Taylor series representation on some neighborhood* $N_r(\xi_0) \subset \mathbb{D}$. *Then* f *is analytic at each point* $\xi \in N_r(\xi_0)$.

▶ **9.31** *Prove the above theorem. (Hint: Consider arbitrary $\xi_1 \in N_r(\xi_0)$. Choose $\rho = \frac{1}{2}(r - |\xi_1 - \xi_0|)$, and consider $N_\rho(\xi_1)$. Show that f has Taylor series representation on $N_\rho(\xi_1)$.)*

Example 3.4 *From the above result we see that the complex function $f : N_1(0) \to \mathbb{C}$ in the previous example given by $f(z) = \frac{1}{1-z}$ is analytic on $N_1(0)$. Likewise, the real function $f : N_1(0) \to \mathbb{R}$ given by $f(x) = \frac{1}{1-x}$ is analytic on $N_1(0) = (-1, 1)$.* ◀

Clearly if f is analytic at ξ_0, it is infinitely differentiable there. But infinite differentiability alone is not enough to ensure analyticity, at least not for real functions. In fact, the following example illustrates that a real function can be infinitely differentiable and even have a Taylor series representation at a point, and yet not be analytic there.

Example 3.5 *Consider the function $f : \mathbb{R} \to \mathbb{R}$ given by*

$$f(x) = \begin{cases} \exp(-1/x^2) & \text{for } x \neq 0 \\ 0 & \text{for } x = 0 \end{cases}.$$

We will show that this function has a Taylor series centered at $x_0 = 0$ that converges for all $x \in \mathbb{R}$. Yet, this Taylor series will only equal the function's value at $x_0 = 0$, and so the function is not analytic there despite being C^∞ there. We begin by finding the Taylor series centered at $x_0 = 0$ for f. Note that $f(0) = 0$. To find $f'(0)$, we use the limit definition to obtain

$$\begin{aligned} f'(0) &= \lim_{h\to 0} \frac{\exp(-1/h^2)}{h} \\ &= \lim_{y\to\pm\infty} \frac{y}{e^{y^2}} && \text{for } y = 1/h, \\ &= 0 && \text{by L'Hospital's rule.} \end{aligned}$$

We leave it to the reader to show that the same strategy yields

$$f^{(j)}(0) = 0 \quad \text{for all } j \geq 0,$$

and therefore the Taylor series for f centered at $x_0 = 0$ is given by

$$\sum_{j=0}^{\infty} \frac{f^{(j)}(0)}{j!} x^j \equiv 0.$$

Clearly this does not equal the function's values at $x \neq 0$. ◀

▶ **9.32** *Show that $f^{(j)}(0) = 0$ for all $j \geq 0$ for the function f of the previous example.*

It might be surprising to learn that such a function as the one in the previous example exists. In fact, in the complex case such functions *don't* exist. As we already know, if a complex function f is differentiable at a point it is infinitely differentiable there. In the next subsection we will discover even more, namely, a complex function f is differentiable on a neighborhood $N_r(z_0)$ if and only if it is analytic on that neighborhood. Differentiable complex functions are very nice functions.

► **9.33** *Suppose $f, g : \mathbb{D} \to \mathbb{F}$ are analytic at $\xi_0 \in \mathbb{D}$. Show the following:*
a) $f \pm g$ is analytic at ξ_0,
b) fg is analytic at ξ_0,
c) f/g is analytic at ξ_0 provided $g(\xi_0) \neq 0$.

3.2 Complex Analytic Functions

The following proposition establishes an important and useful fact, namely, the interchangeability of the terms *differentiable* and *analytic* for complex functions on open connected sets D.

Proposition 3.6 *Consider the complex function $f : D \to \mathbb{C}$. Then f is differentiable on D if and only if f is analytic on D.*

► **9.34** *Prove the above proposition.*

► **9.35** *Show that the function $f : \mathbb{C} \to \mathbb{C}$ given by $f(z) = |z|^2$ is not analytic anywhere in $\mathbb{C}$. Is it differentiable anywhere?*

As a consequence of establishing Proposition 3.6, many results stated earlier in terms of differentiable complex functions can now be restated in terms of analytic complex functions (although we will not do so). The following exercise provides such an opportunity to the reader.

► **9.36** *Show that if $f_n : D \to \mathbb{C}$ for $n \geq 1$ is a sequence of analytic complex functions on the open connected set $D \subset \mathbb{C}$ that converges uniformly to $f : D \to \mathbb{C}$ on D, then f is analytic on D.*

Example 3.7 *Consider the complexified version of the real function from Example 3.5, which we saw was not analytic at the origin, even though it was C^∞ there. The complexified version is $f : \mathbb{C} \to \mathbb{C}$ given by*

$$f(z) = \begin{cases} \exp(-1/z^2) & \text{for } z \neq 0 \\ 0 & \text{for } z = 0 \end{cases}.$$

Clearly f is differentiable for $z \neq 0$, and is therefore analytic on $\mathbb{C} \setminus \{0\}$. If f were also (complex) differentiable at $z = 0$ then f would be analytic there as well, and so f would have a Taylor series representation

$$\sum_{j=0}^{\infty} \frac{f^{(j)}(0)}{j!} z^j,$$

on $N_r(0)$ for some $r > 0$. But this in turn would imply that for $z = x$, the real *version of the function would have Taylor series representation*

$$\sum_{j=0}^{\infty} \frac{f^{(j)}(0)}{j!} x^j,$$

on $N_r(0) = (-r, r) \subset \mathbb{R}$, which would imply that the real version of the function was analytic at $x = 0$. This contradiction tells us that the complex version of the function in this example cannot be differentiable at $z = 0$. In fact, one can show that the complex version of this function isn't even continuous at $z = 0$ since $\lim_{z\to 0} f(z)$ does not exist. ◀

▶ **9.37** *Show that $\lim_{z\to 0} f(z)$ does not exist for the function f of the above example.*

Analytic functions play a particularly important role in complex function theory, so determining their key properties (as related to their Taylor series representation) is fundamental to any development of analysis. Loosely stated, a function that is analytic on a neighborhood behaves much like an infinite-degree polynomial there. As with polynomials, it is often useful to characterize the zeros of analytic functions.

Zeros of Complex Analytic Functions

Recall that the fundamental theorem of algebra says that every nonzero polynomial with complex coefficients has at least one complex root, i.e., for polynomial $p(z) = a_0 + a_1 z + \cdots + a_n z^n$, with $a_n \neq 0$ and $a_j \in \mathbb{C}$ for $0 \leq j \leq n$, there exists $z_0 \in \mathbb{C}$ such that $p(z_0) = 0$. From elementary algebra we may write

$$p(z) = (z - z_0)\, q_1(z),$$

where $q_1(z)$ is a polynomial of degree $n - 1$. If $q_1(z_0) = 0$, then we can write

$$p(z) = (z - z_0)^2\, q_2(z),$$

and so on, until finally we can write

$$p(z) = (z - z_0)^m\, q(z),$$

where $q(z)$ is a polynomial such that $q(z_0) \neq 0$. In this case we say that z_0 is a *root* or *zero of multiplicity* $m \geq 1$. We now extend these ideas to more general complex functions.

Definition 3.8 Let $f : D \to \mathbb{C}$ be differentiable on D. Then,

1. The point z_0 is an **isolated zero** of f if and only if $f(z_0) = 0$ and $f(z) \neq 0$ on some deleted neighborhood $N_r'(z_0) \subset D$.
2. Let $z_0 \in D$ be an isolated zero of f such that there exists $m \in \mathbb{Z}^+$ with $f(z) = (z - z_0)^m\, g(z)$, where $g : D \to \mathbb{C}$ is differentiable on D and $g(z_0) \neq 0$. Then the number m is called the **multiplicity** of

the zero of f at z_0.

Note in part 2 of the above definition that since g is differentiable it is also continuous. Since g is nonzero at z_0 the continuity of g implies the existence of a neighborhood $N_r(z_0)$ on which g is nonzero. In a certain sense, then, the function f, which has an isolated zero of multiplicity m at z_0, behaves much like $(z-z_0)^m$ near z_0.

Theorem 3.9 *Let $f : D \to \mathbb{C}$ be differentiable on D, and let $z_0 \in D$ be an isolated zero of f. Then there exists some integer $m \geq 1$ such that z_0 is a zero of f of multiplicity m.*

PROOF Choose $N_r(z_0) \subset D$. By Taylor's theorem we have

$$f(z) = \sum_{j=0}^{\infty} a_j\,(z-z_0)^j \quad \text{for } z \in N_r(z_0),$$

and since $f(z_0) = 0$ it follows that we may start the series index at $j = 1$. Let a_m be the coefficient corresponding to the smallest $m \geq 1$ such that $a_m \neq 0$. Note that such an m exists, otherwise $a_j = 0$ for all $j \geq 1$ implying $f(z) \equiv 0$ on $N_r(z_0)$, which contradicts that z_0 is an isolated zero of f. Since $a_1 = a_2 = \cdots = a_{m-1} = 0$, we have

$$f(z) = \sum_{j=m}^{\infty} a_j\,(z-z_0)^j = (z-z_0)^m \big[a_m + a_{m+1}\,(z-z_0) + \cdots\big] \quad \text{on } N_r(z_0).$$

Let $g : D \to \mathbb{C}$ be defined by

$$g(z) = \begin{cases} \frac{f(z)}{(z-z_0)^m} & \text{for } z \in D \setminus \{z_0\} \\ a_m & \text{for } z = z_0 \end{cases}.$$

Then $f(z) = (z-z_0)^m\,g(z)$ on D, where $g(z_0) = a_m \neq 0$. We leave it to the reader to show that g is differentiable on D. ♦

► **9.38** *Show that g is differentiable on D.*

Suppose $D \subset \mathbb{C}$ is open and connected. The following theorem establishes that if $f : D \to \mathbb{C}$ is a complex differentiable function that is not identically equal to zero on D, then the zeros of f must be isolated.

Theorem 3.10 *Let $f : D \to \mathbb{C}$ be differentiable on the open connected set $D \subset \mathbb{C}$, and let $Z_f \subset D$ be the set of zeros of f in D. If Z_f has a limit point in D, then $f(z) \equiv 0$ on D.*

PROOF Define the set $A \subset D$ according to

$$A = \left\{ z \in D : f^{(j)}(z) = 0 \quad \text{for all } j \geq 0 \right\}.$$

We will show that the sets A and $D \setminus A$ are both open. To see that A is open, suppose $z_0 \in A$. Then for some neighborhood $N_r(z_0)$,

$$f(z) = \sum_{j=0}^{\infty} \frac{f^{(j)}(z_0)}{j!} (z - z_0)^j = 0 \quad \text{for} \quad z \in N_r(z_0) \subset D.$$

This implies that $N_r(z_0) \subset A$, and therefore A is open. To show that $D \setminus A$ is open, let $z_0 \in D \setminus A$. Then there exists $m \in \mathbb{N}$ such that $f^{(m)}(z_0) \neq 0$. By continuity of $f^{(m)}(z)$ at z_0 there exists some neighborhood $N_r(z_0) \subset D$ on which $f^{(m)}(z)$ is nonzero, and hence $N_r(z_0) \subset D \setminus A$. Therefore $D \setminus A$ is open. Now, since D is connected and $D = A \cup (D \setminus A)$, it follows that either $A = \varnothing$ or $A = D$. Finally, suppose Z_f has a limit point $z^* \in D$. We will show that $z^* \in A$, and therefore $A = D$, proving that $f(z) \equiv 0$ on D. To see that $z^* \in A$, we write

$$f(z) = \sum_{j=0}^{\infty} \frac{f^{(j)}(z^*)}{j!} (z - z^*)^j \quad \text{for} \quad z \in N_r(z^*) \subset D.$$

If $z^* \notin A$ there exists a smallest $m \geq 0$ such that $f^{(m)}(z^*) \neq 0$. This implies that for some $0 < s < r$,

$$f(z) = (z - z^*)^m \, h(z) \quad \text{on} \quad N_s(z^*) \subset N_r(z^*) \subset D,$$

where $h(z) \neq 0$ on $N_s(z^*)$. (Why?) This contradicts the fact that z^* is a limit point of Z_f. Therefore $z^* \in A$, and so $A = D$, completing the proof. ♦

► **9.39** *Answer the (Why?) in the above proof.*

Corollary 3.11 *Suppose $f, g : D \to \mathbb{C}$ are differentiable on the open connected set $D \subset \mathbb{C}$.*

a) If $f(z) = 0$ for all $z \in A \subset D$ where the set A has a limit point in D, then $f(z) \equiv 0$ on D.

b) If $f(z) = g(z)$ for all $z \in A \subset D$ where the set A has a limit point in D, then $f \equiv g$ on D.

► **9.40** *Prove the above corollary.*

Part *b*) of the above corollary is sometimes referred to as *the identity principle.*

Example 3.12 *We will apply the above corollary to show that $1 + \tan^2 z = \sec^2 z$ for all $z \in \mathbb{C}$. To see this, let $f : \mathbb{C} \to \mathbb{C}$ be given by $f(z) = \sec^2 z - \tan^2 z - 1$. Then since $f(x) = \sec^2 x - \tan^2 x - 1 \equiv 0$ for all $x \in \mathbb{R}$, the corollary yields that $f(z) \equiv 0$ for all $z \in \mathbb{C}$.* ◄

► **9.41** *What other trigonometric identities can be derived in this way?*

4 LAURENT'S THEOREM FOR COMPLEX FUNCTIONS

Consider the function $f : \mathbb{C} \setminus \{2\} \to \mathbb{C}$ given by $f(z) = \frac{1}{z-2}$. By Taylor's theorem, f has a Taylor series centered at $z_0 = 0$ with neighborhood of convergence $N_2(0)$. That is,

$$f(z) = \sum_{j=0}^{\infty} \frac{f^{(j)}(0)}{j!} z^j \quad \text{on } N_2(0).$$

But this function f is also defined for $|z| > 2$, so it is natural to ask if f can be represented by some (other) series expansion centered at $z_0 = 0$ and convergent for $|z| > 2$. To explore this, note that

$$f(z) = \frac{1}{z-2} = \frac{1}{z}\left(\frac{1}{1-\frac{2}{z}}\right) = \frac{1}{z}\sum_{j=0}^{\infty}\left(\frac{2}{z}\right)^j = \sum_{j=0}^{\infty} \frac{2^j}{z^{j+1}},$$

and this series converges for $\left|\frac{2}{z}\right| < 1$, i.e., for $|z| > 2$. This type of power series, which involves powers of $\frac{1}{z-z_0}$, or equivalently, negative powers of $(z - z_0)$, is known as a *Laurent series*. More generally, the form of a Laurent series centered at a point z_0 involves both positive and negative powers of $(z - z_0)$, that is,

$$f(z) = \sum_{j=0}^{\infty} a_j (z - z_0)^j + \sum_{j=1}^{\infty} \frac{b_j}{(z - z_0)^j}, \tag{9.21}$$

and so it consists of *two* infinite series. The first series in (9.21) is often referred to as *the analytic part* and the second series in (9.21) is often referred to as *the singular part*, or *the principal part* of the Laurent series. For the Laurent series expansion to exist at a particular $z \in \mathbb{C}$, both the analytic part and the singular part must be convergent at z. In fact, the analytic part consisting of positive powers of $(z - z_0)$ will converge for all z inside some circle of radius R, while the singular part consisting of negative powers of $(z - z_0)$ will converge for all z *outside* some circle of radius r, as in our example. It is the overlap of the two regions of convergence associated with the analytic part and the singular part that comprises the region of convergence of the Laurent series. This region of overlap is typically the annulus centered at z_0 and denoted by $A_r^R(z_0) = \{z \in \mathbb{C} : r < |z - z_0| < R\}$. See Figure 9.3 for an illustration. The situation $R = \infty$ and $r = 0$ corresponds to the case where the Laurent series converges everywhere in $\mathbb{C}$ except possibly at z_0. (In our particular example, we have convergence in the annulus $A_2^{\infty}(0)$.) Note that since the annulus so described *excludes* the point z_0, it is not necessary that f be differentiable at z_0 for it to have a Laurent series expansion centered there. (In fact, the function f need not be differentiable anywhere within $N_r(z_0)$.)

If $0 < r < r_1 < R_1 < R$ we will refer to the annulus $A_{r_1}^{R_1}(z_0) \subset A_r^R(z_0)$ as a *proper subannulus* of $A_r^R(z_0)$. In a manner similar to that of a convergent

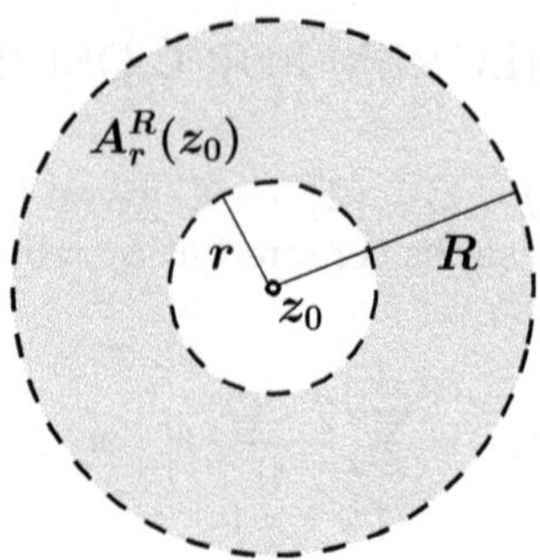

Figure 9.3 *The region of convergence of a Laurent series.*

Taylor series on subsets of its neighborhood of convergence, when a function f has a convergent Laurent series expansion on an annulus $A_r^R(z_0)$, it will converge absolutely on $A_r^R(z_0)$ and uniformly on $\overline{A_{r_1}^{R_1}}(z_0)$ where $A_{r_1}^{R_1}(z_0)$ is any proper subannulus of $A_r^R(z_0)$. To establish this, we must consider the convergence of a given Laurent series a bit more carefully. Since the analytic part of the Laurent series is a power series, it will converge absolutely on its neighborhood of convergence $N_R(z_0)$. Recall that the convergence is uniform on $\overline{N_{R_1}(z_0)}$ for any $0 < R_1 < R$. To analyze the singular part, we define

$$r \equiv \inf\left\{ |z - z_0| : \sum_{j=1}^{\infty} \frac{b_j}{(z - z_0)^j} \text{ converges} \right\}.$$

If r fails to exist then the singular part never converges. If r exists then we will show that the singular part converges absolutely for $(\overline{N_r(z_0)})^C$, and uniformly on $(N_{r_1}(z_0))^C$ for any $r_1 > r$. To see this, choose $r_1 > r$. Then as indicated in Figure 9.4 there exists z_1 such that $r \le |z_1 - z_0| < r_1$ and $\sum_{j=1}^{\infty} \frac{b_j}{(z_1 - z_0)^j}$ converges. Therefore, there exists $M \ge 0$ such that

$$\left| \frac{b_j}{(z_1 - z_0)^j} \right| \le M \quad \text{for all } j \ge 1.$$

Now, for $z \in (N_{r_1}(z_0))^C$ we have

$$\left| \frac{b_j}{(z - z_0)^j} \right| = \left| \frac{b_j}{(z_1 - z_0)^j} \right| \left(\frac{|z_1 - z_0|}{|z - z_0|} \right)^j \le M \left(\frac{|z_1 - z_0|}{r_1} \right)^j.$$

Since $|z_1 - z_0| < r_1$ this establishes via the Weierstrass M-test that $\sum_{j=1}^{\infty} \frac{b_j}{(z - z_0)^j}$ converges absolutely on $(\overline{N_r(z_0)})^C$ and uniformly on $(N_{r_1}(z_0))^C$. Since $r_1 > r$ was arbitrary, this result holds for any $r_1 > r$. We leave it to the reader to show that $\sum_{j=1}^{\infty} \frac{b_j}{(z - z_0)^j}$ diverges on $N_r(z_0)$. As with a Taylor series, the points on the boundary $C_r(z_0)$ must be studied individually.

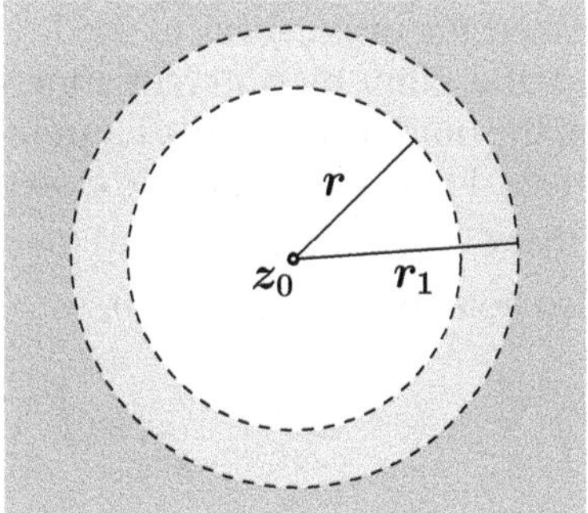

Figure 9.4 *The regions of convergence and divergence of the singular part of a Laurent series.*

► **9.42** *As claimed above, show that* $\sum_{j=1}^{\infty} \frac{b_j}{(z-z_0)^j}$ *diverges on* $N_r(z_0)$.

The above discussion and exercise establish the following result.

Proposition 4.1 *Suppose* $f : D \to \mathbb{C}$ *has a Laurent series expansion* $f(z) = \sum_{j=0}^{\infty} a_j(z-z_0)^j + \sum_{j=1}^{\infty} \frac{b_j}{(z-z_0)^j}$ *on the annulus* $A_r^R(z_0) \subset D$ *(where* $r \geq 0$ *and* R *may be* ∞*). Then for any proper subannulus* $A_{r_1}^{R_1}(z_0) \subset A_r^R(z_0)$ *the given Laurent series expansion for* f *converges absolutely on* $A_{r_1}^{R_1}(z_0)$ *and uniformly on* $\overline{A_{r_1}^{R_1}(z_0)}$.

This result, in turn, implies the following. It will be instrumental in proving part of the key result of this section.

Proposition 4.2 *Let* $\sum_{j=1}^{\infty} \frac{b_j}{(z-z_0)^j}$ *be the singular part of a Laurent series expansion for* $f : D \to \mathbb{C}$ *on* $A_r^R(z_0) \subset D$. *Then* $\sum_{j=1}^{\infty} \frac{b_j}{(z-z_0)^j}$ *represents a continuous function on* $(\overline{N_r(z_0)})^C$, *and for any contour* $C \subset (\overline{N_r(z_0)})^C$ *we have*

$$\int_C \left(\sum_{j=1}^{\infty} \frac{b_j}{(z-z_0)^j} \right) dz = \sum_{j=1}^{\infty} \int_C \frac{b_j}{(z-z_0)^j}\, dz.$$

► **9.43** *Prove the above proposition.*

It is also true, and of great practical value as we will see, that when a Laurent series expansion exists for a function it is unique. To see this, suppose f is differentiable on the annulus $A_r^R(z_0)$ and suppose too that it has a convergent Laurent series there given by

$$f(z) = \sum_{k=0}^{\infty} a_k(z-z_0)^k + \sum_{k=1}^{\infty} \frac{b_k}{(z-z_0)^k}. \tag{9.22}$$

Recall that we established the uniqueness of the Taylor series representation of a function centered at a point by differentiating the series term-by-term; here, we will establish the uniqueness of the Laurent series representation by integrating term-by-term. Let C be any simple closed contour in $A_r^R(z_0)$ with $n_C(z_0) = 1$, and note that $C \subset A_{r_1}^{R_1}(z_0) \subset A_r^R(z_0)$ for some proper subannulus $A_{r_1}^{R_1}(z_0) \subset A_r^R(z_0)$. Also note that for any $k \geq 0$,

$$\begin{aligned}\oint_C \frac{f(\zeta)}{(\zeta - z_0)^{k+1}}\, d\zeta &= \oint_C \sum_{j=0}^{\infty} a_j(\zeta - z_0)^{j-k-1}\, d\zeta + \oint_C \sum_{j=1}^{\infty} \frac{b_j}{(\zeta - z_0)^{j+k+1}}\, d\zeta \\ &= \sum_{j=0}^{\infty} \oint_C a_j(\zeta - z_0)^{j-k-1}\, d\zeta + \sum_{j=1}^{\infty} \oint_C \frac{b_j}{(\zeta - z_0)^{j+k+1}}\, d\zeta \qquad (9.23) \\ &= 2\pi i\, a_k\, n_C(z_0) \\ &= 2\pi i\, a_k.\end{aligned}$$

Note in (9.23) above that the integral of each summand corresponding to a_j for $j \neq k$, and for b_j for $j \geq 1$, vanishes due to the integrand having an antiderivative within $N_{R_1}(z_0)$. Also, in integrating the singular part term-by-term, we have applied Proposition 4.2. This shows that each a_k in (9.22) is uniquely determined by f and z_0. A similar argument can be made for the uniqueness of each b_k for $k \geq 1$ in (9.22) by considering the integral $\oint_C \frac{f(\zeta)}{(\zeta - z_0)^{-k+1}}\, d\zeta$ for any fixed $k \geq 1$. We leave this to the reader.

▶ **9.44** *Establish the uniqueness of the b_k terms in the Laurent series representation (9.22) by carrying out the integral $\oint_C \frac{f(\zeta)}{(\zeta - z_0)^{-k+1}}\, d\zeta$ for any fixed $k \geq 1$.*

With these important facts about an existing Laurent series established, we now state and prove Laurent's theorem, the key result of this section. It is a kind of sibling to Taylor's theorem in complex function theory. It gives conditions under which a function $f : D \to \mathbb{C}$ is guaranteed a Laurent series representation convergent on an annulus $A_r^R(z_0) \subset D$.

Theorem 4.3 (Laurent's Theorem)
Let $f : D \to \mathbb{C}$ be differentiable on the annulus $A_r^R(z_0) = \{z : r < |z - z_0| < R\} \subset D$ (where $r \geq 0$ and R may be ∞). Then $f(z)$ can be expressed uniquely by

$$f(z) = \sum_{j=0}^{\infty} a_j(z - z_0)^j + \sum_{j=1}^{\infty} \frac{b_j}{(z - z_0)^j},$$

on $A_r^R(z_0)$. For any choice of simple closed contour $C \subset A_r^R(z_0)$ with $n_C(z_0) = 1$, the coefficients a_j and b_j are given by

$$a_j = \frac{1}{2\pi i} \oint_C \frac{f(\zeta)}{(\zeta - z_0)^{j+1}}\, d\zeta \quad \textit{for} \quad j \geq 0,$$

and

$$b_j = \frac{1}{2\pi i}\oint_C \frac{f(\zeta)}{(\zeta - z_0)^{-j+1}}\,d\zeta \quad \textit{for} \quad j \geq 1.$$

PROOF As in the proof of Taylor's theorem, we may assume that $z_0 = 0$. Fix $z \in A_r^R(0)$. Choose circles C_1 and C_2 in $A_r^R(0)$ such that both are centered at 0, with radii R_1 and R_2, respectively, satisfying $r < R_2 < |z| < R_1 < R$.

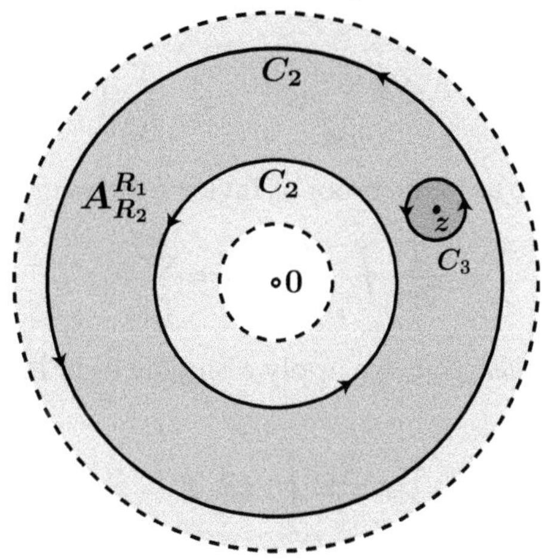

Figure 9.5 *The situation in the proof of Laurent's theorem.*

As indicated in Figure 9.5, choose a third circle C_3, centered at z and having radius R_3, with R_3 small enough that C_3 is contained in $A_r^R(0)$ and does not intersect C_1 or C_2. Then, since $n_{C_1} = n_{C_3} + n_{C_2}$ on $\left(A_r^R(0)\right)^C \cup \{z\}$ (Why?), we have

$$\frac{1}{2\pi i}\oint_{C_1} \frac{f(\zeta)}{\zeta - z}\,d\zeta = \frac{1}{2\pi i}\oint_{C_3} \frac{f(\zeta)}{\zeta - z}\,d\zeta + \frac{1}{2\pi i}\oint_{C_2} \frac{f(\zeta)}{\zeta - z}\,d\zeta \quad \text{by Theorem 3.6, Chap. 8}$$

$$= f(z) + \frac{1}{2\pi i}\oint_{C_2} \frac{f(\zeta)}{\zeta - z}\,d\zeta \quad \text{by Cauchy's formula.}$$

Solving for $f(z)$ gives

$$f(z) = \frac{1}{2\pi i}\oint_{C_1} \frac{f(\zeta)}{\zeta - z}\,d\zeta - \frac{1}{2\pi i}\oint_{C_2} \frac{f(\zeta)}{\zeta - z}\,d\zeta$$

$$= \frac{1}{2\pi i}\oint_{C_1} \frac{f(\zeta)}{\zeta - z}\,d\zeta + \frac{1}{2\pi i}\oint_{C_2} \frac{f(\zeta)}{z - \zeta}\,d\zeta. \tag{9.24}$$

We will show that the first integral on the right-hand side of (9.24) leads to the *analytic part* of the Laurent series expansion for $f(z)$, while the second integral on the right-hand side leads to the *singular part*. The analysis of the first integral proceeds just as in the proof of Taylor's theorem, that is,

$$\frac{1}{2\pi i}\oint_{C_1}\frac{f(\zeta)}{\zeta - z}\,d\zeta = \sum_{j=0}^{N} z^j \frac{1}{2\pi i}\oint_{C_1}\frac{f(\zeta)}{\zeta^{j+1}}\,d\zeta + \frac{1}{2\pi i}\oint_{C_1}\left(\frac{z}{\zeta}\right)^{N+1}\frac{f(\zeta)}{\zeta - z}\,d\zeta.$$

In this case, however, we can no longer expect that $\frac{1}{2\pi i}\oint_{C_1}\frac{f(\zeta)}{\zeta^{j+1}}\,d\zeta = \frac{f^{(j)}(0)}{j!}$, because f is not necessarily differentiable inside C_1. Therefore, we define a_j by $a_j \equiv \frac{1}{2\pi i}\oint_{C_1}\frac{f(\zeta)}{\zeta^{j+1}}\,d\zeta$, for whatever value this integral takes. This yields

$$\frac{1}{2\pi i}\oint_{C_1}\frac{f(\zeta)}{\zeta - z}\,d\zeta = \sum_{j=0}^{N} a_j z^j + \frac{1}{2\pi i}\oint_{C_1}\left(\frac{z}{\zeta}\right)^{N+1}\frac{f(\zeta)}{\zeta - z}\,d\zeta,$$

and letting $N \to \infty$ as in the proof of Taylor's theorem gives the *analytic part*,

$$\frac{1}{2\pi i}\oint_{C_1}\frac{f(\zeta)}{\zeta - z}\,d\zeta = \sum_{j=0}^{\infty} a_j z^j.$$

To obtain the *singular part*, we apply a similar technique. Consider that for $\zeta \in C_2$,

$$\begin{aligned}\frac{f(\zeta)}{z-\zeta} = \frac{f(\zeta)}{z}\left(\frac{1}{1-\frac{\zeta}{z}}\right) &= \frac{f(\zeta)}{z}\left(1 + \frac{\zeta}{z} + \cdots + \left(\frac{\zeta}{z}\right)^N + \left(\frac{\zeta}{z}\right)^{N+1}\frac{1}{1-\frac{\zeta}{z}}\right)\\ &= f(\zeta)\left(\frac{1}{z} + \frac{\zeta}{z^2} + \cdots + \frac{\zeta^N}{z^{N+1}} + \left(\frac{\zeta}{z}\right)^{N+1}\frac{1}{z-\zeta}\right)\\ &= \sum_{j=1}^{N+1}\frac{\zeta^{j-1}}{z^j}f(\zeta) + \left(\frac{\zeta}{z}\right)^{N+1}\frac{f(\zeta)}{z-\zeta},\end{aligned}$$

so that,

$$\frac{1}{2\pi i}\oint_{C_2}\frac{f(\zeta)}{z-\zeta}\,d\zeta = \sum_{j=1}^{N+1}\frac{1}{z^j}\left(\frac{1}{2\pi i}\oint_{C_2}\zeta^{j-1}f(\zeta)\,d\zeta\right) + \frac{1}{2\pi i}\oint_{C_2}\left(\frac{\zeta}{z}\right)^{N+1}\frac{f(\zeta)}{z-\zeta}\,d\zeta.$$

In this case, define $b_j \equiv \frac{1}{2\pi i}\oint_{C_2}\zeta^{j-1}f(\zeta)\,d\zeta$, and take the limit as $N \to \infty$ to obtain

$$\frac{1}{2\pi i}\oint_{C_2}\frac{f(\zeta)}{z-\zeta}\,d\zeta = \sum_{j=1}^{\infty}\frac{b_j}{z^j}.$$

Finally, consider $a_j = \frac{1}{2\pi i}\oint_{C_1}\frac{f(\zeta)}{\zeta^{j+1}}d\zeta$. To show that the conclusion holds for *any* simple closed contour $C \subset A_r^R(z_0)$ with $n_C(z_0) = 1$, consider any such C. Then $n_C = 1$ on $N_r(0)$, and also $n_{C_1} = n_C$ on $\left(A_r^R(0)\right)^C$, obtaining via Theorem 3.6 from Chapter 8,

$$a_j = \frac{1}{2\pi i}\oint_{C_1}\frac{f(\zeta)}{\zeta^{j+1}}d\zeta = \frac{1}{2\pi i}\oint_{C}\frac{f(\zeta)}{\zeta^{j+1}}d\zeta.$$

A similar argument works to show that $b_j = \frac{1}{2\pi i}\oint_C \zeta^{j-1} f(\zeta)\, d\zeta$. Finally, the uniqueness of the Laurent series expansion was established prior to the statement of the theorem. ♦

► **9.45** *Answer the (Why?) question in the above proof. To do so, exploit the fact that the contours are all circles, and consider the cases z, $\overline{\text{Int}(C_r(0))}$, and $\overline{\text{Ext}(C_R(0))}$ separately.*

In a certain sense, the concept of the Laurent series expansion generalizes that of the Taylor series expansion. It allows for a series representation of $f : D \to \mathbb{C}$ in both negative and positive powers of $(z - z_0)$ in a region that excludes points where f is not differentiable. In fact, if the function f is differentiable on all of $N_R(z_0)$ the Laurent series will reduce to the Taylor series of the function centered at z_0.

One practical use of a function's Laurent series representation is the characterization of that function's *singularities*. We consider this topic next.

5 SINGULARITIES

5.1 Definitions

In our motivating example from the last section we discussed the function $f : \mathbb{C} \setminus \{2\} \to \mathbb{C}$ given by $f(z) = \frac{1}{z-2}$. The point $z_0 = 2$ is one where the function f fails to be differentiable, and hence no Taylor series can be found for f centered at $z_0 = 2$. Yet, f is differentiable at points z arbitrarily near to z_0. Such a point z_0 is called a *singularity* of the function f.

Definition 5.1 Suppose $f : D \to \mathbb{C}$ is not differentiable at $z_0 \in \mathbb{C}$. If f is differentiable at some point z in every deleted neighborhood $N_r'(z_0)$, then z_0 is called a **singularity** of f.

There are different types of singularities, and we characterize them now. First, we distinguish between *isolated* and *nonisolated* singularities.

Definition 5.2 Suppose $z_0 \in \mathbb{C}$ is a singularity of $f : D \to \mathbb{C}$. If f is differentiable on some deleted neighborhood $N_r'(z_0) \subset D$ centered at z_0, then z_0 is called an **isolated singularity** of f. Otherwise, z_0 is called a **nonisolated singularity** of f.

Example 5.3 *The function $f : \mathbb{C} \setminus \{0\} \to \mathbb{C}$ given by $f(z) = \frac{1}{z}$ has an isolated singularity at $z_0 = 0$. The function $\text{Log} : \mathbb{C} \setminus \{0\} \to \mathbb{C}$ given by $\text{Log}(z) = \ln|z| + i\,\text{Arg}(z)$ has nonisolated singularities at every point $z = x \in (-\infty, 0]$.* ◄

► **9.46** *Consider the function* $f : \mathbb{C} \to \mathbb{C}$ *given by* $f(z) = \begin{cases} z & \text{if } \operatorname{Im}(z) \geq 0 \\ -\overline{z} & \text{if } \operatorname{Im}(z) < 0 \end{cases}$. *Find the singularities of* f *and characterize them as isolated or nonisolated. What if* $-\overline{z}$ *is replaced by* $\overline{z}$*?*

While we cannot find a Taylor series centered at a singularity of f, we can always find a Laurent series centered at an *isolated* singularity, convergent in an annulus that omits that point. The Laurent series can then be used to further characterize the isolated singularity. There are three subtypes to consider. In the following definition, note that $N_r'(z_0)$ is an annulus $A_0^r(z_0)$ centered at z_0.

Definition 5.4 Let z_0 be an isolated singularity of $f : D \to \mathbb{C}$, and suppose f has Laurent series representation $f(z) = \sum_{j=0}^{\infty} a_j(z - z_0)^j + \sum_{j=1}^{\infty} \frac{b_j}{(z-z_0)^j}$ on $N_r'(z_0) \subset D$.

1. If $b_j = 0$ for all $j \geq 1$, then f is said to have a **removable singularity** at z_0.
2. If there exists $N \in \mathbb{N}$ such that $b_j = 0$ for $j > N$ but $b_N \neq 0$, then f is said to have a **pole of order** N at z_0. In the special case where $N = 1$, the pole is often referred to as a **simple pole.**
3. Otherwise f is said to have an **essential singularity** at z_0.

It is not hard to show that f has an essential singularity at z_0 if and only if $b_j \neq 0$ for infinitely many values of $j \geq 1$ in its Laurent series expansion, that is, if the Laurent series expansion's singular part is an infinite series.

► **9.47** *Prove the above claim.*

According to this definition, if z_0 is an isolated singularity of $f : D \to \mathbb{C}$ one can determine what type of singularity z_0 is by finding the Laurent series representation for f on an annulus $A_r^R(z_0)$ centered at z_0, and scrutinizing its singular part. However, in practice it is often difficult to obtain the Laurent series directly via Laurent's theorem. Finding alternative means for making this determination is therefore worthwhile, and we do so now. The key will be to exploit the *uniqueness* of the Laurent series representation of a given function on a particular annular region. The following example illustrates the idea.

Example 5.5 *Consider the function* $f : \mathbb{C} \setminus \{0\} \to \mathbb{C}$ *given by* $f(z) = \frac{1}{z^3} e^z$. *We leave it to the reader to confirm that* $z_0 = 0$ *is an isolated singularity. To determine what type of isolated singularity* $z_0 = 0$ *is we will find the Laurent series representation of* f *on* $N_1'(0)$. *To this end, note that* f *has the form* $f(z) = \frac{1}{z^3} f_0(z)$ *where the function* f_0 *is differentiable on* $\mathbb{C}$. *Therefore, we can expand* f_0 *in a Taylor series centered at* $z_0 = 0$, *namely,* $f_0(z) = e^z = \sum_{j=0}^{\infty} \frac{1}{j!} z^j$. *We* *then can write,*

$$f(z) = \frac{1}{z^3} \sum_{j=0}^{\infty} \frac{1}{j!} z^j = \sum_{j=0}^{\infty} \frac{1}{j!} z^{j-3} = \frac{1}{z^3} + \frac{1}{z^2} + \frac{1}{2!}\frac{1}{z} + \frac{1}{3!} + \frac{1}{4!} z + \frac{1}{5!} z^2 + \cdots$$

which is convergent on $A_0^\infty(0)$. *By the uniqueness property of Laurent series representations, this must be the Laurent series centered at* $z_0 = 0$ *for* f. *Since* $b_3 = 1$, *and* $b_n = 0$ *for* $n \geq 4$, *we see that* f *has a pole of order 3 at* $z_0 = 0$. ◀

▶ **9.48** *Consider the function* $f : \mathbb{C} \setminus \{0\} \to \mathbb{C}$ *given by* $f(z) = \frac{\sin z}{z}$. *Show that* $z_0 = 0$ *is a removable singularity.*

Example 5.6 *Consider the function* $f : \mathbb{C} \setminus \{-1, 1\} \to \mathbb{C}$ *given by* $f(z) = \frac{1}{z^2-1}$. *We will determine what kind of singularity* $z_0 = 1$ *is. Note that* $f(z) = \left(\frac{1}{z-1}\right)\left(\frac{1}{z+1}\right)$, *and*

$$\frac{1}{z+1} = \frac{1}{2 + (z-1)} = \frac{1}{2}\left(\frac{1}{1 + \frac{z-1}{2}}\right)$$
$$= \frac{1}{2} \sum_{j=0}^{\infty} (-1)^j \left(\frac{z-1}{2}\right)^j = \sum_{j=0}^{\infty} \frac{(-1)^j}{2^{j+1}} (z-1)^j,$$

which converges for $|z - 1| < 2$. *Overall then, we have*

$$f(z) = \left(\frac{1}{z-1}\right)\left(\frac{1}{z+1}\right) = \sum_{j=0}^{\infty} \frac{(-1)^j}{2^{j+1}} (z-1)^{j-1},$$

and this Laurent series converges on the annulus $A_0^2(1)$. *From this we see that* $z_0 = 1$ *is a simple pole.* ◀

▶ **9.49** *For the function in the above example, show that* $z = -1$ *is also a simple pole.*

▶ **9.50** *For* $N \in \mathbb{N}$, *what kind of singularity is* $z = 1$ *for the function* $f : \mathbb{C} \setminus \{1\} \to \mathbb{C}$ *given by* $f(z) = \frac{1}{(z-1)^N}$?

Example 5.7 *Consider the function* $f : \mathbb{C} \setminus \{0\} \to \mathbb{C}$ *given by* $f(z) = e^{1/z}$. *By Taylor's theorem, we know that* $e^z = \sum_{j=0}^{\infty} \frac{1}{j!} z^j$, *which converges for all* $z \in \mathbb{C}$. *Therefore the Laurent series for* f *centered at* $z_0 = 0$ *is given by*

$$f(z) = e^{1/z} = \sum_{j=0}^{\infty} \frac{1}{j!} \frac{1}{z^j},$$

which converges on $A_0^\infty(0)$. *Clearly* $z_0 = 0$ *is not a removable singularity, nor is it a pole of any order. Therefore, it is an essential singularity.* ◀

Finally, we define what it means for a function to have a singularity at infinity.

Definition 5.8 Consider $f : D \to \mathbb{C}$. Let $D^* \equiv \{z \in D : \frac{1}{z} \in D\}$, and define the function $g : D^* \to \mathbb{C}$ by $g(z) = \frac{1}{z}$. Then the function $f \circ g : D^* \to \mathbb{C}$ is given by $(f \circ g)(z) = f(g(z)) = f\left(\frac{1}{z}\right)$. We say that f has a **singularity at infinity** if the function $f \circ g$ has a singularity at $z_0 = 0$. The singularity at infinity for f is characterized according to the singularity at $z_0 = 0$ for $f \circ g$.

Example 5.9 *Consider the function $f : \mathbb{C} \to \mathbb{C}$ given by $f(z) = z^3$. Then the function $g : \mathbb{C} \setminus \{0\} \to \mathbb{C}$ given by $g(z) = \frac{1}{z}$ obtains $f \circ g : \mathbb{C} \setminus \{0\} \to \mathbb{C}$ given by $(f \circ g)(z) = f\left(\frac{1}{z}\right) = \frac{1}{z^3}$, which has a pole of order 3 at $z_0 = 0$, and therefore f has a pole of order 3 at infinity.* ◀

▶ **9.51** *Consider the function $f : \mathbb{C} \setminus \{0\} \to \mathbb{C}$ given by $f(z) = e^{1/z}$. Show that the associated singularity of $f \circ g$ at $z_0 = 0$ is removable, and hence that f has a removable singularity at infinity.*

▶ **9.52** *Consider the function $f : \mathbb{C} \setminus \{0\} \to \mathbb{C}$ given by $f(z) = \frac{\sin z}{z}$. Show that f has an essential singularity at ∞.*

5.2 Properties of Functions Near Singularities

We now prove some results particular to each type of isolated singularity, and thereby reveal something about how a function behaves near each such point.

Properties of Functions Near a Removable Singularity

Suppose z_0 is a removable singularity of the function $f : D \to \mathbb{C}$. Then for some $N_r'(z_0) \subset D$, the function f has Laurent series representation given by

$$f(z) = \sum_{j=0}^{\infty} a_j (z - z_0)^j \quad \text{on } N_r'(z_0),$$

and we may extend f to z_0 by defining $F : D \cup \{z_0\} \to \mathbb{C}$ according to

$$F(z) = \begin{cases} f(z) & \text{if } z \in D, \\ a_0 & \text{if } z = z_0. \end{cases}$$

Clearly the function F is differentiable at z_0. (Why?) This leads us to the following theorem.

Theorem 5.10 *Let z_0 be an isolated singularity of $f : D \to \mathbb{C}$.*

a) The point z_0 is a removable singularity of f if and only if f can be extended to a function $F : D \cup \{z_0\} \to \mathbb{C}$ that is differentiable at z_0 and such that $F(z) = f(z)$ on D.

b) The point z_0 is a removable singularity of f if and only if there exists $N_r'(z_0) \subset D$ such that f is bounded on $N_r'(z_0)$.

PROOF Half of the proof of part *a*) precedes the statement of the theorem, and we leave the remaining half of the proof to the reader. To prove part *b*), suppose f is differentiable and bounded on $N_r'(z_0) \subset D$, and let $g : N_r(z_0) \to \mathbb{C}$ be given by

$$g(z) = \begin{cases} (z - z_0)^2 f(z) & \text{for } z \neq z_0 \\ 0 & \text{for } z = z_0 \end{cases}.$$

Since f is differentiable on $N_r'(z_0)$, the function g must be too. To see that g is also differentiable at z_0, consider

$$g'(z_0) = \lim_{z \to z_0} \frac{g(z)}{z - z_0} = \lim_{z \to z_0} (z - z_0) f(z) = 0,$$

where the last equality holds since f is assumed bounded on $N_r'(z_0)$. Therefore, g is differentiable on $N_r(z_0)$ and hence has Taylor series representation there given by

$$g(z) = \sum_{j=0}^{\infty} a_j (z - z_0)^j = \sum_{j=2}^{\infty} a_j (z - z_0)^j.$$

From this, and the fact that $f(z) = \frac{g(z)}{(z-z_0)^2}$ on $N_r'(z_0)$, we obtain

$$f(z) = \sum_{j=2}^{\infty} a_j (z - z_0)^{j-2} = \sum_{j=0}^{\infty} a_{j+2} (z - z_0)^j \quad \text{on} \quad N_r'(z_0).$$

This last expression must be the Laurent series for f on $N_r'(z_0)$, and hence z_0 is a removable singularity of f. Now assume z_0 is a removable singularity of f. Then f has Laurent series representation of the form

$$f(z) = \sum_{j=0}^{\infty} a_j (z - z_0)^j \quad \text{on} \quad N_R'(z_0),$$

for some $R > 0$. Since this power series is defined and continuous (in fact, it is differentiable) on $N_R(z_0)$, it must be bounded on $\overline{N_r(z_0)}$ for $0 < r < R$. Therefore $f(z)$, which equals this power series on $N_r'(z_0) \subset \overline{N_r(z_0)}$, must be bounded on $N_r'(z_0) \subset \overline{N_r(z_0)}$. ♦

► **9.53** *Complete the proof of part a) of the above theorem.*

► **9.54** *Prove the following corollaries to the above theorem. Let z_0 be an isolated singularity of $f : D \to \mathbb{C}$.*

a) Then z_0 is a removable singularity if and only if $\lim_{z \to z_0} f(z)$ exists.

b) Then z_0 is a removable singularity if and only if $\lim_{z \to z_0} (z - z_0) f(z) = 0$.

Use of the above theorem is illustrated in the following examples.

Example 5.11 *As we have already seen in exercise 9.48, the function $f : \mathbb{C} \setminus \{0\} \to \mathbb{C}$ given by $f(z) = \frac{\sin z}{z}$ has a removable singularity at $z_0 = 0$. We can extend f to $z_0 = 0$ by assigning it the value $a_0 = 1$ from its Laurent series expansion centered at 0. The resulting function $F : \mathbb{C} \to \mathbb{C}$ given by $F(z) = \begin{cases} \frac{\sin z}{z} & \text{if } z \neq 0 \\ 1 & \text{if } z = 0 \end{cases}$ is therefore entire.* ◀

Example 5.12 *Consider the function $f : \mathbb{C} \setminus \{1\} \to \mathbb{C}$ given by $f(z) = \frac{z^2-1}{z-1}$. Since $f(z)$ equals $z + 1$ on $\mathbb{C} \setminus \{1\}$, it is clearly bounded on $N_1'(1)$. From the above theorem, we conclude that $z_0 = 1$ is a removable singularity of f.* ◀

Properties of Functions Near an Essential Singularity

As established by the following theorem, near an essential singularity a function will take values that are arbitrarily close to any fixed value $w_0 \in \mathbb{C}$.

Theorem 5.13 (The Casorati-Weierstrass Theorem)
Let $z_0 \in \mathbb{C}$ be an essential singularity of $f : D \to \mathbb{C}$. Then, for any $w_0 \in \mathbb{C}$ and any $\epsilon, r > 0$, there exists $z \in N_r'(z_0)$ such that $|f(z) - w_0| < \epsilon$.

PROOF [2] We use proof by contradiction. To this end, let $z_0 \in \mathbb{C}$ be an essential singularity of $f : D \to \mathbb{C}$ and assume the negation of the conclusion. Then there exists $w_0 \in \mathbb{C}$ and $\epsilon, r > 0$ such that $|f(z) - w_0| \geq \epsilon$ on $N_r'(z_0)$. Since z_0 is an isolated singularity of f, there exists $\rho > 0$ such that f is differentiable on $N_\rho'(z_0)$. From this we may conclude that $\frac{1}{f(z)-w_0}$ is differentiable and bounded on $N_\rho'(z_0)$. By Theorem 5.10, z_0 is a removable singularity of $\frac{1}{f(z)-w_0}$ and its Laurent series expansion has the form

$$\frac{1}{f(z) - w_0} = \sum_{j=0}^{\infty} a_j (z - z_0)^j \quad \text{on} \quad N_\rho'(z_0). \tag{9.25}$$

Let $m \geq 0$ be the smallest integer such that $a_m \neq 0$. Then (9.25) yields for $z \in N_\rho'(z_0)$,

$$\begin{aligned} \frac{1}{f(z) - w_0} &= \sum_{j=m}^{\infty} a_j (z - z_0)^j \\ &= (z - z_0)^m \sum_{j=m}^{\infty} a_j (z - z_0)^{j-m} \\ &\equiv (z - z_0)^m g(z) \quad \text{on} \quad N_\rho'(z_0), \end{aligned}$$

where $g(z)$ is differentiable on $N_\rho(z_0)$ and $g(z_0) \neq 0$. Also note that there exists $N_s(z_0) \subset N_\rho(z_0)$ such that $g(z) \neq 0$ on $N_s(z_0)$. (Why?) From this, we can write

[2] We follow [Con78].

$$f(z) - w_0 = \frac{1}{(z-z_0)^m} \frac{1}{g(z)} \quad \text{on } N_s'(z_0), \tag{9.26}$$

where $\frac{1}{g(z)}$ is differentiable on $N_s(z_0)$. Therefore, $\frac{1}{g(z)}$ has Taylor expansion

$$\frac{1}{g(z)} = \sum_{j=0}^{\infty} c_j (z-z_0)^j \quad \text{on } N_s(z_0), \quad \text{with } c_0 \neq 0, \quad \text{(Why?)}$$

and (9.26) becomes

$$\begin{aligned} f(z) - w_0 &= \frac{1}{(z-z_0)^m} \frac{1}{g(z)} \\ &= \frac{1}{(z-z_0)^m} \sum_{j=0}^{\infty} c_j (z-z_0)^j \\ &= \sum_{j=0}^{\infty} c_j (z-z_0)^{j-m} \\ &= \frac{c_0}{(z-z_0)^m} + \frac{c_1}{(z-z_0)^{m-1}} + \cdots + \frac{c_{m-1}}{(z-z_0)} + \sum_{j=m}^{\infty} c_j (z-z_0)^{j-m}. \end{aligned}$$

Adding w_0 to both sides of the above equality obtains the Laurent series representation for f centered at z_0 from which we can clearly see that z_0 is a pole of order m. This is a contradiction. ♦

▶ **9.55** *Answer the two (Why?) questions in the above proof.*

The Casorati-Weierstrass theorem is closely related to another result, often referred to as Picard's theorem, which we will not prove here. Picard's theorem states that if f has an essential singularity at z_0, then in any deleted neighborhood of z_0 the function f takes on every complex value with possibly a single exception.

Properties of Functions Near a Pole

We now characterize the behavior of a function near a pole.

Theorem 5.14 *Suppose $f : D \to \mathbb{C}$ has an isolated singularity at z_0. Then z_0 is a pole of order $N \geq 1$ if and only if for some $N_r'(z_0) \subset D$ there exists a function $f_0 : N_r(z_0) \to \mathbb{C}$ differentiable on $N_r(z_0)$ such that $f_0(z_0) \neq 0$ and $f(z) = \frac{f_0(z)}{(z-z_0)^N}$ for all $z \in N_r'(z_0)$.*

PROOF Suppose a function $f_0 : N_r(z_0) \to \mathbb{C}$ exists as described. Since f_0 is differentiable on $N_r(z_0)$, it has a convergent Taylor series

$$f_0(z) = \sum_{j=0}^{\infty} a_j (z-z_0)^j \quad \text{on } N_r(z_0).$$

Then

$$f(z) = \sum_{j=0}^{\infty} a_j (z - z_0)^{j-N}$$

is the Laurent series for f valid on $A_0^r(z_0)$. Since $f_0(0) = a_0 \neq 0$, we see that f has a pole of order N at z_0. Now suppose z_0 is a pole of order $N \geq 1$ associated with the function f. Then there exists a deleted neighborhood $N_r'(z_0) \subset D$ on which f is differentiable and on which f has the Laurent series representation given by

$$f(z) = \sum_{j=0}^{\infty} a_j (z - z_0)^j + \sum_{j=1}^{N} \frac{b_j}{(z - z_0)^j}, \tag{9.27}$$

with $b_N \neq 0$. Define $f_0 : N_r(z_0) \to \mathbb{C}$ according to $f_0(z) = (z - z_0)^N f(z)$. Then clearly f_0 has Laurent series representation on $N_r'(z_0)$ given by

$$\begin{aligned} f_0(z) &= (z - z_0)^N \sum_{j=0}^{\infty} a_j (z - z_0)^j + (z - z_0)^N \sum_{j=1}^{N} \frac{b_j}{(z - z_0)^j}, \quad (9.28) \\ &= \sum_{j=0}^{\infty} a_j (z - z_0)^{j+N} + \sum_{j=1}^{N} b_j (z - z_0)^{N-j}, \\ &= \sum_{j=0}^{\infty} a_j (z - z_0)^{j+N} + b_1 (z - z_0)^{N-1} + b_2 (z - z_0)^{N-2} + \cdots + b_N. \end{aligned}$$

From this we see that f_0 is, in fact, differentiable on $N_r(z_0)$, and that $f_0(z_0) = b_N \neq 0$. Comparing equations (9.27) and (9.28) shows that $f(z) = \frac{f_0(z)}{(z-z_0)^N}$ on $N_r'(z_0)$. ♦

Corollary 5.15 *Suppose $f : D \to \mathbb{C}$ has an isolated singularity at z_0. Then z_0 is a pole if and only if $\lim_{z \to z_0} |f(z)| = \infty$.*

PROOF Suppose z_0 is a pole of order $N \geq 1$ of f. Then there exists a neighborhood $N_r(z_0) \subset D$ and a function $f_0 : N_r(z_0) \to \mathbb{C}$ differentiable on $N_r(z_0)$ such that $f_0(z_0) \neq 0$ and $f(z) = \frac{f_0(z)}{(z-z_0)^N}$ on $N_r'(z_0)$. By continuity there exists a neighborhood $N_s(z_0) \subset N_r(z_0)$ such that $|f_0(z)| \geq c > 0$ on $N_s(z_0)$. Therefore, $|f(z)| \geq \frac{c}{|z-z_0|^N}$ on $N_s'(z_0)$, which implies $\lim_{z \to z_0} |f(z)| = \infty$. Conversely, suppose $\lim_{z \to z_0} |f(z)| = \infty$. Since z_0 is not a removable singularity (Why?), it must either be a pole or an essential singularity. But $\lim_{z \to z_0} |f(z)| = \infty$ implies that for any $M \geq 0$ there is a $\delta > 0$ such that $|f(z)| \geq M$ for $|z - z_0| < \delta$. By the Casorati-Weierstrass theorem, z_0 can't be an essential singularity (Why?), and so it must be a pole. ♦

► **9.56** *Answer the two (Why?) questions in the above proof.*

Note the difference between the behavior of a function near a pole as opposed to its behavior near an essential singularity. While approaching a pole, the magnitude of the function's value grows without bound, whereas approaching an essential singularity causes the function's value to go virtually "all over the place" infinitely often as z gets nearer to z_0. The behavior near an essential singularity is singular indeed!

The following theorem points to yet another way in which poles of a function are "less strange" than essential singularities. In fact, if a function f has a pole of order N at z_0, then near z_0 the function f behaves like $\frac{1}{(z-z_0)^N}$ for the same reason that near an isolated zero of order N a differentiable function behaves like $(z-z_0)^N$. The theorem suggests even more, namely, that the reciprocal of a function with an isolated zero of order N at z_0 will be a function with a pole of order N at that point. Also, a type of converse holds as well. The reciprocal of a function f with a pole of order N at z_0 is not quite a function with a zero of order N at z_0. But it can ultimately be made into one, as the proof shows. Ultimately, the poles of f can be seen to be zeros of the reciprocal of f in a certain sense.

Theorem 5.16

a) Suppose $f : D \to \mathbb{C}$ is differentiable on D and has an isolated zero of order N at z_0. Then there exists $N_r'(z_0) \subset D$ such that the function $g : N_r'(z_0) \to \mathbb{C}$ given by $g(z) \equiv \frac{1}{f(z)}$ has a pole of order N at z_0.

b) Suppose $f : D \to \mathbb{C}$ has a pole of order N at z_0. Then there exists $N_r'(z_0) \subset D$ such that the function $g : N_r(z_0) \to \mathbb{C}$ given by $g(z) \equiv \begin{cases} \frac{1}{f(z)} & \text{for } z \in N_r'(z_0) \\ 0 & \text{for } z = z_0 \end{cases}$ has a zero of order N at z_0.

PROOF To prove part a), note that there exists $h : D \to \mathbb{C}$ differentiable on D with $h(z_0) \neq 0$ such that $f(z) = (z - z_0)^N h(z)$. By continuity of h there exists $N_r(z_0) \subset D$ such that $h(z) \neq 0$ on $N_r(z_0)$. Therefore, the function $g : N_r'(z_0) \to \mathbb{C}$ given by $g(z) = \frac{1}{f(z)} = \frac{1/h(z)}{(z-z_0)^N}$ has a pole of order N at z_0. To prove part b), we may assume there exists a function $f_0 : N_s(z_0) \to \mathbb{C}$ where $N_s(z_0) \subset D$, f_0 is differentiable on $N_s(z_0)$, and $f_0(z_0) \neq 0$ such that $f(z) = \frac{f_0(z)}{(z-z_0)^N}$ on $N_s'(z_0)$. Again, by continuity of f_0 there exists $N_r(z_0) \subset N_s(z_0)$ such that $f_0(z) \neq 0$ on $N_r(z_0)$. Therefore, on $N_r'(z_0)$ we have $\frac{1}{f(z)} = (z - z_0)^N \frac{1}{f_0(z)}$, which implies that z_0 can be defined as a zero of order N for $\frac{1}{f}$. In fact, defining $g : N_r(z_0) \to \mathbb{C}$ by $g(z) = \begin{cases} \frac{1}{f(z)} & \text{for } z \in N_r'(z_0) \\ 0 & \text{for } z = z_0 \end{cases}$ obtains

$g(z) = \begin{cases} (z-z_0)^N \frac{1}{f_0(z)} & \text{for } z \in N_r'(z_0) \\ 0 & \text{for } z = z_0 \end{cases}$. This function g has a zero of order N at z_0 as was to be shown. ♦

It should be noted that in the proof of part *b)* above, something a bit subtle occurs. It is not quite true that the reciprocal of the function f having a pole of order N at z_0 is a function with a zero of order N at z_0. This is because the original function f isn't even defined at z_0, and so the reciprocal of f isn't defined there either, initially. However, it turns out that z_0 is a removable singularity of the reciprocal of f. When assigned the value zero at z_0, the reciprocal of f becomes a differentiable function at z_0 with a zero of multiplicity N there. So while the domain of the function f excludes z_0, we can extend the domain of the function $\frac{1}{f}$ so as to include z_0.

6 The Residue Calculus

6.1 Residues and the Residue Theorem

Let z_0 be an isolated singularity of $f : D \to \mathbb{C}$ and suppose f is differentiable on the annulus $A_r^R(z_0) \subset D$. By Laurent's theorem, f has a Laurent series expansion on $A_r^R(z_0)$,

$$f(z) = \sum_{j=0}^{\infty} a_j (z-z_0)^j + \sum_{j=1}^{\infty} \frac{b_j}{(z-z_0)^j}.$$

The coefficient b_1 in this expansion is especially significant.

Definition 6.1 Let z_0 be an isolated singularity of the function $f : D \to \mathbb{C}$. The coefficient b_1 of the Laurent series representation of f on $A_r^R(z_0) \subset D$ is called the **residue** of f at z_0 and is denoted by $\text{Res}(f, z_0)$.

In fact, if $C \subset A_r^R(z_0)$ is a simple closed contour with $n_C(z_0) = 1$, Laurent's theorem tells us that

$$b_1 = \frac{1}{2\pi i} \oint_C f(z)\, dz.$$

Rearranging this formula obtains

$$\oint_C f(z)\, dz = 2\pi i\, b_1, \tag{9.29}$$

and so the value of b_1 is instrumental in evaluating complex integrals. If f is differentiable for all $z \in \text{Int}(C)$, and in particular at z_0, Cauchy's integral theorem implies that this integral will be zero and so b_1 should be zero as well. In fact, for such a function the whole singular part of the Laurent expansion will vanish, leaving just the analytic part as the Taylor series expansion for

f. But for a function f that is *not* differentiable at z_0, we will find that for a simple closed contour C with $n_C(z_0) = 1$ the only term in the Laurent series expansion for f that contributes a nonzero value to the integral in (9.29) is the b_1 term, hence the name "residue." For these reasons, it is of particular interest to be able to compute the residue b_1 easily.

Example 6.2 *Recall the function $f : \mathbb{C} \setminus \{0\} \to \mathbb{C}$ given by $f(z) = \frac{1}{z^3}\, e^z$ discussed in Example 5.5 on page 494. We found the Laurent series of f centered at $z_0 = 0$ and convergent on $A_0^\infty(0)$ to be*

$$f(z) = \frac{1}{z^3}\sum_{j=0}^{\infty}\frac{1}{j!}z^j = \sum_{j=0}^{\infty}\frac{1}{j!}z^{j-3} = \frac{1}{z^3} + \frac{1}{z^2} + \frac{1}{2!}\frac{1}{z} + \frac{1}{3!} + \frac{1}{4!}z + \frac{1}{5!}z^2 + \cdots .$$

From this, we see that $z_0 = 0$ is a pole of order 3, *and* $\mathrm{Res}(f, 0) = b_1 = \frac{1}{2}$. ◀

Note that in the above example we found the Laurent series expansion of the function $f(z) = \frac{1}{z^3} f_0(z)$ on $A_0^\infty(0)$ by finding the Taylor series for the differentiable function f_0 on a neighborhood of $z_0 = 0$, and then dividing that Taylor series by z^3. The residue b_1 of the resulting Laurent series corresponded to the second coefficient of the Taylor series of f_0, that is, $\mathrm{Res}(f, 0) = \frac{f_0''(0)}{2!} = \frac{1}{2}$. This is so because the coefficient of z^2 in the Taylor series of f_0, when divided by z^3, becomes the coefficient of the $\frac{1}{z}$ term of the final Laurent series. This is true more generally, and is stated in the following important result.

Theorem 6.3 *Suppose $f : D \to \mathbb{C}$ is differentiable on $A_r^R(z_0)$ and let z_0 be a pole of order N of f where $f_0 : N_R(z_0) \to \mathbb{C}$ is the differentiable function on $N_R(z_0)$ such that $f(z) = \frac{f_0(z)}{(z-z_0)^N}$, and $f_0(z_0) \neq 0$. Then*

$$\mathrm{Res}(f, z_0) = \frac{f_0^{(N-1)}(z_0)}{(N-1)!}.$$

Note that if $N = 1$, then $\mathrm{Res}(f, z_0) = f_0(z_0)$. Also recall that if z_0 is a pole of order N for f, Theorem 5.14 on page 499 showed that we may choose f_0 to be given by $f_0(z) = (z - z_0)^N f(z)$ in such a way that f_0 is differentiable at z_0.

▶ **9.57** *Prove Theorem 6.3.*

▶ **9.58** *Find* $\mathrm{Res}(\frac{\sin z}{z}, 0)$.

▶ **9.59** *Find the residues of the function $f : \mathbb{C}\setminus\{\pm 1\} \to \mathbb{C}$ given by $f(z) = \frac{1}{z^2-1}$ at each of its singularities.*

▶ **9.60** *Prove the following: Suppose the functions f and g are differentiable at z_0, $f(z_0) \neq 0$, $g(z_0) = 0$, and $g'(z_0) \neq 0$. Then the function given by $\frac{f(z)}{g(z)}$ has a simple pole at z_0, and* $\mathrm{Res}\left(\frac{f(z)}{g(z)}, z_0\right) = \frac{f(z_0)}{g'(z_0)}$.

Example 6.4 *Consider $f : N'_{\frac{3}{2}}(1) \to \mathbb{C}$ given by $f(z) = \frac{1}{z^2-1}$. We will use equality (9.29) and Theorem 6.3 to evaluate*

$$\oint_C \frac{dz}{z^2-1}, \quad \text{where } C = C_1(1).$$

Note that $z_0 = 1$ is a simple pole of f, and that $f_0 : N_{\frac{3}{2}}(1) \to \mathbb{C}$ given by $f_0(z) = \frac{1}{z+1}$ is differentiable on $N_{\frac{3}{2}}(1)$ with $f_0(1) = \frac{1}{2} \neq 0$, and $f(z) = \frac{f_0(z)}{z-1}$. By Theorem 6.3, Res$(f,1) = f_0(1) = \frac{1}{2}$*, and therefore equality (9.29) yields*

$$\oint_C \frac{dz}{z^2-1} = 2\pi\, i \, \text{Res}(f,1) = \pi\, i.$$

◀

► **9.61** Evaluate $\oint_C \frac{dz}{z^2-1}$ where $C = C_1(-1)$.

► **9.62** Evaluate $\oint_C \frac{dz}{z^2-1}$ where $C = [2i+2, 2i-2, -2i-2, -2i+2, 2i+2]$.

The following important theorem generalizes the idea illustrated in the previous example. It highlights the great practicality in using residues to compute complex integrals.

Theorem 6.5 (The Residue Theorem)
Let $\widetilde{D} \subset \mathbb{C}$ be open and connected and suppose $\{z_1, \ldots, z_M\} \subset \widetilde{D}$. Let $D \equiv \widetilde{D} \setminus \{z_1, \ldots, z_M\}$ and suppose $f : D \to \mathbb{C}$ is differentiable on D. If $C \subset D$ is any closed contour such that $n_C(w) = 0$ for all $w \in \widetilde{D}^C$, then,

$$\oint_C f(z)dz = 2\pi i \sum_{j=1}^{M} n_C(z_j)\text{Res}(f, z_j).$$

PROOF Since f is differentiable on $D \equiv \widetilde{D} \setminus \{z_1, \ldots, z_M\}$, we may choose $r > 0$ small enough such that the deleted neighborhoods $N'_r(z_j) \subset D \setminus C$ for $j = 1, \ldots, M$ are disjoint (see Figure 9.6). Also, choose circles C_j parametrized by $\zeta_j : [0, 2\pi] \to N'_r(z_j)$ with

$$\zeta_j(t) = \tfrac{1}{2}\, r\, e^{i n_j t}, \quad \text{where } n_j \equiv n_C(z_j).$$

Then it follows that

(1) $n_{C_j}(z_j) = n_j = n_C(z_j)$,

(2) $n_C(w) = 0 = \sum_{j=1}^{M} n_{C_j}(w)$ for $w \in \widetilde{D}^C$,

which by Theorem 3.6 of Chapter 8 implies

$$\oint_C f(z)dz = \sum_{j=1}^{M} \oint_{C_j} f(z)dz.$$

But since

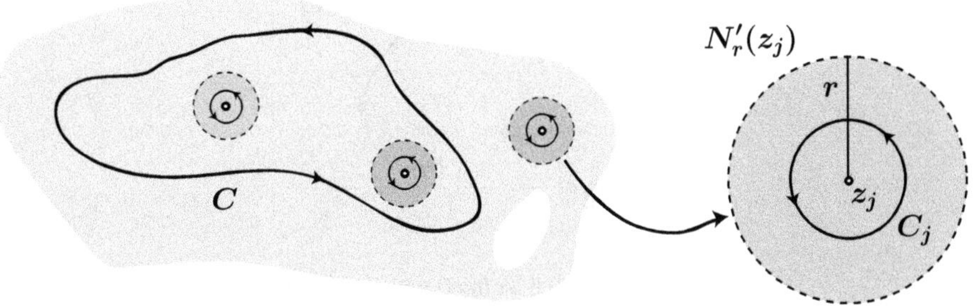

Figure 9.6 *The situation in the proof of the residue theorem.*

$$\begin{aligned}\oint_{C_j} f(z)dz &= n_j \oint_{C_{\frac{r}{2}}(z_j)} f(z)dz \\ &= n_j\, 2\pi i\, \mathrm{Res}(f, z_j) \\ &= 2\pi i\, n_C(z_j)\, \mathrm{Res}(f, z_j),\end{aligned}$$

the theorem follows. ♦

The above theorem can also be thought of as a generalized Cauchy's integral theorem in that, for a function $f : D \to \mathbb{C}$ integrated along contour $C \subset D$, it provides a formula for the value of the integral in the more complicated case where there are finitely many isolated singularities in $D \setminus C$. When no such singularities are present, it reduces to Cauchy's integral theorem.

Example 6.6 *We will evaluate the integral* $\oint\limits_C \frac{dz}{z^2-1}$ *for the contour* C *shown in Figure 9.7.*

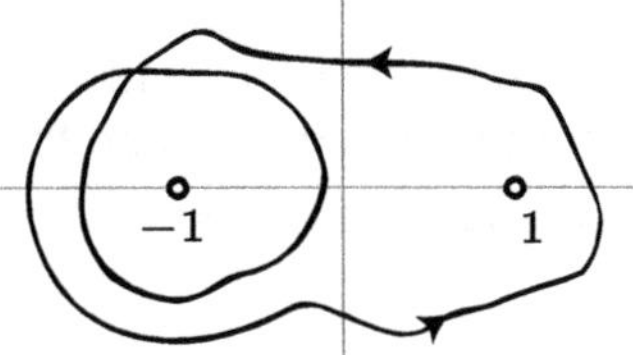

Figure 9.7 *The contour C of Example 6.6.*

Note that $f : \mathbb{C} \setminus \{\pm 1\} \to \mathbb{C}$ *given by* $f(z) = \frac{1}{z^2-1}$ *is differentiable on* $D \equiv \mathbb{C} \setminus \{\pm 1\}$. *Since* $z = \pm 1 \notin C$, *and* $n_C(1) = 1$ *and* $n_C(-1) = 2$ *can be easily verified using the visual inspection technique justified in the Appendix of Chapter 8, we have*

$$\begin{aligned}\oint\limits_C \frac{dz}{z^2-1} &= 2\pi i \Big[n_C(-1)\mathrm{Res}\Big(\tfrac{1}{z^2-1}, -1\Big) + n_C(1)\mathrm{Res}\Big(\tfrac{1}{z^2-1}, 1\Big)\Big] \\ &= 2\pi i \Big[2\big(-\tfrac{1}{2}\big) + 1\big(\tfrac{1}{2}\big)\Big] = -\pi i.\end{aligned}$$

◄

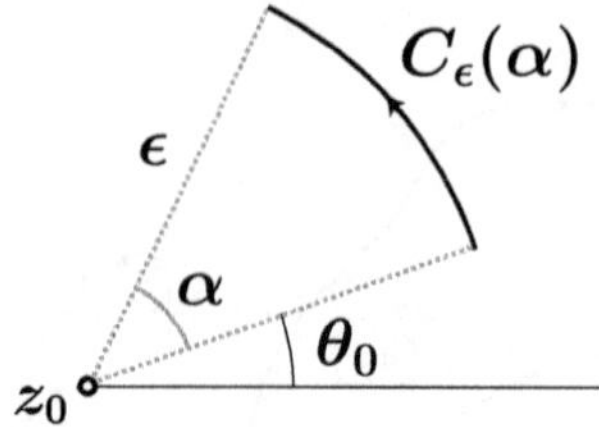

Figure 9.8 *A fractional circular arc.*

The Fractional Residue Theorem

Suppose z_0 is a simple pole of the function $f : D \to \mathbb{C}$. Let θ_0 be some fixed angle, and suppse $\epsilon > 0$. Let $C_\epsilon(\alpha) \equiv \{z_0 + \epsilon e^{it} : \theta_0 \le t \le \theta_0 + \alpha\}$ be the circular arc subtending α radians of the circle of radius ϵ centered at z_0 as shown in Figure 9.8. We will sometimes need to evaluate integrals of the form

$$\lim_{\epsilon \to 0} \left(\int_{C_\epsilon(\alpha)} f(z)\, dz \right). \tag{9.30}$$

In such cases, the following result will be useful.

Theorem 6.7 *Suppose z_0 is a simple pole of the function $f : D \to \mathbb{C}$, and let $C_\epsilon(\alpha)$ be defined as above. Then,*

$$\lim_{\epsilon \to 0} \left(\int_{C_\epsilon(\alpha)} f(z)\, dz \right) = i\alpha \operatorname{Res}(f, z_0).$$

Note that if $\alpha = 2\pi$ this result is consistent with the residue theorem. And yet, the proof of this result requires much less than that of the residue theorem.

PROOF [3] Since z_0 is a simple pole, we know that on some neighborhood $N_r(z_0)$ we have

$$f(z) = \frac{b_1}{z - z_0} + a_0 + a_1(z - z_0) + \cdots = \frac{b_1}{z - z_0} + g(z),$$

where g is differentiable on $N_r(z_0)$. If $\epsilon > 0$ is small enough, then $C_\epsilon(\alpha) \subset N_r(z_0)$, and

$$\int_{C_\epsilon(\alpha)} f(z)\, dz = b_1 \int_{C_\epsilon(\alpha)} \frac{dz}{z - z_0} + \int_{C_\epsilon(\alpha)} g(z)\, dz. \tag{9.31}$$

Note that

$$\int_{C_\epsilon(\alpha)} \frac{dz}{z - z_0} = \int_{\theta_0}^{\theta_0 + \alpha} \frac{i\epsilon e^{it}}{\epsilon e^{it}}\, dt = i\alpha,$$

[3] We follow [MH98].

and so all that remains is to show that the second integral on the right of (9.31) vanishes as $\epsilon \to 0$. To establish this, note that $g(z)$ must be bounded on $C_\epsilon(\alpha)$, and so

$$\left| \int\limits_{C_\epsilon(\alpha)} g(z)\, dz \right| \le M\, L_{C_\epsilon(\alpha)} \quad \text{for some } M \ge 0.$$

From this we obtain $\lim_{\epsilon \to 0}\Big(\int_{C_\epsilon(\alpha)} g(z)\, dz \Big) = 0$, and the theorem is proved. ♦

6.2 Applications to Real Improper Integrals

The residue theorem is an especially useful tool in evaluating certain real improper integrals, as the following examples serve to illustrate.

Example 6.8 *We begin by applying the technique to a real improper integral whose value we already know, namely,* $\int_{-\infty}^{\infty} \frac{dx}{x^2+1} = \pi$. *Traditionally this result is obtained by simply recognizing* $\tan^{-1} x$ *as the antiderivative to* $\frac{1}{x^2+1}$. *Here, we'll begin by defining the contour* $C = \{C_1, C_2\}$, *where* $C_1 = [-R, R]$ *for* $R > 1$ *and* C_2 *is the semicircle going counterclockwise from* $(R, 0)$ *to* $(-R, 0)$ *as illustrated in Figure 9.9.*

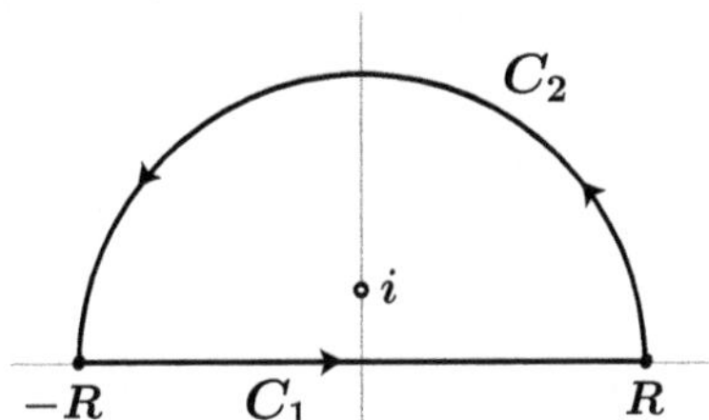

Figure 9.9 *The contour of Example 6.8.*

We will evaluate

$$\oint_C \frac{dz}{z^2+1} = \int_{C_1} \frac{dz}{z^2+1} + \int_{C_2} \frac{dz}{z^2+1}.$$

By the residue theorem, we know

$$\oint_C \frac{dz}{z^2+1} = 2\pi\, i \mathrm{Res}\Big(\frac{1}{z^2+1}, i\Big).$$

To find $\mathrm{Res}\big(\frac{1}{z^2+1}, i\big)$, *we write*

$$\frac{1}{z^2+1} = \frac{\frac{1}{z+i}}{z-i},$$

which gives us $\text{Res}\left(\frac{1}{z^2+1}, i\right) = \frac{1}{2i}$. Therefore, $\oint_C \frac{dz}{z^2+1} = 2\pi i \left(\frac{1}{2i}\right) = \pi$, and

$$\pi = \oint_C \frac{dz}{z^2+1} = \int_{C_1} \frac{dz}{z^2+1} + \int_{C_2} \frac{dz}{z^2+1}.$$

If we parametrize C_1 by $z_1 : [-R, R] \to \mathbb{C}$ given by $z_1(t) = t$ and C_2 by $z_2 : [0, \pi] \to \mathbb{C}$ given by $z_2(t) = Re^{it}$, we can write

$$\pi = \int_{-R}^{R} \frac{dt}{t^2+1} + \int_0^{\pi} \frac{iRe^{it}dt}{R^2e^{2it}+1}. \tag{9.32}$$

We will take the limit of this expression as $R \to \infty$. In doing so, the second integral on the right-hand side of (9.32) will vanish. To see this, note that

$$\begin{aligned} \left| \int_0^{\pi} \frac{iRe^{it}}{R^2e^{2it}+1} dt \right| &\leq \int_0^{\pi} \frac{R}{|R^2e^{2it}+1|} dt \\ &\leq \int_0^{\pi} \frac{R}{R^2-1} dt = \frac{R\pi}{R^2-1}, \end{aligned}$$

and from this we obtain

$$\lim_{R\to\infty} \int_0^{\pi} \frac{iRe^{it}}{R^2e^{2it}+1} dt = 0.$$

Given this, we conclude that

$$\pi = CPV \int_{-\infty}^{\infty} \frac{dt}{t^2+1} = \int_{-\infty}^{\infty} \frac{dt}{t^2+1}.$$

The last equality follows since the integrand is an even function of t, and therefore the Cauchy principal value and the "regular" value are equal in this case. ◀

Example 6.9 In this example, we will evaluate $\int_0^{\infty} \frac{dx}{x^3+1}$. To do so, we consider the integral $\oint_C \frac{dz}{z^3+1}$, where $C = (C_1, C_2, C_3)$, and where $\alpha = \frac{2\pi}{3}$ as shown in Figure 9.10.

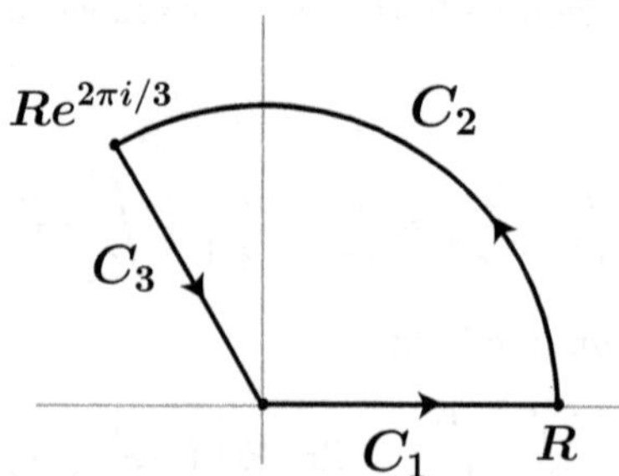

Figure 9.10 *The contour of Example 6.9.*

We will first find the singularities of the function $f(z) = \frac{1}{z^3+1}$. We will show *that they are poles and find the residue for each. To this end, we set* $z^3+1=0$ *and solve for the singularities* $z_1 = e^{\frac{i\pi}{3}}, z_2 = -1, z_3 = e^{-\frac{i\pi}{3}}$. *Note that* $z^3+1 = (z-z_1)(z-z_2)(z-z_3)$. *(Why?) Now it follows that each singularity of*

$$f(z) = \frac{1}{z^3+1} = \frac{1}{(z-z_1)(z-z_2)(z-z_3)}$$

is a pole of order 1 *and that*

$$\text{Res}\left(\frac{1}{z^3+1}, z_1\right) = \frac{1}{(z_1-z_2)(z_1-z_3)} = \frac{1}{i\sqrt{3}\left(\frac{3}{2}+i\frac{\sqrt{3}}{2}\right)},$$

$$\text{Res}\left(\frac{1}{z^3+1}, z_2\right) = \frac{1}{(z_2-z_1)(z_2-z_3)} = \frac{1}{3}, \quad \text{and,}$$

$$\text{Res}\left(\frac{1}{z^3+1}, z_3\right) = \frac{1}{(z_3-z_1)(z_3-z_2)} = -\frac{1}{i\sqrt{3}}\left(\frac{1}{\frac{3}{2}-i\frac{\sqrt{3}}{2}}\right).$$

There is only one singularity of $\frac{1}{z^3+1}$ *inside* C *(which one?), and therefore,*

$$\oint_C \frac{dz}{z^3+1} = 2\pi i \text{Res}\left(\frac{1}{z^3+1}, z_1\right) = \frac{2\pi i}{i\sqrt{3}}\left(\frac{1}{\frac{3}{2}+i\frac{\sqrt{3}}{2}}\right) = \frac{2\pi}{\sqrt{3}}\left(\frac{1}{\frac{3}{2}+i\frac{\sqrt{3}}{2}}\right).$$

We now parametrize as follows:

a) C_1 *by* $z_1 : [0,R] \to \mathbb{C}$ *given by* $z_1(t) = t$,

b) C_2 *by* $z_2 : [0, \frac{2\pi}{3}] \to \mathbb{C}$ *given by* $z_2(t) = Re^{it}$, *and*

c) C_3 *by* $z_3 : [0,R] \to \mathbb{C}$ *given by* $z_3(t) = (R-t)e^{i\frac{2\pi}{3}}$,

and calculate

$$\begin{aligned}
\oint_C \frac{dz}{z^3+1} &= \int_{C_1} \frac{dz}{z^3+1} + \int_{C_2} \frac{dz}{z^3+1} + \int_{C_3} \frac{dz}{z^3+1} \\
&= \int_0^R \frac{dt}{t^3+1} + \int_0^{\frac{2\pi}{3}} \frac{iRe^{it}dt}{R^3e^{3it}+1} + \int_0^R \frac{-e^{i\frac{2\pi}{3}}dt}{(R-t)^3+1} \\
&= \int_0^R \frac{dt}{t^3+1} + \int_0^{\frac{2\pi}{3}} \frac{iRe^{it}dt}{R^3e^{3it}+1} - e^{i\frac{2\pi}{3}} \int_0^R \frac{dt}{t^3+1} \\
&= (1-e^{i\frac{2\pi}{3}}) \int_0^R \frac{dt}{t^3+1} + \int_0^{\frac{2\pi}{3}} \frac{iRe^{it}dt}{R^3e^{3it}+1}.
\end{aligned} \tag{9.33}$$

We will show that the last integral on the right of (9.33) *is zero. To do this,* we will allow R to get arbitrarily large. First, note that

$$\left|\int_0^{\frac{2\pi}{3}} \frac{iRe^{it}dt}{R^3e^{3it}+1}\right| \leq \int_0^{\frac{2\pi}{3}} \frac{Rdt}{|R^3e^{3it}+1|}$$

$$\leq \int_0^{\frac{2\pi}{3}} \frac{Rdt}{R^3-1} = \left(\frac{R}{R^3-1}\right)\frac{2\pi}{3}.$$

This implies that

$$\lim_{R\to\infty} \int_0^{\frac{2\pi}{3}} \frac{iRe^{it}dt}{R^3e^{3it}+1} = 0.$$

Putting this all together, and letting $R \to \infty$, we obtain

$$\frac{2\pi}{\sqrt{3}}\left(\frac{1}{\frac{3}{2}+i\frac{\sqrt{3}}{2}}\right) = \oint_C \frac{dz}{z^3+1} = (1-e^{i\frac{2\pi}{3}})\int_0^\infty \frac{dt}{t^3+1},$$

which implies

$$\int_0^\infty \frac{dt}{t^3+1} = \frac{2\pi}{\sqrt{3}}\left(\frac{1}{\frac{3}{2}+i\frac{\sqrt{3}}{2}}\right)\left(\frac{1}{1-e^{i\frac{2\pi}{3}}}\right) = \frac{2\pi}{3\sqrt{3}}.$$

◀

Example 6.10 *Consider the integral*

$$\oint_C e^{1/z}dz,$$

where the contour C is shown in Figure 9.11.

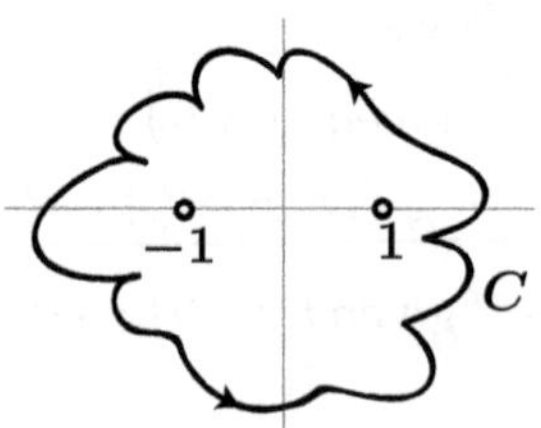

Figure 9.11 *The contour of Example 6.10.*

Since $e^w = \sum_{j=0}^{\infty} \frac{w^j}{j!}$ converges for every $w \in \mathbb{C}$, it follows that

$$e^{1/z} = \sum_{j=0}^{\infty} \frac{1}{j!z^j} \quad \text{converges for} \quad z \neq 0.$$

But this series is the Laurent series for $e^{1/z}$, from which we obtain $\text{Res}(e^{1/z}, 0) = 1$ and hence,

$$\oint_C e^{1/z}dz = 2\pi i.$$

◀

Example 6.11 *Consider the integral* $\int_0^\infty \frac{\sin x}{x}dx$. *Since*

$$\int_0^\infty \frac{\sin x}{x}dx = \mathrm{Im}\left(\int_0^\infty \frac{e^{ix}}{x}dx\right),$$

we will evaluate $\oint_C \frac{e^{iz}}{z}dz$, *where* $C = (C_1, C_2, C_3, C_4)$, *as shown in Figure 9.12.*

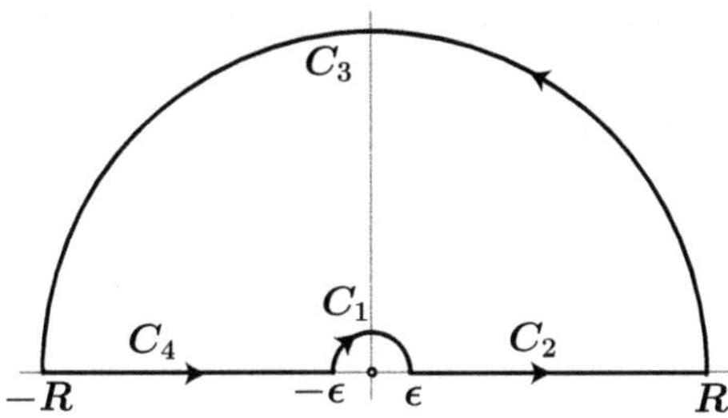

Figure 9.12 *The contour of Example 6.11.*

To evaluate each of these integrals, we note that:

$$\int_{C_1} \frac{e^{iz}}{z}dz = -\int_{C_\epsilon(\pi)} \frac{e^{iz}}{z}dz,$$

$$\int_{C_2} \frac{e^{iz}}{z}dz = \int_\epsilon^R \frac{e^{it}}{t}dt,$$

$$\int_{C_3} \frac{e^{iz}}{z}dz = \int_0^\pi \frac{e^{iRe^{it}}}{Re^{it}}iRe^{it}dt = i\int_0^\pi e^{iRe^{it}}dt, \qquad \text{and}$$

$$\int_{C_4} \frac{e^{iz}}{z}dz = \int_{-R}^{-\epsilon} \frac{e^{it}}{t}dt.$$

Since $\oint_C \frac{e^{iz}}{z}dz = 0$ *(Why?), it follows that*

$$\begin{aligned} 0 &= -\int_{C_\epsilon(\pi)} \frac{e^{iz}}{z}dz + i\int_0^\pi e^{iRe^{it}}dt + \int_\epsilon^R \frac{e^{it}}{t}dt + \int_{-R}^{-\epsilon} \frac{e^{it}}{t}dt \\ &= -\int_{C_\epsilon(\pi)} \frac{e^{iz}}{z}dz + i\int_0^\pi e^{iRe^{it}}dt + \int_\epsilon^R \frac{e^{it} - e^{-it}}{t}dt \\ &= -\int_{C_\epsilon(\pi)} \frac{e^{iz}}{z}dz + i\int_0^\pi e^{iRe^{it}}dt + \int_\epsilon^R \frac{2i\sin t}{t}dt. \end{aligned}$$

From this we obtain

$$2i\int_0^\infty \frac{\sin t}{t}dt = \lim_{\epsilon\to 0}\left(\int_{C_\epsilon(\pi)} \frac{e^{iz}}{z}dz\right) - \lim_{R\to\infty} i\int_0^\pi e^{iRe^{it}}dt. \tag{9.34}$$

To evaluate the limits in the above, we note that

$$\left|\int_0^\pi e^{iRe^{it}}\,dt\right| = \left|\int_0^\pi e^{iR\cos t}e^{-R\sin t}\,dt\right| \le \int_0^\pi e^{-R\sin t}\,dt,$$

the last of which $\to 0$ *as* $R\to\infty$ *due to Jordan's inequality (see Exercise 7.24 on page 358 in Chapter 7). Also, by the fractional residue theorem,*

$$\lim_{\epsilon\to 0}\left(\int_{C_\epsilon(\pi)}\frac{e^{iz}}{z}\,dz\right) = i\pi\,\mathrm{Res}\left(\frac{e^{iz}}{z},0\right) = i\pi.$$

All of this allows us to rewrite (9.34) as

$$2i\int_0^\infty \frac{\sin t}{t}\,dt = i\pi,$$

which obtains

$$\int_0^\infty \frac{\sin t}{t}\,dt = \frac{\pi}{2}.$$

◀

7 Supplementary Exercises

1. *Find the power series expansion and radius of convergence for* $f:\mathbb{C}\setminus\{-2\}\to\mathbb{C}$ *given by* $f(z)=\frac{1}{2+z}$ *in powers of* z*. How about in powers of* $z-i$*? In powers of* $z+2$*?*

2. *Find the radius of convergence for each of the following:*

a) $\sum_{j=0}^\infty \frac{j^3}{(j+3)4^j}z^j$ b) $\sum_{j=0}^\infty \frac{j!x^j}{e^{2j}}$ c) $\sum_{j=1}^\infty \left(\frac{2^j+5^j}{j^5}\right)z^j$ d) $\sum_{j=1}^\infty \frac{(z-\pi)^j}{j^5}$

e) $\sum_{j=1}^\infty \frac{2^j(z+i)^j}{j(j+1)}$ f) $\sum_{j=1}^\infty \frac{(z+1)^j}{j^{2j}}$ g) $\sum_{j=1}^\infty \frac{5^j(x-1)^j}{j^2}$

3. *Use the series expansion* $\frac{1}{1-z}=\sum_{j=0}^\infty z^j$ *to find the series expansion of* $\frac{1}{(1-z)^2} = \left(\sum_{j=0}^\infty z^j\right)\left(\sum_{j=0}^\infty z^j\right)$*. What is the radius of convergence?*

4. *Use the expansions* $f(z)=\frac{1}{1-z}=\sum_{j=0}^\infty z^j$ *and* $g(z)=\frac{1}{1-z^2}=\sum_{j=0}^\infty z^{2j}$ *to find the expansion for* $\frac{f(z)}{g(z)}$*. Does this expansion reduce to* $1+z$ *for appropriate values of* z*?*

5. *Consider the function* $f:\mathbb{R}\setminus\{0\}\to\mathbb{R}$ *given by* $f(x)=\frac{1}{x}$*. Find the Taylor series expansion for* f *in powers of* $x-1$*. Where does it converge?*

6. *Find the Taylor series expansion for the function* $f:\mathbb{C}\setminus\{i\}\to\mathbb{C}$ *given by* $f(z)=\frac{e^z}{z-i}$ *in powers of* $z+1$*. Where does it converge?*

7. *Confirm that the product of the Taylor series representation for* e^x *with that of* e^{iy} *yields that of* e^z*, as it must.*

8. *Differentiate the Taylor series expansion for* $\sin z$ *centered at* $z_0 = 0$ *term-by-term to obtain the Taylor series for* $\cos z$ *centered at* $z_0 = 0$.

9. *Find the Taylor series representation for* $f(z) = \frac{1}{z}$ *centered at* $z_0 = -1$. *Differentiate it term-by-term to find the Taylor series for* $g(z) = \frac{1}{z^2}$.

10. *Use multiplication and division of power series to find the power series representations centered at the origin for the following:*

a) $\frac{\ln(1+x)}{1+x}$ *b)* $\cos^2 x$ *c)* $e^x \sin x$ *d)* $\frac{e^z}{1-z^2}$ *e)* $\sin 2x = 2\sin x \cos x$

11. *Let* $\frac{x}{e^x-1} = \sum_{j=0}^{\infty} \frac{B_j}{j!} x^j$. *The coefficients* B_j *for* $j \geq 1$ *are called the* Bernoulli numbers. *Show that* $B_3 = B_5 = B_7 = \cdots = 0$, *and calculate* B_1, B_2, B_4, *and* B_6.

12. *Find the Taylor series expanded around* $x_0 = 0$ *for the following functions:*
a) $F(x) = \int_0^x \frac{\sin t}{t}\,dt$ *b)* $G(x) = \int_0^x \left(\frac{\sin t}{t}\right)^2 dt$.

13. *Consider the function* $f : \mathbb{C}\backslash\{\pm 3\} \to \mathbb{C}$ *given by* $f(z) = \frac{1}{z^2-9}$.

a) *Find the Taylor series for* $f(z)$ *that is valid for* $|z| < 3$.

b) *Find the Laurent series for* $f(z)$ *that is valid for* $0 < |z-3| < 6$ *and* $\text{Res}(f(z), 3)$.

c) *Find the Laurent series for* $f(z)$ *that is valid for* $|z| > 6$ *and* $\text{Res}(f(z), 3)$.

14. *Find the Laurent expansion for the function* $f : \mathbb{C}\backslash\{\pm 1\} \to \mathbb{C}$ *given by* $f(z) = \frac{1}{z^2+1}$ *in powers of* $z-1$. *Where does it converge? How about in powers of* $z+1$? *Where does it converge now?*

15. *Find the Laurent expansion for the function given by* $f(z) = \frac{\cos z}{z^2+(1-i)z-i}$ *that converges for* $0 < |z-i| < \sqrt{2}$. *How about for* $|z-1| < \sqrt{2}$?

16. *Find the Laurent expansion for the function given by* $f(z) = \frac{1}{z(z-1)(z-2)}$ *that converges on* $A_0^1(0)$. *How about on* $A_1^2(0)$? *How about on* $A_2^\infty(0)$?

17. *Find the Laurent series for the following:*
a) $\csc z$ *on* $A_0^\pi(0)$ *b)* $\frac{1}{e^z-1}$ *on* $A_0^{2\pi}(0)$ *c)* $\frac{e^z}{z\,(z^2+1)}$ *on* $A_0^1(0)$.

18. *Classify the indicated singularities:*
a) $f(z) = \frac{z^2-2z+3}{z-2}$ *at* $z_0 = 2$ *b)* $f(z) = \frac{\sin z}{z^4}$ *at* $z_0 = 0$ *c)* $f(z) = \frac{e^z-1}{z}$ *at* $z_0 = 0$.

19. *Classify the singularities at* $z_0 = 0$ *for the following:*
a) $f(z) = z \sin\frac{1}{z}$ *b)* $f(z) = z \cos\frac{1}{z}$ *c)* $f(z) = \frac{\text{Log}(z+1)}{z^2}$.

20. *Show that a complex function* f *has a removable singularity at* ∞ *if and only if* f *is constant.*

21. *Show that a complex function* f *is entire with a pole at* ∞ *of order* m *if and only if* f *is a polynomial of degree* m.

22. *Show that* $\int_0^{2\pi}\frac{dt}{2+\cos t}=\frac{2\pi}{\sqrt{3}}$. *To do so, express* $\cos t$ *in terms of the complex exponential function to show that the given integral expression is equivalent to* $\frac{2}{i}\oint\limits_{C_1(0)}\frac{dz}{z^2+4z+1}$. *Then use residues to obtain the value* $\frac{2\pi}{\sqrt{3}}$. *For* $a>b>0$, *show more generally that* $\int_0^{2\pi}\frac{dt}{a+b\,\cos t}=\frac{2\pi}{\sqrt{a^2-b^2}}$.

23. *Evaluate:* $a)\ \oint\limits_{C_1(0)}\frac{e^z}{4z^2+1}\,dz$ $\ b)\ \oint\limits_{C_2(0)}\frac{z^3+1}{(z+1)(z^2+16)}\,dz$ $\ c)\ \oint\limits_{C_2(0)}\frac{5z-2}{z\,(z-1)}\,dz$ $\ d)\ \oint\limits_{C}\frac{dz}{(z^2-1)^2+1}$ *where* $C=[2,2+i,-2+i,-2,2]$.

24. *Evaluate the following real integrals.*

$a)\ \int_0^\infty\frac{dx}{(1+x^2)^2}$ $\quad b)\ \int_0^\infty\frac{x^2+3}{(1+x^2)(4+x^2)}\,dx$ $\quad c)\ \int_{-1}^1\frac{dx}{(2-x^2)\sqrt{1-x^2}}$ $\quad d)\ \int_0^\infty\frac{x^2+1}{x^4+5x^2+4}\,dx$

$e)\ \int_0^\infty\frac{\cos x}{1+x^2}\,dx$ $\quad f)\ \int_0^\infty\frac{dx}{(1+x)\sqrt{x}}$ $\quad g)\ \int_0^\infty\frac{\cos x}{1+x^4}\,dx$ $\quad h)\ \int_0^\infty\frac{\ln x}{x^4+1}\,dx$ $\quad i)\ \int_0^\infty\frac{\ln x}{x^2+1}dx$

$j)\ \int_0^\infty\left(\frac{\sin x}{x}\right)^2dx$ $\quad k)\ \int_0^\infty\frac{dx}{\sqrt{x}(x^2+1)}$ $\quad l)\ \int_{-\infty}^\infty\frac{dx}{x^2+4x+2}$ $\quad m)\ \int_0^\infty\frac{x\,\sin 2x}{x^4+1}\,dx$

$n)\ \int_0^\infty\frac{\cos x-1}{x^2}\,dx$ $\quad o)\ \int_{-\infty}^\infty\frac{dx}{x^2+x+1}$ $\quad p)\ \int_{-\infty}^\infty\frac{e^{x/2}}{1+e^x}\,dx$

25. *What kind of singularity does the function* $f:\mathbb{C}\setminus\{0\}\to\mathbb{C}$ *given by* $f(z)=\left(e^z-1\right)^{-1}$ *have at* $z_0=0$? *Find three nonzero terms of the Laurent series expansion of this function centered at* $z_0=0$. *Then, find the largest value of* $R>0$ *such that the Laurent series converges on* $0<|z|<R$.

26. *Show that the function* $f:\mathbb{C}\setminus\{0\}\to\mathbb{C}$ *given by* $f(z)=\frac{z}{e^z-1}$ *has a removable singularity at* $z_0=0$. *Find the first two terms in the Laurent series expansion for* f, *and find the radius of convergence.*

27. *For each of the following, find the residue and determine the type of singularity:*

$a)\ \mathrm{Res}\big(ze^{1/z},0\big)$ $\qquad b)\ \mathrm{Res}\big(\frac{1-e^{2z}}{z^4},0\big)$ $\qquad c)\ \mathrm{Res}\big(\frac{e^{2z}}{(z-1)^2},1\big)$

$d)\ \mathrm{Res}\big(\frac{z}{\cos z},\frac{\pi}{2}\big)$ $\qquad e)\ \mathrm{Res}\big(\frac{e^z}{z^2+\pi^2},i\pi\big)$ $\qquad f)\ \mathrm{Res}\big(\frac{e^z}{z^2+\pi^2},-i\pi\big)$

28. *Let the function* $f:\mathbb{C}\to\mathbb{C}$ *be entire such that* $\lim\limits_{z\to\infty}f(z)=\infty$. *Show that* f *must be a polynomial.*

29. *True or false? Let* $f:\mathbb{C}\to\mathbb{C}$ *be an entire function.*

a) If $f(x)=0$ *on* $(0,1)$, *then* $f(z)=0$ *for all* $z\in\mathbb{C}$.

b) If $f(x)$ *is bounded on* $\mathbb{R}$, *then* $f(z)$ *is constant on* $\mathbb{C}$.

30. *Find the radius of convergence for* $\sum_{n=1}^\infty n^{2/n}z^n$ *and find a point on the circle of convergence where the series diverges.*

31. *Fix* $\alpha,\beta\in\mathbb{C}$. *Find the radius of convergence for* $\sum_{n=1}^\infty\left(\alpha^n+\beta^n\right)z^n$.

10

COMPLEX FUNCTIONS AS MAPPINGS

Mathematicians do not study objects, but relations between objects. Thus, they are free to replace some objects by others so long as the relations remain unchanged. Content to them is irrelevant: they are interested in form only.

Jules Henri Poincaré

1 THE EXTENDED COMPLEX PLANE

In this section we will establish a convenient correspondence between the complex plane and the surface of a sphere known as the *Riemann sphere.* A consequence of this mapping will be a well-defined notion of *the point at infinity* in what is referred to as *the extended complex plane* $\widehat{\mathbb{C}} \equiv \mathbb{C} \cup \{\infty\}$.

Stereographic Projection and the Riemann Sphere

Consider the unit sphere $\mathbb{S} \equiv \{(x_1, x_2, x_3) \in \mathbb{R}^3 : x_1^2 + x_2^2 + x_3^2 = 1\} \subset \mathbb{R}^3$, and let $N = (0, 0, 1)$ be the "north pole" of $\mathbb{S}$. Consider the plane $\mathbb{R}^2 \subset \mathbb{R}^3$ intersecting the sphere $\mathbb{S}$ along the equator as shown in Figure 10.1, and define a map f according to the following rule. For the ray emanating from N and intersecting the sphere $\mathbb{S}$ at (x_1, x_2, x_3) and the plane at (x_1', x_2') define the map $f : \mathbb{S} \setminus \{N\} \to \mathbb{R}^2$ by $f(x_1, x_2, x_3) = (x_1', x_2')$. To develop an algebraic formula for f, we consider the line in $\mathbb{R}^3$ given by

$$\big(tx_1, tx_2, 1 + t(x_3 - 1)\big) \quad \text{for } t \in \mathbb{R}.$$

This line contains the point N and the point with coordinates $(x_1, x_2, x_3) \in \mathbb{S}$. To find the coordinates of the point in the plane that also lies on the line, set the third coordinate of the line to zero and obtain $t = \frac{1}{1-x_3}$. This yields $(x_1', x_2') = \left(\frac{x_1}{1-x_3}, \frac{x_2}{1-x_3}\right)$, and therefore $f(x_1, x_2, x_3) = \left(\frac{x_1}{1-x_3}, \frac{x_2}{1-x_3}\right)$. Clearly the real plane can be replaced by the complex plane, and doing so lets us define $F : S \setminus \{N\} \to \mathbb{C}$ given by

$$F(x_1, x_2, x_3) = \frac{x_1}{1 - x_3} + i\frac{x_2}{1 - x_3}. \tag{10.1}$$

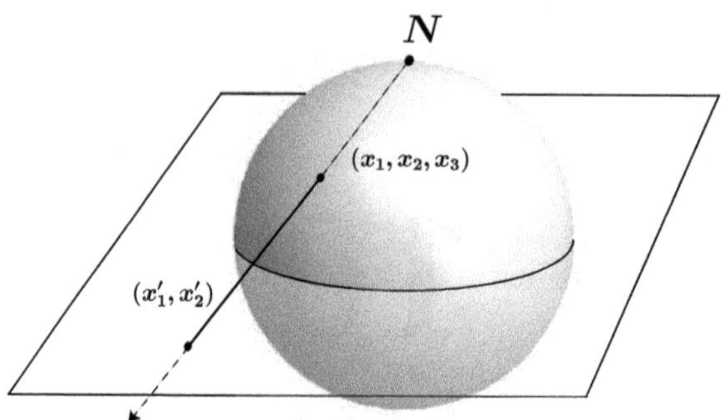

Figure 10.1 *Stereographic projection.*

We will now show that F is a one-to-one correspondence. To see that F is one-to-one, consider (x_1, x_2, x_3) and (y_1, y_2, y_3) both in $\mathbb{R}^3$ and such that $F(x_1, x_2, x_3) = F(y_1, y_2, y_3)$. This implies that $\frac{x_1+ix_2}{1-x_3} = \frac{y_1+iy_2}{1-y_3}$, which in turn yields $\frac{x_1^2+x_2^2}{(1-x_3)^2} = \frac{y_1^2+y_2^2}{(1-y_3)^2}$. Since $x_1^2 + x_2^2 + x_3^2 = y_1^2 + y_2^2 + y_3^2 = 1$, this obtains $\frac{1-x_3^2}{(1-x_3)^2} = \frac{1-y_3^2}{(1-y_3)^2}$, which in turn yields that $x_3 = y_3$. (Why?) It is then easy to see that $x_1 = y_1$ and $x_2 = y_2$, and therefore that F is one-to-one. To see that F is onto, let $w = \phi + i\psi \in \mathbb{C}$ be arbitrary. We will find $(x_1, x_2, x_3) \in S \setminus \{N\}$ such that $F(x_1, x_2, x_3) = \phi + i\psi$. In fact, simple algebra obtains

$$x_1 = \frac{2\phi}{1+\phi^2+\psi^2}, \quad x_2 = \frac{2\psi}{1+\phi^2+\psi^2}, \quad x_3 = \frac{\phi^2+\psi^2-1}{1+\phi^2+\psi^2},$$

and so F is onto. Since F is one-to-one and onto, it has an inverse. We leave it to the reader to show that $F^{-1} : \mathbb{C} \to \mathbb{S} \setminus \{N\}$ is given by

$$F^{-1}(z) = \left(\frac{z+\overline{z}}{|z|^2+1}, \frac{-i(z-\overline{z})}{|z|^2+1}, \frac{|z|^2-1}{|z|^2+1} \right). \tag{10.2}$$

► **10.1** *Verify that F^{-1} is the inverse function to F.*

We may extend the function F^{-1} and thereby extend the complex plane to include an ideal *point at infinity*, denoted by ∞, in the following way. Note that there is no element of $\mathbb{C}$ that gets mapped by F^{-1} to N. In fact, we can show that as (x_1, x_2, x_3) approaches N, the magnitude of $F(x_1, x_2, x_3)$ becomes unbounded. To see this, consider that

$$|F(x_1, x_2, x_3)|^2 = \frac{x_1^2+x_2^2}{(1-x_3)^2} = \frac{1-x_3^2}{(1-x_3)^2} = \frac{1+x_3}{1-x_3},$$

from which it follows that $\lim_{(x_1,x_2,x_3)\to N} |F(x_1, x_2, x_3)|^2 = \infty$. Motivated by this, we define the *point at infinity*, denoted by ∞, and adjoin it to the complex plane $\mathbb{C}$. We then define the extension $\widehat{F}^{-1}$ of F^{-1} to be the one-to-one correspondence $\widehat{F}^{-1}:\mathbb{C}\cup\{\infty\}\to\mathbb{S}$ given by

$$\widehat{F}^{-1}(z)=\begin{cases}F^{-1}(z) & \text{if } z\in\mathbb{C}\\ N & \text{if } z=\infty\end{cases}. \tag{10.3}$$

Of course, we can similarly extend F to all of $\mathbb{S}$ by defining $\widehat{F}:\mathbb{S}\to\mathbb{C}\cup\{\infty\}$ according to

$$\widehat{F}(x_1,x_2,x_3)=\begin{cases}F(x_1,x_2,x_3) & \text{if } (x_1,x_2,x_3)\in\mathbb{S}\setminus\{N\}\\ \infty & \text{if } (x_1,x_2,x_3)=N\end{cases}. \tag{10.4}$$

Then the functions $\widehat{F}^{-1}$ and $\widehat{F}$ are inverses of each other. This discussion motivates the following definition.

Definition 1.1 The **point at infinity**, denoted by ∞, is defined as the unique preimage of the point $N\in\mathbb{S}$ under the mapping $\widehat{F}^{-1}:\widehat{\mathbb{C}}\to\mathbb{S}$ corresponding to the **stereographic projection** described above. The set $\widehat{\mathbb{C}}\equiv\mathbb{C}\cup\{\infty\}$ is called **the extended complex plane**.

Note that ∞ is not a point in $\mathbb{C}$. Yet it will be convenient to sometimes refer to points in $\mathbb{C}\subset\widehat{\mathbb{C}}$ that are "near" ∞. To do this properly, we must develop a consistent notion of "distance" from points in $\mathbb{C}\subset\widehat{\mathbb{C}}$ to ∞. This might not appear possible since the distance from any point $z_0\in\mathbb{C}\subset\widehat{\mathbb{C}}$ to ∞ would seemingly be infinite, no matter how large $|z_0|$ might be. The trick is to define a new distance function (called a metric) between any two points in $\widehat{\mathbb{C}}$ in terms of a well-defined distance between the corresponding points *on the sphere* $\mathbb{S}$. Then, with this new distance function, we can say that a point z_0 in $\mathbb{C}\subset\widehat{\mathbb{C}}$ is "close" to ∞ exactly when the corresponding point $\zeta_0=\widehat{F}^{-1}(z_0)$ on the sphere is "close" to N. One such distance function defined on the sphere is just the Pythagorean distance in $\mathbb{R}^3$. Define the distance function $d:\widehat{\mathbb{C}}\times\widehat{\mathbb{C}}\to\mathbb{R}$ for any $z,z'\in\widehat{\mathbb{C}}$ by

$$d(z,z')\equiv\left|\widehat{F}^{-1}(z)-\widehat{F}^{-1}(z')\right|.$$

Then it can be shown that

$$d(z,z')=\begin{cases}\frac{2|z-z'|}{\sqrt{(1+|z|^2)(1+|z'|^2)}} & \text{for } z,z' \text{ in } \mathbb{C}\\ \frac{2}{\sqrt{1+|z|^2}} & \text{for } z \text{ in } \mathbb{C} \text{ and } z'=\infty\end{cases},$$

and that $d(z_1,z_2)\geq 0$ with equality if and only if $z_1=z_2$, $d(z_1,z_2)=d(z_2,z_1)$, and $d(z_1,z_2)\leq d(z_1,z_3)+d(z_2,z_3)$ for all z_1,z_2, and z_3 in $\widehat{\mathbb{C}}$. These properties are the necessary ones of any well-defined metric or distance function in Euclidean space.

► **10.2** *Verify the claims made above about the distance function* d.

► **10.3** *Using the new distance function on $\widehat{\mathbb{C}}$ described above, show that for any given ϵ satisfying $0 < \epsilon < 2$, any $z \in \widehat{\mathbb{C}}$ will be within ϵ of ∞ if $|z| > \sqrt{\frac{4}{\epsilon^2} - 1}$. What if $\epsilon \geq 2$?*

Continuity of the Stereographic Projection

We may now establish that the stereographic projection mapping $\widehat{F}^{-1}$ is continuous on $\mathbb{C}$. Continuity is a "metric dependent" property, and so we need to check it under each of the metrics we now have on $\mathbb{C}$. To see continuity in the old metric, note that $\widehat{F}^{-1} = F^{-1}$ on $\mathbb{C}$, and simply consider (10.2). To establish continuity for the d metric, fix $z' \in \widehat{\mathbb{C}}$ and let $\epsilon > 0$ be given. We must find $\delta > 0$ such that $d(z, z') < \delta$ implies $|\widehat{F}^{-1}(z) - \widehat{F}^{-1}(z')| < \epsilon$. But $d(z, z') \equiv |\widehat{F}^{-1}(z) - \widehat{F}^{-1}(z')|$, and so taking $\delta = \epsilon$ does the trick. In fact, the d metric can measure distances to ∞ as well, and therefore continuity of $\widehat{F}^{-1}$ also holds at ∞. Because of this, we may hereafter think of $\mathbb{S}$ as a representation of $\widehat{\mathbb{C}}$ and we will refer to $\mathbb{S}$ as the *Riemann sphere*. Through this special identification we can relate convergence in the complex plane to convergence on $\mathbb{S}$ and vice versa. In fact, convergence in $\mathbb{C}$ as previously defined (with the usual distance function) is preserved under the new distance function d. That is, $\lim z_n = z_0$ in the usual metric if and only if $\lim d(z_n, z_0) = 0$. To see this, note that $\lim z_n = z_0$ if and only if $\lim \widehat{F}^{-1}(z_n) = \widehat{F}^{-1}(z_0)$ by continuity of $\widehat{F}^{-1}$. This in turn holds if and only if $\lim |\widehat{F}^{-1}(z_n) - \widehat{F}^{-1}(z_0)| = 0$, which is true if and only if $\lim d(z_n, z_0) = 0$. Also, using d we can extend the idea of convergence in $\mathbb{C}$ to include sequences $\{z_j\} \subset \mathbb{C} \subset \widehat{\mathbb{C}}$ that converge to ∞ in the following way. Let $\{z_j\}$ be a sequence with $z_j \in \mathbb{C} \subset \widehat{\mathbb{C}}$ for $j \geq 1$. Then, denoting the traditional symbol of unboundedness by $+\infty$ (rather than ∞, which now has a different meaning in this chapter), we have

$$\lim_{j \to +\infty} d(z_j, \infty) = \lim_{j \to +\infty} \frac{2}{\sqrt{1 + |z_j|^2}} = 0 \text{ if and only if } \lim_{j \to +\infty} |z_j| = +\infty.$$

We have established the following result.

Proposition 1.2 *Let $\{z_j\}$ be a sequence with $z_j \in \mathbb{C} \subset \widehat{\mathbb{C}}$ for $j \geq 1$. Then $\lim_{j \to +\infty} z_j = \infty$ if and only if $\lim_{j \to +\infty} |z_j| = +\infty$.*

The point at infinity is said to "compactify" the complex plane, since with it and the new metric d the set $\widehat{\mathbb{C}} = \mathbb{C} \cup \{\infty\}$ is the image of the compact set $\mathbb{S}$ under the continuous mapping $\widehat{F}$. As we will see, the point at infinity can be a convenience when discussing unbounded sets in the complex plane and mappings of them. For example, we will adopt the following rather typical convention: Let A be any subset of $\mathbb{C}$. Since $\mathbb{C} \subset \widehat{\mathbb{C}}$, the set A may also be viewed as a subset of $\widehat{\mathbb{C}}$. Consider $\widehat{F}^{-1}(A)$ as a subset of $\mathbb{S}$. Then A is unbounded when considered as a subset of $\mathbb{C}$, and $\infty \in A$ when A is considered as a subset of $\widehat{\mathbb{C}}$, if and only if $N \in \widehat{F}^{-1}(A)$. That is, *any subset A of $\mathbb{C}$ is unbounded if and only if when A is considered as a subset of $\widehat{\mathbb{C}}$ it contains the point ∞.* As a particular example, any line in $\mathbb{C}$ is unbounded, and hence when considered as a subset of $\widehat{\mathbb{C}}$ it will contain the point ∞.

One last detail is necessary in order to complete our understanding of the extended complex plane. We must define an arithmetic on $\widehat{\mathbb{C}}$. To do this, we must accommodate the new element ∞ within our already established arithmetic on $\mathbb{C}$. To this end, let z_0 be any element in $\mathbb{C}$. Then define $z_0 \pm \infty \equiv \infty$. For any $z_0 \neq 0$ in $\mathbb{C}$, define $z_0 \times \infty \equiv \infty$. Finally, for any z_0 in $\mathbb{C}$ we define $\frac{z_0}{\infty} \equiv 0$. We cannot, however, define $0 \times \infty$, $\frac{\infty}{\infty}$, or $\infty - \infty$, and so these expressions will remain indeterminate in our extended complex plane. When all the symbols are defined, the usual commutative and associative laws hold: for instance, $z_1(z_2 + z_3) = z_1 z_2 + z_1 z_3$ if all the sums and products are defined and $z_1, z_2, z_3 \in \widehat{\mathbb{C}}$. This is a simple verification that we leave to the reader.

► **10.4** *Verify that the commutative and associative laws hold for $z_1, z_2, z_3 \in \widehat{\mathbb{C}}$ when all the operations are defined.*

2 Linear Fractional Transformations

A *linear fractional transformation*, hereafter denoted LFT (and sometimes also referred to as a *Mobius transformation*), is a special kind of mapping from the extended complex plane to itself.

Figure 10.2 *Translation.*

2.1 Basic LFTs

Translation

For example, the transformation $T_1(z) = z + 2i$ translates every point in $\mathbb{C}$ two units in the direction of the positive imaginary axis. Consistent with our notion of extended arithmetic on $\widehat{\mathbb{C}}$, we see clearly that $T_1(\infty) = \infty$. In fact, it can be shown that if A is a line in $\widehat{\mathbb{C}}$ (which must contain the point at infinity), then $T_1(A)$ is also a line in $\widehat{\mathbb{C}}$. A typical translation is illustrated in Figure 10.2.

► **10.5** *Verify that T_1 described above maps lines to lines in $\widehat{\mathbb{C}}$. Verify too, that $T_1(A)$ is a circle, if $A \subset \mathbb{C} \subset \widehat{\mathbb{C}}$ is a circle.*

Rotation and Scaling (Dilation)

Fix nonzero $z_0 \in \mathbb{C}$, and let $T_2 : \widehat{\mathbb{C}} \to \widehat{\mathbb{C}}$ be given by

$$T_2(z) = z_0 z.$$

Clearly $T(0) = 0$ and $T(\infty) = \infty$. For any other $z \in \mathbb{C} \subset \widehat{\mathbb{C}}$ we may employ polar form with $z_0 = r_0 e^{i\theta_0}$ and $z = re^{i\theta}$ to obtain

$$T_2(z) = r_0 r e^{i(\theta+\theta_0)}.$$

From this we see that the point $z = re^{i\theta}$ gets mapped to a new point whose radius is rescaled by r_0 and whose argument is increased by θ_0. Therefore a mapping of the form $T_2(z) = z_0 z$ has the effect of rotating by θ_0 and scaling by a factor of r_0. In the case where the angle θ_0 is negative, the rotation is clockwise of magnitude $|\theta_0|$, whereas if θ_0 is positive the rotation is counterclockwise. If r_0 is greater than 1 the scaling effect is a stretching, whereas if r_0 is less than 1 the scaling effect is a shrinking. As an explicit example, consider $T_2(z) = iz$. Then

$$T_2(z) = iz = e^{i\pi/2} re^{i\theta} = re^{i(\theta+\frac{\pi}{2})},$$

and it is easy to see that in this case T_2 rotates points counterclockwise through an angle of $\frac{\pi}{2}$ radians (see Figure 10.3). Again, consistent with our notion of

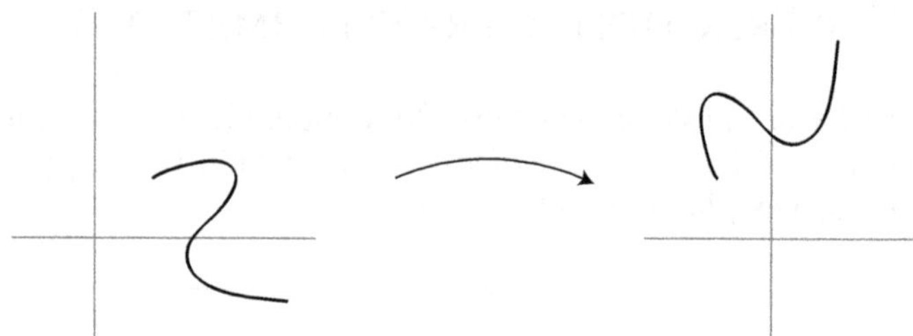

Figure 10.3 *Rotation by 90 degrees.*

extended arithmetic on $\widehat{\mathbb{C}}$, we see clearly that $T_2(\infty) = \infty$. As in the case of translations, it can be shown that T_2 takes lines to lines, and circles to circles.

► **10.6** *Show that if $A \subset \widehat{\mathbb{C}}$ is a line, and if $B \subset \mathbb{C} \subset \widehat{\mathbb{C}}$ is a circle, then $T_2(A)$ is a line and $T_2(B)$ is a circle.*

Inversion

Let $T_3 : \widehat{\mathbb{C}} \to \widehat{\mathbb{C}}$ be given by

$$T_3(z) = \frac{1}{z}.$$

Clearly, $T(0) = \infty$ and $T(\infty) = 0$. To determine the image of any other point in $\mathbb{C} \subset \widehat{\mathbb{C}}$, note that in polar coordinates we have

$$T_3(z) = \frac{1}{z} = \frac{1}{re^{i\theta}} = \frac{1}{r}e^{-i\theta}.$$

From this we see that T_3 rescales z by a factor of $\frac{1}{r^2}$, and rotates z through an angle -2θ. This rotation amounts to a reflection across the real axis.

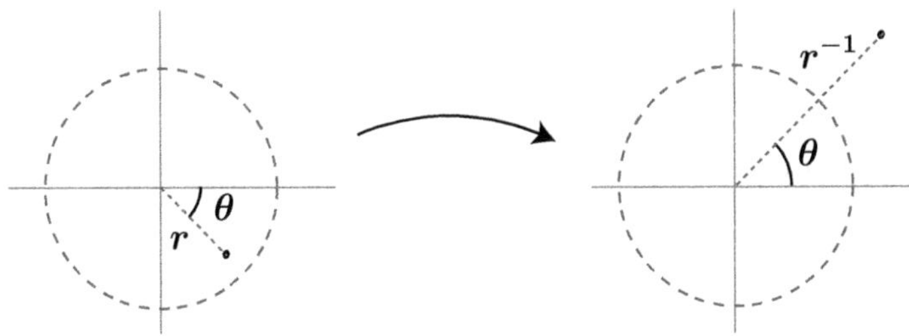

Figure 10.4 *Inversion.*

Owing to the rescaling,[1] any point inside the unit circle will be mapped by T_3 to the exterior of the unit circle, and vice versa (see Figure 10.4). Points on the unit circle will be mapped back to the unit circle.

► **10.7** *Let $A \subset \widehat{\mathbb{C}}$ be a line that passes through the origin. Show that $T_3(A)$ is also a line. Explain why the image of a line that does not pass through the origin cannot be a line. In fact, show that such a line is mapped to a circle.*

► **10.8** *Find the image of the square $A = [0, 1, i, 0] \subset \mathbb{C} \subset \widehat{\mathbb{C}}$ under T_3.*

► **10.9** *Let $B \subset \mathbb{C} \subset \widehat{\mathbb{C}}$ be a circle. Show that $T_3(B)$ is a circle if B does not pass through the origin, but is a line if B does pass through the origin.*

2.2 General LFTs

We now note the following important results regarding LFTs.

Proposition 2.1 *Let $T : \widehat{\mathbb{C}} \to \widehat{\mathbb{C}}$ be an LFT given by $T(z) = \frac{az+b}{cz+d}$. Then,*

a) The mapping T is continuous on $\widehat{\mathbb{C}}$ and is differentiable on

$\mathbb{C} \setminus \{-\frac{d}{c}\}$ *if* $c \neq 0$, *and on* $\mathbb{C}$ *if* $c = 0$.

b) The mapping T is a one-to-one correspondence from $\widehat{\mathbb{C}}$ to itself.

c) The mapping T has an inverse T^{-1}, which is also an LFT.

► **10.10** *Prove the above. Begin by computing the derivative of $T(z)$.*

We will now show that the special-case LFTs T_1, T_2, and T_3 from the last subsection are the building blocks of every LFT.

[1]If one were to also consider more general inversions that mapped points on a circle of radius R to itself, rather than just the inversions that map points of the unit circle to itself as we do here, it would in fact be advisable to characterize the scaling differently. For a discussion of this more general inversion see [Nee99].

Proposition 2.2 *Every LFT can be written as a composition of mappings of the type T_1, T_2, and T_3.*

PROOF There are two cases.

Case 1: $c = 0$. In this case, $d \neq 0$ and $a \neq 0$, and we have

$$T(z) = \frac{a}{d}z + \frac{b}{d} = (T_1 \circ T_2)(z),$$

where $T_1(z) = z + \frac{b}{d}$ and $T_2(z) = \frac{a}{d}z$.

Case 2: $c \neq 0$. In this case,

$$T(z) = \frac{a}{c} - \left(\frac{ad - bc}{c}\right)\frac{1}{cz + d}.$$

From this we may write

$$T(z) = (\widehat{T}_1 \circ T_2 \circ T_3 \circ T_1 \circ \widehat{T}_2)(z),$$

where $\widehat{T}_2(z) = cz$, $T_1(z) = z + d$, $T_3(z) = \frac{1}{z}$, $T_2(z) = -\left(\frac{ad-bc}{c}\right)z$, and $\widehat{T}_1(z) = \frac{a}{c} + z$. ♦

It can be shown that LFTs map lines to lines or circles, and circles to lines or circles. We leave this to the reader.

► **10.11** *Suppose T is an LFT. If $A \subset \widehat{\mathbb{C}}$ is a line, show that $T(A)$ is a line or a circle. When will it be a circle? If $B \subset \mathbb{C} \subset \widehat{\mathbb{C}}$ is a circle, show that $T(B)$ is a line or a circle. When will it be a line?*

The simple proof of the following result is left to the reader.

Proposition 2.3 *If S and T are both LFTs, then $S \circ T$ is also an LFT.*

► **10.12** *Prove the above.*

Example 2.4 *We will specify an LFT that maps the points $-1, 0$, and 1 to the points $-i, 1$, and i, respectively. To do so, let $T(z) = \frac{az+b}{cz+d}$. We will determine a, b, c, and d satisfying the above. Note that since $T(0) = 1$, we must have $b = d$. Also, $T(\pm 1) = \pm i$, which implies $\frac{a+b}{c+b} = i$ and $\frac{-a+b}{-c+b} = -i$. Solving these two equations for a and c in terms of b obtains $a = ib$ and $c = -ib$. Overall we have*

$$T(z) = \frac{ibz + b}{-ibz + b} = \frac{iz + 1}{-iz + 1}.$$

◄

► **10.13** *Find an LFT that maps $0, 1$, and ∞ to $1, 0, 2i$, respectively.*

The above example illustrates that one can specify an LFT by specifying its action on three distinct points. This might be surprising, since the formula for an LFT involves four complex numbers a, b, c, and d. In fact, we will show that specifying the images of three distinct points z_1, z_2, and z_3 in $\widehat{\mathbb{C}}$ is sufficient to characterize an LFT completely. One way to see this [Nee99] is to first consider what are called the *fixed points* of an LFT. The point z_0 is a fixed point of the LFT given by $T(z) = \frac{az+b}{cz+d}$ if $T(z_0) = z_0$. Simple algebra reveals that *an LFT that is not the identity mapping can have at most two such fixed points,* since the equality $T(z_0) = z_0$ amounts to a quadratic polynomial in z_0.

► **10.14** *Show that if $c \neq 0$ then the fixed points of T lie in $\mathbb{C}$. Show too that if $c = 0$ then ∞ is a fixed point, and if $a = d$ as well, so that $T(z) = z + b$, then ∞ is the only fixed point.*

Given the above, if T is an LFT taking the distinct points z_1, z_2, z_3 in $\widehat{\mathbb{C}}$ to the points w_1, w_2, w_3, then it is unique. To see this, suppose S is another LFT that also maps z_1, z_2, and z_3 to w_1, w_2, and w_3. Then S^{-1} is an LFT, and $S^{-1} \circ T$ is an LFT that takes each of z_1, z_2, and z_3 to itself. This implies that $S^{-1} \circ T$ is the identity mapping, and hence $S = T$. Finally, note too that for any three distinct points z_1, z_2, and z_3 in $\widehat{\mathbb{C}}$ and three distinct image points w_1, w_2, and w_3 in $\widehat{\mathbb{C}}$ we can always find an LFT that maps each z_i to each respective w_i for $i = 1, 2, 3$. This is left to the reader in the following exercise.

► **10.15** *Show that the equation $\frac{(w-w_1)(w_2-w_3)}{(w-w_3)(w_2-w_1)} = \frac{(z-z_1)(z_2-z_3)}{(z-z_3)(z_2-z_1)}$ implicitly defines an LFT given by $w = T(z)$ that maps z_1, z_2, and z_3 to w_1, w_2, and w_3, respectively. The expression on each side of the equality implicitly defining w in terms of z is called a* **cross ratio**.

► **10.16** *Use the above discussion to revisit Example 2.4.*

► **10.17** *Consider the distinct points z_1, z_2, and z_3 in $\mathbb{C}$ as shown in Figure 10.5. Find*

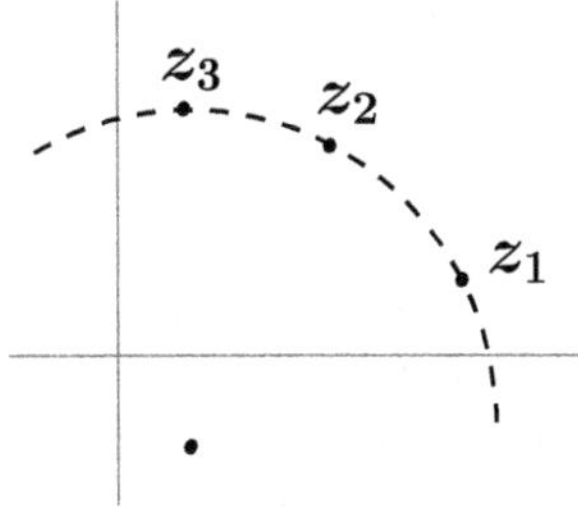

Figure 10.5 *The situation in Exercise 10.17*

an LFT that takes z_1, z_2, and z_3 to w_1, w_2, and ∞, respectively, with $w_1 < w_2$ real. If T is the LFT so found, show that $\text{Im}(T(z)) = 0$ if and only if $z \in C$ where C is the circle containing z_1, z_2, and z_3. If z_0 is any point inside the circle C show that $T(z_0)$ is in the upper half-plane. Likewise, show that points outside the circle C get mapped to the lower half-plane. Can you show that Int(C) *is mapped onto the upper half-plane? Can you use this to find a map from the upper half-plane to the unit disk?*

Mapping the Right Half-Plane to the Unit Disk

We now show that we can map, via LFT, the right half-plane to the unit disk with the imaginary axis mapped to the unit circle. To this end, consider the LFT given by $T(z) = \frac{az+b}{cz+d}$, and suppose we wish to map the points $-i, 0, i$ to the points $i, 1, -i$, respectively. Since $f(0) = 1$, we have $d = b$, and since $f(\pm i) = \mp i$ we have

$$\frac{ai+b}{ci+d} = -i \quad \text{and} \quad \frac{-ai+b}{-ci+d} = i.$$

Solving for a, b, c, and d, we obtain $c = b$ and $d = -a$. Therefore we may take as our mapping

$$f(z) = \frac{1-z}{1+z}.$$

Note that multiplying a, b, c, d by a constant does not change the linear transformation. (This is a clue, in fact, as to why only three distinct points need be specified in order to characterize the LFT.) One can check that this LFT maps the imaginary axis to the unit circle. To see that the right half-plane is mapped to the interior of the unit disk, note that 1 is in the right half-plane. If z is in the right half-plane too, then $(-z)$ is in the left half-plane. Remembering that absolute values measure distance in the traditional way, we note that $|1-z| < |1+z| = |1-(-z)|$, and therefore $|f(z)| = \frac{|1-z|}{|1+z|} < 1$. We leave it to the reader to show that the mapping is onto.

► **10.18** *Show that the mapping constructed above is onto the open unit disk. Also, show that the imaginary axis is mapped onto the unit circle.*

► **10.19** *Find an LFT that maps the upper half-plane to the unit disk. Do this by first mapping the upper half-plane to the right half-plane and then map the result to the unit disk via the LFT given above. You must apply Proposition 2.3.*

► **10.20** *Can you use Exercise 10.15 on page 523 to find a map from the right half-plane to the unit disk with the imaginary axis mapped to the unit circle?*

3 Conformal Mappings

LFTs are a special subclass of important mappings that have the property that they "preserve angles." Such mappings are called *conformal mappings*, and the following motivation for our formal definition of this class of mappings will make this angle-preservation property more precise. For convenience in this section, we restrict our attention to $\mathbb{C}$ (not $\widehat{\mathbb{C}}$).

3.1 Motivation and Definition

Consider a complex function $f : D \to \mathbb{C}$ that is differentiable on D, and let z_0 be in D. We wish to investigate conditions under which the function f, when acting as a mapping, will preserve angles having their vertex at z_0. To make this idea more precise, consider any two smooth curves in D parametrized

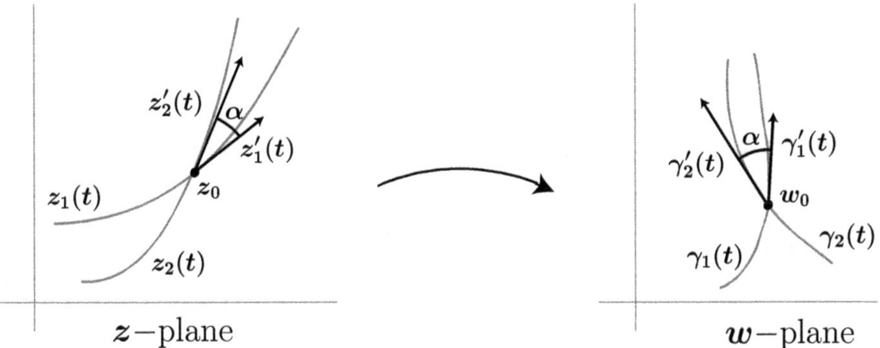

Figure 10.6 *Conformality.*

by $z_i : [a,b] \to \mathbb{C}$ for $i = 1,2$, such that for both $i = 1,2$,

(i) $z_i(t_0) = z_0$ for $a < t_0 < b$,

(ii) $z_i'(t)$ is continuous and nonzero for $a \le t \le b$.

Geometrically, $z_i'(t_0)$ can be thought of as a tangent vector to the curve parameterized by $z_i(t)$ at the point $z_i(t_0)$. Denote the angle between $z_1'(t_0)$ and $z_2'(t_0)$ by α, where $\alpha \in [0, 2\pi)$. Then if $\theta_i(t)$ is a continuous argument of $z_i'(t)$ note that $\alpha = \theta_2(t_0) - \theta_1(t_0)$. We wish to see how α transforms under the mapping f. (It is useful to think of f as taking points $z = x + iy \in D$ from the "z-plane" to points $w = u + iv$ in the "w-plane" as shown in Figure 10.6.) To this end, let $\gamma_i(t) = f\left(z_i(t)\right)$ be the image of the curve $z_i(t)$ under the mapping f. The chain rule yields

$$\gamma_i'(t) = f'\left(z_i(t)\right) z_i'(t), \tag{10.5}$$

which has a continuous argument $\Theta_i(t)$ if $f'(z) \neq 0$ on D. It follows that

$$\begin{aligned}\Theta_2(t_0) - \Theta_1(t_0) &= \left[\arg\left(f'(z_2(t_0))\right) + \arg\left(z_2'(t_0)\right)\right] \\ &\quad - \left[\arg\left(f'(z_1(t_0))\right) + \arg\left(z_1'(t_0)\right)\right] \\ &= \theta_2(t_0) - \theta_1(t_0) = \alpha,\end{aligned}$$

and so the angle between the curves at z_0 has been preserved. Also note that if we think of $z_i'(t_0)$ as the tangent vector to the curve $z_i(t)$ at z_0, and $\gamma_i'(t_0)$ as the tangent vector of the image curve at $w_0 = f(z_0)$, then the differentiable mapping f has the effect according to equation (10.5) of acting on the tangent vector $z_i'(t_0)$ by multiplying it by $f'\left(z_i(t_0)\right) = f'(z_0) \neq 0$. Recall that $f'(z_0)$ is just a complex number, and as such can be represented in polar form by $|f'(z_0)|e^{i\arg(f'(z_0))}$. Seen this way, the mapping f then has the effect of stretching or shrinking the tangent vector $z_i'(t_0)$ by the factor $|f'(z_0)|$ and rotating it by $\arg(f'(z_0))$. The key point here is to note that the smooth curve $z_i(t)$ passing through z_0 was arbitrary, and for any choice of such a smooth curve, the mapping f has the same local effect on the tangent vector to the curve at the point z_0, namely, a stretching or shrinking by the same constant factor $|f'(z_0)|$ and a rotation by the same constant angle $\arg(f'(z_0))$. We refer to this property of the scaling factor of f by saying that it *scales uniformly in all directions* at z_0. This and the uniformity of action by f on the angles subtended by vectors at z_0 are what make it special, and we call such maps *conformal*. This motivates the following definition.[2]

Definition 3.1 Let $f : D \to \mathbb{C}$ be differentiable on D and suppose $z_0 \in D$. If $f'(z_0) \neq 0$ then we say that f is **conformal at** z_0. If f is conformal at each point of D we say that f is **conformal on** D.

We give some examples.

Example 3.2 *Consider the mapping $f : \mathbb{C} \setminus \{1\} \to \mathbb{C}$ given by $f(z) = \frac{1}{z-1}$. This map is conformal on its domain since $f'(z)$ exists and is nonzero there. If $w = f(z) = u + iv$, then we may solve for x and y in terms of u and v to obtain*

$$x = \frac{u^2 + v^2 + u}{u^2 + v^2}, \qquad y = -\frac{v}{u^2 + v^2}.$$

The circle $x^2 + y^2 = 1$ and the line $y = x$ in the z-plane are orthogonal at their point of intersection $z_0 = \left(\frac{1}{\sqrt{2}}, \frac{1}{\sqrt{2}}\right)$. The image in the w-plane of the line $y = x$

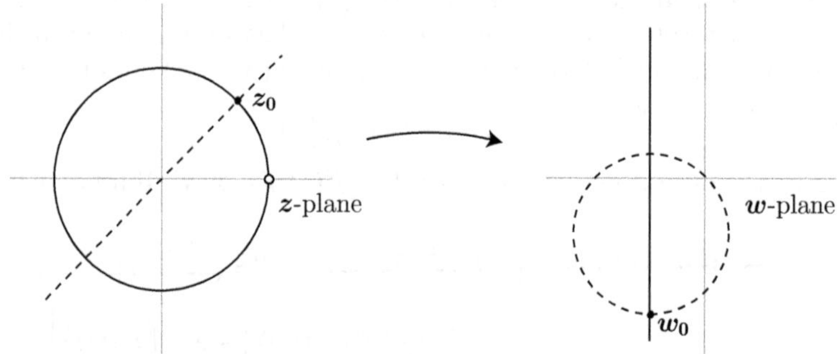

Figure 10.7 *The situation in Example 3.2.*

is the circle $\left(u + \frac{1}{2}\right)^2 + \left(v + \frac{1}{2}\right)^2 = \frac{1}{2}$, and the image of the circle $x^2 + y^2 = 1$ (minus the point $(x, y) = (1, 0)$) is the line $u = -\frac{1}{2}$. The point corresponding to $z = 1$ clearly gets mapped to the point at infinity. The image of the point of intersection is $w_0 = \left(-\frac{1}{2}, -\frac{1}{2\sqrt{2}-2}\right)$, and it can be seen in Figure 10.7 that the image curves are also orthogonal at w_0. ◀

[2]Some authors define a complex mapping $f : D \to \mathbb{C}$ to be conformal at a point $z_0 \in D$ if the mapping preserves angles at the point. This alone need not imply uniform scaling near the point. However, upon consideration of open connected sets mapped by f, the geometric property of angle preservation can be shown to obtain precisely when f is differentiable with f' nonzero on the set. The uniform scaling property then comes along in the bargain; see [Nee99].

Example 3.3 *Let $f : \mathbb{C} \to \mathbb{C}$ be given by $f(z) = e^z$. Then f is conformal on $\mathbb{C}$ since $f'(z) = e^z \neq 0$ there. Consider the lines $y = 0$ and $x = 0$ in the complex plane. These lines are orthogonal at their point of intersection $z_0 = 0$. The reader can easily verify that the image of the line $x = 0$ under f is the circle $u^2 + v^2 = 1$ and the image of the line $y = 0$ is the ray from the origin along the positive real axis. The image of the point of intersection $z_0 = 0$ is $w_0 = f(z_0) = 1$. It is not hard to verify that the tangent to the circle is orthogonal to the ray at the point w_0.* ◀

▶ **10.21** *Show that any LFT is conformal on $\mathbb{C}$ if $c = 0$, or on $\mathbb{C} \setminus \{\frac{-d}{c}\}$ if $c \neq 0$.*

It is convenient that conformality is preserved under compositions.

Proposition 3.4 *Suppose $f : D_f \to \mathbb{C}$ is conformal at $z_0 \in D_f$ and that $g : D_g \to \mathbb{C}$ is conformal at $f(z_0)$, where $D_g \supset \mathcal{R}_f$. Then $g \circ f : D_f \to \mathbb{C}$ is conformal at $z_0 \in D_f$.*

▶ **10.22** *Prove the above proposition.*

Points at which conformality of a mapping fail are called *critical points* of the mapping. For example, the function $f : \mathbb{C} \to \mathbb{C}$ given by $f(z) = z^n$ for $n \geq 2$ has a critical point at $z = 0$. In fact, angles are multiplied by a factor of n there. To see this, consider a point $z_\epsilon = \epsilon e^{i\theta}$ on the circle of radius $\epsilon < 1$ centered at $z = 0$. Then $w = f(z_\epsilon) = \epsilon^n e^{in\theta}$ so that the vector corresponding to the point z_ϵ has been "shrunk" by a factor of ϵ^{n-1} and rotated by an angle $(n-1)\theta$. That is, the image point's argument is n times the preimage point's argument. A circle that is parametrized to go around $z = 0$ once in the z-plane will have its image go around $f(0) = 0$ in the w-plane n times. This means that each point on the w-circle has n preimages on the z-circle. Near the critical point $z = 0$ then, the function is not one-to-one. Also note that near the point $z = 0$ the mapping $f(z) = z^n$ for $n \geq 2$ has the effect of contracting, since a disk centered at 0 and of radius $\epsilon < 1$ is mapped to a disk centered at 0 of radius $\epsilon^n < \epsilon < 1$.

3.2 More Examples of Conformal Mappings

In this section we explore some other mappings of the complex plane to itself.

Example 3.5 *Fix α such that $0 < \alpha \leq \pi$, and consider the open sector given by $A = \{z \in \mathbb{C} : z = re^{i\theta}$ with $r > 0$ and $0 < \theta < \alpha\}$. Let the mapping $f : \mathbb{C} \setminus (-\infty, 0] \to \mathbb{C}$ be given by $f(z) = z^{\pi/\alpha}$. We will find the image $f(A)$ in the w-plane (see Figure 10.8). Note that we may write*

$$f(z) = z^{\pi/\alpha} = e^{\frac{\pi}{\alpha}\mathrm{Log} z} = e^{\frac{\pi}{\alpha}(\ln r + i\theta)} = e^{\frac{\pi}{\alpha}\ln r} e^{i\frac{\pi\theta}{\alpha}} = r^{\frac{\pi}{\alpha}} e^{i\frac{\pi\theta}{\alpha}}.$$

From this we see that the ray $z = re^{i\theta}$ for fixed $0 < \theta < \alpha$ gets mapped

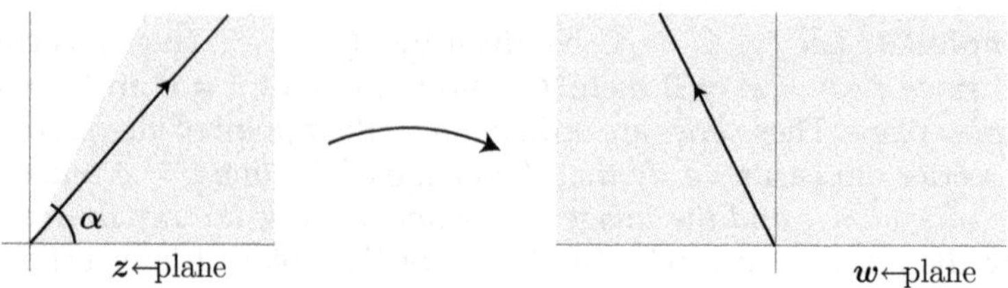

Figure 10.8 *The region A in Example 3.5.*

into the ray $w = r^{\frac{\pi}{\alpha}} e^{i\frac{\pi\theta}{\alpha}}$, which emanates from the origin of the w-plane and makes an angle $\frac{\pi\theta}{\alpha}$ as measured from the positive u-axis. It is now easy to see that the sector A gets mapped into the upper half-plane. The boundary ray $\theta = 0$ stays fixed while the other boundary ray $\theta = \alpha$ gets mapped to the negative u-axis. ◀

▶ **10.23** *Show that the mapping f is conformal. Also show that it maps its domain onto the upper half-plane.*

▶ **10.24** *Fix α as before so that $0 < \alpha \leq \pi$. Also, fix $\theta_0 \in \mathbb{R}$. Consider the sector $A = \{z \in \mathbb{C} : z = re^{i\theta}$ with $r > 0$ and $\theta_0 < \theta < \theta_0 + \alpha\}$. Using composition of mappings, find a map from the sector A onto the upper half-plane. Now for any fixed $z_0 \in \mathbb{C}$, do the same for the sector given by $A_{z_0} = \{z \in \mathbb{C} : z = z_0 + re^{i\theta}$ with $r > 0$ and $\theta_0 < \theta < \theta_0 + \alpha\}$. Can you find a map of A_{z_0} to the unit disk?*

Example 3.6 Consider $f : \mathbb{C} \to \mathbb{C}$ given by $f(z) = e^z$, and suppose $A = \{z \in \mathbb{C} : \text{Im}(z) = y_0\}$, a horizontal line in the z-plane. Then for $z \in A$, we have that $w = f(z) = e^x e^{iy_0}$. As x varies from $-\infty$ to $+\infty$ with y_0 fixed, we see (as illustrated in Figure 10.9) that $0 < e^x < +\infty$, and hence the points on the line in the z-plane are mapped to the points on a ray emanating from the origin at angle y_0 in the w-plane. If instead we consider the horizontal strip

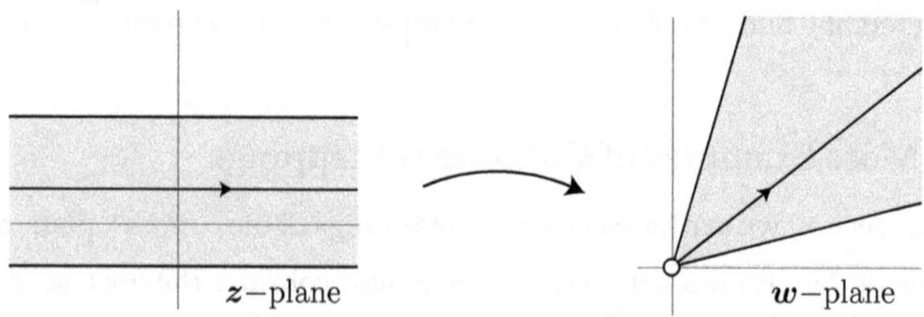

Figure 10.9 *The region A in the latter part of Example 3.6.*

$B = \{z \in \mathbb{C} : y_1 \leq \text{Im}(z) \leq y_2\}$ where $y_2 - y_1 \leq 2\pi$, we see that B is mapped into the sector $\{w \in \mathbb{C} : w = \rho e^{i\phi}$ with $\rho > 0$ and $y_1 \leq \phi \leq y_2\}$. ◀

▶ **10.25** *Fix an angle α. If B_α is the rotated strip $\{e^{i\alpha}z \in \mathbb{C} : y_1 \leq \mathrm{Im}(z) \leq y_2\}$, where $y_2 - y_1 \leq 2\pi$, find a map of B_α into the sector $\{w \in \mathbb{C} : w = \rho e^{i\phi}$ with $\rho > 0$ and $y_1 \leq \phi \leq y_2\}$. Can you find a map of the strip B_α to the unit disk?*

Example 3.7 *Consider the mapping $f : \mathbb{C} \to \mathbb{C}$ given by $f(z) = \cos z = \cos x \cosh y - i \sin x \sinh y$, and let A be the "infinite" rectangular region shown in Figure 10.10. We will find $f(A)$. To this end, let $z = x + iy$ and consider the*

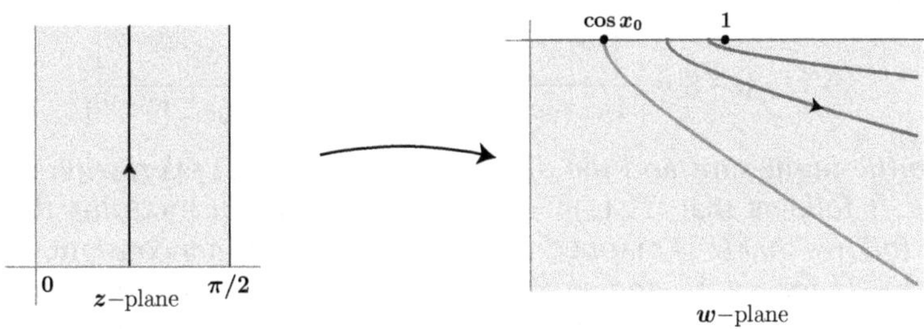

Figure 10.10 *The region A in Example 3.7.*

ray given by $x = x_0 \in [0, \frac{\pi}{2}]$ where $y \geq 0$. If $w = f(z) = u + iv$ is the image of the ray, then $u = \cos x_0 \cosh y$ and $v = -\sin x_0 \sinh y$, which implies that for $x_0 \in (0, \frac{\pi}{2})$ we have

$$\frac{u^2}{\cos^2 x_0} - \frac{v^2}{\sin^2 x_0} = 1.$$

Since $y \geq 0$, we have $v \leq 0$, i.e., we are on the bottom half of a hyperbola. If $x_0 = 0$, then $u = \cosh y$, $v = 0$, which is the ray $[1, \infty)$ on the u-axis. Finally if $x_0 = \frac{\pi}{2}$ we have $u = 0$, $v = -\sinh y$, which is the ray $(-\infty, 0]$ on the v-axis. From this it is easy to see that $f(A) = \{w \in \mathbb{C} : \mathrm{Re}(w) \geq 0,\ \mathrm{Im}(w) \leq 0\}$. ◀

Example 3.8 *Consider the mapping $f : \mathbb{C} \setminus (-\infty, 0] \to \mathbb{C}$ given by $f(z) = \mathrm{Log}(z)$. We will show that the domain $A = \mathbb{C} \setminus (-\infty, 0]$ is mapped to the strip $B = \{w \in \mathbb{C} : -\pi < \mathrm{Im}(w) \leq \pi\}$. To see this, let $z = re^{i\Theta}$ for $-\pi < \Theta \leq \pi$. Then $w = f(z) = \ln r + i\Theta$ with $\Theta \in (-\pi, \pi]$ and $-\infty < \ln r < +\infty$. Fixing $\Theta = \Theta_0 \in (-\pi, \pi]$ and allowing r to increase through $(0, +\infty)$ obtains $w = u + iv = \ln r + i\Theta_0$, and so the ray emanating from the origin in the z-plane gets mapped to the horizontal line in the w-plane corresponding to $v = \Theta_0$ and $-\infty < u < +\infty$.* ◀

▶ **10.26** *Show that the above map is onto the strip B.*

▶ **10.27** *Fix angle α and let $\mathcal{R}_\alpha$ be the ray emanating from the origin at angle α in the z-plane. Let $g : \mathbb{C} \setminus \mathcal{R}_\alpha \to \mathbb{C}$ be given by $f(z) = \log(z) = \ln r + i\theta$ where $\theta \in (\alpha, \alpha + 2\pi]$. If $A = \mathbb{C} \setminus \mathcal{R}_\alpha$, find the image $g(A)$ in the w-plane.*

Our next example provides a nontrivial conformal mapping of the unit disk to itself.

Example 3.9 *We define the unit disk to be $U \equiv N_1(0)$. For $z_0 \in U$, define*

$$T_{z_0}(z) \equiv \frac{z - z_0}{1 - \overline{z_0} z}.$$

Then T_{z_0} is differentiable on U (since $|z_0| < 1$). We leave it to the reader to verify that T_{z_0} is conformal on U. We will now show that T_{z_0} maps the unit circle to itself. To see this, let $z_0 = r_0 e^{i\theta_0}$ with $r_0 < 1$, and suppose $z = e^{i\theta} \in \partial U$. Then

$$\left|T_{z_0}(e^{i\theta})\right| = \frac{\left|e^{i\theta} - r_0 e^{i\theta_0}\right|}{\left|1 - r_0 e^{-i\theta_0} e^{i\theta}\right|} = \left|e^{i\theta}\right| \frac{\left|1 - r_0 e^{i(\theta_0 - \theta)}\right|}{\left|1 - r_0 e^{-i(\theta_0 - \theta)}\right|}.$$

Since the numerator and the denominator in the last expression are conjugates, it follows that $|T_{z_0}(z)| = 1$. By the maximum modulus theorem, it then follows that U is mapped into itself since T_{z_0} is nonconstant. Therefore $T_{z_0}(z) \in U$ for $z \in U$, and so T_{z_0} is a nontrivial conformal mapping of U to itself. The reader can verify that the inverse transformation is given by

$$T^{-1}(z) = T_{-z_0} = \frac{z + z_0}{1 + \overline{z_0} z},$$

and note that the inverse is of the same form with z_0 replaced by $-z_0$. Therefore, T_{-z_0} is a conformal mapping of U into itself too. Now, T_{z_0} and T_{-z_0} being inverses, we have:

$$U = T_{z_0}\left(T_{-z_0}(U)\right) \subset T_{z_0}(U) \subset U$$

This shows that $T_{z_0}(U) = U$. The reader can verify that the mapping T_{z_0} is one-to-one, and therefore it is a one-to-one correspondence of the unit disk with itself. ◀

► **10.28** *Show that T_{z_0} is conformal on U. Also, verify that T_{-z_0} is the inverse mapping for T_{z_0}, and that T_{z_0} is one-to-one.*

► **10.29** *Suppose $a \in \mathbb{R}$, and consider $T \equiv e^{ia} T_{z_0}$. Show that T is also a conformal one-to-one correspondence of U with itself.*

3.3 The Schwarz Lemma and the Riemann Mapping Theorem

We begin with a technical definition.

Definition 3.10 An open set $D \subset \mathbb{C}$ is called **simply connected** if for every closed contour $C \subset D$ and $w \notin D$,

$$n_C(w) = 0.$$

Intuitively, a simply connected subset of $\mathbb{C}$ (or of $\mathbb{R}^2$) is one that contains no "holes." This is also sometimes described as a set within which any closed

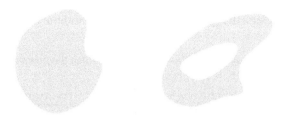

Figure 10.11 *Examples of a simply connected region (left) and a region that is not simply connected (right).*

contour can be "shrunk down" to a point without ever leaving the set (see Figure 10.11).

Example 3.11 *The interior of a simple closed contour C' is a simply connected open set. In fact, if $C \subset \text{Int}(C')$, then $n_C(w)$ is constant on the connected set $\overline{\text{Ext}(C')} = C' \cup \text{Ext}(C')$ by Theorem 2.7 of Chapter 8. Since that set is unbounded, $n_C(w)$ is zero there.* ◀

▶ **10.30** *Let $D \subset \mathbb{C}$ be a simply connected open set. If $f : D \to \mathbb{C}$ is differentiable and $C \subset D$ is a closed contour, show that $\oint_C f(z)dz = 0$.*

▶ **10.31** *Show that a star-shaped region is simply connected.*

The Schwarz Lemma

The following is a significant result in the theory of complex functions as mappings. Its proof is a consequence of the maximum modulus theorem.

Theorem 3.12 (The Schwarz Lemma)
Let $U = \{z \in \mathbb{C} : |z| < 1\}$ be the unit disk. Suppose $f : U \to \mathbb{C}$ is differentiable and satisfies $f(0) = 0$ and $f(U) \subset \overline{U}$, i.e., $|f(z)| \leq 1$ for $z \in U$. Then

$$|f(z)| \leq |z| \quad \text{for all } \; z \in U.$$

If equality holds at some $z \neq 0$, then $f(z) = e^{ia}z$ for some $a \in \mathbb{R}$.

Note that the final condition and its particular conclusion imply that if $|f(z)| = |z|$ for some nonzero z in U, then the map f amounts to a rigid rotation.

PROOF Note that, since $f(0) = 0$, the function $g(z) \equiv \frac{f(z)}{z}$ is differentiable on the unit disk if it is defined to be $f'(0)$ at $z = 0$. Fix $r < 1$. On the circle $C_r = \{z : |z| = r\}$,

$$|g(z)| = \left|\frac{f(z)}{z}\right| \leq \frac{1}{r} \tag{10.6}$$

by the assumption on f. By the maximum modulus theorem, the inequality holds for all z satisfying $|z| \leq r$. Take the limit as $r \to 1^-$ and we find that $|g(z)| \leq 1$ for all z, with equality implying g is constant. (See also the exercise below.) This means that $|f(z)| \leq z$ for all $z \in U$, and the final assertion follows as well. ♦

► **10.32** *Use the maximum modulus theorem to finish the proof of the above lemma. Prove also that* $|f'(0)| \leq 1$.

A consequence of the Schwarz lemma is that any conformal one-to-one mapping of U to itself must be of a particular form.

Proposition 3.13 *If $f : U \to U$ is a conformal one-to-one correspondence of U with itself, then there exists $z_0 \in U$, $a \in \mathbb{R}$, so that*

$$f(z) = e^{ia} T_{z_0}(z) = e^{ia} \frac{z - z_0}{1 - \overline{z_0} z}.$$

PROOF Take z_0 to be the unique root of f and consider $g : U \to U$ given by $g(z) = (f \circ T_{-z_0})(z)$. Then $g(0) = 0$, since $T_{-z_0}(0) = z_0$, so that the Schwarz lemma applies. Therefore $|g(z)| \leq |z|$. Now g is also a one-to-one correspondence of the disk with itself, since T_{-z_0} and f are. We may consider its inverse $h : U \to U$ given by $h(z) = g^{-1}(z)$. Then $h(0) = 0$ too, and we apply the Schwarz lemma to h. We have that $|h(z)| \leq |z|$ for all $z \in U$. Therefore,

$$|z| = |h(g(z))| \leq |g(z)| \leq |z|,$$

which means that $|g(z)| = |z|$ for all $z \in U$. Thus g is a rotation, $g(z) = e^{ia} z$ for some $a \in \mathbb{R}$. Since $g = f \circ T_{-z_0}$, it then follows that $f = g \circ T_{-z_0}^{-1} = g \circ T_{z_0}$. Thus $f(z) = e^{ia} T_{z_0}(z)$. ♦

► **10.33** *Suppose $f_1, f_2 : D \to U$ are differentiable and are conformal one-to-one correspondences. Show by considering the composite mapping $f_2 \circ f_1^{-1}$ and applying the above proposition, that for some $a \in \mathbb{R}$, $z_0 \in U$, f_2 is given by $f_2(z) = e^{ia} \frac{f_1(z) - z_0}{1 - \overline{z_0} f_1(z)}$.*

The Riemann Mapping Theorem

The Riemann mapping theorem is a celebrated result in complex analysis. Originally stated and proved in Riemann's doctoral thesis, the proof he provided was dependent upon a result he called Dirichlet's principle that was later found to not necessarily apply to all the types of sets Riemann considered.[3] Subsequent refinements by a series of mathematicians (Weierstrass, Hilbert, Osgood, Carathéodory) firmly established the result. The proof is rather technical and involves certain concepts and techniques we have not developed, and we therefore omit it.[4] Yet, it provides a nice culmination to the present chapter, and an enticement to further reading.

[3]In particular, the nature of the boundary of the set is critical. In fact, the behaviors of functions and solutions to equations at and near the boundaries of the domains on which they are applied are often complicated and/or difficult to establish.

[4]See [Ahl79].

Theorem 3.14 (The Riemann Mapping Theorem)
Let $D \subset \mathbb{C}$ be an open, connected, and simply connected proper subset of the complex plane. Then there exists a differentiable function $f : D \to \mathbb{C}$ such that $f'(z) \neq 0$ on D, and so that f establishes a conformal one-to-one correspondence between D and $U = N_1(0)$.

► **10.34** *In the statement of the above theorem, why can't the set D be the whole complex plane? What would that imply?*

► **10.35** *Suppose $D_1, D_2 \subset \mathbb{C}$ are open, connected, and simply connected. Use the Riemann mapping theorem to show that there exists a conformal one-to-one correspondence $f : D_1 \to D_2$.*

Note that the result is relatively simple to state, and is schematically illustrated in Figure 10.12. Its consequences are significant, in that if one is working on a problem in a complicated domain $D \subset \mathbb{C}$, even with a somewhat complicated boundary, one can map the problem to the much nicer environment that is the unit disk. However, while the existence of such a mapping is guaranteed by the theorem, much as with the Tietze extension theorem of Chapter 4, there is no algorithm suggested for finding the map that does the job. Even the proof of the theorem does not provide a practical means for obtaining the desired map, much in the same way the Tietze extension theorem failed to provide such details. It is also worth pointing out that there is no analogous result for mappings from higher-dimensional Euclidean spaces to themselves. Because of the Riemann mapping theorem, it is sometimes joked that those who study and understand the analysis of differentiable complex functions on the unit disk U know all there is to know about the subject. But we suggest you consult the bibliography to learn even more.

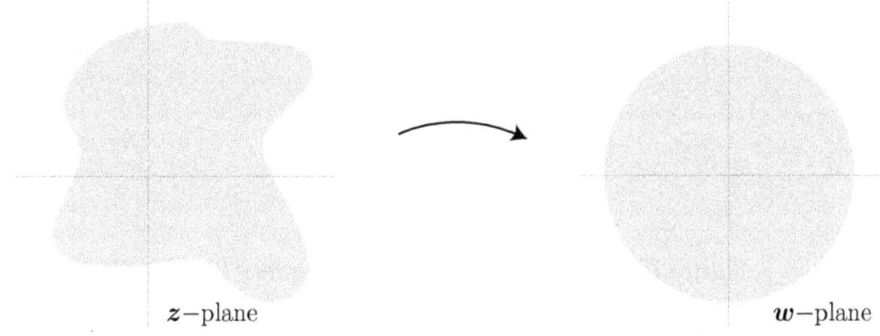

Figure 10.12 *The Riemann Mapping Theorem.*

4 Supplementary Exercises

1. *Find an LFT that maps* $1, 0, -1$ *to* $i, \infty, 1$*, respectively.*

2. *Find an LFT that maps* $\infty, i, 0$ *to* $0, i, \infty$*, respectively.*

3. *A fixed point of a function* f *is an element* z *such that* $f(z) = z$*.*
a) *Show that an LFT that is not the identity has at most two fixed points in* $\widehat{\mathbb{C}}$*.*
b) *Find the fixed point or points of* $w = f(z) = \frac{z-i}{z+i}$*.*

4. *Let* $f(z) = \frac{1}{z}$*, and let* A *be the line* $\{z : \text{Im}(z) = 1\}$*. Let* B *be the line* $\text{Re}(z) = \text{Im}(z)$*. Let* C *be the circle* $|z| = 1$*. Find* $f(A), f(B)$*, and* $f(C)$*.*

5. *Let* $f(z) = \frac{1}{z}$*. Let* $A = \{z : 0 < \text{Im}(z) < \frac{1}{2}\}$*. Find* $f(A)$*.*

6. *Let* $f(z) = \frac{1}{z}$*. Let* A *be the disk of radius* r *centered at the origin. Find* $f(A)$*. If* B *is the disk of radius* r *centered at* $z = r$*, find* $f(B)$*.*

7. *Let* $f(z) = z + \frac{1}{z}$*, and let* A *be the top half of the unit disk. Find* $f(A)$*. To what does the boundary of* A *get mapped?*

8. *Let* $f(z) = iz + 1$*. Let* A *be the region* $\{z : \text{Re}(z) > 0, 0 < \text{Im}(z) < 2\}$*. Find* $f(A)$*.*

9. *Let* $f(z) = iz - 3$ *and* $A = [0, 2, 2+i, 2i, 0]$*,* $B = [0, 2, 2+i, 0]$*. Find* $f(\overline{\text{Int}(A)})$ *and* $f(\overline{\text{Int}(B)})$*.*

10. *Let* $f(z) = z^2$ *and* A *be the region shown in Figure 10.13, the set of points* $z = x + iy$ *such that* $1 \le xy \le 2$ *and* $1 \le x^2 - y^2 \le 2$*. Find* $f(A)$*.*

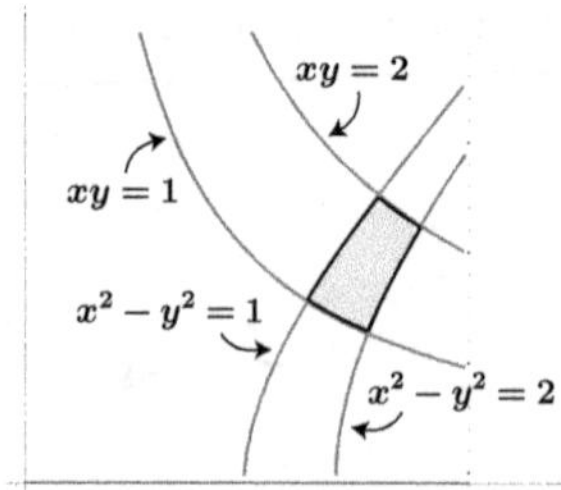

Figure 10.13 *The region A in Supplementary Exercise 10.*

11. *For* $f : \mathbb{C} \to \mathbb{C}$ *given by* $f(z) = \sin z$ *find* $f(A)$ *for the following:*
a) $A = \{z \in \mathbb{C} : \text{Im}(z) = c > 0, \ -\frac{\pi}{2} \le \text{Re}(z) \le \frac{\pi}{2}\}$*, where* $c \in \mathbb{R}$
b) $A = \{z \in \mathbb{C} : -\frac{\pi}{2} \le \text{Re}(z) \le \frac{\pi}{2}, 0 \le \text{Re}(z) \le \text{Im}(z) \le 1\}$

12. *Consider the set* $A = \{z \in \mathbb{C} : -\frac{\pi}{2} \le \text{Re}(z) \le \frac{\pi}{2}, \text{Im}(z) \ge 0\}$*. Show that* $f(A)$ *where* $f : \mathbb{C} \to \mathbb{C}$ *is given by* $f(z) = \sin z$ *is the upper half-plane with the boundary of* A *mapped to the real axis.*

13. *Let* $f(z) = \sin z$ *and* $A = \{z : -\pi \leq x \leq \pi, 1 \leq y \leq x\}$, *where as usual* $z = x + iy$. *Find* $f(A)$.

14. *Let* $f(z) = \sin z$, *and let* A *be the region shown in Figure 10.14. Find* $f(A)$.

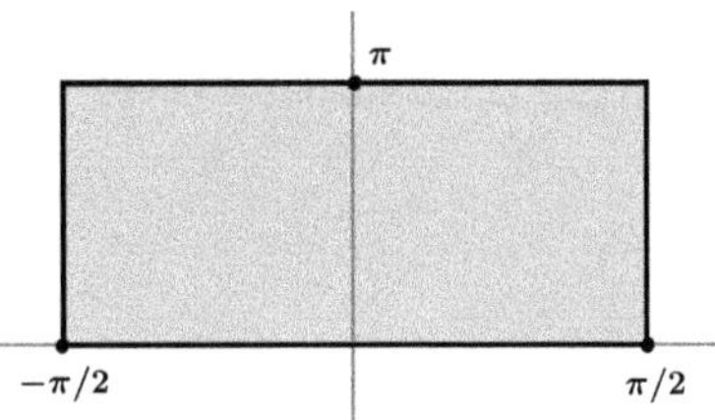

Figure 10.14 *The region A in Supplementary Exercise 14.*

15. *Define conformality at* ∞ *in terms of angle preservation at* $N \in \mathbb{S}$. *For any two curves* $C_1, C_2 \subset \mathbb{C}$ *that are unbounded, consider their images under stereographic projection* $\Gamma_1, \Gamma_2 \subset \mathbb{S}$, *where* Γ_1 *and* Γ_2 *intersect at* N.

16. *Show that the line* $y = mx$ *for* $m \neq 0$ *in the* z*-plane is mapped to a spiral in the* w*-plane by the mapping* $f(z) = e^z$.

17. *Consider the set* $A = \{z \in \mathbb{C} : 1 \leq \text{Re}(z) \leq 3,\ \frac{\pi}{2} \leq \text{Im}(z) \leq \pi\}$. *Show that* $f(A)$ *where* $f : \mathbb{C} \to \mathbb{C}$ *is given by* $f(z) = e^z$ *is a quarter annulus in the second quadrant of the complex plane.*

18. *Find a conformal map that maps* $A = \{z \in \mathbb{C} : \text{Im}(z) > 1,\ |\text{Re}(z)| < b\}$ *where* $b \in \mathbb{R}$ *onto the upper half-plane such that* $b + i$ *and* $-b + i$ *are mapped to* 1 *and* -1, *respectively.*

19. *Find a conformal map that maps* $A = \{z \in \mathbb{C} : |z| \geq 1,\ \text{Im}(z) < 0\}$ *onto* $B = \{w \in \mathbb{C} : |w| < 1\}$.

20. *Let* $w = f(z)$ *be defined by* $\frac{w-1}{w+1} = \left(\frac{z-i}{z+i}\right)^2$. *Find the domain in the* z*-plane that is mapped onto* $\{w \in \mathbb{C} : \text{Im}(w) > 0\}$.

21. *Find the image of the set* $\{z \in \mathbb{C} : |z| < 1\}$ *under the mapping* $w = \frac{z}{1-z}$.

22. *Find a conformal map that maps* $\{z \in \mathbb{C} : |z| < 1,\ \text{Re}(z) < 0\}$ *onto* $\{w \in \mathbb{C} : \text{Im}(w) > 0\}$ *such that* -1 *is mapped to* 1.

23. *Find a conformal map that maps* $\{z \in \mathbb{C} : 0 < \text{Im}(z) < \pi\}$ *onto* $\{w \in \mathbb{C} : |w| < 1\}$.

24. *Consider the mapping* $w = \frac{a}{2}\left(z + z^{-1}\right)$ *where* $a > 0$. *Find the domain in the* z*-plane that gets mapped onto* $\{w \in \mathbb{C} : \text{Im}(w) > 0\}$.

25. *Find a conformal mapping that maps* $\mathbb{C} \setminus \{z \in \mathbb{C} : \text{Re}(z) \leq -1\}$ *onto the interior of the unit disk such that the origin is mapped to itself.*

BIBLIOGRAPHY

Real Analysis References

[Bar76] Robert G. Bartle. *The Elements of Real Analysis*. Wiley, 1976.

[Buc03] R. Creighton Buck. *Advanced Calculus*. Waveland Pr, Inc., 2003.

[Cro94] Michael J. Crowe. *A History of Vector Analyis: The Evolution of the Idea of a Vectorial System*. Dover, 1994.

[Fit95] Patrick Fitzpatrick. *Advanced Calculus*. Brooks Cole, 1995.

[Ful78] Watson Fulks. *Advanced Calculus: An Introduction to Analysis*. Wiley, 1978.

[GO64] Bernard L. Gelbaum and John M. H. Olmstead. *Counterexamples in Analysis*. Holden-Day, Inc., 1964.

[HH01] John H. Hubbard and Barbara B. Hubbard. *Vector Calculus, Linear Algebra, and Differential Forms: A Unified Approach*. Prentice Hall, 2001.

[Hof07] Kenneth Hoffman. *Analysis in Euclidean Space*. Dover, 2007.

[Kos04] Witold A. J. Kosmala. *A Friendly Introduction to Analysis: Single and Multivariable*. Pearson/Prentice Hall, 2 edition, 2004.

[KP02] Steven G. Krantz and Harold R. Parks. *The Implicit Function Theorem: History, Theory, and Applications*. Birkhauser Boston, 2002.

[Lan94] Serge Lang. *Differentiable Manifolds*. Springer, 1994.

[Rud76] Walter Rudin. *Principles of Mathematical Analysis*. McGraw-Hill, 1976.

[Rud87] Walter Rudin. *Real and Complex Analysis*. McGraw-Hill, 1987.

[Sch85] Norman Schaumberger. More applications of the mean value theorem. *The College Mathematics Journal*, 16(5):397–398, 1985.

[Spi65] Michael Spivak. *Calculus on Manifolds*. W. A. Benjamin, 1965.

[Thi03] Johan Thim. Continuous nowhere differentiable functions. Master's thesis, Lulea University of Technology, 2003.

[Zyg77] Antoni Zygmund. *Trigonometric Series*. Cambridge University Press, 1977.

Complex Analysis References

[Ahl79] Lars V. Ahlfors. *Complex Analysis*. McGraw-Hill, 1979.

[AN07] Robert Ash and W.P. Novinger. *Complex Variables*. Dover, 2007.

[BC03] James W. Brown and Ruel V. Churchill. *Complex Variables and Applications*. McGraw-Hill, 2003.

[Con78] John Conway. *Functions of One Complex Variable*. Springer, 1978.

[Gam03] Theodore W. Gamelin. *Complex Analysis*. Springer, 2003.

[GK06] Robert E. Greene and Steven G. Krantz. *Function Theory of One Complex Variable*. AMS, 3 edition, 2006.

[Hil62] Einar Hille. *Analytic Function Theory*. Ginn and Company, 1962.

[Kra99] Steven G. Krantz. *Handbook of Complex Variables*. Birkhauser, 1999.

[Mar05] A. I. Markushevich. *Theory of Functions of a Complex Variable*. American Mathematical Society, 2005.

[MH98] Jerrold E. Marsden and Michael J. Hoffman. *Basic Complex Analysis*. W. H. Freeman, 1998.

[Nee99] Tristan Needham. *Visual Complex Analysis*. Oxford University Press, 1999.

[Rud87] Walter Rudin. *Real and Complex Analysis*. McGraw-Hill, 1987.

[Tit39] Edward C. Titchmarsh. *The Theory of Functions*. Oxford University Press, 1939.

Miscellaneous References

[Ful97] William Fulton. *Algebraic Topology: A First Course*. Springer, 1997.

[Hat02] Allen Hatcher. *Algebraic Topology*. Cambridge University Press, 2002.

[HK71] Kenneth Hoffman and Ray Kunze. *Linear Algebra*. Prentice Hall, 1971.

[KD03] Helmut Kopka and Patrick W. Daly. *A Guide to LaTeX*. Addison-Wesley, 2003.

[MGB+04] Frank Mittelbach, Michel Goossens, Johannes Braams, David Carlisle, and Chris Rowley. *The LaTeX Companion*. Addison-Wesley, 2004.

[Mun00] James Munkres. *Topology*. Prentice Hall, 2000.

Index

Milton Keynes UK
Ingram Content Group UK Ltd.
UKHW021928071024
449327UK00022B/1732